3rd International Conference on Mechanical, Automotive and Aerospace Engineering 2016

IOP Conference Series: Materials Science and Engineering
Volume 184

Kuala Lumpur, Malaysia
25 – 27 July 2016

ISBN: 978-1-5108-3897-0
ISSN: 1757-8981

Printed from e-media with permission by:

Curran Associates, Inc.
57 Morehouse Lane
Red Hook, NY 12571

Some format issues inherent in the e-media version may also appear in this print version.

Printed by Curran Associates, Inc. (2017)

For permission requests, please contact the Institute of Physics
at the address below.

Institute of Physics
Dirac House, Temple Back
Bristol BS1 6BE UK

Phone: 44 1 17 929 7481
Fax: 44 1 17 920 0979

techtracking@iop.org

Additional copies of this publication are available from:

Curran Associates, Inc.
57 Morehouse Lane
Red Hook, NY 12571 USA
Phone: 845-758-0400
Fax: 845-758-2633
Email: curran@proceedings.com
Web: www.proceedings.com

Table of contents

Volume 184

3rd International Conference on Mechanical, Automotive and Aerospace Engineering 2016

25–27 July 2016, Kuala Lumpur, Malaysia

Accepted papers received: 28 February 2017
Published online: 3 April 2017

Preface

Aerospace Engineering

Automotive Engineering

Mechanical Engineering

IOP Conf. Series: Materials Science and Engineering **184** (2017) 011001 doi:10.1088/1757-899X/184/1/011001

Innovative Technologies for a Sustainable Future

Program Book and Abstracts Proceeding

3rd International Conference on Mechanical, Automotive and Aerospace Engineering 2016 (ICMAAE'16)

25 – 27 July 2016
Kulliyyah of Engineering, IIUM

IIUM Engineering Congress 2016

IOP Conf. Series: Materials Science and Engineering **184** (2017) 011001 doi:10.1088/1757-899X/184/1/011001

3rd International Conference on Mechanical, Automative and Aerospace Engineering 2016 2016

FOREWORD FROM THE CHAIRMAN OF ICMAAE 2016

PROF. DR. MEFTAH HRAIRI

Chairman of ICMAAE'16

Head of Mechanical Engineering Department

International Islamic University Malaysia

The 3rd International Conference on Mechanical, Automotive and Aerospace Engineering (ICMAAE'16) was held in the International Islamic University Malaysia (IIUM), Kuala Lumpur during July 25-27, 2016. ICMAAE'16 was organized as part of IIUM Engineering Congress 2016 (IEC'16) which also comprised the 4th International Conference on Biotechnology Engineering (ICBioE'16) and the 6th International Conference on Computer and Communication Engineering (ICCCE'16).

The main objective of organizing this conference is to provide an international technical forum for engineers, academicians, scientists and researchers to present result of on-going and completed research in the areas of Mechanical, Automotive and Aerospace Engineering. The primary focus of the conference is to create an effective medium for institutions and industries to meet, present and discuss the latest research developments, challenges and trends in mechanical, automotive and aerospace engineering. In addition, ICMAAE'16 aims at providing an opportunity to renew old friendships and make new acquaintances.

The success and prosperity of ICMAAE'16 depended completely on the effort, talent, and energy of researchers who have written and submitted papers on a variety of topics in the field of mechanical, automotive and aerospace engineering. We were indeed glad at the favorable response received from the scientific community around the world. Praise is also deserved for all of those who supported ICMAAE'16. Especially, we would like to thank the organizing committee and reviewers who have invested significant time in analyzing and assessing multiple papers and who held and maintained a high standard of quality for this conference. My thanks go to all the sponsors and all participants in making this conference a success.

These proceedings, which covers 3 chapters: Aerospace, Automotive, Mechanical, are a compilation of the accepted papers and represent an interesting outcome of the conference. It is hoped that the proceedings will serve as an important research source of reference and knowledge.

Prof. Dr. Meftah Hrairi

Head of Mechanical Engineering Department
Kulliyah of Engineering
International Islamic University Malaysia

IOP Conf. Series: Materials Science and Engineering **184** (2017) 011001 doi:10.1088/1757-899X/184/1/011001

3rd International Conference on Mechanical, Automotive and Aerospace Engineering 2016

2016

ICMAAE'16 ORGANIZING COMMITTEES

Chairman :
Meftah Hrairi

Co-Chairman:
Ahmad Faris Ismail
Waqar Asrar

Secretary
Sanisah Saharin
Hanan Mokhtar

Technical Committee
AKM Mohiuddin – Chairman
Mohammed Abdelrahman – Co-chair

Sany I Ihsan
MNA Hawlader
Waleed Faris
Qasim H Shah
Yulfian Aminanda
Moumen Idris
Mazharul Islam
Sanisah Saharin
Hanan Mokhtar
Nabilah Ramli
Syed Noh Syed Abu Bakar
Muhammad Saifuddin Rehan
Mohamed Okasha
Erwin Sulaeman
JS Mohammed Ali
Kassim A Abdullah
Meftah Hrairi
Ari Legowo
Fadly J Darsivan
Mohd Sultan Ibrahim
M. Ataur Rahman
Sher Afghan
Waqar Asrar
Hilmi Hela Ladin
Zahir Hanouf

Keynote/Invited speakers
Hanan Mokhtar – Chairman
Ari Legowo - Co-chairman
Kassim A. Abdullah

Financial Sponsorship, Food and Souvenirs
Ari Legowo – Chairman
Saifuddin Co-chairman
Azweeda Dahlan
Azliza Embong

Publication in Journals
Yulfian Aminanda – Chair
Erwin Sulaeman - Co-chair

Waqar Asrar
Waleed Faris
M Hawlader

Registration
JS Mohammad Ali – Chairman
Syed Noh - Co-chair
Mohammad Azizudeen
Syazwani Mahmoodin
Nooryani Haji Ali

Hospitality, and Transportation
Sanisah Saharin - Chair
Saifuddin Rehan
Firdauz Hakem
Nor Hafez Adnan
M Farid

Publicity
Moumen Idres – Chair
Mohamed Okasha - Co-chair

Mohd Sultan
M Abdelrahman
MNA Hawlader
M Ataur Rahman

Qasim Shah
Waleed Faris
Yulfian Aminanda
Waqar Asrar

Internal Advisory Board
Waqar Asrar - Chair
Ari Legowo
Yulfian Aminanda
Ahmad Faris Ismail

Web Master
Fadly J Darsivan (Chair)
Mohd Sultan – Co Chair
Zahratu Hillal

Proceedings and Programme book
Nabilah Ramli - Chair
M Abdelrahman - Co-Chair
Yulfian Aminanda
Erwin Sulaeman

IOP Conf. Series: Materials Science and Engineering **184** (2017) 011001 doi:10.1088/1757-899X/184/1/011001

3rd International Conference on Mechanical, Automative and Aerospace Engineering 2016

2016

ADVISORY BOARD

IOP Conf. Series: Materials Science and Engineering **184** (2017) 011001 doi:10.1088/1757-899X/184/1/011001

3rd International Conference on Mechanical, Automative and Aerospace Engineering 2016

2016

CONFERENCE TOPICS

Modeling and Simulation

Experimental Techniques and Measurements

Dynamics and Controls

Structures and Materials

Thermo-fluids

Aerodynamics and Aeroelasticity

Aerospace Dynamics and Controls

Aerospace Propulsion

Aerospace Structures

Aerospace Design, Testing, and Performance

Engine and its Systems

Alternative Fuels

Combustion and Emission control

Vehicle Structures and Crashworthiness

Power train

Vehicle Dynamics and Control

Alternative Power Systems and Management

Vehicle Aerodynamics

Noise Vibration & Harshness

Experimental Methods and Measurements

IOP Conf. Series: Materials Science and Engineering **184** (2017) 011001 doi:10.1088/1757-899X/184/1/011001

REVIEWERS

Dr Mohamed Abd Rahman
Dr. Salmiah Ahmad
Prof. Ahsan Ali Khan Ali Khan
Prof Yulfian Aminanda
Prof. Waqar Asrar
Dr. Fadly Jashi Darsivan
Mohammad Nurul Alam Hawlader
Prof. Meftah Hrairi
Dr. Moumen Idres
Dr Sany Izan Ihsan
Dr. Mazharul Islam
Prof. Ahmad Faris Ismail
Dr Mohamed Ali Jaffar Syed
Dr. Nor Khairusshima Khairussaleh
Dr. S A Khan
Prof. Md. Raisuddin Khan
Dr. Ari Legowo
Prof Md Abdul Maleque
Prof. Dr. AKM Mohiuddin
Dr. Hanan Mokhtar
Prof. A.N. Mustafizul Karim
Prof. A.K.M. Nurul Amin Nurul Amin
Dr. Mohamed Okasha
Dr. Md Mozasser Rahman
Prof. Dr Ataur Rahman
Dr. Nabilah Ramli
Dr. Mahbubur Rashid
Dr Sanisah Saharin
Dr. Muhammad Saifuddin
Dr. Quasim Hussain Shah
Dr. Mohd. Sultan Ibrahim Shaik Dawood
Dr Erwin Sulaeman
Dr. Suryanto Suryanto
Dr. Syed Noh Syed Abu Bakar

3rd International Conference on Mechanical, Automative and Aerospace Engineering 2016

2016

Peer review statement

All papers published in this volume of *IOP Conference Series: Materials Science and Engineering* have been peer reviewed through processes administered by the proceedings Editors. Reviews were conducted by expert referees to the professional and scientific standards expected of a proceedings journal published by IOP Publishing.

Slew maneuver control of flexible spacecraft for vibration suppression

Okubo. H. [1,a]*
[1]Department of Mechanical Engineering
Kanagawa Institute of Technology, Japan
1030 Shimo-ogino, Atsugi, Kanagawa, Japan

E-mail: [a]okubo@me.kanagawa-it.ac.jp

Abstract. This paper presents the input shaping technique for reducing the vibrations of flexible structures under attitude control inputs applied to the flexible spacecraft. The input shaping profile is investigated for the effective suppression of transient vibrations of modal responses. It is applied to the input commands to minimize the residual vibrations and suppress the overshoot of the modal responses. The results of numerical simulations, using a simple dynamic model of a flexible spacecraft, show that the input shaping technique is useful for suppressing the residual vibrations caused by attitude maneuvers.

1. Introduction

One of the problems faced during the attitude control of a spacecraft is unsatisfactory precision control caused by vibrations of the flexible structures. It is very important to suppress the residual vibrations quickly to meet the mission requirements. The input shaping technique proposed by Smith [1], is one of the most effective methods for suppressing the transient vibrations of a flexible structure and it has been used in many applications [1]-[4]. This technique has also been applied to the concurrent design with feedback control of nonlinear coupled vibration systems such as the trajectory tracking control of a flexible manipulator [5] and the slewing maneuver of a flexible spacecraft [6].

The main disadvantage of the input shaping is that the equation used to reduce the residual vibrations requires the estimates of the natural frequencies and damping ratios. If these structural parameters are different from the real values, the method does not work and the effectiveness of the input shaping in vibration suppression is degraded. Furthermore, the input shaper is designed for suppressing a single vibration mode and it is not effective for other vibration modes.

This paper introduces an input shaping technique based on Nil-Mode-Exciting profiler (NME). The NME profiler has been studied for shaping the input command to attenuate the vibration of the spacecraft appendage structures [7]-[8]. The input command is designed such that not to excite all modes of vibration higher than the designated natural frequency of the flexible structure.

The Input Shaping technique and the NME profiler are studied in this paper for the suppression of residual vibrations due to the rest-to-rest maneuver of the spacecraft and they are integrated to achieve a superior performance in vibration damping.

2. Input shaping method for vibration suppression

Figure 1 shows the block diagram of a servo control system consisting of a feedback loop for pursuing the control objective, such as the attitude maneuver of a flexible satellite, and an input shaper for suppressing the excitation of vibrational responses.

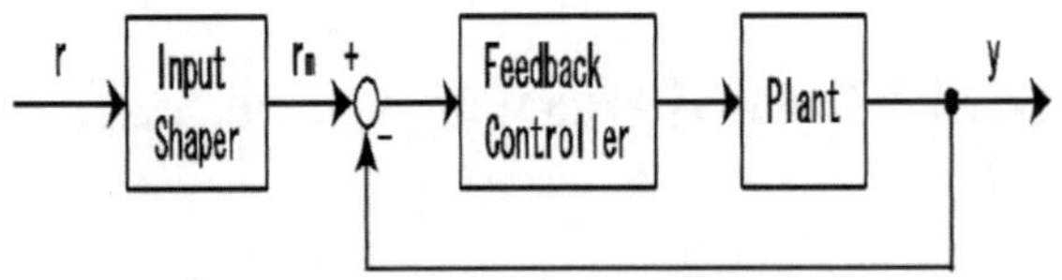

Figure 1. Block diagram of input shaper with a feedback controller.

The amplitude of vibration response when N impulses are fed to a one-degree-of-freedom vibration system is given as follows.

$$A_{amp} = \sqrt{\left(\sum_{j=1}^{N} B_j \cos \omega_d t_j\right)^2 + \left(\sum_{j=1}^{N} B_j \sin \omega_d t_j\right)^2} \tag{1}$$

where

$$B_j = A_j \frac{\omega_0}{\sqrt{1-\zeta^2}} exp\left(-\zeta\omega_o(t_N - t_j)\right) \tag{2}$$

In addition, A_j denotes the amplitude of the jth impulse, t_j is the time at which the impulse occurs, ω_0 is the natural frequency of the system, and ζ, is the damping ratio of the system. To eliminate the vibrational response after the input has been applied, the expression for A_{amp} should be zero at time t_N; when the input ends. In other words, both the squared terms in Eq. (1) should be equal to zero.

$$\sum_{j=1}^{N} A_j \, exp\left(-\zeta\omega_o(t_N - t_j)\right) \cos \omega_d t_j = 0 \tag{3}$$

$$\sum_{j=1}^{N} A_j \, exp\left(-\zeta\omega_o(t_N - t_j)\right) \sin \omega_d t_j = 0 \tag{4}$$

For example, consider two impulse inputs at times t_1 and t_2 with amplitudes A_1 and A_2. The responses to the two impulse inputs cancel each other, and the vibrations after time t_2 vanish, only if the natural frequency and damping ratio of the system are known exactly. Figure 2 demonstrates the input shaping technique using two impulses.

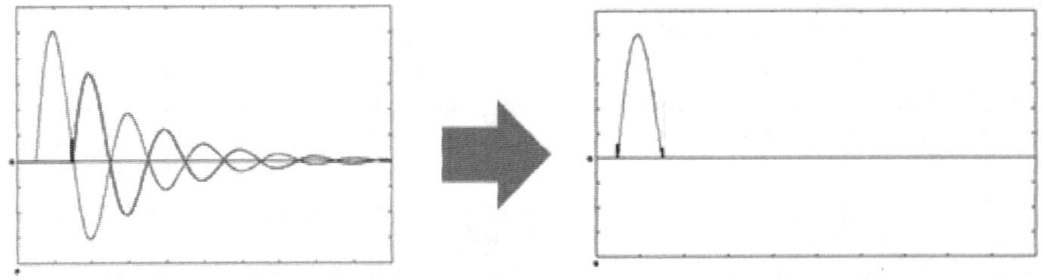

Figure 2. Basic idea of the input shaping technique.

For the successful application of the input shaping technique in most real systems, the constraint equations, i.e., Eqs. (3) and (4) must ensure robustness to modeling errors. Singer and Seering [2] have developed a method of robust input shaping by using the condition that the derivative of Eqs. (3) and (4) with respect to the frequency of the residual vibrations equals zero. The resulting shaper is called a zero vibration and derivative (ZVD) shaper. Further details may be obtained by referring to their study.

3. Nil-Mode-Exciting profiler

The concept of Nil-Mode-Exciting profiler is based on the use of the sinc function

$$sinc(t) = \frac{\sin(\omega_0 t)}{\omega_0 t} \tag{5}$$

A Fourier transform of the sinc function shows the characteristics of an ideal low-pass filter with a cutoff frequency of ω_0 rad/s. The frequency response function and time history of the sinc function are shown in Fig. 3.

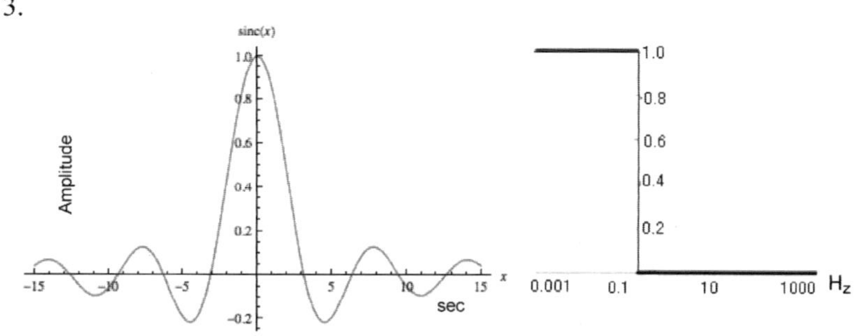

Sinc function Frequency response

Figure 3. Characteristics of a sinc function.

An NME profiler consists of a couple of sinc functions with opposite signs as follows:

$$A = \frac{\sin\{\omega_a(t-t_{offset1})\}}{\omega_a(t-t_{offset1})} - \frac{\sin\{\omega_a(t-t_{offset2})\}}{\omega_a(t-t_{offset2})} \tag{6}$$

where

$$t_{offset2} - t_{offset1} = \frac{2\pi}{\omega_a} \tag{7}$$

Equation (6) represents the sum of the accelerating and decelerating functions with time offsets. A typical synthesized sinc function is shown in Figure 4.

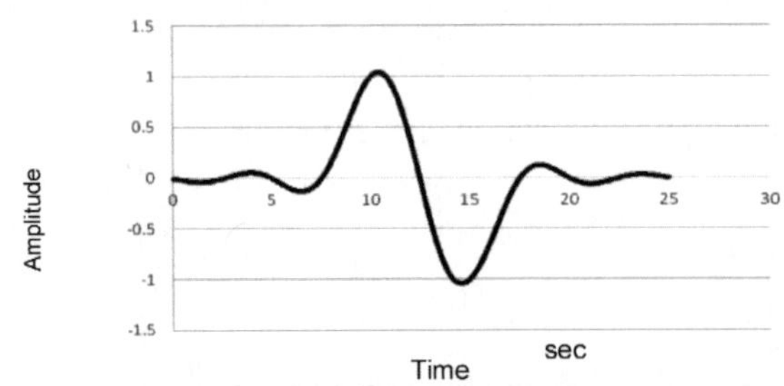

Figure 4. A synthesized sinc function for NME.

Practically, to modify the above infinite sinc function into a finite-time function, it is necessary to introduce a time window function such as a Hamming function:

$$w(x) = 0.54 - 0.46\cos 2\pi x \quad (\text{if } 0 \le x \le 1) \tag{8}$$

4. Integration of Input Shaping and Nil-Mode-Exciting profiler

The third method is introduced in this paper and shown in Fig. 5, where a single NME profile for the torque input (solid line) is divided into three inputs (dotted lines) according to the sequence of three inputs for the Input Shaping.

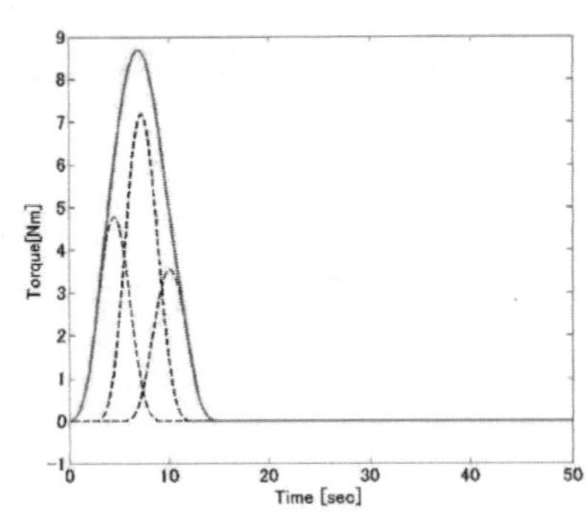

Figure 5. Integrated Input Shaping and NME profiler.

5. Dynamics of a flexible spacecraft

A simplified dynamic model of a satellite with flexible appendages is shown in Figure 6. The model of the flexible structure is considered as a one-degree-of-freedom-vibration system, as defined in literature [4].

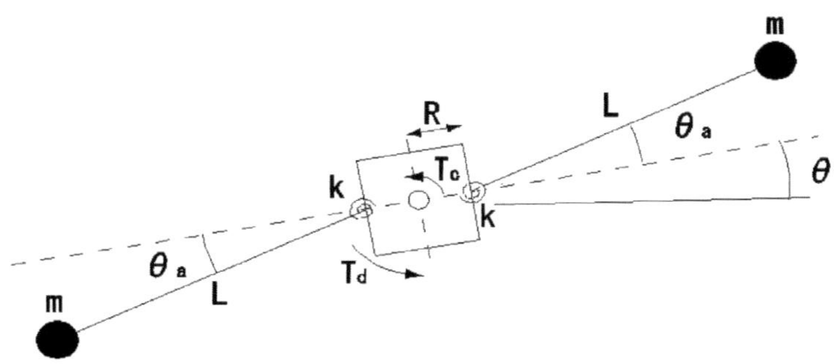

Figure 6. Dynamic model of a flexible spacecraft.

where J_b is the moment of inertia of the satellite body, m is the mass of the adjunct, R is the mounting position from the center of the satellite, θ is the attitude angle, θ_a is the angle of the arm, k is the spring constant of torsion, T_c is the torque of the reaction wheel, and T_d is the torque of the thrusters.

The dynamics of the attitude of the satellite with a flexible structure is derived by Lagrange's equations of motion as follows.

$$[J_b + 2m(R+L)^2]\ddot{\theta} + 2m(R+L)L\ddot{\theta}_a = T_d + T_c \quad (9)$$

$$2\mathrm{m}(\mathrm{R}+\mathrm{L})\mathrm{L}\ddot{\theta} + 2\mathrm{m}L^2\ddot{\theta}_a = -2k\theta_a \quad (10)$$

where the angular velocity of the satellite body is ignored, because it is very small when compared to the angular velocity of the reaction wheel.

The torque of the thruster is considered as a pulse in one direction, and it functions with unit amplitude, changing sign at a specified time sequence. The impulses of the thruster torque are shown in Table 1, where A_j denotes the normalized amplitude of a single impulse, t_j is the switching time; t_1 = 0, because the first impulse is applied at the initial time.

Table 1. Amplitudes and switching times of thruster torque

A_j	t_j
1	t_1
-1	t_2
1	t_3
-1	t_4
⋮	⋮
-1	t_n

In order to apply the ZVD, Eqs. (3) and (4), and their derivatives with respect to the natural frequency are used for the constraint equations. The number of switching times, n, of the impulses to suppress the tangent vibration with frequency robustness is determined by solving the following equations.

$$\sum_{j=1}^{N}(-1)^{j+1} sin\omega_d t_j = 0 \quad (11)$$

$$\sum_{j=1}^{N}(-1)^{j} cos\omega_d t_j = 0 \quad (12)$$

$$\sum_{j=1}^{N}(-1)^{j+1} t_j sin\omega_d t_j = 0 \quad (13)$$

$$\sum_{j=1}^{N}(-1)^{j+1} t_j cos\omega_d t_j = 0 \quad (14)$$

We consider the torque of the reaction wheel. The attitude angle of the satellite is controlled by a simple proportional-integral-derivative (PID) controller.

$$T_c = -I_w \ddot{\theta}_w = -k_p \theta - k_i \int_0^t \theta d\tau - k_d \dot{\theta} \quad (15)$$

whereI_w is the moment of inertia of the reaction wheel and $\ddot{\theta}_w$ is the angular acceleration of the reaction wheel.

6. Result of numerical simulation

The following numerical values are used in simulation.

J_b = 12800 kg$\cdot m^2$, m = 25 kg, R = 2 m, L = 10 m, k = 5685 Nm

The amplitudes and switching times of the thruster are obtained by using the damped natural frequency ω_d. These are listed in Table 2, where the time for the ignition of the thruster is set to $t_1 = 0$ sand $t_4 = 100$ s.

Table 2 Amplitudes and switching times of thruster torque.

A_i	$t_i(s)$
1	0
-1	0.55
1	1.11
-1	100
1	100.55
-1	101.11

The results of numerical simulations are shown in Figures 7-11. Figure 7 shows the time response of the flexible structure when input shaping is not used, and Figure 8 shows the response with input shaping. The vertical axis representsθ_a, which is the amplitude of the angular displacement of the flexible structure, and the horizontal axis represents time. These figures show that the convergence time is shorter, and the maximum amplitude of displacement is smaller, when input shaping is applied for the input torque.

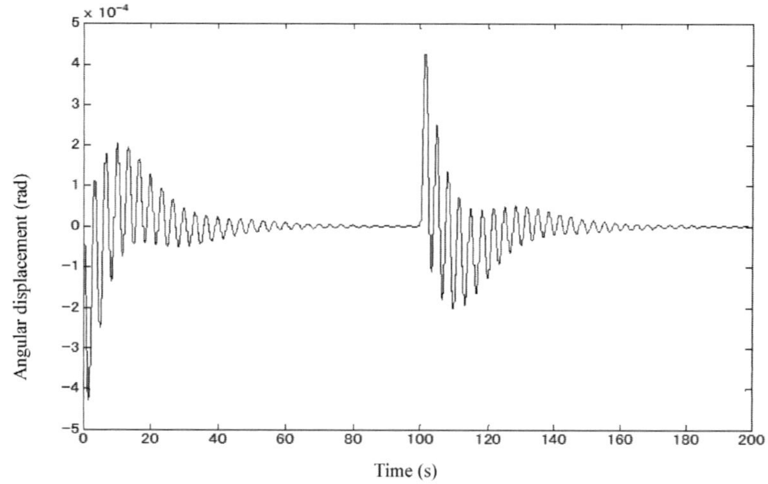

Figure 7. Displacement of flexible structure without input shaping.

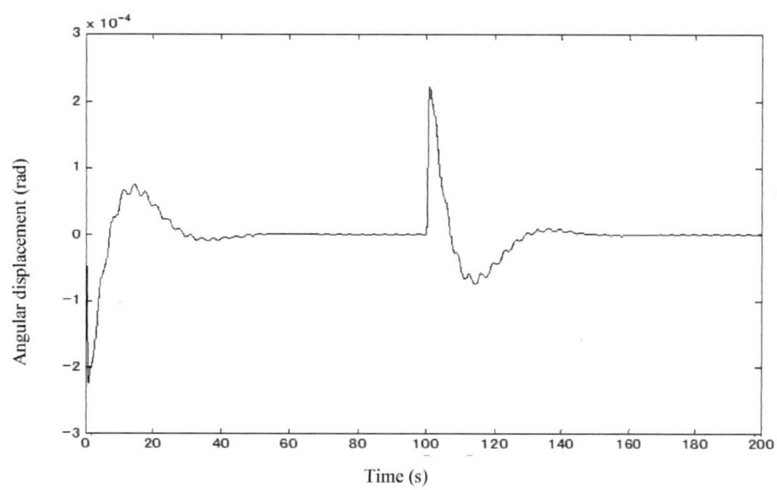

Figure 8. Displacement of flexible structure with input shaping.

The input torque T_d of the thruster with input shaping, and the PID feedback control torque $T_c$of the reaction wheel are shown in Figure 9. Figures 10 and 11 show the effect of zero vibration and derivative (ZVD) shaper for enhancing the robustness against the parameter errors. The vibrational response is excited due to the 15% error in the natural frequency of the flexible structure (Figure 10), whereas it is remarkably improved by using the robust input shaping (Figure 11). Figure 12 shows the input torque corresponding to the robust input shaping. It is observed from these numerical simulations that the input shaping technique and the ZVD version are very effective in suppressing the residual vibrations of the flexible structures excited by the attitude control torque.

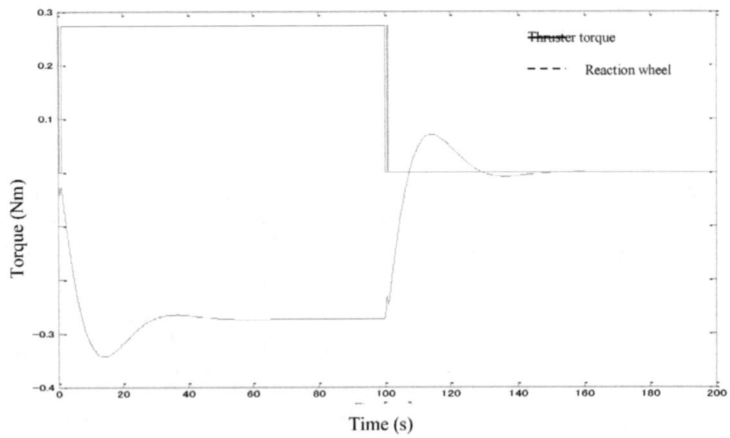

Figure 9. Torque of the reaction wheel and thruster with input shaping.

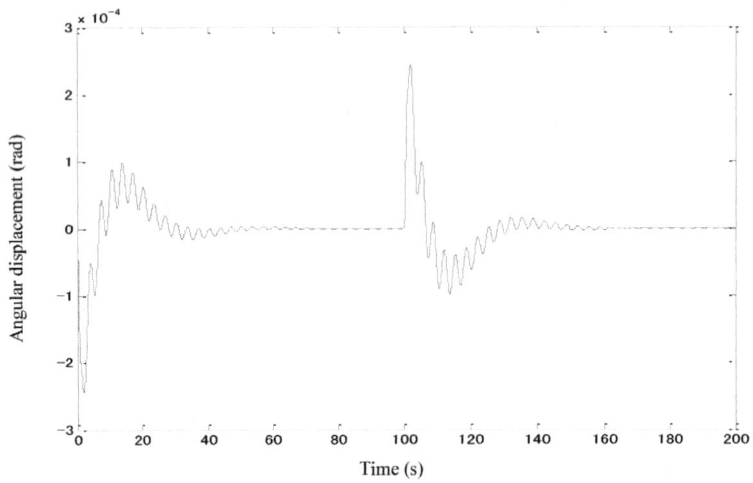

Figure 10. Displacement with ordinary input shaping (15% error in natural frequency).

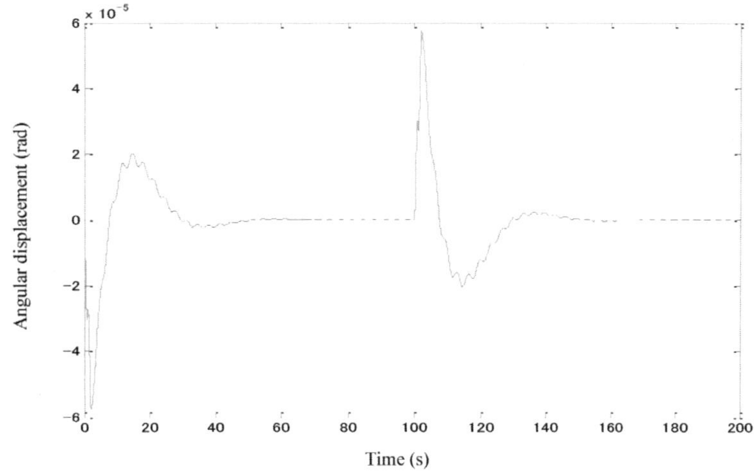

Figure 11. Displacement with robust input shaping (15% error in natural frequency).

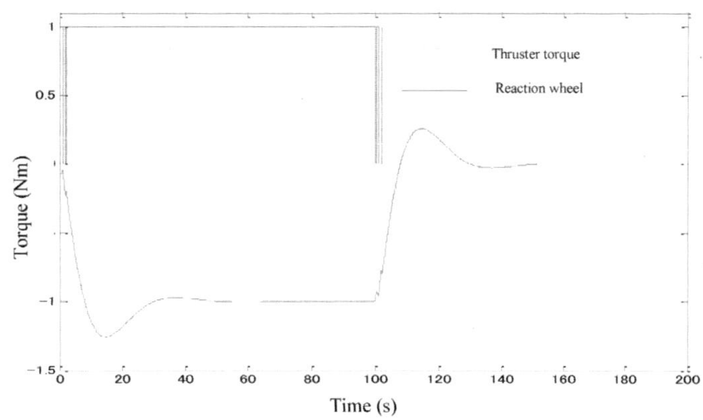

Figure12. Torque of the reaction wheel and thruster with robust input shaping.

7. Conclusion

Input shaping is very effective in suppressing the residual vibrations of flexible structures. The results of the numerical simulation demonstrate that the input shaping method is highly effective for agile attitude maneuvers of a flexible spacecraft. Future work involves comparing the input shaping method with other command shaping methods such as Nill-Mode Excitation (NME) profiler, and experimental demonstration of the input shaping method with a simple laboratory model.

References

[1] Smith, O.J.M. 1958 Feedback Control Systems *McGraw-Hill, New York* p.338.
[2] Singer, N.C., Seering, W.P. 1990 Preshaping command inputs to reduce system vibration *J. Dynamics Systems, Measurement and Control 112, 1*, 76-82.
[3] Singhose, W.E. 1996 Extra-Insensitive Input Shaper for Controlling Flexible Spacecraft *Journal of Guidance, Control and Dynamics, Vol. 19, No. 2* pp. 385–391.
[4] Singhose, W.E., Banerjee, A.K., and Seering, W.P. 1997 Slewing Flexible Spacecraft with Detection-Limiting Input Shaping *Journal of Guidance, Control and Dynamics, Vol. 20, No. 2* pp. 291–298.

[5] Hasegawa, Y. and Okubo, H. 2008 Input Shaping for Fast Trajectory racking Control of a Flexible Manipulator *Proceedings of the MOVIC, Paper 327* 10 pages.

[6] Kashima, M. 2010 Attitude Control of Flexible Spacecraft Using Control Momentum Gyros *Master Thesis, Graduate School of Engineering, Osaka Prefecture University*.

[7] Kamiya, T., Maeda, K., Ogura, N., Hashimoto, T., and Sakai, S. 2007 Preshaping Profiler for Flexible Spacecraft Rest-to-Rest Maneuvers *JAXA-ISAS, 229-8510,* 6 pages.

[8] Okubo, H. and Kuwamoto, S. 2014 Integrated Feedback/Feedforward Control of Flexible Spacecraft for Agile Attitude Maneuver and Vibration Suppression *25th International Conference on Adaptive Structures and Technologies*, The Hague, The Netherlands, Oct. 6-8, 12 pages.

Experimental Investigation of the Influence of a Reverse Delta Type Add-on Device on the Flap-tip Vortex of a Wing

A Altaf[1,a], T B Thong[1,b], A A Omar[2,c*] and W Asrar[3,d]

[1]School of Engineering and Computer Sciences, New York Institute of Technology, P.O. Box 5464, Abu Dhabi, United Arab Emirates

[2]Mechanical Engineering Department, University of Tripoli, P.O. Box 81507, Tripoli, Libya

[3]Mechanical Engineering Department, International Islamic University Malaysia, P.O. Box 10, 50728 Kuala Lumpur, Malaysia

Email: [a]aafaq@nyit.edu, [b]tan.boon.thong@monash.edu, [c]aao@aerodept.edu.ly, [d]waqar@iium.edu.my

Abstract
Particle Image Velocimetry was used in a low speed wind tunnel to investigate the effect of interactions of vortices produced by an outboard flap-tip of a half wing (NACA 23012 in landing configuration) and a slender reverse delta type add-on device, placed in the proximity of the outboard flap-tip, on the upper surface of the half wing. This work investigates the characteristics of the vortex interactions generated downstream in planes perpendicular to the free stream direction at a chord-based Reynolds number of $Re_c = 2.74\times10^5$. It was found that the add-on device significantly reduces the tangential velocity magnitude and enlarges the vortex core of the resultant vortex by up to 36.1% and 36.8%, respectively.

Nomenclature

b	= wing span length (m)
Re_c	= Reynolds number based on chord length
r	= radius, m
r_c	= core radius, m
V_θ	= tangential velocity, m/s
V_∞	= free stream velocity, m/s
x	= stream-wise coordinate, m
y	= span-wise coordinate, m
z	= transverse coordinate, m
α	= angle of attack, deg.

1. Introduction

The study and control of swirling vortex flows is vital towards developing ways to suppress vortices generated by commercial aircraft which leads to a reduction in the hazard posed to trailing aircraft [1, 2].

Vortices created by aircrafts are an inevitable consequence of the creation of lift. Vortices persist for many kilometers and wake-vortex encounters pose a grave hazard to trailing aircrafts that fly in close proximity near the airport runway especially during takeoff and landing [3] because the wake vortex circulation is at a maximum. Wake vortex and turbulence generated by large aircraft are strong and can be unsafe to the following aircraft which can experience motion anywhere from sudden up-wash to downwash to uncontrollable rolls to sudden loss of altitude, depending on the aircraft's position and orientation with respect to the wake [4]. Incidents have been recorded during landing approaches, when wake turbulence has resulted in fatal accidents because of insufficient time and altitude for pilots to regain full control of their aircraft after being severely affected by the strong vortices.

Vortex wakes are very slow to dissipate naturally and therefore there is a need to find ways of increasing the dissipation rate. One of the most promising concepts of vortex dissipation enhancement involve modification of the wing by means of winglets or wing fences attached to the suction surface of the wing. These wing devices either create countersign vorticity that subsequently diminishes the intensity of the rolled up vortex or cause the vortex core dimension to be increased, which also hastens vortex dissipation [5].

Some of the most promising concepts of vortex modification make use of multiple vortex pair systems [6, 7, 8]. The addition of a strong vortex pair counter-rotating to the basic tip vortices can lead to more diffused vortices whose strength is the sum of the co-rotating and counter-rotating vortices.

Researchers are optimistic about solving wake hazard passively in the near-wake field. Numerous experimental investigations have been conducted with devices of various shapes and sizes to alter the vortex rollup process and to create instability in the wake of the aircraft so that the resultant vortex is enlarged and more diffused. Tests have been conducted with a spoiler mounted on the wing tip [9], splines mounted downstream of the wing tip [10] spoilers of delta type plan-form deployed in the area of the outboard flap [11], a tip-mounted slender half delta wing [12] and a tip-mounted reverse half delta wing [13].

Recent investigations have pointed out that reverse delta type add-on devices may have the capability to be used in vortex alleviation [2, 11, 13, 14]. Reverse delta type add-on device vortices may excite some instability (countersign vorticity) into the wingtip and flap-tip vortices and modify the vortex roll up process [2]. The interaction of the vortices would then tend to create a weaker resultant vortex with a reduction in peak vorticity, maximum tangential velocity and core circulation. These may lead to a diffused resultant vortex which can enhance wake vortex decay and thus, lead to wake vortex alleviation. The timings between consecutive landings and take-offs, and the separation distances between aircrafts can be reduced significantly by the use of a reverse delta type add-on device. This will directly increase the airports aircraft handling capacity and increase its profitability.

The purpose of this paper is to investigate the influence of the interaction of the vortices of a reverse delta type add-on device and the outboard flap-tip vortex.

2. Experimental Setup

The experiment was carried out in the closed-loop low-speed wind tunnel at the International Islamic University Malaysia (IIUM). The wind tunnel test section dimensions are 2.30 m (width), 1.50 m (height) and 6.00 m (length). The wind tunnel has a turbulence intensity of less than 0.11%.

A half wing model (NACA 23012 in landing configuration; slat extension 15° and flap extension 20°), shown in Fig. 1, is used in this investigation along with a reverse delta type add-on device, shown in Fig. 2. A reverse delta type add-on device is bounded by a leading edge and by a pair of trailing edges extending from the ends of the leading edge towards a trailing apex point. The reverse delta type add-on device has a bevel angle of 20° and is attached to the half wing model by a 20 mm long movable joint.

The velocity components in planes perpendicular to the stream-wise direction at four locations in the downstream direction are investigated. The experimental setup is shown in Fig. 3. Multiple laser positions are chosen so as to differentiate between their flow characteristics in terms of roll up, tangential velocity, vorticity and circulation.

PIV measurements are taken only at the half model outboard flap-tip with/without the reverse delta add-on device attached. The CCD camera is focused at the half model wingtip at all downstream planes.

The free stream velocity was set to 12 m/s corresponding to a $Re_c = 2.74 \times 10^5$. The flow was seeded by particles of a mean diameter of 1 μm. Important PIV data acquisition parameters such as time interval (Δt= 60 μs) and laser sheet thickness (2 mm) were fixed. A light sheet from a Nd:YAG laser system of wavelength 532 nm was used to illuminate the flow. A Flow Sense M2 8 bit CCD camera with Micro-Nikkor 60 mm camera lens was placed perpendicular to the laser sheet, downstream of the wing.

PIV data was recorded for the High Lift Configuration (HLC) at α=7.7° and HLC with a reverse delta type add-on device at α=9.7°. At 12 m/s, the HLC at an angle of attack 7° yields a target lift coefficient C_L=1.06. When an add-on device was attached to the half wing model at HLC, it was found that increasing the half wing model angle of attack by 1°- 2° recovers the target lift coefficient that was achieved at HLC without the add-on device. Hence, the half wing model with the add-on device is tested at α=9.7°. The reverse delta type add-on device angle of attack was fixed at α=30°.

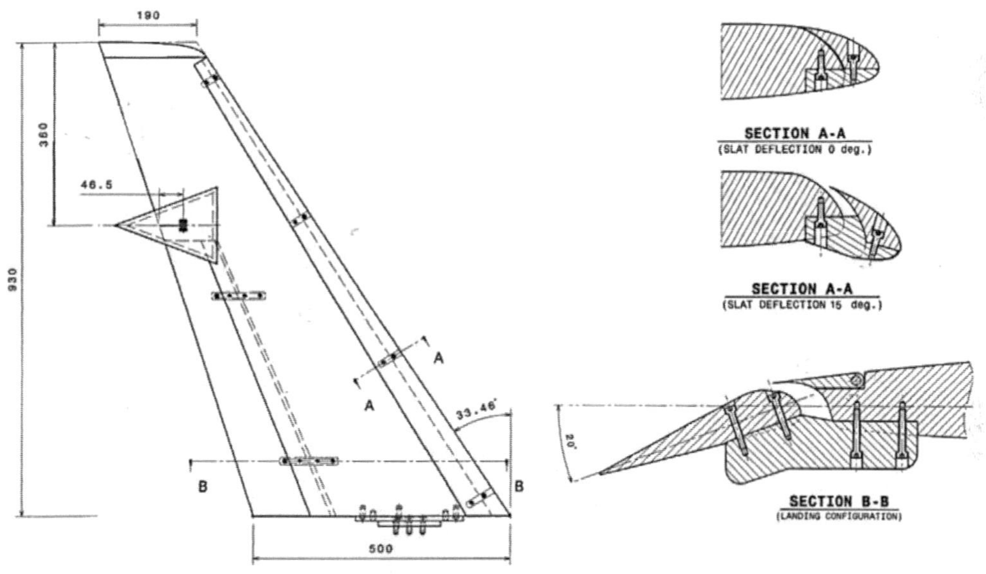

Figure 1. Schematic of the Half-Span Wing model. Dimensions are in millimetres [mm].

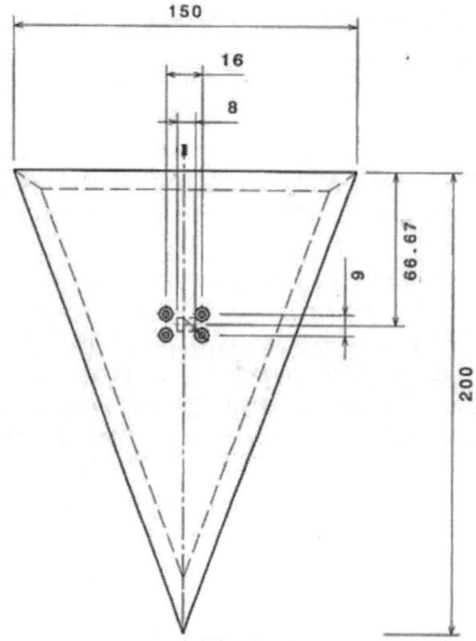

Figure 2. Schematic of the reverse delta type add-on device.
Dimensions are in millimetres [mm].

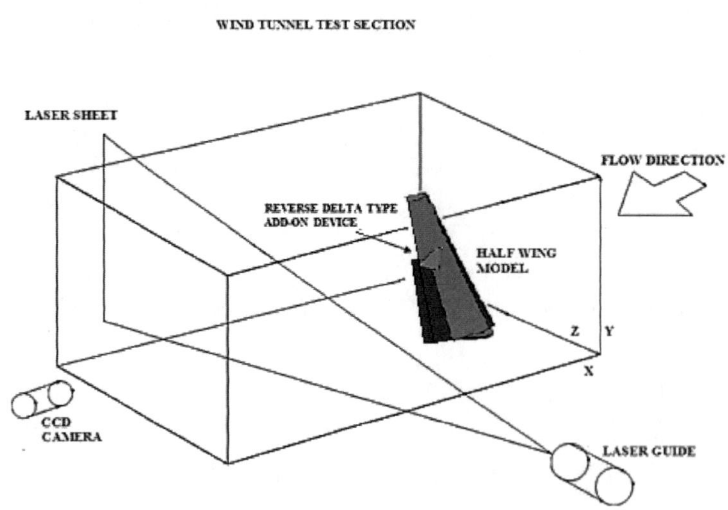

Figure 3. Schematic of the experimental setup.

3. Results and Discussions

3.1 Velocity Vectors and Vorticity Contours

Vortices are very slow to dissipate naturally and may take several miles to dissipate downstream of a large aircraft. Vortex dissipation rate can be accelerated if the vortex core dimension is increased significantly by using vortex modification techniques such as using wing devices that can alter the vortex rollup process. A more diffused vortex core at the same downstream location implies a weaker vortex.

Figs. 4a-4h show the velocity vectors of a half wing model at HLC with/without a reverse delta type add-on device attached at a location near the half model outboard flap-tip. The results are obtained at four downstream planes; x/(b/2)= 0.021, 0.548, 1.075 and 2.387. This investigation was carried out to discover if the resultant vortex formed after the merging of the reverse delta type add-on device vortices and the flap-tip vortex yields a more diffused vortex.

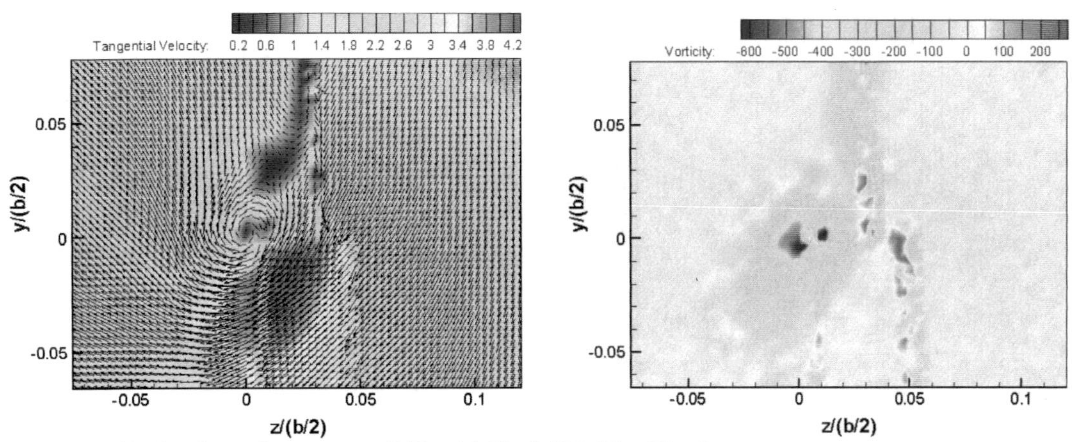

a) High Lift Configuration case, α=7.7°, x/(b/2)=0.021, V_∞=12 m/s.

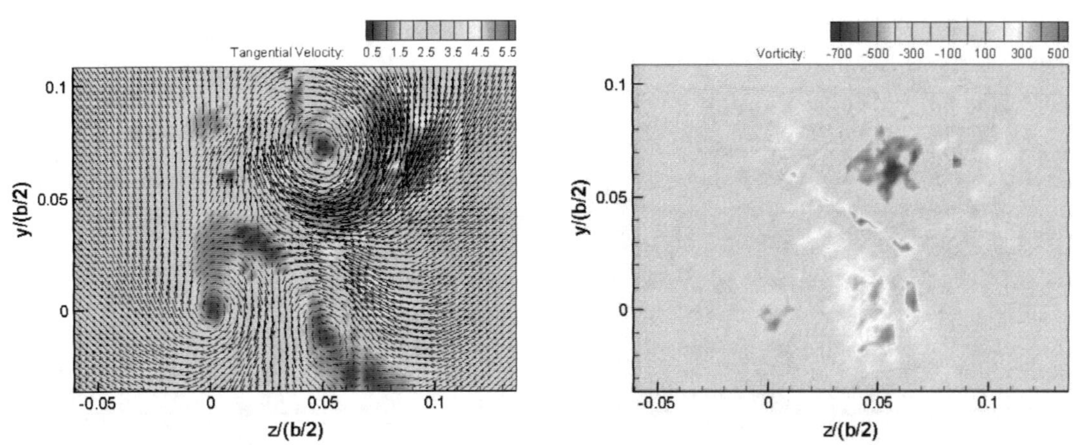

b) High Lift Configuration case with reverse delta type add-on device, α=9.7°; x/(b/2)=0.021, V_∞=12 m/s.

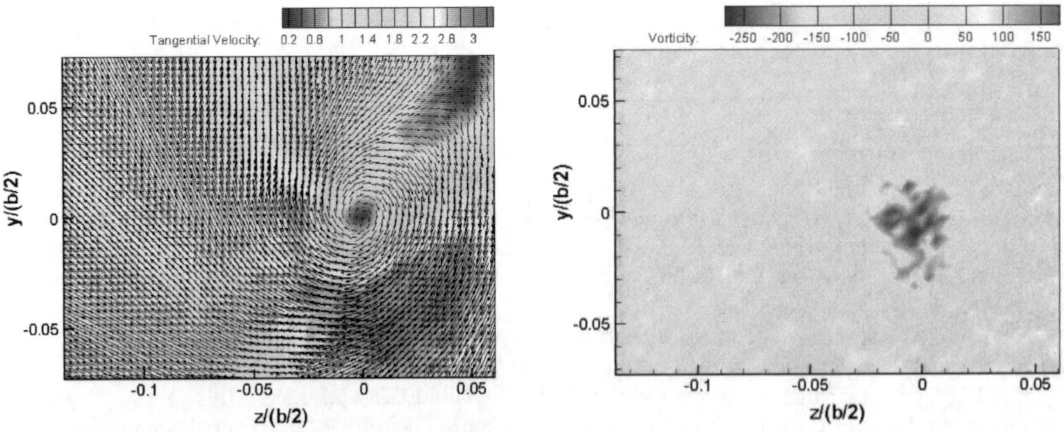

c) High Lift Configuration case, α=7.7°, x/(b/2)=0.548, V_∞=12 m/s.

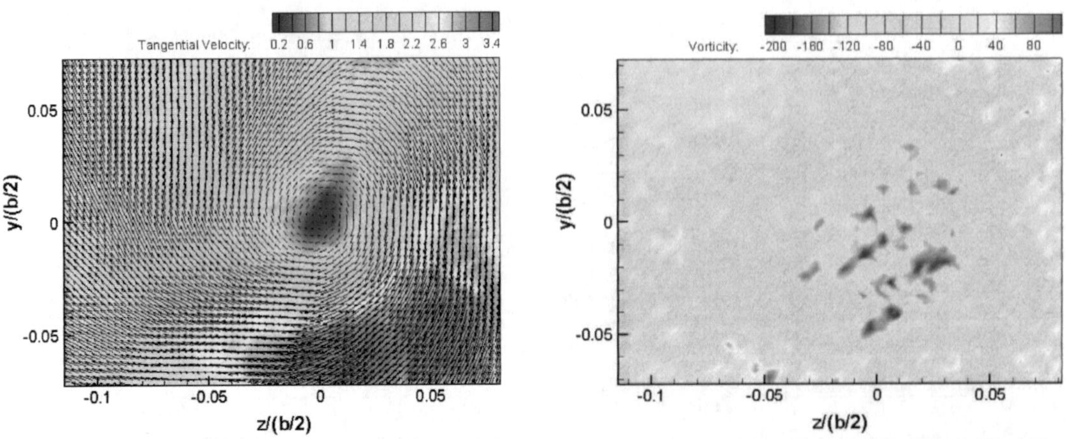

d) High Lift Configuration case with reverse delta type add-on device, α=9.7°; x/(b/2)=0.548, V_∞=12 m/s.

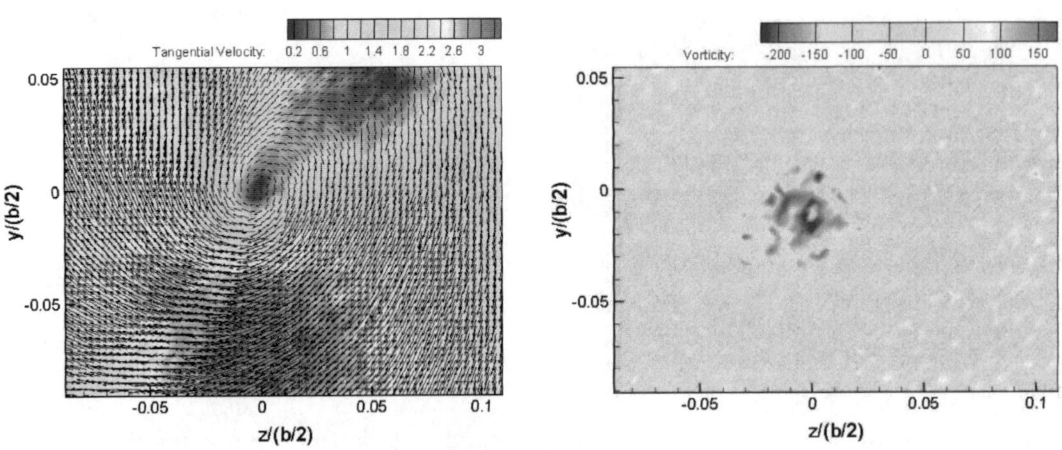

e) High Lift Configuration case, α=7.7°, x/c=1.075, V_∞=12 m/s.

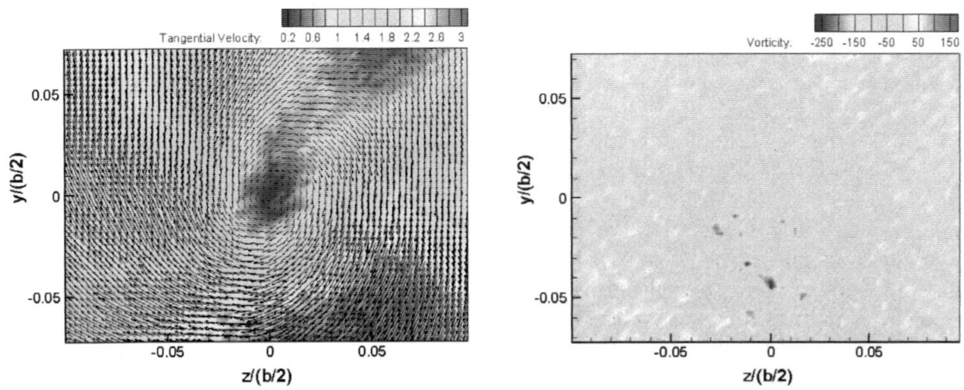

f) High Lift Configuration case with reverse delta type add-on device, α=9.7°; x/(b/2)=1.075, V_∞=12 m/s.

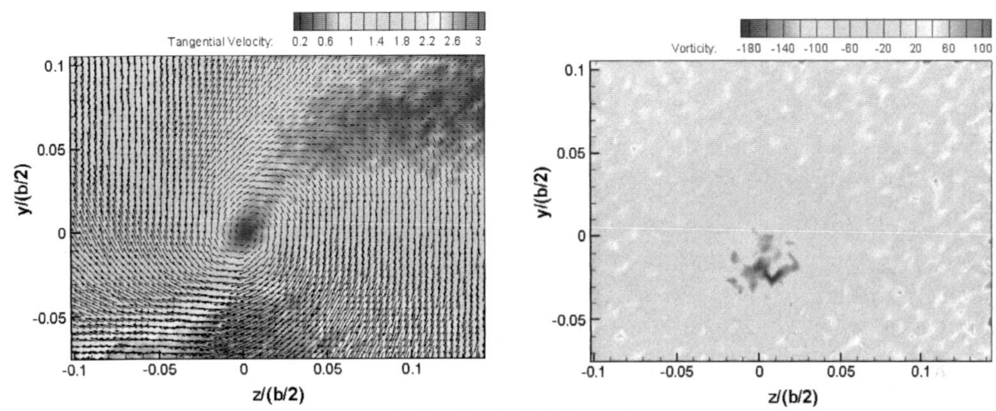

g) High Lift Configuration case, α=7.7°, x/c=2.387, V_∞=12 m/s.

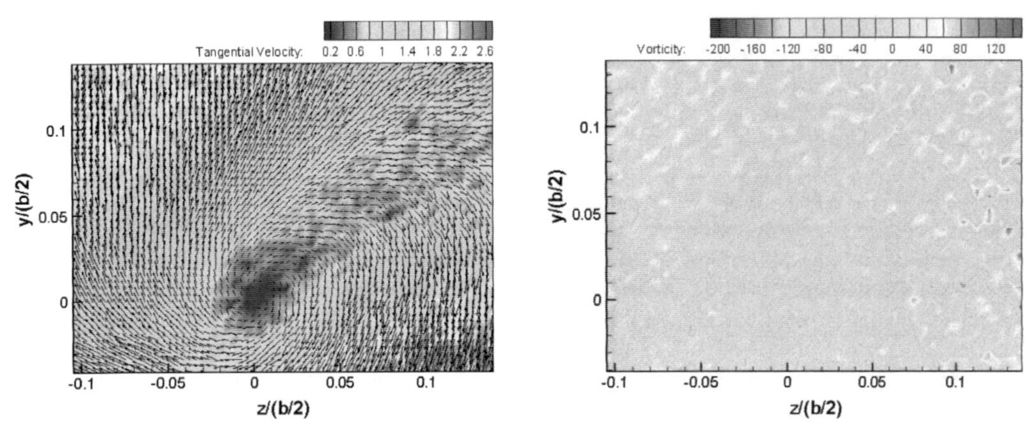

h) High Lift Configuration case with reverse delta type add-on device, α=9.7°; x/(b/2)=2.387, V_∞=12 m/s.

Figure 4. Tangential Velocity Magnitude and Vorticity Contours for High Lift Configuration α=7.7° and High Lift Configuration α=9.7° with reverse delta type add-on device at x/(b/2)=0.021, 0.548, 1.075 and 2.387.

Figs. 4a, 4c, 4e and 4g show the velocity vectors, tangential velocity magnitude and vorticity contours of the HLC flap-tip vortex. It is observable that the flap-tip vortex is not very strong. This is because the flap deflection is set to δ=20° only. A flap deflection of δ=20° will yield a weaker vortex than a flap deflection of δ=40° as less pressure differential exists. It is clearly noticeable that the flap-tip vortex is well rolled up at downstream plane x/(b/2)=0.548 but the vortex rollup is weaker (less compact vortex) at a farther downstream plane. The tangential velocity magnitude is seen to decrease slightly between downstream planes. Reynolds [15] attributed the reduction in the tangential velocity to the growth of the vortex in order to conserve momentum. As the vortex propagates through the fluid its volume increases with time due to the entrainment of the surrounding fluid.

The flap-tip vortex of the HLC without the add-on device exhibits a better rollup than the HLC with the add-on device, as shown in Figs. 4c, 4e and 4g. The flap-tip vortex of the HLC has diffused (enlarged) from x/(b/2)=0.548 to x/(b/2)=1.075 by 12.1% and from x/(b/2)=1.075 to x/(b/2)=2.387 by 20.8%. This increase in vortex core radius within 2 half-span lengths is significantly high. This indicates that the flap-tip vortex of the HLC is not very strong and hence, it has significantly diffused without any instabilities being introduced. When the reverse delta type add-on device is used, as shown in Figs. 4b, 4d, 4f and 4h, the vortex core radius is found to have diffused more than the HLC case. At x/(b/2)=0.548, the resultant vortex core radius compared to the flap-tip vortex core radius of the HLC has increased by 31.9% for the reverse delta type add-on device case. At x/(b/2)=1.075, the resultant vortex core radius compared to the flap-tip vortex core radius of the HLC has increased by 36.8% for the reverse delta type add-on device case. At x/(b/2)=2.387, the resultant vortex core radius compared to the flap-tip vortex core radius of the HLC has increased by 33.4% for the reverse delta type add-on device case. The significant increase in resultant vortex core radius indicates that the reverse delta type add-on device is capable of alleviating wake vortex strength.

When a reverse delta type add-on device is used the tangential velocity magnitude of the resultant vortex is seen to decrease slightly with respect to the HLC flap-tip vortex. It can be said that the add-on device reduces the tangential velocity magnitude of the resultant vortex which enables the resultant vortex to diffuse (weaken and enlarge) more rapidly than the HLC case.

For all investigated cases, the vorticity decreased gradually from a maximum at the centre to nearly zero at the outer region of the vortices. At x/(b/2)= 0.021, closer vorticity contours at the centre of the vortex are recorded, as shown in Figs. 4a-b. It can be noticed that the vortex cores are distinguishable. At x/(b/2)= 0.548, the vortex core for the HLC only is distinguishable (Fig. 4c), whereas the vorticity contours for the add-on device case (Figs. 4d, 4f and 4h) are broken into small vorticity patches. These small patches of vorticity have high individual vorticity magnitudes but they do not pose a hazard as they are too tiny to cause any significant circulation of the entire vortex core. At farther downstream locations, the tiny vorticity patches increase in number for the add-on device cases. This suggests that the vortex core is mostly diffused/scattered/distributed and hazard from such vortices is minimal. A rapidly diffusing vortex rejects vorticity from the vortex core. The rejection of vorticity reduces the circulation of the vortex and its strength [16]. For the HLC, the vortex core is only largely diffused at x/(b/2)= 2.387. Thus, it can be said that when the add-on device is used, the resultant vortex formed diffuses earlier and more rapidly than the HLC case.

The direct interaction of the vortices generated by the add-on device with the flap-tip vortex creates a resultant vortex with a weaker rollup, lower tangential velocity and larger vortex core. The vortex system with co-rotating and counter-rotating vortices experiences a lower tangential velocity due to instabilities created by the direct interaction of the co-rotating and counter-rotating vortices within the vortex system which lead to an enlarged and weaker resultant vortex [2]. The flap-tip vortex and the reverse delta type add-on device vortex in close proximity of the flap-tip are co-rotating vortices, as shown in Fig. 4b. These co-rotating vortices are very close to each other and at the point where the vortices meet; they have opposite direction of rotation. There is also a counter rotating vortex of the reverse delta type add-on device present. The interaction between these three vortices lowers the tangential velocity of the resultant (merged) vortex and causes it to diffuse rapidly, as shown in Figs. 4a-4h. It can be concluded that the use of a reverse delta type add-on device causes the resultant vortex to diffuse rapidly.

3.2 Tangential Velocity

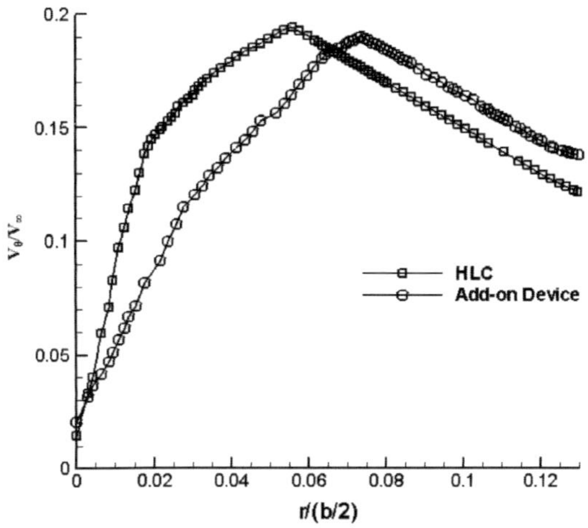

(a)

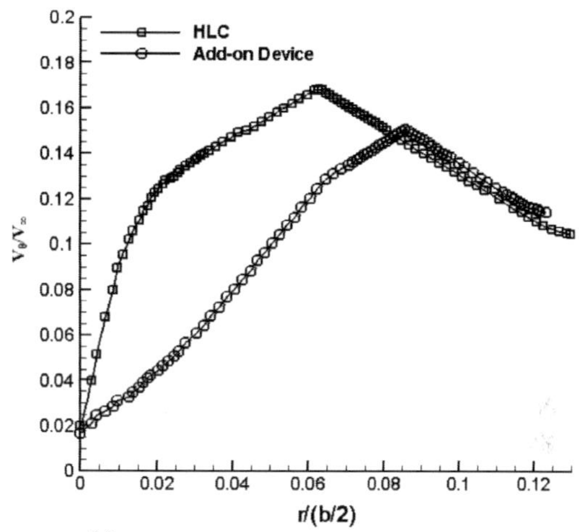

(b)

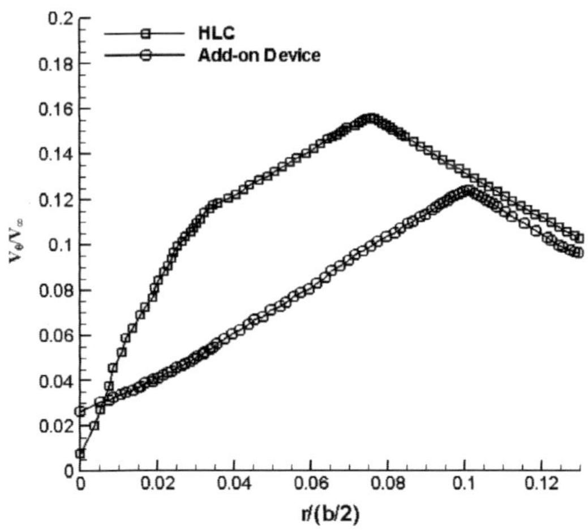

(c)

Figure 5. Non-dimensional tangential velocity distributions of HLC, and HLC with L-rdw at (a) x/(b/2)=0.548, (b) x/(b/2)=1.075 and (c) x/(b/2)=2.387.

The tangential velocity around the vortex centerline, V_θ, is calculated as $v_\theta(r) = \frac{1}{k}\sum_{i=1}^{k} v_{\theta,i}(y,z)\big|_{r=\sqrt{y^2+z^2}}$ where k is the number of points for each radius; $r = \sqrt{y^2+z^2}$; V_θ is normalized by the free stream velocity V_∞ and plotted versus the radial distance from the vortex centerline r, normalized by the half span (b/2); and r_c is defined as the core radius of the tip vortex, where maximum tangential velocity occurs [17].

From Figs. 5a-c, the reduction of tangential velocity at the vortex core for the HLC from x/(b/2)=0.548 to x/(b/2)=2.387 is 19.6%. Whereas, the tangential velocity reduction of the resultant vortex core from x/(b/2)=0.548 to x/(b/2)=2.387 for the reverse delta type add-on case is 36.1%. There is a 16.5% higher reduction in tangential velocity magnitude when the reverse delta type add-on device is used. This significant reduction in tangential velocity magnitude causes the vortex to diffuse more rapidly and hence, alleviate wake vortex strength.

From Figs. 5a-c, it is evident that the resultant vortex core radius has increased significantly when the reverse delta type add-on devices are used.

4. Conclusion

Reverse delta type add-on device was attached to a half wing model in the proximity of the outboard flap-tip at HLC and its resultant downstream vortex structures were obtained using Particle Image Velocimetry (PIV) in the IIUM closed-loop low-speed wind tunnel.

The vortex structures showed that there was a significant increase in the resultant vortex core radius and a significant reduction in the tangential velocity magnitude of the resultant vortex compared to the flap-tip vortex core radius at HLC.

The maximum tangential velocity reduction recorded between downstream plane 2 and downstream plane 4 for the HLC was 19.6% whereas the maximum tangential velocity reduction for the reverse delta type add-on device case was 36.1%.

The reverse delta type add-on device case is more favourable as its tangential velocity magnitude is lower and the resultant vortex is more diffused (weaker and larger vortex core radius) than the HLC case.

From the above observations, it can clearly be seen that the reverse delta type add-on device has several benefits. The add-on device results in a significant reduction in tangential velocity magnitude and a significant increase in the resultant vortex core radius which ensures that the vortex strength is significantly weakened. The weak resultant vortex will continue to decay rapidly and this will avoid trailing aircrafts to encounter strong swirling vortical flows. This permits aircraft spacing to be reduced, time between consecutive landings and take-offs to be reduced and aircraft capacity at airports to be increased.

Acknowledgement

The authors would like to thank the International Islamic University Malaysia for supporting this work through the Ministry of Science, Technology and Innovation (MOSTI) E-Science research grant no. SF13-001-005. The authors wish to acknowledge the support of the Ministry of Education, Malaysia, through the Fundamental Research Grant Scheme (FRGS/1/2014/TK01/MUSM/02/1). The authors would also like to thank Mr. Mohd Norhafiz Adnan and Mr. Firdauz Hakem Marof for their kind assistance.

References

[1] Coustols E (2006). An overview of European projects on wake vortices, Proceedings of the Workshop on "*Principles of Wake Vortex Alleviation Devices*", ONERA Toulouse, France, 2006.

[2] Coustols E, Jacquin L and Shrauf G (2006). Status of wake vortex alleviation in the frame work of European collaboration: validation attempts using tests and CFD results, *ECCOMAS CFD conference*, Technical University of Delft, The Netherlands, 2006.

[3] Arndt R E A, Arakeri V H and Higuchi H (1991). Some observations of tip vortex cavitation, *Journal of Fluid Mechanics*, Vol. 229, pp. 269-289, 1991.

[4] Ortega J M, Bristol R L and Savas O (2002). Wake Alleviation Properties of Triangular-Flapped Wings, *AIAA Journal*, 40(4), pp. 709–721.

[5] Rossow V (1991). Prospects for Alleviation of Hazard posed by Lift-generated Wakes, Proceedings of the *FAA Wake Vortices Conference*, Washington, D.C., 1991.

[6] Fabre D, Jacquin L and Loof A (2002). Optimal Perturbations in a Four-Vortex Aircraft Wake in Counter-Rotating Configuration, *Journal of Fluid Mechanics*, Vol. 451, pp. 319-328.

[7] Durston D A, Walker S M, Driver D M, Smith S C and Savas O (2005). Wake-Vortex Alleviation Flowfield Studies, *Journal of Aircraft*, 47(4), pp. 894-907.

[8] Savas O (2005). Experimental Investigations on Wake Vortices and their Alleviation, *Comptes Rendus Physique*, Academie des Sciences, Paris, 6(4-5), pp. 415-429.

[9] Corsiglia V R, Jacobsen R A and Chigier N (1971). An experimental investigatation of trailing vortices behind a wing with a vortex dissipater, *Aircraft wake turbulence and its detection*, Plenum Press, pp. 229-242, 1971.

[10] Patterson J C (1975). Vortex Attenuation Obtained in the Langley Vortex Research Facility, *Journal of Aircraft,* 12(9), pp. 745-749, 1975.

[11] Breitsamter C (2011). Wake vortex characteristics of transport aircraft, Progress in *Aerospace Sciences*, 47(10, 89-134, 2011.

[12] Lee T and Su Y Y (2012). Wingtip vortex control via the use of a reverse half-delta wing, *Experiments in Fluids*, DOI: 10.1007/s00348-012-1274-8, 2012.

[13] Lee T and Pereira J (2013). Modification of Static-wing tip vortex via a slender half-delta wing, *Journal of Fluids and Structures*, 2013.

[14] Ludin H, Omar A A and Asrar W (2015). Engineering Applications of Computational Fluid Dynamics, Advanced Structured Materials, Chapter No. 7, "Numerical Investigation of the Flow over Delta Wing and Reverse Delta Wing", Editor, Ku Zilati Ku Shaari and Mokhtar Awang, Springer International Publishing, pp85-101. ISBN: 978-3-319-02835-4, 2015.

[15] Reynolds O (1876). "On the Resistance Encountered by Vortex Rings and Relation between the Vortex Rings and the Streamlines of a Disk", *Nature*, Vol. 14, pp. 477-479.

[16] Maxworthy T (1972). The Structure and Stability of Vortex Rings, *Journal of Fluid Mechanics*, Vol. 51, pp. 15-32.

[17] Saffman P G (1978). The Number of Waves on Unstable Vortex Rings, *Journal of Fluid Mechanics*, Vol. 84, pp. 625-639.

Estimation of Stability Derivatives in Pitch for an Oscillating Wedge in Hypersonic Flow

Sher Afghan Khan[1a], Mohammed Asadullah[1b]

[1]Department of Mechanical Engineering,
Kulliyyah of Engineering, International Islamic University Malaysia,
53100 Kuala Lumpur, Malaysia

Email: [a]sakhan@iium.edu.my, [b]shiblibhai@gmail.com

Abstract. A similitude has been obtained for an oscillating two-dimensional wedge in pitch with attached shock at high angle of incidence in hypersonic flow. A strip theory, in which flow at infinite span wise location is two dimensional and independent of each other is being used. Further this unites the similitude with piston theory to give one dimensional piston theory with large flow deflection. Closed form solution is obtained for stiffness and damping derivatives in pitch. The present theory is valid when the shock wave is attached with the nose of the wedge. With the increase in semi vertex angle of the wedge the stiffness and the damping derivatives are found to increase progressively, and with the increase in the Mach number both stiffness and damping derivatives are found to decrease then with increase in Mach number it becomes independent of Mach number. From the amalgamation of theory developed, we can evaluate the stiffness and damping derivatives for in pitch for a wide range of Mach numbers, for various pivot positions, and angle of attack. Significantly the same results are being validated with the analytical results of Liu and Hui [4] and Lighthill's [3] theory for some cases.

1. Introduction

The analysis of hypersonic flow over wedge, cones or flat plate over wide range of angle of incidence and Mach number is of current interest in aerospace and aeronautical programs. At low hypersonic speeds, the molecular bonds vibrate, which changes the magnitude of the forces generated by the air on the aircraft. At higher hypersonic speeds, the molecules break apart producing electrically charged plasma around the aircraft. Large variations in air density and pressure occur because of shock waves, and expansion waves. Linearized theory was developed for the supersonic flow with the assumption of trailing edge and sharp leading edge at low angles of incidence. But these assumptions break down for similar bodies at hypersonic speeds. The hypersonic small disturbance theory is connected with classical hypersonic similitude but fails to provide hypersonic flow solution Binoy [5]. The idea of hypersonic similitude is due to Tsien [5], who analysed the 2-D and axisymmetric irrotational equation of motion. Lighthill [3] linked the unsteady problem to that of a gas flow in a tube driven by piston but gave solutions without considering the effects of secondary wave reflection. Lighthill developed a theory of oscillating airfoils at high Mach number. Here the pitching oscillation is taken into

IOP Conf. Series: Materials Science and Engineering **184** (2017) 012003 doi:10.1088/1757-899X/184/1/012003

consideration. Appleton [6] and McInthosh [7] have considered the effects of secondary wave reflection and considered the hypersonic flow past a slender wedge. Hui [8] analytically studied and gave exact formulae for stability of wedge of any thickness with attached bow shock for hypersonic flow. Sychev [1] considered 3-D hypersonic flow past bodies whose transverse dimensions are substantially smaller than their length and developed a number of general conclusions regarding the properties of 3D hypersonic flow past bodies at high angle of incidence. Ericsson's [9] theory covers viscous and elastic effects for airfoil with large flow deflection. Orlik–Ruckemann [10] has included viscous effect and Mandl [11] has addressed small surface curvature effect for oscillating thin wedges. Ghosh's [2] similitude and piston theory for the infinite span case with large flow deflection is valid for airfoil with planar or non-planar surfaces whereas Hui's [9] theory is for plane wedges. The method to analyze the aerodynamic derivatives for stiffness and damping of oscillating hypersonic delta wings with curved leading edge is developed by Crasta et al [12].

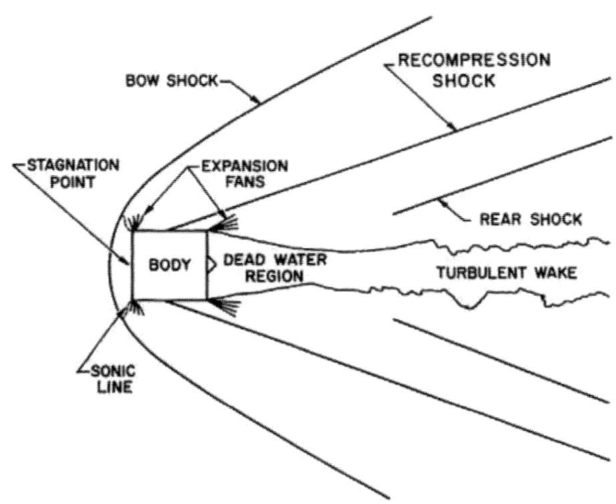

Fig. 1. Sketch of Hypersonic flow field on flat surface at M = 5.

This study is derive the mathematical expressions to enable us determine the stiffness and damping derivative for 2-D wedge. From the study, strip theory is used which combines with aerodynamic similitude to give the piston theory. With piston theory, we can evaluate the stiffness and damping derivatives in pitch for a range of hypersonic Mach number for various pivot positions and angle of incidence.

2. Experimental

2.1. Strip and Piston Theory

Ghosh's [2] similitude and piston theory for the infinite span case with large flow deflection is valid only for aerofoils with planar surfaces. In the present work the Ghosh similitude has been extended for supersonic flows past a 2D wedge. This method can give approximate result to determine the stability derivatives but is only valid for attached shock wave case. In the present study an attempt has been made to calculate the stability derivatives in pitch for an oscillating wedge in hypersonic flow for wide range of Mach number and angle of incidence. The same theory can be extended for supersonic flow as well.

$$\frac{P_1}{P_2} = 1 + \gamma\frac{U_p}{a_1} + \frac{\gamma(\gamma+1)}{4}\left(\frac{U_p}{a_1}\right)^2 + \frac{\gamma(\gamma+1)}{12}\left(\frac{U_p}{a_1}\right)^3 \quad (1)$$

The airfoil geometry and motion gives piston velocity U_p which is related to pressure p. since the piston Mach number M >>1, instead of the approximate expression of Lighthill [3] , the exact expression is used which can be written in quadratic form in pressure ratio,

$$\frac{P}{P_\infty} = 1 + AM_1^2(\sin\alpha)^2 + AM_1\sin\alpha\sqrt{B + M_1^2(\sin\alpha)^2} \quad (2)$$

Where, $A = \frac{\gamma(\gamma+1)}{4}$, $B = \left(\frac{4}{\gamma+1}\right)^2$

The stiffness and damping derivatives are, respectively:

$$-C_{m_\alpha} = \frac{1}{\frac{1}{2}\rho_1 U_1^2 L^2}\left(-\frac{\partial m}{\partial\alpha}\right) \quad (3)$$

and

$$-C_{m_q} = \frac{1}{\frac{1}{2}\rho_1 U_1 L^3}\left(-\frac{\partial m}{\partial q}\right) \quad (4)$$

Evaluated at $= \theta$ and $q = 0$, piston Mach number is equal to:

$$M_p = \frac{1}{\alpha_\infty}[U_\infty(\sin\alpha)^2 + (x - x_o)q] \quad (5)$$

Assuming that two flat plates have formed a sharp wedge, close form formulae for stiffness and damping coefficients have been obtained for it. Viscous effects and wave reflections are not taken into account in this study. Results are obtained for hypersonic flow of a perfect gas over the wedges of different semi-vertex angles, Mach numbers, and pivot positions.

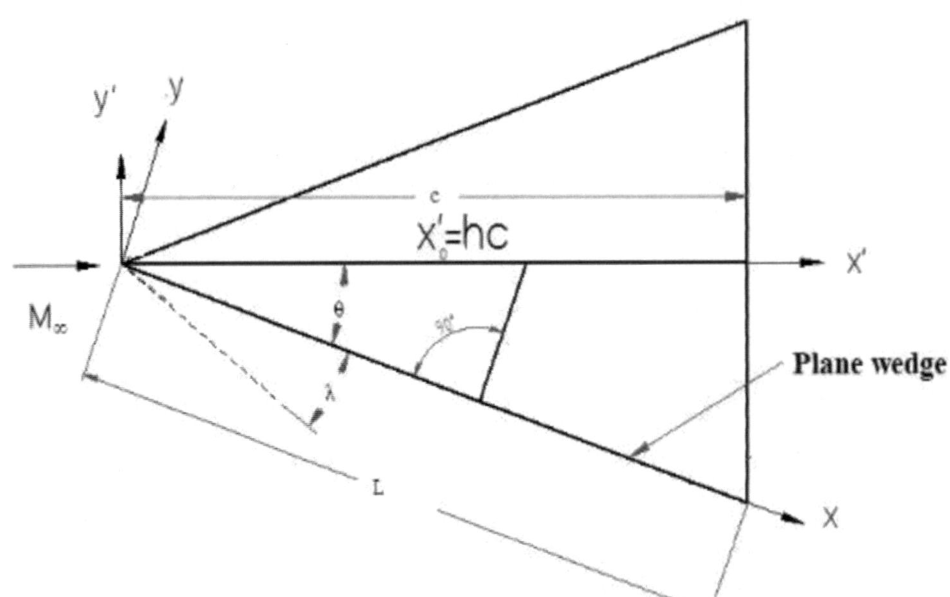

Fig. 2. Geometry of the wedge.

So in the flow past an airfoil at high Mach number, the perturbation and gradients are much larger in the lateral direction than those in the axial direction. Consider the flat plate aerofoil with the length

L at mean angle of incidence θ and oscillating in pitch with small amplitude about a pivot point 01 at a distance x0.

$$\frac{P}{P_\infty} = 1 + AM_\infty^2 \tan^2\alpha + AM_\infty \tan\alpha \sqrt{B + M_\infty^2 \tan^2\alpha} \tag{6}$$

where P_∞ is free stream pressure.
The aerodynamic stiffness is the coefficient of α and damping is of q in the nose down moment,

$$-m = \int_0^L (x - x_0)\, p dx \tag{7}$$

of the load distribution about the axis $x = x_o$. p is the pressure on the windward surface.
From equation (3), substituting the value of p into equation (4) and obtain

$$-m = \int_0^L (x - x_0)\, p_\infty \left[1 + AM_\infty^2 \tan^2\alpha + AM_\infty \tan\alpha \sqrt{B + M_\infty^2 \tan^2\alpha}\right] dx \tag{8}$$

Final stiffness derivative can be written as,

$$-Cm_{\alpha_0} = \left[\frac{(\gamma+1)}{M_\infty \cos\alpha_0 \cos\varphi}\right] F(S_1)\left(\frac{1}{2} - h\cos^2\alpha_0\right) \tag{9}$$

Final Damping derivative can be written as,

$$-Cm_q = \frac{(\gamma+1)}{M_\infty}\left[\frac{F(S_1)}{\cos\varphi \cos^3\alpha_0}\right]\left[\frac{1}{3} - h\cos^2\alpha_0 + h^2\cos^4\alpha_0\right] \tag{10}$$

To relate the piston velocity with pressure on the face of piston velocity with pressure on the face of piston, Light Hill suggested the use of three terms in the isentropic expression for the pressure on a piston as a power series in its velocity. So a condition that the piston velocity should be less than or equal to free stream sound velocity is imposed to satisfy isentropic assumption. This is consistent with hypersonic small disturbance theory on which Light Hill's (3) piston analogy is based.

3. Results and discussion

3.1. Comparison of Lighthill's Theory with the present theory

The Fig. and Fig. for Mach number 12 and 17 it is clearly shown to be valid up to angle of incidence of 15 degrees. The figure shows that the *Lighthill's* curve is rapidly moves away from the oblique shock results whereas the present theory shows a good accuracy throughout.

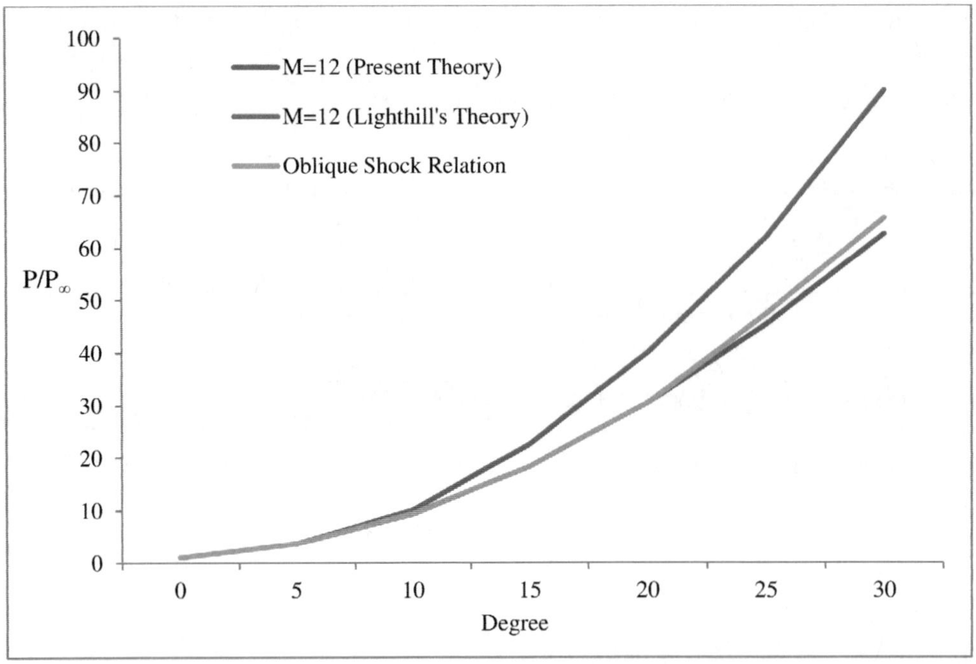

Fig. 3: Comparison graph of pressure ratio vs. angle of incidence for M = 12.

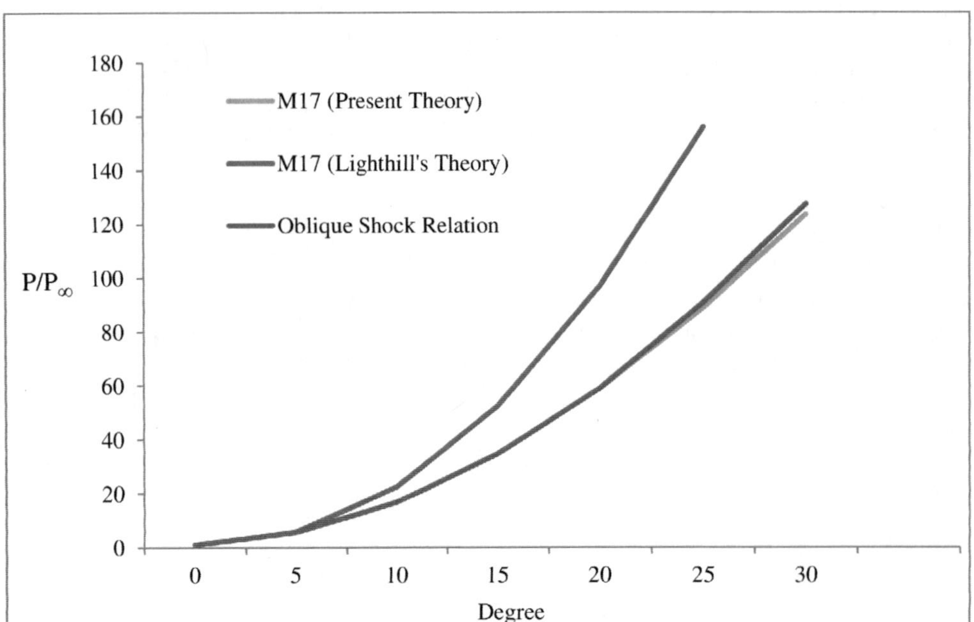

Fig. 4: Comparison graph of pressure ratio vs. angle of incidence for M = 17.

The stiffness coefficients and damping derivatives were also compared with the results of the oblique shock theory as shown in Fig. and Fig. and the present theory is found to be in a good agreement with oblique shock at M = 17.

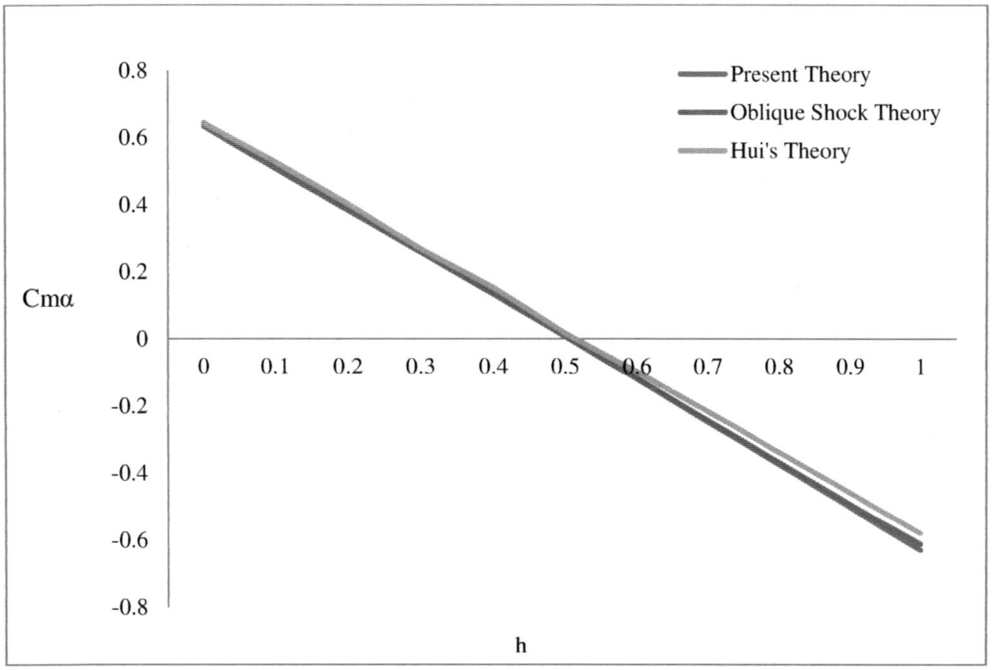

Fig. 5: Comparison graph of stiffness derivative vs. angle of incidence for M = 17.

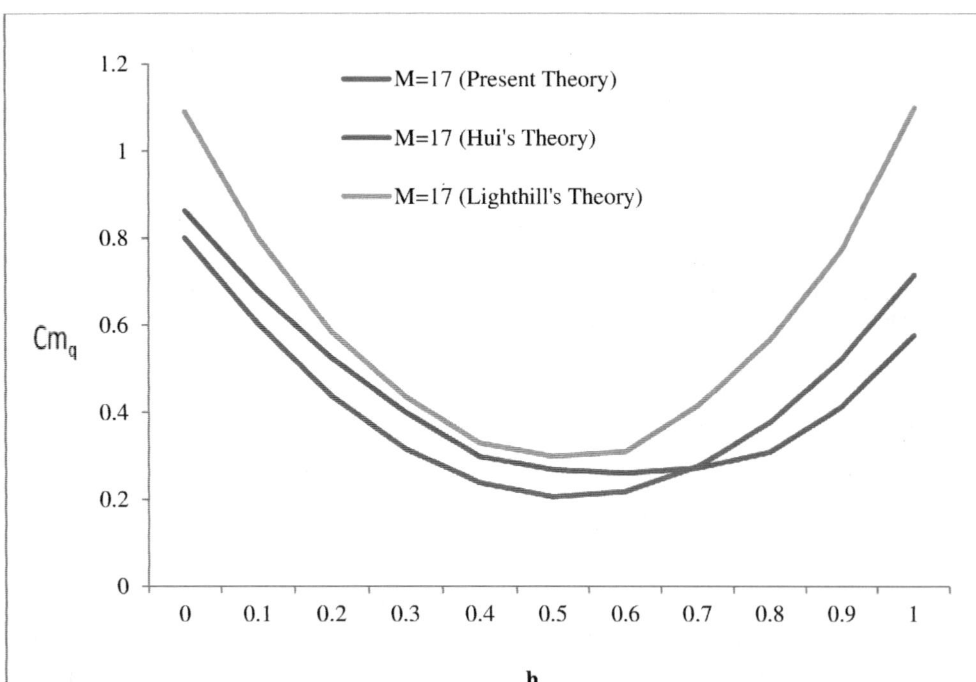

Fig. 6: Comparison graph of damping derivative vs. angle of incidence for M = 17.

4. Conclusion

Based on the analytical method, it is evident that with the increase in angle of attack at h = 0.0 there is a continuous increase in the values of the damping derivatives. It is also observed that the stiffness derivative decreases linearly with pivot position. In addition, it is seen that the stiffness derivatives are linearly increases with the increase of angle of incidence which is very true and the trend is on the expected lines. Further, from the graph another observation is made that with the increase in the angle of incidence there is a continuous shift of the centre of pressure towards the aft position of the wedge. Due to this phenomenon, the method of continuous increasing the angle of incidence could be used to stabilize the aerodynamic vehicle for some special cases; the requirement of large stabilizing surface could be avoided. The present theory is valid only when the shockwave is attached with the nose of the wedge. Meanwhile, it is observed that the pressure distribution is significantly large at h = 0 and it is increasing abruptly as the Mach number increased up to 20. At this stage, aerodynamic heating is the main concern as extreme pressure could lead to extreme temperature.

References

[1] Sychev V V 1960 Three Dimensional Hypersonic Gas Flow past Slender Bodies at High Angles of Attack *Journal of Applied Mathematics and Mechanics* **24.2** 296-306.

[2] Ghosh K 1968 Hypersonic Large Deflection Similitude for Oscillating Delta Wings *Aeronautical Journal* **88.858** 357-361.

[3] Light Hill M J 1953 Oscillating Aerofoil at High Mach Numbers *Journal of Aeronautical Sciences* **20.6** 402-406.

[4] Lui D D and Hui W H 1977 Oscillating Delta Wings with Attached Shock Waves *AIAA Journal* **15.6** 804-812.

[6] Tsien H S 1946 Similarity Laws of Hypersonic Flow Journal of *Math. Physics* 247-251.

[7] Appleton J P 1964 Aerodynamics Pitching Derivatives of a Wedge in Hypersonic Flow *AIAA Journal* **2.11** 1524-1530.

[8] Mc Inthosh S C Jr 1956 Studies in Unsteady Hypersonic Flow Theory *Ph.D. Dissertation Stanford University California*.

[9] Hui W H 1969 Stability of Oscillating Wedges and Caret Wings in Hypersonic and Supersonic Flows *AIAA Journal* **7.8** 1524-1530.

[10] Ericsson L E 1977 Viscous and Elastic Perturbation Effects on Hypersonic Unsteady Airfoil Aerodynamics *AIAA Journal* **15.10** 1481-1490.

[11] Orlik-Ruckemann, K J 1966 Stability Derivatives of Sharp Wedge in Viscous Hypersonic Flow *AIAA Journal* **4.6** 1001-1007.

[12] Mandl P 1970 Effect of Small Surface Curvature on Unsteady Hypersonic Flow over an Oscillating Thin Wedge *C. A. S. I. Transactions* 45-47.

[13] Asha Crasta and. Khan S A 2012 High Incidence Supersonic Similitude For Planar Wedge **21** 468–471.

Numerical simulation of non-isothermal thin liquid film flow on inclined plane using an implicit finite difference scheme

Mohammad Ameer Hamza[1*], Ahmad Tariq Jameel[2], Waqar Asrar[1] and Asif Hoda[3]

[1]Dept. of Mechanical Engineering, [2]Dept. of Biotechnology Engineering,
International Islamic University Malaysia, P.O. Box 10,
Gombak, 50728 Kuala Lumpur, Malaysia

[3]Dept. of Mechanical Engineering, Jubail University College, P. O. Box 10074
Jubail Industrial City 31961, Kingdom of Saudi Arabia

Corresponding author's e-mail address: mahamza2000@gmail.com

Abstract. The classical problem of the stability and dynamics of thin liquid films on solid surfaces has been studied extensively. Particularly, thin liquid films subjected to various physico-chemical effects such as thermocapillarity, solutal-Marangoni and evaporative instabilities at the film surface has been the focus of research for more than two decades. Various flow configurations of thin film such as thin film on plane, inclined, and wavy surfaces has been the subject of recent investigations. An inclined film compared to a horizontal film, also experiences the gravity force which may significantly influence the nonlinear dynamics of the film coupled with other forces. In this research, we attempt to study the stability and dynamics of thin liquid films subjected to thermocapillarity and evaporative instabilities at the free surface besides instability owing to ubiquitous van der Waals attraction, using numerical simulations. For a Newtonian liquid, flow in thin liquid film on a planar support and bounded by a passive gas, is represented by Navier-Stokes equation, equation of continuity and appropriate boundary conditions. The external effects are incorporated in the body force term of the Navier-Stokes equation. These governing equations are simplified using the so called long-wave approximation to arrive at a nonlinear partial differential equation, henceforth called equation of evolution (EOE), which describes the time evolution of the interfacial instability in the film caused by internal and/or external effects. Efficient numerical method is required for the solution of the equation of evolution (EOE) in order to comprehend the nonlinear dynamics of the thin film. Here we present the results of our numerical simulation using Crank-Nicholson implicit finite difference scheme applied to the thin film model incorporating instabilities owing to gravity, evaporation and thermo-capillarity. Comparison of our results with those obtained from Spectral method, show remarkable agreement for most of the cases investigated.

Keywords: Thin liquid film flow; Nonlinear Dynamics; Implicit Finite Difference; Spectral Method.

1. Introduction

The stability and dynamics of thin liquid films on solid substrates has been studied extensively for the past three decades. Particular attention has been focused on the case of thin liquid films subjected to various physical-chemical effects such as thermocapillarity, solutal Marangoni and evaporative instabilities at the surface. Furthermore, if the film is inclined at an angle to the horizontal, it experiences the gravitational force which might influence the nonlinear dynamics of the film coupled to other forces. Heat transfer in thin liquid film flow on an inclined plane occurs in numerous chemical and biochemical processes such as solid coating, food processing and in systems with evaporative cooling as well as high throughput exchange devices.

In the case of Newtonian fluid flow of thin liquid film on a solid support and bounded by a passive gas, the dynamics are represented by Navier Stokes equations and equation of evolution together with appropriate boundary conditions. The external effects are generally incorporated in the body force term of the Navier Stokes equations. These governing equations can then be simplified using the so called long wave approximation to arrive at fourth order nonlinear partial differential equation, henceforth called the equation of evolution which describes the time evolution of the interfacial instability caused by internal and/or external effects. The details of the derivation of the equation of evolution are available in the literature.

The linear stability characteristics can be obtained by the solution of the linearized equation of evolution. However, for the accurate simulations of the nonlinear dynamics and surface morphology of thin-film, we require efficient numerical methods for the solution of the equation of evolution (EOE). The extent of nonlinearity and the stiffness of the resulting differential equation depend upon the nature of various physico-chemical effects incorporated in the thin-film model [1-11]. Numerous works on numerical simulations of various liquid thin-film models have been reported in the literature [e.g., 12, 13, 14, 15]. Burelbach et al. (1988) [14] has solved the problem of evaporating/condensing liquid film on horizontal plane using an implicit finite difference scheme, Crank-Nicholson mid-point rule, by employing centered difference for space and forward difference for time derivatives. The resultant nonlinear difference equations are solved by Newton-Raphson iterations. Ali et al. (2005) [13] have employed similar discretization scheme for their isothermal liquid thin-film on inclined plane, however IMSL subroutines were employed for the solution of resulting difference equations which essentially uses an adoption of Newton-Raphson method. Fourier spectral method was used by Joo et al. (1991) [12] for numerical simulations of heated falling films under the influence of thermocapillarity and evaporative effects at the surface, however, they have ignored van der Waals effects in their formulation. There are many other works that have been reported in the literature.

The present paper illustrates some results from the numerical simulation of the flow of the thin liquid layer on inclined plane and subjected to evaporative and thermocapillarity instabilities. An implicit time averaged Crank Nicholson finite difference scheme formulation of the EOE has been used for numerical simulations. To validate our numerical codes of our implicit Crank Nicholson finite difference scheme we apply them to the solution of the equation of evolution of long-wave instabilities of heated falling film discussed by Joo et al. (1991) [12]. The results generated using our code are compared with the results reported by Joo et al. (1991) [12] which uses Fourier spectral method. The results from our implicit finite difference code closely match with the Fourier spectral thereby proving the viability of the implicit scheme as a reliable alternative to an explicit finite difference method. The validated solution procedure has thereafter been employed to study the time evolution of surface instabilities in a non-isothermal thin-film by incorporating the thermocapillarity and evaporation effects.

2. Thin film model

We assume a thin viscous Newtonian liquid film. The film is bounded above by its vapour and below by a uniformly heated rigid plane inclined at an angle θ to the horizontal. The film is sufficiently thick so that a continuum theory of the liquid is valid. The liquid film consists of an incompressible Newtonian fluid with constant material properties and is laterally unbounded. Evaporation occurs at the layer resulting in mass loss, momentum transfer and energy consumption. Plate has a fixed constant temperature T_H. Liquid temperature T_F on the free surface is controlled by losses to the passive gas above. The surface tension σ_o depends on temperature so that the thermocapillarity effects are present.

We consider a 2-D Cartesian coordinate system as shown in figure 1. The vapor=liquid interface is located at z=h(x,t). Liquid is assumed to evaporate in a direction normal to the interface. The Navier-Stokes and continuity equations in a non-dimensional Cartesian co-ordinate system (x, z) defined as in figure 1 are:

$$x: u_t + uu_x + wu_u = -p_x - \phi_x + Gsin\theta + u_{xx} + u_{zx} \quad (1)$$

$$z: w_t + uw_x + ww_z = -p_z - Gcos\theta + w_{xx} + w_{zz} \quad (2)$$

$$EOC: u_x + w_z = 0 \quad (3)$$

Where, the spatial coordinates are scaled by mean film thickness h_0; time by viscous time scale h_0^2. u and w are the components of velocity in x- and z- directions, respectively and p is the pressure. The subscripts denote the derivatives.

2.1 Equation of Evolution

The length scale of the initial instability in a thin film is much larger than the mean film thickness h_0. For flows with a characteristic length in the x-direction typically proportional to the disturbance wavelength and much larger than the thickness of the liquid layer, the long wave analysis of Benney (1966) [15] can be applied to the problem at hand. Following the derivation in Joo et. al. (1991) [2], the long wave approximation of our system of equations result in an equation of evolution (in space & time for the film interface h(x, t) in original coordinates (x, z) given by

$$\begin{aligned} h_t &+ \frac{E}{h+K} + Gh^2h_x sin\theta \\ &+ [\frac{2}{15}G^2h^6h_x sin^2\theta + \frac{KM}{P}\frac{h^2h_x}{(h+K)^2} + \frac{E^2}{D}\frac{h^3h_x}{(h+K)^3} - \frac{1}{3}Gh^3h_x cos\theta + \frac{A}{h}h_x \\ &+ Sh^3h_{xxx}]_x + \frac{5}{24}GE(\frac{h^4}{h+K})_x sin\theta \\ &+ EP(\frac{h}{h+K})^3\left[\frac{E}{3(h+K)} + \frac{G}{120}(7h-15K)hh_x sin\theta\right] = 0 \end{aligned} \quad (4)$$

2.2 EOE by Joo et. al. (1991)

The EOE of Joo et. al. is [12]

$$\begin{aligned} h_t &+ \frac{\bar{E}}{h+K} + Gh^2h_\xi sin\theta \\ &+ \varepsilon[\frac{2}{15}G^2h^6h_\xi sin^2\theta + \frac{KM}{P}\frac{h^2h_\xi}{(h+K)^2} + \frac{E^2}{D}\frac{h^3h_\xi}{(h+K)^3} - \frac{1}{3}Gh^3h_\xi cos\theta + \frac{A}{h}h_\xi \\ &+ \bar{S}h^3h_{\xi\xi\xi}]_\xi + \frac{5}{24}\varepsilon GE(\frac{h^4}{h+K})_\xi sin\theta \\ &+ \varepsilon EP(\frac{h}{h+K})^3\left[\frac{E}{3(h+K)} + \frac{G}{120}(7h-15K)hh_\xi sin\theta\right] = 0 \end{aligned} \quad (5)$$

The following scale factors are used: ξ=εx, ζ=z, τ=εt

IOP Conf. Series: Materials Science and Engineering **184** (2017) 012004 doi:10.1088/1757-899X/184/1/012004

$(E, D, S) = (\varepsilon\bar{E}, \varepsilon^2 D, \varepsilon^{-2}\bar{S})$. Joo et. al. have employed ε=0.2 and $\bar{S}$=0.1 throughout their simulations.

3. Numerical method

We have employed an implicit finite difference numerical discretization scheme discussed below.

3.1 Implicit Finite Difference Scheme-Crank Nicholson Mid Point Rule

We use the conservative form of the rescaled equation of evolution. This equation of evolution is discretized using forward difference in time and central difference in space with time averaging, resulting into a set of difference equations for the film thickness h. These equations are then solved as a set of nonlinear coupled algebraic equations by an iterative procedure of the type of Newton's method. Let subscript i denotes space discretization and superscripts n and n + 1 denote values of variables at nth and (n + 1)th times. Thus, Eq. (5) after discretization appears as:

$$
\begin{aligned}
&\frac{h_i^{n+1}}{\Delta\tau}+\frac{S}{2A^2}\left[\left(\frac{E}{h_i+k}\right)^{n+1}+\left(\frac{E}{h_i+k}\right)^{n}\right]+(\frac{S}{A})^{1/2}\frac{Re}{A}\frac{1}{4\Delta x}[\{{h_i}^2(h_{i+1}-h_{i-1})\}^{n+1}\\
&\quad+\frac{1}{2x}[\frac{2}{15}\frac{Re^2}{A}\frac{1}{4\Delta x}[\{h_{i+1}^6(h_{i+2}-h_i)-h_{i-1}^6(h_i-h_{i-1})\}^{n+1}\\
&\quad+\{h_{i+1}^6(h_{i+2}-h_i)-h_{i-1}^6(h_i-h_{i-1})\}^{n}]\\
&\quad+\frac{KM}{PA}\frac{1}{4\Delta x}[\left\{\frac{h_{i+1}^2}{(h_{i+1}+K)^2}(h_{i+2}-h_i)-\frac{h_{i-1}^2}{(h_{i-1}+K)^2}(h_i-h_{i-2})\right\}^{n+1}\\
&\quad+\left\{\frac{h_{i+1}^2}{(h_{i+1}+K)^2}(h_{i+1}-h_i)-\frac{h_{i-1}^2}{(h_{i-1}+K)^2}(h_i-h_{i-2})\right\}]^{n}\\
&\quad+\frac{E^2}{DA}\frac{1}{4\Delta x}\left[\left\{\frac{h_{i+1}^3(h_{i+2}-h_i)}{(h_{i+1}+K)^3}-\frac{h_{i-1}^3(h_i-h_{i-2})}{(h_{i-1}+K)^3}\right\}^{n+1}\right.\\
&\quad\left.+\left\{\frac{h_{i+1}^3(h_{i+2}-h_i)}{(h_{i+1}+K)^3}-\frac{h_{i-1}^3(h_i-h_{i-2})}{(h_{i-1}+K)^3}\right\}^{n}\right.\\
&\quad-\frac{1}{3}\frac{Gcos\theta}{A}\frac{1}{4\Delta x}\left[\{h_{i+1}^3(h_{i+2}-h_i)-h_{i-1}^3(h_i-h_{i-2})\}^{n+1}\right.\\
&\quad\left.+\{h_{i+1}^3(h_{i+2}-h_i)-h_{i-1}^3(h_i-h_{i-2})\}^{n}\right]\\
&\quad+\frac{1}{4(\Delta x)^3}\left[\{h_{i+1}^3(-h_{i-1}+2h_i-2h_{i+2}+h_{i+3})\right.\\
&\quad-h_{i-1}^3(-h_{i-3}+2h_{i-2}-2h_i+h_{i+1})\}^{n+1}\\
&\quad\left.+\{h_{i+1}^3(-h_{i-1}+2h_i-2h_{i+2}+h_{i+3})-h_{i-1}^3(-h_{i-3}+2h_{i-2}-2h_i+h_{i+1})\}^{n}\right]\\
&\quad+\frac{1}{4\Delta x}[\{(h_{i+1})^{-1}(h_{i+2}+h_i)-(h_{i-1})^{-1}(h_i+h_{i-2})\}^{n+1}\\
&\quad\left.-\{(h_{i+1})^{-1}(h_{i+2}+h_i)-(h_{i-1})^{-1}(h_i+h_{i-2})\}^{n}]\right]\\
&\quad+\frac{5}{24}\left(\frac{S}{A^3}\right)^{1/2}\frac{Re}{4\Delta x}[\left\{\frac{h_{i+1}^4}{h_{i+1}+K}-\frac{h_{i-1}^4}{h_{i-1}+K}\right\}^{n+1}+\left\{\frac{h_{i+1}^4}{h_{i+1}+K}-\frac{h_{i-1}^4}{h_{i-1}+K}\right\}^{n}\\
&\quad+\left(\frac{S}{A^3}\right)^{1/2}\frac{EP}{2}\left[\left\{\left(\frac{S}{A}\right)^{1/2}\frac{Eh_i^3}{3(h_i+K)^4}+\frac{Re}{120}\frac{(7h_i-15K)h_i^4}{(h_i+K)^3}\left(\frac{h_{i+1}-h_{i-1}}{2\Delta x}\right)\right\}^{n+1}\right.\\
&\quad\left.+\left\{\left(\frac{S}{A}\right)^{1/2}\frac{Eh_i^3}{3(h_i+K)^4}+\frac{Re}{120}\frac{(7h_i-15K)h_i^4}{(h_i+K)^3}\left(\frac{h_{i+1}-h_{i-1}}{2\Delta x}\right)\right\}^{n}\right]=0
\end{aligned}
\tag{6}
$$

4. Results and discussions

The results of the numerical solution of the evolution equation of Joo et. al. (1991) using implicit method is compared with the numerical results of Joo et. al. (1991) obtained using Fourier spectral method. We have incorporated the various thermophysical effects from Joo et. al. (1991) model in our thin film model, with the exception of van der Waals effects. Figure (2) shows the evolution of film interface for an isothermal film. Our implicit finite difference method gives almost identical results with Joo et. al. (1991) with only a slight deviation. Figure (3) shows the evolution of liquid layers with thermocapillarity in the absence of gravity. Significant deviation is observed both in the film rupture time as well as film profile.

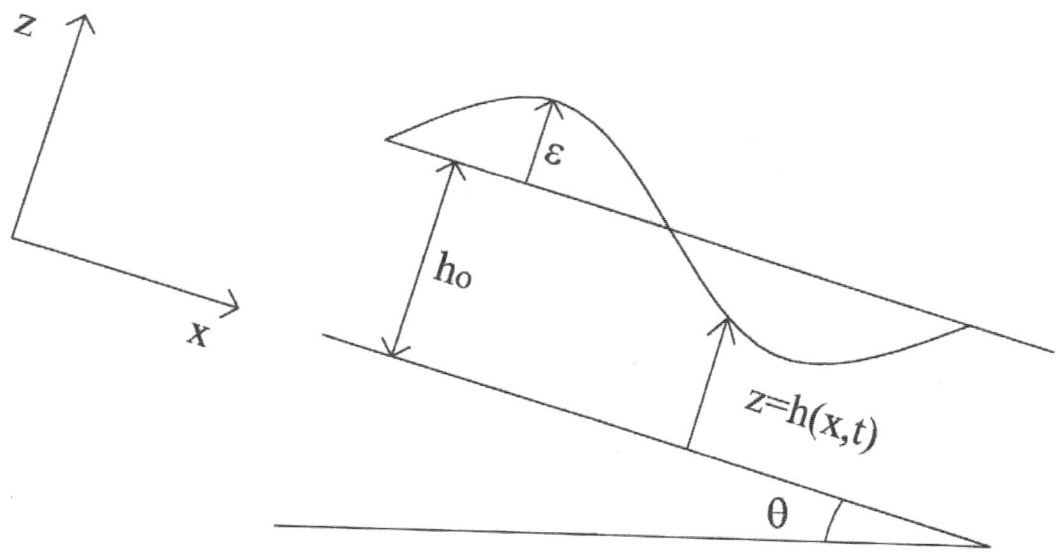

Figure 1

IOP Conf. Series: Materials Science and Engineering **184** (2017) 012004 doi:10.1088/1757-899X/184/1/012004

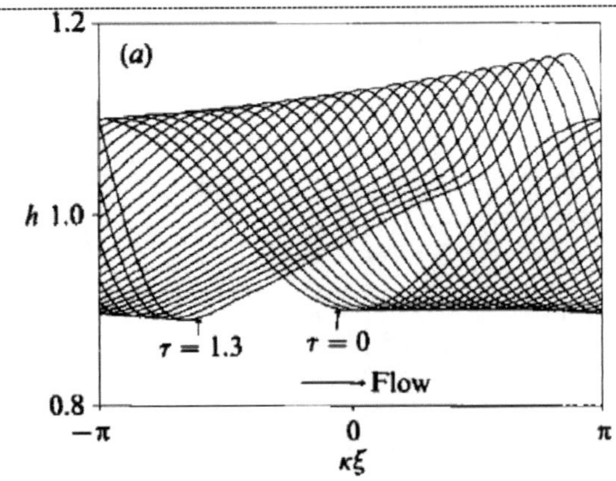

Figure 2(a):Spectral method (Joo et. al. 1991)

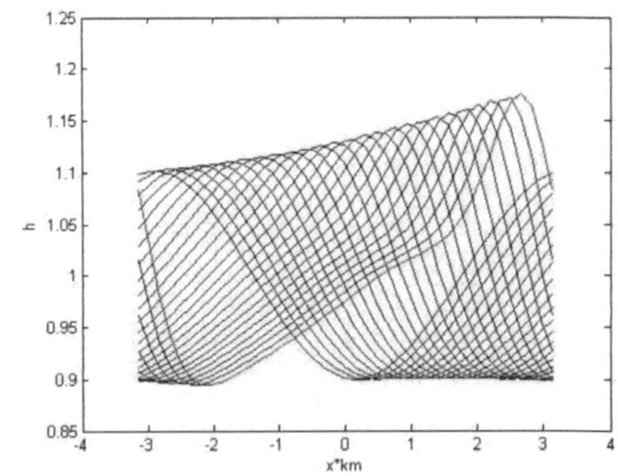

. Figure 2(a): Implicit FDM

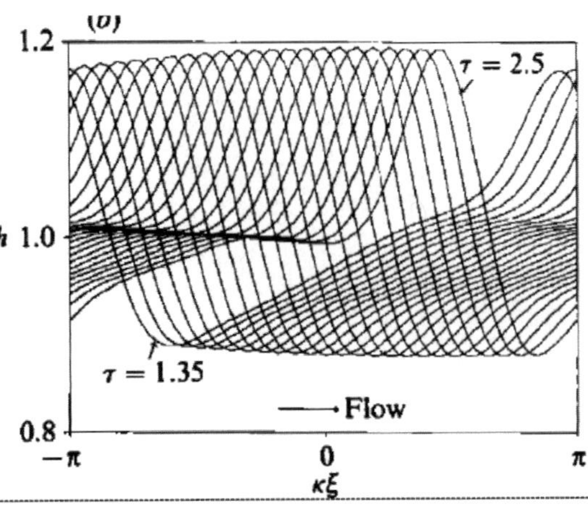

Figure 2(b): Spectral method (Joo et. al. 1991)

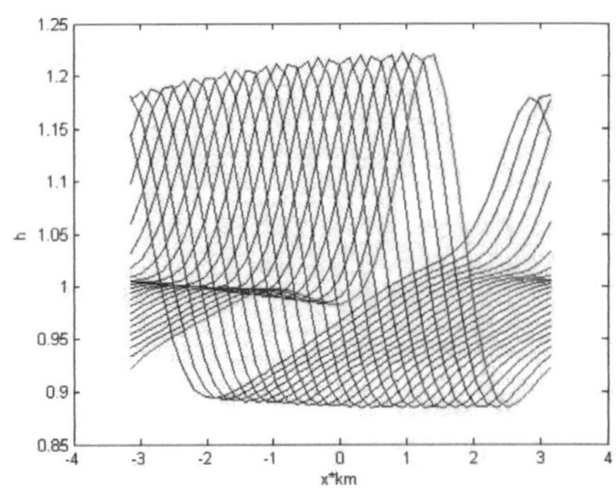

Figure 2(b): Implicit FDM

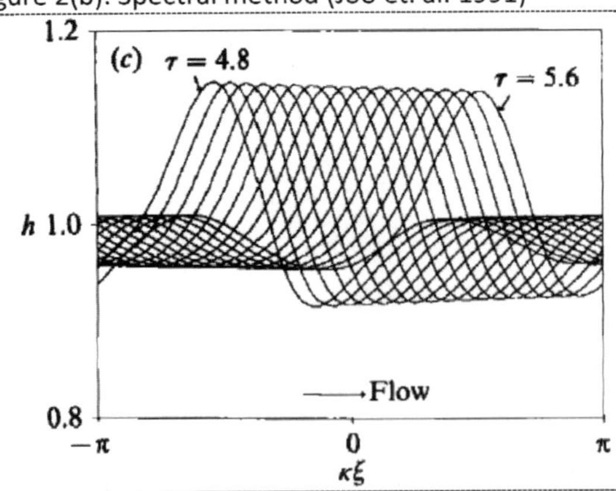

Figure 2(c):Spectral method (Joo et. al. 1991)

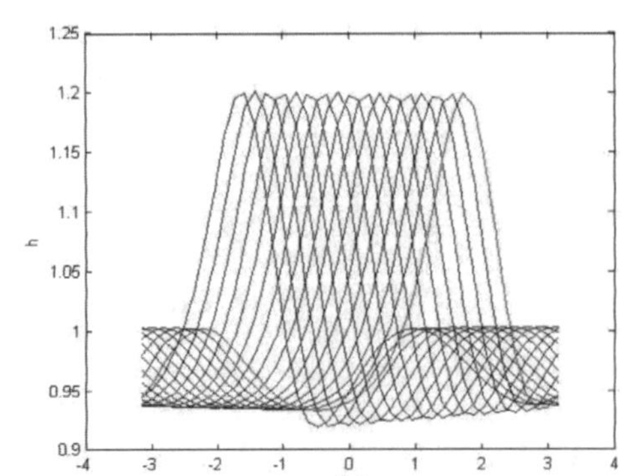

Figure 2(c): Implicit FDM

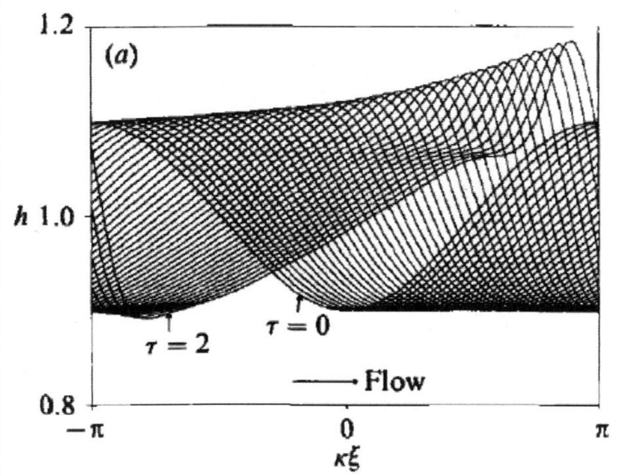

Figure 3: Spectral method (Joo et. al. 1991)

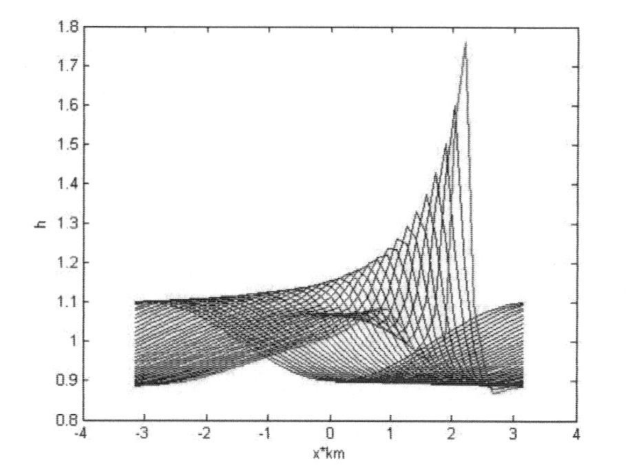

Figure 3: Implicit FDM

5. Conclusion

The results obtained by our implicit Crank Nicholson finite difference method demonstrate a close conformity to the results obtained by Fourier spectral method of Joo et. al. (1991), with some slight deviation in some cases and significant deviation in one of the cases investigated. This indicates that the implicit finite difference method is a reliable method for investigating the highly nonlinear thin film model to a high degree of accuracy.

Figure Captions

Figure 1: The physical configuration of thin film flowing down an inclined plane

Figure 2: Surface wave instability for an isothermal layer with k=1.11, G=5, $\beta=45^0$, $\bar{S}=0.1$. (a) Free surface configuration for $0 \leq \tau \leq 1.3$ with $\Delta\tau=0.05$; (b) free surface configurations for $1.35 \leq \tau \leq 2.5$ with $\Delta\tau=0.05$; (c) free surface configurations for $4.8 \leq \tau \leq 5.6$;

Figure 3: Free surface configuration of gravitational instability for an isothermal layer with k=0.7, G=5, $\beta=45^0$ and $\bar{S}=0.1$ with $\Delta\tau=0.05$ up to $\tau=2$.

References

[1] Huppert, H. E., and J. E. Simpson, "The slumping of gravity currents," 1980, J. Fluid Mech. **99,** 785-789

[2] Huppert, H. E., "The propagation of two-dimensional and axisymmetric viscous gravity currents over a rigid horizontal surface," 1982a, J. Fluid Mech. **121,** 43-58.

[3] Grotberg, J. B., "Pulmonary flow and transport phenomena," 1994, Annu. Rev. Fluid Mech. **26,** 529-571.

[4] Sharma, A., and E. Ruckenstein, "An analytical nonlinear theory of thin film rupture and its application to wetting films," 1986a, J. Colloid Interface Sci. **113,** 456-479.

[5] Wong, H., I. Fatt, and C. J. Radke, 1996, "Deposition and thinning of the human tear film," J. Colloid Interface Sci. **184,** 44-51.

[6] Gaines, G. L., 1966, *Insoluble monolayers at liquid-gas interfaces* (Interscience, New York).

[7] Bikerman, J. J., 1973, *Foams* (Springer, New York).

[8] Edwards, D. A., H. Brenner, and D. T. Wasan, 1991, *Interfacial Transport Phenomena* (Butterworth Heineman, Boston).

[9] Schramm, L. L., and F. Wassmuth, 1994, "Foams: basic principles," in *Foams: Fundamentals and Applications in the PetroleumIndustry*, edited by L. L. Schramm (American Chemical Society, Washington), pp. 3-46.

[10] Wasan, D. T., K. Koczo, and A. D. Nikolov, 1994, "Mechanisms of aqueous foam stability and antifoaming action with and without oil: a thin film approach," in *Foams: Fundamentalsand Applications in the Petroleum Industry*, edited by L. L. Schramm (American Chemical Society, Washington), pp. 47-114.

[11] Wong, H., C. J. Radke, and S. Morris, "The motion of long bubbles in polygonal capillaries. Part 1. Thin films," 1995, J. Fluid Mech. **292,** 71-94.

[12] S. W. Joo, S. H. Davis, S. G. Bankoff, "Long wave instabilities of heated falling films: Two-dimensional theory of uniform layers," 1991, J. Fluid. Mech. **230,** 117-146.

[13] M. A. Ali, A. T. Jameel, F. R. Ahmadun, "Stability and rupture of nano-liquid film (NLF) flowing down an inclined plane," 2005, Computers and Chemical Engineering, **29**, 2144–2154.

[14] J. P. Burelbach, S. G. Bankoff, "S. H. Davis, Nonlinear stability of evaporating/condensing liquid films," 1988, J. Fluid Mech. **195**, 463–494.

[15] M. B. Williams, S. H. Davis, "Nonlinear theory of film rupture," 1982, Journal of Colloid and Interface Science. **90**, 220-228.

Investigation of effect of process parameters on suddenly Expanded flows through an axi-symmetric nozzle for different Mach Numbers using Design of Experiments

Jaimon D. Quadros 1, S. A. Khan2, Antony A. J.3

[1]Department of Mechanical Engineering, Birla Institute of Technology, Offshore campus, Ras- Al-Khaimah, UAE

[2]Department of Mechanical Engineering, International Islamic University Malaysia, Kuala Lampur, Malaysia

[3]Department of Mechanical Engineering, P. A. College of Engineering, Mangalore, India

Email: jaimonq@gmail.com

Abstract. Experiments are conducted to determine base pressure variation through micro-jets from a suddenly expanded axisymmetric passage. Four micro-jets having an orifice diameter of 1mm are situated at 90^0 interims along the base at 6.5 mm from the geometrical axis of the main jet. The Mach number and L/D ratios were the process parameters employed in the study. The flow stream was extended all of a sudden into an axi-symmetric duct of 4.84 cross-sectional area for all the Mach numbers and L/D ratios respectively. The test Mach numbers used in the study was 2, 2.5 and 3; length-to-diameter ratios selected for the sudden expansion tube were 4, 6 and 8. The jets were operated at an overexpansion level of (Pe/Pa = 0.277). The experiments are conducted as per Taguchi design of experiments. From this investigation, one will be able to identify the enlargement length to diameter ratio resulting in maximum increasing or decreasing base pressure. Mathematical models are also developed for base pressure based on Mach number and L/D ratio with a maximum error of ± 10%.

1. Introduction

In the area of Space flights and missile technology, base flows are deemed to be a core area of research due to high Reynolds numbers. Following these, the concern has been drifted towards hypersonic speed regime in prospect of heat transfer at the base and near-wake structure. However the understanding of base flows is incomplete due to the presence of strong pressure gradient and insufficient turbulence knowledge. Activated fundamentally by the necessities prevailing in technological developments, various inquisitive studies have been proclaimed in literature dedicated towards lowering the base drag penalty by implementing both active and passive techniques. Practically, these aspire in controlling the near-wake flow-field for the development of base pressure. The experimental studies involving sudden expansion with internal flows have various benefits when compared to external flows. The volume of air supply needed is reduced by discarding the requirement for tunnel with a cross-section considerable enough such that flow over the model is not disturbed by wall interference. The static and surface pressure measurements are made not only through the entrance section; but also through the wake region. Because of such extensive applicability, suddenly expanded flows have been studied broadly [1]. Many researchers have made an attempt to control the pressure in wake area with passive control. However, to the best of the authors' knowledge there has been no work reported with active control of base pressure as per Taguchi design of experiments for high inertia levels and L/D ratio as control variables. Therefore, the present study involves control of base pressure control by active and passive means in the micro jets form under favorable pressure gradient as per Taguchi design of experiments.

Axi-symmetric expanded flow is a complicated aspect identified by the flow separation, flow recirculation and reattachment. This type of flow may be classified into two regions separated by a shear layer namely i) region of flow recirculation and ii) region of main flow [2]. The point where the segregated streamline touches the wall is called as reattachment point. The salient features of the flow field have been illustrated in Fig. 1. There is abundant data available in the case of sudden expansion problem which has been registered previously. The boundary layer effect was studied by experimentally by Wick [3], where the boundary layer was considered as the source of mass flow for corner flow. It was recognized that the pressure developed at the expansion corner was directly relevant to the expansion thickness upstream and boundary layer. However Hoerner [4] concluded that the layer of the boundary acts as insulating air thereby reducing the efficacy of the jet as pump. He finally concluded that the phenomenon of base pressure can be experimentally studied by use of internal flow. Khan et. al. [5-9] carried out experiments in order to study the response of micro jets influencing over, under and correctly expanded nozzles for controlling base pressure in axisymmetric ducts that were suddenly expanded. It was concluded that the micro jets do not adversely affect wall pressure distribution. Furthermore, it was also concluded that the NPR has a definitive role in controlling the variation the base pressure for both cases i.e. with and without control. Durst et. al. [10] concluded that symmetric flows exist in a 2-D plane for a limited range of Reynolds numbers. The higher Reynolds number generated minute disturbances at the tip of the sudden expansion which become amplified between the main flow and recirculation flow. Rathakrishnan et. al. [11] studied flows in pipe that were subjected to sudden enlargement. It was concluded that non-dimensional base pressure is an active function of expanded area ratios, NPR and L/D ratios of ducts. Thus for a given nozzle and enlargement area ratio, the duct has to have a minimum value in order to obtain a minimum value for base pressure. Thus for flows with sudden expansion, the total pressure also has to be taken into account along with base pressure minimization. From previous literature as stated above, it is evident that suddenly expanded flows with control can be useful in understanding many practices. It can be of incredible help for rocket and space program applications for thoroughly understanding the behavior of base pressure and a methodology which can be formulated for controlling base pressure. This is because; control of base pressure will result in the development of either increment or decrement of base pressure. Thus base pressure is minimized in case of a combustion chamber for maximizing the mixing, and maximized for the case of rockets and missiles for the purpose of base drag reduction.

2. Experimental set-up

Fig. 2 depicts the experimental setup used for the present investigation. Eight holes are marked at the exit rim of the nozzle. Out of these four are marked ‘c’ utilized for blowing; four are marked ‘m’ and utilized for measuring base pressure. The pressure at the base region was controlled by blowing through ‘c’ marked gapes by means of pressure that was generated in the settling chamber. The experiments were administered for Mach numbers 2.0, 2.5 and 3.0. Since active controls are used in the present study, L/D ratios greater than or equal to 10 can be employed. However for this particular study L/D ratios of 4, 6 and 8 have been employed. The results presented here are for an over expanded case of ($P_e/P_a = 0.277$) and therefore the corresponding NPR’s for Mach number 2.0, 2.5 and 3.0 are 2.16, 4.64 and 10 respectively. A PSI system 2000 pressure transducer was implemented for base pressure measurement. It consisted of 16 channels with optimum pressure ranging from 0 to 300 psi. It displayed reading by averaging 250 samples per second.

3. Plan of experiments

Taguchi design of experiments for two factors at three levels was used for exercising the experiment. The orthogonal array selected was L9 (23) as shown in Table 1. The experiments were carried out considering Mach number and L/D ratio for a common expansion area ratio of 4.84. Each experiment has been repeated for two times and average values of base pressure have been recorded.

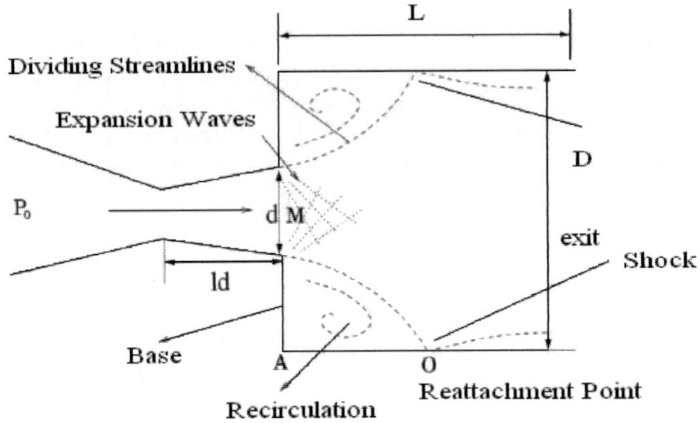

Fig. 1. Sudden expansion flow field.

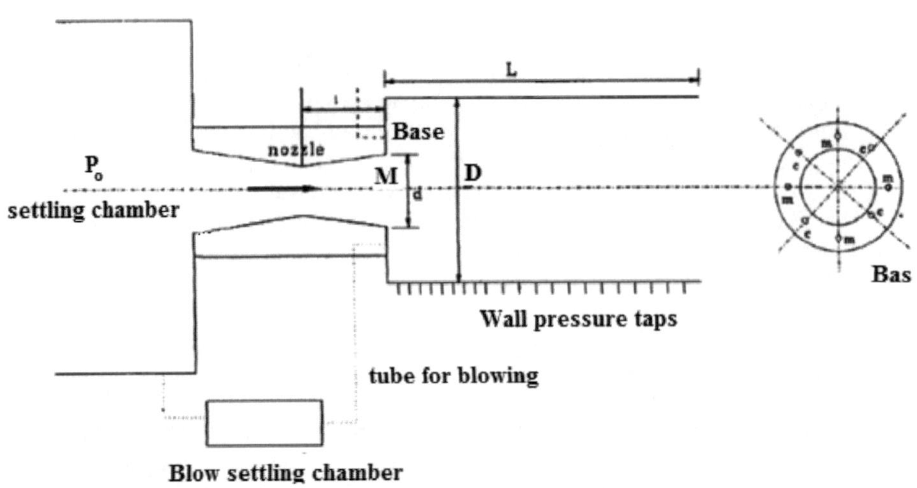

Fig. 2. Experimental Set up.

Table 1. Allocation of factors an $L_9(2^3)$ Orthogonal Array.

Experiment No.	Mach number (M)	L/D Ratio	Non-dimensional base pressure without control (P_b/P_a)	Non-dimensional base pressure with control (P_b/P_a)
1	2.0	4	0.808	0.806
2	2.0	6	0.795	0.793
3	2.0	8	0.802	0.802
4	2.5	4	0.68	0.675
5	2.5	6	0.64	0.64
6	2.5	4	0.637	0.637
7	3.0	4	0.575	0.577
8	3.0	6	0.505	0.506
9	3.0	8	0.484	0.486

4. Results and Discussions

The obtained data consisted measured base pressure *(P_b)*, across the length of the augmented pipe and nozzle pressure ratio (NPR) defined as the proportion of stagnation pressure *($P_{Stagnation}$)* to the back pressure *(P_{atm})*. The registered pressure values were non-dimensionalized by subdividing them with the surrounding ambient pressure. Fig. 3(a) presents results of base pressure variation with respect to L/D ratios for Mach number M = 2.0. It is observed that base pressure continues to decrease with L/D attaining a minimum value at L/D = 6. For an L/D greater than 6, it tends to increase again. This minimum base pressure obtained at L/D = 6 are compatible with the results Khan and Rathakrishnan [5]. However the influence of control is only marginal over base pressure. Fig. 3(b) presents results of base pressure variation Mach number M = 2.5. The behavior is distinctive when compared to M=2.0. Once again in this case, control marginally influences the base pressure. Furthermore, for an L/D = 8; did observe minimum base pressure. Fig. 4 illustrates the results for Mach number M = 3.0. Here again an L/D upto 8 only plays a compelling effect on minimizing base pressure. Thus for all three cases control marginally influences base pressure. This minimum base pressure achieves the reattachment length for the corresponding Mach number. Thus for Mach number 2.0, the reattachment length is L/D = 6 and Mach numbers 2.5 and 3, the reattachment length is L/D = 8. This is because the quality of the vortex at the base area which is high for higher Mach numbers prompted for bigger suction at the base. For higher Mach numbers i.e. 2.5 and 3.0, a decrement in the base pressure is observed as compared to Mach 2.0; for which the physical reason is the impacted oblique stun or development fan that is positioned at the spout lip [12]. The base pressure values for Mach 3.0 show distinct behavior when control is applied thus making control effective. The values for control are imperceptibly higher when compared to without control which is vice-versa otherwise in the case 2.0 and 2.5. In the existing study the position of microjets are fixed. Therefore, for a particular value of area ratio, the microjets are fixed to a specific distance away from the base corner. Hence, for higher Mach numbers, the microjets will tend to produce a shock effect which in turn contributes to increase in the base pressure values more effectively. Hence, the vortex at the base will generate more suction at the base. This appears to be the purpose for control becoming more effective at Mach number 3.0 than Mach numbers 2.5 and 2.0.

4.1. Experimental Model for Failure Index

A Statistical model based on multiple linear regression equations were developed for non- dimensional base pressure (P_b/P_a) using the experimental parameters for both cases of with and without control are given below. The standard commercial statistical software MINITAB 17 has been used to derive. The polynomial model disposed below represents the non-dimensional base pressure (P_b/P_a) as a function of Mach number (M) and L/D ratio. The regression equations are

$$P_b/P_a \text{ (With Control)} = 1.4078 - 0.2803\,(M) - 0.01167\,(L/D) \quad (1)$$

$$P_b/P_a \text{ (Without Control)} = 1.4178 - 0.2773\,(M) - 0.01108\,(L/D) \quad (2)$$

Where M=Mach number; L/D= length to diameter ratio.

4.2. Validation of Experimental Results

To validate the base pressure results, for non- dimensional base pressure (P_b/P_a) was computed theoretically by standard Eq. 1 and Eq. 2 developed by for arbitrary experimental conditions and validated with modeling base pressure results. It was found from Table 2 and Table 3 that; the percentage of error is within 9.56% for the case of with control and 6.07% for the case of without control respectively. The instrument errors in measurement system and lack of surface finish at the base region produce uneven rate of change base pressure. This may be the plausible reason for the error, which is experimentally reasonable.

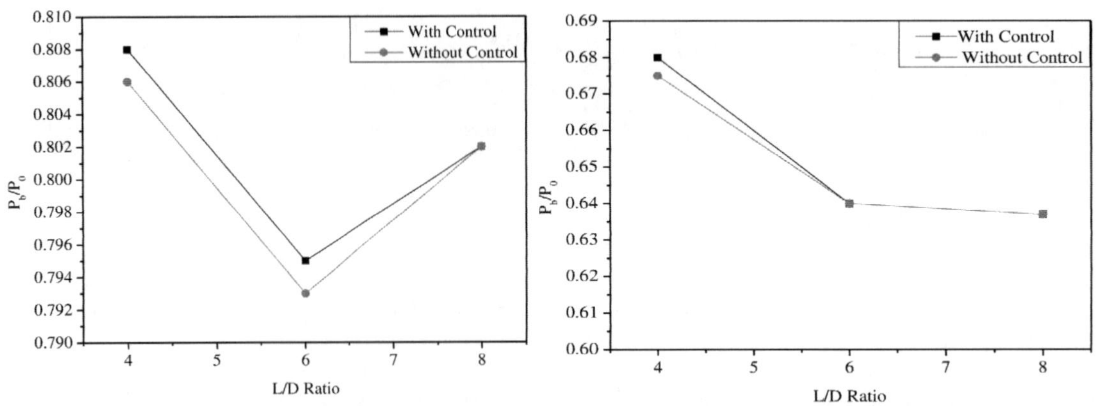

Fig. 3. Variation in Base Pressure with L/D Ratio for (a) M=2.0, P_e/P_a=0.277; (b) M=2.5, P_e/P_a=0.277.

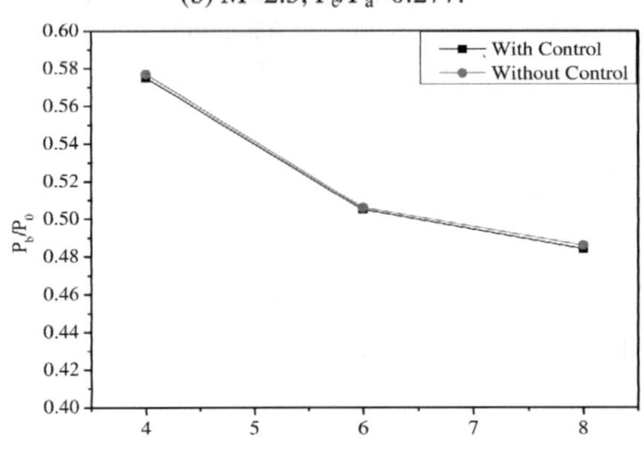

Fig. 4. Variation in Base Pressure with L/D Ratio for a) M=3.0, P_e/P_a=0.277.

Table 2. Validation for non-dimensional base pressure results with control.

Mach number (M)	L/D Ratio	Experimental Non-dimensional base pressure (P_b/P_a)	Predicted Non-dimensional base pressure (P_b/P_a)	% Error
2.0	4	0.808	0.801	0.86
2.0	6	0.795	0.778	2.13
2.0	8	0.802	0.754	5.98
2.5	4	0.68	0.660	2.94
2.5	6	0.64	0.637	0.46
2.5	8	0.637	0.614	3.6
3.0	4	0.575	0.520	9.56
3.0	6	0.505	0.497	1.58
3.0	8	0.484	0.474	2.07

Table 3. Validation for non-dimensional base pressure results without control.

Mach number (M)	L/D Ratio	Experimental Non-dimensional base pressure (P_b/P_a)	Predicted Non-dimensional base pressure (P_b/P_a)	% Error
2.0	4	0.806	0.819	1.60
2.0	6	0.793	0.797	0.50
2.0	8	0.802	0.775	3.37
2.5	4	0.675	0.680	0.74
2.5	6	0.64	0.658	2.81
2.5	8	0.637	0.636	0.16
3.0	4	0.577	0.542	6.07
3.0	6	0.506	0.520	2.77
3.0	8	0.486	0.497	2.26

4.2 Analysis of Variance (ANOVA)

ANOVA helps in formally testing the implication of the two factors i.e., Mach number and L/D ratio by correlating the mean square against the approximate calculation of the base pressure errors at specific confidence levels. The objective of analysis of variance is to inspect the factors which considerably affect the variation of base pressure. Table 4 and Table 5 illustrate the analysis of variance for non-dimensional base pressure results for the case of with and without control respectively. It is observed that Mach number has the highest significance of 95.91 % for the case of with control and 95.21% for the case of without control respectively. This is further followed by L/D ratio having less significance of 2.63% and 2.43% for both cases of without and with control respectively. The dependence of base pressure on Mach number observes that in the Mach numbers in the supersonic regime have a substantial influence on base pressure when compared to L/D ratio. This is due to the account of these flows which are highly overexpanded; the shock at the nozzle exit becomes stronger. These shocks possess a larger shock angle owing to smaller flow deflection angles. Thus, the base pressure is managed by the recycling flow stream instead of the flow behind the shock [5]. Thus, these waves are to be considered while analyzing base flows with Mach numbers in the supersonic regime which become the prime aspect for influencing base pressure. It is apparently clear from the outcome that, the L/D has a lesser effect on control of base pressure when compared to Mach number. It can be declared that, the pressure at base is predominated by reattachment length; defined as span of flow from its inception at the expansion to the location where in the nozzle's shear layer adheres with the wall duct. This can be plausible only by placing a duct of definitive length. However for supersonic Mach numbers employed in the current study, the minimum length for reattachment as per experimental results was L/D = 6. Thus supersonic Mach numbers associate themselves with higher L/D ratios thus minimizing their significance in controlling the expanded flows.

Table 4. ANOVA results for non-dimensional base pressure with control.

Source	DF	Adj SS	Adj MS	F-Value	P-Value	%P
A	1	0.117880	0.117880	263.97	0.000	95.91
B	1	0.003267	0.003267	7.32	0.035	2.63
Error	6	0.002679	0.000447			1.46
Total	8	0.12382				100

Table 5. ANOVA results for non-dimensional base pressure without control.

Source	DF	Adj SS	Adj MS	F-Value	P-Value	%P
A	1	0.115371	0.115371	242.96	0.000	95.21
B	1	0.002948	0.002948	6.21	0.047	2.43
Error	6	0.002849	0.000475			2.36
Total	8	0.121168				100

5. Conclusions

Controls of base flows for various process parameters as per Taguchi design of experiments have been studied. The following conclusions have been made:

- The reattachment length for Mach number 2.0 is L/D=6 and for Mach numbers 2.5 and 3, the reattachment length is L/D=8. This is mainly due to the high strength vortex at the base region which is high for higher Mach numbers leading to larger suction at the base.
- The control becomes effective for Mach number 3.0 since the microjets fixed to a specific distance away from the base corner tend to produce a shock effect in turn contributing to increase in the base pressure for higher Mach numbers.
- Linear polynomial models representing the non-dimensional base pressure (P_b/P_a) as a function of Mach number (M) and L/D ratio were developed for both the cases of with and without control. The models agreed well with the experimental results with percentage error of 9.56% for the case of with control and 6.07% for the case of without control respectively.
- Analysis of variance is also conducted to inspect the factors which considerably affect the variation of base pressure. It is observed that Mach number has the highest significance of 95.91 % and 95.21% in cases of with control and without control respectively. This is further followed by L/D ratio having less significance of 2.63% and 2.43% for both cases of with and without control respectively.

References

[1] Korst H A 1956 Theory of base pressure in transonic and supersonic flow *Journal of Applied Mechanics* **23.4** 593-600.
[2] Hall W B and Orme E M 1956 Flow of a compressible fluid through a sudden enlargement

in a pipe *Proceedings of the Institute of Mechanical Engineers*, **169** 1007-1022.
[3] Wick R S 1953 The effect of boundary layer on sonic flow through an abrupt cross-sectional area change *Journal of Aeronautical Science* **20** 675-682.
[4] Hoerner S F 1950 Base drag and thick trailing edges *Journal of the Aeronautical Science* **17** 622-628.
[5] Khan S A and Rathakrishnan E 2002 Active control of suddenly expanded flows from over expanded nozzles *International Journal of Turbo and Jet Engines* **19.1-2** 119- 126.
[6] Khan S A and Rathakrishnan E 2003 Control of suddenly expanded flows with micro jets, *Journal of Turbo and Jet Engines* **20.2** 63-81.
[7] Khan S A and Rathakrishnan E 2004 Active control of suddenly expanded flow from under expanded nozzles *International Journal of Turbo and Jet Engines* **21.4** 233-253.
[8] Khan S A and Rathakrishnan E 2004 Control of suddenly expanded flow from correctly expanded nozzles *International Journal of Turbo and Jet Engines* **21.4** 255-278.
[9] Khan S A and Rathakrishnan E 2006 Control of suddenly expanded flows Aircraft *Engineering and Aerospace Technology: An International Journal* **78.4** 293-309.
[10] Durst F, Meiling A and Whitelaw J H Low Reynolds number flow over a plane symmetric sudden expansion *Journal of Fluid Mechanics* **12.6** 111-128.
[11] Rathakrishnan E and Sreekanth A K 1984 Flow in pipes with sudden enlargement. *Proceedings of the 14th International Symposium on Space Technology and Science* Tokyo, Japan, 491-496.
[12] Rehman S and Khan S A 2008 Control of base pressure with micro jets: part I *Aircraft engineering and Aerospace Technology: An International Journal* **80.2** 158-164.

Development of Sense and Avoid system based on multi sensor integration for unmanned vehicle system

Ari Legowo[1], M. Faiz Bin Ramli[2a], Syariful Syafiq Shamsudin[2b]
[1] Department of Mechanical Engineering, Kulliyyah of Engineering, International Islamic University, Malaysia, P.O.Box 10, 50728 Kuala Lumpur, Malaysia.
[2] Faculty of Mechanical and Manufacturing Engineering, Universiti Tun Hussein Onn Malaysia.

E-mail: [1]legowo@iium.edu.my

Abstract. Unmanned Aerial Vehicles (UAV) have a great potential to replace manned aircraft in many tasks or missions. However, in order to allow these UAVs to execute their mission appropriately and especially to deploy them in national or civil airspace, UAVs must have a proper and reliable collision avoidance system. This avoidance system is known as a Sense and avoidance (SAA) system. The purpose of the system is to detect any obstacles either in static or moving condition and respond with proper avoidance maneuvers in order to maintain minimum separation distances. Therefore the research project deals with the development of a unique SAA system which is based on multisensory integration. In particular, it focuses on sensors, processing logic and hardware required on the UAVs to acquire situational awareness. Dissimilar kind of sensors which is optical and laser type is proposed to be employed, so that weaknesses from each individual sensor can be tolerated. Proper logics or algorithm related to the fusion of output from each sensor will also be studied.The combination of Extended KalmanFilter (EKF) and Differential Evolution(DE) method is used for the purpose of development of detection and tracking system which considered to be the main challenges in SAA system development. Subsequently, once the algorithms are put together, including maneuvers avoidance technique algorithm, the developed SAA system will be experimentally tested in order to validate and evaluate the overall capability of the system.

1. Introduction

Unmanned Aerial Vehicle (UAV) as the name implies, it is an aircraft without a pilot on-board or attached to it. Hypothetically, UAVs have a great potential to replace manned aircraft in many tasks or missions (e.g. surveillance, rescue mission, attack mission, collecting data, and etc.).In order to let these UAVs execute their mission appropriately and especially to deploy them in national or civil airspace, UAVs need to have a proper and reliable collision avoidance system.

In the field of UAVs, the term Sense and Avoid (SAA) is used for collision avoidance system. SAA system is an ability of the UAVs to detect traffic or obstacles (Static/moving) and respond with proper avoidance maneuvers in order to maintain minimum separation distances.Under requirement of Part 91 General operating and flight rules (Federal Aviation administration FAA), the key requirement for any vehicle to enter an airspace, that they must have 'See and avoid" technique incorporated in their vehicle [1].

In general, any SAA system must consists of five key stages or processes which are Surveillance, identification of risk, determination of appropriate maneuvers, maneuvers execution, and return to course [2]. Surveillance or close observation of the SAA system highly depends on the type of the sensors being used. There are two types of sensors which are cooperative and non-cooperative sensor (see table 1). Cooperative surveillance uses intruders information (e.g position, velocity, trajectory, and etc) constantly by employ a transponding method by which the intruders transmit its information. On the other hand, Non-cooperative sensor obtains intruders information via indirect method either by passive or active technique [3]. The ability to deploy UAVs across a multitude of operating environments and mission scenarios has rendered them exceedingly useful for both defense and commercial applications and this has fuelled their rapid development in the last decade.

In Korea, Chungnam National University has developed SAA system for the agriculture purposes using only Light Detection and Range (LIDAR) sensor system. The sensor is seletected because this agriculture UAVs only hover or fly near the ground and do not require any communication between other aircraft [4]. Other sensing methods that are available for SAA system is vision sensing typesystem. Pazmany Peter Catholic University used five camera vision system in order to develop collision warning system and they have selected five pieces of WVGA (752x480) cameras (MBSV034M-FFC) to cover the required resolution [5]. Northrop Grumman Global Hawk the United States surveillance UAVs also utilizes a vision system which by adopting three camera system to detect intruder aircraft [6]. There is also research and studies of SAA system using unique sensor which is Doppler radar. These sensor radars require relative motion between the vehicle and target to detect obstacles. Pulsed Doppler radar can be used to give intruders velocity, range and in some cases angular location with respect to observer. By measuring the airspeed of the UAVs, it is possible to acquire the relative speed between the UAVs and target [7].

This paper focused on proposing a unique enhancement solution to develop UAVs SAA system by using the integration of dissimilar type of sensors that capable to efficiently and accurately detect intruders or obstacles and ultimately evade the collision.

2. Challenges

A lot of research and investigation pertaining to SAA system are still undergoing. When it comes to development of SAA system for UAVs, every researcher must considers properly two important aspects related to system itself which is sensor characteristic and algorithm method for tracking and avoidance system. Field of Regard (FOR), Field of vision (FOV), and range of the sensors have to be properly considered, so that the SAA system being develops is accurate and reliable. Furthermore, weight and size of the sensors attached must be suitable with the physical size and power produced by the UAVs. Most of the sensors only sense their target within their own capabilities. For example, an UAVs equipped with optical sensors, will only detect their target by bearing attitude and not by range. Therefore, the detection reliability of the SAA system will certainly not favourable.

The complexity of the SAA system also lies on the development of the detection and tracking algorithm (filtering/estimation method) because it has several methods and each methods has its own advantage and disadvantage. These two aspects are closely related to each other, because each selection of sensors can have a direct impact on the complexity of the algorithm. Thus, when the algorithm become too complex, the tracking system may not be accurate enough and might as well fail.

Table 1. Sensor characteristics

Name	*Type*	*Range*	*Bearing Azimuth*	*Bearing Elevation*
A/C transponder	Cooperative	Accurate	Calculated	Calculated
ADS-B	Cooperative	Accurate	Calculated	Calculated
Optical	Non-Cooperative	Not sensed	Accurate	Accurate
Thermal	Non-Cooperative	Not sensed	Accurate	Accurate
LIDAR	Non-Cooperative	Accurate	Narrow	Narrow
Radar	Non-Cooperative	Accurate	360 Degrees	360 Degrees
Acoustic	Non-Cooperative	Accurate	360 Degrees	360 Degrees

3. Proposed solution

As stated above, the proposed solution of the research project to enhance the level of accuracy and reliability of the SAA system is by the employment of multiple sensors that have a different type of sensing capabilities into the system. Apart from that, we will also introduce a combination of Extended Kalman filter (EKF) and Differential Evolution (DE) as our state estimation or filtering algorithm.

3.1 Multi sensors integration.

There are several techniques to encounter the disadvantages of sensor types or single sensor. One could introduce multiple and similar type of sensors or multiple sensor types. In order to integrate multiple sensors into one system that produce a single output, one must properly combine all the data coming from all sensors (similar/dissimilar). This combination of data is called as a data fusion. Data fusion techniques combine data from multiple sensors, and related information from associated databases to achieve improve accuracies and more specific inferences that could be achieved by the use of single sensor alone [8]. In this project, dissimilar sensors will be employed which LIDAR and Optical sensors.

Combination of these sensors can wipe off the disadvantages from each individual sensor, because LIDAR can detect accurately the range and the distance of the obstacle while not for optical sensors. Conversely, Optical sensor have the capability to detect the bearing of the obstacles while this capability is not the case for LIDAR. Since data coming from these sensors are different, it is considered as a non-commusurate data, therefore, these data must be fused at decision level and cannot be used at raw level [9].

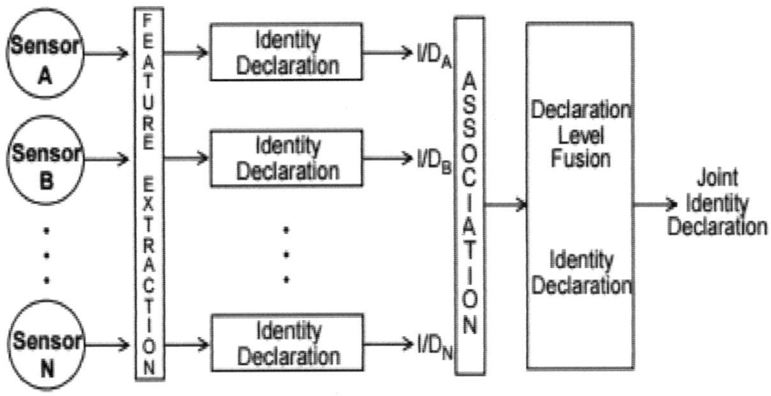

Figure 1. Fusion Architecture

3.2 Detection Method.

The project will uses EKF with combination the ofDE to make the tracking system computationally efficient and capable of running real time on-board of the UAVs. Since computing all data (e.g source image) or coefficients at every position is computationally expensive and consumes a lot of time, it is more efficient and cost-effective to predict the position of the interest obstacles and searching only a region around the predicted position. The position of the obstacles is estimated using the measurement from the obstacles detection and the process covariance matrix is updated and optimized by DE algorithm method[1,10,11]. These are the few of the governing equations:

predicted state

$$\boldsymbol{X_k} = \boldsymbol{f}(\boldsymbol{X_{k-1}}, \boldsymbol{U_{K-1}}, \boldsymbol{W_{k-1}}) \quad (3.1)$$

Measurement input

$$\boldsymbol{Z_k} = \boldsymbol{f}(\boldsymbol{X_k}, \boldsymbol{V_k}) \quad (3.2)$$

Position and velocity

$$\boldsymbol{X_k} = \begin{bmatrix} \overline{\boldsymbol{X}}_k \\ \dot{\boldsymbol{X}}_k \end{bmatrix} \quad (3.3)$$

Process covariance matrix

$$P_k = (I - KH)P_{K_p} \quad (3.4)$$

These are general steps in implementing the technique:

i. **Initialization**

In this stage the whole image is searched for the object due we do not know previously the obstacle position. Process covariance matrix is also initialized.

ii. **Prediction**

The state of the obstacle is predicted using the motion model of the kalman filter for the next frame at next time step.

iii. **Correlation**

In this stage, the obstacles is detected within the search window (measurement Z_k) and the State position and covariance matrix is updated with the measurement data.

4. Result and Discussion

Research project methodology starts with analysing related literature reviews on SAA systems including regulations or requirement, sensors characteristics, sensor detection algorithms, maneuver avoidance method and others. After extensive reviews have been done, type of sensors will be chosed (e.g size, weight, range, bearing angle and etc).

Each sensor chosed will be tested or simulated to determine what kind of data can be captured. After the sensors simulation, proper detection and tracking algorithm including data fusion algorithm will be developed. The combination of EKF and DE method is used for the purpose of filtering the data (Optimized) to develop the detection and tracking system. Subsequently, once the algorithms are put together, the system will be experimentally tested.

When detection and tracking system is accomplished, maneuver avoidance method for the SAA system will be developed. Finally, this SAA system will going through several experiment in order to validate and evaluate the overall capability of the system.

5. Conclusion

By using a unique UAVs SAA system based on multisensory will be developed. The system is expected to succeed in detecting and tracking the obstacles either in static or moving condition. Ultimately, the UAVs must be able to perform safe maneuver when risk of collision detected. The SAA system will give a contribution (e.g multisensory technique, sense and detection method, sensor fusion method and manoeuvre avoidance technique) on the matters regarding the operational possibilities of UAVs in the national or civil airspace.

References

[1] United States Department of Defense (2004), "Airspace Integration Plan for Unmanned Aviation"

[2] Andrew R. Lancher, David R Maroney, Dr Andrew D. Zeitlin (2007) "Unmanned Aircraft Collision Avoidance Technology Assessment and Evaluation Methods", The MITRE Corporation, McLean, VA, USA.

[3] Hung Pham, Scott A. Smolka, Scott D. Stoller, Dung Phan, Junxing Yang (2015) "A survey on Unmanned Aerial vehicle Collision Avoidance Systems". Department of Computer Science, Stony Brook University, Stony Brook, NY, USA.

[4] Jingu Kim, Sua Song, Seungkeun Kim, Jiyoung Souk (2014),"Collision Avoidance Sytem for Agriculture Unmanned Helicopter Using LIDAR Sensor", Asia Pacific International Symposium On Aerospace Technology (APISAT).

[5] AkosZarandy, Zoltan Nagy, BalintVanek, TamasZsedrovits, Andra Kiss, Mate Nemeth, (2011),"A five camera vision system for UAV visual attitude calculation and collision warning", Springer-Verlag Berlin Heidelberg.

[6] J. Daniel Griffith, Mykel J. Kochenderfer, James K. Kuchar (2018)," Electro Optical System Analysis for Sense and Avoid", AIAA Guidance, Navigation and Control Conference and Exhibit.

[7] AndrewViquerat, Lachlan Blackhall, Alistair Reid, Salah Sukarrieh, Graham Brooker, (2017),"Reactive Collision Avoidance For Unmanned Aerial Vehicles Using Doppler Radar", 6th International Conference on Field and Service Robotics.

[8] David L.Hall, James Llinas (1997), " An introduction to multisensor Data Fusion", Proceeding of the IEEE, Vol-85, No-1. Appl. Res. Lab., Pennsylvania State Univ., University Park, PA, USA

[9] David L.Hall, James Llinas (2001), "Introduction to multisensor Data Fusion", CRC Press LLC.

[10] Ashraf Qadir, William Semke, Jeremiah Neubert, (2011),"Implementation of an onboard visual tracking system with Small unmmaned Aerial Vehicle (UAV)", International Journal of Innovative Technology & Creative Engineering, Vol.1 No.10.

[11] Robert H. Chen, Arthur Gevorkian, Alex Fung, Won Zon Chen, (2011),"Multisensor Data Integration for Autonomous Sense and Avoid", AIAA 2011-1479.

CFD Studies on Triangular Micro-Vortex Generators in Flow Control

Yashodhar V[1], Humrutha G[2], M Kaushik[3], and S A Khan*[4]

[1,2,3]Department of Aerospace Engineering, Indian Institute of Technology Kharagpur-721302, India
[4]Department of Mechanical Engineering, International Islamic university, Kualalumpur, Malaysia

E-mail: [3]mkaushik@aero.iitkgp.ernet.in, [4*]sakhan@iium.edu.my

Abstract. In the present study, the flow characteristics of the commercially used S809 wind turbine airfoil controlled with triangular counter-rotating micro-vortex generators at stall angle of attack of 15 degrees and 10 m/s, 15 m/s and 20 m/s (speed range used in the wind turbine applications) had been computationally investigated. In addition to the controlled airfoil, an uncontrolled airfoil was also studied for the comparison. The modelling and analysis had been carried out using incompressible, Reynolds Averaged Navier Stokes equation using Spalart-Allmaras one equation turbulence model. The numerical computations were performed with SIMPLE algorithm. The velocity profiles at different locations on the suction surface were plotted for both uncontrolled and controlled airfoils. The shear stresses exerted on the upper surface of the airfoil in both the configurations were also compared. It is found that the controlled airfoil, the shear stress distribution was greatly increased near to trailing edge of the airfoil revealing the superiority of vortex generators in increasing the efficiency of wind turbine by delaying boundary layer separation. The qualitative results of flow visualization in the spanwise direction also support the quantitative findings of velocity profiles and shear stress distribution.

Keywords: Micro-vortex generators, Boundary layer separation, Stream-wise vortices, Critical shear stress, Flow visualization

Nomenclature

MVG	=	Micro-vortex generator
h	=	Height of micro-vortex generator
δ	=	Boundary layer thickness
α	=	Vertex angle of triangular micro-vortex generator
β	=	Angle of attack
Z	=	Length between adjacent micro vortex generators
c	=	Chord of the airfoil
x	=	X-axis (measured from the leading edge of the airfoil)
U_∞	=	Free-stream velocity

τ	=	RMS values of surface shear stress
Tc	=	Critical shear stress
RANS	=	Reynolds Averaged Navier Stokes

1. Introduction

In aircrafts, the surface roughness or placement of micro-vortex generators over wing delays boundary layer separation and hence increases lift thereby increasing stall angle. The optimized MVG-controlled flow over airfoil increases its aerodynamic efficiency. The boundary layer separation is the most critical problem for the wind turbines which reduces its performance the necessity of flow control arises. Micro-vortex generators are small trapezoid shaped fin-like devices, which are used in aircrafts to improve their aerodynamic efficiency. Since long they are considered as a cost effective solution due to their ease in manufacturing and installation over wings. They greatly influence the aerodynamic forces and stall characteristics by controlling the flow over the wind turbine rotors as well. Generally wind turbines can be classified into two types, pitch controlled turbines which alter the pitch of the rotor and Stall controlled turbines which stall the entire turbine, if the RPM exceeds a certain value. The MVGs are best suited for the stall controlled wind turbines. Khalfallaha and Koliub (2007) studied the advantages of micro-vortex generators in flow control and dust removal in dusty regions where power losses can aggravate up to 57% in the span of nine months **Error! Reference source not found.**. They performed their experimental study to see the effect of dust accumulation on rotor blades of a stall regulated 300 kW wind turbine for a period of one year. They found that the location of dust deposition is dominated near the stagnation points. They have also concluded that the area of dust deposition and dust particle size increases with lapse of time. These dust particles were found to affect the aerodynamic loads on the rotors which has led to decrease in lift and increase in drag. Further, it was reported that the roughness effects resulting in degradation, depends on the roughness length relative to Boundary layer thickness, Reynolds number and Airfoil type. Nianxin and Jinping (2009) performed a computational study on the flow past a wind turbine blade to investigate the effect of roughness size and roughness area covered on the lift and drag coefficients **Error! Reference source not found.**. It is found that airfoils are more sensitive to surface roughness if they are located in the former 50% chord than the latter half. The lift coefficient experienced a decrease of about 30% in the extreme case at the roughness location of 25% chord. It was also seen that the roughness at the critical locations promote premature turbulence and flow separation. It is known that the vortex generators promote turbulence which leads to losses in the wake region of the airfoil. Hence to minimize these losses, the dimensions of the vortex generators must be precise. Bray (1998) has performed his studies on the efficiency of vortex generators in delaying flow separation and reducing boundary layer thickness. He found that the flow control through manipulation of the boundary layer either in preventing or delaying the separation has significant positive effects on the lift and performance of aircraft **Error! Reference source not found.**In the recent past, many studies have been performed on various geometries of the vortex generators under adverse pressure gradient. Gad-el-Hak (2000) has predicted the point of boundary layer separation using various low profile vortex generators. He concluded that the introduction of high momentum flow into the boundary layer is one of the most efficient methods in delaying the flow separation which eventually enhances the performance **Error! Reference source not found.**. The micro-vortex generators generally fetch turbulence into the wake of the airfoil which eventually leads to losses, therefore its shape and size must be properly designed. Lin (2002) has performed his studies on variable size micro-vortex generators **Error! Reference source not found.**. It was found that the height of the vortex generators must be less than the boundary layer thickness (δ). It was seen that when vortex generator's height (h) was kept 62.5% of the boundary layer thickness then it showed a higher performance than the case when $h= \delta$. Even when the height of MVGs lies in the range of $0.1 < h/\delta < 0.2$, it provides a sufficient moment transfer into the Boundary layer. Anand et al (2010) have shown in their studies that the optimum MVG geometry and configuration may also depend on the airfoil shape **Error! Reference source not found.**. It is known that the different airfoil shapes has different stall characteristics and they also have different transition points along the surface. Hence they concluded that a particular MVG configuration optimized for one airfoil profile may not be optimal for another profile of different thickness, camber, or leading-edge radius. The test conducted in low-turbulence pressure wind tunnel at NASA Langley Research Centre (2000) showed that the MVGs dramatically increased the performance of the aircraft **Error! Reference source not found.**. In this study it was also found that the generated noise gets attenuated while landing. It has been shown as one of the cost effective means to improve the performance of an aircraft. Shan et al (2008) have found that the MVGs are advantageous in many aspects as compared to other high lift augmenting devices such as flaps and slats **Error! Reference source not found.**It is seen that for wings or any other lifting surface; the fully attached flow not only leads to increased lift but also minimizes aeroacoustics noise levels. For wind turbine blades these advantages manifest in higher aerodynamic torque and power output for given wind conditions.

From the above, it is evident that most of the studies are concentrated in understanding the flow physics behind the vortex generators applications such as, the formation of separation bubbles, delay of flow separation and boundary layer thinning effects. Although, some investigations have been done on the shape optimization of these vortex generators to improve airfoil performance but the effects on the surface shear stress at high angle of attacks have been hardly looked into. The shear stress distribution is critical for the wind turbine blades. This is because when the blade surface has roughness then it is prone to an early boundary layer separation. From the literature, it is understood that the separation can be delayed by the deployment of the vortex generators over airfoil surface. However, all these vortex generators shape do not improve the wind turbine's performance. Hence, there is a need to optimize MVGs geometry as well as their placement over the wind turbine rotor blade. Keeping view on this objective, an innovative design named as triangular counter rotating micro-vortex generator by combining the vortex manipulation advantages of triangular shape and counter rotating shapes, has been studied as flow control. Thus, the present investigation aims at evaluating and comparing the velocity profiles of the triangular counter rotating micro-vortex generators controlled airfoil with that of plain airfoil configuration. To investigate the MVGs efficacy in delaying boundary layer separation the shear stress was also compared for both uncontrolled and controlled configurations. The commercially used S809 airfoil in wind turbine blades was chosen for the study.

2. Methodology

The micro-vortex generators were used to control the flow over airfoil. In determining the height of vortex generators the thickness of boundary layer at a point about 1 cm upstream of the separation point was calculated. This boundary layer thickness was also used in determining the dimensions of the MVGs. The flow over MVG controlled airfoil configuration had been simulated at the stall angle of attack and the velocity profiles along with shear stress were computed at three different flow speeds. The results obtained were compared with their uncontrolled counterpart under similar flow conditions.

The micro-vortex generators were placed in the mid-span of airfoil just upstream of the separation point. In the present investigation, a counter-rotating configuration and a triangular configuration were merged together in obtaining triangular counter-rotating micro-vortex generators as shown in Figs 2.1(a)-(c).

The three dimensional models were developed using the SOLIDWORKS designing package. These models were imported into the ANSYS design modeler and a flow domain was chosen for the analysis. The control volume was meshed in the ANSYS mesh modeler. The discretization of the domain was done using unstructured grid consisting of approximately 1350000 tetrahedral and hexahedral elements. A fine mesh in the vicinity of the airfoil model was adapted. The cross-sectional views of uncontrolled airfoil mesh and MVG controlled airfoil mesh were shown in the Figs 2.2(a)-(b). It can be seen that the mesh was slightly inflated near the airfoil surface in both uncontrolled and controlled configurations. This was purposefully done to have very fine mesh in the proximity of the suction surface of the airfoil to obtain more accurate flow characteristics. Sufficiently large number of elements was issued at sharp edges and low curvature points such as near the micro-vortex generators, leading edge and trailing edge of the airfoil.

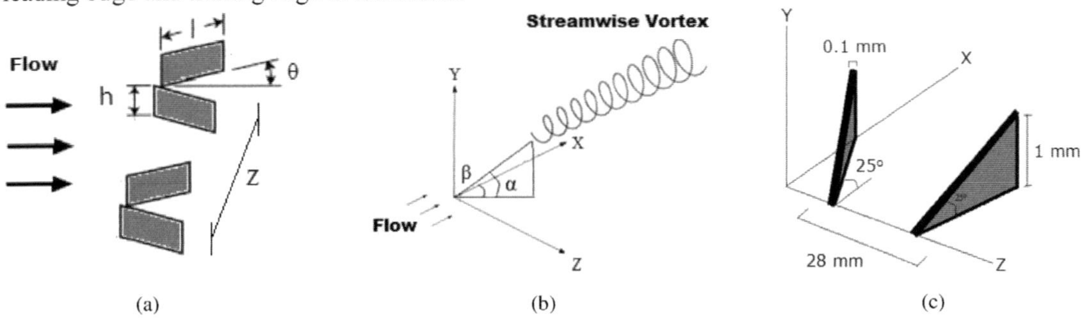

Fig 2.1 Schematic diagrams of micro-vortex generators geometry, (a) Counter-rotating vortex generators, (b) Triangular vortex generators (c) Triangular counter-rotating micro-vortex generators.

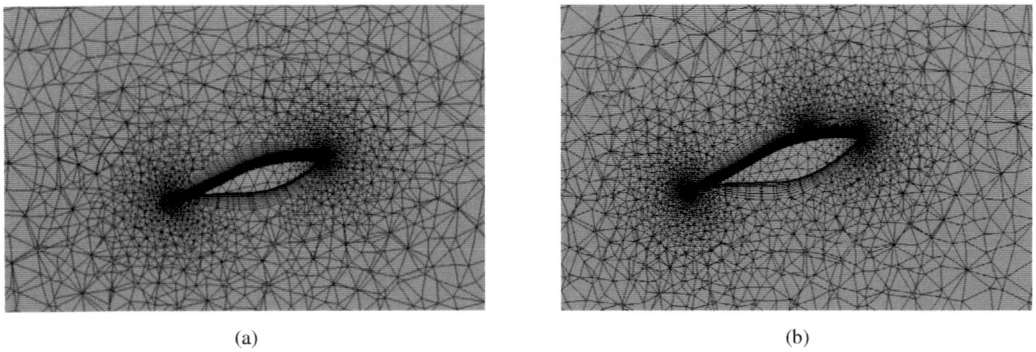

Fig 2.2 Cross-sectional views of the meshed airfoils, (a) uncontrolled airfoil (b) controlled airfoil with MVGs.

The computational flow analysis was done using the FLUENT software package. With air as flowing fluid at incompressible speeds the 3D models were analyzed using RANS equations. The Spalart-Allmaras turbulence model was used to model the Reynolds stresses in the Navier Stokes equations. This turbulence model was a one-equation model that solves a modelled transport equation using the kinematic viscosity. From the literature, it is found that the Spalart-Allmaras model provides accurate results especially when the boundary layer is subjected to an adverse pressure gradient. Also, this model is relatively stable and less sensitive to the grid resolution than the two-equation model **Error! Reference source not found.**. However, the Spalart-Allmaras model suffers from a deficiency common to transport type models that, it fails to predict the reduction in shear layer growth rate with increasing jet Mach number. The model will predict the incompressible shear layer growth rate regardless of the jet Mach number. The SIMPLE scheme was used for pressure-velocity coupling, and pressure and momentum equations are approximated using the second order upwind scheme. Based on chord of the airfoil taken as characteristics length, the Reynolds number were found to be 2×10^5, 3×10^5 and 4×10^5 for the free-stream velocities of 10 m/s, 15 m/s, and 20 m/s, respectively. A no-slip boundary condition was applied on the surface of the airfoil. The inlet turbulence intensity was set to 10% which had closely simulated the atmospheric boundary layer exposure to industrial wind turbine blades.

3. Results and Discussion

From the literature, it is understood that increasing surface roughness delays the boundary layer separation. One such technique to increase the roughness is the deployment of boundary layer submerged vortex generators over the upper surface of airfoil. These MVGs induce shear stress greater than the critical shear stress required for separation, hence the boundary layer remains attached and thereby increasing the aerodynamic performance of wind turbine blades. Further, since not all the micro-vortex generators shapes improve the aerodynamic performance, thus their design and installation over airfoil surface must be optimized. With this objective, the present computational study was carried out to investigate the efficacy of an innovative design of micro-vortex generators named as, triangular counter rotating micro-vortex generators deployed over the wind turbine S809 airfoil. To quantify the MVGs effects on aerodynamic performance the streamwise velocity profiles were plotted for both uncontrolled and controlled configurations. In addition, the shear stress distributions were also compared to investigate the MVGs effectiveness in delaying boundary layer separation. For qualitative study, the visualization images of velocity magnitude contours were also analyzed.

3.1. Streamwise Velocity Profiles

The streamwise velocity profiles at two different axial locations at free stream velocities of 10 m/s, 15 m/s and 20 m/s are shown in Figs 3.1(a)-(b). The axial locations studied are 0.5c (near field location), and 0.83c (far downstream location). The angle of attack was kept constant at 15 degrees, which is the stall angle for uncontrolled S809 airfoil. The local velocity (u) is plotted against the transverse distance (y) from the surface of the airfoil to obtain the velocity profiles. These plots were not non-dimensionalized purposefully to optimize their presentation.

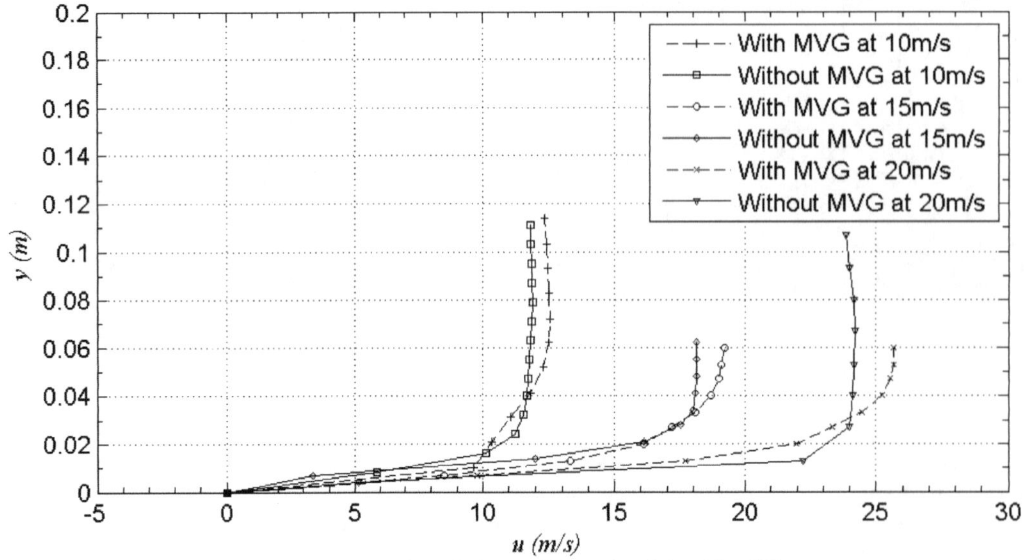

Fig 3.1(a) Streamwise velocity profiles at x/c = 0.5 and β = 15°.

The velocity at x/c = 0.5 and β = 15° for both uncontrolled airfoil and the airfoil controlled with triangular counter rotating micro-vortex generators are shown in Fig 3.1(a). It is seen that the velocity profiles for the controlled airfoil become shallower than the uncontrolled one at all the flow speeds investigated. In other words, there exists a large velocity gradient in transverse direction (y-direction) inside the boundary layer in comparison to the plain airfoil configuration. Since, the height of micro-vortex generators was less than the boundary layer thickness (h/δ = 0.5), thus it induces larger shear stress on the airfoil surface which delays the boundary layer separation thereby increasing aerodynamic performance by minimizing flow separation losses. Thus, the superiority of controlled airfoil configuration over uncontrolled one is established in this case. Further, it is also seen that at the flow speed of 10 m/s both the curves are approaching each other. With the increase of flow speed to 15 m/s, performance of controlled airfoil is slightly inferior to its uncontrolled counterpart. However, when the flow speed is further increased to 20 m/s the controlled airfoil is again exhibiting its superiority in increasing the velocity gradient and hence the shear stress over the airfoil surface. Interestingly, the boundary layer is still attached over the controlled airfoil and thus increasing the stall angle beyond 15°.

When the location is increased to 0.83c (i.e. near to trailing edge of the airfoil), the larger back-flow region at β = 15° is seen in all the uncontrolled cases (Fig 3.1(b)). It is obvious because this zone is highly vortex dominated. At all the flow speeds the back-flow region is seen and the boundary layer exhibits the tendency of separation at 10 m/s and 15 m/s free stream velocities. That is, there exists a zero-pressure gradient (du/dy =0) at this location for these flow speeds. Interestingly, again at U_∞= 20 m/s the MVG controlled airfoil has maintained its superiority over the uncontrolled airfoil. Hence, at stall angle the boundary layer is completely separated leading to decreased aerodynamic performance. However, the magnitude of the maximum backflow velocity and the width of the back-flow region are greatly reduced in the case of controlled configuration. It again shows the superiority of micro-vortex generators controlled airfoil in improving aerodynamic performance and delaying boundary layer separation than the uncontrolled airfoil.

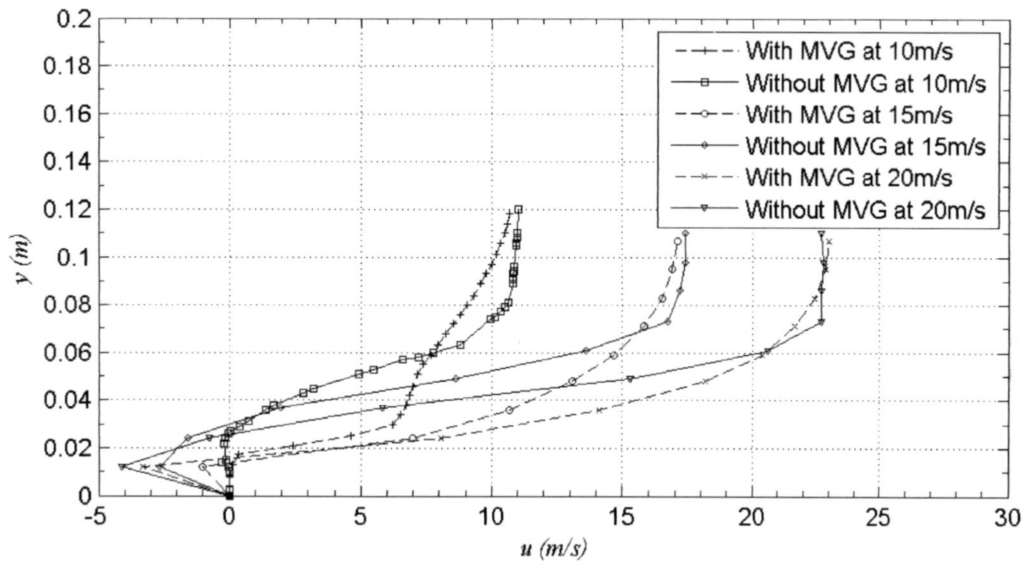

Fig 3.1(b) Streamwise velocity profiles at x/c = 0.83 and β = 15°.

From the above discussions, it is evident that the velocity profiles of the airfoil controlled with the micro-vortex generators have a larger velocity gradient than the plain airfoil. It clearly indicates the formation of shear layers along the surface of the airfoil. Although, the backflow region is reduced in some of the cases but the boundary layer thickness is still greater in case of controlled airfoils. This may be because of the gradual change in velocity due to significant number of shear layers formed.

3.2. Shear Stress Distribution

The shear stress distribution is characterised by the measured root mean square (RMS) stress fluctuations in three dimensions as shown below.

$$\tau = \frac{1}{2}\sqrt{\left(\overline{(\tau_x')^2} + \overline{\left(\tau_y'\right)^2} + \overline{(\tau_z')^2}\right)} \qquad \text{(Eq. 1)}$$

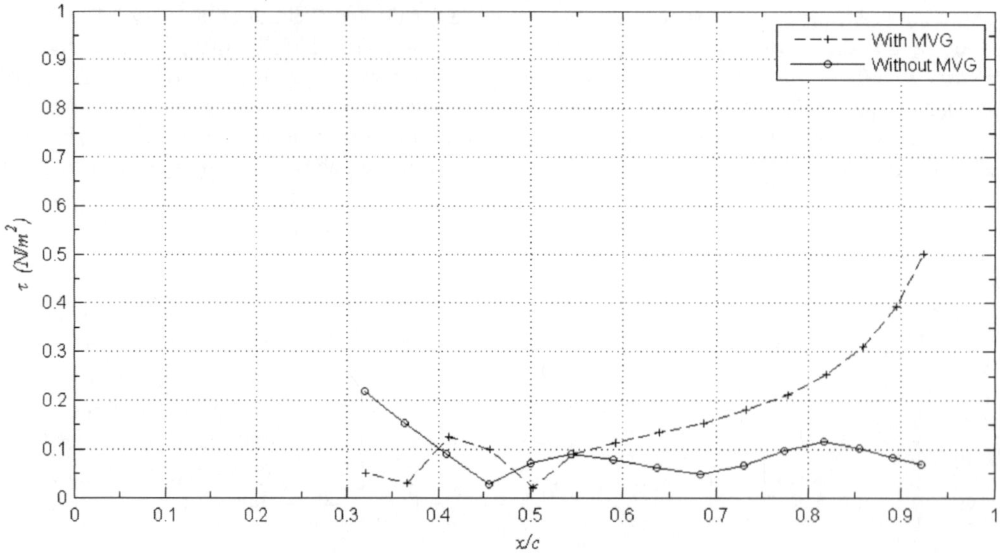

Fig 3.2 Shear stress distribution at $\beta = 15^{\circ}$ and U_{∞}= 15 m/s.

The shear stress distribution at the free-stream velocity of 15 m/s at $\beta = 15^{\circ}$ for both uncontrolled airfoil and the airfoil controlled with micro-vortex generators are presented in Fig 3.2.

It was seen that at the stall angle of attack, both uncontrolled and controlled airfoils are possessing almost similar shear stress distribution along the surface. However, in the case of airfoil without controls the shear stress shows a dip at 0.5c , which could be due to the formation of a small bubble or recirculation zone. In the vicinity of this location the flow is unsteady due to frequent attachment or detachment of the boundary layer and this unsteadiness owe to the acoustic waves which travels upstream are generated through the vortices downstream. Ahead of this location, the curve starts increasing and then decreases again reaching to its minimum at 0.7c. This may be because of the boundary layer separation taking place at the trailing edge of airfoil. However, when the micro-vortex generators are installed over the suction surface the shear stress is continuously increasing till the trailing edge except at 0.36c. The increased shear stress on the controlled airfoil surface denotes the larger skin friction. Hence, it has greater ability in delaying the flow separation especially in the far downstream locations.

3.3. *Variation of Pressure Coefficient*

The spanwise average of the coefficient of pressure varying with the chord of the wind turbine airfoil at $\beta = 15^{\circ}$ and U_{∞}= 15 m/s is shown in the Fig 3.3. The pressure coefficient distribution exhibits the considerable increase in negative pressure on the suction side as the boundary layer remains attached due to the deployment of micro-vortex generators. However, a slight increase in the positive pressure on the lower surface is also observed. This distribution suggests the increase in lift on the controlled airfoil compared to the plain one.

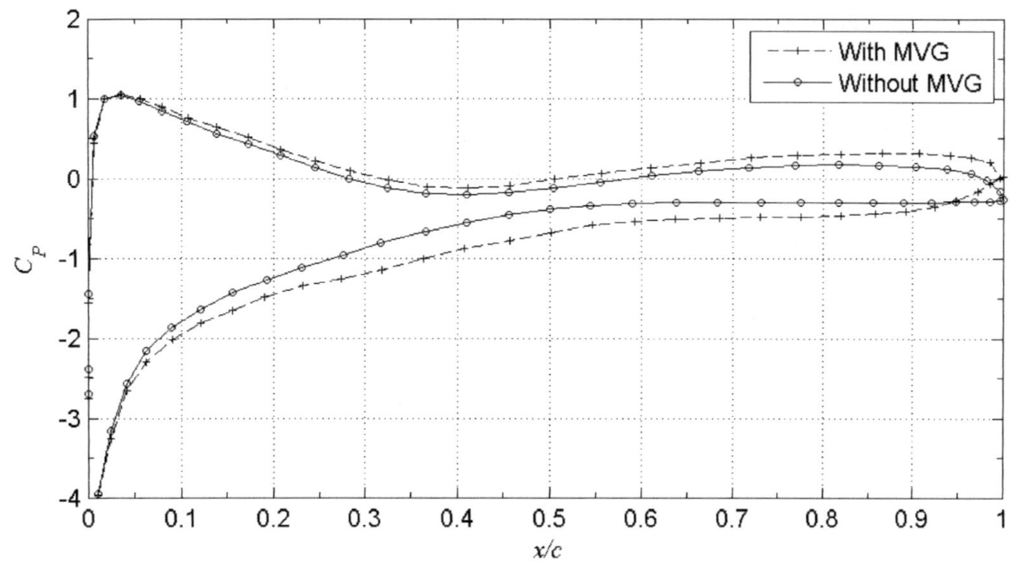

Fig 3.3 C_P distribution for airfoils at and U_∞= 15 m/s.

The experimental results of Somers (1997) are found to be in close agreement with the computed C_P values **Error! Reference source not found.**.

3.4. Flow Field Visualization

In investigating the effectiveness of micro-vortex generators, the flow field was visualized along the spanwise direction for both uncontrolled and controlled airfoil configurations at airfoil stall angle (β = 15°) and free stream velocity of 15 m/s as shown in the Figs 3.4(a)-(b). It can be seen that a large recirculation zone does exist over uncontrolled airfoil near trailing edge due to boundary layer separation at stall angle (Fig 3.4(a)). These trailing edge vortices induce boundary layer separation near to leading edge by sending acoustic information in upstream direction. Therefore, there exists a large recirculation zone near to leading edge.

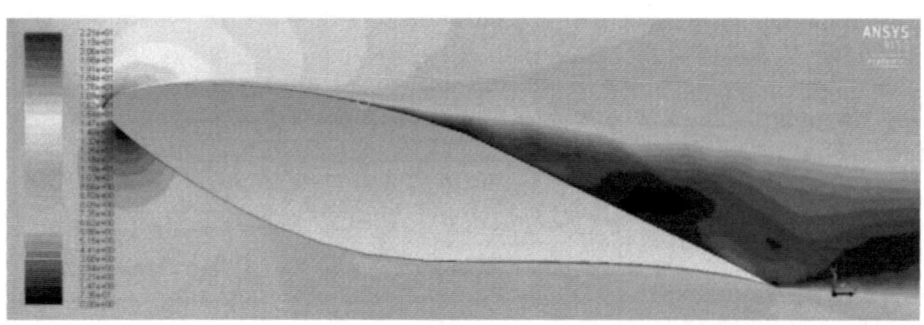

(a)

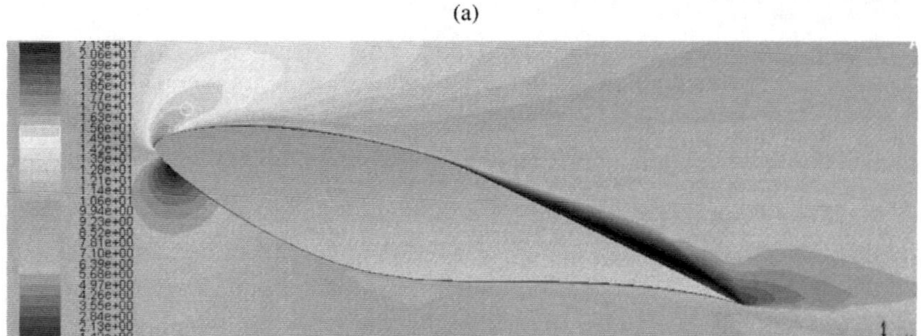

(b)

Fig 3.4 Spanwise flow field visualization of airfoils β = 15° and U_∞= 15 m/s, (a) uncontrolled airfoil, (b) controlled airfoil.

However, once micro-vortex generators were deployed over the suction surface of the airfoil the flow field was seen as given in Fig 3.4(b). It is clearly seen that, the vortices generated by the MVGs delays the boundary layer separation and hence the recirculation zone is suppressed. Hence, due to decreased losses, the performance output of the wind turbine will be increased. These visualization results clearly support the findings obtained in the sections on velocity profiles and shear stress distribution.

4. Conclusions

The quantitative and qualitative results on the efficacy of the proposed triangular counter rotating micro-vortex generators show that the deployment of micro-vortex generator reduces the boundary layer thickness in the flow past them and delays separation. Due to supressed flow recirculation zones on the controlled airfoil surface, the shear stress was found to be linearly increasing. It will lead to further increase in stall angle for controlled configurations beyond the existing 15° stall angle of uncontrolled airfoil. In addition, the images obtained by flow visualization along the spanwise direction show the advantages of micro-vortex generators in improving the aerodynamic performance. Although, the increase in drag may be expected but the drag analysis along with L/D ratio will be performed in the future studies.

References

[1] Khalfallaha M G and Koliub A M 2007 Effect of dust on the performance of wind turbines *Desalination* **209** 209–220.

[2] Nianxin R and Jinping Ou 2009 Dust effect on the performance of wind turbine airfoils *J. Electromagnetic Analysis & Applications* **1** 102–107.

[3] Bray 1998 A parametric studies of vane and air-jet vortex generators (PhD thesis Cranfield University).

[4] Gad-el-Hak M 2000 Flow control, passive, active, and reactive flow management (Cambridge University Press).

[5] Lin J C 2002 Review of research on low-profile vortex generators to control boundary-layer separation *Progress in Aerospace Sciences* **38** 389–420.

[6] Anand et al 2010 Passive flow control over NACA0012 airfoil using vortex generators *Conference on Fluid Mechanics and Fluid Power* (FMFP10 – FP – 12).

[7] NASA Langley Research Centre 2000 Micro-vortex generators enhance aircraft performance (NASA Fact Sheet FS-2000-06-52-LaRC).

[8] Shan H et al 2008 Numerical study of passive and active flow separation control over a NACA0012 airfoil *Computers & Fluids* **37** 975–992.

[9] Yaras M I and Grosvenor A D 2003 Evaluation of one- and two-equation low-Re turbulence models Part II—Vortex-generator jet and diffusing S-duct flows *International Journal for Numerical Methods in Fluids* **42** 1321–1343.

[10] Somers D 1997 Design and experimental results for the S809 airfoil *National Renewable Energy Laboratory, USA* (NREL/SR-440-6918).

Generation of atmospheric boundary layer in the IIUM low speed wind tunnel

Khalid Aldin Bin Za'aba, Waqar Asrar*, and Mohamad Al Dheeb

Dept. of Mechanical Engineering, International Islamic University Malaysia
PO Box 10, 50728 Kuala Lumpur, Malaysia

waqar@iium.edu.my

Abstract. The purpose of this paper is to describe an attempt made to simulate the atmospheric boundary layer (ABL) in the Low Speed Wind Tunnel (LSWT) at IIUM. This was performed through modifications of the inlet and surface conditions of the test section. Devices such as spires, fences and surface roughness were used. The ASCE 7-10 standard was used as a reference to validate the results between the wind tunnel data and the full-scale ABL flow characteristics. The velocity profile and turbulence intensity were measured using a one-dimensional hotwire (CTA) probe. The data obtained show good agreement with ASCE 7-10 velocity profiles for exposures B and C while using a scale of 1:488 to match the wind tunnel data. The turbulence intensity profiles do not show a good agreement.

1. Introduction

The atmospheric boundary layer is the air close to the earth's surface. The atmospheric wind has had a robust impact on human beings and their surroundings. The interest to study the effect of wind include economic loss and loss of life cause by wind devastation, problems that constantly grow due to the increase in world population and urbanization. For engineering and design purposes, it is important to understand the wind flow of atmospheric boundary layer and the characteristics of turbulent flow associated with it.

Most of the wind tunnels are usually designed for aeronautical and/or automotive testing purposes. Typically, the flow in the test section is nearly laminar or of low turbulence intensity. The boundary layer in such wind tunnels is as thin as possible on the walls in the test section. On the other hand the Atmospheric Boundary Layer (ABL) has much higher turbulence intensity and the boundary layer in the tests section is very thick. The test section of IIUM low speed wind tunnel is too short (6 m long) when compared to an environmental wind tunnel test section. Thus, the creation of the ABL in this wind tunnel requires certain devices to make the boundary layer thicker and have much higher turbulent intensity in a relatively short distance.

For this study, the generated ABL in the IIUM wind tunnel will be compared with recommended standard values in international wind loading standards and codes.

The surface of the earth can be categorized into three distinct terrain types; urban and suburban, rural, and smooth surface. The ASCE 7-10 [1]standard categorizes each type of terrain according to the type of 'exposure' as follows:

- **Exposure B**

"Urban and suburban areas, wooded areas, or other terrain with numerous closely spaced obstructions having the size of single family dwellings or larger. This exposure is representative of terrain in the upwind direction for a distance of at least 460 m (1,500 ft.) or 10 times the height of the building or other structure, whichever is greater".

- **Exposure C**

"Open terrain (rural terrain) with scattered obstacle having heights generally less than 9.1 m (30 ft.). This category includes flat open country, grasslands and shorelines in hurricane prone regions".

- **Exposure D**

"Flat, no obstacle areas exposed to wind flowing over open water for a distance of at least 1.61 km (1 mile). Shorelines in this exposure include inland waterways, the Great Lakes and coastal areas such as California. Exposure D extends inland from the shoreline a distance of 460 m (1500 ft.) or 10 times the height of the building or structure, whichever is greater".

Numerous researchers have obtained the ABL in a wind tunnel (for example, [2, 3]). They and others have contributed in many ways to construct the atmospheric boundary layer in the wind tunnel. This simulation of the atmospheric boundary layer in a long test section wind tunnel is performed by simulating a long stretch of terrain upstream of the model so that vertical shear and turbulence will develop naturally. [4] remarks that, there must be target characteristics which should be met. [5]have defined that "the characteristics are; the variation of the mean wind speed and turbulence intensities with height".

2. ExperimentalTechniques

To generate ABL, a combination of spires, fence and surface roughness are needed. To alter the external flow, spires are used that extend the bottom wall into the oncoming flow. Design formulae have been developed by [6]for spires. These are used to determine the height and width of the spires used in this work. The surface roughness elements were also designed by using the design formulae of [6]

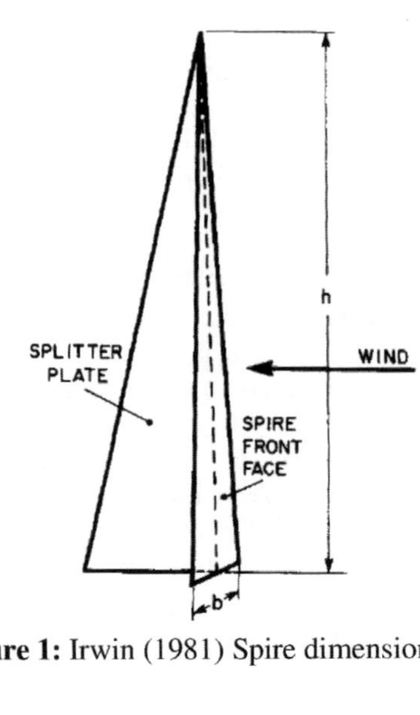

Figure 1: Irwin (1981) Spire dimensions

[4] evaluated three types of roughness surfaces. The first type consisted of wooden strip panels attached to removable thick floor boards. The strip panels lay cross stream to the flow and characterized a "rough terrain".

A later modification added blocks to the floor boards in order to increase the effective roughness of the strip panels. The square blocks were applied in a diamond-shape pattern on the floor board next to the wooden strip panels. Carpet was utilized as the third type of roughness fetch to characterize a smoother type terrain. The carpet extended from the downstream fence barrier to the boundary layer test section area.

In order to avoid wake effects on the model, [7]felt that the last row of the roughness elements should not be closer to the model than ten times their dimension. Also without any further floor roughness between the last row of large roughness elements and model location, an inner boundary layer with a typical smooth-surface velocity profile will develop. In order to retard the development of such a smooth surface profile, seven rows of smaller cubic roughness elements are placed at regular intervals just upstream from the model location.

2.1. Thickening devices/Barriers

The role of the barrier is to give an initial momentum deficit and depth to the layer which is mixed into the developing layer by turbulence generated by the mixing-device. The flow is tricked by the barrier into believing the fetch of roughness to be longer, and by the mixing-device that the barrier is not there at all. In the ideal case, the flow in the test area near the downstream end of the surface roughness should have the characteristics of a boundary layer grown naturally over a much longer fetch of the same surface roughness, without any additional characteristics imposed by the barrier. The intended role of the barrier is therefore to establish the boundary-layer height h. There is often interaction between the barrier and the mixing-device.

[8]has made the arrangement of a castellated barrier wall, vortex generators and surface roughness elements in the wind tunnel test section. The height (127 mm) and form of the castellated barrier wall and vortex generators, as well as the layout of surface roughness elements, were tuned to generate rural, suburban and urban ABL models.

[9] have made use of elliptic generators. They have followed the procedure and the design made by [10]but with a wedge angle of 5.7^0.

2.2. Standard Mean velocity and turbulence intensity profile Mean velocity profiles

The ASCE standard mean velocity profile for the ABL in rural and suburban areas are shown in Fig. 2

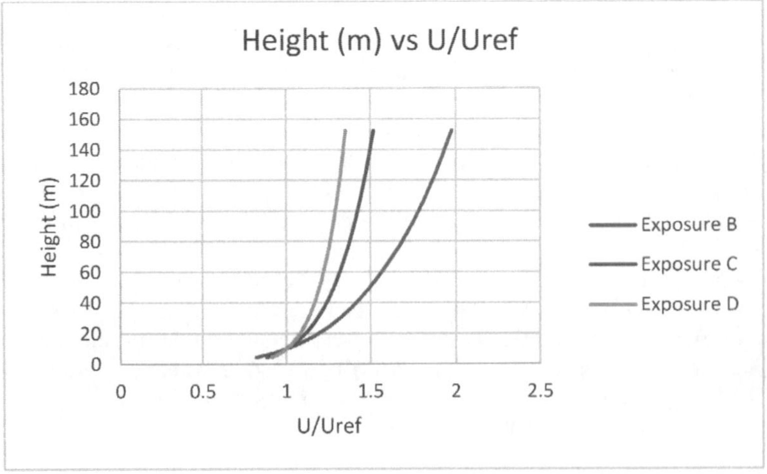

Figure 2: Velocity profile for rural and suburban (ASCE-7, 2010)

Using the following formula to plot the graph

$$\frac{\overline{V}_{z,Tc}}{\overline{V}_{10.Tc}} = \left(\frac{z}{10}\right)^{\overline{\alpha}_{Tc}} \tag{1}$$

Where α values are 0.13, 0.15, and 0.18 for exposure D, C, and B respectively. Meanwhile, V is the mean velocity and z is the height. Subscript Tc stands for terrain category in the ASCE standard.

Turbulence intensity profiles

The ASCE standard turbulence intensity profiles for the ABL in exposures B, C, and D are shown in Figure 3.

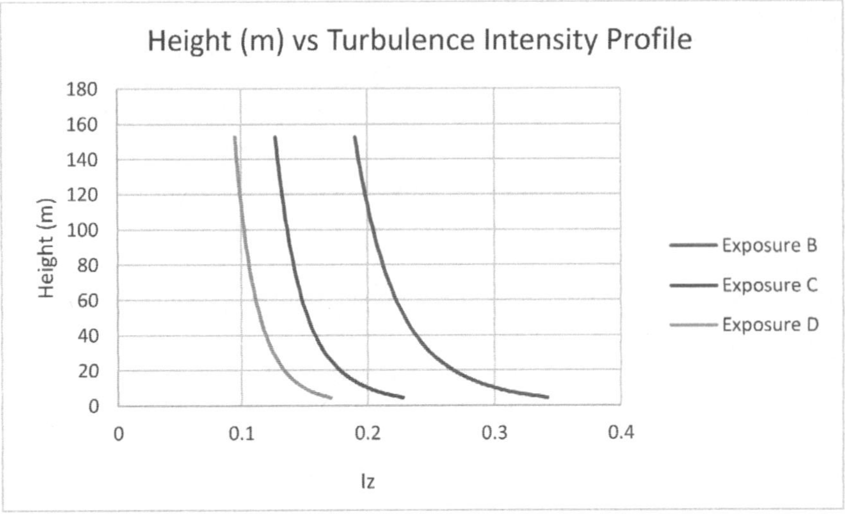

Figure 3: Turbulence intensity profiles (ASCE-7,2010)

Using the following formula to plot the graph

$$I_{z,Tc} = C_{Tc}\left(\frac{10}{\bar{z}}\right)^{\frac{1}{6}} \quad (2)$$

C_{Tc} is a constant for each exposure; 0.3, 0.2 and 0.15 for exposure B, C and D respectively.

2.3. Scaling the profile

Most of the previous studies simply assume the scale for modelling the atmospheric boundary layer in the wind tunnel[11] has defined the scale for simulation in wind tunnel to be 1/500 when the height of the atmospheric boundary layer produced in the wind tunnel is 2m by assuming the real typical boundary layer as 1000 m. Another researcher, [12] also has assumed, the scale to be 1:3000 when he simulates the atmospheric boundary layer is generated in a small wind tunnel with the height of 22.1cm and compare it with the height of 700 m atmospheric boundary layer in the real world.

3. Experiment Setup

The atmospheric boundary layer is generated in the International Islamic University Malaysia Low Speed Wind Tunnel. The IIUM LSWT is a closed-loop type wind tunnel having a test section of 1.5 m x 2.3 m x 6 m with maximum speed of 50 m/s.The arrangement of the devices has been classified into three cases. The details of the cases are as follows:

3.1. Case A:

In this case a combination of spires and blocks as surface roughness are used to generate the atmospheric boundary layer as shown in Figure 4.

Figure 4: Spires and surface roughness installed (Case A)

3.2. Case B:

In this case a combination of spires, fence and blocks as surface roughness are used to generate the atmospheric boundary layer as shown in Figure 5.

Figure 5: Spires, fence and surface roughness installed (Case B)

3.3. Case C:
In Case C, fence and blocks as surface roughness are used to generate the atmospheric boundary layer as shown in Figure 6.

Figure 6: Fence and surface roughness (Case C)

3.4. Measurement device
A constant temperature anemometer (CTA) – also known as a hotwire was used to measure the mean and turbulent velocity as well as the turbulence intensity. A 1-D CTA probe was used. The tests were performed at free stream speeds of 15m/s, 20m/s and 30m/s.

4. Results
Empty wind tunnel test data: Figures 7 show the velocity profiles and turbulence intensity of the empty test section at various speeds. The measurements were carried out near the end of the test

IOP Conf. Series: Materials Science and Engineering **184** (2017) 012008 doi:10.1088/1757-899X/184/1/012008

section (5.21 m from the beginning of the test section) using a 1D CTA probe. The results prove the thin boundary layer and low turbulence intensity even at the end of the test section.

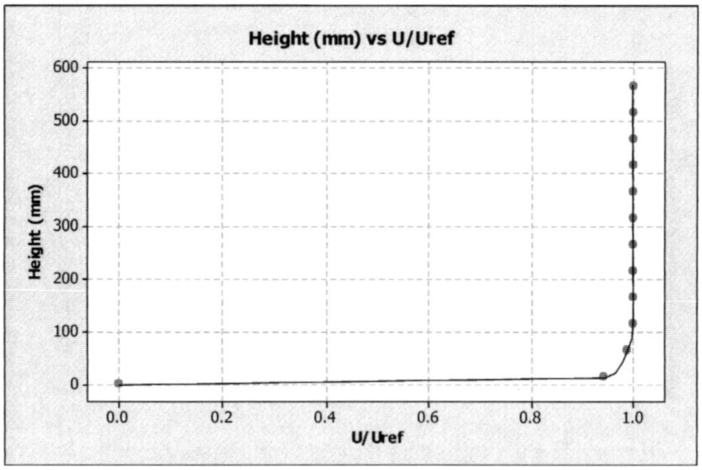

Figure 7: Free Stream Velocity profile for 15m/s

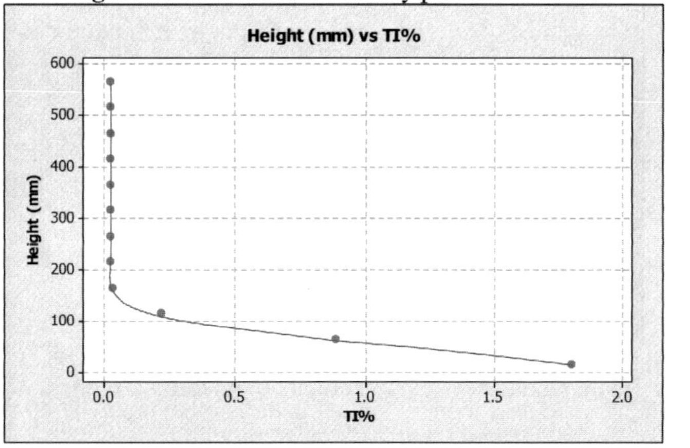

Figure 8: Free Stream Turbulence intensity for 15m/s

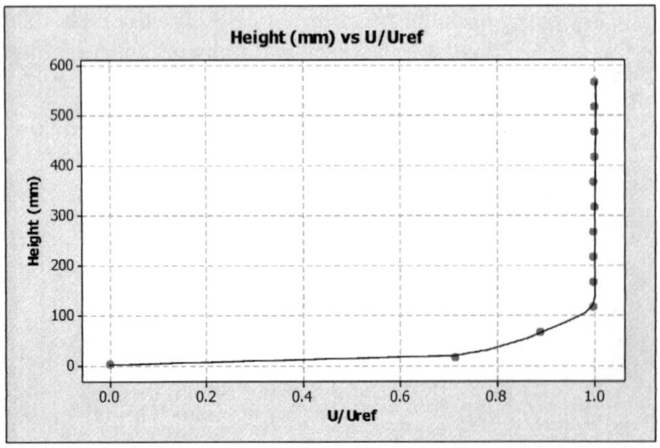

Figure 9: Free Stream Velocity profile for 20m/s

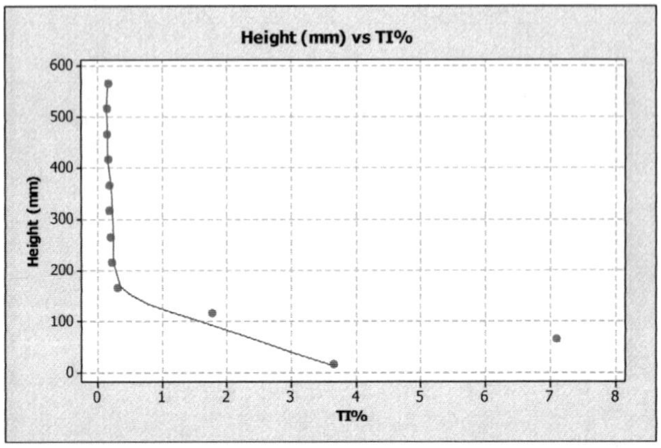

Figure 10: Free stream Turbulence intensity for 20m/s

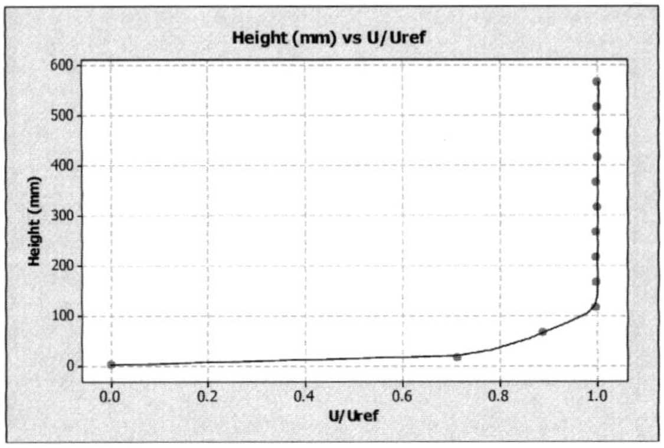

Figure 11: Free steam velocity profile 30m/s

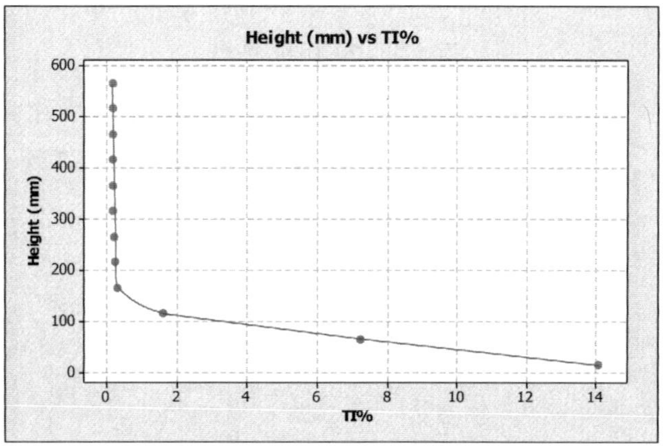

Figure 12: Free stream turbulence intensity for 30m/s

4.1. Analysis of results

The measurements were made at a location 5.21m from the beginning of the test section. The tests have been divided into three cases that are; Case A, Case B and Case C for free stream speeds of 15m/s, 20m/s and 30m/s. Only for case A, a velocity of 30m/s could be attained. For case A, the installation of spires at height of 0.8m and blocks as the surface roughness produced a boundary layer of 750mm height. This method proves that, any desired height of boundary layer, can be produced by using spires of nearly the same height as the required boundary layer height. In all cases, the boundary layer height producedwas nearly of the same height as of the spires used for all speeds tested. However, the turbulence intensity produced varied with the speed. Turbulence intensities of 5.88%, 11.65% and 16.133% were produced, near the floor at speeds of 15m/s, 20m/s and 30m/s respectively, in Case A.

For case B, all three devices were installed together that are, spires, fence and blocks as surface roughness. The data for velocity profile fluctuates more at low level of height that is below 200mm, than in case A. There is no change of height of boundary layer which is the same as in case A. However, there is a difference in turbulence intensity near the floor especially ata speed of 20m/s. The increase in turbulence intensity was substantial at 27.68%. For case C, only the fence and the blocks as the surface roughness were used. From the graph, velocity profile for velocity of 15m/s and 20m/s shows an increment until the height of 700mm and start decrease at the height of 750mm. This happens due to the absence of spires. At a speed of 15m/s, the turbulence intensity is smaller than at 20m/s. The percentage near the floor are 8.36% and 21.31% respectively. Therefore, we can say that, the turbulence intensity near the floor increases as the velocity increases, as well as when the fences are installed. The fence proves to be great device to increase the turbulence intensity as well as slow the air stream.

The experimental data is compared to the ASCE standard. From the comparison it seems that the best combination to simulate the ABL profile of 'Exposure B' is fence and blocks as the surface roughness at a tunnel speed of 20m/s. To simulate the ABL for 'Exposure C', the best combination should be spires, fence and blocks as the surface roughness at the tunnel speed of 20m/s. This combination is achieved by scaling the wind tunnel data to match with the ASCE standard. From the standard, the nominal height of atmospheric boundary layer is 365.76m. Thus, a scale of 1:488 is proposed to match with the velocity profiles of the ASCE 7-10 standard.

The measured turbulence intensity is lower than expected. The major cause seems to be the use of a 1-D probe measuring only the x-component of the turbulence intensity. The actual turbulence intensity would be higher if a 3D CTA probe is used.

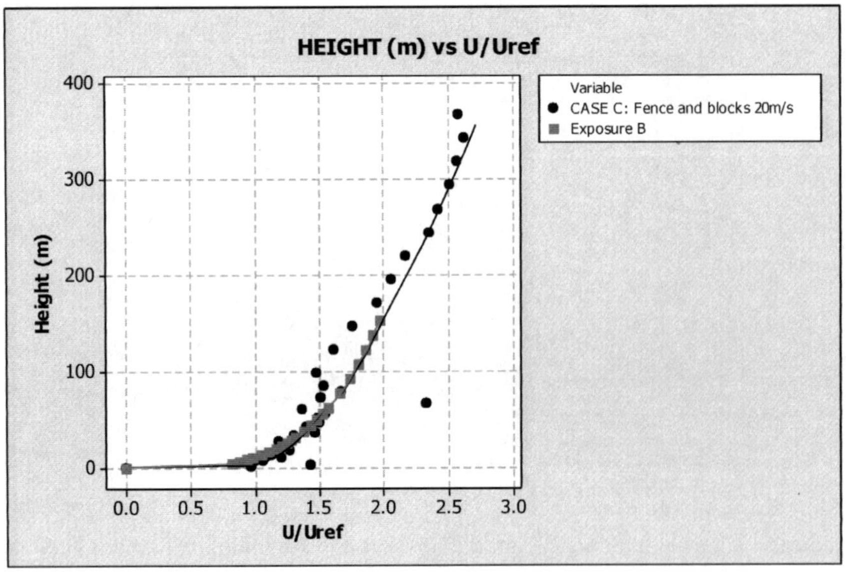

Figure 13: The most suitable comparison between experimental and ASCE standard suburbun velocity profile

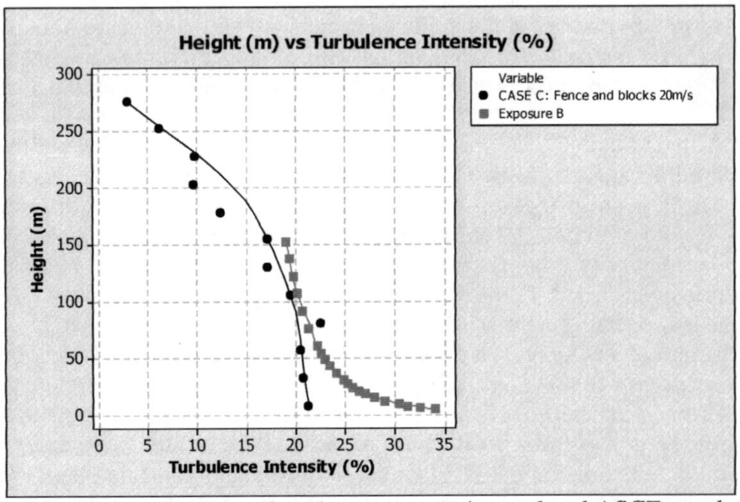

Figure 14: The most suitable comparison between experimental and ASCE standard suburban turbulence intensity profile

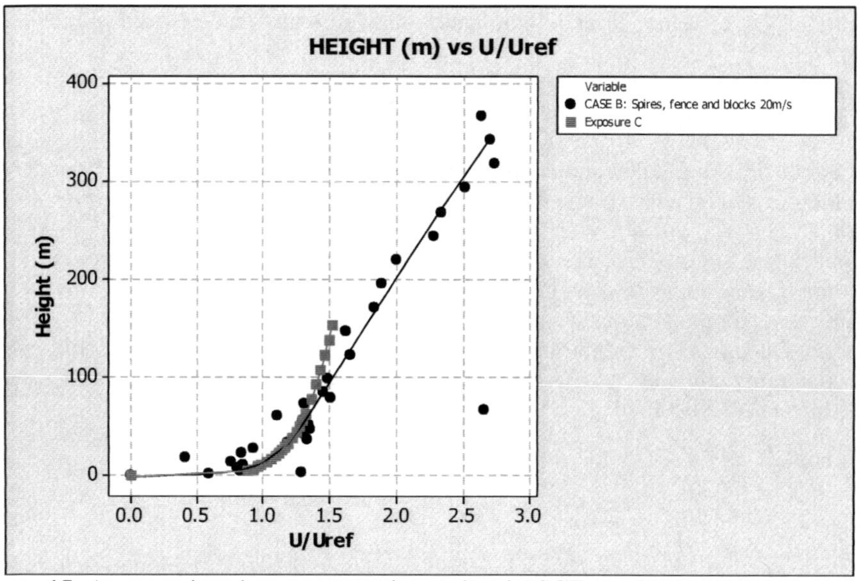

Figure 15: A comparison between experimental and ASCE standard rural velocity profile

5. Conclusion

The results prove that by utilizing spires we can increase the height of the boundary layer and also increase its turbulence intensity. Also by adding fence upstream of the spire we can significantly increase the turbulence intensity. The blocks seem to have little effect on the turbulence intensity and velocity profile. The main function of the blocks is to retard air flow on the lower portion of the boundary layer. In conclusion, all the devices tested prove to be effective in altering the flow properties and simulating an ABL for 'Exposures B and C'.

The suburban and rural atmospheric boundary layer was generated in the IIUM LSWT using the configuration of fence, spires and blocks as the surface roughness. The scale of the simulated ABL is 1:488 after unifying and finding a compromise between the scale of the velocity profile and the turbulence intensity. Further measurements should be made using a 3D CTA probe to obtain better results.

References

[1] ASCE 7-10., Minimum design loads for buildings and other structures. 2010, American Society of Civil Engineers, Reston, VA.

[2] Counihan, J., An improved method of simulating an atmospheric boundary layer in a wind tunnel. *Atmospheric Environment* (1967), 1969. **3**(2): p. 197-200.

[3] Cermak, J.E., Laboratory simulation of the atmospheric boundary layer. *AIAA Journal,* 1971. **9**(9): p. 1746-1754.

[4] Burton, W.V., *Wind tunnel simulation of an atmospheric boundary layer*. 2001, Texas Tech University.

[5] Simiu, E. and R.H. Scanlan, *Wind effects on structures*. 1996: Wiley.

[6] Irwin, H., The design of spires for wind simulation. *Journal of Wind Engineering and IndustrialAerodynamics,* 1981. **7**(3): p. 361-366.

[7] Tieleman, H.W., T.A. Reinhold, and R.D. Marshall, On the wind-tunnel simulation of the

atmospheric surface layer for the study of wind loads on low-rise buildings. *Journal of Wind Engineering and Industrial Aerodynamics,* 1978. **3**(1): p. 21-38.

[8] Kozmar, H. Natural wind simulation in the TUM boundary layer wind tunnel. in *Proceedings of the 5thEuropean-African Conference on Wind Engineering, Florence, Italy*. 2009.

[9] Barbosa, P.H., M. Cataldi, and A. Silva Freire. Simulation of Atmospheric Boundary Layer Flows in Short Wind Tunnels ". in *Anais do XI Congresso Brasileiro de Meteorologia*. 2000.

[10] Armitt, J. and J. Counihan, The simulation of the atmospheric boundary layer in a wind tunnel. *Atmospheric Environment* (1967), 1968. **2**(1): p. 49-71.

[11] Holmes, J.D., *Wind loading of structures*. 2015: CRC Press.

[12] Hobson-Dupont, M., *The Development of a Small Scale Wind Tunnel Simulating the Atmospheric Boundary Layer.* Master's Theses and Graduate Research, San Jose State University (2015).

Investigation of Malaysian Low Earth Orbits for Remote Sensing Applications

M Okasha[1], **I Shuib**[2] **and M Idres**[3]

[1,2,3] Dept. of Mechanical Engineering, International Islamic University Malaysia (IIUM), Kuala Lumpur 50725, Malaysia

E-mail: mokasha@iium.edu.my

Abstract. This paper explores the orbit design procedures of a low Earth satellite orbit for remote sensing mission. The mission objective is to gather images over Malaysia as an area of interest for a variety of applications such as land monitoring and digital mapping. The design procedures are mainly driven by the payload performance, lighting conditions and the coverage requirements. Standard CubeSat platform with a NanoCam is considered for the mission. Numerical investigations show the trade studies of various satellite orbital parameters and their impact on the performance. The orbital parameters are optimized to meet the mission design requirements.

1. Introduction

"Remote sensing is the practice of deriving information about the Earth's land and water surfaces using images acquired from an overhead perspective, using electromagnetic radiation in one or more regions of the electromagnetic spectrum, reflected or emitted from the Earth's surface" [1]. It is the process of gathering information about an object without direct physical contact. The photographic images taken by a camera, is the most common remote sensing method for collecting of information.

Recently "CubeSat" satellites are widely popular for a wide range of remote sensing applications. These include not only educational but commercial applications such as mapping, real estate and construction, and oil and gas monitoring as defined by the NASA web site. Orbit design is a crucial process for remote sensing applications. It involves the selection of the orbital parameters to satisfy the mission objectives successfully. This process is driven basically by the payload performance, lighting conditions, access and coverage requirements and the satellite lifetime [2-4].

In this paper, the process of selecting the orbital parameters are presented. Malaysia is considered as an area of interest. Standard CubeSat platform with a NanoCam is selected for this mission [5]. Numerical investigations and trade studies are performed using MATLAB [6] and AGI/STK [7] software packages to define the optimum orbital parameters to meet the design objectives. The paper is arranged as follow. Section 2 presents a brief description of the orbital parameters and defines the mission requirements. Section 3 shows the impact of various orbital parameters on the satellite performance through trades studies. Finally, Section 4 provides conclusions and recommendations.

2. Orbital parameters and mission requirements

The six classical Keplerian orbital elements can be used to define the satellites orbit as shown in Figure 1. These elements are: semi-major axis a, eccentricity e, inclination i, ascending node Ω, argument of perigee ω and true anomaly ν. The main goal of orbit design is to find the optimal set of these parameters to meet the mission objectives (see Figure 2).

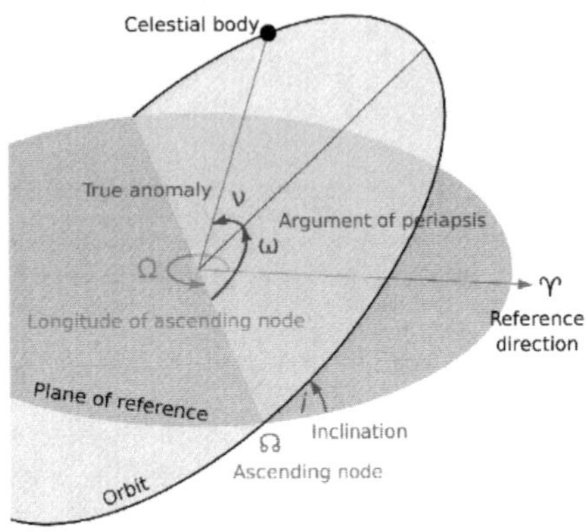

Figure 1. The Six Keplerian Elements.

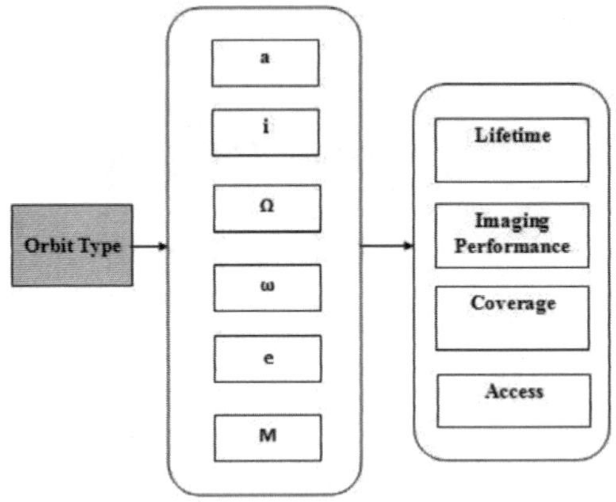

Figure 2. The Orbit Design Flow.

The mission primary objective is defined to gather images over Malaysia with the following requirements:

- 100 % coverage of the entire territory within two weeks time period;
- At least, 10 minutes daily access with the Malaysian Space Centre ground station;
- At maximum, 100 meters ground sample resolution;
- All the captured images are during the direct Sun light;
- At least 5 years satellite lifetime.

NanoEye CubeSat platform is selected with a NanoCam as shown in Figure 3 and Figure 4 [5]. This type of satellite provides a standard design for picosatellites which helps in reducing cost and development time, as well as enable small organizations with limited budget to get involved in space exploration activities. This system is designed specifically for low cost Earth observation missions with effective resolution better than 100 meters per pixel.

To better understand how the low Earth satellite orbit and payload parameters impact on access, coverage and lifetime mission requirements, a series of parametric studies is considered. For each parametric study, a single design parameter is varied through a range of values and at each value the design requirements are analysed. MATLAB and AGI/STK software are used through numerical investigations to select the orbital parameters that compromise the fulfilment of all the mission requirements.

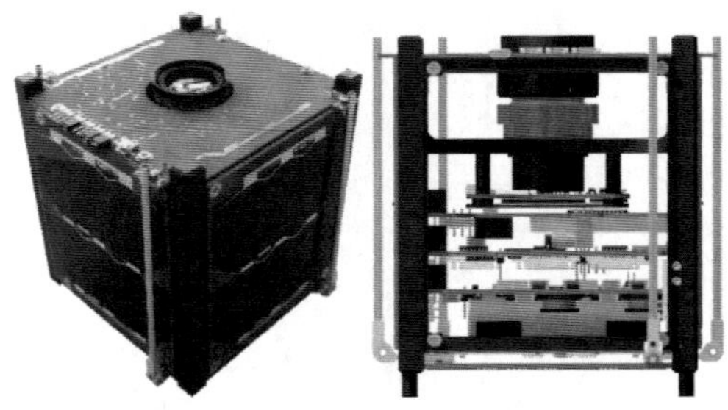

Figure 3. NanoEye CubeSat Platform [5].

Figure 4. NanoCam C1U [5].

3. Numerical studies

Table 1 shows the basic properties of the STK scenario. It includes the properties of the scenario, satellite, payload, and ground station. In this scenario, a CubeSat satellite is launched to monitor the territories of Malaysia and to scan its surface. Data and images that collected from the sensors are then transferred to Malaysian Space Centre ground station for processing during the communication sessions as shown in Figure 5.

Table 1. STK Scenario setup.

	Item	Value
Scenario Properties	Analysis Period	Start: 22 Jun 2016 04:00:00.000 UTCG Stop: +15 days
	Epoch Time	22 Jun 2016 04:00:00.000 UTCG
Satellite Properties (CubeSat)	Propagator	High-Precision Orbit Propagator (HPOP)
	Orbital Parameters	Default parameters
	Mass, drag area, radiation area	Mass = 1 kg Drag Area = Radiation Area =0.02 m^2
Payload Properties (NanoCam C1U)	Sensor Type	Rectangular
	Rectangular	Vertical Half Angle = 5.4 deg Horizontal Half Angle = 4.1 deg
	Resolution	Focal length = 0.035 m Detector Pitch = 0.0032 mm
	Constraints	Resolution/Ground Sample Distance/Max = 100 m
Ground Station (Malaysian Space Centre)	Position	Latitude = 2.784 deg Longitude = 101.508 deg Altitude = 0.005 km
Coverage	Area of interest	Malaysia

Figure 6 and Figure 7 show the basic configuration of the imaging sensor (NanoCam C1U). It is a rectangular CMOS sensor with 35 mm focal length and 2048 x 1536 pixels. The formulas in Figure 6 are used to compute the sensor detector pitch (DP) and vertical and horizontal half angles. This sensor can provide 60 m resolution at 650 km altitude.

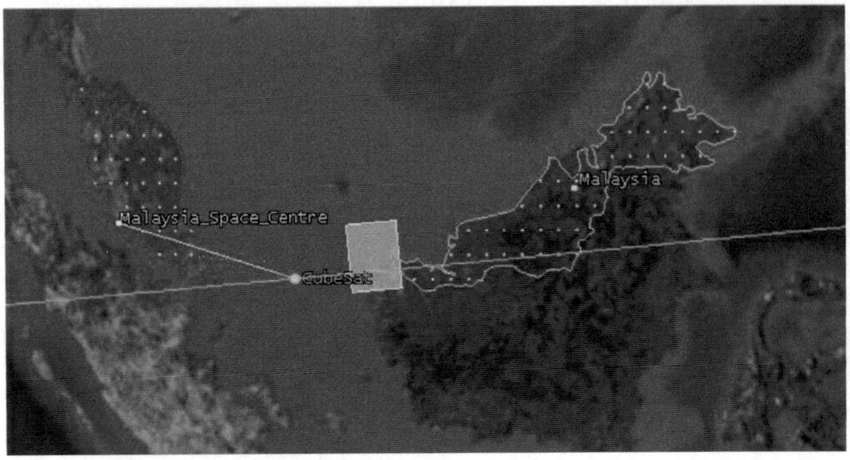

Figure 5.STK Scenario Elements.

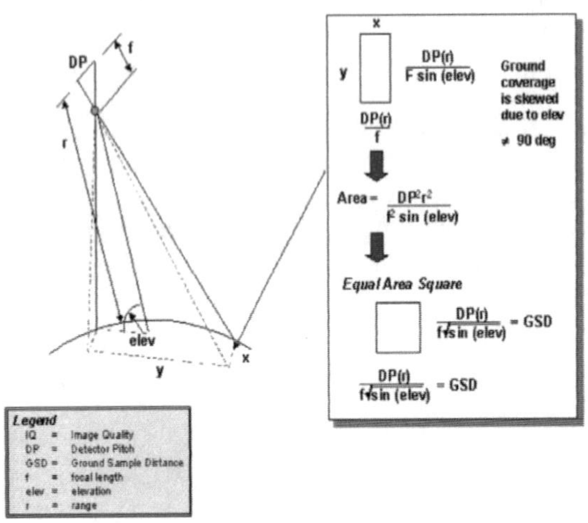

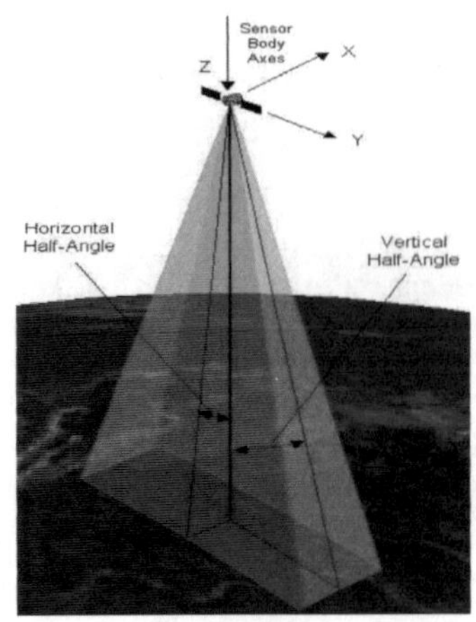

Figure 6. Sensor Resolution.

Figure 7. Rectangular Sensor.

It is required here to specify the orbital parameters and senor configuration to meet the mission requirements mentioned in the previous section. To do that, a series of parametric studies will be run to better understand how the satellite orbit and sensor parameters impact coverage capabilities. For each parametric study, a single design parameter will be run through a sweep of values and coverage statics on each grid shown in Figure 5 will be collected using the Figure of Merit object in STK. The coverage per day statics of the Figure of Merit will be used to determine the minimum, maximum and average values for each grid during the scenario time period. The goal here is to find the optimal configuration of these parameters to best cover the territories of Malaysia.

To determine the impact of inclination of coverage, Figure 8 shows the inclination trade study. Results from this study are expected, the closer from the country latitude, the better coverage capabilities will be. This study gives a better understanding on how to assure maximum coverage of the whole country. The inclination region of interest between 5 and 12 degrees will assure at lease few seconds coverage per day. An inclination of 5.5 degree will be selected form this study.

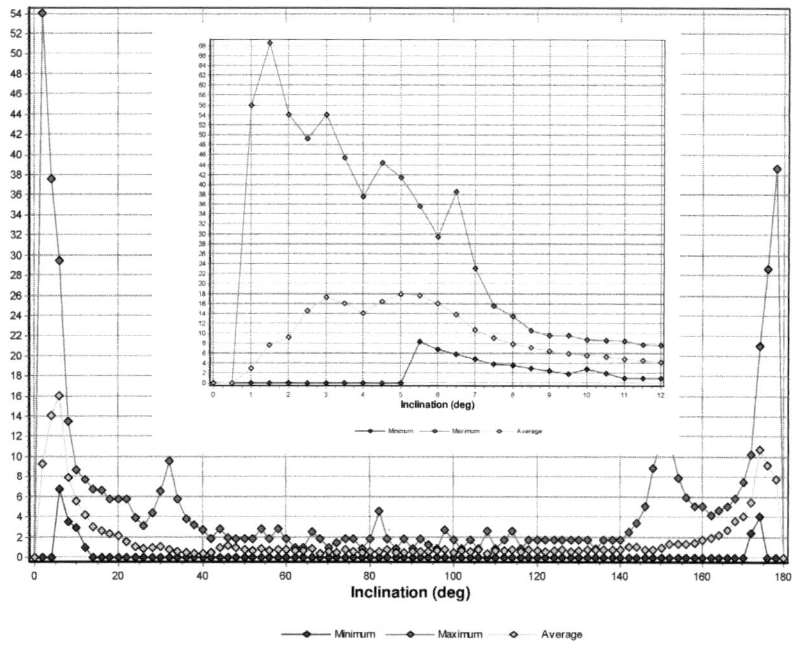

Figure 8. Inclination Trade Study Including Zoom In Part In The Region of Interest.

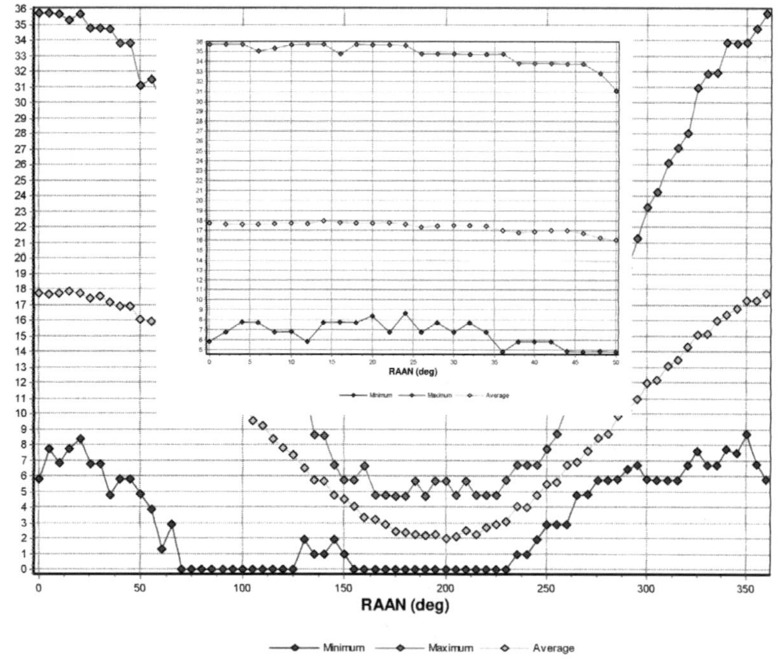

Figure 9. RAAN Trade Study Including Zoom In Part In The Region of Interest.

Another parameter that might impact the coverage capabilities is the RAAN. Figure 9 shows the RANN trade study. From this study, two regions are interested. One is between 0 and 50 degrees and the other is between 300 and 360 degrees. A 24 degree RAAN is selected from this study to maximize the grid minimum coverage per day.

Altitude or alternatively the classical orbital parameter, semi-major axis, is another parameter that impacting coverage. For this study, semi-major axis is varied form 6,300 km to 7,500 km as shown in Figure 10. As expected, coverage increases rapidly with semi-major axis until certain altitude is reached at which point there is an equally sharp drop off. This due to a resolution constraint of 100 metres. Beyond 7450 km semi-major axis, the sensor can no longer resolve 100 meter objects.

The final orbital parameters considered in this study that might impact coverage capabilities are eccentricity and argument of perigee. Figure 11 and Figure 12 show the trade study of these parameters respectively. From these figures, the data appears to be insensitive to eccentricity and argument of perigee. In summary, the orbital parameters listed in Table 2 are selected to be the initial guess of the influenced parameters that expected to achieve all the mission requirements.

Orbital parameters and sensor parameters have a large impact on coverage capabilities. In order to optimize these parameters, an optimization tool will be employed. Although, each parameter trends are clearly shown on previous analysis, guessing all parameters will be difficult because multiple parameters have a wide range. On previous studies, one parameter was changed at a time. To solve this complex problem, optimization techniques can be very useful. In these optimization algorithms, all the parameters are systematically modified until some goals are achieved. AGI/STK [7] optimizer is an automated tool that utilizes mathematical calculations about a design problem and incrementally attempts to find an optimal solution. The intelligent SEQOPT algorithm in this tool uses surrogate models to accelerate the optimization process. It systematically modifies variables in a Scenario until some objective is met.

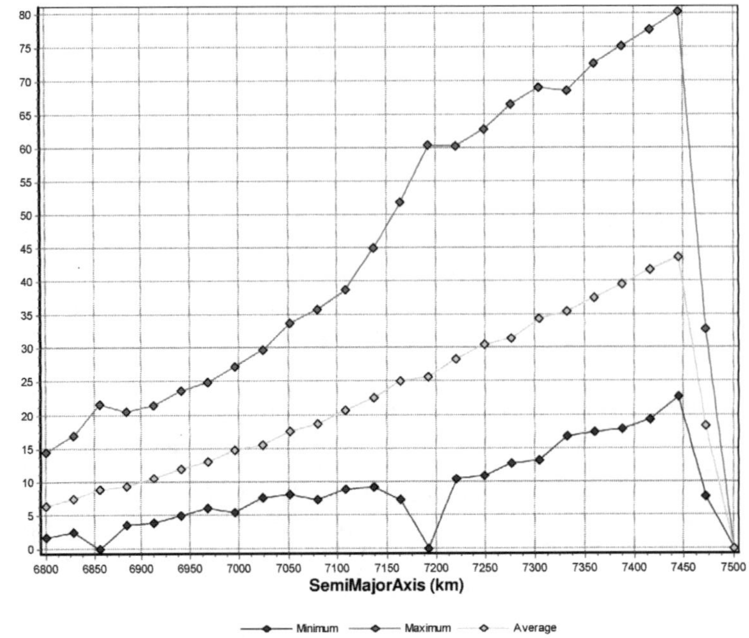

Figure 10. Semi-major Axis Trade Study.

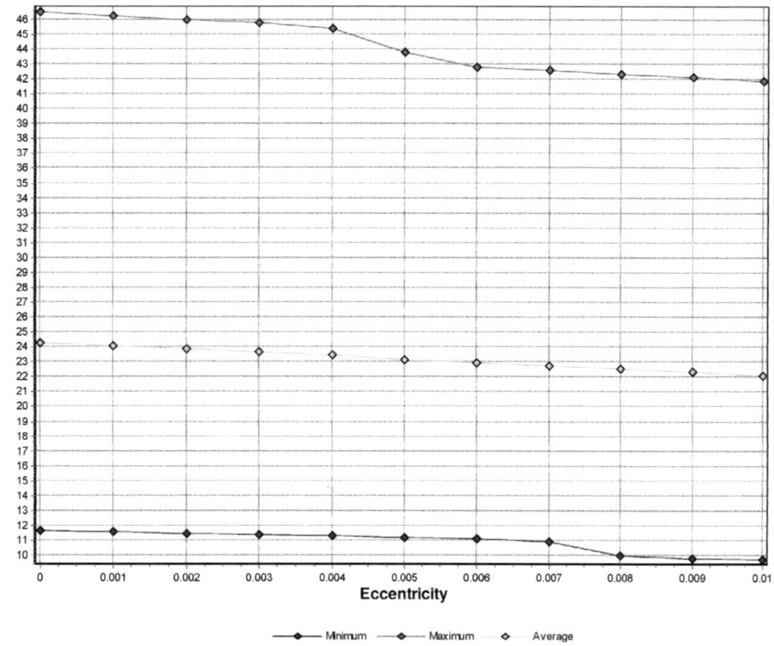

Figure 11. Eccentricity Trade Study.

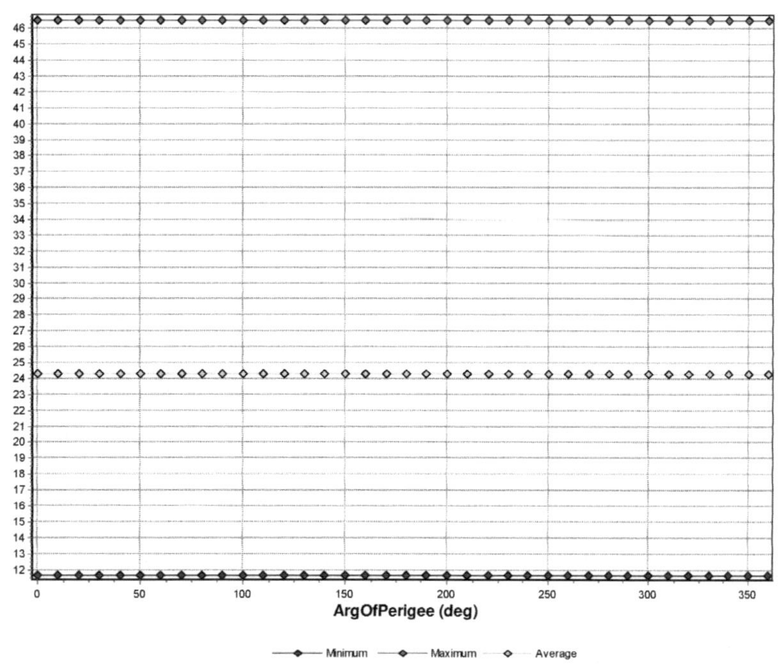

Figure 12. Argument of Perigee Trade Study.

Table 2. Initial Orbit Parameters.

Parameter	Value
Inclination	5.5 deg
Semi-major axis	7140 km
RAAN	24 deg
Argument of Perigee	0 deg
Eccentricity	0 deg

Two optimization cases are considered here. The objective for both is to maximize the minimum coverage per day to monitor each grid point on the map considering lighting and resolution constraints. In the first case, the satellite considered to be nadir pointing without attitude manoeuvres capabilities. while in the second one, the satellite has the ability to perform the attitude manoeuvres. The optimal results of both cases are presented in Table 3. It is observed that the attitude manoeuvres capabilities dramatically improve the grid minimum coverage. In both cases, the satellite can achieve more than 10 min communication access time with the ground station. In addition, the 5 years lifetime requirements are met.

Table 3. Nadir and Off Nadir Optimized Parameters.

Parameter	Nadir/Value	Off Nadir/Value
Inclination	5.8 deg	5.4 deg
Semi-major axis	7245.1km	7250 km
RAAN	38.9 deg	10.5 deg
Grid Minimum Converge	13.9 sec	104.4 sec
Mean Access duration	14.1 min	14.2 min
Satellite Lifetime	> 5 years	> 5 years
Attitude Capabilities	None	≈28 deg

4. Conclusion

In summary, the orbital parameters design trades are presented for a low Earth orbit considering Malaysia as an area of interest. The orbital parameters have been selected to meet all the mission requirements. Trade studies show that altitude, RAAN and inclination have the most impact on these mission requirements. Numerical optimization techniques are employed to optimize the coverage capabilities considering realistic sensor parameters and constraints. These studies show that attitude manoeuvres greatly improve coverage requirements.

References

[1] James B. Campbell and Randolph H. Wynne "Introduction to Remote Sensing," The Cuilford Press New York, NY, 2001.

[2] J. R. Wertz, D. F. Everett and J. J. Puschell, editors, "Space Mission Engineering: The New SMAD," Microcosm Press, Hawthorne, CA, 2011.

[3] L. Qiao, C. Rizos and A. Dempster, "Satelllites Orbit Design and Determination for the

Australian Garada Project," Proceedings of the 24th International Technical Meeting of The Satellite Division of the Institute of Navigation (ION GNSS 2011), Portland, Oregon, September 20-23, 2011.

[4] Karim Kamalaldin and Mohamed Okasha "Low Inclination Circular Orbits for Remote Sensing Satellites," Applied Mechanics and Materials, Vol. 629, pp. 298-303, 2014.

[5] GomSpace's systems for cubesats: http://www.gomspace.com/index.php?p=products

[6] Using MATLAB, TheMathworks Company, Inc., Natick, Massachusetts, Version 15.

[7] Satellite Tool Kit, Analytical Graphics, Inc., Malvern, PA.

Aeroelastic passive control optimization of supersonic composite wing with external stores

E Sulaeman[1], N A Abdullah[1,2] and S.M. Kashif[1,3]

[1]Department of Mechanical Engineering, International Islamic University Malaysia
[2]Department Aerospace Engineering, University of Sheffield, UK
[3]Presently at SABIC, the Netherland

E-mail: esulaeman@iium.edu.my

Abstract. This paper provides a study on passive aeroelastic control optimization, by means of aeroelastic tailoring, of a composite supersonic wing equipped with external stores. The objective of the optimization is to minimize wing weight by considering the aeroelastic flutter and divergence instability speeds as constraints at several flight altitudes. The optimization variables are the composite ply angle and skin thickness of the wing box, wing rib and its control surfaces. The aeroelastic instability speed is set as constraint such that it should be higher than the flutter speed of a metallic base line model of supersonic wing having previously published. A finite element analysis is applied to determine the stiffness and mass matric of the wing and its multi stores. The boundary element method in the form of doublet lattice method is used to model the unsteady aerodynamic load. The results indicate that, for the present wing configuration, the high modulus Graphite/Epoxy composite provides a desired higher flutter speed and lower wing weight compare to that of Kevlar/Epoxy composite as well as the base line metallic wing materials. The aeroelastic boundary thus can be enlarged to higher speed zone and in the same time reduce the structural weight which is important for a further optimization process.

1. Introduction

The importance to optimize structural weight has been the goal since the beginning of air vehicle design concept. The effort has been conducted through structural optimization by implementation of structural design concept such as semi- monocoque design, improvement on structural analysis using finite element method to accurately estimate stress and displacement level of highly redundant complicated structure, enhancement on manufacturing capability to produce just necessary structural thickness, and advancement on material technology such as to use composite and more recently smart material.

On the other hand, reducing structural weight may induce some adverse effects that relate to comfort and safety aspects. One important adverse effect is that the air vehicle may become sensitive to vibration or flutter where, at this point, aeroelasticity plays important part. The interaction between structural weight and stiffness under steady and unsteady aerodynamic loading is thoroughly investigated in aeroelasticity[1]. For the reason of safety issue, aeroelasticity becomes necessary procedure in aircraft design. All civil and military aircraft standard regulations include aeroelastic stability and response assessment as part of the requirement [2 – 4]. A huge number of studies have been devoted to increase the accuracy and efficiency of aeroelastic estimation through analytical, numerical and experimental investigations [1].

Advancement in composite technology itself opens a new branch in aeroelasticity called aeroelastic tailoring which is defined as, according to Ref. 5, a technique to passively control structural deformation by altering the structure directional stiffness to improve its aeroelastic performance. The first implementation of aeroelastic tailoring is perhaps as early as 1949 where Munk design a propeller blade made by wood [6]. Realizing that the wood has a non-isotropic material properties, Munk rotated the fiber orientation of the blade wood material in order to gain a beneficial effect of the blade deformation under its selected operation condition. Aeroelastic tailoring is also called passive aeroelastic control [6] as the composite fiber orientation can be used to passively control aeroelastic mode shapes which can improve critical flutter and divergence speed. Without aeroelastic tailoring, the swept forward wing concept may not be accomplished into a safe high speed aircraft [5].

Many articles have contributed important review on the direction of aeroelastic tailoring [5 - 7]. Recently, Ref. 6 presents a comprehensive state of the art and promising technology of passive aeroelastic control that may lead to a new structural design. Some of previous optimization studies investigate the wing composite laminate with fiber orientations of 0°, ±45° and 90° arrangements[6] or, in other word, the variation on the fiber orientation between laminates is restricted to be 0°, 45° or 90°. In the present work, the optimization is performed in such a way that the variation of the fiber orientation can be as small as 5°. With this more relax restriction, it is expected that the aeroelastic performance can be further improved.

2. Aeroelasticity and Optimization

2.1. *Structural model*

The present work performed the aeroelastic analysis based on two numerical simulation models: structural model and aerodynamic model. The structural model of the wing is constructed based on finite element method where combination of quadrilateral shell elements and bar elements are used. The in-plane membrane and lateral bending characteristics of the quadrilateral elements are exploited to model the composite laminate for each layer with different material orientation direction. The launcher of the external stores is modeled as bar element where its attachment to the wing box is carefully set to simulate accurate boundary conditions. In the present work the composite materials used is Kevlar/Epoxy. The materials data is taken from [12] as shown in Table 1. Note that the Kevlar/Epoxy is selected since its modulus elasticity E_l is near the baseline model material which is Aluminum, but with lighter density.

2.2. *Aerodynamic model*

The aerodynamic model of the wing for the supersonic region is based on the so-called ZONA51 method, a boundary element method for unsteady aerodynamics of multiple lifting surfaces oscillating in supersonic flow [13]. For the subsonic region, the doublet lattice method of MSC/Nastran is used [13]. The wing lifting surface is divided in chordwise and spanwise directions into a number of trapezoidal panel elements. The parallel side of the trapezoidal element is set in line with the incoming flow. It is noted that even though the present wing is designed for the supersonic region, the flutter analysis is required to be conducted at the subsonic region as well to ensure the instability does not occur in the complete flight design envelope. To ensure compatibility between the structural finite element and aerodynamic boundary element model, an infinite surface spline method is used such that the 6 degrees of freedom structural deformations at each point is related to 6 degrees of freedom aerodynamic forces of each trapezoidal panel [13].

2.3. *Aeroelastic Solution*

Prior to the aeroelastic solution, a modal analysis is performed using SOL 103 module of MSC/Nastran to calculate the vibration frequencies and mode shapes. SOL 145 module of MSC/Nastran is then used to solve the aeroelastic solution, where in the present case, the British PK method of the flutter solution is selected. The results of SOL 145 module are in the form of a set of frequency and structural damping for each speed. The flutter speed and frequency are then obtained by interpolating the structural damping to zero. This aeroelastic procedure is repeated for several altitudes covering the required operational flight zone. Note that a certain negative altitude for aeroelastic analysis is required in the military and civil standard regulations [2 - 4] to take into account the uncertainty of atmospheric properties among others as discussed in detailed in [9].

Table 1. Composite material properties (adapted from [12])

	S-glass/ epoxy	Kevlar/ epoxy	HM Graphite/ epoxy
Elastic Properties:			
E_1, GPa	55	80	230
E_2, GPa	16	5.5	6.6
G_{12}, GPa	7.6	2.1	4.8
ν_{12}	0.26	0.31	0.25
Tensile Strengths:			
σ_1, MPa	1800	2000	1100
σ_2, Mpa	40	20	21
σ_{12}, MPa	80	40	65
Compressive Strengths:			
σ_1, MPa	690	280	620
σ_2, MPa	140	140	170
Physical Properties:			
α_1, $10^{-6}/°C$	2.1	−4.0	−0.7
α_2, $10^{-6}/°C$	6.3	60	28
Volume fraction	0.7	0.54	0.7
Thickness, mm	0.15	0.13	0.13
Density, Mg/m^3	2.0	1.38	1.63

2.3. Optimization

The optimization is performed with the objective to minimize the wing weight by considering the flutter speed as constraints. The constraint is set such that it should be higher than the flutter speed at the flight altitudes of the supersonic Aluminum base line wing model of Ref. 8. The optimization variables are the composite ply angle, skin thickness and material of the wing box, wing rib and its control surfaces. In the present work, the optimization is performed first to find the optimum material orientation for each layer of the wing skin. The increment of the angle orientation of each layer is selected as 5°. The second step is to find the optimum composite layer thickness with the increment of $^1/_8$ of the thickness of the base line model. The third step is to find the optimum wing rib thickness that it is found to be sensitive to the flutter speed result. The reduction of the rib thickness is performed by considering the spanwise distance from the wing root. In this case, the wing is divided into four spanwise zones where its zone has the same wing rib thickness.

3. Validation of the Flutter Analysis Procedure

To demonstrate the present aeroelastic procedure, comparison with two benchmark cases are conducted. The first case is to check the flutter speed at subsonic flow by referring to the well-known BAH wing model. A more detailed data of the BAH wing can be found in Ref. 14. The second case is to check the flutter speed at supersonic flow by comparing to the Hedgepeth wing model [15]. The results of both cases are presented in Tables 2 and 3 which show a good agreement with those of the benchmark models. Note that the critical eigenvalue in Table 3 is another non-dimensional representation of flutter frequency.

Table 2. Comparison of subsonic flutter solution procedure results

Parameter	Bisplinghoff et al.[14]	Present procedure result
Flutter speed,V (m/s)	386.69	339.85
Flutter frequency, f (Hz)	2.960	2.976

Table 3 Comparison of supersonic flutter solution procedure results

Parameter	Hedgepeth [15]	Present procedure result
Critical eigen value	492	493.96
Flutter speed, V (m/s)	570.36	577.59

4. Baseline Configuration

To demonstrate the enhancement of the present aeroelastic passive control, the results are compared with a baseline configuration which is a metallic wing with multiple external stores presented in [8]. The baseline model has a wing span of 7.5 m, aspect ratio of 5, taper ratio of 0.5 and leading edge swept-back angle of 30°. The wing section has a diamond shape airfoil with the wedge angle of 10°. The wing box portion in is 65% of chord-length as shown in figure 1, whereas the leading edge and trailing edge are 15% and 20%, respectively. The AMRAAM AIM-120A external sores are positioned at two locations along the wing span and Sidewinder AIM-9M at wing tip. A detailed wing sizing has been conducted in [8] for this configurations with a flight load factor of n_z = 5.5 and safety factor of 1.5. With this load factor, the wing sizing was performed resulting to the total wing and external store weight of 811 kg [8]. The flutter analysis performed at five altitudes shows that the aircraft flutter Mach numbers are M=3.01, M=2.42, M=1.98, M=1.65 and M=1.44 at the altitudes of 30 kft, 20 kft, 10 kft, sea level and -7.9 kft, respectively.

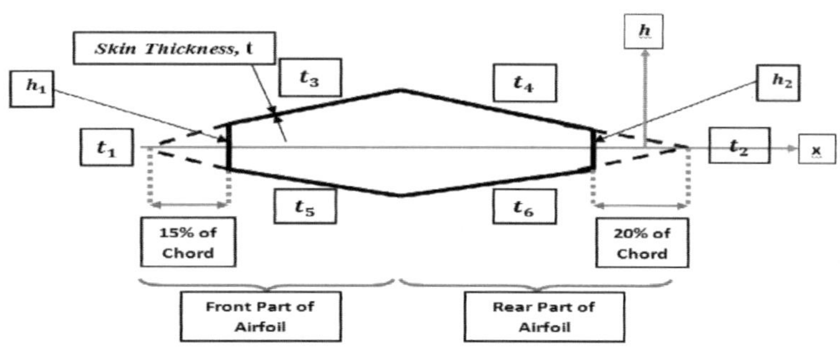

Figure 1: Double wedge airfoil (Taken from [8] with permission)

5. Kevlar/Epoxy Laminate Configuration

A preliminary work using Kevlar/Epoxy composite laminate for the surface skins of wing box and control surfaces has been conducted in [10]. Ref. 10 basically performed the first step of the present optimization procedure, where the optimum fiber orientation layer are found to be $[45/20/0]_s$ for the wing skin and $[90/60/45]_s$ for flap and aileron control surfaces. The second step is then performed in the present work to find the optimum thickness of the composite layer. The thickness of the skin layer is decreased with the increment of $^1/_8$ of the thickness of the base line model. Some of important iteration results are shown in Table 4. Note that r_{wing} in Table 4 indicates the wing skin thickness ratio with respect to that of baseline model. Similarly, $r_{surface}$ is the control surface skin thickness ratio with respect to that of baseline model. The clean wing weight is to represent the wing weight without external stores and its launchers.

Table 4. Flutter Mach numbers for the Step 2 iteration

Case	r_{Wing}	$r_{Surface}$	Flutter Mach number					Flutter mode	Clean Wing Weight (kg)
			-8 kft	0 kft	10 kft	20 kft	30 kft		
Baseline*	1	1	1.437	1.651	1.983	2.420	3.008	4	289.1
1#	1	1	1.741	2.003	2.412	2.949	3.668	5	181.3
2#	1	1/2	1.751	2.015	2.427	2.967	3.693	5	162.8
3	3/4	1	1.603	1.844	2.210	2.716	3.381	5	164.7
4	3/4	1/2	1.616	1.859	2.241	2.740	3.411	5	146.3
5	3/4	3/8	1.618	1.862	2.244	2.744	3.415	5	141.7

In Table 4, the result of Baseline is adapted from [8], and the results of Case 1 and 2 are from Ref. 10. Cases 3 until 5 (and a number of iterations not showing in Table 4) are the optimization of the present work. Case 1 until 5 results indicates a changes between the flutter mode of the baseline and composite wing. This is as expected that the composite laminate alter the bending and torsional modes in such a way that increase the flutter boundary with a less weight. Case 2 gives the desired highest flutter speed as reported in Ref. 10, however the wing weight is *not* the lowest one. It is Case 5 that gives the lowest weight and in the same time satisfies the constraint that its flutter speed is higher than the baseline configuration. As shown in Table 4, the flutter boundary of Case 5 in all altitudes is enlarge by approximately 113% with the wing weight reduction by 51%. A more detailed result for Case 5 are presented in figure 2 and Table 5 where the first 8 modes are shown. The results show that there are almost no pure bending or torsional modes. Obviously this is due to the influence of the external store along the wingspan and the composite laminates.

Table 5. Natural frequency and mode description of Case 5 configuration

Mode	Frequency, f (Hz)	Mode Description
1	2.223	First wing bending
2	3.753	Combination of Wing bending and torsion
3	5.389	Comb. of Wing torsion and Station 2 external store pitch
4	5.817	Comb. of Wing torsion and bending
5	8.437	2nd Wing bending
6	10.276	2nd Wing torsion and station 3 external store
7	12.297	Comb. of Wing torsion and control surface deflection
8	17.266	Comb. of Wing bending and station 1 external store

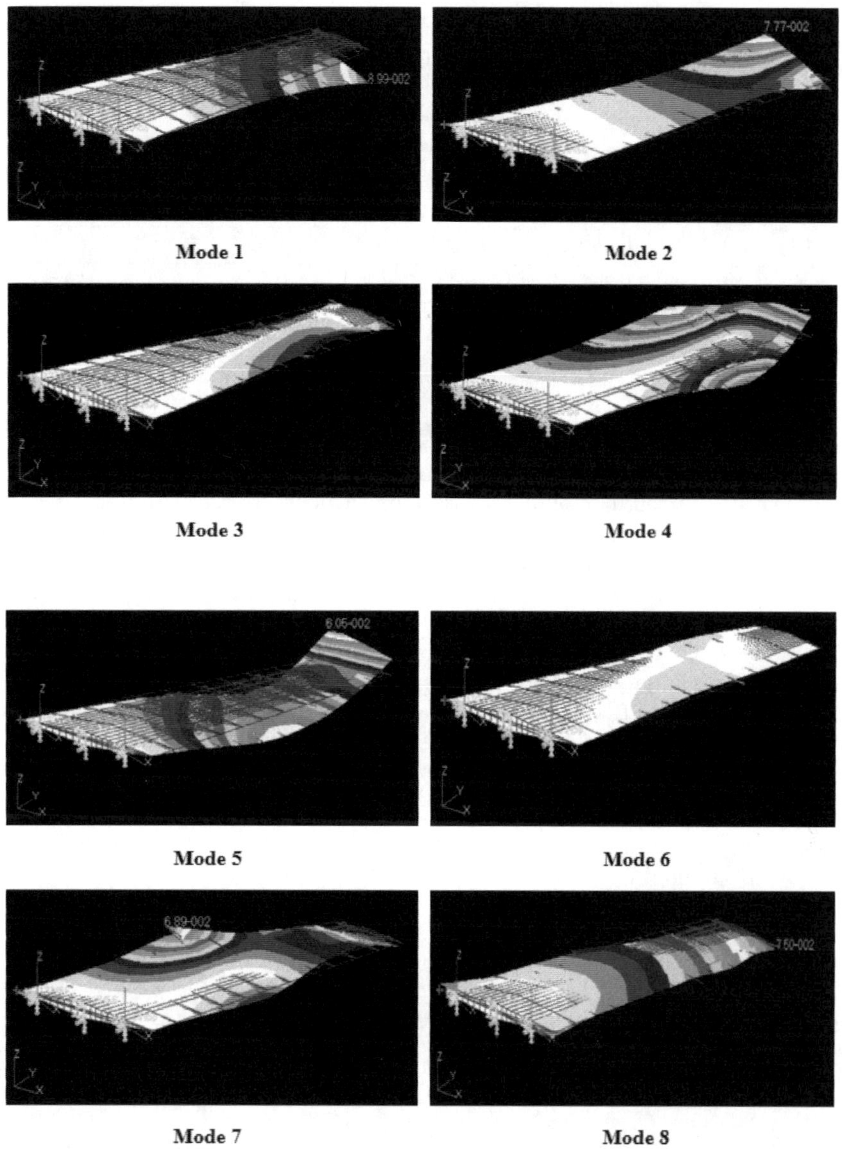

Figure 2. Vibration Mode Shapes of Case 5 Configuration

The third step of the optimization procedure is conducted in the present work by varying the thickness of the wing rib. Started from Case 5 configuration, the wing rib is reduced by taking into account the four spanwise zone along the wingspan. The thickness is reduced for each zone with the incremental reduction ratio of $^1/_8$ with respect to the baseline configuration. A number of iterations are conducted where three cases of the results are presented in Table 6.

Table 6. Flutter Mach numbers for the Step 3 iteration

Case	Flutter Mach number					Flutter mode	Clean Wing Weight (kg)
	-8 kft	0 kft	10 kft	20 kft	30 kft		
Baseline*	1.437	1.651	1.983	2.420	3.008	4	289.1
5	1.618	1.862	2.244	2.744	3.415	5	141.7
5.1	1.618	1.862	2.244	2.744	3.415	5	136.2
5.2	1.607	1.849	2.228	2.725	3.392	6	131.8
5.3	1.628	1.874	2.258	2.761	3.437	6	127.0

Table 6 shows that Case 5.3 is perhaps the optimum configuration for the present wing configuration where the wing weight is the lowest and in the same time the flutter boundary is enlarge by more than 112.4%. A further reduction may be difficult to perform since the strength and manufacturing requirement should be satisfied. The flutter plot for the structural damping and frequency for sea level is shown in figure 3 which shows that Mode 6 is the dominant modes that induced the destructive flutter mode. There is a possible hump flutter mode represented by Mode 3 that may induced a limit cycle oscillation. Figure 4 shows the flutter mode for each altitude. It shows that the flutter mode is very dangerous since the structural damping curve is very stiff near the flutter speed that indicates a very rapid changing of the structural damping with the increase of flight speed. This information would be useful to design the velocity range of the aircraft to assure a safe flight mission profile. In addition, for an explosive type flutter, the flight flutter test should be conducted with a well prepared planning, sufficient real time monitoring of aeroelastic stability and fast, accurate flutter speed predictions.

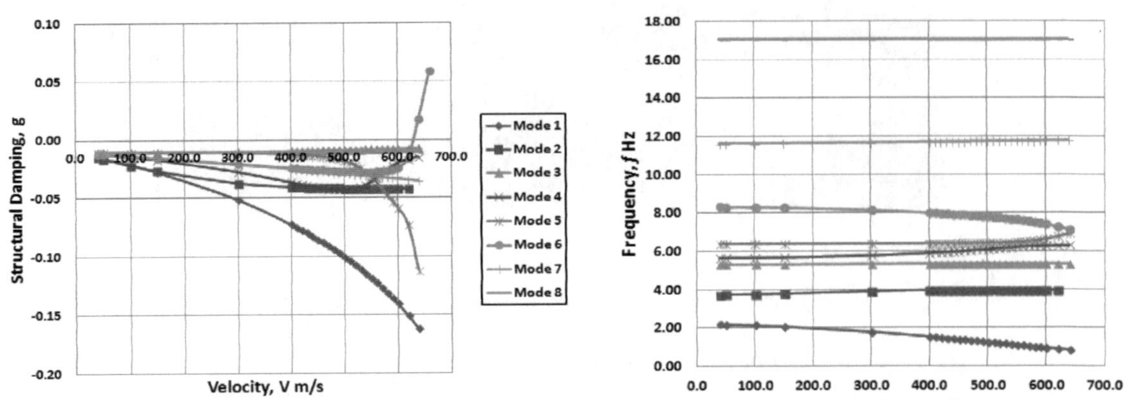

(a) V – g plot (b) V – f plot

Figure 3. Structural damping g and frequency plot of Case 5.3at sea level

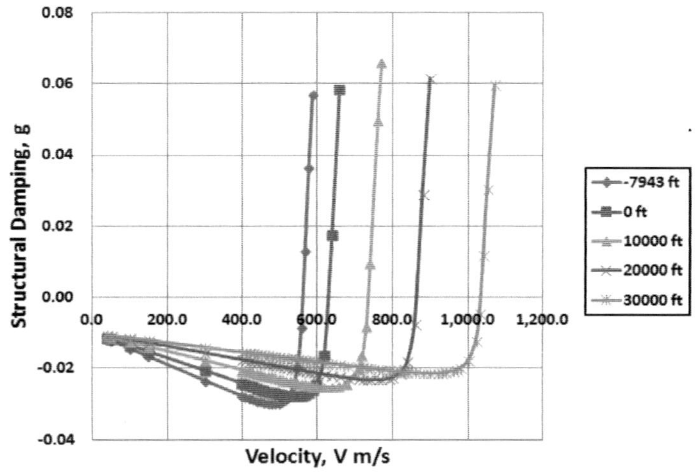

Figure 4. V-g graph of flutter mode of Case 5.3 at several altitudes

6. Conclusion

The present work shows that Kevlar/Epoxy laminate can be used to reduce the wing structural weight and enlarge the aeroelastic stability boundary. The results show that the external store and the anisotropic behavior of the composite laminate influence the vibration mode shapes and therefore the flutter mode characteristic. By exploiting the laminate thickness and fiber orientation of the wing skin, control surface and rib, the lowest wing weight that can be achieved is 56% less than the baseline model with the flutter speed improved to more than 12%. The optimum configuration has an explosive flutter type that needs careful arrangement during flight flutter test.

Acknowledgment

The supports of the Malaysian Ministry of Higher Education (MOHE) under the grant FRGS 13-020-0261 are gratefully acknowledged. Special thanks to the Science and Technology Research Institute for Defense (STRIDE) Malaysia for allowing the present authors to use MSC Nastran and MSC Patran in their facilities.

7. References

[1] Wright J R and Cooper J E 2007 Aircraft Aeroelasticity and Loads (West Sussex: Wiley)
[2] Anon. 1993 Military Specification: Airplane Strength and Rigidity Vibration, Flutter, Divergence. MIL-A-8870C(AS) (US Naval Air Warfare Center Aircraft Division)
[3] Anon. 2009 UAV System Airworthiness Requirement (USAR) for NATO Military UAV System - Standard Agreement STANAG 4671 (NATO)
[4] Anon. 2006 Federal Aviation Regulation, Part 25- Airworthiness Standards: Transport Category Airplanes (US Department of Transportation, Amendment 117)
[5] Shirk M, Hertz T and Weisshaar T 1986 J. Aircraft 23 6-18
[6] Jutte C V and Stanford B T 2014 Aeroelastic Tailoring of Transport Aircraft Wings: State-of-the-Art and Potential Enabling Technologies (TM-2014-218252) (NASA: Langley)
[7] Livne E and Weisshaar T 2003 *J. Aircraft* **40** pp 1047-65

[8] Abdullah N A and Sulaeman E 2014 *IIUM Engineering J* **15** pp 1-11
[9] Sulaeman E 2012 *Appl Mechanics and Materials* **225** pp. 397-402
[10] Abdullah N A and Sulaeman E 2013 *Appl Mechanics and Materials* **464** pp 110-5
[11] Chen P C and Sulaeman E 2003 *AIAA J* **41** pp 1060-8
[12] Roylance D 2000 *Laminated Composite Plates* http://www.saylor.org/courses/ME102 MIT OpenCourseWare
[13] Johnson E and Rodden W H 1994 *Aeroelastic Analysis User's Guide* (Los Angeles: MSC Software Corporation)
[14] Bisplingfoff R L, Ashley H and Halfman R L 1996 *Aeroelasticity* (New York: Dover)
[15] Hedgepeth J M 1957 *J. Aero. Sci.* **24** pp 563-73

Effect of Temperature and moisture dependent material properties on the Bending behavior of Laminated Cylindrical Shell under Hygrothermal Loads

Saleh Alsubari.[1,a] *, J. S. Mohamed Ali[1,b], Yulfian Aminanda[1,c] and M.S.I.Shaik Dawood[1,d]

[1]Department of Manufacturing and Material Engineering
Kulliyyah of Engineering, International Islamic University, Malaysia
P.O.Box 10, 50728 Kuala Lumpur, Malaysia

E-mail : [a]snsubari@gmail.com,[b]jaffar@iium.edu.my, [c]yulfian@iium.edu.my, [d]sultan@iium.edu.my

Abstract. The effect of temperature and moisture dependent material properties on the bending behavior of simply supported angle-ply composite laminated shells has been investigated. A 13 term accurate higher order shear deformation theory with zigzag function is used in this analysis in which the effects of transverse shear deformation are taken into account. A number of examples are solved to illustrate the numerical results concerning bending response of simply supported angle-ply composite shell subjected to hygrothermal effects. It is suggested that temperature-dependent and/or moisture-dependent material properties ought to be used in the analysis of laminated composite shell subjected to hygrothermal loads.

1. Introduction

Laminated composite shell structures are widely used in civil, mechanical, aerospace, and other engineering applications. These materials are becoming popular because of their high specific strength and high specific stiffness. Structures used in the above fields are very often exposed to high temperatures as well as moisture during their service life. The varying environmental conditions due to moisture absorption and temperature seem to have an adverse effect on the stiffness and strength of the structural composites.

Considerable work has been done recently on composite laminated shell which can be found from a complete review of static analysis of shells by Qatu [1]. Very recently numerous works have been reported on the hygrothermal analysis of composite laminates. Salvatore [2] analyzed hygrothermal elastic analysis of multilayered composite and sandwich shells, by means of the Carrera Unified Formulation (CUF). Also Jafari and Sobhani [3] presented the thermo-elastic response of a 2-D functionally graded open cylindrical shell with temperature-dependent material properties based on third-order shear deformation plate theory of Reddy. Brischetto and Carrera [4] studied the static analysis of multilayered smart shells subjected to mechanical, thermal and electrical loads using higher order models. Similar to these works, in this paper the higher order theory presented by Alsubari et al [5] and Ali et al[6,7] for analysis of thermal hygrothermal loads has been extended to include the temperature/ moisture dependent material properties effects on shells. Thus, the aim of this study is to investigate the temperature/ moisture dependent material properties behavior of simply supported, composite laminated cylindrical shell strip, using higher order shear deformation theory.

2. Governing Equation

Consider an infinitely long laminated cylindrical shell panel. The panel coordinate system is such that $0\leq x\leq \infty$, $0\leq \theta \leq \theta_m$ and $-h/2\leq z \leq h/2$. The radius of the shell strip is considered as R and the load is uniform along the x axis so that the shell undergoes cylindrical bending (ε_x=0). The displacement field used in this problem is [5-7]:

$$(u,v)=(u_0,v_0)+z(\emptyset_x,\emptyset_\theta)+z^2(\gamma_x,\gamma_\theta)+z^3(\aleph_x,\aleph_\theta)+\varphi(S_x,S_\theta) \qquad (1)$$
$$w=w_0+z\emptyset_z+z^2\gamma_z$$

where $u(x,\theta,z)$, $v(x,\theta,z)$ and $w(x,\theta,z)$ are the displacement at any point in the laminate. The parameters, u_0, v_0 are the in-plane displacements, and w_0 is the transverse displacement of a point (x,θ) on the middle plane respectively. The functions $\emptyset_x$, $\emptyset_\theta$ are the rotations of the normal to the middle plane about x and θ respectively. The other parameters, γ_θ γ_x, $\aleph_x$, $\aleph_\theta$ and $\emptyset_z$ are unknown higher-order terms which are function of x and θ only. φ is zigzag function term and defined by:

$$\varphi=2(-1)^k z_k/h_k \qquad (2)$$

where z_k is a local transverse coordinate with its origin at the center of the kth layer and h_k is the corresponding layer thickness.

The stress-stain law for the kth layer with respect to the x,θ coordinate is given by:

$$\begin{Bmatrix}\sigma_x\\ \sigma_y\\ \sigma_z\\ \tau_{yz}\\ \tau_{xz}\\ \tau_{xy}\end{Bmatrix}=\begin{bmatrix}C_{11} & C_{12} & C_{13} & 0 & 0 & C_{16}\\ C_{12} & C_{22} & C_{23} & 0 & 0 & C_{26}\\ C_{13} & C_{23} & C_{33} & 0 & 0 & C_{36}\\ 0 & 0 & 0 & C_{44} & C_{45} & 0\\ 0 & 0 & 0 & C_{45} & C_{55} & 0\\ 0 & 0 & 0 & 0 & 0 & 0\end{bmatrix}^k\begin{Bmatrix}\varepsilon_{xx}-\Delta T\alpha_x-\Delta C\beta_x\\ \varepsilon_{yy}-\Delta T\alpha_y-\Delta C\beta_y\\ \varepsilon_{zz}-\Delta T\alpha_z-\Delta C\beta_z\\ \tau_{yz}\\ \tau_{xz}\\ \tau_{xy}-\Delta T\alpha_{xy}-\Delta C\beta_{xy}\end{Bmatrix}^k \qquad (3)$$

where the C_{IJ}^k are the transformed elastic coefficients and $\Delta T=T-T_0$, $\Delta C=C-C_O$ in which, T(x,y,z) and C(x,y,z), are the temperature and moisture distribution loads respectively, whereas T_0 and $C_O$are the reference temperature and moisture concentration respectively. α_x, α_θ, $\alpha_z$and $\alpha_{x\theta}$ are the thermal expansion coefficients and β_x, β_θ, β_z and $\beta_{x\theta}$ are the moisture expansion coefficients.

Expression for the principle of virtual work is

$$\iiint(\sigma_x\,\delta\varepsilon_x+\sigma_\theta\,\delta\varepsilon_\theta+\sigma_z\,\delta\varepsilon_z+\tau_{xz}\,\delta\gamma_{xz}+\tau_{x\theta}\,\delta\gamma_{\theta x}+\tau_{\theta z}\,\delta\gamma_{\theta z})\,dzdA-\iint q\delta w dA=0 \qquad (4)$$

where q is the load per unit area on the mid-surface of the shell.

Integrating the displacement gradient given in Eq. 4 by parts and by setting the coefficient of δu_0, δv_0, δw_0, $\delta\emptyset_x$, $\delta\emptyset_\theta$, $\delta\gamma_x$, $\delta\gamma_\theta$, $\delta\gamma_z$, $\delta\aleph_x$, $\delta\aleph_\theta$, $\delta\emptyset_z$, δS_x and δS_θ to zero separately, we obtain the following equilibrium equations with associated boundary conditions.

$$N_{x\theta,\theta}=0 \qquad H_{\theta,\theta}/R-H_{\theta z}=0 \qquad M_{\theta z,\theta}-M_\theta-RN_z=0$$
$$N_{\theta,\theta}+N_{\theta z}=0 \qquad N_{\theta z,\theta}/R-N_\theta+Rq=0 \qquad K_{\theta,\theta}-T_{\theta z}=0 \qquad (5)$$
$$M_{x\theta,\theta}/R-N_{xz}=0 \qquad K_{\theta z}-K_\theta-2RM_z=0 \qquad H_{x\theta,\theta}/R-H_{xz}=0$$
$$M_{\theta,\theta}/R-N_{\theta z}=0 \qquad J_{\theta,\theta}/R-J_{\theta z}=0 \qquad K_{x\theta}-2RM_{xz}=0$$
$$Jx_{\theta,\theta}/R-3K_{xz}=0$$

Boundary conditions are given as
At θ=constant: one from each of the following bracketed quantities should be specified:

$(N_{x\theta}, u_0)$, (N_θ, v_0), $(M_{x\theta}, \emptyset_x)$,$(M_\theta, \emptyset_\theta)$, $(K_{x\theta}, \gamma_x)$, $(K_\theta, \gamma_\theta)$, $(J_{x\theta}, \aleph_x)$, $(J_\theta, \aleph_\theta)$, $(H_{x\theta}, S_x)$, (H_θ, S_θ), $(N_{\theta z}, w_0)$, $(M_{\theta z}, \emptyset_z)$ and $(K_{\theta z}, \gamma_z)$

The assumed solution for Eq. 5 which satisfies the above boundary conditions is

$$(u_0, \emptyset_x, \gamma_x, \aleph_x, S_x)= (p_1, p_3, p_5, p_7, p_9) \cos(\pi\theta/\theta_m)$$
$$(v_0, \emptyset_\theta, \gamma_\theta, \aleph_\theta, S_\theta)= (p_2, p_4, p_6, p_8, p_{10})\cos(\pi\theta/\theta_m) \quad (6)$$
$$(w_0, \emptyset_z, \gamma_z)=(p_{11}, p_{12}, p_{13}) \sin(\pi\theta/\theta_m)$$

From Eq. 1, Eq. 3, Eq. 5, and Eq. 6, the stress resultants can be expressed in terms of displacements, and finally a system of linear algebraic equations will be formed as:

$$[L]\{\delta\}=\{f\} \quad (7)$$

where [L] is a matrix of differential operators with respect to x and y, $\{\delta\}$ represent the displacement vector $[u_0\ v_0\ w_0\ \emptyset_x\ \emptyset_\theta\ \gamma_x\ \gamma_\theta\ \gamma_z\ \aleph_x\ \aleph_\theta\ \emptyset_z\ S_x\ S_\theta]$, and $\{f\}$ is the load vector that depends on the mechanical and hygrothermal load. Using Eq. (7) unknown coefficient $\{\delta\}$ can be obtained readily, and subsequently in-plane stresses and transverse stresses can be obtained. Although the transverse shear stresses can be calculated from the constitutive relations, these stresses may not satisfy the continuity conditions at the interface between layers, hence transverse shear stresses are obtained by integrating three dimensional equilibrium equations of elasticity.

3. Results And Discussion

In previous studies [5-7], the material properties of the composite laminate are assumed to be independent of temperature and moisture concentration. However the elastic moduli of laminate in general degenerate with the elevation of temperature and moisture concentration. Thus, in this section, we will illustrate the effect of temperature and moisture on the material properties, and hence its effect on deflections and stresses

3.1 Effect of temperature variation

In this section, we will investigate the static response of composite cylindrical shell laminate by taking into account the change of material properties due to temperature variation. This example will examine the effect of temperature variation on the deflections and stresses for four layered angle-ply composite laminate $(45/15)_2$. All laminae are assumed to have same thickness and made of the same orthotropic materials. The material properties used are as given in Table 1 [8]. The load composite laminated cylindrical shells considered to be subjected to a sinusoidal temperature load that is varying through thickness [9]

$$\Delta T = (T-t)\,z(1-4/3\,z^2/h)\sin(\pi\theta/\theta_m) \quad (8)$$

The results obtained expressed in dimensionless terms given as $\bar{w}=w/(h\,\alpha_L \bar{T}_0 S^2)$, $\bar{v}=v/(h\,\alpha_L \bar{T}_0 S)$ and $(\bar{\sigma}_{ij}, \bar{\tau}_{ij}) =(\sigma_{ij}, \tau_{ij})/(E_T \alpha_L \bar{T}_0)$ with $S=R/h$.

The results of deflections and stresses are presented for two different cases. The case of dependent material properties (DMP) on temperature variation and independent material properties of temperature (IMP) are given in Table 2 and Figure 1.It is to be noted that the IMP means that material properties are considered at T=300 K.

Table 2 shows that, when the effects of temperature on material properties are considered, the deflections and displacement decreased, while in-plane and shear stresses increased. Also Figure 1 shows that, as temperature rise, the difference between IMP and DMP results increased.

It is to be noted that, transverse shear stresses results for the case of DMP are less affected by temperature rise than the corresponding one for the case of IMP. In general one can conclude that, as the temperature rises, the difference between DMP and IMP results increased.

Table 1. Elastic properties at different temperature $(\alpha_1 = -0.3, \alpha_2 = 28.1)$

	Temperature T [K]					
Elastic moduli [GPa]	300	325	350	375	400	425
E_2	9.5	8.5	8.0	7.5	7.0	6.75
G_{12}	6.0	6.0	5.5	5.0	4.75	4.5

Note : $\boldsymbol{E_1 = 130}$**GPa**, $\boldsymbol{E_3 = E_2,\ G_{13} = G_{12},\ G_{23} = 0.5G_{12},\ v_{12} = v_{23} = v_{13} = 0.3}$

Table 2. Normalized deflections and stresses for the $(\mathbf{45/15})_2$ cylindrical shell for (R/h=4)

	T=350 K		T=375 K		T=400 K		T=425 K	
Results	IMP	DMP	IMP	DMP	IMP	DMP	IMP	DMP
$\bar{w}(-1/2)$	-0.93	-0.91	-1.39	-1.37	-1.86	-1.81	-2.32	-2.27
$\bar{v}(-1/2)$	-0.59	-0.58	-0.89	-0.87	-1.19	-1.15	-1.483	-1.44
$\bar{\sigma}_x(1/2)$	238.79	265.71	231.75	268.62	224.71	272.19	217.67	269.57
$\bar{\sigma}_\theta(1/2)$	234.33	260.96	225.01	261.72	215.79	262.88	206.51	258.12
$\bar{\tau}_{x\theta}(1/2)$	122.21	149.29	116.64	153.79	111.06	158.76	105.49	157.67
$\bar{\tau}_{xz}(0)$	7.8329	9.73	7.36	9.85	6.89	10.06	6.43	9.84
$\bar{\tau}_{\theta z}(0)$	24.28	24.69	23.22	23.65	22.14	22.61	21.09	21.47

3.1 Effect of moisture variation

In this section, we will investigate the effect of moisture variation on the material properties and the static hygrothermal bending response of the simply-supported $(45/15)_2$ cylindrical shells. The composite laminate cylindrical shell assumed to be subjected to sinusoidal moisture and thermal distribution across the thickness of the laminate as follows [9]

$$\Delta T = (T - t)\, z(1 - 4/3\, z^2/h)\sin(\pi\theta/\theta_m) \quad (9a)$$

$$\Delta C = C\, z(1 - 4/3\, z^2/h)\sin(\pi\theta/\theta_m) \quad (9b)$$

The relationship between the moisture concentration and material properties are as given in Table 3 [8]. In this section, "T" and" t" given in Eq. 9 are assumed to be 300 K and 375 K respectively. Different values of moisture concentration are considered. In this example displacements and the stresses of a four-layer $(45/15)_2$ cylindrical shell strip subjected to hygrothermal effect are reported in Table 4 with R/h=4. Similarly displacements of the four-layer $(30/-30)_2$ laminate for (R/h=4) are plotted in Figure 2. Once again, Table 4 shows that all displacements and stresses results are increasing with the increase of the moisture concentration. In addition, Figure 2 shows displacements for the shell strip for different moisture concentration. It can be noted that the transverse displacement $\bar{w}$ increases as moisture concentration increase.

Table 3. Elastic properties at different moisture concentration (E_1 =130 GPa and G_{12} =6 GPa)

	Moisture concentration C %						
Elastic moduli [GPa]	0.00	0.25	0.50	0.75	1.00	1.25	1.5
E_2	9.5	9.25	9.0	8.75	8.5	8.5	8.5

Note: $\boldsymbol{E_3 = E_2,\ G_{13} = G_{12},\ G_{23} = 0.5G_{12},\ v_{12} = v_{23} = v_{13} = 0.3}$
and $\beta_1 = 0$, β_2=0.44, $\alpha_1 = -0.3 * 10^{-6}$, $\beta_2 = 28.* 10^{-6}$

Table 4. Normalized deflections and stresses for the $(45/15)_2$ cylindrical shell for (R/h=4)

	C %					
Results	0.25	0.5	0.75	1.0	1.25	1.5
$\bar{w}(-1/2)$	-2.13	-2.85	-3.57	-4.29	-5.02	-5.76
$\bar{v}(-1/2)$	-1.37	-1.83	-2.30	-2.76	-3.23	-3.70
$\bar{\sigma}_x(1/2)$	246.53	284.03	310.58	339.55	362.32	385.09
$\bar{\sigma}_\theta(1/2)$	256.95	269.70	292.25	316.96	335.92	354.87
$\bar{\tau}_{x\theta}(1/2)$	129.84	144.82	159.20	175.85	186.52	197.19
$\bar{\tau}_{xz}(0)$	8.14	9.06	9.93	10.98	11.57	12.17
$\bar{\tau}_{\theta z}(0)$	25.05	26.90	28.74	30.59	32.36	34.13

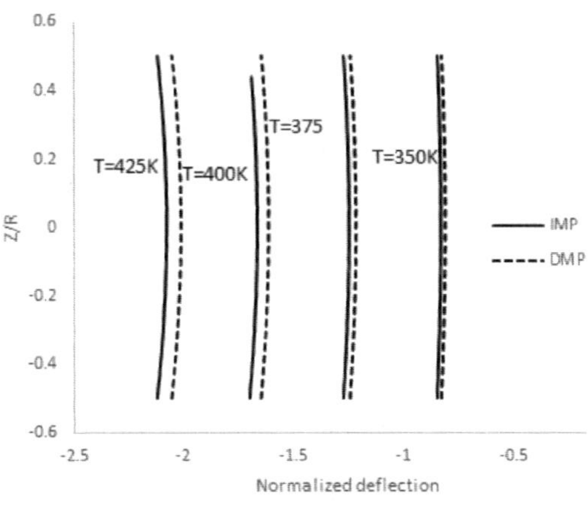

Fig 1. Normalized deflection $\bar{w}$ for $(45/15)_2$ cylindrical shell with and without the effect of temperature on material properties (R/h=4)

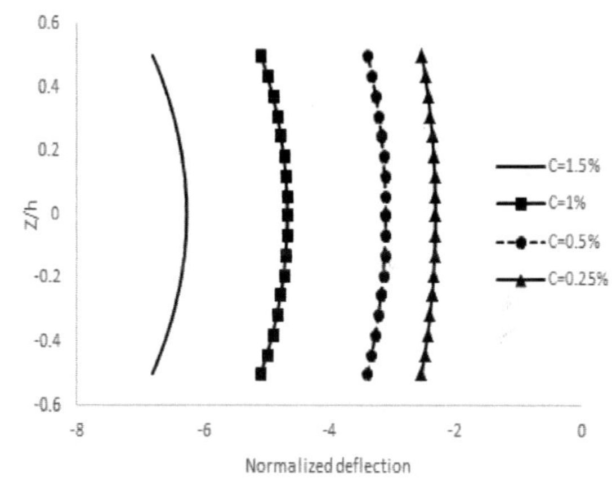

Fig 2. Normalized deflection $\bar{w}$ for $(30/-30)_2$ cylindrical shell strip at different moisture concentration

4. Conclusion

A higher order shear deformation theory is proposed to study the hygrothermal analysis of composite laminated shells including the effect of temperature and moisture dependent material properties. The present results show that the transverse displacements and shear stresses are sensitive to the variation due to change on material properties caused by the rise of temperature and moisture concentration. Moreover, numerical results shows that, the hygrothermal responses taking into account the hygrothermal effect into material properties is quite different from those ignoring hygrothermal effect on material properties, thus in the accurate stress analysis the change in material properties with temperature and moisture variation should be considered when such a laminate is subjected to hygrothermal loading.

Acknowledgments
The authors are grateful to the support of the Ministry of Higher Education Malaysia through the Grant FRGS 12-073-0222.

References

[1] Qatu, M.S., E. Asadi, and W. Wang 2012 Review of recent literature on static analyses of composite Shells: 2000-2010. eview of recent literature on static analyses of composite shells: 2000-2010. *Journal of Composite Materials.* **2**(3) : p. 61-86.
[2] Brischetto, S. 2013 Hygrothermoelastic analysis of multilayered composite and sandwich shells. *Journal of Sandwich Structures and Materials,* **15**(2) : p. 168-202.
[3] Mehrabadi, S.J. and B.S. Aragh 2013, On the thermal analysis of 2-D temperature-dependent Functionally graded open cylindrical shells. *Composite Structures*. 96: p. 773-785.
[4] Brischetto, S. and E. Carrera 2013 Static analysis of multilayered smart shells subjected to mechanical, Thermal and electrical loads. *Meccanica*. 48(5): p. 1263-1287.
[5] Alsubari, S., J.S. Mohamed Ali, and Y. Aminanda 2015 Hygrothermoelastic analysis of anisotropic cylindrical Shells. *Composite Structures*. 131: p. 151-159.
[6] Mohamed Ali, J.S., S. Alsubari, and Y. Aminanda. 2015 A Higher Order Theory for Bending of Cross Ply Laminated Cylindrical Shell under Hygrothermal Loads. Advanced Materials Research. 1115 : p. 509-512.
[7] Mohamed Ali, J. S., S. Alsubari, and Y. Aminanda 2016 Hygrothermoelastic Analysis of Orthotropic Cylindrical Shells. *Latin American Journal of Solids and Structures*. **13**(3): p. 573-589.
[8] Patel, B., M. Ganapathi, and D. Makhecha 2002 Hygrothermal effects on the structural behaviour of Thick composite laminates using higher-order theory. *Composite Structures,* **56**(1): p. 25-34.
[9] Zenkour, A.M. 2014 Simplified theory for hygrothermal response of angle-ply composite plates. *AIAA Journal*. **52**(7): p. 1466-1473.

IOP Conf. Series: Materials Science and Engineering **184** (2017) 012012 doi:10.1088/1757-899X/184/1/012012

Optimal location of piezoelectric patches for active vibration control

Mohammad F. Labanie[1,a], **J S Mohamed Ali**[1,b], **M.S.I Shaik Dawood**[1,c],
[1]Department of Mechanical Engineering
Kulliyyah of Engineering, International Islamic University, Malaysia
P.O.Box 10, 53100 Kuala Lumpur, Malaysia

E-mail: [a]mflabanie@gmail.com , [b]jaffar@iium.edu.my , [c]sultan@iium.edu.my

Abstract. This paper focuses on finding the optimal location for a piezoelectric patch for minimizing the settling time of an excited isotropic and orthotropic plate. COMSOL Multiphysics has been used to design and model the plate with PID controller. Classical Optimization tool called Parametric Sweep has been used to achieve the objective of the experiment. Five different stacking sequences were used in the study of orthotropic plate. The results obtained by the FEA software indicated that by placing the piezoelectric patches at the optimal location, the settling time of a plate can decrease by 40% compared to placing it at the centre of the fixed end.

1. Introduction

The use of piezoelectric in the field of active vibration control has gained a lot of attention in the recent years due to their ability to minimize the settling time. Active vibration control is important in space application and in structures where it has low stiffness and high flexibility. In space application, the density of the atmosphere is almost zero which require the structure longer time to settle. In low stiffness structures, the damping effect will be too low to supress the structure in short time. The implementation of active vibration control has shown a good results to restore the structure to its equilibrium state in short time. The piezoelectric patches can be placed in different configurations, either bonded to the upper and lower surface or imbedded between layers of the material. The optimal location depends of the objective of the setup. Kumar et al [1] considered a flexible beam using LQR to control his model. Guadenzi et al [2] studied the effect of open and close loop circuit effect on a cantilever composite Beam. Bruant et al [3] studied a flexible beam and defined different optimal characteristics for the sensor and actuator patches. In his model, the optimal criterion for the sensor was to maximize the energy output from the patch, while, for the actuator the optimal location was based on minimize mechanical energy in the plate. Hac et al [4] considered the optimal location for the piezoelectric patches based on the observability and controllability degree of the structure. Biglar et al [5] tried to find the optimal location placement of a single patch of sensor and actuator over a cantilever steel plate. He implemented the genetic algorithms to increase the controllability and observability of the cantilever plate. Roy et al [6] proposed an improved genetic algorithm to be applied, in order to maximize the controllability of the structure. Kim et al [7] considered the optimal location based on minimizing the sound level generated by a vibrating beam. Nor et al [8] used ant colony optimization method to determine the optimal placement of the actuator and sensor based on minimizing the energy of the plate.

2. Mathematical Modeling

In this work a cantilever plate made of isotropic/orthotropic material was considered with the dimension a, b and thickness t, as illustrated in Fig. 1

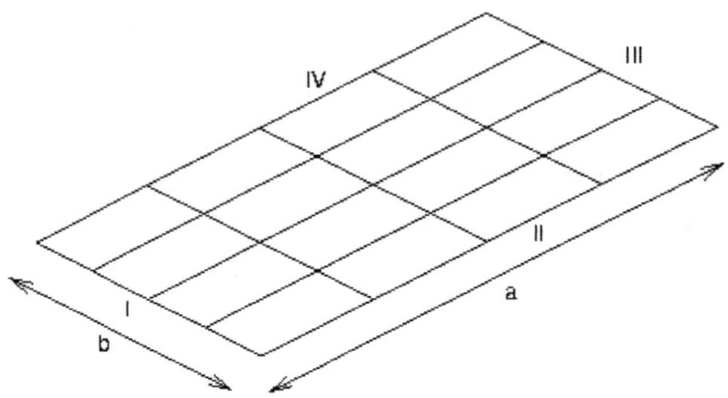

Figure 1. Isotropic plate model

According to Guedenzi [9], the governing equations that describe the piezoelectric effect are

$$\epsilon = S\,\sigma + d^T\,E \quad \ldots (2.1)$$

$$D = \epsilon\,E + d\,\sigma \quad \ldots (2.2)$$

Where $\boldsymbol{\sigma}$ is the stress vector, $\boldsymbol{S}$ is the compliance matrix, $\boldsymbol{\epsilon}$ the dielectric permittivity at zero stress, $\boldsymbol{D}$ is the electrical displacement and E is electrical field vector. For an isotropic plate the equation that relates the stress to the voltage applied can be expressed as:

$$\begin{bmatrix}\sigma_x\\ \sigma_y\\ \tau_{xy}\end{bmatrix} = \begin{bmatrix}\frac{E_p}{1-v_p^2} & \frac{v_pE_p}{1-v_p^2} & 0\\ \frac{v_pE_p}{1-v_p^2} & \frac{E_p}{1-v_p^2} & 0\\ 0 & 0 & \frac{E_p}{2(1-v_p)}\end{bmatrix}\begin{bmatrix}\epsilon_x\\ \epsilon_y\\ \epsilon_{xy}\end{bmatrix} - \begin{bmatrix}\frac{E_{pz}}{1-v_{pz}^2} & \frac{v_{pz}E_{pz}}{1-v_{pz}^2} & 0\\ \frac{v_{pz}E_{pz}}{1-v_{pz}^2} & \frac{E_{pz}}{1-v_{pz}^2} & 0\\ 0 & 0 & \frac{E_{pz}}{2(1-v_{pz})}\end{bmatrix}\begin{bmatrix}d_{31}\\ d_{32}\\ d_{33}\end{bmatrix}\frac{V(t)}{t_{pz}} \quad \ldots (2.3)$$

where $\boldsymbol{E_p}, \boldsymbol{v_p}, \boldsymbol{E_{pz}}$ and $\boldsymbol{v_{pz}}$ represent the modulus of elasticity, possion's ratio of the plate, modulus of elasticity of the piezoelectric patch and possion's ratio of the piezoelectric patch respectively.

In case of composite material, a different representation is required due to the fiber orientation. The governing equation in the case of composite plate is given as:

$$\begin{bmatrix}\sigma_x\\ \sigma_y\\ \tau_{xy}\end{bmatrix} = \begin{bmatrix}Q_{11} & Q_{12} & Q_{13}\\ Q_{21} & Q_{22} & Q_{23}\\ Q_{31} & Q_{32} & Q_{33}\end{bmatrix}\begin{bmatrix}\epsilon_x\\ \epsilon_y\\ \epsilon_{xy}\end{bmatrix} - \begin{bmatrix}\frac{E_{pz}}{1-v_{pz}^2} & \frac{v_{pz}E_{pz}}{1-v_{pz}^2} & 0\\ \frac{v_{pz}E_{pz}}{1-v_{pz}^2} & \frac{E_{pz}}{1-v_{pz}^2} & 0\\ 0 & 0 & \frac{E_{pz}}{2(1-v_{pz})}\end{bmatrix}\begin{bmatrix}d_{31}\\ d_{32}\\ d_{33}\end{bmatrix}\frac{V(t)}{t_{pz}} \quad \ldots (2.4)$$

The Q matrix is the reduced stiffness matrix that depends of the fibre orientation in the lamina.

In this paper, the optimal location of a single isotropic sensor and actuator patches to minimize the settling time of a cantilever orthotropic and isotropic plate. Proportional integral derivative controller has been adopted for this model. The gain values for the PID controller were calculated based on Ziegler–Nichols method theory of control. Six cases will be presented in this paper, Aluminium plate, Graphite/Epoxy plate with the following stacking sequences, $[0/0]_s$, $[90/90]_s$, $[45/45]_s$, $[0/90]_s$ and $[45/-45]_s$. The plates are modelled under the plane stress assumption i.e. $\sigma_z = \tau_{zx} = \tau_{zy} = 0$

3. Benchmark model

3.1 Controller model

The selected controller was PID. The governing equation of the PID controller is given by:

$$V_{applied} = K_p(V_{Measured} - V_{set}) + K_i \int_0^t (V_{Measured} - V_{set})\, dt + K_d \frac{\partial}{\partial t}(V_{Measured} - V_{set}) \quad \text{... (3.1)}$$

where $\boldsymbol{K_P}, \boldsymbol{K_I}$ and $\boldsymbol{K_D}$ are the gain values for the PID controller, respectively. $\boldsymbol{V_{set}}$ is set to equal zero because the aim of the controller is to return the structure to its equilibrium condition. The following figure illustrates the relation between the generated voltage by the model and the applied controller.

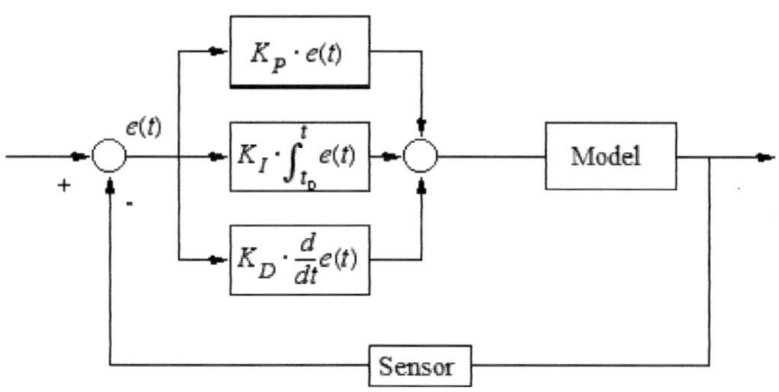

Figure 2. PID controller block diagram.

The gain values used in this simulation were:

Table 1: PID Gain Values used

	K_p	K_i	K_d
Gains	1.023	0.85	0.1

The gains are calculated based on the initial position which was centre position near the fixed end.

3.2 Validation

Before optimization of location of patches on isotropic and orthotropic plates using the COMSOL software, the software's capability in handling piezoelectric patches and composite laminates was first tested and validated against the following 2 standard benchmark case studies results
Case 1: The model of the Nechibvute et al [10] in which it evaluated the analytical voltage produced due to a force applied at the tip of a cantilever beam as shown in Figure 3. An Aluminium cantilever beam was considered with PZT-5H attached to the upper face of the cantilever beam. Using COMSOL Multiphysics, this problem was modelled and validated against Nechibvute et al results as shown in Table 2. It can be seen the results are very closer.

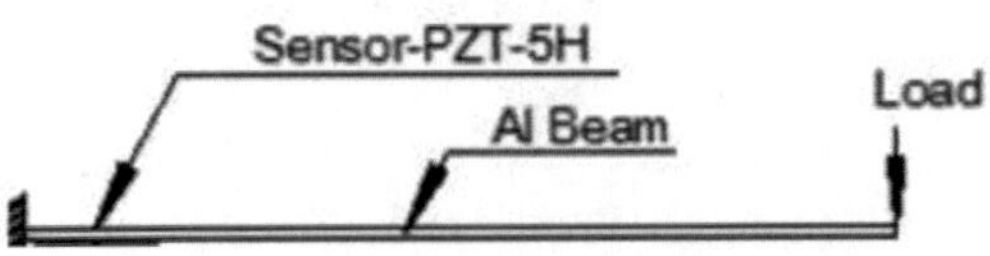

Figure 3. Composite Plate Geometry and Boundary Condition [10].

Table 2: Comparison between Nechibvute et al result and simulation results.

Amplitude of AL Beam tip (mm)	Open Circuit Voltage (V) Nechibvute et al Experimental	Open Circuit Voltage (V) Simulated
0.33	3.19	3.24
0.44	4.10	4.23
0.58	5.80	5.94
0.86	8.11	8.23
1.04	9.87	10.04
1.27	11.99	12.17
1.50	14.22	14.33

Case 2 : Reddy [11] model for finding the centre deflection for an orthotropic laminated plate under uniform pressure as shown in Fig. 4.

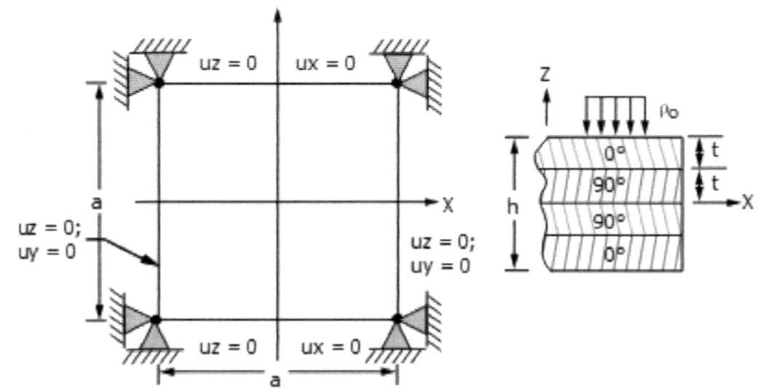

Figure 4. Composite Plate Geometry and Boundary Condition.

Composite material are defined by the orientation of the fibre with respect to the initial frame. In order to simulate the composite material in COMSOL we need to define a base vector system that is rotated with the desired orientation. The results that were obtained by COMSOL Multiphysics for the above problem is validated against the Reddy [11] results. The displacement at the centre of $[0/90]_s$ plate was evaluated by COMSOL and it was very close to the theoretical results of Reddy [11] with the error of 0.12966 %.

3.3 Optimization technique

In this model, classical optimization method called parametric sweep is adopted. Parametric sweep method is where the FEA software will work out on the optimal patch location for minimizing the settling time. For all plates, a force is applied at the free end for 5 seconds, which will ensure the structure is fully deflected, and then released to measure the time needed to reach settling condition. The isotropic plate, was divided into 16 location, for sensor and actuator to be tested, where x_1, y_1 and x_2, y_2 are the coordinates of the centre of the piezoelectric sensor and actuator, respectively.

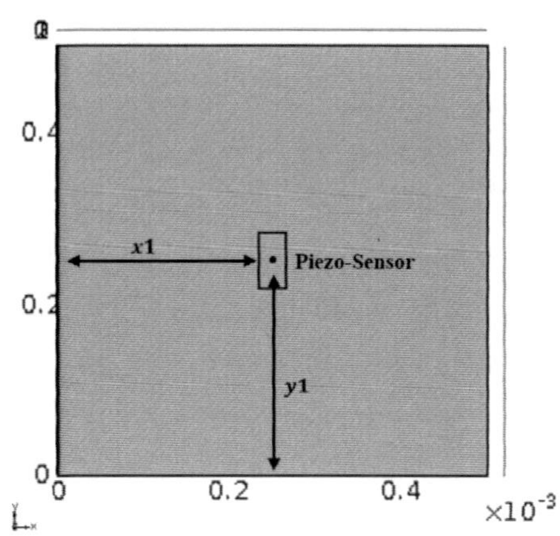

Figure 5. Top view. Measurement of x_1, y_1 to the centre of the piezoelectric sensor patch.

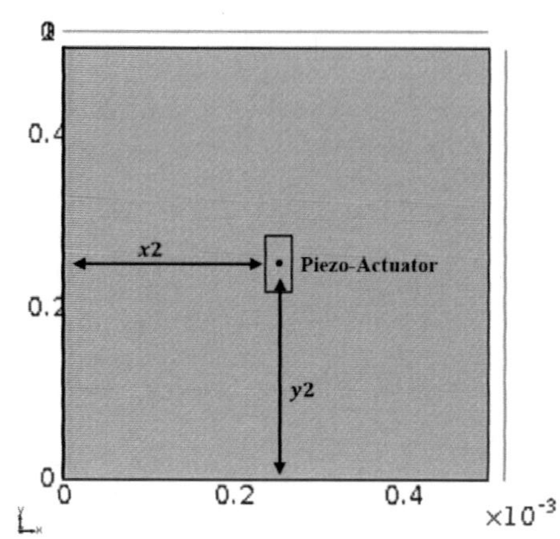

Figure 6. Bottom view. Measurement of x_2, y_2 to the centre of the piezoelectric actuator patch.

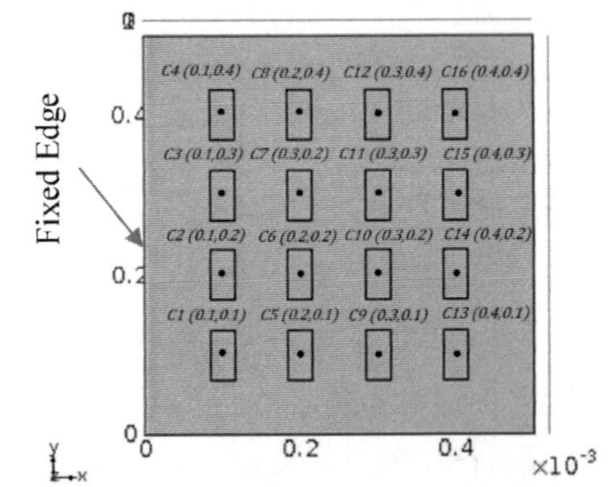

Figure 7. Top View. Location of PZT Sensor to be tested on the isotropic plate.

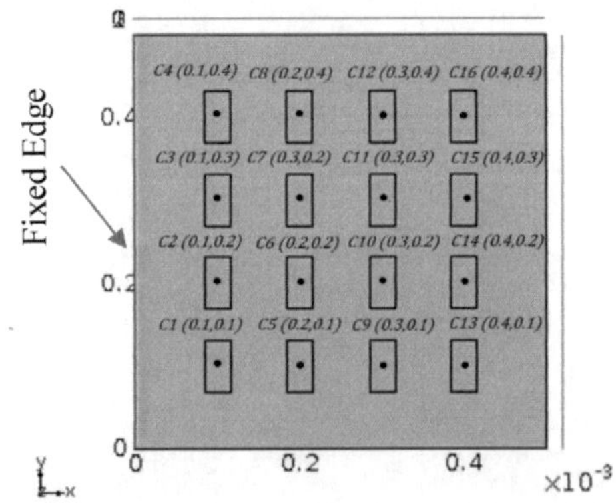

Figure 8. Bottom View. Location of PZT Actuator to be tested on the isotropic plate.

The orthotropic plate is defined in the same manner, but, the dimensions of the plate was changed so that the number of possible location combination of sensor and actuator patches is reduced. In the isotropic 265 location combination was tested to determine the best location to minimize the settling time. In the orthotropic, the number of location combination reduced to 81 per case, as this model will test five different cases for the orthotropic plate.

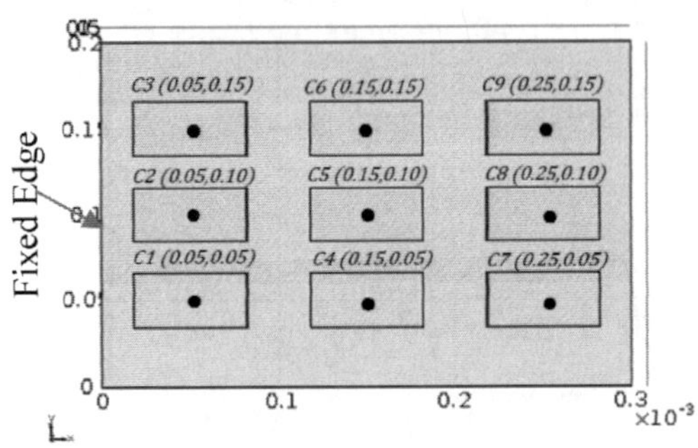

Figure 9. Top View. Location of PZT Sensor to be tested on the orthotropic plate.

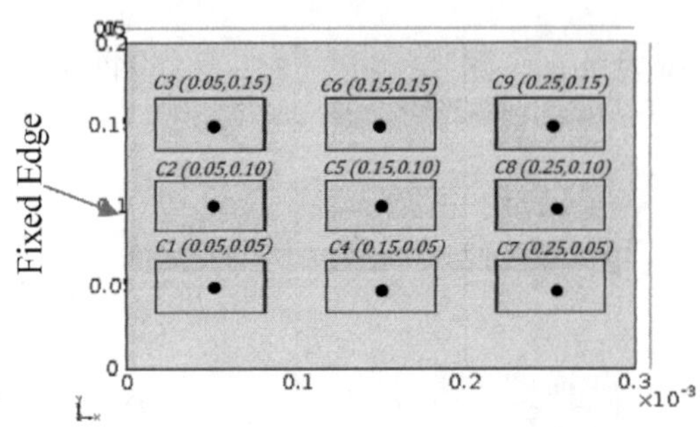

Figure 10. Bottom View. Location of PZT Actuator to be tested on the isotropic plate.

4. Results

4.1 Plate model

The following Tables 3 & 4 show the material properties of aluminium plate, piezoelectric (PZT-5H) patch and graphite/epoxy composite used in this analysis.

Table 3: Material Property of Al Plate and PZT-5H Patch

Parameter	Plate (Aluminium)	Piezoelectric (PZT-5H)

Length	0.5 [m]	31.8 [mm]
Width	0.5 [m]	63.5 [mm]
Thickness	2 [mm]	0.86 [mm]
Density	2700 Kg/m^3	7500 Kg/m^3
Young's Modulus	69 GPa	70 GPa
Passion's ratio	0.33	0.31
Piezoelectric Constant, d_{31}	-	-2.74e-010 [C/N]
Piezoelectric Constant, d_{32}	-	-2.74e-010 [C/N]
Loss factor	0.06	0.06

Table 4: Material Property of Graphite/Epoxy Plate

	Length	Width	Thick-ness	E_l	E_t	v_{12}	G_{lt}	ρ	α	β
Gr/Ep	0.3 [m]	0.2 [m]	0.25 [mm]	181 [GPa]	10.3 [GPa]	0.28	7.17 [GPa]	1600 Kg/m^3	1.0	5×10^{-5}

4.2 Isotropic Aluminium plate

A force of 10 N in applied at the free end for 5 seconds to allow full deflection of the plate and then released to measure the settling time. The results for the aluminium plate indicated that the best location is near the fixed end that would enable the actuator to produce a counter moment to supress the oscillation. The optimal location was to place the sensor is at $\boldsymbol{C_1}$ in Fig. 5 while the actuator at $\boldsymbol{C_4}$ in Fig. 6 . The following Fig. 11 shows the difference in settling time between optimal location compared to if it was placed mid plane near to the fixed end.

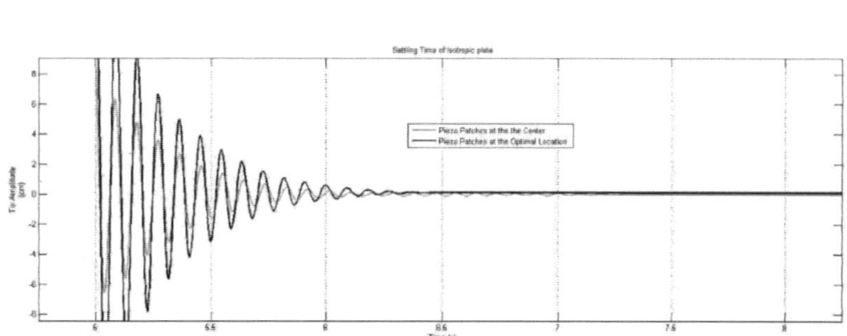

Figure 11. Comparing initial guess and optimal location for isotropic plate.

4.3 Orthotropic plate

In this section the results obtained for the orthotropic plate will be presented. All plates are made of 4 lamina with different stacking sequence. The initial guess for all of the cases was by placing the patches at $\boldsymbol{C_2}$ as illustrated in Fig. 7 for both sensors and actuators. A force of 10 N in applied at the free end for 5 seconds to allow full deflection of the plate and then released to measure the settling time.

4.3.1 $[0/0]_s$

The unidirectional 0 degree plate, indicated that placing the patches at the optimal location would decrease the settling time required for the plate by 0.83 seconds. The 0 degree oriented fiber plate, needed more time to settle due to the lower density and internal damping as compared to the isotropic plate. The optimal location of the patches for this plate turned out to be $\boldsymbol{C_1}$ for the sensor patch and $\boldsymbol{C_3}$ for the actuator patch.

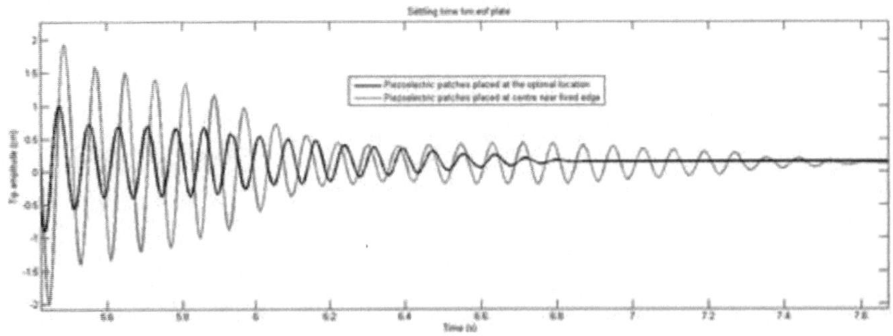

Figure 12. Comparing initial guess and optimal location for 0 degree plate.

4.3.2 $[90/90]_s$

The 90 degree laminate, required the longest time to settle. The reason that the moment being applied from the actuator isn't in the fiber direction. Placing the patches at the optimal location reduced the settling time to 2.76 second compared to placing the patches at the initial guess. The optimal location for this plate was also $\boldsymbol{C_1}$ for sensor and $\boldsymbol{C_3}$ for the actuator. Switching the location of the sensor and actuator gave similar results.

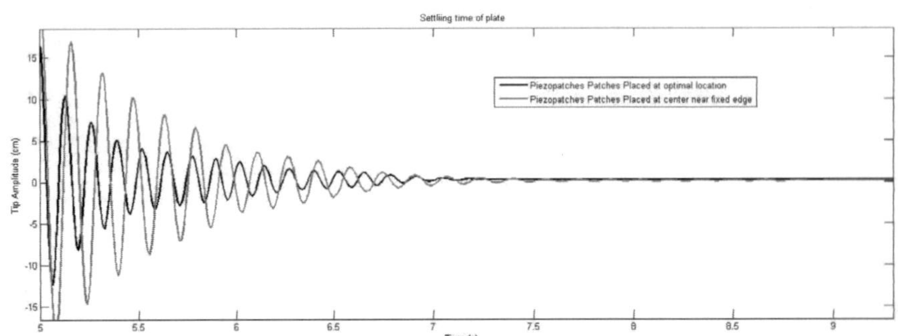

Figure 13. Comparing initial guess and optimal location for 90 degree plate

4.3.3 $[45/45]_s$

In the 45 degree laminate, the moment being applied from the actuator isn't in the fiber direction. Placing the patches at the optimal location reduced the settling time by 0.96 seconds compared to placing the patches at the initial guess. The optimal location for this plate was also $\boldsymbol{C_1}$ for sensor and $\boldsymbol{C_3}$ for the actuator. Switching the location of the sensor and actuator gave the same results for settling time.(Fig. 14)

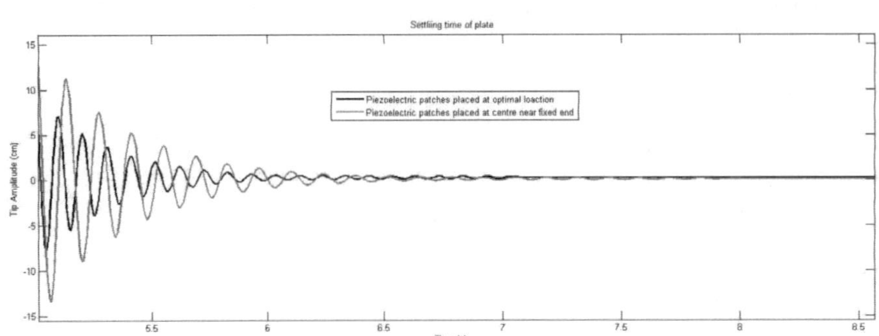

Figure 14. Comparing initial guess and optimal location for 45 degree plate

4.3.4 $[0/90]_s$

Vibration of the cross ply plate settled in shorter time by placing it at the optimal location. The optimal location for this plate turned out to be $C_1$for sensor and C_3 for actuator. The settling time reduction between the optimal location and initial guess turned out to be around 45%. The optimal location allowed the structure to settle within 1.23 second from being released.

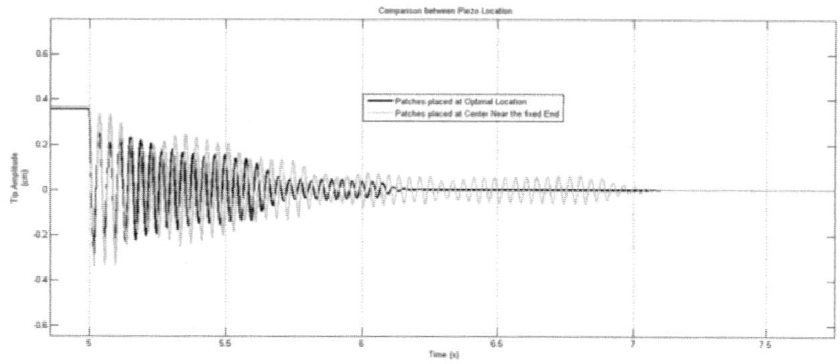

Figure 15. Comparing initial guess and optimal location for cross-ply plate

4.3.5 $[45/-45]_s$

The angle ply laminate, required more time to settle compared to the cross ply laminate. The plate required 2.54 seconds to settle even after placing the patches in the optimal location. The optimal location turned out to be, $C_1$for sensor and C_3 for actuator. While switching the location of the patches gave the same output results.

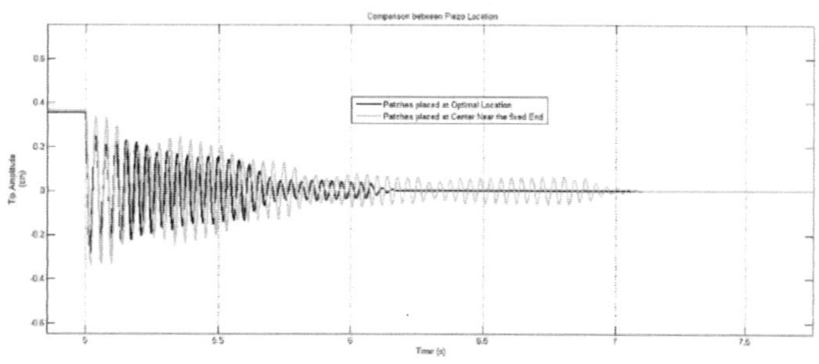

Figure 16. Comparing initial guess and optimal location for $[45/-45]_s$ plate

5. Conclusion

This study focused on finding the optimal location of patches to reduce the settling time of isotropic plate and orthotropic plate with different stacking. This study indicated that the settling time of the structure can be minimized by placing the patches at the optimal location. By reducing the settling time, the structure faces less internal force and shear stress which will increase the reliability of the structure during its life cycle. The optimal location for both actuator and sensor patch was near the fixed edge as the actuator will produce moment to counter the tip oscillating.

6. References

[1] Kumar, K.R., Narayanan, S.: Active vibration control of beams with optimal placement of piezoelectric sensors/actuator pairs. Smart Mat. Struct. 17, 055008 (2008)

[2] Gaudenzi, P., Carbonaro, R., & Barboni, R. (1997). Vibration control of an active laminated beam. Composite Structures, 38 (1-4), 413-420. doi:10.1016/s0263-8223(97)00076-7

[3] Bruant, I., Coffignal, G., Le'ne', F., Verge', M.: A methodology for determination of piezoelectric actuator and sensorlocation on beam structures. J. Sound Vib. 243, 862–882 (2001)

[4] Hac, A., Liu, L.: Sensor and actuator location in motion control of flexible structures. J. Sound Vib. 167, 239–261 (1993)

[5] Biglar, M., Gromada, M., Stachowicz, F., & Trzepieciński, T. (2015). Optimal configuration of piezoelectric sensors and actuators for active vibration control of a plate using a genetic algorithm. Acta Mechanica Acta Mech, 226(10), 3451-3462. doi:10.1007/s00707-015-1388-1

[6] Roy, T., & Chakraborty, D. (2009). Optimal vibration control of smart fiber reinforced composite shell structures using improved genetic algorithm. Journal of Sound and Vibration, 319(1-2), 15-40. doi:10.1016/j.jsv.2008.05.037

[7] Kim, J. and Ko, B. 1998. "Optimal Design of a Piezoelectric Smart Structure for Noise Control," Smart Materials and structures, 7:801–808

[8] Nor, K. A., Muthalif, A. G., & Wahid, A. N. (2014). Optimization in Active Vibration Control: Virtual Experimentation Using COMSOL Multiphysics - MATLAB Integration. 2014 5th International Conference on Intelligent Systems, Modelling and Simulation. doi:10.1109/isms.2014.71

[9] Gaudenzi, P. (2009). Smart structures: Physical behaviour, mathematical modelling and applications. Chichester, U.K.: Wiley.

[10] Nechibvute, A., Akande, A., & Luhanga, P. (2011). Modeling of a PZT Beam for Voltage Generation. 19(2), 259-271.

[11] J. N. Reddy, "Exact Solutions of Moderately Thick Laminated Shells", ASCE Journal Engineering Mechanics, Vol. 110 No. 5, 1972, pg. 805.

Fe-C-Si ternary composite coating on CP-titanium and its tribological properties

M. A. Maleque[a*], W. Saffina[b] , A. S Ahmed[c] and M. Y. Ali[d]

[a]Department of Manufacturing and Materials Engineering
Kuliyyah of Engineering, International Islamic University Malaysia, P.O. Box 10, 50728 Gombak, Kuala Lumpur, Malaysia.
[c]Department of Mechanical and Manufacturing Engineering, Faculty of Engineering
Universiti Malaysia Sarawak, 94300 Kota Samarahan, Sarawak

Email: [a*]maleque@iium.edu.my

Abstract. This study focused on the development of ternary composite coating through incorporation of Fe-C-Si ternary powder mixtures on CP-Ti substrate and characterizes the microstructure, hardness and wears behavior in presence of Jatropha oil. In this work, the surface of commercial purity titanium (CP-Ti) was modified using a tungsten inert gas (TIG) surface melting technique. The wear behavior of coated CP-titanium was performed using pin-on-disk machine. The results showed that the melt track has dendritic microstructure which was homogenously distributed throughout the melt pool. This Fe-C-Si ternary composite coating enhanced the surface hardness of CP-Ti significantly from 175 HV for the untreated substrate to ~800 HV for the Fe-C-Si coated CP-Ti due to the formation of intermetallic compounds.. The wear results showed that less wear volume loss was observed on the composite coated CP-Ti in presence of Jatropha-biodiesel compared to uncoated CP-Ti. The achievement of this hard Fe-C-Si composite coating on the surface of CP-Ti can broadened new prospect for many engineering applications that use biodiesel under different tribological variables.

1. Introduction

Titanium is one of the few allotropic metals that exist in two different crystallographic forms. At room temperature, it has a close-packed hexagonal structure, designated as the alpha phase whereas, at around 884 °C, the alpha phase transforms into a body-centred cubic structure, known as beta phase which is stable up to titanium's melting point of 1677 °C. Alloying elements promote the formation of one or the other of the two phases. Carbon, for example, stabilizes the alpha phase which means it raises the alpha to the beta transformation temperature. Iron, copper, chromium and vanadium are beta stabilizer which lowers the transformation temperature, therefore allowing the beta phase to remain stable at lower temperatures, and even at room temperature. Thus, titanium's mechanical properties are closely related to these allotropic phases. For example, the beta phase will be much stronger but more brittle than alpha phase. Commercially pure titanium (CP-Ti) is unalloyed titanium characterized by an alpha phase hexagonal close-packed crystal structure. Commercially pure titanium is over

99.6% titanium with traces of carbon, iron, nitrogen, oxygen, and hydrogen [1]. However, CP-Ti has poor tribological properties due to lower hardness and high friction coefficient [2]. Therefore, it is utmost important to improve the tribological properties of CP-Ti especially under biodiesel lubricated condition through surface modification to increase the performance of the engineering components and extend their service life.

TIG welding was introduced as a surface alloying and modification technique in the late 19th century as an effective nitriding technique [3]. It was used as a reliable nitriding technique for the improvement of wear resistance of Ti-6Al-4V alloy. Later, several researches have been conducted addition research on the surface modification by cladding or surface alloying using this concept for ferrous material substrate [4]. Eroglu and Ozdemir [6] investigated the TIG surface alloying with preplaced graphite, chromium and high-carbon ferro-chromium powders on SAE 1020 steel. The result showed remarkable enhancement in wear resistance and corrosion resistance without impairing bulk properties.

Biodiesel is chemically comprised of alkyl mono-esters of fatty acids originally derived from vegetable oils or animal fats [7, 8]. "Biodiesel" is well known chemically as the mono-alkyl esters of long-chain fatty acids and is produced from several types of conventional and non-conventional vegetable oils and animal fats including those of used oils from the frying industry, soybean oil, rapeseed oil, tallow, rubber seed oil and palm oil [9, 10]. Biodiesel have better lubrication behaviour rather than diesel, which lubrication behaviour plays vital roles in pumping and injecting the fuel. It acts as an alternative green fuel which offers some technical advantages. The advantages of using biodiesel as a diesel engine fuel because it has the biodegradability, higher cetane number, and higher flashpoint and reduced greenhouse gases (GHG) emission [11]. It is very important to study the effect of biodiesel on the tribological behaviour of composite coated CP-Ti which is not available in literature. Therefore, the main aim of this work is to form a ternary composite coating layer via surface modification process under TIG torch melting and study the wear behaviour in jatropha biodiesel oil.

2. Processing route for hard surface layer

2.1. Materials

Commercially purity titanium (CP-Ti) plate was used with the dimension of this specimen was 50 mm x 40 mm x 12 mm. The specimen was abraded with coarse emery paper to remove the oxide layer as well as to obtain an even surface finish and degreased by using acetone clean from any dirt such as oil and dust. The Fe-C-Si ternary powder mixture was prepared using polyvinyl acetate (PVA). The biodiesel used in this research was synthesized from crude jatropha curcas oil (JCO) that was supplied by UNMAS, Malaysia. The chemical compositions of the free fatty acids in the crude jatropha oil.

2.2 Powder Preparation

Iron (Fe), silicon (Si) and carbon (C) ternary mixture were used with the composition 94%, 4% and 2% respectively. After measuring the powders (Fe, Si, and C) according to their compositions, it was gone through ball milling process. These powders were mixed for one hour of a speed of 200 rpm in order to make it homogenous mixture. The amount of Fe-C-Si powder mixture to be placed on the surface of CP-Ti was according to the ratio 1 mg/mm2. After the calculation, the amount of powder need to be placed on sample is 2g. For the pre-placement of powder, the binder PVA, distilled water and alcohol (ethanol) were used. The powder was first placed on the sample and makes slurry. Then, the powder was mixed with water and ethanol and binder (PVA). The binder is used in order to prevent the powder from blown away during the TIG melting process. The binder was mixed thoroughly with the powder until a paste like mixture was obtained. The specimen then dried in the oven at 80°C for 60 minutes.

2.3 Coating with Tungsten Inert Gas (TIG) Melting

In this experiment, TIG torch or arc was used as a heat source to melt the powder mixture in order to form a composite coating layer on the CP-ti substrate. TIG uses a non-consumable tungsten electrode to produce the torch. The melt area is protected from oxidation by a shielding gas. The shielding gas used is argon at a rate of 20 l/min.. The samples position was adjusted at 1 mm below the electrode tip. The arc voltage and current were, 30 V and 100 A respectively. Once the arc is struck, the torch in a small circle to create a melting pool, the size of which depends on the size of the electrode and the current. Then, the molten single track is allowed to be solidify at room temperature. The energy input for melting was calculated using an expression: V=voltage (V), I=current (I), S=speed (mm/s), Heat Input Energy, $H = \eta \ VI/S$ where η is heat absorption efficiency equal to 0.45.

2.4 Material Surface Preparation

The CP-titanium plates were cut into required size and dimension using an EDM wire cut machine (model: W11FX2 K).A standard metallographic technique of grinding, polishing, and etching was used to prepare the samples for microstructure investigation and coating. Final polishing was done using diamond polishing paste. Then the cross sections of the coated materials were chemically etched for revealing their microstructure. The etching solution was prepared using 50% hydrofluoric acid (HF), 30% nitric acid (HNO3) and 20% of distilled water. The surface grinding process was done on the coated CP-Ti.

3. Results and Discussion

3.1 Microstructural Morphology

Figure 1 shows the microstructure of the coated layer which showed dendrites and are strongly bonded to the substrate after melting the Fe-C-Si particles which re-precipitated after being solidified. The melt microstructure also consisted of same unmelted Fe-C-Si particles (refer to Figure 1 (b)) along with the dendrites precipitates (Fig. 1 (a). The presence of particles always present at the bottom of the melt pool due to the mechanism of heating, stirring, and solidification [2].

Table 1 shows the matrix of the advantages and disadvantages of melting and diffusion processes on materials of the surface modification. It is important to understand of the process and to choose the best method in the application.

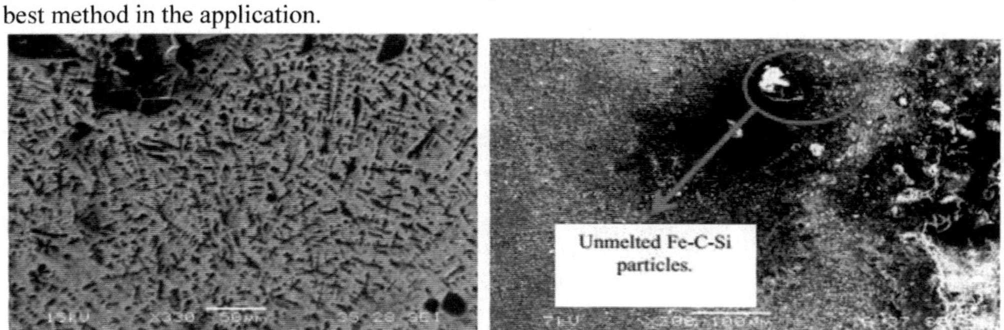

Figure 1. (a) Dendrites microstructure observed by SEM; (b) formation of unmelted Fe-C-Si particles (x200 magnification)

3.2 Microhardness of the Coated CP-titanium

The depth hardness profile was measured from the top surface of composite coating to the bottom of the composite coating layer and the hardness values are plotted against melt depth as shown in

Figure 2. It can be seen from the profiles that the hardness through the pool is not stable. It is increasing and decreasing due to the different distribution of Fe-C-Si ternary powders. However, it can be seen that the hardness of the coated layer increased in comparison with that of the substrate. The diagram also shows that maximum hardness value in profile is near the surface and they gradually decrease towards the substrate. This is related to the concentration of the phases present in the modified layers. According to hardness profile, the maximum hardness was more than 800Hv. This result showed that the surface hardness was significantly enhanced from 175 HV for the untreated substrate to 800 HV for the coated surface, due to the formation of intermetallic compound such as TiC and Fe3C in the modified layers.

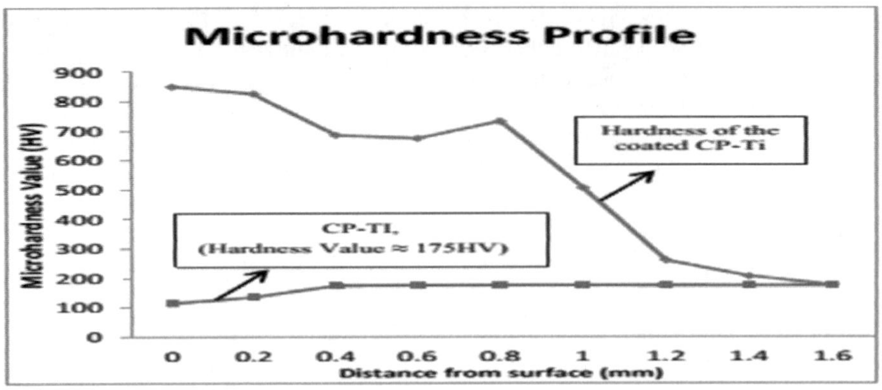

Figure 2. Microhardness profile across the coated layer via replacing powder of Fe-C-Si

3.3 Wear of Coated CP-Ti with and without Jatropha biodiesel

In this work, wear test was done for 10 minutes at a speed of 2.36 cm/s and load of 25N using wear testing machine. After the wear testing, the width and radius of the wear track for the sample were measured and recorded. Table 1 shows the wear test parameters and test results with and without jatropha biodiesel.

Table 1. Wear test parameters for the samples with and without lubricants

Experimental Runs	Speed (cm/s)	Load (N)	Sliding distance (m)	Wear track radius (mm)	Wear tracks width (μm)	Wear volume loss (mm^3)
Without lubricant	2.36	25	14	1.5	0.91	727
With lubricant	2.36	25	14	1.5	0.59	469

From the Table 1, the difference of wear track width for each sample can be observed clearly. Moreover, when measured using Mitutoyo Surface Test SJ-400, there is a difference in wear track width between samples. It can be concluded that as for the sample under lubricated conditions, the wear track width is smaller than the one without lubricant. Figure 3 shows the depth of penetrations (Rz value) and average surface roughness (Ra value) measured by Mitutoyo Surface Test SJ-400 machine of coated CP-Titanium samples were tested under biodiesel lubricated condition and without the biodiesel lubricants.

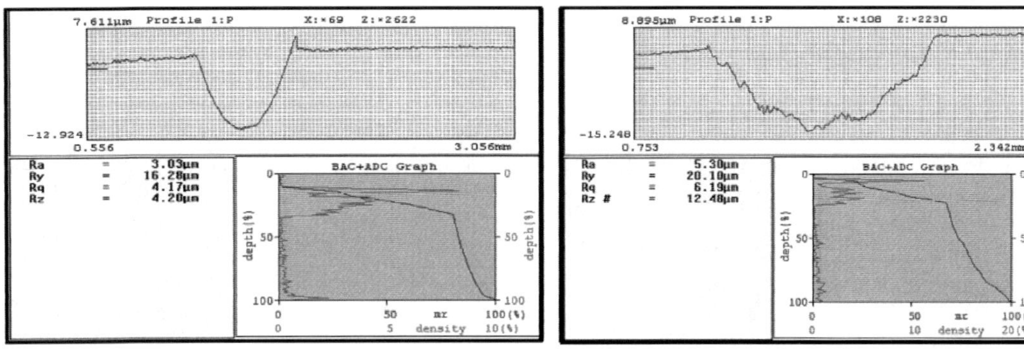

(a) (b)

Figure 3. Comparison depth of penetration (Rz value) and surface roughness (Ra value) for (a) sample under lubricated conditions (Rz =4.20 um, Ra = 3.03 um) and (b) sample without lubricants (Rz =12.48um, Ra = 5.30 um).

The sample under the lubricated condition showed smaller Rz and Ra indicated that the presence of lubricants reduce the wear of the sample. Without lubricants, average surface roughness is larger than the one with the lubricants. The higher the Rz and Ra values leading to higher wear volume loss.

4. Conclusions

The CP-Titanium has been modified successfully by surface coating with ternary powder mixtures of Fe-C-Si using TIG melting technique in argon gas environment. This Fe-C-Si composite coating has increased surface hardness from 175 HV for the untreated substrate to more than 800 HV for the coated composite Fe-C-Si layers, due to the formation of intermetallic compound such as TiC and Fe3C in the modified layers. Then, the coated CP-Ti samples were tested for wear test under biodiesel lubricated conditions. It shows that less wear volume loss was observed on the composite coated CP-Ti in presence of Jatropha-biodiesel. Since, Fe-C-Si composite coating samples experienced less wear; it is preferable to use this material in many engineering applications that use biodiesel under different tribological variables.

Reference

[1] Adeleke S A and Maleque, M A, Thin surface layers of iron-based alloys deposited by TIG hardfacing, Tribology Online, vol. 10, (6), 2015,pp.434-440.-263.

[2] Adeleke S A and M.A. Maleque, Tungsten Inert Gas Surface Alloying of Commercial Purity Titanium (CP-Ti) with Fe-C-Si Ternary Mixtures, Advanced Materials Research, 1024, 2014, pp.207-210.

[3] Khan, T. I., A Review of Method Used For Surface Melting and Gaseous Alloying. Pakestan Acad Sci, 63-72, 2007.

[4] Mridha, S. Idriss, A. N. M., Maleque, M. A., Suryanto, & Souad, A. (2011), 22-24 November 2011). Effect of voltage on the consolidation of TiC particulates on steel substrate fues by TIG welding arc. Paper presented at the regional tribology conference bayview hotel, langkawi island, Malaysia

[5] Eroglu, M. Ozdemir, N. (2002). Surf Coat Technol, 154:209.

[6] Carraretto, C., Macor, A., Mirandola, A., Stoppato, A., and Tonon S. (2004). Biodiesel as alternative fuel: Experimental analysis and energetic evaluations, Energy, vol. 29, pp. 2195–2211.

[7] Maleque, M. A, HH Masjuki, M Ishak, Bio-fuel-contaminated lubricant and hardening effects on the friction and wear of AISI 1045 steel, Tribology transaction, Vol 41, Issue 1, 1998, pg. 155-159.

[8] Maleque M A, Ghazal B A, Ali M Y, Hayyan M and Ahmed A S 2015 Wear behaviour of TiC coated AISI 4340 steel produced by TIG surface melting, *Materials Science Forum* **819** pp 76-80.

[9] Shah, S., Sharma, S. and Gupta, M.N. (2004). Biodiesel preparation by lipase-catalyzed transesterification of Jatropha oil. Energy Fuel, 18: 154-159.

[10] Ramadhas, A S, Jayaraj, S. and Muraleedharan, C. (2005). Biodiesel production from FFA rubber seed oil. Fuel, 84: 335-340.

[11] Reddy, ANR, Abu Ahmed Saleh, M Saiful Islam, Sinin Hamdan, and Md Abdul Maleque, Biodiesel production from crude jatropha oil using a highly active heterogeneous nano–catalyst by optimizing transesterification reaction parameters, Energy & Fuels, 30(1), 2016, pp.334-343.

Acknowledgement
Special thanks to the RMC whom has funded this research project under Research Matching Grant Scheme RMGS12-007-0020. Authors also acknowledged the support from Universiti Malaysia Sarawak.

Analysis of the internal temperature of the cells in a battery pack during SOC balancing

R Mizanur[1], M M Rashid[1], A Rahman[2], A H M Zahirul Alam[3], S Ihsan[2] and M S Mollik[1]

[1]Department of Mechatronics Eng., International Islamic University Malaysia, Malaysia
[2]Department of Mechanical Eng., International Islamic University Malaysia
[3]Department of Electrical and Computer Eng., International Islamic University Malaysia

Abstract. Lithium-ion batteries are more suitable for the application of electric vehicle due to high energy and power density compared to other rechargeable batteries. However, the battery pack temperature has a great impact on the overall performance, cycle life, normal charging-discharging behaviour and even safety. During rapid charge transferring process, the internal temperature may exceed its allowable limit (46^0C). In this paper, an analysis of internal temperature during charge balancing and discharging conditions is presented. Specific interest is paid to the effects of temperature on the different rate of ambient temperature and discharging current. Matlab/Simulink Li-ion battery model and quasi-resonant converter base balancing system are used to study the temperature effect. Rising internal temperature depends on the rate of balancing current and ambient temperature found in the simulation results.

Keywords: SOC, temperature, quasi-resonant converter, EV.

1. Introduction

The battery technology is widely employed in the different applications, namely space vehicles, traction, hybrid powertrain, electric vehicle, and electronic utilities as the standard power source. Especially, in electric vehicle, it has a great impact to run long time and to reduce cost and fuel consumption which causes co_2 emission [1]. The configuration and type of the battery technology are based on the power and energy demand. In the present market, there are different types of rechargeable batteries available, for example, NiMH, Li-ion, $LiFeO_4$, sodium-nickel chloride, nickel-cadmium (NiCd) and lead acid. The terminal voltage of a single cell is limited such as 2V for lead acid battery, 1.2V for NiMH, 3.6V for Li-Ion and 3.3V for $LiFePO_4$. To meet the high energy and power demand, battery cells are connected in series or parallel and both combinations. Among these types of batteries, Li-ion base batteries are more preferable for EV because of low self-discharge rate (6%- 10%), higher power density and energy capacity.

Ideally, each battery cell has unique characteristics, for example, voltage and current, so they can provide same power and energy. However, in reality, the characteristics of individual cell differ from cell to cell because of ambient temperature variation, manufacturing process, internal impedance and chemical degradation [2-4]. A battery pack always undergoes the charging and discharging process. During charging and discharging, battery capacity and lifecycle decrease gradually over time. Especially, high chemical reaction occurs throughout the discharging process (while a large amount of current drawing by the traction motor) and increases the battery cell internal temperature. Accordingly, state of charge (SOC) of the individual cell in a battery pack becomes unbalanced. Once this problem happens, some cells are charged faster than the other cells during charging and led to overcharging (explosion and catch fire). Similarly, during the discharging process, cells are under discharged which led to permanent damage. It is indeed necessary that a charge balancing system is required to solve the unbalancing problem. During the quick and efficient charge balancing process, cell internal temperature is required to consider. Since, battery pack plays an important role for EV, the temperature of the battery pack which has a significant impact on the overall performance is a critical parameter. The operating temperatures of Li-ion battery are 0-46^{0}C for charging and -20^{0}C to 60^{0}C for discharging [6]. Usually, at room temperature of 25^{0}C, batteries are designed. Lower and higher temperatures than room temperature both have the effect on the power, energy and lifecycle of the battery cell.

Along with excellent properties, Li-ion cell has low mass density leads to lower thermal capacity, which allows rapid internal temperature rising during exothermic chemical reaction and joule heating (resistive heating, I^2R). Joule heating takes place while charging and discharging process and chemical reaction occurs during overcharge and under-discharge condition [7]. The performance of battery cell is influenced by temperature variation. Excessive temperature variation increases internal resistance, self-discharge rate and voltage. If cell internal temperature continues to increase, the eventual consequence may be an explosion and fire. Hence, many Li-ion manufacturer companies have implemented a wide range of techniques to ensure safety against rupture and fire at the point of view. They used conventional circuit breakers, thermal fuses and current interrupted device even the latest technology of thermal cooling system to manage the problem discussed. Nevertheless, it is undoubtedly necessary to understand the impact of internal temperature during charging and discharging which will be a constructive way of improving the design of the quick charge equalization system. Therefore, in this work, an investigation of the battery internal temperature during the charge balancing process and discharging process has been performed. Battery model and balancing system has been discussed as well to analyse the above mentioned problem for designing a quick charge balancing system.

2. Battery Model

In terms of defining and studying of SOC, voltage, current, and temperature, a battery model is required. There are several battery models proposed in the preceding literatures [8-10]. The models are made of the equivalent circuit of combination of resistance and capacitance. For example, the RC model, internal resistance model, resistance capacitance model, thevenin equivalent circuit model, Shepherd model and the PNGV model [11-12]. In this study, the nonlinear battery model has been applied which is shown in figure 1. According to that model, figure 2 shows a typical battery discharging model.

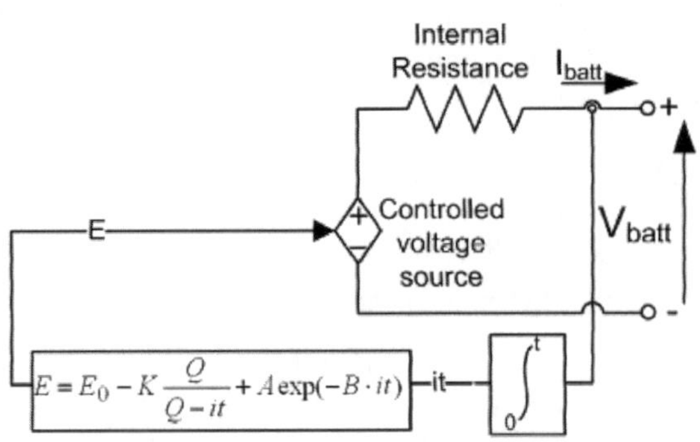

Figure 1. Nonlinear battery model [12].

The following parameters are found while defining this battery model.
Discharging (i*>0)

$$V_{batt} = E_0 - K.\frac{Q}{Q-it}i* - K.(\frac{Q}{Q-it}).it + A.\exp(-B.it) \quad (1)$$

Charging (i*<0)

$$V_{batt} = E_0 - K.\frac{Q}{Q-0.1.Q}i^* - K.(\frac{Q}{Q-it}).it + A.\exp(-B.it) \quad (2)$$

$$V_{batt(full)} = E_0 - R.i + A \quad (3)$$

$$V_{batt(\exp)} = E_0 - K.\frac{Q}{Q-Q_{\exp}}.(Q_{\exp} + i) - R.i + A.\exp(\frac{-3}{Q_{\exp}}Q_{\exp}) \quad (4)$$

$$V_{batt(nom)} = E_0 - K.\frac{Q}{Q-Q_{nom}}.(Q_{nom} + i) - R.i + A.\exp(\frac{-3}{Q_{\exp}}Q_{nom}) \quad (5)$$

Where,
E is no load voltage,
E_O battery constant voltage,
K polarization voltage,
Q battery capacity (Ah),
R internal resistance,
i battery current,
V_{Batt} nominal voltage (V),
exp(s) Exponential zone dynamics (V).

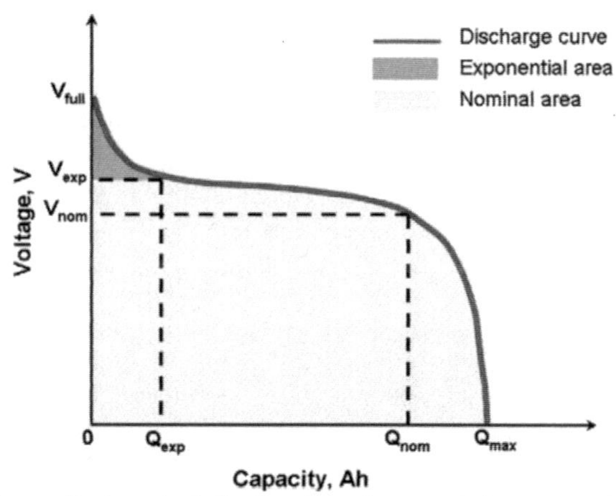

Figure 2. A typical discharging characteristic curve [13].

3. SOC Balancing Methods

Ideally, every cell in a battery pack has the same chemical reaction but in reality each cell has different chemical reaction due to manufacturing process, internal resistance, self-discharge rate, degradation and temperature. These various chemical reaction causes unbalance the SOC of the cell in a battery pack. For safety and long time performance, every cell needs to be the same SOC level in a battery stack. Based on charging, discharging and inoperative condition, various charge equalization methods have been designed, developed and implemented. The methods can be divided into two main methods: namely energy consuming or passive method and energy transferring or active methods. In the energy consuming method, resistor components are used across the battery cell with higher charge. On the contrary, inductor, capacitor and diode are used in energy transferring methods because of the non-consuming element. According to the use of non-consuming components, inductor base, capacitor base and different kinds of converter base equalization system have been presented for example buck-boost converter, flyback converter, quasi-resonant converter, resonant converter etc. [14-20]. In this study, the quasi- resonant LC converter base charge equalization system for two series connected battery cell has been considered and shown in figure. 3.

Battery pack in an electric vehicle is made of sub modules to acquire high voltage and current mentioned before. Based on the above methods, the charge balancing operation is performed from module to cell, module to module, cell to module and cell to cell. The module to cell balancing system can charge a weakest cell efficiently but in cell to module, strong cell can charge a module inefficiently [21].

Most of the researches have been led on balancing with inoperative condition due to the most straightforward approach to transfer energy from strong cell or module to weak cell or module in a battery pack. If the cells are connected in series and both series- parallel combination, the strong and weak cells are found using efficient algorithm. Converter base equalization system with a high switching frequency holds the energy from strongest cell and releases to weakest cell for a while. In the charging and discharging condition, there is a need to think about bypass current to protect the overcharging and undercharging.

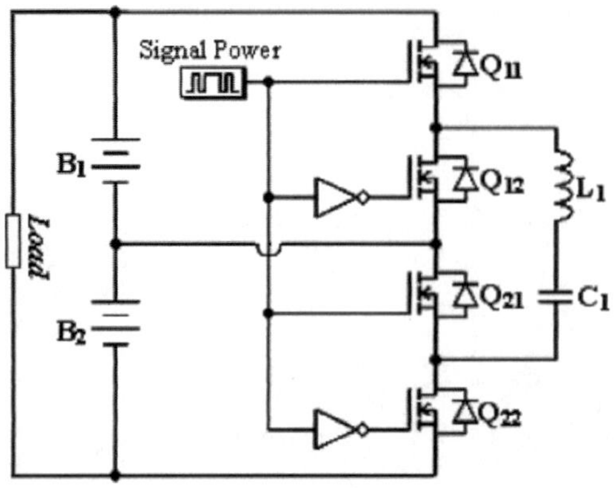

Figure 3. Equalization circuit based on quasi-resonant converter [19].

4. Temperature analysis

During charging and discharging cycle of charge balancing, cell temperature rises dramatically and may exceed its allowable limit (46°C). Battery cell performance depends on internal and ambient temperature. According to joule heating, temperature is generated inside a battery due to I^2R losses when current flows through its internal resistance shown in Eq. (7). Another two heat generation sources are the change of entropy and surrounding temperature of a battery [22]. Dynamic battery temperature is defined by a general energy balance equation shown in Eq. (6).

$$\rho C_{cell}\frac{dT_{cell}}{dt} = Q_{gen} - Q_p - Q_s \qquad (6)$$

where

$$Q_{gen} = i(E_0 - V_{batt} + T\frac{dU}{dT}) + i^2 R$$

$$Q_p = hA_s(T_{cell} - T_{amb})$$

$$Q_s = E\sigma_{sb}(T_{cell} - T_{amb})$$

The internal temperature (T) can be found at any given time (t) and shown as follows

$$T(t) = L^{-1}(\frac{Q_{gen}R_{th} + T_{amb}}{1 + S.t_C}) \qquad (7)$$

where C_{cell} is the heat capacity of a cell, t time, T_{cell} the battery temperature, Q_{gen} the overall heat generation inside the cell, i charging/Discharging current, T_{amb} ambient temperature, Q_p heat generation due to chemical reaction, AS the total surface area, h the heat temperature coefficient, Qs heat generation due to entropy changes, R_{th} thermal resistance, cell to ambient (°C/W) and t_C thermal time constant, cell to ambient (s).

5. Simulation Results

To demonstrate the investigation of battery cell temperature with respect to time in second, a Li-ion (5.4AH, 3.7V) battery model and quasi-resonant converter base balancing circuit were chosen. The simulation time for discharging and the charge balancing operation is different due to limitation of simulation tool. The proposed work is simulated by Matlab-Simulink 2015b with the consideration of wide ambient temperature of 25-40°C, constant discharging current of 1-4Amp and SOC difference of 0.5%. In essence, the cell temperature is generated due to ambient temperature and current flowing through it during charging and discharging. Therefore, initially the ambient temperature and discharging current are assumed to be 25°C as a room temperature and 1 Amp respectively.

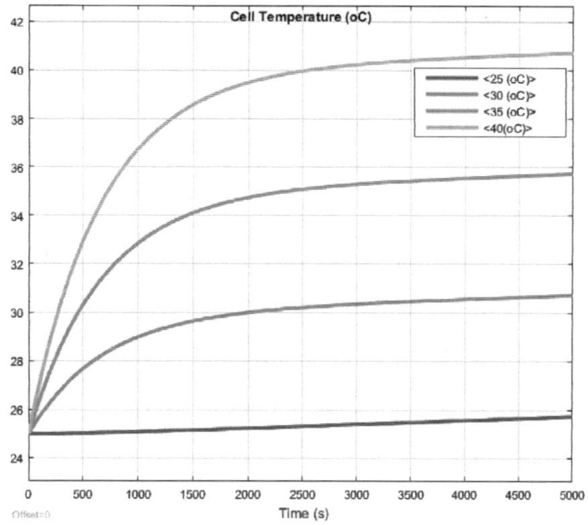

Figure 4. Internal temperature on different ambient temperature and constant discharging current (1A).

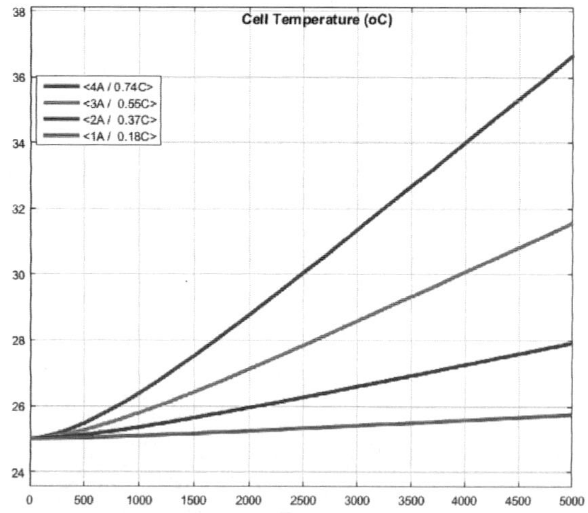

Figure 5. Internal temperature based on constant ambient temperature (25°C) and discharging current variation (1-4A).

Ambient temperature and constant discharging current variation have been considered to reflect the internal cell temperature. Cell temperature was rising as varying ambient temperature from 25-40°C where the discharging current were constant (1 Amp) shown in figure 4. From the figure, it can be seen that the temperature reached at 26°C after a certain time of 5000sec when the ambient temperature was 25°C. Again, ambient temperature was changed from 20°C to 25°C, 30°C and 35°C respectively to observe the effect of internal cell temperature. Accordingly, it is noticed that the rate of temperature rising was significant and proportional. When the ambient temperature was settled on 40°C, cell temperature changed from 25°C to around 41°C which was close to the maximum allowable operating range of 46°C. If the temperature exceeds the maximum range, Li-ion battery is unable to perform its normal behaviour during charging and discharging.

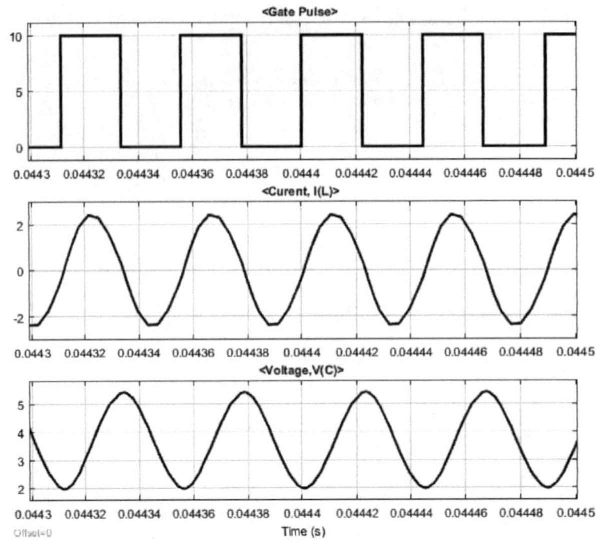

Figure 6. Resonant current and voltage curve in LC network.

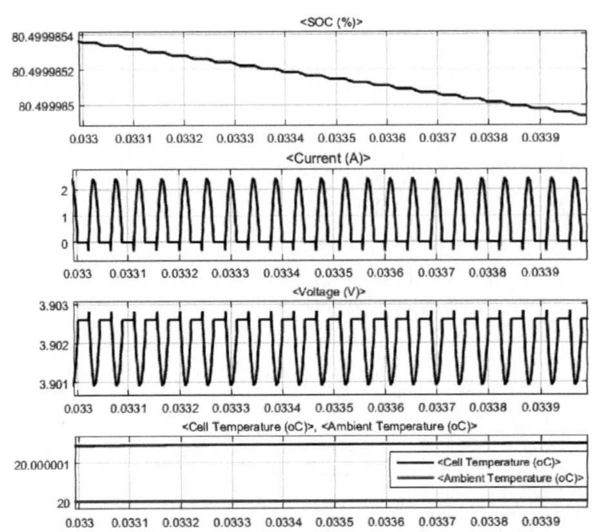

Figure 7. Battery cell (B1): SOC, Voltage, Current and internal temperature during balancing

Usually, temperature rising of the battery cell depends on battery C rate or discharging current and AH (ampere hour) rating. It is suggested that, C rate should be less than 1C to get the best performance [23-25]. In figure 5, discharging current was varied and ambient temperature were constant. Maximum current for discharging were 4Amp/ 0.74C in this case. The figure shows that the cell temperature gradually increased to nearly 26oC at a certain time of 5000sec when discharging current were 1Amp. In the similar manner, temperature reached to almost 37oC at a discharging current of 4Amp.

During the charge balancing operation, both charging and discharging is occurring. The charge balancing circuit for two series connected cells, which has been exposed in the figure 3 where one cell is assumed to be a strong cell and another to be a weak cell. In addition, the charge transferring process was performed by dint of the LC network. This LC network was controlled by at a constant frequency of 22KHz and 50% duty ratio. The resonant current and voltage in the LC network was shown in figure 6. During the charge equalization process, higher cell was discharged from 80.5% and lower cell was charging from 80%. For both discharging and charging condition, cell temperatures were increasing from 20°C to 20.0002°C for B1 and 20.001°C for B2 respectively presented in figure7 and 8 respectively.

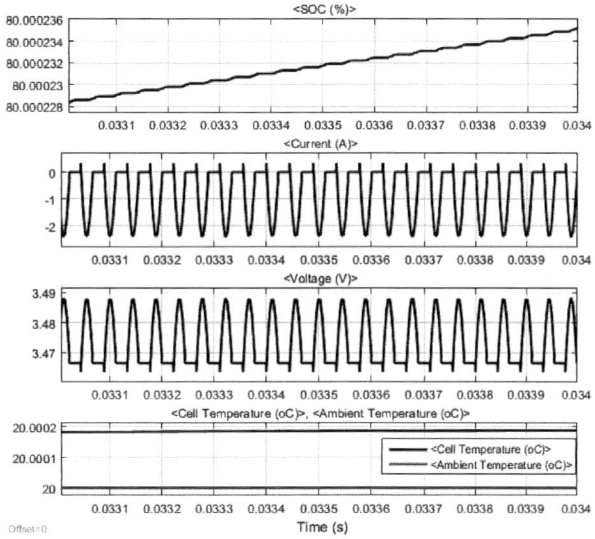

Figure 8. Battery cell (B2): SOC, Voltage, Current and internal temperature during balancing

The rising temperature depends on SOC difference, charging-discharging current and ambient temperature. The simulation time was setup around 1 sec and 2 secs for 2% and 5% respectively due to the limitation of simulation tools. Figure 9 shows the result of SOC charge equalization with temperature where the initial temperature was nearly 25oC and SOC for two cells were almost 80% and 80.2% individually. After 2sec, both SOCs become equal to 80.125% and the temperature increased to just over 20.1°C. At this moment, it seems that rising temperature is not so much. In this case, if the SOC difference is large, growing temperature will be high. From the Figure 10, it can be said that, internal temperature increased more than the 2% difference.

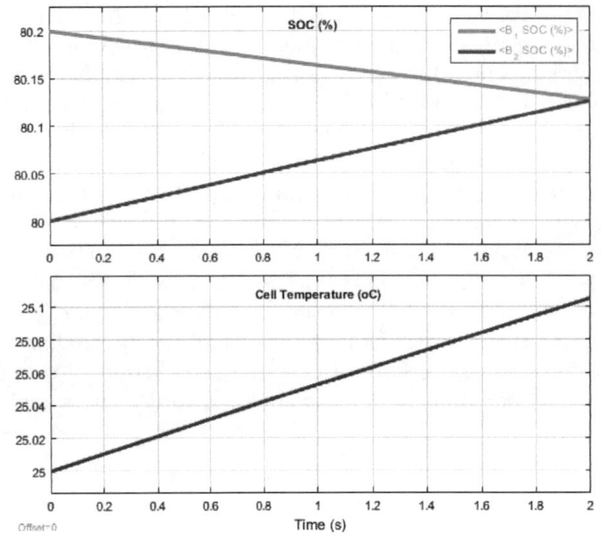

Figure 9. Battery cell (B1 and B2): SOC of 2% variation and internal temperature during balancing.

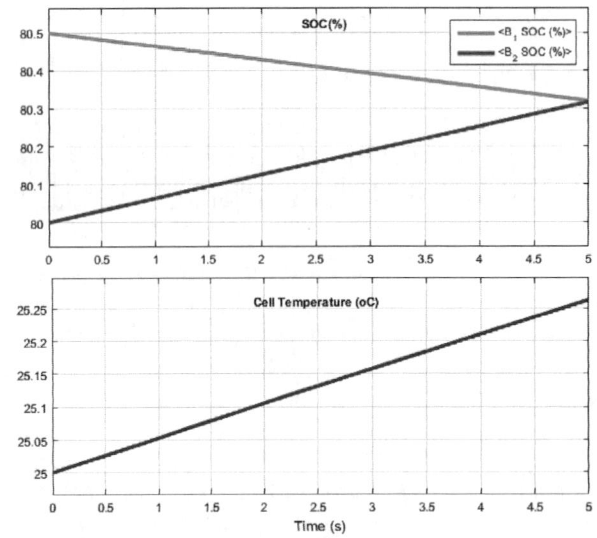

Figure 10. Battery cell (B1 and B2): SOC of 5% variation and internal temperature during balancing.

6. Conclusion

This paper has presented the analysis of the internal temperature of the individual cell during the charge balancing process. Among the charge equalization methods, quasi-resonant base charge balancing method has been applied for analysing the internal temperature. The analysis clearly shows that the rate of increasing internal temperature is influenced by ambient temperature and charging-discharging current (C-rate). In addition, it is obviously true that the highest peaks of the internal temperature of the battery cells depend on SOC difference. During the charge balancing process, the internal temperature may exceed its allowable limit if the SOC difference of the cells is 60%-80% (one is almost full and another is almost empty). In future, experimental work on different balancing methods to analyse battery cell and pack temperature and to design an efficient charge equalization system.

References

[1]. Gallardo-Lozano, Javier, et al. 2014 *Battery equalization active methods*. Journal of Power Sources 246 pp 934-949.

[2]. Imtiaz, Abusaleh M., and Furqan H. Khan. 2013 *"Time Shared Flyback Converter" Based Regenerative Cell Balancing Technique for Series Connected Li-Ion Battery Strings*. Power Electronics, IEEE Transactions on 28.12 pp 5960-5975.

[3]. Kularatna, Nihal. 2011 *Rechargeable batteries and their management*. Instrumentation & Measurement Magazine, IEEE 14.2 pp 20-33.

[4]. Panday, Aishwarya, Hari Om Bansal, and Pramod Srinivasan. 2016 *Thermoelectric Modeling and Online SOC Estimation of Li-Ion Battery for Plug-In Hybrid Electric Vehicles*. Modelling and Simulation in Engineering 2016.

[5]. Quinn, D. Dane, and Tom T. Hartley. 2013 *Design of novel charge balancing networks in battery packs*. Journal of Power Sources 240 pp 26-32.

[6]. Leng, Feng, Cher Ming Tan, and Michael Pecht. 2015 *Effect of Temperature on the Aging rate of Li Ion Battery Operating above Room Temperature*. Scientific reports 5.

[7]. Srinivasan, Rengaswamy, et al. 2011 *Instantaneous measurement of the internal temperature in lithium-ion rechargeable cells*. Electrochimica Acta 56.17 pp 6198-6204.

[8]. Lee, S.; Kim, J.; Lee, J.; Cho, B.H. 2008 *State-of-charge and capacity estimation of lithium-ion battery using a new open-circuit voltage versus state-of-charge*. J. Power Sources, 185, pp 1367–1373.

[9]. Gao, Lijun, Shengyi Liu, and Roger A. Dougal. 2002 *Dynamic lithium-ion battery model for system simulation*. IEEE transactions on components and packaging technologies 25.3 pp 495-505.

[10]. Johnson, V.H. 2002 *Battery performance models in ADVISOR*. J. Power Sources, 110, pp 321–329.

[11]. Idaho National Engineering & Environmental Laboratory. 2010 *Battery Test Manual for Plug-In Hybrid Electric Vehicles*; Assistant Secretary for Energy Efficiency and Renewable Energy (EE) Idaho Operations Office: Idaho Falls, ID, USA, 2010

[12]. Tremblay, Olivier, and Louis-A. Dessaint. 2009 *Experimental validation of a battery dynamic model for EV applications*." World Electric Vehicle Journal 3.1 pp 1-10.

[13]. Saw, L. H., et al. 2014 *Electro-thermal analysis of Lithium Iron Phosphate battery for electric vehicles*. Journal of Power Sources 249 pp 231-238.

[14]. Einhorn, Markus, Werner Roessler, and Juergen Fleig. 2011 *Improved performance of serially connected Li-ion batteries with active cell balancing in electric vehicles*. Vehicular Technology, IEEE Transactions on 60.6 pp 2448-2457.

[15]. Pascual, Cesar, and Philip T. Krein. 1997 *Switched capacitor system for automatic series battery equalization*. Applied Power Electronics Conference and Exposition, 1997. APEC'97 Conference Proceedings 1997., Twelfth Annual. Vol. 2. IEEE, 1997.

[16]. M Rahman, MM Rashid, A Rahman, S Khan 2016. *Resonant-Boost LC converter base charge balancing system for electric vehicle* ARPN Journal of Engineering and Applied Sciences pp 362-367

[17]. Lee,Yuang-Shung, and Guo-Tian Cheng. 2006 *Quasi-resonant zero-current-switching bidirectional converter for battery equalization applications*. Power Electronics, IEEE Transactions on 21.5 pp 1213-1224.

[18]. Baronti, Federico, et al. 2013 *High-efficiency digitally controlled charge equalizer for series-connected cells based on switching converter and super-capacitor*. Industrial Informatics, IEEE Transactions on 9.2 pp 1139-1147.

[19]. Baughman, Andrew C., and Mehdi Ferdowsi. 2008 *Double-tiered switched-capacitor battery charge equalization technique*. Industrial Electronics, IEEE Transactions on 55.6 pp 2277-2285.

[20]. Yuanmao, Ye, K. W. E. Cheng, and Y. P. B. Yeung.2012 *Zero-current switching switched-capacitor zero-voltage-gap automatic equalization system for series battery string*. Power Electronics, IEEE Transactions on 27.7 pp 3234-3242.

[21]. Stuart, Thomas A., and Wei Zhu. 2011 *Modularized battery management for large lithium ion cells*." Journal of Power Sources 196.1 pp 458-464.

[22]. Zhu, Cong, et al. 2013 *Development of a theoretically based thermal model for lithium ion battery pack*. Journal of Power Sources 223 pp 155-164.

[23]. Dearborn, Scott. 2005 *Charging Li-ion batteries for maximum run times*." Power Electron. Technol. Mag 31, no. 4 pp 40-49.

[24]. Rahman, A., Nur, F., Hawlader, M.N.A. and Afroz, R., 2015. *Fuzzy controlled evaporative battery thermal management system for EV/HEV*. International Journal of Electric and Hybrid Vehicles, 7(1), pp.22-39.

[25]. Wang, H., Chunyu, D.U. and Wang, C. 2009 *Study of low temperature performance of li-ion battery*, Battery Bimonthly, Vol. 139, No. 14, pp.208–210.

Numerical Investigation of Aerodynamic Characteristics of High Speed Train

J.S. Mohamed Ali [1,a], Ashraf Ali Omar[2], Muhammad 'Atif B. Ali[1] and Abdul Rahman Bin Mohd Baseair[1]

[1]Department of Mechanical Engineering, Kulliyyah of Engineering, International Islamic University Malaysia (IIUM)
P.O.Box 10, 50728 Kuala Lumpur, Malaysia
[2] Department of Aeronautical Engineering, Tripoli University, Libya

Email : [a]jaffar@iium.edu.my

Abstract. In this work, initially the effect of nose shape on the drag characteristics of a high speed train is studied. Then the influence of cross winds on the aerodynamics and hence the stability of such modern high speed trains is analyzed. CFD analysis was conducted using STAR-CCM+ on trains with different features and important aerodynamic coefficients such as the drag, side force and rolling moment coefficients have been calculated for yaw angles of crosswinds ranging from 0° to 90°. The results show that the modification on train nose shape can reduce the drag up to more than 50%. It was also found that, bogie faring only reduces small percentage of drag but significantly contributed to higher rolling moment and side force coefficient hence induced train instability.

1. Introduction

Many of the current generation high speed trains such as the French TGV Duplex and Japanese Shinkansen E6 reach speeds of 300km/h in regular operation. At these speeds aerodynamic forces and moments get amplified and hence it greatly influences the performance of the train. In addition to the high forward speed, strong cross wind may affect the stability and riding comfort of the vehicles and it may cause the train to overturn. Therefore, understanding of train stability under crosswinds is a topic of safety concern in the railway community of every country. Thus crosswind stability of rail vehicles has been a research topic during the last few decades [1-6]. Moreover, to achieve high speeds, the aerodynamic drag of the rail vehicles was decreased using various design features, such as adopting streamlined shape head [7] and also by covering the train bogies using fairing [8], etc. These design features may have an influence on the crosswind stability of the bogies. Thus in this work, the effect of both, bogie fairings and nose shape, on the aerodynamic characteristics was studied together in normal and cross winds conditions. The aim of this study is to modify the train nose into a streamlined shape and also add bogie fairings for the purpose of aerodynamic drag reduction and to investigate the effect of the bogie fairings and nose shape on the aerodynamic characteristics of high speed train under turbulent cross winds.

2. Methodology

The CFD analysis on train models with different configuration were simulated in CFD software STAR-CCM+, to study the aerodynamic characteristic of the train such as, drag, lift, moment and side force coefficient. For the study of the cross wind effect on train, a large computational domain is created simulating the wind tunnel working section. In the computational domain, the air flow direction can be changed to meet the required yaw angle for simulation. The physics of the simulation were set up according to that of Biadgo et al [6]. The numerical simulation scenario consists of a stationary train exposed to a constant cross wind of 70m/s at different yaw angles ranging from 0° to 90°. Implicit Unsteady-RANS Method was used with turbulent and K-Epsilon model. On the outlet, a uniform Neumann boundary condition is applied, meaning that the pressure gradient equals zero. The inflow turbulence intensity and length scale were set to be 3% and 0.3m respectively. All runs were performed with a time step of 0.08 sec. Fig 1 shows the side view of the 3-D meshed train with the polyhedral volume mesh around the train.

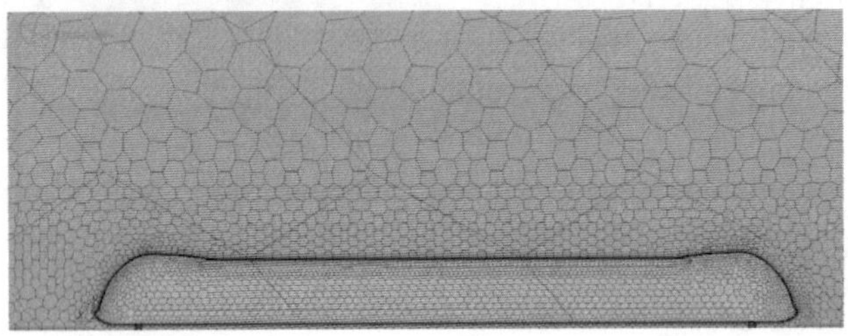

Figure 1. Side view of the meshed train and the volume mesh around train with prism layers

3. Results and Discussion

3.1 *Effect of Nose Shapes on Drag and Lift Coefficient*

Initially the effect of different nose shape on the aerodynamic performance of the high speed train is studied. Models resembling modern high speed trains with different nose shape were considered for CFD analysis as shown in Fig 2. Baseline train model dimension is chosen based on the Desiro ET425M Train also known as KLIA Express. The major dimensions of Desiro ET 425 M is 50m x 4m x4.6m. In addition, other train models with optimized nose shape is also chosen for analysis.

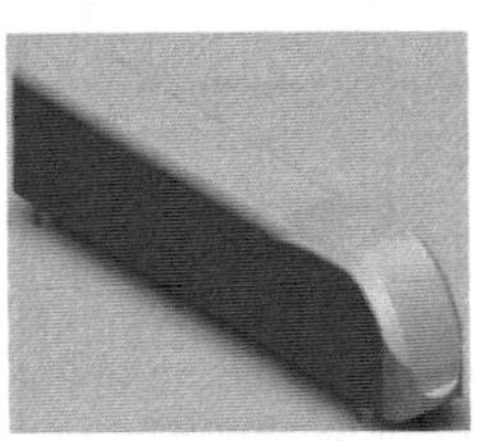

Baseline

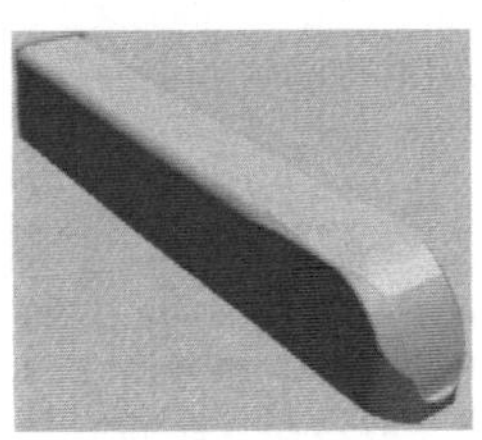

Model 1 (M1)

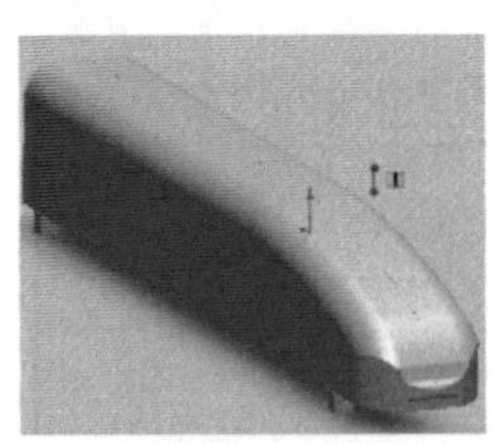

Model 2 (M2)

Model 3 (M3)

Figure 1. Train models with different nose shape.

Table 1. Drag and Lift coefficient of different models

Model	Drag Coefficient, C_D	% change in C_D	Lift Coefficient, C_L	% change in C_L
Baseline	0.532	-	0.471	-
M1	0.544	+2.25%	0.299	+36.5%
M2	0.353	-33.6%	0.263	-44.2%
M3	0.230	-56.8%	0.074	-84.3%

*negative sign '-' indicating decrement while positive sign '+' indicating increment.

The results obtained from the CFD analysis for different nose shapes are presented in Table 1. We can see that, the M1 has the largest drag coefficient, followed by baseline, M2 and lastly M3. The lift coefficient is highest for baseline model and followed by M1, M3 and lowest M3. The lift coefficient for M3 has been reduced drastically which is more than 50% less than the baseline model.

3.2 *Effect of Bogie Fairings and Nose shape on the Aerodynamic Characteristics of High Speed Train under Cross winds*

The following three models as shown in Figures 3-5 are considered for study on the effect of cross winds at different yaw angles. The CFD results obtained for side force coefficient and rolling moment coefficient for different yaw angles are presented in Figure 6 and Figure 7.

Figure 3. Standard model (Without fairing)

Figure 4. Train with fairing

Figure 5. Sharp nose train without fairing

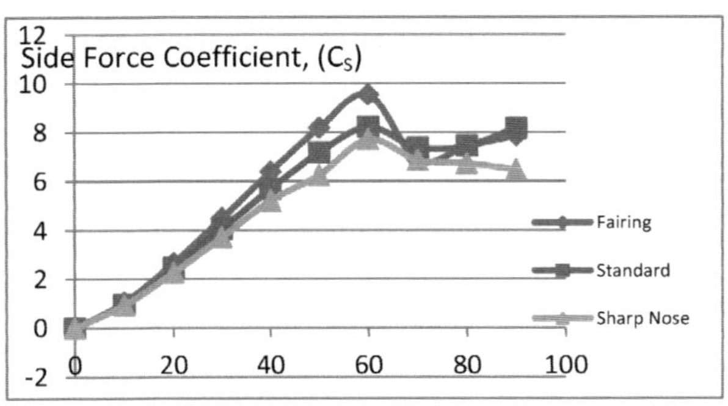

Figure 6. Side force coefficient vs yaw angle for train with fairing

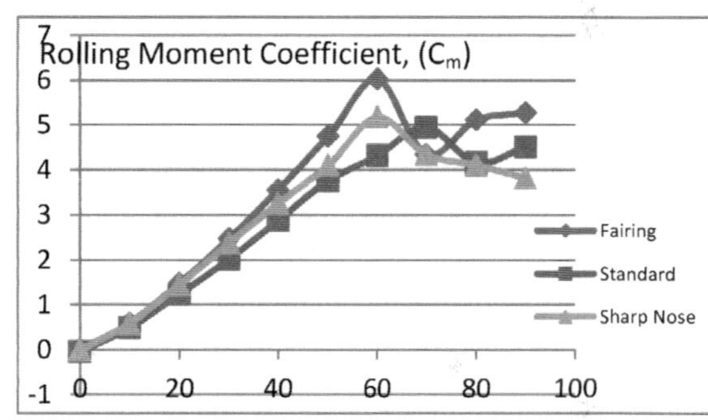

Figure 7. Rolling moment coefficient vs. yaw angle

From figure 6, it can be seen that the side force coefficient for standard and train model with fairing increases steadily with yaw angle until 60° before it starts to decrease and increase again after 70° to 90°. From 0° to 60°, the side coefficient value of model with fairings is slightly higher compared to the standard model. However at 70° and 90°, the value of side force coefficient for standard model is slightly higher compared to the model with fairings. On the other hand, the side force coefficient for sharp nose model is lowest compared to other models at every yaw angle.

As can be seen from figure 7, the rolling moment coefficient varies almost in a similar fashion to the side force except that the rolling moment coefficient of standard model is higher than model with fairing only at 70°. The rolling moment coefficient of sharp nose model is considerably lower compared to model with fairing but higher than standard model at every yaw angle. At yaw angle of 90°, the rolling moment coefficient for sharp nose is lowest compared to other models.

In general, the variation of the side force and rolling moment coefficients of the models with and without faring is in a good agreement with the numerical and experimental results presented by Biadgo et al [6]. It can be seen from Fig 7 that there is a significant rolling moment resulting from the lift and side forces that can lead to overturning of the train with overloading of the railway track as well.

To understand the effect of fairings on the aerodynamic performance of the train, the three models (Figs 3-5) were analyzed at different speeds and the results for the drag force are as given in Fig 8.

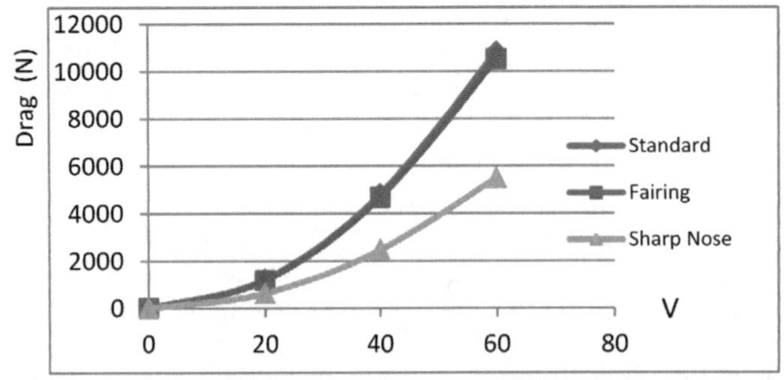

Figure 8. Drag force vs. speed of train

It is obvious that the drag curve for the configuration with fairings stands below the curve of the standard configuration and the drag curve for sharp nose model is way below the base model. It was found that the drag reduction by use of fairings is only about 44.6 N (3.6%) at the speed 20m/s and 316.4N (2.85%) at 60m/s.

The drag coefficient (C_D) for sharp nose model was lowest at around 0.26 compared to the other two models at 0.5. Thus the fairings at the train bogie reduce the drag of the train by a very small decrement. However, the modification on the nose gives significant drag reduction.

4. Flow Structure

Results have been generated for flow around the train for different yaw angles from 0° to 90° and typical flow structure around the train for a yaw angle of 60° is shown in Figure 9. The flow structures of the trains were taken along the train's cross section at different cutting plane (front, mid and rear end).

Cross Section	Total Pressure	Velocity Vector
1	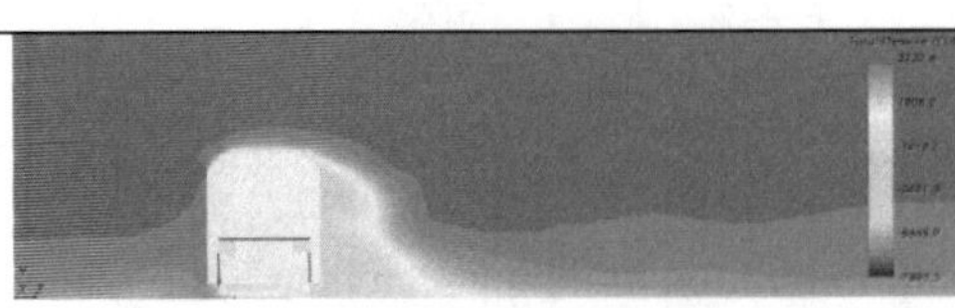	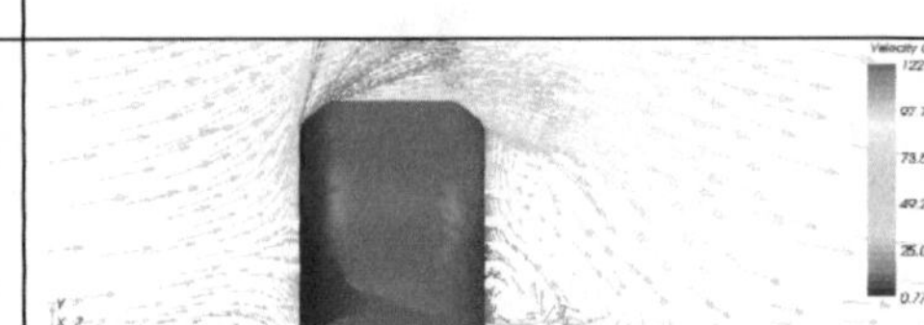
2	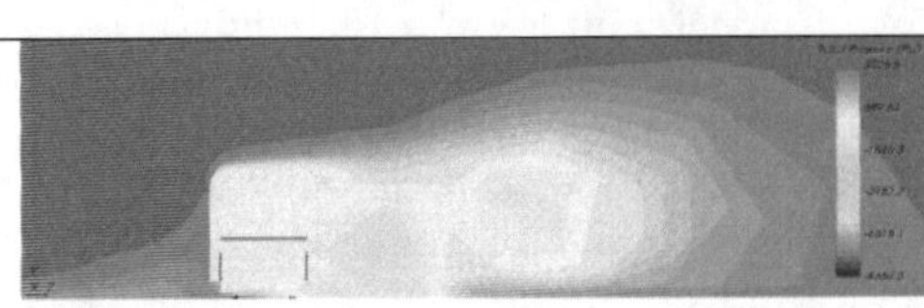	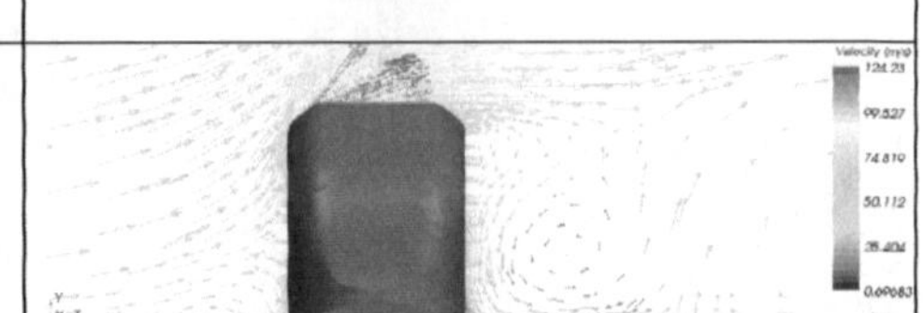
3	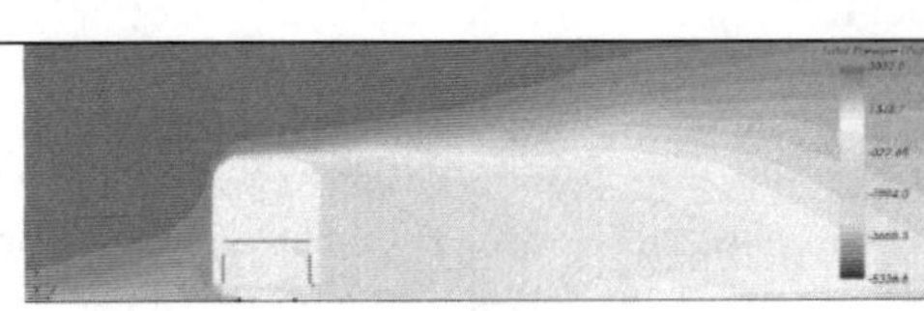	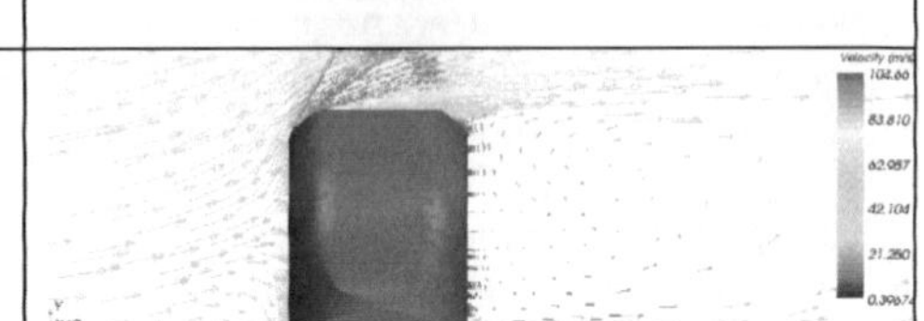

Figure 9. Velocity vectors, colored by velocity magnitude (m/s) and total pressure contour (Pa) along the standard train's cross section planes for yaw angle 60°.

As can be seen from the figure 9, flow separation takes place on both the lower and upper edges of the bogie. Contours of velocity vectors and total pressure distributions have also been computed at different cross-sections from the nose of the train along its length for different yaw angles. As can be seen in Fig 9, there is a high velocity on the top of the train i.e flow accelerates on the trains' top surface and also the presence of vortex creates a low pressure region. The existence of the lower pressure region explains the increased side force and roll moment. The results obtained are very similar to that of Biadgo et al [6].

5. Conclusion

Based on the findings, it can be concluded that, Baseline Model and Model 1 designs can be considered as bluff body design compared to Model 2 and 3. The results of drag coefficient for Baseline and Model 1 are significantly large compared to streamlined shape of Model 2 and 3. A change in train nose design can significantly improve the aerodynamics of high speed train. The study has also shown that, bogie fairing only reduce drag in small percentage but contribute to higher rolling moment and side force coefficient. Thus the bogie fairing can reduce small amount of drag at the cost of higher tendency for the train to be unstable and overturn. A sharper and streamlined train nose will contribute to both drag reduction and give better crosswind stability compared to other models.

References

[1] Baker, C.J. 2002, The wind tunnel determination of crosswind forces and moments on a high speed train, Notes on Numerical Fluid Mechanics, Vol. 79, pp. 46–60.

[2] Baker C. J., Measurements of the cross wind forces on trains 2004, Journal of Wind Engineering and Industrial Aerodynamics, Vol 92, pp 547-563.

[3] Carrarini, A. 2007, Reliability based analysis of the cross wind stability of railway vehicles, , Journal of Wind Engineering and Industrial Aerodynamics, Vol 95, pp 493-509.

[4] Cheli, F, Ripamonti, F., Rocchi, D. and Tomasini, G. 2010, Aerodynamic behaviour investigation of the new EMUV250 train to crosswind, Journal of Wind Engineering and Industrial Aerodynamics, Vol 98, pp 189-201.

[5] Guilmineau, E., Chikhaoui, O., Deng, G. and Visonneau, M. 2013, Cross wind effects on a simplified car model by a DES approach, Computers and Fluids, Vol 78, pp 29-40.

[6] Asress Mulugeta Biadgo, Aleksandar Simonovic, Jelena Svorcan, Slobodan Stupar, 2014, Aerodynamic Characteristics of High Speed Train under Turbulent Cross Winds: a Numerical Investigation using Unsteady-RANS Method, FME Transtions, Vol 42, No 1, pp 10-18.

[7] Munoz-Paniagua, J., García, J., Crespo, A. and Laspougeas, F., 2015, Aerodynamic Optimization of the Nose Shape of a Train Using the Adjoint Method, Journal of Applied Fluid Mechanics, Vol. 8, No. 3, pp. 601- 612.

[8] Mancini, G., Malfatti, A., Violi, A. and Matschke, G. 2001 Effects of experimental bogie fairings on the aerodynamic drag of the ETR 500 high speed train. Proceedings of the World Congress on Railway Research (WCRR 2001), Cologne, Germany.

Implementing Lean Manufacturing in Malaysian Small and Medium Startup Pharmaceutical Company

[1]Wan Mohd Khairi bin Wan Ibrahim, [2]Mohamed Abdul Rahman, [3]Mohd Rushdi bin Abu Bakar

[1,2]Department of Manufacturing and Materials Engineering, International Islamic University (IIUM), Kuala Lumpur 53100 Malaysia, [3]IKOP Sdn. Bhd., Jalan Sultan Ahmad Shah, Bandar Indera Mahkota, 25200 Kuantan Pahang

wmkwi79@gmail.com, mrahman@iium.edu.my, mohd.rushdi.abu.bakar@gmail.com

Abstract. Domestic pharmaceutical industry has been identified by the Malaysian government as an industry to be developed under its 11th economic development plan. Most homegrown pharmaceutical companies fall under the category of small and medium enterprises (SME) and therefore need to be highly efficient in their operations to compete with the multinationals. Though lean manufacturing is a well-known methodology to achieve an efficient operation, only a small percentage of the local SMEs implement it. The study aims to determine the real success factors in lean implementation through systematic review of relevant literature on lean manufacturing implementation in local companies, onsite observation of a selected SME company, Global Factor Sdn. Bhd. (GFSB), that successfully implemented lean manufacturing followed by actual implementation of lean project at IKOP Sdn. Bhd., a small startup pharmaceutical company. Lean tools like Gemba, value stream map (VSM) and spaghetti diagram were used to analyze and improve a process at IKOP Sdn. Bhd. The literature review showed that the implementation of lean manufacturing at Malaysian SMEs involved in pharmaceutical industry is at its infancy. Study at GFSB indicated that successful implementation of lean manufacturing stems from management support, employee's commitment, government support and knowledge on lean among employees. Application of lean tools in IKOP Sdn. Bhd. to improve the process cycle efficiency of hand sanitizer, i-Hand 4.0, has shown that the GMP guidelines are not jeopardized. The Kaizen improvement project resulted in 46.3% reduction in lead time. It may be concluded that implementing lean manufacturing in any small local startup pharmaceutical company is beneficial in reducing operational costs and increasing the efficiency and effectiveness and does not conflict with the existing GMP guidelines.

1. Introduction

The cost of developing drugs is very high as pharmaceutical companies are required to invest approximately 3-4 times more than other manufacturing companies [1]. This high cost is due to higher failure rates, longer times required for clinical trials, and increasing drug complexity [2]. To offset these costs, pharmaceutical companies are now facing increasing pressure to reduce manufacturing costs through improving operational efficiencies. Lean manufacturing is a manufacturing system that aims to minimize production cost and maximize the profit by eliminating waste or non-value added activities [3]. Derived largely from the Toyota Production System, the practices have been implemented widely in many industries [4].

Pharmaceutical industry, however, has been slow in adopting LM [5,6,7]. And this may be more so in small and medium enterprises (SME) and especially so in the Malaysian context [8,9,10]. The domestic pharmaceutical industry has been identified by the Malaysian government as an industry to be developed under its new 11th economic development plan. Most of the homegrown companies within the industry fall under small and medium enterprise (SME) and therefore need to be highly efficient in their operations in order to compete with the multinationals. Therefore, this study aimed to achieve better understanding of the success factor(s) of LM implementation in Malaysian startup pharmaceutical SMEs through a systematic review of relevant literatures, an onsite observation at a local SME which successfully implemented LM and a case study implementation at another local small pharmaceutical startup company.

2. Literature Review

One approach to achieve an efficient operation is through implementing lean manufacturing. Exploratory inquiry however indicated that relatively a small percentage of the local SME companies implements lean manufacturing as also found by previous studies on this issue [10,11]. This is true for pharmaceutical companies due to their focus on establishing themselves in the initial market since for many drugs, the first to market captures the majority market share [6]. The complexity of designing operational procedures to incorporate LM into the existing system highly controlled and regulated by the current Good Manufacturing Practices (cGMP) may also be an issue.

Barriers and difficulties in the implementation of lean in pharmaceutical industry may come from the cGMP guidelines. The cGMP requires a lot of paper works; documentations and standard operation procedures (SOP) that affects the lead time [12]. cGMP are the guidelines formulated by the US Food and Drug Administration (FDA) that pharmaceutical industries in the entire world should follow in the drug production. At times, the rules and regulation in cGMP may not be applicable and become waste in pharmaceutical production process [13]. Lean manufacturing is a manufacturing system that has aimed to minimize production cost and maximize the profit by eliminating waste or non-value added activities [3]. This system was invented and implemented by Toyota Production System successfully and the practices have since been implemented widely in many industries [4].

According to [8], lean implementation is still new to SMEs. This is true in the Malaysian context; the implementation is still low within them [9,10]. Malaysian SMEs are categorized based on two categories: (i) manufacturing sectors and (ii) services and other sectors [14]. A company in manufacturing sector that operates with less than 200 full time workers and obtains annual income of RM50 million is considered as an SME. In services and other sectors, for a company to be considered as an SME, the number of full time workers must be less than 75 and the annual income must be less than RM20 million. There are 60,158 companies listed by SME Corp Malaysia as SMEs. SME Corp Malaysia is the authorized government agency entrusted to improve, support and advice SMEs in competing in the global business environment. The agency has been recommending SME to apply lean manufacturing concept due to its contribution towards company's efficiency and competency [15].

Although Toyota shared the knowledge of lean concept, companies need to run and implement it using their own style [16]. The implementation of lean without right understanding will create more waste and decrease in employees' morale [17]. The level of success of lean implementation was directly determined by the understanding of lean concept by top management responsible in the organization [18]. To succeed in implementing lean, the organization must understand the lean concept really well through actual implementation [19,20]. Researchers [10,15,21,22,23,24], found that, the main barriers to the successful lean implementation in Malaysian SMEs are lack of understanding of lean concept and the employees' attitude.

3. Methodology

Multiple case study concept [25] was used to collect the data. Review of literature relevant to the implementation of LM in Malaysia was carried out. Special focus was given to the LM implementation in Malaysian SMEs and pharmaceutical companies. The main aim was to gain a better understanding of the level of lean implementation in Malaysia as well as establish the factors behind the successes and failures of LM implementation.

An onsite observation was carried out at a small manufacturing company; GF Sdn. Bhd. (GFSB) which successfully implemented lean manufacturing. Data collection was done through interviews, process observation and document analysis [26] in order to evaluate the real factors behind the successful implementation of lean manufacturing in GFSB.

Subsequently, onsite observation was done at a small local startup pharmaceutical company, IKOP Sdn. Bhd. to identify the benefits of implementing LM in this company and whether or not the cGMP poses a hindrance to successful LM implementation. Two (2) lean tools namely, value stream map (VSM) and spaghetti diagram were used in a lean project at a selected process. The implementation process at IKOP Sdn. Bhd. began with gathering data to develop the current VSM (cVSM) of the production process. Spaghetti diagram was utilized to represent flow paths of items and activities through a stream in order to identify opportunities to expedite process flow. Future VSM (fVSM) was later developed to identify improvement opportunities.

4. Results and Discussion

Exploratory study in the initial phase showed that GFSB was the one and only company which implemented lean manufacturing system in east coast region as recorded from interview with a Malaysian Productivity Corporation personnel [11]. Through systematic literature review, a total of 39 papers relevant to the lean implementation within the Malaysian context were collected. It was found that the lean implementation in the Malaysian pharmaceutical sector was in the infancy. Figure 1 indicates the percentage of reviewed papers by industries in which pharmaceutical industry was included in the "Others" category, with only one (1) paper published in 2015.

Data were collected in two SME companies, one in manufacturing (GFSB) and another one in pharmaceutical (IKOP). The lean implementation activities were successfully implemented in GFSB while in IKOP lean manufacturing was to be implemented. Table 1 describes the two main lean tools used namely, VSM and Kaizen, and success factors identified at GFSB. Top management support was found to be the starting point of this achievement as they were responsible to provide the direction of the company. Employees' attitude was found to be the second factor behind the successful lean implementation. Only with positive attitudes of the workforce can improvement program such as lean manufacturing be successful. The third factor came in the form of training prior to the implementation; it was found that the employees were given an intensive training by an appointed lean consultant. With this knowledge, they could implement the concept correctly. Lastly, an external factor in the form of government support was the ignition point for this company to implement lean. The government authority such as SME Corporation, that offers manufacturing grants, requires the company to implement lean manufacturing system. This indirectly introduced the company to lean implementation. The analysis of this observation was used in the case of a newly established pharmaceutical company in term of success factors, effects on the implementation and performance of lean manufacturing system for the company.

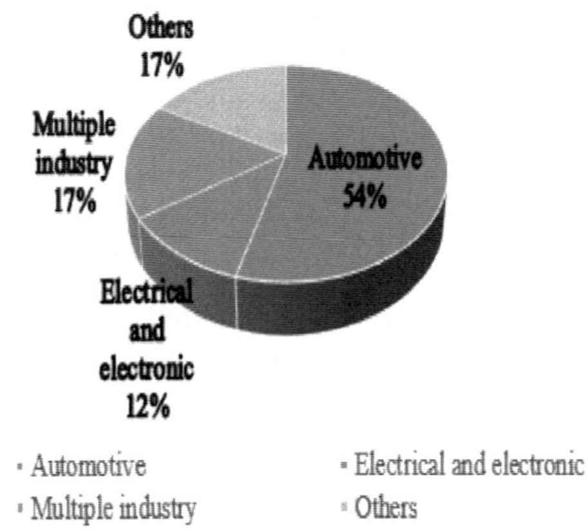

Figure 1. Papers distribution by sector.

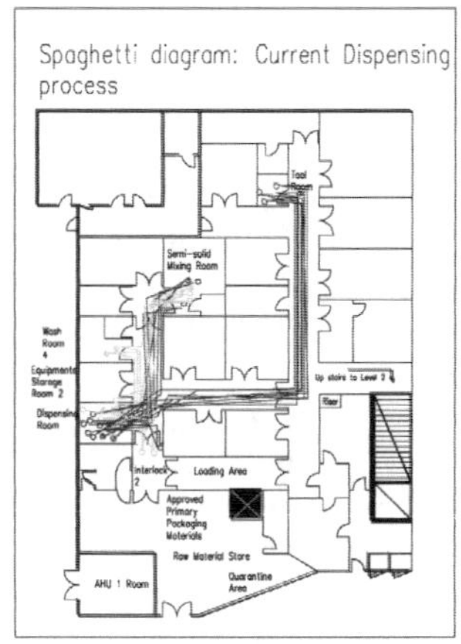

Figure 2. Current spaghetti diagram.

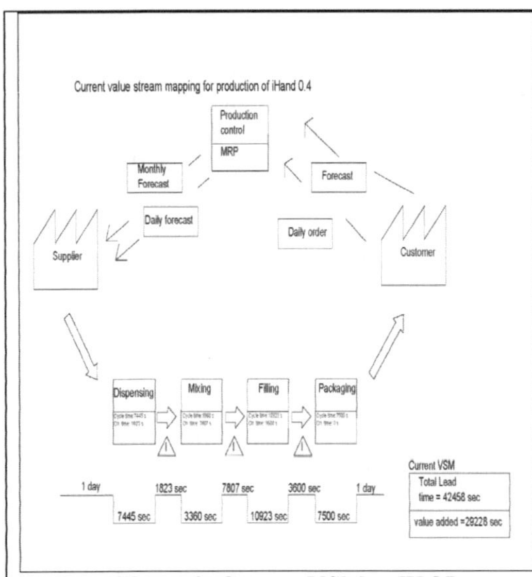

Figure 3. Current VSM at IKOP.

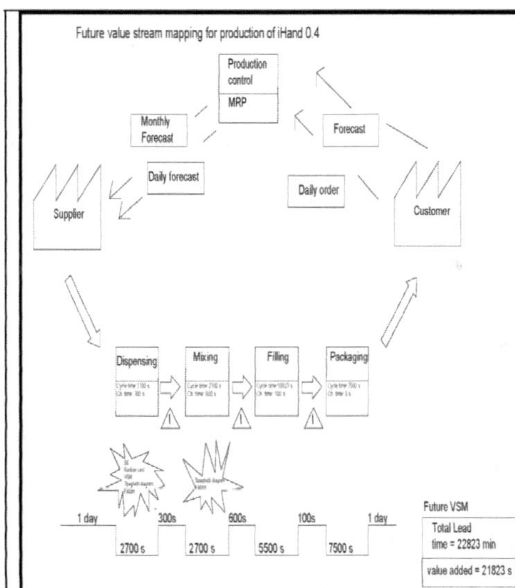

Figure 4. Future VSM at IKOP.

Figure 2 and Figure 3 show the current spaghetti diagram at dispensing process and current VSM for the whole production of i-Hand 4.0 respectively. The current lead time recorded was 42,458 seconds or more than 11 hours. Figure 4 shows the suggested future VSM with the application of lean tools such as 5S, VSM, spaghetti diagram and Kaizen. Kanban card application was to be implemented. With this integration, the future lead time was expected to be reduced to 22,823 seconds or less than 7 hours.

Table 1. Lean implementation program in GFSB.

Item	Description
Lean tools	VSM and Kaizen were used. cVSM was developed and studied. Future VSM then produced and implemented. The implementation required small cost with the highest amount was RM50.
Success factor	1. Top management support 2. Employees' attitude 3. Implementation knowledge 4. Government support

Table 2. Lean implementation in IKOP.

Item	Description
Lean tools	VSM, spaghetti diagram and Kaizen was used. cVSM was developed and studied. Future VSM then produced.
Success factor	1. Top management understanding 2. Employees' attitude 3. Implementation knowledge

Table 2 shows the lean implementation in IKOP. In this implementation, three lean tools were used; VSM, spaghetti diagram and Kaizen. A spaghetti diagram could visualize the overall motion of employees in their activities. From the results three factors were identified. Firstly, top management understanding through briefing and explanation session which resulted in the top management agreeing to lean implementation with certain conditions such as no conflict with current good manufacturing practices (cGMP) and minimum cost required in the implementation. The second factor was employees' attitude in which they believed this system can generate better results thus may increase the company's income through better understanding the benefits of lean manufacturing system. Lastly, the implementation knowledge; after a few training sessions on lean and VSM, they understood their functions in the organization.

5. Conclusions

It may be concluded from the study that the implementation of lean manufacturing at Malaysian SMEs involved in pharmaceutical industry is at its infancy. Successful implementation of lean manufacturing stems from management support, employee's commitment, government support and knowledge on lean among employees. Application of lean tools in IKOP Sdn. Bhd. to improve the process cycle efficiency of hand sanitizer, i-Hand 4.0, has shown that the GMP guidelines are not jeopardized with the Kaizen improvement project resulted in 46.3% reduction in lead time. It may be further concluded that implementing lean manufacturing in any small local startup pharmaceutical company is beneficial in reducing operational costs and increasing the efficiency and effectiveness and does not conflict with the existing GMP guidelines.

Acknowledgement
The authors would like to express their thanks to the management of GF Sdn. Bhd. and IKOP Sdn. Bhd. for giving the permission to carry out observations and interviews at their premises.

References

[1] F.J. Cohen. Macro Trend in Pharmaceutical Innovation, Nature Review: Drug Discovery, January 2005, Volume 4. Retrieved from www.nature.com/reviews/drugdisc
[2] M.E. Lawler, Improving Shop Floor Visualization and Metrics. (MSc. Thesis), 2010. Massachusetts Institute of Technology.
[3] J.P. Womack, D.T. Jones, and D. Ross, The Machine that Changed the World, Rawson Associates, New York, NY, 1990.
[4] M. Holweg, The Genealogy of Lean Production, J. Operation Management. 25 (2007) 420 – 437.

[5] A. Greene, and D. O'Rourke. Lean Manufacturing Practice in a cGMP environment. (2006). Retrieved from Pharmaceutical Technology Europe: http://www.ptemagazine.com

[6] A.D., Villa. Lean Transformation Methodology and Implementation in Biopharmaceutical Operation. (MSc. Thesis) 2008. Massachusetts Institute of Technology.

[7] V. Jaiganesh, and J.C. Sudhahar, Sketching Out the Hidden Lean Management Principles", International Journal of Scientific and Research Publications. 3(2) (2013).

[8] P. Achanga, E. Shehab, R. Roy and G. Nelder, Critical Success Factors for Lean Implementation within SMEs, Journal of Manufacturing Technology Management Vol. 17 No. 4 (2006) 460-471.

[9] A.M.N. Rose, B. Md. Deros, M.N. Ab. Rahman, and N. Nordin, Lean Manufacturing Best Practices in SMEs, Proceedings of the 2011 International Conference on Industrial Engineering and Operations Management Kuala Lumpur, Malaysia, January 22 – 24, 2011

[10] Y.C. Wong, K. Y. Wong, and A. Ali, A Study on Lean Manufacturing Implementation in the Malaysian Electrical and Electronics Industry, European Journal of Scientific Research ISSN 1450-216X Vol.38 No.4 (2009), 521-535.

[11] Sharifah Mazina Syed Mohamed, Interviewed by Author, Kuantan, Pahang, 15 October 2014.

[12] M. Khlat, A.H. Harb, A. Kassem. Lean Manufacturing: Implementation and Assessment in the Lebanese Pharmaceutical Industry. International Journal of Computing and Optimization, 1, (2014) 47 – 62.

[13] B.V. Chowdary, and D. George, Improvement of Manufacturing Operations at a Pharmaceutical Company, Journal of Manufacturing Technology Management, Vol. 23, (2011) 56 – 75.

[14] Information on www.smecorp.gov.my

[15] N. Nordin, B. Md. Deros, and D. Abd. Wahab, A Survey on Lean Manufacturing Implementation in Malaysian Automotive Industry, International Journal of Innovation, Management and Technology, 1(4), (2010) 374 – 380.

[16] J.K. Liker, The Toyota Way – 14 Management Principles from the World's Greatest Manufacturer, McGraw-Hill, New York, NY. 2004.

[17] J.H. Marvel, and C.R. Standridge, A Simulation-enhanced Lean Design Process, Journal of Industrial Engineering & Management, Vol. 2 No. 1, (2009) 90-113.

[18] R. Shah, P.T. Ward, Lean Manufacturing: Context, Practice Bundles, and Performance, Journal of Operations Management 21, (2010) 129-149.

[19] A.D. Rymaszewska, The Challenges of Lean Manufacturing Implementation in SMEs, Benchmarking: An International Journal, Vol. 21 Iss 6 (2014) 987 – 1002.

[20] Flinchbaugh, "Beyond lean: building sustainable business and people success through new ways of thinking", available at: http://mhlnews.com/archive/beyond-lean-building-sustainable-business-and-people-success-through-new-ways-thinking-2 (2005).

[21] E.N. Roslin and S.A.M. Shahadat,. A Conceptual Model for Full-blown Implementation of Lean Manufacturing System in Malaysian Automotive Industry. Proceedings of the 2014 International Conference on Industrial Engineering and Operations Management. (2014) 1309 – 1315.

[22] A.R. Rahani, M. al-Ashraf, (2012) Production Flow Analysis through Value Stream Mapping: A Lean Manufacturing Process Case Study. Procedia Engineering 41, (2012) 1727 – 1734.

[23] N. Nordin, B. Md. Deros, M.N. Ab. Rahman. Lean Manufacturing Implementation in Malaysian Automotive Industry: An Exploratory Study. Proceedings of the 2011 International Conference on Industrial Engineering and Operations Management. 4(1), (2011) 1979 – 3871

[24] R. Muslimen, S.M. Yusof, and A.S.Z. Abidin, Lean Manufacturing Implementation in Malaysian Automotive Components Manufacturer: A Case Study. Proceedings of the World Congress on Engineering 2011 Vol I WCE 2011, 1. (2011)

[25] R.K. Yin, Case Study Research: Design and Methods, 2nd Edition. (1994) Thousand Oaks, CA:

Sage.
[26] D.E. Gray, Doing Research in the Real World (2004) London: Sage Publication

Finite element analysis of a composite crash box subjected to low velocity impact

M S I Shaik Dawood[1,2], A L Ahmad Ghazilan[1] and Q H Shah[1]

[1]Department of Mechanical Engineering, Faculty of Engineering, International Islamic University Malaysia, 50728 Kuala Lumpur, Malaysia

[2]sultan@iium.edu.my

Abstract. In this work, finite element analyses using *LS-DYNA* had been carried out to investigate the energy absorption capability of a composite crash box. The analysed design incorporates grooves to the cross sectional shape and E-Glass/Epoxy as design material. The effects of groove depth, ridge lines, plane width, material properties, wall thickness and fibre orientation had been quantitatively analysed and found to significantly enhance the energy absorption capability of the crash box.

1. Introduction

Crash tubes serve primarily as energy absorbers and are used in many applications to absorb energy particularly in the event of impact. In automotive application, these tubes are mainly used in absorbing impact energy during the frontal crash event. The tubes are required to be designed to absorb highest possible energy with minimum component weight. Lighter materials such as fiber composites and aluminum provide enhanced strength, durability and lower weight compared to the steel counterpart. Finite element analysis is one of the preferred virtual test method employed in the development of a new component design as analysis can be performed relatively faster and with less expense.

In the event of collision, car manufacturers require the passenger cabin not to deform excessively [1]. Therefore, it is necessary for the frontal part of the car to absorb as much energy as possible over a longer period of time to minimize occupant injury. Crash boxes are normally mounted at the front end of the car frame. In the event of frontal impact, these boxes are intended to collapse and absorb impact energy in order to reduce deformation of other vehicle components. Metallic crash tubes absorb impact energy via plastic deformation and the process can be further enhanced with the inclusion of foam material [2].

The fundamental process of an axial crash is briefly described in [3]. The process involves initiation of compressive strain accumulation at certain area of ridge lines followed by initial plastic buckling [3]. This collapses the area between the ridge lines and later initiates a series of plastic buckling events followed by multiple collapse of area between the subsequent ridge lines causing the load to fluctuate throughout the deformation process [3]. The energy absorption can be improved by ensuring high buckling load at the ridge lines and minimizing the buckling cycle time and load fluctuation [3]. Metallic crash tubes with annular grooves filled with polyurethane foam are reported to offer better energy absorption capacity by almost twofold compared to empty tubes [4]. In another study, experimental results of single and multi-cell aluminum thin wall tubes with different cross sections under quasi static loading showed that the energy absorption of the multi-cell tube was greater [5].

An innovative cross sectional shape as part of a new crash box design was developed in a study by [3]. The box was designed with four grooves in the direction of impact with as many as 24 ridge lines enabling formation of finely detailed wrinkles and absorption of twice the impact energy as well as about 30% weight reduction compared to the conventional crash box. The ridge lines affected the buckling cycle time and the plane width was found to control the number of buckling generated during the crash in which a shorter plane width created more number of buckling [3].

Though metallic tubes offer high stiffness and strength, the setback is their weight. Fiber composite materials are known to have superior specific strength and specific stiffness compared to metallic counterparts. This directly means crash tubes designed using composite materials are lighter and can be efficient energy absorbers compared to the metallic tubes. The composite tubes can be made to collapse in a progressive manner in which energy absorption is dominated by fiber and matrix failure as well as delamination [6]. The mechanism of failure depends on the geometry of the tube, material properties, laminate orientation and impact speed. Layups made of (±85), (±75) and (±65) were found to fail in a progressive manner compared to (±55), (±45) and (±35) layups [6].

The objective of this work is to analyze the energy absorption characteristics of a composite crash box following a similar design approach proposed by [3] for metallic box. In the present work the analyses was carried out using finite element method, employing *LS-DYNA* as the analysis tool.

2. Method of investigation

This section provides a brief description on the fundamentals of crashworthiness parameters for an axially loaded thin walled member, outline on the method of investigation, validation of results for subsequent *FEA* investigation and results and discussions

2.1. Crashworthiness parameters of axially loaded thin walled member

The energy absorption characteristics of an axially loaded member can be determined from the load-displacement curve. It allows the determination of the peak impact load, the mean load, the crushing distance, energy absorbed, and specific energy absorption.

The peak load, P_{max}, must be kept small or reduced as it relates to the sudden deceleration in the event of impact, which is not desirable. The mean load, P_{mean}, and the crushing distance, D_s, are closely related to the energy absorbed by the tube. For a tube to absorb more energy it is required to either have the mean load to be higher or longer crushing distance. The area under the load-displacement curve represents the total energy absorbed, $E_{absorbed}$, during the crash event whereas specific energy absorption is defined by total energy absorbed per unit mass, m, or more commonly abbreviated as *SEA*.

$$E_{absorbed} = \int P_{mean} D_s \quad (1)$$

$$SEA = \frac{E_{absorbed}}{m} \quad (2)$$

2.2. Finite element modelling: result validation

As a first step to the *FE* analysis, a test case obtained from the literature [7] was validated. The case considered was a square E-glass/epoxy tube (25mm × 25mm), with 150mm length and 2.5mm wall thickness, hit by a 100kg impactor with an initial velocity 7.9m/s. The investigation was carried out using *LS-DYNA*, an explicit finite element solver. The composite material was modeled using *MAT 54/55*.

The result of the simulation, impact load vs. displacement, is shown in Figure 1. The peak impact load and energy absorbed obtained from the current simulation are 34.13kN and 1596.45kJ respectively. The results are very close to the ones reported in [7] which were 33.42kN for the peak impact load and 1560.10kJ for the total energy absorbed.

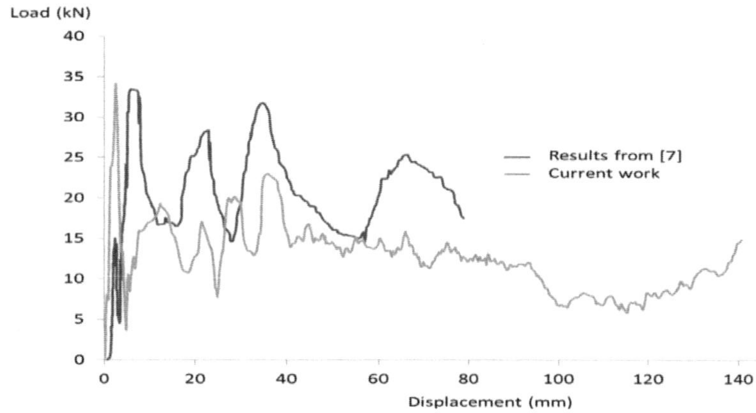

Figure 1. Load-displacement curve

2.3. Finite element modelling, results and discussion

In this section, the effects of different design parameters on the energy absorption of composite crash boxes will be presented. The cross section of the crash box is shown in Figure 2 with total cross sectional area of 3750mm^2 and height of 150mm.

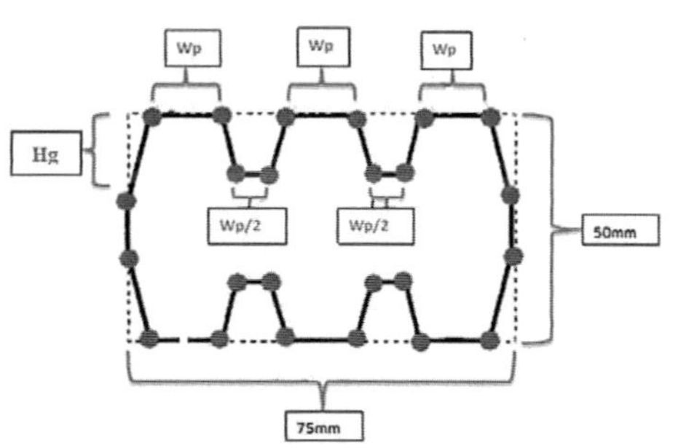

Figure 2. Cross sectional shape of crash box design, adopted from [3]

a. 20 ridge lines b. 24 ridge lines

Figure 3. *FEA* model of the crash box

The attributes of the sample design and geometry are shown on Figure 3 and 4. The geometrical effects of grooves depth, H_g, plane width, W_p, and different material properties were studied. The dimensions of each model are summarized in Table 1.

Table 1. Modeling properties

Simulation Number	Material	Thickness (mm)	Dimension (mm)		Fiber orientation
			Plane Width	Groove Depth	
1	E-Glass/Epoxy	2.5	15	10	[45/-45]
2	E-Glass/Epoxy	2.5	15	12.5	[45/-45]
3	E-Glass/Epoxy	2.5	15	15	[45/-45]
4	E-Glass/Epoxy	2.5	11.5	15	[45/-45]
5	E-Glass/Epoxy	2.5	12.5	15	[45/-45]
6	E-Glass/Epoxy	2.5	13.5	15	[45/-45]
7	Steel	2.5	11.5	15	-
8	Aluminum	2.5	11.5	15	-
9	E-Glass/Epoxy	2.5	11.5	15	[0/90/0]s
10	E-Glass/Epoxy	2.0	11.5	15	[45/-45]
11	E-Glass/Epoxy	2.5	11.5	15	[45/-45]
12	E-Glass/Epoxy	3.0	11.5	15	[45/-45]
13	E-Glass/Epoxy	2.5	11.5	15	[45/-45]
14	E-Glass/Epoxy	2.5	11.5	15	[0/45/90/-45]
15	E-Glass/Epoxy	2.5	11.5	15	[0/30/90/-30]
16	E-Glass/Epoxy	2.5	11.5	15	[0/90]

2.3.1. Effect of groove depth. Table 2 shows the effect of groove depth, where the highest peak load and mean load are observed for groove depth of 15mm. As the groove depth is reduced, the values of the peak and mean loads are reduced as well. Although higher groove depth will lead to higher energy absorption, it will also increase the peak load which is not desirable.

Table 2. Effect of groove depth on crashworthiness parameters

Simulation Number	Peak Load (kN)	Mean Load (kN)	Deformation (mm)	Absorbed Energy (kJ)	Specimen Mass (kg)	(SEA) (kJ/kg)
1	88.95	32.72	115.00	3.763	0.216	17.4
2	92.85	35.20	115.00	4.048	0.229	17.6
3	100.60	37.44	115.00	4.306	0.243	17.7

2.3.2. Effect of ridge lines. Table 3 shows the effect of ridge lines on the peak and mean loads as well as on energy absorbed. The highest peak load is observed for the case 20 ridge lines whereas increasing the ridge lines to 24 resulted in reduction of peak load but higher mean load. The *SEA* is also higher in the case of 24 ridge lines. Although higher groove depth can increase the energy absorption, increasing the number of ridge lines is seen as a better option since it does not increase the peak load.

Table 3. Effect of ridge lines on crashworthiness parameters

Simulation Number	Peak Load (kN)	Mean Load (kN)	Deformation (mm)	Absorbed Energy (kJ)	Specimen Mass (kg)	(SEA) (kJ/kg)
3	100.60	37.44	115.00	4.306	0.243	17.7
4	87.35	42.72	115.00	4.913	0.133	36.9

2.3.3. Effect of plane width. Table 4 shows the effect of plane width on the peak and mean loads as well as on energy absorbed. From the results it can be seen that shorter plane width leads to higher mean load and better *SEA*. However, it also increases the peak load.

Table 4. Effect of plane width on crashworthiness parameters

Simulation Number	Peak Load (kN)	Mean Load (kN)	Deformation (mm)	Absorbed Energy (kJ)	Specimen Mass (kg)	(SEA) (kJ/kg)
4	87.35	42.72	115.00	4.913	0.133	36.9
5	86.15	29.03	115.00	3.338	0.137	24.4
6	86.88	27.38	115.00	3.149	0.140	22.5

2.3.4. Effect of material properties. Table 5 shows the effect of material properties on the crashworthiness parameters. As can be seen from the table, steel tube results in highest peak and mean loads. A comparison between aluminium and $[0/90/0]_s$ composite shows, composite results in higher peak and lower mean loads compared to aluminium but the benefits can be observed in terms of *SEA*.

Table 5. Effect of material properties on crashworthiness parameters

Simulation Number	Peak Load (kN)	Mean Load (kN)	Deformation (mm)	Absorbed Energy (kJ)	Specimen Mass (kg)	(SEA) (kJ/kg)
7	445.30	156.11	52.61	8.213	0.581	14.1
8	183.14	92.94	102.59	9.535	0.200	47.7
9	230.74	76.15	101.72	7.750	0.133	58.1

2.3.5. Effect of wall thickness. Table 6 shows the effects of wall thickness on the crashworthiness parameters. The results show that the peak and mean load increased significantly with the increase in wall thickness. Since increasing the wall thickness results in higher mean load, this means, greater energy will be absorbed by the crash box thus improving protection. Improvement in terms *SEA* is also observed with the use of higher wall thickness.

Table 6. Effect of wall thickness on crashworthiness parameters

Simulation Number	Peak Load (kN)	Mean Load (kN)	Deformation (mm)	Absorbed Energy (kJ)	Specimen Mass (kg)	(SEA) (kJ/kg)
10	66.80	27.35	122.72	3.360	0.107	31.3
11	87.35	42.72	115.00	4.913	0.133	36.9
12	110.02	47.90	116.36	5.580	0.159	35.0

2.3.6. Effect of fibre orientation. Table 7 shows the effects of fibre orientation on the crashworthiness parameters. As can be seen from the table, quasistatic layup provides improvements in terms of mean load and *SEA* compared to other layups as it strengthens the performance of the tube in axial direction. The [0/90] orientation is also seen here to improve the energy absorption compared to the [45/45] layup.

Table 7. Effect of fibre orientation on crashworthiness parameters

Simulation Number	Peak Load (kN)	Mean Load (kN)	Deformation (mm)	Absorbed Energy (kJ)	Specimen Mass (kg)	(SEA) (kJ/kg)
13	87.35	42.72	115.00	4.910	0.133	36.9
14	135.16	67.85	98.17	6.660	0.133	50.0
15	142.71	58.05	105.88	6.140	0.133	46.1
16	119.64	43.18	114.57	4.950	0.133	37.1

3. Conclusions

A finite element analyses was carried out to investigate the effects of groove depth, ridge lines, plane width, material properties, wall thickness and fibre orientation on the energy absorption characteristics

IOP Conf. Series: Materials Science and Engineering **184** (2017) 012017 doi:10.1088/1757-899X/184/1/012017

of a composite crash box. The analyses concluded the use of 15 mm groove depth, 24 ridges lines and 11.5 mm plane width significantly improves the specific energy absorption capability as compared to the conventional crash boxes. It is also observed that the use of higher wall thickness and proper fibre layup greatly affects the energy absorption of the crash box. The study concludes that proper selection of design parameters can be used to improve the energy absorption capability of crash boxes made of composite materials and they can be made comparable in performance to that of convention metal tubes.

References

[1] Du Bois P, Chou C C, Fileta B B, Khalil T B, King A I, Mahmood H F, Mertz H J, Wismans J, Prasad P, Belwafa J E 2004 *American Iron and Steel Institute* p 29

[2] Seitzberger M, Rammerstorfer F G, Degischer H P and Gradinger R 1997 *Acta Mechanica* **125(1-4)** pp 93-105.

[3] Nakazawa Y, Tamura K, Yoshida M, Tagaki K and Kano M 2005 *VIII International Conference on Computational Plasticity* (Barcelone) pp 1-4

[4] Darvizeh A, Darvizeh M, Ansari R and Meshkinzar A 2013 *Thin-Walled Structures* **71** pp 81-90

[5] Nia A A and Parsapour M 2014 *Thin-Walled Structures* **74** pp 155-65

[6] Hull D 1991 *Composites science and technology* **40** pp 377-421

[7] Shetty S K 2006 Finite element study of energy absorption characteristics of a hybrid structure-composite wrapped on a square metal tube (Doctoral dissertation, Wichita State University)

IOP Conf. Series: Materials Science and Engineering **184** (2017) 012018 doi:10.1088/1757-899X/184/1/012018

Abrasive wear response of TIG-melted TiC composite coating: *Taguchi approach*

M A Maleque[1], K A Bello[1, 2], A A Adebisi[1] and A Dube[3]
[1] Department of Manufacturing and Materials Engineering, International Islamic University Malaysia.Darul Salam, 53100 Gombak, Selangor, Malaysia.
[2] Department of Metallurgical and Materials Engineering, Ahmadu Bello University Zaria, Nigeria
[3] Ducom Instruments Ltd., Peenya Industrial Area Bangalore, KA-560 058, India.

E-mail: maleque@iium.edu.my

Abstract. In this study, Taguchi design of experiment approach has been applied to assess wear behaviour of TiC composite coatings deposited on AISI 4340 steel substrates by novel powder preplacement and TIG torch melting processes. To study the abrasive wear behaviour of these coatings against alumina ball at 600° C, a Taguchi's orthogonal array is used to acquire the wear test data for determining optimal parameters that lead to the minimization of wear rate. Composite coatings are developed based on Taguchi's L-16 orthogonal array experiment with three process parameters (welding current, welding speed, welding voltage and shielding gas flow rate) at four levels. In this technique, mean response and signal-to-noise ratio are used to evaluate the influence of the TIG process parameters on the wear rate performance of the composite coated surfaces. The results reveal that welding voltage is the most significant control parameter for minimizing wear rate while the current presents the least contribution to the wear rate reduction. The study also shows the best optimal condition has been arrived at A_3 (90 A), B_4 (2.5 mm/s), C_3 (30 V) and D_3 (20 L/min), which gives minimum wear rate in TiC embedded coatings. Finally, a confirmatory experiment has been conducted to verify the optimized result and shows that the error between the predicted values and the experimental observation at the optimal condition lies within the limit of 4.7 %. Thus, the validity of the optimum condition for the coatings is established.

1. Introduction

Arc welding is a major joining process which has found wide application in modern manufacturing processes. Among the various types of arc welding, tungsten inert gas (TIG) arc welding technology is being increasingly used as a suitable heat source for surface treatment of steel alloy components due to its attractive features such as flexible power requirement, limited heat inputs, highly autogenous [1-3]. Unlike conventional high-energy beam heat sources (such as laser and electron beams technology), the TIG fusion welding process is a low energy technique that guarantees quality weld surface with minimum defect and production cost. Due these inherent benefits, TIG arc surface melting and alloying is gaining popularity among the current trend in hardfacing (liquid phase surface) treatment of steels via incorporation of ceramic composite powder of desirable composition [4, 5].

TIG Ceramic-embedded coatings are being applied on engineering components employed in services where abrasive wear mostly occurs. A full understanding of the effects of TIG processing variables on the wear response is required to be established for the choice of optimal coating variable to reduce wear. A lower wear rate is one the main target of ceramic embedded coatings by TIG arc torch. Despite the technological simplicity of TIG glazing process, the number of processing parameters is too large and the factor-response correlation are not always known [6-8]. The optimization of the influencing parameters is therefore necessary from the point of view of obtaining a desirable functional deposited layer exhibiting lower wear rate. In most current coating development, the desired TIG processing parameters is determined by a trial-and-error procedure based on experience and handbook values. However, this is time-consuming, costly and does not insure that the selected input variables will result in optimal performance characteristics. In order to ease the optimization process, Taguchi design of experiment (DoE) approach which avoids the great number of experiment needed for a full factorial design could be very useful. Taguchi's method uses signal-to-noise ratio (S/N) to identify best parameter combinations and their levels, with the least variability, for optimization [7, 9]. In the present work, the Taguchi DoE approach was applied to study the TIG torch deposition of titanium carbide (TiC) hard coatings on near surface layer of AISI 4340 low alloy steel. The optimum combination of the control parameters (welding current, welding speed, welding voltage and argon gas flow rate) for obtaining minimum wear rate was investigated based S/N ratio.

2. Experimental procedure

2.1. Materials and specimen preparation
The ceramic material used for coating in this work was commercially available structural TiC, supplied by Sigma Aldrich USA (average size, 40µm; 99.9 wt. % purity). A commercial grade AISI 4340 low alloy steel sheet machined to nominal dimensions of 100 mm x 40 mm x 15 mm were acquired as the substrate material. The chemical compositions of the steel substrates are 0.42 percent C, 0.65 percent Si, 0.16 percent Mn, <0.05 percent S, <0.05 percent P, 0.75 percent Cr, 1.76 percent Ni and 0.24 percent Mo. The surfaces of the machined specimens were made flat by grinding to a surface roughness of about R_a = 0.4 µm using SiC emery paper. This process was followed by degreasing in acetone and ethanol under ultrasonic waves for 20 minutes to remove any impurities such as grease, dust, and oxide layer from the substrate surface.

2.2. Composite coating deposition
The composite coatings based on TiC_p powder reinforcement were prepared by powder preplacement and TIG torch melting processes. In preplacing the powder, TiC powder with approximate content of 1.0 mg/mm^2 was mixed with a PVA (polyvinyl acetate) organic binder, distilled water and alcohol to form a ceramic paste. The paste was evenly preplaced on the cleaned substrate surface using a flat plastic spatula. After 30 minutes, the paste was dried in an electric oven at a lower temperature of 60 ^oC for 1 hour. The melting of the preplaced surfaces was achieved under the torch of a DCEN TIG arc welding machine (Model: Miller-Telwin TIG 165) to produce continuous coating tracks with overlapping ratio of 50 % for ball on disc tribo-wear tests. The surface coating was performed by traversing the preplaced specimen at 1 to 2 mm/s speed beneath the TIG arc with varying currents of 70-100 A and voltages of 20-35 V. A 3.2 mm diameter tungsten electrode was used to produce a stable arc length of about 2mm. Pure argon shielding gas was channelled at a flow rate of 10-25 l/min to prevent surface oxidation during irradiation. All the tracks were produced based on the plan of Taguchi L_{16} OA design. The essential features of TIG melting process are shown figure 1.

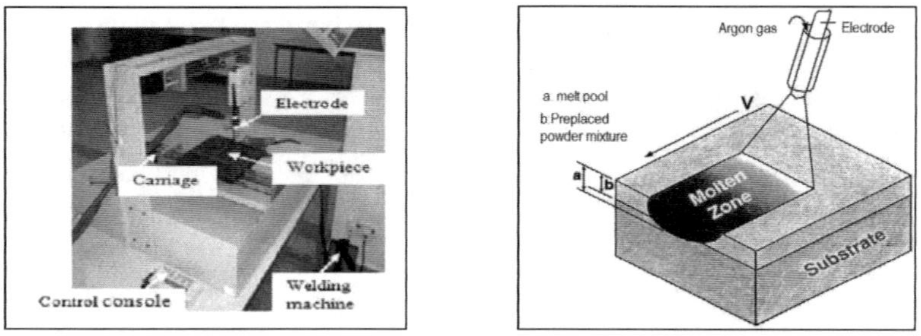

(a) (b)

Figure 1. TIG welding process (a) photo of the machine (b) schematic of TIG glazing features.

2.3. Abrasive wear tests

The wear test was conducted using a DUCOM high-temperature tribometer (TR-20-PHM-CHM-600 model) with ball and disc contact geometry in rotating sliding motion. The rotating disc specimens for wear tests were machined to block with specified size of 26 mm x 10 mm x 10 mm. Alumina ball bearings with hardness of about 1800 Hv and maximum temperature of 1500-1800 °C were used as the static ball partner during the test. The test uses normal force of 10 N, rotating speed 0.2 m/s and a sliding distance of 500m. The sliding experiment was performed without lubrication at high temperature of 600 °C. After the test, the surface wear properties for each specimen were evaluated in terms of wear rate which was assessed by monitoring wear volume loss (V) using a standard formula [2]:

$$V=2\pi R\{r^2\sin^{-1}(d/2r)-d/4\sqrt{(4r^2-d)}\} \quad (1)$$

where R is the wear track radius (mm), d the wear track width (mm) measured with the aid of 3D non-contact profilometer, and r is the radius of alumina ball (mm). Thus, the wear rate (w) mm^3/N was calculated from Eq. 2 [2]:

$$w=V/F_n l \quad (2)$$

where F_n represents the normal load applied to the samples and l is the sliding distance.

2.4. Taguchi design of experiment

The In Taguchi DoE, the most important stage is the selection of control parameters. Based on the extensive review of literature, parameters such as welding current, welding speed, welding voltage and argon gas flow rate were selected in this TIG hardfacing experiment and their effects were studied each specimen wear rate. The parameters and their corresponding levels are given in Table 1.

Table 1. Levels of the parameter used in the experiment.

Control Factor	Symbol	Units	Levels			
			1	2	3	4
Welding current	A	A	70	80	90	100
Welding Speed	B	mm/s	1.0	1.5	2.0	2.5
Welding voltage	C	V	20	25	30	35
Argon flow rate	D	L/min	10	15	20	25

A modified Taguchi array of L_{16} was used to plan the experiment. The details of the L_{16} (4^5) OA with five four-level factors is reported in the column 2, 3, 4 and 5 of Table 2. Each experimental run was repeated two times and the average value of the wear rate response was recorded. Furthermore, Taguchi S/N ratio analysis was used to evaluate the experimental data. Taguchi employs the S/N ratio to measure the quality characteristics deviating from the desired value. Based on the objective of this study, wear rate characteristic corresponds to the lower-the-better (LB) criterion and the S/N ratio (η_w) for wear rate is defined by [7, 9]:

$$\eta_w = -10\log(M.S.D) \quad (3)$$

where M.S.D is the mean square deviation for the output wear rate. Therefore, the M.S.D for the LB characteristic can be expressed as:

$$M.S.D = \frac{1}{n}\sum_{i=1}^{n} y_i \quad (4)$$

where n *is* the number of experiments and y_i is the value of experimental result of the *i*th experiment. The results of S/N ratio for minimum wear rate are determined based on Eq. 3 and Eq. 4 using MINITAB 17 statistical software.

Table 2. Experimental design using L_{16} OA for wear rate of TiC composite coatings

Test run	Welding current, A: (A)	Welding peed, B: (mm/s)	Welding voltage C: (V)	Argon flow rate D: (L/min)	Wear rate, (mm^3/N)	S/N ratio (dB)
1	90.00	2.50	25.00	20.00	9.8072E-06	-19.8309
2	100.00	1.00	35.00	20.00	1.0825E-05	-20.6888
3	100.00	2.50	20.00	15.00	8.3717E-06	-18.4563
4	100.00	2.00	25.00	10.00	2.5134E-05	-28.0051
5	70.00	2.00	30.00	20.00	8.5910E-06	-18.6808
6	70.00	1.00	20.00	10.00	1.2182E-05	-21.7142
7	80.00	2.00	35.00	15.00	1.1809E-05	-21.4439
8	80.00	2.50	30.00	10.00	1.2262E-05	-21.7714
9	100.00	1.50	30.00	25.00	9.7859E-06	-19.8121
10	90.00	1.00	30.00	15.00	1.1206E-05	-20.9892
11	80.00	1.50	20.00	20.00	1.2645E-05	-22.0387
12	90.00	1.50	35.00	10.00	7.9527E-06	-18.0103
13	70.00	2.50	35.00	25.00	1.2795E-05	-22.1406
14	80.00	1.00	25.00	25.00	1.7089E-05	-24.6542
15	70.00	1.50	25.00	15.00	1.4426E-05	-23.1828
16	90.00	2.00	20.00	25.00	1.9787E-05	-25.9277

3. Experimental procedure

3.1. Abrasive Test Results and Taguchi Analysis.

In Table 2, the sixth and seventh column shows the experimental results for average wear rate and S/N ratio. The S/N ratio was computed and analysed using MINITAB 17 software. A greater S/N ratio value indicates better quality characteristics. As such, the optimal level of the control parameters is the level with the greatest S/N ratio. The level average responses from the S/N ratio data for wear rates of TiC embedded surface are summarized in Table 3. Accordingly, the difference between the maximum and the minimum values (delta) is a decisive determinant of the influence of each parameter. The higher the delta value, the more influential the parameter appears. Therefore, the response table gives the significant trend with which each control factor affects the wear rate of TIG embedded coating when tested at 600 °C. It can be noticed from Table 3 that the welding voltage 'A' and the welding speed 'B' are the most important factor affecting the wear response. The argon gas flow rate 'C' has a less relevant effect, while the welding current shows the lowest effect among those factors.

Table 3. S/N response table for wear rate of TiC composite coatings

Factor		Level 1	Level 2	Level 3	Level 4	delta	Rank
Welding current	A	-21.43	-22.48	-21.19	-21.74	1.29	4
Welding Speed	B	-22.01	-20.76	-23.51	-20.55	2.96	2
Welding voltage	C	-22.03	-23.92	-20.31	-20.57	3.60	1
Argon flow rate, AFR	D	-22.38	-21.02	-20.31	-23.13	2.82	3

The main effects plots for wear rates of the three coating systems are presented as shown in figure 2. Analysis of the results from the response plots leads to the conclusion that factor combination of A_3 (90 A), B_4 (2.5 mm/s), C_3 (30 V) and D_3 (20 L/min) gives minimum wear rate for TiC embedded coatings.

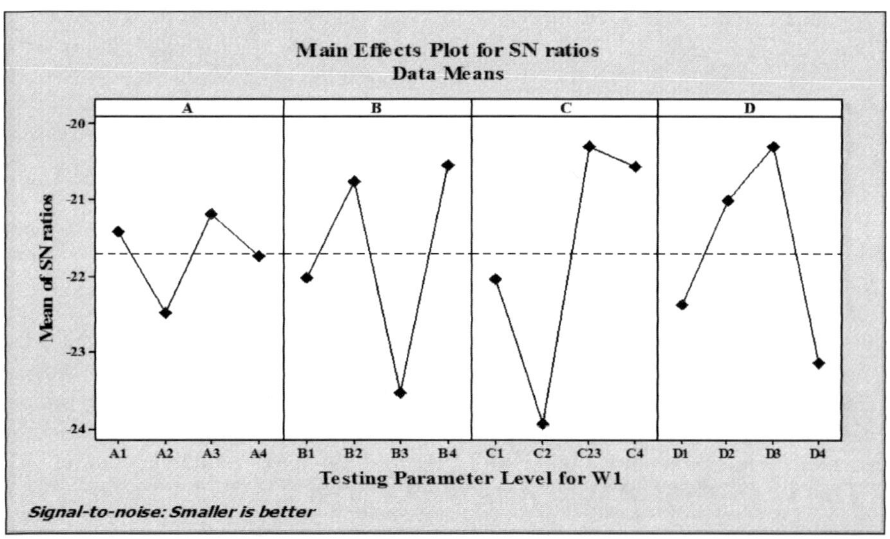

Figure 2. Influence of control factors on average S/N ratio for wear rate of TiC coatings.

3.2. Confirmation Experiment

The final stage of the Taguchi analysis is to conduct confirmation experiment by comparing the predicted value and the experimental results. In this process, a new set of experiment was performed using the optimal levels of the process parameters ($A_3B_4C_3D_3$) to predict the wear rate. The predicted S/N ratio for wear rate can be estimated with the aid of following predictive equation [9]:

$$\bar{\eta}_w = \bar{\eta}_m + \sum_{i=1}^{k} \left(\eta_i - \bar{\eta}_m\right) dB \quad (5)$$

where $\overline{\eta}_w$ is the predicted S/N ratio for wear rate response, $\overline{\eta}_m$ is the overall experimental average, η_i is the mean response for parameters at the optimal levels and k is the number of the control factors. Using predictive Eq. 5, the predicted S/N ratio ($\overline{\eta}_w$) for the wear rate was obtained from MINITAB 17 software and found to be $\overline{\eta}_w = -16.3069$. The experimental observations are then compared with the results obtained from the predictive equations. The percentage errors observed for the mathematical models are computed using the experimental value as reference. It is found that the percentage error observed for wear rate in TIG embedded TiC coating is 4.6 % for the S/N ratio value. Therefore, the resulting Taguchi model seems to be valuable for predicting wear rate to a reasonable accuracy.

4. Conclusion

This study shows that TIG torch can be used to produce TiC embedded composite coatings on LAS substrate. It is noticed that the wear rate of the ceramic-embedded layer is greatly affected by the processing parameter level of the TIG torch. Taguchi experimental design has been applied to analyze the influence of the control parameters on the wear rate of the coatings. Welding voltage and the welding speed are found to be the most influencing parameters for minimizing the wear rate. Argon gas flow rate and welding current are identified to have less relevant effect as far as the minimization of wear rates is concerned. The predictive based on Taguchi approach has been demonstrated to be consistent with the experimental observations. Therefore, the obtained results are satisfactory to a reasonable accuracy and can be used to investigate wear rate of any ceramic embedded composite coatings.

Acknowledgement

The authors acknowledge the financial assistance from Research Management Center of International Islamic University Malaysia under the grant no RMGS-12-007-0020.

References

[1] Buytoz S 2006 Microstructural and microhardness characteristics of gas tungsten are synthesized Fe-Cr-C coating on AISI 4340 *Surf Coat Technol.* **200** 3734-42.

[2] Mirdha S, Idriss A N M, Maleque M A, Suryanto S and Souad A 2012 Effect of voltage on the consolidation of TiC particulates on steel substrate fused by TIG welding arc *Int. J. Mech. Mater. Eng.* **7** 48-53.

[3] Adeleke S A and Maleque M A 2014 Tungsten inert gas surface alloying of commercial purity titanium (CP-Ti) with Fe-C-Si ternary mixtures *Adv. Mater. Res.* **1024** 207-210.

[4] Hojjatzadeh S M H, Halvaee A and Heydarzadeh M 2012 Surface Alloying of AISI 1045 steel in a nitrogen environment using a gas tungsten arc process *J. of Mater Process Tech.* **213** 2496-2504.

[5] Maleque M A, Bello K A, Idriss A N M and Mirdha S 2013 Processing of TiC-CNT hybrid composite coating on low alloy steel using TIG torch technique *Appl. Mech. Mater.* **378** 259-64.

[6] Pal S, Malviya S K, Pal S K and Samantaray A K 2009 Optimization of quality characteristics parameters in a pulsed metal inert gas welding process using grey-based Taguchi method *Intl J Adv. Manuf. Techn.* **44** 1250-60.

[7] Juang S C, Tarng Y S 2002 Process parameter selection for optimizing the weld pool geometry in the tungsten inert gas welding of stainless steel *J. Mater Proces Tech.* **122** 33-37.

[8] Bello K A, Maleque M A, Zuraida A and Mirdha S 2015Optimization of hardness behaviour of TIG modified ceramic coating using the Taguchi approach *Adv. Mater. Res.* **115** 238-42.

[9] Peng D X 2012 The effects of welding parameters on wear performance of clad layer with TiC ceramic *Ind. Lubr. Tribol.* **64** 303-311.

Investigation of Parametric Influence on the Properties of Al6061-SiC$_p$ Composite

A A Adebisi[1], M A Maleque[1] and K A Bello[1]

[1] Department of Manufacturing and Materials Engineering,
International Islamic University Malaysia
53100, Kuala Lumpur, Malaysia

E-mail: maleque@iium.edu.my

Abstract. The influence of process parameter in stir casting play a major role on the development of aluminium reinforced silicon carbide particle (Al-SiC$_p$) composite. This study aims to investigate the influence of process parameters on wear and density properties of Al-SiC$_p$ composite using stir casting technique. Experimental data are generated based on a four-factors-five-level central composite design of response surface methodology. Analysis of variance is utilized to confirm the adequacy and validity of developed models considering the significant model terms. Optimization of the process parameters adequately predicts the Al-SiC$_p$ composite properties with stirring speed as the most influencing factor. The aim of optimization process is to minimize wear and maximum density. The multiple objective optimization (MOO) achieved an optimal value of 14 *wt%* reinforcement fraction (RF), 460 *rpm* stirring speed (SS), 820 °*C* processing temperature (PT$_{emp}$) and 150 *secs* processing time (PT). Considering the optimum parametric combination, wear mass loss achieved a minimum of 1 x $10^{-3}g$ and maximum density value of 2.780g/mm^3 with a confidence and desirability level of 95.5%.

1. Introduction

Light weight composite materials possesses the requirements for structural applications, with aluminium matrix composite (AMC) as the most attractive due to its inherent properties. The major attraction of AMC resulted due to its reduction in mass; improve wear resistance and material properties such as high strength and stiffness. Moreover, AMC has attracted much attention by researchers owing to its low density, high thermal conductivity, low melting point and ability to be reinforced by a wide variety of reinforcement phases such as SiC, Al_2O_3, TiC, B_4C and fly ash [1-3].

Cost effective stir casting process is one of the potential techniques to develop AMC due to its simplicity, flexibility. The properties of cast AMC from stir casting process is majorly influenced by the processing parameters. Sozhamannan et al. [4] examined the influence of PT$_{emp}$ and stirring time (ST) on the mechanical properties of Al-SiC$_p$. It was observed that the hardness value increases linearly between 750-800 °*C* at 20 min processing time. Balasivanandha et al. [5] reported that higher SS and ST influences the hardness of AMC. Furthermore, Naher et al. [6] studied the effect of different SS on uniform distribution of particle in simulated fluids. It is concluded from the study that SS has a significant effect on the particle distribution. Based on majority of studies conducted in the literatures, the processing parameters are examined independently of each other, although a simultaneous evaluation of these parameters is required because of their influence on the processing condition. As a result, several experimental design techniques maybe recommended to investigate the effect of these processing factors on the properties of AMC. The response surface methodology (RSM) of central composite design (CCD) has been a preferred method for experimental plan because it is a very efficient in reducing the number of experiments with a large number of factors and levels [7]. CCD is also capable of achieving the optimum conditions required to attain the best characteristic properties. Therefore, the aim of this study is to examine the influence of parametric optimization on wear and density properties of AMC using the stir casting technique.

2. Experimental details

2.1. Materials and method

AA6061 alloy is used as the matrix material and silicon carbide particle (SiC_p) as the reinforcement material. Table 1 shows the properties of the matrix alloy and reinforcement materials.

Table 1. Properties of matrix (AA6061) and reinforced SiC_p phase.

Property	Unit	Al (6061)	SiC_p
Density	g/cm^3	2.7	3.22
Melting point	°C	660	2973
Coefficient of thermal expansion	µm/m°C	23.4	4
Thermal conductivity	W/mK	166	126

The composite is developed using the liquid metallurgy route through the bottom pouring stir casting process. The melting and preheating furnace chamber has a maximum heating capacity of 1000 °*C* which is sufficient to conduct all the experimental design runs. The SiC_p powder is treated in the preheating furnace to 900 °*C* for 2 hrs to make their surfaces oxidized [8]. After melting of the aluminium alloy, dross formation on the surface is extracted using a preheated coated skimmer. Prior to incorporation of the preheated SiC_p powder, 1*wt%* magnesium (Mg) is added to improve the wettability between the matrix alloy and reinforcement phase. The presence of Mg not only has the beneficial effect of alloying but also reduces the surface tension of the melt and decreases the solid-liquid interfacial energy. The preheated SiC_p was gradually injected into the melt as the coated stainless steel stirrer is automatically lowered into the furnace for stirring purpose. This process is conducted for each of the experimental run considering the process parameters in the design plan.

2.2. Wear test condition

High frequency reciprocating wear testing machine is used to conduct the wear test according to ASTM D 6079-97. All AMC samples developed based on the design plan were evaluated at a constant operating condition of 50 N load at 600 *rpm* for 30 *min* using a 52100 chrome steel ball as the counter surface. The ball has a diameter of 6mm and hardness value of 60-67 RC. The difference in mass before and after the test is used to evaluate the WML.

2.3. Experimental design

The experiments were planned based on CCD method. It utilized four factor five level design scheme as shown in Table 2. The factors are reinforcement fraction (RF), stirring speed (SS), processing temperature (PT_{emp}), and processing time (PT).

Table 2. Design plan for factors and their levels in CCD

Factors		Levels				
		-2	-1	0	+1	+2
Reinforcement fraction *(wt %)*	X_1	5	10	15	20	25
Stirring speed *(rpm)*	X_2	200	300	400	500	600
Processing temperature *(°C)*	X_3	700	750	800	850	900
Processing time *(secs)*	X_4	60	90	120	150	180

The CCD consist of $2^k + 2k + n$ runs, where k is the number of factors, 2^k is the number of the factorial points at the corners of the cube, $2k$ is the number of the axial points. Thus, the number of experimental runs in this study is 16+8+6=30. The experimental outcome is presented in Table 3

Table 3. Experimental design plan and response based on CCD

No	a: Reinforcement Fraction *(wt %)*	b: Stirring speed *(rpm)*	c: Processing Temperature*(°C)*	d: Processing Time *(secs)*	Wear mass loss *(g)*	Density *(g/mm³)*
1	15.00	400.00	800.00	120.00	0.001600	2.7793
2	20.00	300.00	850.00	150.00	0.001725	2.7650
3	15.00	400.00	700.00	120.00	0.002025	2.7550
4	10.00	300.00	850.00	90.00	0.002395	2.7630
5	10.00	300.00	850.00	150.00	0.001750	2.7695
6	25.00	400.00	800.00	120.00	0.001250	2.7600
7	10.00	500.00	850.00	90.00	0.001350	2.7743
8	20.00	300.00	750.00	90.00	0.002050	2.7540
9	15.00	400.00	800.00	120.00	0.001250	2.7785
10	10.00	500.00	750.00	150.00	0.001100	2.7770
11	15.00	400.00	800.00	120.00	0.001150	2.7743
12	15.00	400.00	800.00	120.00	0.001100	2.7800
13	15.00	400.00	800.00	120.00	0.001100	2.7808
14	10.00	300.00	750.00	150.00	0.001950	2.7553
15	20.00	500.00	750.00	90.00	0.001650	2.7498
16	10.00	500.00	750.00	90.00	0.001825	2.7603
17	20.00	300.00	750.00	150.00	0.001425	2.7538
18	15.00	200.00	800.00	120.00	0.002310	2.7500
19	20.00	500.00	750.00	150.00	0.001025	2.7650
20	15.00	400.00	800.00	120.00	0.000900	2.7733
21	20.00	300.00	850.00	90.00	0.001375	2.7440
22	15.00	400.00	800.00	60.00	0.001800	2.7580
23	5.00	400.00	800.00	120.00	0.002200	2.7575
24	10.00	500.00	850.00	150.00	0.000850	2.7823
25	15.00	400.00	800.00	180.00	0.001025	2.7645
26	15.00	600.00	800.00	120.00	0.001500	2.7658
27	20.00	500.00	850.00	90.00	0.001525	2.7550
28	15.00	400.00	900.00	120.00	0.001200	2.7655
29	10.00	300.00	750.00	90.00	0.002450	2.7483
30	20.00	500.00	850.00	150.00	0.001200	2.7663

3. Results and discussion

3.1. Model development

The mathematical relationship between the response and processing factors is expressed in equation 1 by the second order polynomial equation [9].

$$Y = \beta_0 + \sum_{i=1}^{k} \beta_i X_i + \sum_{i=1}^{k} \beta_{ii} X_i^2 + \sum_{i=1}^{k} \sum_{j \geq i}^{k} \beta_{ij} X_i X_j + \varepsilon \tag{1}$$

where Y represents the response, X_i and X_j are the equation value of the factors, β_0 is the constant, β_i, β_j and β_{ij} are linear, interaction and quadratic end coefficient respectively, and k is the number of the factors.

3.2. *Parametric influence on wear mass loss (WML)*

The relationship between the processing parameters and WML is evaluated by regression analysis. It is observed that a quadratic model is suggested as shown in Table 4. The model is associated with p values of < 0.0045, which is < 0.05, indicating that the model is considered to be statistically significant.

Table 4. Test for significance of WML

Source	SS	df	MS	F Value	p-value	
Mean	7.070E-005	1	7.070E-005			
Linear	3.615E-006	4	9.037E-007	9.02	0.0001	
2FI	5.743E-007	6	9.571E-008	0.94	0.4888	
Quadratic	*1.184E-006*	*4*	*2.960E-007*	*5.94*	*0.0045*	*Suggested*
Cubic	3.858E-007	8	4.823E-008	0.93	0.5424	Aliased
Residual	3.614E-007	7	5.163E-008			
Total	7.682E-005	30	2.561E-006			

The adequacy and validity of the model is ascertained using ANOVA in Table 5 by confirming the significant terms with p values < 0.05. However, model terms with p values > 0.05 are considered insignificant. The numerical analysis for WML model equation is therefore obtained based on the significant terms. Moreover, the difference between the adjusted R^2 (0.7591) and predicted R^2 (0.6171) is > 0.2, which establishes a reasonable agreement as required.

Table 5. Test result of ANOVA for WML

Source	SS	df	MS	F Value	p-value	
Model	5.053E-006	8	6.316E-007	12.43	< 0.0001	*significant*
A-Reinforcement Fraction	5.385E-007	1	5.385E-007	10.59	0.0038	
B-Stirring speed	1.609E-006	1	1.609E-006	31.66	< 0.0001	
C-Processing Temperature	3.638E-007	1	3.638E-007	7.16	0.0142	
D-Processing time	1.103E-006	1	1.103E-006	21.70	0.0001	
AB	3.150E-007	1	3.150E-007	6.20	0.0213	
A^2	3.934E-007	1	3.934E-007	7.74	0.0112	
B^2	7.487E-007	1	7.487E-007	14.73	0.0010	
C^2	2.288E-007	1	2.288E-007	4.50	0.0459	
Residual	1.067E-006	21	5.083E-008			
Lack of Fit	7.941E-007	16	4.963E-008	0.91	0.6023	*not significant*
Pure Error	2.733E-007	5	5.467E-008			
Cor Total	6.120E-006	29				

Based on the confirmation, model equations were developed for WML. Equation 2 is used to recreate the response for any given level of each processing parameters in the experiment.

$$W_m\ (g) = +\ 0.034060\ -2.84440E\text{-}004a\ -\ 1.98811E\text{-}005b\ -6.03196E\text{-}005c\ -\ 7.14583E\text{-}006d\ +\ 2.80625E\text{-}007ab + 4.74107E\text{-}006a^2 + 1.63527E\text{-}008b^2 + 3.61607E\text{-}008c^2 \quad (2)$$

Moreover, equation 3 is used to identify the relative impact of the processing parameters by comparing their coefficients. It is concluded from the equation that the process parameter with the most influencing impact on WML is SS as illustrated in Figure 1.

$$W_m\ (g) = +\ 1.237E\text{-}003\ -\ 1.498E\text{-}004a\ -\ 2.590E\text{-}004b\ -\ 1.231E\text{-}004c\ -\ 2.144E\text{-}004d + 1.403E\text{-}004ab + 1.185E\text{-}004a^2 + 1.635E\text{-}004b^2 + 9.040E\text{-}005c^2 \quad (3)$$

From the quadratic influence, it is observed from Figure 2a that WML decreases as both SS and RF gradually increases. However, the WML did not show any obvious decrease beyond 500 *rpm* and 18 *wt%*, rather a further tendency of an increase in WML may be observed. Higher SS beyond 500 *rpm* initiates an increase in WML due to vigorous stirring which gives rise to vortex phenomenon leading to porosity formation as a result of gas entrapment in the melt. Therefore, the RF shows little or no obvious effect on the WML beyond 18 *wt%* due to the influence of further increase in the SS.

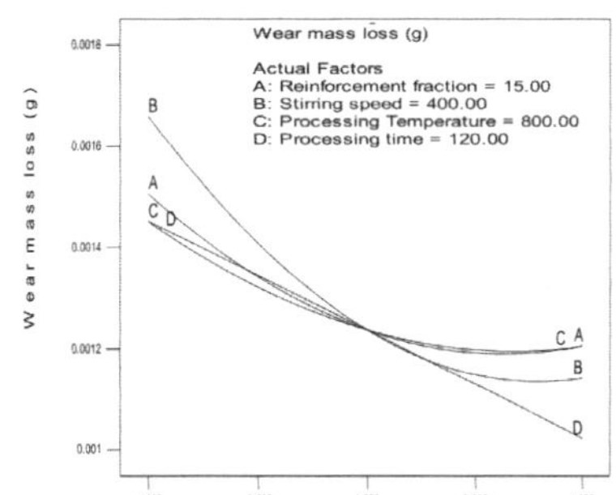

Figure 1. Perturbation plot for wear mass loss

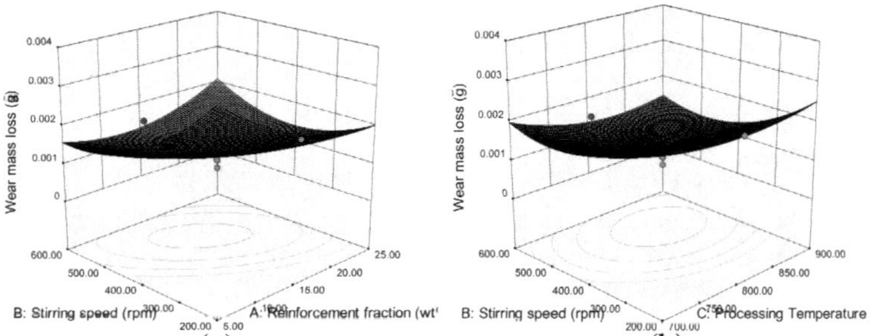

(a) (b)

Figure 2. Surface plot profile of (a) SS and RF (b) SS and PT_{emp} on WML

Figure 2b shows the simultaneous influence of SS and PT_{emp} on WML. It is observed that both factors attained minimum WML at 500 *rpm* and 830 *°C* respectively. Though, beyond these processing conditions an increase in WML is experienced due to undesirable situation of porosity formation and particle settling due to higher SS and PT_{emp}.

3.3. Parametric influence on density property

A quadratic relationship is suggested between the process parameters and density as indicated in Table 6. The model has a p values of 0.0002 which is < 0.05 (i.e. $\alpha = 0.05$ or 95% confidence) indicating that the model is considered to be statistically significant.The adequacy and validity of the developed model is tested through ANOVA as shown in Table 7. The significant model terms are considered to generate the model and are identified with p values < 0.05. However, model terms with p value > 0.05 are regarded as insignificant. Insignificant lack of fit implies that the model fits the data adequately. The difference between the adjusted R^2 (0.7257) and predicted R^2 (0.5755) is within the acceptable limit < 0.2 in order to establish a reasonable agreement.

Table 6. Test for significance of density

Source	SS	df	MS	F Value	p-value	
Mean	229.22	1	229.22			
Linear	1.358E-003	4	3.395E-004	4.12	0.0107	
2FI	2.235E-004	6	3.726E-005	0.39	0.8794	
Quadratic	*1.382E-003*	*4*	*3.456E-004*	*11.37*	*0.0002*	*Suggested*
Cubic	3.696E-004	8	4.620E-005	3.75	0.0491	Aliased
Residual	8.615E-005	7	1.231E-005			
Total	229.22	30	7.64			

Based on the validity and adequacy through ANOVA, two model equation developed for density. Equation 4 determines the density value for any given level of each processing parameters in the experiment.

$$\rho = + 1.49158 + 4.70000E\text{-}003a + 4.16042E\text{-}004b + 2.65083E\text{-}003c + 1.14792E\text{-}003d - 1.76667E\text{-}004a^2 - 4.63542E\text{-}007b^2 - 1.61667E\text{-}006c^2 - 4.21296E\text{-}006d^2 \quad (4)$$

Moreover, equation 5 defines the relative impact of the processing parameters on density.

$$\rho = + 2.78 - 3.000E\text{-}003a + 4.521E\text{-}003b + 3.208E\text{-}003c + 4.104E\text{-}003d - 4.417E\text{-}003a^2 - 4.635E\text{-}003b^2 - 4.042E\text{-}003c^2 - 3.792E\text{-}003d^2 \quad (5)$$

The perturbation plot in Figure 3 shows that there is interaction among the process parameters and it also describes the SS as the most influential factor on density. The interactions indicate that the factors influences the density property due to their quadratic effect. Figure 4 shows the effects of the process parameters on density and it is observed that density increases as the factors increases. However, an optimum condition is attained as density shows a reverse trend due to further increase in the parameters. In Figure 4(a) and (b), the SS, RF and PT achieved optimum at 500 *rpm*, 14 *wt%* and 160 *secs* respectively. Moreover, in Figure 4(c) the PT_{emp} attained an optimum of 835 *°C*. With PT_{emp} above 835 °*C* the reinforced particles gain higher energy which facilitates faster movement and settling of the particles in the melt. This phenomenon of particle settling at higher temperature hinders the effective distribution of the reinforcement thereby influencing an increase in WML.

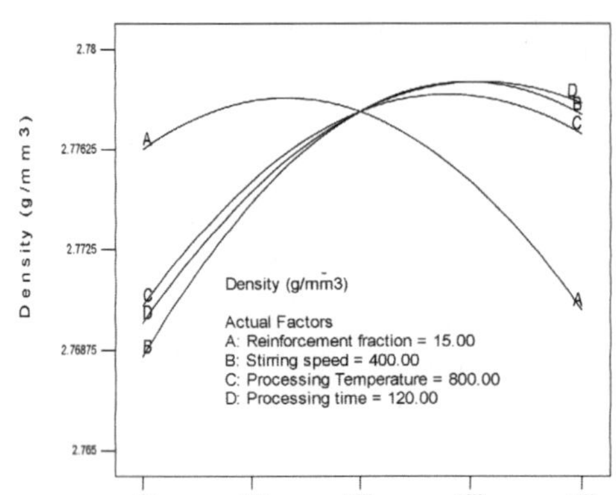

Figure 3. Perturbation plot for density

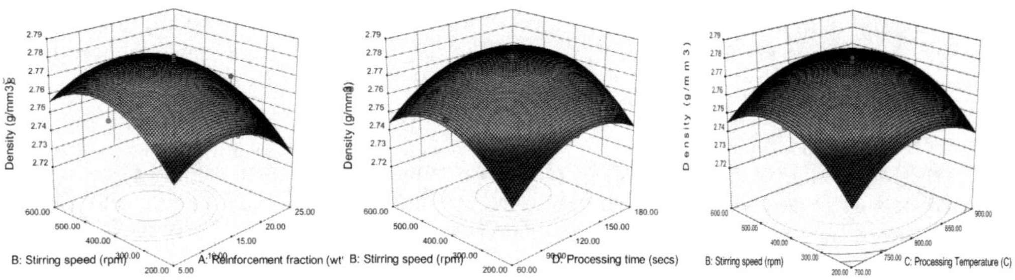

(a) (b) (c)

Figure 4. Surface plot profile of (a) SS and RF, (b) SS and PT and (c) SS and PT_{emp} on density

4. Multiple Objective Optimization Analysis

The multiple objective optimization analysis generates a set of condition that simultaneously optimizes both responses (WML and density) through optimal solutions. Based on the solution analysis in Table 8, the optimum values of input process parameters is achieved with experiment No 1. The conditions are reinforcement fraction 14 *wt%*; stirring speed 460 *rpm*; processing temperature 820 °*C* and processing time 150 *secs* with a confidence and desirability level of 95.5%.

Table 8. Optimal solution generated for the response

No	Reinforcement fraction	Stirring speed	Processing Temperature	Processing time	Wear mass loss	Density	Desirability	
1	*14.030*	*462.562*	*821.590*	*150.000*	*0.001*	*2.780*	*0.955*	*Selected*
2	14.428	464.695	824.793	149.997	0.001	2.780	0.955	
3	14.444	464.457	824.277	149.997	0.001	2.780	0.955	
4	14.413	464.853	824.280	149.998	0.001	2.780	0.955	
5	14.457	464.020	824.266	149.999	0.001	2.780	0.955	

Figure 5 shows the desirability histogram of WML and density generated for the solutions and multiple (combined) optimum for the stir casting process condition. It is observed that MOO improved the density response as indicated from the desirability histogram.

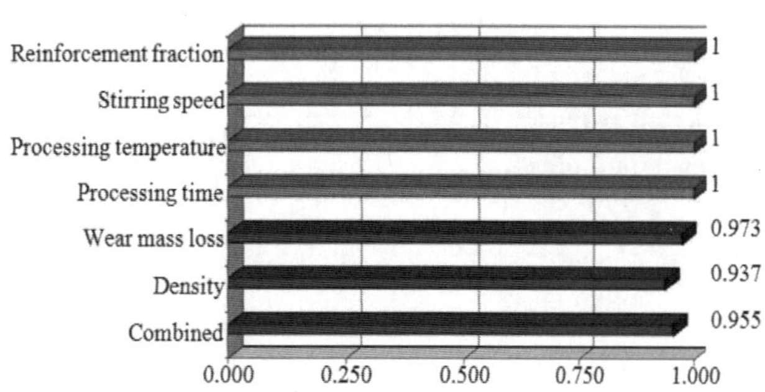

Figure 5. Desirability histogram of MOO for stir casting process parameters

5. Conclusion

Optimum parametric condition for processing Al-SiC_p composite was successfully achieved using the bottom pouring stir casting technique. Models have been developed based on CCD. The developed models has the potential to evaluate WML and density under various process parameter conditions. The influence of different process parameters was exhibited through the perturbation and surface plots.

It is described from the response surface plot that WML decreases with an increase in the processing parameter. However, the reinforcement fraction reaches an optimum condition where there is little or no obvious change in WML as other processing parameter initiates an increase in WML at higher processing conditions. For density, the surface plot exhibits an increase in density as the processing parameter increases. Though, an optimum condition is attained after which the density shows a reverse trend as the processing parameter further increases. From the MOO analysis, the set of conditions that simultaneously optimizes both WML and density was found as RF 14 *wt%*; SS 460 *rpm*; PT_{emp} 820 °*C* and PT 150 *secs*. As a result the WML achieved a minimum of 1 x $10^{-3}g$ and a maximum density value of 2.780g/mm^3 with a confidence and desirability level of 95.5%.

Acknowledgement
The authors acknowledge the support of the Ministry of Education and International Islamic University Malaysia for financial support to conduct this research work under Project No.: PRGS12 002-0002.

References

[1] Hariprasad T, Varatharajan K and Ravi S 2015 Wear characteristics of B_4C and Al_2O_3 reinforced with Al-5083 metal matrix based hybrid composite, *Procedia Eng.* **97** 925-929.
[2] Adebisi A A, Maleque M A and Shah Q H 2014 Performance assessment of aluminium composite material for automotive brake rotor, *Int. J. Veh. Syst. Model. Test.* **9**(3-4) 207-217.
[3] Adebisi A A, Maleque M A and Ali M Y 2015 Wear characteristics of multiple particle size silicon carbide reinforced aluminium composite, *Adv. Mat. Res.* **1115** 174-177.
[4] Sozhamannan G G, Balasivanandha S P and Venkatagalapathy V S K 2012 Effect of processing parameters on metal matrix composites: stir casting process, *J. Surf. Eng. Mater. Adv. Technol.* **2** 11-15.
[5] Balasivanandha S P, Karunamoorthy L, Kathiresan S and Mohan B 2006 Influence of stirring speed and stirring time on distribution of particles in cast metal matrix composite, *J. Mater. Process. Tech.* **171** 268–273.
[6] Naher S, Brabazon D and Looney L 2003 Simulation of the stir casting process, *J. Mater. Process. Tech.* **143** 567-571.
[7] Niranjan K and Lakshminarayanan P 2013 Optimization of process parameters for in situ casting of Al/TiB_2 composites through response surface methodology, *Trans. Nonferrous Met. Soc. China.* **23**(5) 1269-1274.
[8] Adebisi A A, Maleque M A, Ali M Y and Bello K A 2016 Effect of variable particle size reinforcement on mechanical and wear properties of 6061 Al-SiC_p composites. *Compos Interfaces.* **23**(6) 533-547.
[9] Myers R H, Montgomery D C and Anderson-Cook C M 2009 Response surface methodology: process and product optimization using designed experiments. *John Wiley & Sons*, New York.

IOP Conf. Series: Materials Science and Engineering **184** (2017) 012020 doi:10.1088/1757-899X/184/1/012020

Investigation on the Maximum Power Point in Solar Panel Characteristics Due to Irradiance Changes

M. A. Abdullah[1,a] *, Siti Fauziah Toha[1,b], Salmiah Ahmad[1,c],
[1]Department of Mechatronics
Kulliyyah of Engineering, International Islamic University, Malaysia
P.O.Box 10, 50728 Kuala Lumpur, Malaysia

E-mail: [a]adlirajin@gmail.com, [b]tsfauziah@iium.edu.my, [c]salmiah@iium.edu.my

Abstract. One of the disadvantages of the photovoltaic module as compared to other renewable resources is the dynamic characteristics of solar irradiance due to inconsistency weather condition and surrounding temperature. Commonly, a photovoltaic power generation systems consist of an embedded control system to maximize the power generation due to the inconsistency in irradiance. In order to improve the simplicity of the power optimization control, this paper present the characteristic of Maximum Power Point with various irradiance levels for Maximum Power Point Tracking (MPPT). The technique requires a set of data from photovoltaic simulation model to be extrapolated as a standard relationship between irradiance and maximum power. The result shows that the relationship between irradiance and maximum power can be represented by a simplified quadratic equation. The first section in your paper

1. Introduction

Solar energy is one of the most important renewable energy sources. Solar energy is clean, inexhaustible and free compared to conventional non-renewable resources. It offers an option for clean (pollution free) energy source, with almost no running and maintenance cost. Besides, for remote systems such as radio communications, satellite earth stations, or at sites that are far away from a conventional power system, the hybrid systems have been considered as attractive and preferred alternative sources [1].

In the recent past years, PV power generation system has attracted more attention due to the energy crisis and environment pollution. One of the main renewable energy sources available in nature is the sun, usable for the direct production of electricity through photovoltaic (PV) systems [2]. However, Solar PV electricity output is also highly sensitive to shading and the crucial design involves the energy storage mechanism [3]. When even a small portion of a cell, module, or array is shaded, while the remainder is in sunlight, the output falls dramatically [4]. Besides, Photovoltaics has nonlinear internal characteristics. The voltage-power characteristics of the PV panel is varied which depends upon insolation and temperature. Considering the high initial installation cost of the PV system, it is always necessary to operate PV at its Maximum Power Point (MPP) [5]. Therefore, the dynamic condition of solar power characteristics due to some external factors such as solar irradiance and temperature requires the researchers to improve the charging control in order to optimise the solar power.

Tracking the maximum power point (MPP) of a photovoltaic (PV) array is usually an essential part of a PV system. As such, many MPP tracking (MPPT) methods have been developed and implemented. The methods vary in complexity, sensors required, convergence speed, cost, range of effectiveness, implementation hardware, popularity, and in other respects [6]. The conventional MPPT methods are generally categorized into the following groups according to the control strategies, that are 1) Perturbation and observation (P&O) methods, 2) Incremental conductance methods, 3) Constant current or constant voltage and 4) Miscellaneous (e.g., fuzzy control and voltage- based scheme among them P & O method has drawn much attention due to its simplicity. But the oscillation problem is unavoidable [7].

The traditional prediction of PV energy production is based on the classical approach of time series forecasting of solar energy and weather conditions, which are used to calculate the electrical energy of PV systems: Autoregressive models, Moving Average and Autoregressive Moving Averages are often used to model linear dynamic structures [2]. However, the investigation of the photovoltaic power characteristic in variance of irradiance met some challenges as the conventional MMPT methods rely solely on voltage, current and temperature. Therefore, the objective of this paper is to investigate the maximum power point tracking based on the actual irradiance level. Using a complete simulation model in Matlab Simulink, some sampling has been done to represent the model. Finally some characteristic equations has been proposed as the representation of a simple Power versus Irradiance linearized equation of the sampled model for the use of further research and implementation.

2. System Architecture

The overall block diagram of Solar Intelligent Control is shown in Figure 1 where it consists of the integration of three main subsystems which are MPPT Controller, Buck Converter and Power Monitoring.

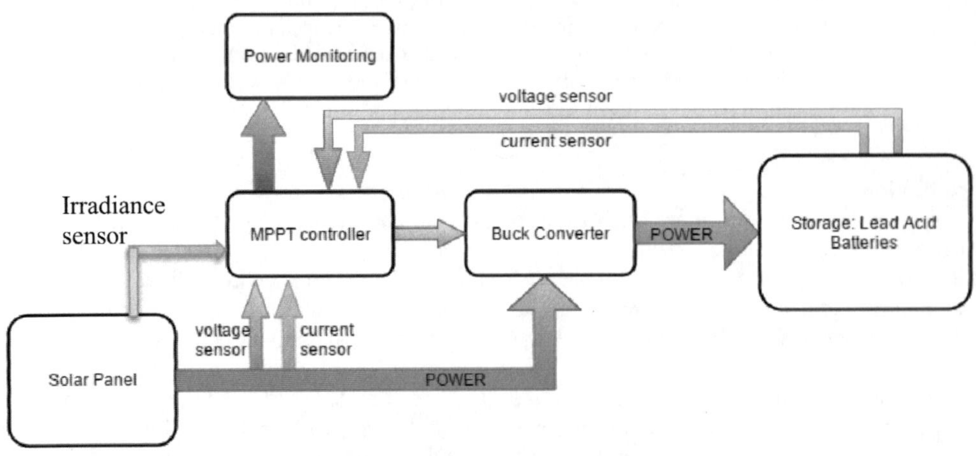

Figure 1: Block Diagram of Solar Intelligent Control

2.1. *Maximum Power Point Tracking (MPPT)*

Maximum power point tracking (MPPT) is a technique used in solar systems to maximize the output power. Some parameters from solar panel are used as references to the power optimiser. In order to perform a more sustainable charging, the condition of the storage unit is taken into consideration to make sure the lifetime of the battery can be preserved.

IOP Conf. Series: Materials Science and Engineering **184** (2017) 012020 doi:10.1088/1757-899X/184/1/012020

Therefore, using simple linearization formula, the system evaluates four main variables that are Voltage, Current and Temperature to utilize the output current and voltage according to solar power characteristics.

2.2. Buck Converter

A buck converter is a step down DC-DC power converter which steps down the voltage rating while stepping up current from solar panel to the storage unit. The buck converter will make sure the solar panel draw power within specific value of voltage and current based on its highest power points.

2.3. Power Monitoring

A power monitoring subsystem is integrated into the Solar Intelligent Control to monitor the condition of solar panel and storage unit. The subsystem will monitor the degradation or defect on the solar panels and the State of Charge (SOC) of the storage unit.

3. Result and Discussion

3.1 Photovoltaic Simulink Block Diagram

The model of photovoltaic was constructed in Matlab/Simulink. A fundamental understanding about the behaviour of solar is achieved by using Simulink application in Matlab2014b as shown in Figure 2.

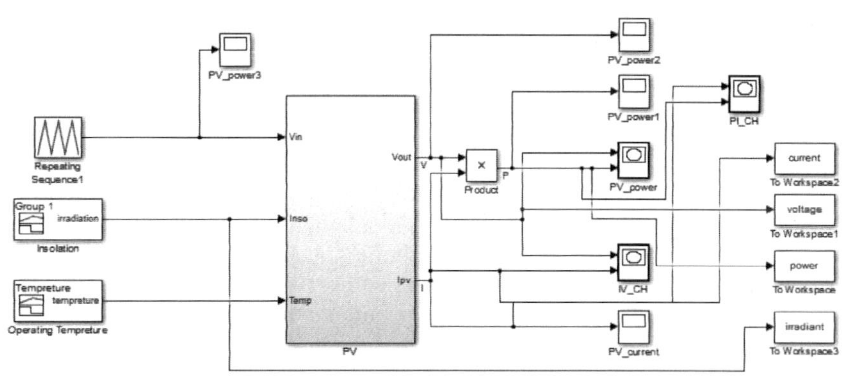

Figure 2: Solar Equivalent Simulink Model

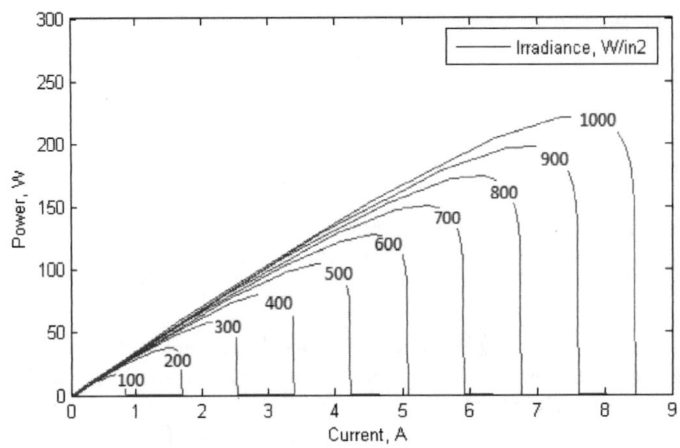

Figure 3: Solar Irradiance Variant

The characteristics of Power(W), Current(I), Voltage(V), and Irradiance(W/in^2); which is the flux of radiant energy per unit area normally measured in Watt per unit area of inch square. The parameters was analysed to propose the most suitable charger controller as the key to optimize the use of solar energy. The preliminary performance measurement are the voltage, current and power behaviour based on different value of solar irradiance. The irradiance value is the input value from the simulation diagram. Some sample irradiance values is taken into consideration starting from 0 W/in^2, 100 W/in^2, 200 W/in^2, 300 W/in^2, 400 W/in^2, 500 W/in^2, 600 W/in^2, 700 W/in^2, 800 W/in^2, 900 W/in^2, and 1000 W/in^2.

Figure 3 shows the current and power relationship in solar panel. The model was simulated based on different value of irradiance ranging from 100 W/in^2 to 1000 W/in^2. Based on the graph, it is clear that the rated current increased when the irradiance level increases. The rated current for maximum irradiance is 8.5 A while the maximum power is 240 W.

Figure 4 shows the voltage and power relationship in solar panel. The model simulates behaviour based on different value of irradiance ranging from 100 W/in^2 to 1000 W/in^2. For 400W/in^2, the optimum Voltage reference is 27.5V and the maximum output power is 65W while for 1000W/in^2, the optimum Voltage reference is 29.5V and the maximum output power is 220W.

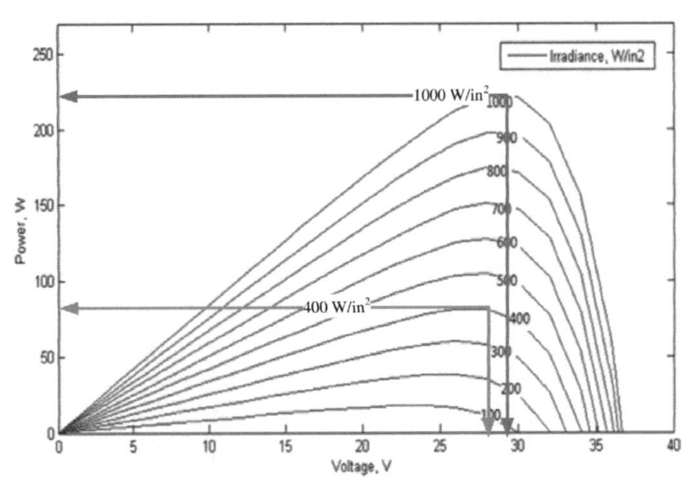

Figure 4: Voltage vs. Power in Solar

IOP Conf. Series: Materials Science and Engineering **184** (2017) 012020 doi:10.1088/1757-899X/184/1/012020

3.2 Linearizing the formula

In order to obtain the maximum power related to the Solar Irradiance, some sampling representing the highest possible power of the system is obtained as shown in Figure 5. Based on the power characteristics of solar panel, the relationship between Power and Irradiance has been simplified into a discrete samples to find out the standard relationship between Power and Irradiance for an optimised Maximum Power Point Tracking (MPPT) control. Using data cursor in Matlab, a tabulated data is performed in Table 1 for the use of linearization technique.

Considering the sets of data sampling, the linearized equations between Maximum Power and Irradiance is identified based on the standard equation stated in (1) and (2).

$$P = mR + y \quad (1)$$

$$P = aR^2 + bR + c \quad (2)$$

Where *m, y, a, b* and *c* are the coefficients of the equation, *P* is the maximum power, *R* is the Solar Irradiance. Using extrapolation technique, the equation (1) and (2) is plotted in Figure 6 to represent the relationship between maximum power and irradiance. The result shows that the quadratic equation and linear equation is in a very small proximity. Therefore to acquire more accurate control is better to use the quadratic equation while to prefer a simpler control it is better to use the linear equation.

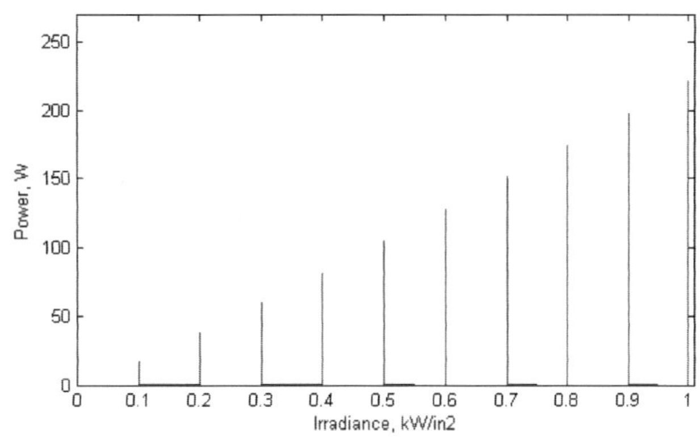

Figure 5. Maximum Power vs. Solar Irradiance

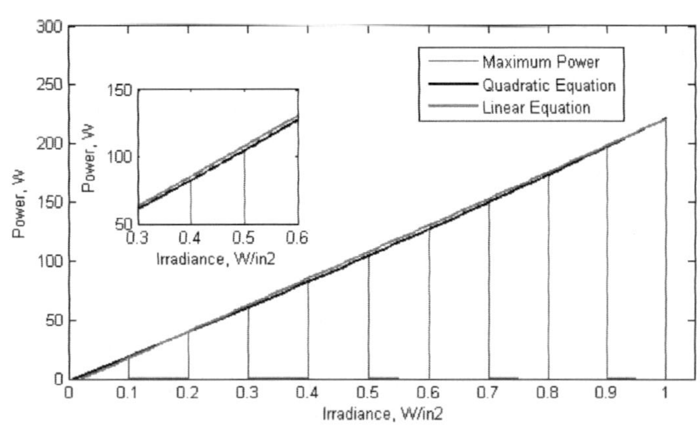

Figure 6: Equation for Maximum Power Point vs Irradiance

Table 1: Maximum Power vs Irradiance

Sample No.	1	2	3	4	5	6	7	8	9	10
Irradiance, W/in^2	100	200	300	400	500	600	700	800	900	1000
Maximum Power, W	17.52	35.85	57.98	78.11	104.55	127.8	151.1	174.3	197.6	220.8

4. Conclusion

An extrapolation approach to investigate the relationship between maximum power point and irradiance level has been presented in this paper. By using a complete Photovoltaic Simulation model in Matlab Simulink, some sampling has been done according to various irradiance level in order to represent an optimized equation. Finally a quadratic equation has been proposed as the representation of a simple Power versus Irradiance. The result suggest the extrapolation technique is applicable to represent a simplified control of Maximum Power Point Tracking (MPPT) for the use of further research and implementation.

Acknowledgements

Authors would like to express their gratitude to International Islamic University Malaysia (IIUM) for providing facilities to conduct this research. This research was funded under Knowledge Transfer Programme (KTP). Grant No: KTP15-010-0011.

References

[1] Z. M. Salameh, "71. 1996 Methodology for Optimally Sizing the Combination of a Battery Bank and PV Array in a Wind□PV Hybrid System.pdf," vol. 11, no. 2, 1996.

[2] R. De Leone, M. Pietrini, and a. Giovannelli, "Photovoltaic energy production forecast using support vector regression," *Neural Comput. Appl.*, pp. 1955–1962, 2015.

[3] V. R. V. Raghunathan, a. Kansal, J. Hsu, J. Friedman, and M. S. M. Srivastava, "Design considerations for solar energy harvesting wireless embedded systems," *IPSN 2005. Fourth Int. Symp. Inf. Process. Sens. Networks, 2005.*, vol. 00, no. C, 2005.

[4] H. Kim, B. Parkhideh, T. D. Bongers, and H. Gao, "Reconfigurable solar converter: A single-stage power conversion PV-battery system," *IEEE Transactions on Power Electronics*, vol. 28, no. 8. pp. 3788–3797, 2013.

[5] R. Ramaprabha and B. Mathur, "Design and Modeling of Standalone Solar Photovoltaic Charging System," *Faculty.Mu.Edu.Sa*, vol. 18, no. 2, pp. 41–45, 2011.

[6] T. Esram and P. L. Chapman, "Comparison of Photovoltaic Array Maximum Power Point Tracking Techniques," *IEEE Trans. Energy Convers.*, vol. 22, no. 2, pp. 439–449, 2007.

[7] R. Sridhar, D. Jeevananathan, N. T. Selvan, and S. Banerjee, "Modeling of PV Array and Performance Enhancement by MPPT Algorithm," *Int. J. Comput. Appl.*, vol. 7, no. 5, pp.

35–39, 2010.

Battery Cell Balancing Optimisation for Battery Management System

M.S. Yusof[1], S.F. Toha[2, a,] N.A Kamisan[3], N.N.W.N. Hashim[4], M. A. Abdullah[5]
[1,2,4,5]Department of Mechatronics, Faculty of Engineering, International Islamic University Malaysia, Kuala Lumpur, Malaysia.
[3]Pumar Marketing Sdn. Bhd., No. 57, Jalan Telawi, Bangsar Baru, Kuala Lumpur, Malaysia.

E-mail: [a] tsfauziah@iium.edu.my

Abstract. Battery cell balancing in every electrical component such as home electronic equipment and electric vehicle is very important to extend battery run time which is simplified known as battery life. The underlying solution to equalize the balance of cell voltage and SOC between the cells when they are in complete charge. In order to control and extend the battery life, the battery cell balancing is design and manipulated in such way as well as shorten the charging process. Active and passive cell balancing strategies as a unique hallmark enables the balancing of the battery with the excellent performances configuration so that the charging process will be faster. The experimental and simulation covers an analysis of how fast the battery can balance for certain time. The simulation based analysis is conducted to certify the use of optimisation in active or passive cell balancing to extend battery life for long periods of time.

1. Introduction

Battery management system is one of the crucial part in various electrical and electronic systems, e.g., commercial electronics device and electric vehicles that help to monitors and reports the state-of-charge (SOC), state-of-health (SOH) and remaining useful life for every rechargeable multi-unit batteries cell [1]–[4]. The battery management system be able to manage and adapt towards the changing of batteries characteristic over time since the applications require the parallel or series attachment of multiple unit battery cells [5]–[7]. A study conducted in [4], shows that battery management system can tolerate with the batteries at the temperature variation range of -30°C to 50°C. This means that the battery still need cell balancing in order to help the management system and help to tolerate with power of the battery.

A research that have been done by [8] show that they are using active cell balancing for their system. The research are conducted from 10 cells of battery and make the comparison between with balancing and without balancing. When they discharge the cells without balancing, one of the cell reach to the discharge voltage limit (DVL) compare to others cell. Even only one of the cell reach the

DVL, the battery is stop working in order to care about the battery life. From that point, it shown that the battery is not fully utilized and make the cycle life shorter. The experiment with balancing show that the discharge voltage limit time is increased by 15%.

In the paper [9], they have done with the balancing algorithm based on outlier detection. The method they been used is by calculate each cell outlier value. Moreover, the cell been balanced if the threshold value of each cell is not less than the threshold assigned. From the result, it shown the frequency of on/off switching significantly reduced from 1150 to 2 and also reduce the voltage cut-off. The algorithms is extremely good, however, it make some disturbing in the cut-off voltage value.

This paper will concentrates on the cell balancing and discuss several proposed balancing method from different point and method by using MATLAB/Simulink. This method is useful based on final voltage information. The value for each cell will be calculated. After get each value, it will compare between others. If the threshold value between maximum and minimum voltage is not over the set value, then the algorithm will proceed with the final-voltage algorithm.

The research work is organized that started with balancing algorithm which explain the type of algorithms. At the second part will discuss about the methodology for this system. In the final section is about the simulation experiments that been done and followed by conclusion.

2. Balancing Algorithm

2.1 Voltage based

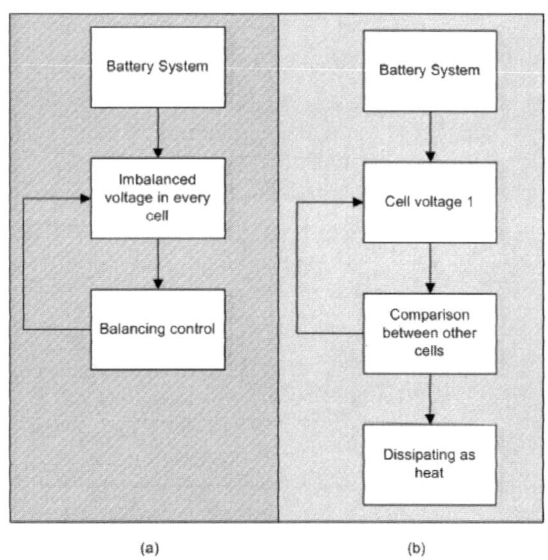

Fig. 1. (a) Voltage based flowchart, (b) Final voltage based flowchart

Figure 1 shows the voltage based algorithm which includes two modules that are imbalanced voltage and balancing control module. Basically, the cell will be charge and discharge up to certain point [5]. However, during the charging process, each cell is charged at different rate resulting of voltage difference between each individual cell. With the different value in voltage reading at each cell, the cell balancing circuit will make an attempt to balance out the voltage reading of individual cell [6], [7]. Cells with abnormal voltage reading will automatically give a feedback information to the previous cell. However, if the voltage reading is under load, it will depend only on the internal cell resistance of every cell.

2.2 Final Voltage based

The abnormal cell will then send feedback information to the previous cell in series next to it. The balancing circuit will be activated when the cell reach an acceptable voltage range. It will then compare its voltage reading to others cell and starts to preform battery balancing operation. The excess power will be dissipated through the resistor and as heat dissipation [3]. The voltage reading will remain around the predetermined nominal voltage value. The voltage reading from the first cell will be compared to the second cell as to validate the value of voltage [10].

2.3 SOC History based

The algorithm attempting to match SOC with cells history based. The balancing will occur at every time intervals [10]. The advantages of this algorithms is it can balance with less current since it occur at any time. However, it requires more power in order to save the memory in the store for every cell.

3. Methodology

In this paper, the system consists of electronic and software parts. It contain two main stages which are designing the control algorithm and electronic implementation in order to measure its performance in the real world environment. Here is the common form of calculation related to the SOC algorithms.

$$\mathrm{SOC}_c(t) = \mathrm{SOC}_c(0) - \left({}^{1}/_{Q}\right) \int_0^t t\, I(t)dt \quad (1)$$

The SOC that shown in Equation 1, use the integration of all charge and having left the cell by the amount of the current [11]. Some battery are extremely sensitive towards the overcharge or discharge. In order to prevent from overcharge or discharge, the additional equation is required.

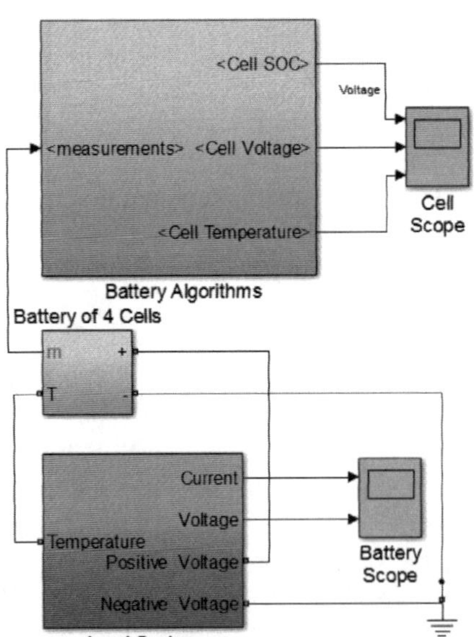

Fig. 2: Full Overview of the Design Software

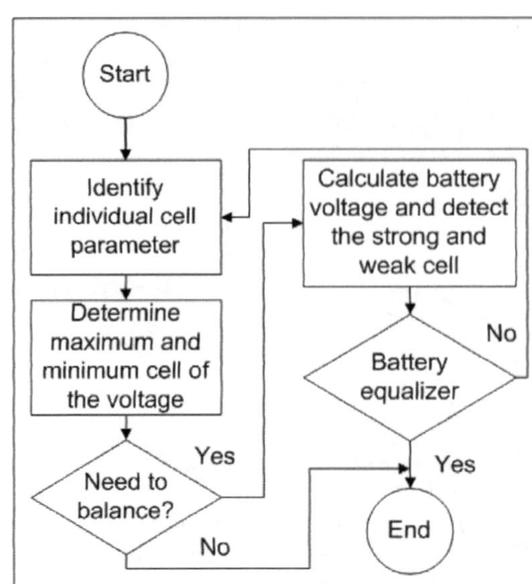

Fig. 3: Cell balancing flowchart

Equation 2 shows the basic equation for comparison of each other cell. V is representative of the voltage of each cell. The voltage been charge or discharge, it will continuously deduct with the voltage equation. If voltage value after operation is the same with the battery cell, the excess charge will be dissipated through the resistor while discharging will stop working.

$$v_{Cell,n} = v_{charge/discharge} - v_{eq,n} \tag{2}$$

Figure 2 shows the overall view of the system. The battery pack consists of 4 cells and the load comes from the current. The current is converted in term of positive and negative voltage. The measurement and balancing calculation is achieved within the battery algorithms. The software used is Matlab 2015 which includes the Simulink and Simscape. Simscape acts as a circuit design which is easy to design the balancer circuit. While the Simulink act as virtual connection between hardware. The using of Matlab give a great result since it already stable and also can be written in the C / C++ language. The software is divides into three parts which arre mechanical part, battery and balancing algorithms. However, this paper is focused only on the algorithms. Figure 3 depicts the flowchart of the cell balancing to works.

4. Results and discussion

In order to compare the efficiency of the balancing algorithms, the constant-current charging-discharging model been used to get the result. In general, most of the parameter that are hard to set; however, the model need to be simplified for the real-time BMS application (Ahmed et al. 2015). The battery pack is modelled in in both Simulink and the circuit in Simscape environments. Table 1 shows the parameters of the battery cell which is $LIFePO_4$. This battery been categories as slight power density compared to the higher power density which is better, but the higher power density is typically more pricy [3].

Table 1. Parameters of the battery

Parameters	*Value*	*Unit*
Nominal voltage	3.5	V
Fully charge voltage	4.45	V
Rated capacity	40	Ah

4.1 Constant Current Charging-Discharging Model and simulation

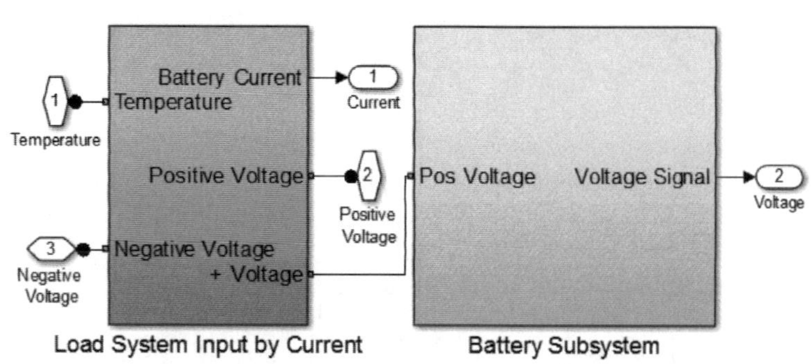

Figure 4: Load operating in the Load System

Figure 4 shows the system of cells balancing. The balancing system is based on the battery pack modularization [5]. The subsystem box contained the algorithm for the balancing. The graph will be shown at the Cell Scope. The constant charging and discharging current progress will be stopped when one of the terminals has reached at below discharging voltage limit or exceeds the charging voltage limit [6]. The discharge voltage level in this paper is 3.23v and charge voltage level is 4.46v.

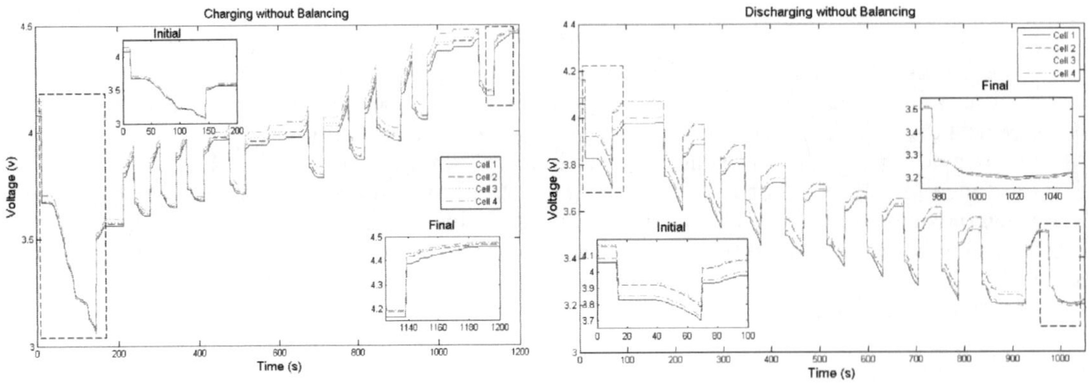

a) b)

Figure 5: (a) Charging process without balancing, (b) Discharging process without balancing

Figure 5 shows the voltage of 4 battery cell during one cycle of charging and discharging. At the charging point, when the cell near 4.45v, the charging process will not continue. The early voltage of each cell is different as the value of initial state is not same. It had been shown at all figure as to show the result of balancing. When the discharging process happen, the battery will not been used anymore when the voltage reach 3.23v. The cell charge and discharge is not balance until the end. Even the gap value is near, but it still needs a balancer to make it more stable. The earlier voltage is started at different value and converges at the same value. After the balancing at certain value, the end of the graph show the value between cells is almost same. It shows that the cell balancing is activated at the end of set value. For Figure 6(a) it shows that the graph of 4 cells battery charging been balanced at some point while Figure 6(b) show the graph of battery discharging been balanced. Each cell compare between each other in order to get the best value.

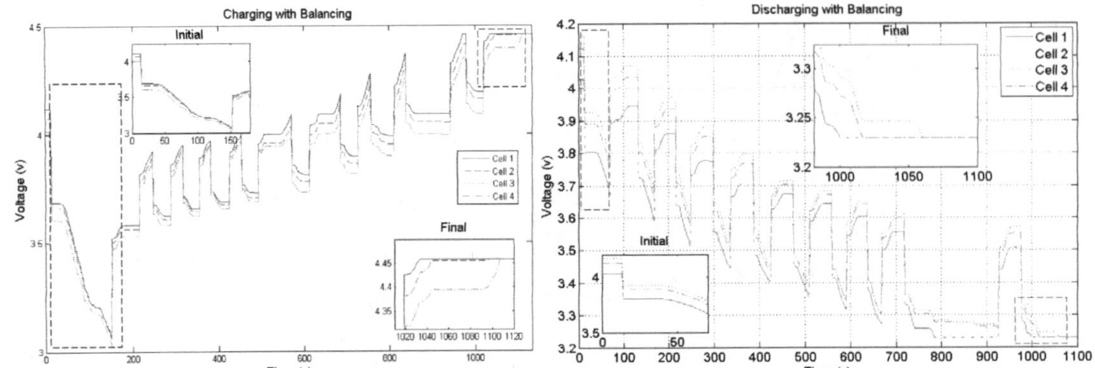

a) b)

Figure 6: (a) Charging with balancing (b) Discharging with balancing

5. Conclusion

The objective of this paper which is to obtain battery cell equalization state during both charging and discharging process are achieved. This results will be further useful for the development of cell balancing circuit and will be tested on Lithium Iron Phosphate ($LIFePO_4$) to reduce both charging and discharging time as well as prolonging battery shelf life in a long run. The balancing circuit implemented within battery stack is vita not only to optimise battery life-time but also will reducing overall system cost as well as the process operation time. The battery balancing must be applied to the electronic and electrical vehicle to achieve a longer lifetime as well as reliability and also safety.

Acknowledgement

This work was funded by the Ministry of Science, Technology and Innovation (MOSTI) Malaysia under the E-Science Grant Scheme: SF15-013-0063.

References

[1] Ahmed, R. et al., 2015. Model-Based Parameter Identification of Healthy and Aged Li-ion Batteries for Electric Vehicle Applications, *Elsevier.*

[2] Bragard, M. et al., 2010. The Balance of Renewable Sources and User Demands in Grids: Power Electronics for Modular Battery Energy Storage Systems. *IEEE Transactions on Power Electronics*, **25.2** 3049-3056.

[3] Cao, J., Emadi, A. & Member, S., 2012. A New Battery / UltraCapacitor Hybrid Energy Storage System for Electric , Hybrid , and Plug-In Hybrid Electric Vehicles, *IEEE Transactions on Power Electronics*, **27.1** 122-132.

[4] Daowd, M. et al., 2014. Battery Management System-Balancing Modularization Based on a Single Switched Capacitor and Bi-Directional DC/DC Converter with the Auxiliary Battery. **Energies**, 2897-2937.

[5] Daowd, M. et al., 2011. *Passive and Active Battery Balancing comparison based on MATLAB Simulation*, IEEE Vehicle Power and Propulsion Conference (VPPC), Chicago USA, 6-9 Sept 2011,

[6] Einhorn, M. et al., 2011. Improved Performance of Serially Connected Li-Ion Batteries With Active Cell Balancing in Electric Vehicles. *IEEE Transactions On Vehicular Technology*, **60.6** 2448-2457.

[7] Javier Gallardo-Lozano, Lateef, A., Romero-cadaval, E. & Milans-montero, M.I., 2013. Active Battery Balancing for Battery Packs. *Electrical, Control and Communication Engineering*, **2.1** 40-46.

[8] Liu, G. et al., 2014. *A comparative study of equivalent circuit models and enhanced equivalent circuit models of lithium-ion batteries with different model structures*. 2014 IEEE Conference and Expo Transportation Electrification Asis-Pacific, 1-6.

[9] Piao, C. et al., 2015. Lithium-Ion Battery Cell-Balancing Algorithm for Battery Management System Based on Real-Time Outlier Detection. *Mathematical Problems in Engineering*, 2015.

[10] Plett, G.L., 2004. Extended Kalman filtering for battery management systems of LiPB-based HEV battery packs Part 2 . Modeling and identification. *Power Sources*, **134**, 262-276.

[11] Products, H.A. & Quarter, F., 2009. Analog Applications. *Analog Applications Journal.*

[12] Qian, H., 2011. A High-Efficiency Grid-Tie Battery Energy Storage System *IEEE Transactions on Power Electronics*, **26.3** 886-896.

[13] Iqbal Husain 2003. Electric and Hybrid Vehicles- Design Fundamental, Second Edition, *CRC Press.*

[14] Wen, B.S., 2009. Cell balancing buys extra run time and battery life. *Power Management*, Texas Instruments Incorpporated.

IOP Conf. Series: Materials Science and Engineering **184** (2017) 012022 doi:10.1088/1757-899X/184/1/012022

Throughput Enhancement of Car Exhaust Fabrication Line by Applying MOST

E. A. H. Hanash[1,a], A. N. M. Karim[1,b], SaravananTanjong Tuan[2], A. K. M. Mohiuddin[3]

[1]Department of Manufacturing and Materials Engineering,

[3]Department of Mechanical Engineering,

Faculty of Engineering, International Islamic University, Malaysia

P.O.Box 10, 50728 Kuala Lumpur, Malaysia

[2]S Automotive Industry Sdn Bhd, Lot 9, Jalan Puchong, Shah Alam, Malaysia

E-Mail: [1,a]ebrahim.hanash@yahoo.com, [1,b]mustafizul@iium.edu.my, [2]saravanantt@toyota.com.my, [3]mohiuddin@iium.edu.my

Abstract. In the fiercely competitive world market of today, manufacturers are facing increasingly tougher challenges and are compelled to find ways for productivity enhancement wherever possible in the whole supply chain. Nevertheless there are many facets in business process which can be explored for possible improvement, an immediate focus goes for the involved processes to re-engineer the activities in different workstations for an efficient and balanced assembly or fabrication line. In this paper an industrial case on fabrication line of a car exhaust system is presented to illustrate the scope of improvement by applying the MOST (Maynard's Operation Sequence Technique) in streamlining the activities followed by assembly line balancing (ALB). The whole process of conducting various tasks is investigated to find out the lapses or wastes, to search for better option and to set the standard times for the tasks. Then individual workstation time is worked out by summing up the standard times of the involved tasks or activities and the concept of ALB is attempted to balance the fabrication line. So by possible reduction or elimination of the identified wastes or lapses workstation times including the bottleneck station are lowered. As a result the throughput of car exhaust systems is enhanced. According to the current practice, the Takt time is set at 3 minutes. However, upon an analysis through use of the MOST, the bottleneck station time is found to be as low as 1.27 minutes. Thus an opportunity of meeting the current level of demand with significantly lower workforce (with 2 operators instead of 5) is revealed. Alternatively, if necessary, an increased workload can be assigned for the current level of workforce. Moreover, with proper distribution of activities among the workstations using the concept of ALB, the line efficiency is found to be improved. So the line balance loss in the current setup of production line is also possible to be largely reduced. Thus daily production of car exhaust, if modified with suggested changes with the current workforce, could be more than double compared to the current daily output. Hence, effectiveness of the MOST followed by ALB applications to expose and remove the operational wastes in the work flow is reiterated with enhanced throughput.

1. Introduction

Fierce competition among business operators in the current world market is due to the change of customer behaviour towards product features and advancement in technology which has led to manifold challenges to the manufacturers [1]. Customers value different products that have comparable features with high quality but lower cost [2, 3]. Companies, therefore, start to introduce new products with different features and reasonable price. However, to sustain under such conditions, manufacturers have to adopt a low cost and/or a differentiation strategy to find ways to increase the throughput and overall productivity of the organization [4]. Thus they find ways to increase the output for enhancing the profitability and maintaining competitive edge [5].

There are many different methods to increase and enhance the productivity, for example, Value Stream Mapping (VSM), Just in Time (JIT), 5's, Lean Manufacturing, Workplace Organization, Work Measurement, and Assembly Line Balancing [6, 7]. Productivity improvement refers to the increase in output per work-hour or time expended. Among the methods for increasing the productivity is the use of predetermined time study (PTS) under the work measurement category. This method is similar to other lean manufacturing methods in removing non-value added activities from the process [8, 9]. An operator can be seen busy in the workplace but when measuring the standard time for certain activity, the wasted time in non-added value activity might be significant.

The predetermined study not only used in setting the standard time in manufacturing sector but it has been applied in the agriculture sector [10]. MTM is found to be more applicable than other methods, except for mental works. Hence, by realizing the significance, a number of researchers have engaged themselves to further the methods and ended up with different techniques. These are predetermined motion time system (PMTS) methods, time estimating methods etc. Moreover, on the basis of data type, motion variations and time unit, several PMTS methods such as Methods Time Measurement (MTM), Work factor, Maynard Operation Sequence Technique (MOST) and Modular Arrangement of Predetermined Time Standard (MODAPTS) are reported in literature [11-14]. The MOST is very popular in setting the standard time of an operation by dividing the process into small elements and calculating the time in Time Measurement Unit (TMU). This method has been widely used by many researchers due to its good accuracy and ease in implementation [14-16]. The MOST system is used to calculate the cycle time for an operation based on Pre-determined time study data.

A typical MOST work sequence code would look like this: an example [17]: The operational element is described as 'Apply pressure and push the chassis trolley about 5-7 steps to the body side panel bonding bay'. There are general and controlled moves and the index values are assigned as follows:

A1 B0 G3 A16 B0 P6 M16 X0 I0

Step 1: Add up all the subscript numbers. Thus, 1+ 0+ 3 + 16 + 0 + 6 + 16 + 0 + 0 = 42 (the subscript is the MOST index value)

Step 2: Multiply the sum of the index by 10. This answer is 42 x 10 = 420 TMU

Step 3: Thus, the time in TMU is converted as 420 TMU * 0.036 seconds = 15.12 seconds

The case as presented in this paper is based on a study conducted in a local auto company to explore the scope of possible improvement in productivity of a car exhaust fabrication line for which excessive time is consumed with the existing level of workforce. Standardization of the task times and revision of work methods have been made to show excess capacity. To do so, the MOST is adopted to set the standard times and to eliminate or reduce the non-value added activities (NVA) and to detect the bottlenecks correctly. The whole assembly line currently consists of five workstations, each is supported by an operator and equipped with necessary facilities.

2. Predetermined Motion Time System

Predetermined motion time studies (PMTS) is one of the work measurement methods which can be applied to set the standard time of the process. PTMS such as methods time measurement (MTM), MOST, and WORK FACTOR, have primarily been used for predicting and quantifying the time it should take to assemble a product [18]. The PMTS methods used to determine the standard time that can be taken to perform certain process. It can be used to divide large tasks into small elements in order to balance the work between workers and make the flow more efficient. Therefore, as recommended by [19] all workstation activities or items have to be organized in an efficient manner.

The PMTS application steps are as follows [20]:

- Synthesize the method that would be used by a worker to perform the task. The method should be described in terms of the basic motions involved in the task based on a predefined workplace layouts and set of tools.
- Retrieve the normal time value for each motion element, based on the work variables and conditions under which the work element is performed. Sum the normal times for all motion elements to determine the normal time for the task.
- Evaluate the method to make improvements by eliminating motions, reducing distances, introducing special tools, introducing simultaneous right and left hand motions and so on.

However, having the variation in level and scope of applications some common PMTS methods which are repeatedly referred in literature [11] for solving the intricate operational issues of automotive industries are described.

2.1 Maynard Operation Sequence Technique (MOST)

Maynard Operation Sequence Technique (MOST) is a predetermined motion time system that is used primarily in industrial settings to set the standard time in which a worker should perform a task [21]. It is a powerful analytical tool that helps increase productivity, improve methods, facilitate planning, establish workloads, estimate labor costs, improve safety and maximize resources [22]. It uses as same time units as MTM; TMU developed around 1967 under the direction of Kjell Zandin [23]. It has been introduced by H. B. Maynard and Company, Inc. in the United States in the year of 1960 [24]. Though industrial application of MOST started from 1967 in the form of Basic MOST, the Basic MOST was reformulated in 1970 and renamed as Clerical MOST for performing the administrative and the clerical work of the production and service industries.

MOST focuses on the movement of an object, as a result it was identified that there are repetitive movement pattern such as, reach, grasp, move, position the object etc. To compute the standard time by using MOST, a task is broken down into individual motion elements. After that, numerical time value in units known as time measurement units (TMUs) is assigned to each of the individual motion element.

In general, there are only two ways by which the object can be transferred and moved, either they are picked and moved freely through space, or they are moved while maintaining contact with another surface. There are three types of MOST, which are Basic MOST (for the activities between 20 s to 2 min.), Mini Most (for the activities shorter than 20 s) and Maxi Most (for the activities above 2 min.) as stated by Jamil et al. [21]. The literatures revealed that, MOST has superior characteristics over other time study methods. Some important advantages of MOST are in terms of its fastness compared to other work measurement techniques, scope of calculating the time in advance, easiness in learning, universal application and economy.

In this project Basic MOST is used for determining the standard time and improving processes and productivity enhancements.

2.2.1 Basic MOST

Basic MOST system, compared to other PMTS systems, concentrates on the movement of 'objects' rather than the basic motions of 'worker'[13]. This criterion made Basic MOST more accurate, fast, economical and simple to implement compared to other PMTS systems. It is found that, Basic MOST is more accurate than the conventional time study method in determining the relevant true time [25]. It consists of sequence models representing two basic activities to measure work which are General Move, Controlled Move AND Tool Use. Table 1 illustrates the different types of moves.

Table 1: Types of Basic Moves in Basic MOST

Activity	Sequence Model	Sub-activities
General Move	ABG ABP A	A – Action Distances
		B – Body Motion
		G – Gain control
		P – place
Controlled Move	ABG MXI A	M – Move Control
		X – Process Time
		I – Align
Tool Use	ABG ABP * ABPA	*F – fasten
		*L – Loosen
		*C – Cut
		*S – Surface treat
		*R- Record
		*M – Measure

The time unit used in calculating and analyzing the work activities by using MOST is TMU where 1 TMU is equal to 0.036 second, or 0.0006 minute or 0.00001 hour.

2.2.1.1 Basic MOST Calculation

There is a sequence of steps to be followed in filling the calculation sheet of the Basic MOST method in order to set the standard time. These steps include:

1. Indicate at the top of the form: Area of work, Operation name, Department, Analyzed by.
2. Document the method to be analyzed by dividing it into a number of successive steps corresponding to the "natural" breakdown of the activity. Number each step in chronological order.
3. Select an appropriate sequence model for each method step.
4. Determine the type of move for that certain small element.
5. Indicate the correct index value for each parameter within each sequence model.
6. Add parameter index value together, multiply by 10 and insert the result in right hand column to arrive at the time for the sequence model in TMU.
7. For the total activity time in TMU, add all sequence times together and insert the result in the bottom right-hand corner. If desired, these time values may be converted to hours, minutes, or seconds in the bottom left-hand corner of the sheet.

As illustrated in **Table 2,** the time-estimation procedure by the MOST, a task 'Welding the flanges to CKD (complete knock-down) pipe' is chosen from the operation at W/S–1 of the undertaken assembly line as an example.

Table 2: Four task elements for 'welding the flanges to CKD pipe', sequence models and index values

SN	Description of task element	Sequence Model	Index value
1	Pick up flanges and clamp on jig.	General Move	A3 B6 G3 A0 B0 P3 A3
2	Pick up CKD pipe and clamp on jig.	General Move	A3 B6 G3 A0 B0 P1 A0
3	Pick up bracket and clamp on jig.	General Move	A3 B6 G3 A0 B0 P0 A0
4	Take up welder and mask, tack weld flanges (3 point).	Tool Use	A6 B6 G3 A0 B3 P3 A0B0P0 A0

In Fig. 1, 2, and 3 the index value tables for each of the three types of moves are presented.

ABG (Get) ABP (Put) A (Return) — General Move

Index x 10	A Action Distance	B Body Motion	G Gain Control	P Placement	Index x 10
0	≤ 2 in. (5 cm.)			Pickup Toss	0
1	Within Reach		GET GRASP: Light Object Light Objects Simo	PUT: Lay Aside Loose Fit	1
3	1 - 2 Steps	Sit or Stand Bend and Arise 50% occ.	GET GRASP: Light Objects Non-simo Heavy or Bulky Blind or Obstructed Disengage Interlocked Collect	PLACE: Loose Fit Blind or Obstructed Adjustments Light Pressure Double Placement	3
6	3 - 4 Steps	Bend and Arise		POSITION: Care or Precision Heavy Pressure Blind or Obstructed Intermediate Moves	6
10	5 - 7 Steps	Sit or Stand with Adjustments			10
16	8 - 10 Steps	Stand and Bend Bend and Sit Climb On or Off Through Door			16

A Action Distance Extended Values

Index	Steps	Distance (ft.)	Distance (m.)
24	11-15	38	12
32	16-20	50	15
42	21-26	65	20
54	27-33	83	25
67	34-40	100	30
81	41-49	123	38
96	50-57	143	44
113	58-67	168	51
131	68-78	195	59
152	79-90	225	69
173	91-102	255	78
196	103-115	288	88
220	116-128	320	98
245	129-142	355	108
270	143-158	395	120
300	159-174	435	133
330	175-191	478	146

Figure 1: Index values for the General Move in Basic MOST

ABG (Get) MXI (Move/Actuate) A (Return) — Controlled Move

Index x 10	M Move Controlled		X Process Time			I Alignment	Index x 10
	Push/Pull/Turn	Crank	Seconds	Minutes	Hours		
1	≤ 12 in. (30 cm.) Button Switch Knob		.5 sec.	.01 Min.	.0001Hr.	1 Point	1
3	>12 in. (30 cm.) Resistance Seat or Unseat High Control 2 Stages ≤ 24 in. (60 cm) Total	1 Rev.	1.5 sec.	.02 Min.	.0004 Hr.	2 Points ≤ 4 in. (10 cm.)	3
6	2 Stages >24 in. (60 cm.) Total 1 - 2 Steps	2 - 3 Revs.	2.5 sec.	.04 Min.	.0007 Hr.	2 Points > 4 in. (10 cm.)	6
10	3 - 4 Stages 3 - 5 Steps	4 - 6 Revs.	4.5 sec.	.07 Min.	.0012 Hr.		10
16	6 - 9 Steps	7 - 11 Revs.	7.0 sec.	.11 Min.	.0019 Hr.	Precision	16

M Push or Pull Extended Values

Index	Steps
24	10-13
32	14-17
42	18-22
54	23-28
67	29-34

Crank Extended Values

Index	Revs.
24	12-16
32	17-21
42	22-28
54	29-36

I Alignment of Machining Tools

Index	Align to
3	Workpiece
6	Scale
10	Indicator Dial

Alignment of Non-typical Objects

Index	Positioning Method
0	Against Stop(s)
3	1 Adjustment to Stop
6	2 Adjustments to Stop(s) 1 Adjustments to 2 Stops
10	3 Adjustments to Stop(s) 2 – 3 Adjustments to Linemark

Non-typical Object Characteristics
Flat, Large, Flimsy, Sharp, Difficult to Handle

Figure 2: Index values for the Controlled Move in Basic MOST

ABG Get Tool · ABP Put Tool · * Tool Action · ABP Aside Tool · A Return

Tool Use

F L Fasten or Loosen

Index x 10	Finger Action	Wrist Action PowerTool				Arm Action						Index x 10
	Spins	Turns	Strokes	Cranks	Taps	Turns		Strokes	Cranks	Strikes	Screw Diam	
	Fingers, Screw-driver	Hand, Screwdriver, Ratchet, T-Wrench	Wrench	Wrench, Ratchet	Hand, Hammer	Ratchet	T-Wrench 2-Hands	Wrench	Wrench, Ratchet	Hand, Hammer	Power Wrench	
1	1	-	-	-	1	-	-	-	-	-	-	1
3	2	1	1	1	3	1	-	1	-	1	1/4 in. (6mm)	3
6	3	3	2	3	6	2	1	-	1	3	1 in. (25 mm)	6
10	8	5	3	5	10	4	-	2	2	5		10
16	16	9	5	8	16	6	3	3	3	8		16
24	25	13	8	11	23	9	6	4	5	12		24
32	35	17	10	15	30	12	8	6	6	16		32
42	47	23	13	20	39	15	11	8	8	21		42
54	61	29	17	25	50	20	15	10	11	27		54

P Tool Placement

Tool	Index
Hammer	0 (1)
Fingers or Hand	1 (3)(6)
Knife	1 (3)
Scissors	1 (3)
Pliers	1 (3)
Writing Instrument	1
Measuring Device	1
Surface Treating Device	1
Screwdriver	3
Ratchet	3
T-Wrench	3
Fixed End Wrench	3
Allen Wrench	3
Power Wrench	3
Adjustable Wrench	6 (3)

Figure 3: Index values for the Tool Use Action Move in Basic MOST

3. Methodology

As mentioned earlier the purpose of the study is primarily to set the standard time for the operations and to find ways for improvement of the whole process by minimizing the waste and/or automate some parts of operations. The Basic MOST has been applied for standard time and Assembly Line Balancing (ALB) is used for balancing the production line by making the local cycle time of all workstations equal or closer to the Takt time.

Basic MOST follows certain procedures by starting to identify the main tasks in the fabrication line. After that, the tasks are divided into small elements. Video recorder is also needed to capture the process, how workers do the job and how the operators move when doing the task. The next step is to study all the movements of a task element and to define them whether in terms of general move, controlled move or a tool use. Then index values are assigned to each move. The sub-numbers of the moves are then added and multiplied by 10 to find the time in TMU to be taken by the operator. To get the time in seconds, the TMU number is multiplied by 0.036. Therefore, by adding all the times of the elements to get the time in each workstation. If the time in a workstation is not equal compared to other workstations then ALB is adopted to rearrange the activities in all workstation to make the cycle time close to the Takt time. The following formulas are used:

$$\text{Takt time}=\frac{total\ work\ time\ available}{Units\ required} \tag{1}$$

$$\text{Workers required}=\frac{\text{Total operation time required}}{\text{Takt time}} \tag{2}$$

$$\text{Efficiency}=\frac{\text{Total operation time}}{(\text{actual number of workstations})*(\text{cycle time})} \tag{3}$$

$$\text{Line balance rate (LBR)}=\frac{Total\ net\ time\ of\ all\ processes}{(longest\ process\ time)(number\ of\ workers)}(100\%) \tag{4}$$

$$\text{Line balance loss rate (LBLR)} = 100\% - \text{LBR} \tag{5}$$

$$\text{Cycle Time}=\frac{\text{production time available per day}}{\text{units required per day}} \tag{6}$$

4. Case Study

As stated earlier this study is conducted in a local company in Malaysia within a supply chain of automotive parts manufacturers. The assembly line engaged in producing car exhaust system in the company is painstakingly investigated. The aim is to set the standard time and also to improve the productivity and performance of the fabrication line. As indicated in **Figure 4**, there have been five workstations in the assembly line and all the manual tasks are currently done by five workers or operators. A general overview of the assembly line and the procedure in evaluating the standard times has been presented in the following sections:

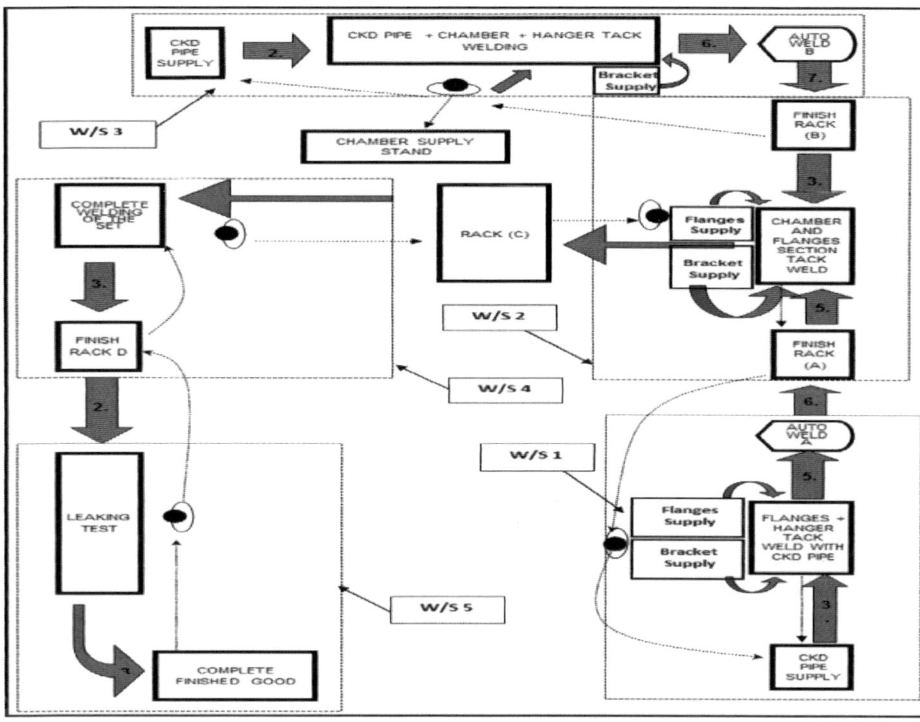

Figure 4: Current layout plan

4.1 Overview of the Undertaken Assembly Line

Five workstations engaged in performing five major operations are
(1) Workstation -1 (W/S-1): Flanges, CKD pipe and Bracket tack weld,
(2) Workstation - 2 (W/S-2): CKD pipe, Chamber and Hanger tack weld,
(3) Workstation - 3 (W/S-3): Flanges, Chamber set and Hanger tack weld,
(4) Workstation - 4 (W/S-4): Complete Welding of Chamber, Pipe and Hanger Bracket and
(5) Workstation - 5 (W/S-5): Leaking Test.

All activities in the workstations are shown in the layout plan in Figure 4 for the whole assembly of car exhaust system. A proper layout plan is meant to improve the line efficiency by arranging the equipment or distributing the workloads according to its function. The whole production line should ideally be designed to eliminate waste in material flows, inventory handling and management. Summary of time for the current operations for car exhaust assembly is presented in Table 3.

Table 3: Current activity time summary for car exhaust assembly

No	Operation Name	Code	On-line				Remark
			TMU	Seconds	With 12% allowance		
					Seconds	Minute	
1	W/S 1	M	1638.9	59.0	66.1	1.10	"M" in column 'code' represent manual assy'
2	W/S 2	M	1555.6	56.0	62.7	1.05	
3	W/S 3	M	1388.9	50.0	56.0	0.93	
4	W/S 4	M	1888.9	68.0	67.2	1.27	
5	W/S 5	M	805.6	29.0	32.5	0.54	
	Total		7277.8	262	293.4	4.90	

Summary of the five workstations' operation times are presented in Table 4. The Table includes the time in time measurement unit (TMUs), seconds and in minutes. An allowance is 12% as decided by the management is included. This allowance is given to the employees for their personal needs, fatigue, etc.

4.2 Process Chart

A process chart is prepared to illustrate the task elements performed in a workstation. A brief description of the elemental process, distance to be covered and the time required in performing the elemental tasks are recorded and presented in a tabular form. Table 4 shows the Process chart for workstation, W/S -1 as per earlier procedure.

Table 4: The process chart for the W/S-1

DISTANCE IN mm	TIME IN SECONDS	CHART SYMBOLS	PROCESS DESCRIPTION
300	6.02	○ ⇨ □ D ▽	Pick up the flanges and clamp on the jig.
300	5.50	○ ⇨ □ D ▽	Pick up the CKD pipe and clamp on the jig.
300	4.20	○ ⇨ □ D ▽	Pick up the bracket and clamp on the jig.
600	7.30	● ⇨ □ D ▽	Take up welder and mask, tack weld flanges (3 point).
20	7.30	● ⇨ □ D ▽	Tack weld hanger bracket and put down the mask and the welder.
200	4.90	● ⇨ □ D ▽	Open the entire clamp and turn the CKD pipe.
250	6.50	● ⇨ □ D ▽	Pick up welder and mask, tack weld other portion of bracket, put down mask and welder
800	5.30	○ ⇨ □ D ▽	Go to auto-weld m/c and pickup previous CKD pipe and place in rack A.
300	12.00	○ ⇨ D ▽	Replace the unfinished CKD pipe, press start & go back & take the flanges
300	20.90	● ⇨ □ D ▽	Press Auto weld on (Simo Factor).
	59.00	5 8 - - -	TOTALS
Value added time = Operation time/Total time = (59.00/78.92)= 74.78% (excluding Offline Job)			

4.3 Takt Time of Car Exhaust Assembly (Present Situation)

Takt time is calculated by dividing the available working time in a day with the daily customer demand for the items. Thus this parameter determines the required ability of a production line to meet the customer orders. To satisfy the demand, the bottleneck station cycle time in the line should not exceed the Takt time. Otherwise it would not be possible to meet the set target of customer order. Thus by evaluating detecting the required production output, necessary resources may be mobilized. The Takt time calculation procedure is shown in Table 5.

Table 5: Takt Time calculation

Total working time	9.0 hours	**Total fabrication time**	4 minutes
Fixed loss	1.5 hours	**Longest Workstation time**	1.27 minutes
Actual working time	8.0 hours	**Required quantity per day**	160 pieces

The formula (1) is used for computing the Takt time is given in following computation:

$$\frac{\text{Actual Working Time}}{\text{Required Quantity}} = \frac{8\,hrs}{160\,units} = 0.05\,\frac{hrs}{pcs} = 3\,\frac{min}{pcs} \quad (1)$$

Takt time represents the ability of the production line to satisfy the demand of the customers. If any workstation time is longer than the Takt time, then it is indication that the production line is not able to satisfy the customer order. On the other hand if each workstation time is less than the Takt time, then this implies that the production line can satisfy the order. However, in this situation the production line is not fully utilized.

By analyzing the current situation for the exhaust production line, it is found that there is no problem in satisfying the customer demand. The Takt time is 4.3 minutes and the workstation time is much less than the takt time. Thus with the current setup, the car exhaust production line has over capacity and operators are to remain idle for considerable time.

4.4 Analysis of current fabrication line capability

The current setup of fabrication line has enough capability to meet the customer daily demand of 160 units. The Takt time in this case is 3 min and the workstations processing times is far less than the Takt time. The longest operation time as observed in workstation 4 (bottleneck) is only 1.27 minutes which should be considered as the cycle time for the line.

Under this situation, the operators are not to be adequately busy and as a result they may be engaged for some other additional activities or a few of them may be laid off.

5. Proposed Modification

Since there is an excess capacity of the fabrication line with the current situation, meeting the demand does not appear to be an issue for the management. However, there should be initiatives for continual improvement to secure competitive advantage. In this respect some improvement strategies are proposed to enhance the performance of the line as well as to improve resource utilization for the company.

When the Takt time is shorter than the bottleneck operation time (cycle time), the assembly line has to be improved by introducing more efficient methods or incorporating additional resources only. Otherwise production target based on customer demand is not possible to be met. But for a situation when takt time is longer than the bottleneck cycle time of the line as prevailing in the car exhaust fabrication line, is a sign of lower productivity of the system. To avoid such situation actions need to be taken either to divert some operators to other activities or lay off them. Alternatively production target should be increased to meet additional demand to be created through proper marketing for the product or product family so that the existing resources can be utilized properly.

5.1 Reducing number of workers maintaining current demand level

Under the current situation there is an excess of workers as can be seen in Figure 5. The assembly line requires less number of operators to assemble for meeting the current demand. In fact, excess workers means extra cost for the company in the form of salaries, wages, bonuses and other privileges to be paid to workers. This cost for the idle time of the operators is a waste. It is necessary to prevent this waste so that the company can benefit from the money they spend and become competitive and sustainable. One of the options to decrease the cost and increase the productivity for the assembly line as well as for the whole company is to reduce the number of operators from the car exhaust production line either by rotating them into other jobs or laying them off.

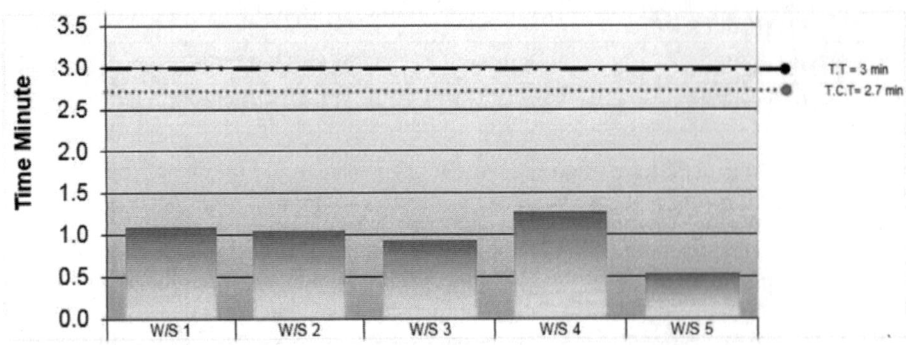

Figure 5: Current line balance loss for car exhaust assembly line

Minimization of the number of workers in the production line by engaging them in any other works can be another option. By this arrangement a company can benefit and increase its productivity. However, first it is necessary to determine the number of workers ideally required for the production line. This evaluation is to be followed for other departments also so that a holistic picture becomes evident for the company in the context of manpower requirement. Thus in case there is lack or shortage of manpower in any department they can make use of the excess workers found in other departments. For the car exhaust assembly line, the calculation procedure of manpower requirement is given below.

We know from Eq. (1), the Takt time $= \frac{\text{Total work time available}}{Units\ required} = \frac{478\ \text{min}}{160} = 3\ \text{min}$

The required number of operators can be found from Eq. (2),

Workers required $= \frac{\text{Total operation time required}}{\text{Takt time}} = \frac{4.9\ \text{min}}{3\ \text{min}} = 1.63 = 2$ workers.

The efficiency of the current situation of the production line can be assessed by using Eq. (3):

Efficiency $= \frac{\text{Total operation time}}{(\text{actual number of workstations}) * (\text{cycle time})} = \frac{4.9\ \text{min}}{(5) * (1.27)} = 77\ \%$

The second option to decrease the number of workers in the car exhaust assembly line is by laying off the excess workers. This option is not usually desirable by the management because of the negative consequences and the cost associated with the firing off labors. Compensation is to be given to the workers laid off while his/her contract not finished yet. Therefore it is often a better option to create opportunities for them to work in another department or production line. This approach appears to be more appropriate for a company to be more conducive and productive.

After properly assessing the required number of workers, the management can decide how many workers to engage in other logistic or production works or to lay off. However, this decision depends on the necessity for the excess workers. In this case the management can make use of 3 workers from the production line and engage them in other works. This proposed solution can help the manufacturer to increase the overall productivity of the company as the 3 excess workers are to work in other best suited operations so that they can be more productive.

For the car exhaust assembly line, number of workers is minimized and workstations are reduced to two from five by merging the first two operations into one and the remaining three operations into another. Thus for the changed situation, number of workstations is reduced to two for which two operators are to be engaged. So with the new arrangement the efficiency of the production line would increase from 77% to 89% as computed by eq. (4).

$$\text{Efficiency} = \frac{\text{Total operation time}}{(\text{actual number of workstations}) * (\text{cycle time})} = \frac{4.9 \text{ min}}{(2) * (2.74)} = 89.4\ \%$$

The changed production line arrangement is depicted in Figure 6 where some of the workstations combined together to be run by only 2 workers instead of 5 workers.

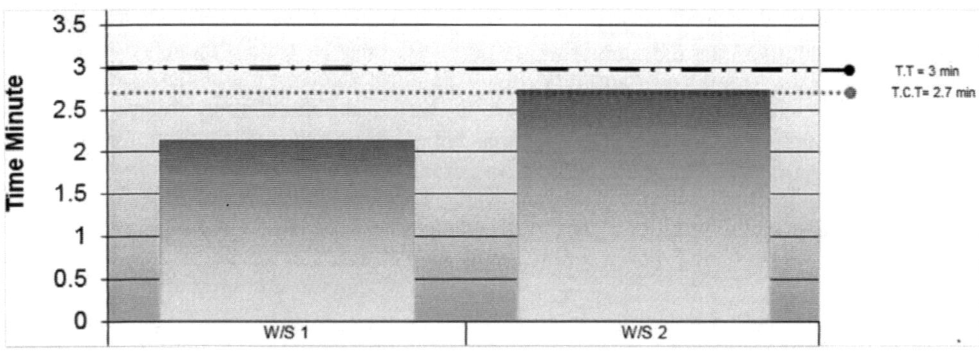

Figure 6: Scenario of car assembly line with two operators in combined activities of workstations

5.1.2 Utilizing excess capacity for increased production

The second solution for the current situation would be to increase production either by adding a similar product along or increasing production of the current car exhaust system. Thus the number of units produced per day can be increased and hence lower the Takt time. When the Takt time is lowered and becomes close to the cycle time this will increase the productivity of the assembly line. As the number of units to be produced increases, the Takt time is to be low until it reaches the optimal level when the Takt time equals to the cycle time. Let us assume a case for doubling the number of units produced i.e. 300 units / day and see the improvement in the production line.

$$\text{The Takt time} = \text{Takt time} = \frac{\text{Total work time available}}{\text{units required}} = \frac{478 \text{ min}}{300 \text{ units}} = 1.60 \text{ minutes}$$

This is to minimize the idle time and increase the efficiency of the assembly line. However, the optimal demand that will make the production line 100% efficient and minimize waste time to zero level. This demand can be found by setting the Takt time equals to the cycle time and this demand is to be 376 units with the takt of 1.30 minutes as shown below.

$$\text{Takt time} = \frac{\text{Total work time available}}{\text{units required}} = \frac{478 \text{ min}}{376 \text{ units}} = 1.30 \text{ minute}$$

However, reduction of Takt time may be possible by facilitating the processes in the workstations with improved methods, tools and equipment.

6. Conclusion

In this endeavor, methods such as MOST and ALB have been used for improving productivity of the assembly line dedicated for production of car exhaust system. The project is initiated to study whole assembly line operations and to identify if there is any scope of operational improvement. Some interesting observations were made with the Takt time being considerably longer than the bottleneck station cycle time. This provided an indication of excess capacity of the assembly line. However, this critical observation would not have been revealed without the investigation. The fact that the line is underutilized with the existence of considerable amount of idle time of the operators opens up an opportunity for revamping the whole production system to improve efficiency. Activities or tasks involved in each of the five workstations were divided into elemental tasks and subsequently their standard times were set by using the MOST. The huge gap between the Takt time and the bottleneck station cycle time is observed which has driven us to search for alternative action plan. It has been found that the same output of 160 set of car exhaust system is possible to be achieved by engaging even less than half of the manpower currently involved. Five workers involved now in five workstations. However, with suggested modification in the assembly line by rearranging the workstations and distribution of activities it would be possible to achieve the same target of production by engaging only two operators. So effectiveness of the MOST and ALB application can be recognized by setting the standard times for the activities as well as balancing of the workstations times of an assembly. A manufacturing company can enhance their resource utilization and achieve operational sustainability by becoming more economic and competitive.

Acknowledgement

This study was conducted under the FRGS project (FRGS14-102-0343) funded by Ministry of Higher Education (MOHE) Malaysia. The authors are grateful to MOHE and RMC, IIUM for their support.

References

[1] Rosnah, M., M.M. Ahmad, S. Sulaiman, and Z. Mohamad. 2003. Increasing competitiveness through advanced manufacturing technologies. *International journal of manufacturing technology and management*, **5**(4), 371-379.

[2] Lapierre, J. 2000. Customer-perceived value in industrial contexts. *Journal of Business & Industrial Marketing*, **15**(2/3), 122-145.

[3] Farahani, R.Z., S. Rezapour, T. Drezner, and S. Fallah. 2014. Competitive supply chain network design: An overview of classifications, models, solution techniques and applications. *Omega*, **45**, 92-118.

[4] Robbins, S., T.A. Judge, B. Millett, and M. Boyle. 2013. Organisational behaviour: Pearson Higher Education AU.

[5] Ilgin, M.A. and S.M. Gupta. 2010. Environmentally conscious manufacturing and product recovery (ECMPRO): a review of the state of the art. *Journal of environmental management*, **91**(3), 563-591.

[6] Dal, V., E. Akçagün, and A. Yilmaz. 2013. Using Lean Manufacturing Techniques to Improve Production Efficiency in the Ready Wear Industry and a Case Study. *Fibres & Textiles in Eastern Europe*, **21**(4), 100.

[7] Chauhan, P., S. Rangrej, K. Samvatsar, and J. Patel. 2015. A Review on Scope, Study & Need of Setup Time Reduction for Conveyor Pulley Manufacturing. *International Journal of Advance Research in Engineering, Science & Technology (IJAREST), ISSN (O)*, 2393-9877.

[8] Pattanaik, L. and B. Sharma. 2009. Implementing lean manufacturing with cellular layout: a case study. *The International Journal of Advanced Manufacturing Technology*, **42**(7-8), 772-779.

[9] Sheth, P.P., V.A. Deshpande, and H.R. Kardani. 2014. Value stream mapping: a case study of automotive industry. *International Journal of Research in Engineering and Technology*, **3**(1), 310-314.
[10] Park, H.S. 2015. A Study on the Applicability of PTS to Establish Standard Time for Agricultural Work of Korea. *대한인간공학회지*, **34**(2), 145-149.
[11] Razmi, J. and M. Shakhs-Niyaee. 2008. Developing a specific predetermined time study approach: an empirical study in a car industry. *Production Planning and Control*, **19**(5), 454-460.
[12] Whitmore, D.A. 1987. Work measurement: Butterworth-Heinemann.
[13] Tuan, S.T., A. Karim, H.E. Kays, A. Amin, and M. Hasan. 2014. Improvement of Workflow and Productivity through Application of Maynard Operation Sequence Technique (MOST).
[14] Yadav, T.K. 2013. Measurement Time Method for Engine Assembly Line with Help of Maynard Operating Sequencing Technique (MOST). *International Journal Of Innovations In Engineering And Technology (IJIET)*, **2**(2).
[15] Belokar, R., Y. Dhull, S. Nain, and S. Nain. 2012. Optimization of Time by Elimination of Unproductive Activities through 'MOST'. *International Journal of Innovative Technology and Exploring Engineering (IJITEE)*, **1**(1), 77-80.
[16] Briggs, J., Y. Jin, M. Price, and R. Burke. 2013. *Mixed-model production system design for aircraft assembly.* Paper presented at the Proceedings of the 11th International Conference on Manufacturing Research (ICMR2013), Cranfield University, UK, 19th.
[17] Vikram, K., D. Shivappa, and J. Sangamesha. 2012. *Establishing Time Standards for Fixing Body Side Panel to the Chassis in Assembly Line Using MOST.* Paper presented at the Proceedings of the National Conference on Trends and Advances in Mechanical Engineering, YMCA University of Science & Technology.
[18] Konz, S. and S. Johnson. 2004. Work design: occupational ergonomics: Holcomb Hathaway, Publishers.
[19] Helander, M. 2005. A guide to human factors and ergonomics: Crc Press.
[20] Sivaloganathan, S. and R. Yanis. 2015. Design for Method Study—Work Measurement: Do We Need It? *ICoRD'15–Research into Design Across Boundaries Volume 2* (pp. 313-326): Springer.
[21] Jamil, M.G., A. Saxena, and M. Mr Vivek Agnihotri. 2013. Optimization of Productivity by Work Force Management through Ergonomics and Standardization of Process Activities using MOST Analysis-A Case Study. *Global Journal of Researches In Engineering*, **13**(6).
[22] Thakre, A.R., D.A. Jolhe, and A.C. Gawande. 2009. *Minimization of Engine Assembly Time by Elimination of Unproductive Activities through'MOST'.* Paper presented at the Emerging Trends in Engineering and Technology (ICETET), 2009 2nd International Conference on.
[23] Zandin, K.B. 2002. MOST work measurement systems: CRC press.
[24] Gupta, M.P.K. and M.S.S. Chandrawat. 2012. To improve work force productivity in a medium size manufacturing enterprise by MOST Technique. *IOSR Journal of Engineering (IOSRJEN)*, **2**(10), 08-15.
[25] Bures, M. and P. Pivodová. 2013. *Comparison of the predetermined time systems MTM-1 and BasicMOST in assembly production.* Paper presented at the Industrial Engineering and Engineering Management (IEEM), 2013 IEEE International Conference on.

IOP Conf. Series: Materials Science and Engineering **184** (2017) 012023 doi:10.1088/1757-899X/184/1/012023

Modelling of an advanced charging system for electric vehicles

Abdul Hassan Jaafar[1], Ataur Rahman[1], A K M Mohiuddin[1] and Mahbubur Rashid[2]

[1] Department of Mechanical Engineering, International Islamic University Malaysia, 50728 KL, Malaysia
[2] Department of Mechatronics Engineering, International Islamic University Malaysia, 50728 KL, Malaysia

Abstract - Climate Change is recognized as one of the greatest environmental problem facing the World today and it has long been appreciated by governments that reducing the impact of the internal combustion (IC) engine powered motor vehicle has an important part to play in addressing this threat. In Malaysia, IC engine powered motor vehicle accounts almost 90% of the national greenhouse gas (GHG) emissions. The need to reduce the emission is paramount, as Malaysia has pledged to reduce 40% of CO_2 intensity by 2020 from 2005 level by 25% of improvement in average fuel consumption. . The introduction of electric vehicles (EVs) is one of the initiatives. However in terms of percentage, the electric vehicles have not been commonly used by people nowadays and one of the reasons is lack in charging infrastructure especially when cars are on the road. The aim of this study is to simulate and model an advanced charging system for the charging infrastructure of EVs /HEVs all over the nation with slow charging mode with charging current 25 A, medium charging mode with charging current 50 A and fast charging mode with charging current 100 A. The slow charging mode is proposed for residence, medium charging mode for office parking lots, and fast charging mode is called fast charging track for charging station on road. With three modes charger topology, consumers could choose a suitable mode for their car based on their need. The simulation and experiment of advanced charging system has been conducted on a scale down battery pack of nominal voltage of 3.75 V and capacity of 1020 mAh. Result shows that the battery could be charging less than 1 hour with fast charging mode. However, due to limitation of Tenaga Nasional Berhad (TNB) power grid, the maximum 50 A current is considered to be the optimized passive mode for the EV's battery charging system. The developed advanced charger prototype performance has been compared with the simulation result and conventional charger performance, the maximum variation has been found 15%, this closed agreement between the advanced charger prototype, simulation model and conventional charger validate the prototype model. Furthermore, based on the result presented in this report, the battery to be charged up to 85% of its rated capacity by constant current mode only rather than continue with constant voltage, which could shorten the battery charging time by 16% and prolong the battery life by 10%.

1. INTRODUCTION

Environmental issues and increasing prices of non-renewable fossil fuels had triggered researchers to find an alternative for internal combustion vehicles. Although exhaust from the vehicles have received increasing attention as a source of air pollution around the globe since 1970, more than 90% of vehicles on the road nowadays still using gasoline and diesel fuels as their source of energy [1]. With improvement in the battery of electric vehicles in terms of size, price, internal resistance, and energy density, it give a hope that the electric vehicles will be widely used by the people. Electric vehicles have advantages over conventional internal combustion vehicles in terms of operating cost, less emission, and energy efficiency. According to Nissan [2], their popular car Nissan Leaf 30 kWh have a travel range up to 155 miles per full charge based on the New European Driving Cycle (NEDC). According to the US Department of Transportation, about 78% of the population drives average 40 miles or less per day [3]. Thus, with a single charging, most of the consumers could get to and from the workplace easily without extra charging needed.

Nissan leaf [2] and i-MiEV [4] electric cars are equipped with 3.3 kW or 6.6 kW on-board charger which will take about 4 to 7 hours for fully charging the batteries at 240 V voltage supply. According to SAE hybrid committee [5], the charging power for Level 2 on-board charger with 240 V supply is allowed up to 19.2 kW. The higher the charging power means the shorter the charging time.The main goal of this study is to simulate and model an advanced charger for electric vehicles whereby the consumers could choose charging power level according to their needs. The general system design of the proposed charging system is presented in figure 1. Through a user interface panel, consumers are able to set the desired charging current accordingly. The integrated management system, which is a computer based control unit, will then set a reference current to the on-board charger. The on-board charger consists of a dc-to-dc converter circuit will be able to control the voltage and current output depends on the pulse widthe modulation (PWM) input signal. Due to the error, i.e. the different between the reference and the actual current level, the pulse width modulator will generate and feed a proper switching signal value to the dc-to-dc converter to provide a proper output current. At the same time, the management system also monitor the condition of the battery in terms of their voltage, current, and temperature in order to control the charging power rate. Besides, there is a switching mechanism to swap between the battery charging mode and the discharging mode.

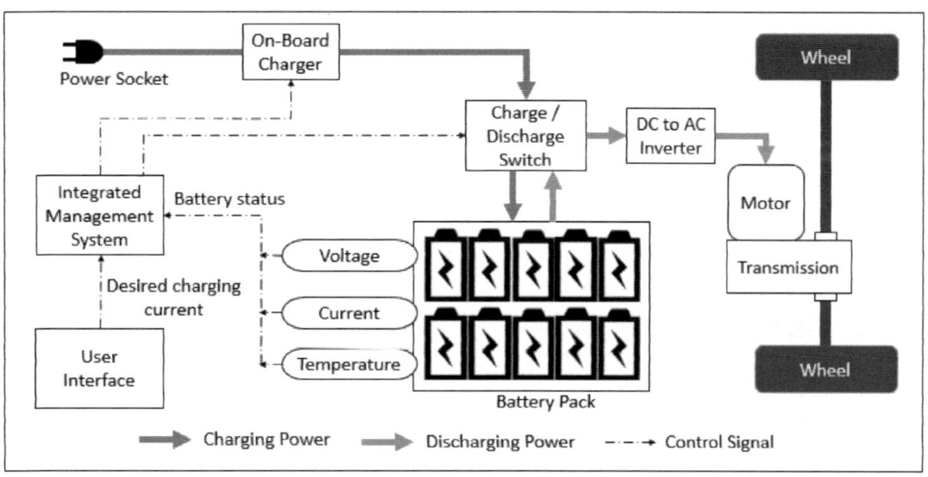

Figure 1. The proposed charging system design

2. METHODLOGY

The propulsion power of an electric car has been estimated based on the vehicle weight, size, road gradient, and speed. An electric car weight of 9.81 kN , wheel base of 2.2 m, CGx location from front wheel of 1.2 m and rear wheel of 1.0 m, CG height of 0.55 m, projected frontal area of 2.86 m^2 is considered to estimate the vehicle propulsion power for the different road gradients and speed. The estimated propulsion power was used to select the traction motor and the battery pack. Vehicle traction torque and propulsion power are estimated by using the equation by Wong [6] and Rahman et al. [7] with taking account the vehicle weight, wheelbase, wheel radius, CG location, rolling motion resistance, aerodynamic resistance, slope angle, air density, drag coefficient and travelling speed.

2.1 *Propulsion power requirement and battery pack design*

An electric car of weight 9.81 kN is used to estimate the vehicle power requirement to maitain the vehicle motion in different road conditions. The traction torque required by a vehicle to propel forward at a desired velocity can be described as follows:

$$T_t = F_t \times r_{wheel} \tag{1}$$

where T_t is traction torque in Nm, F_t is traction force in N and r_{wheel} is radius of the drive wheel in m. The traction force, F_t can be calculated as follows:

2.1.1 *Traction force for initial condition (during stand still to start moving).*

$$F_{t(initial)} = m_a g\left(\mu \frac{l_{fr} + f_r h / L_{wb}}{1 + \mu h / L_{wb}} + \sin\theta_g\right) \tag{2}$$

where m_a is the mass of the vehicle in kg, g is the gravitational acceleration constant equal to 9.81 m/s^2, μ the adhesion coefficient of the road, l_{fr} distance from the front wheel to CG in m, f_r the coefficient of rolling motion resistance, h is the height of CG in m, L_{wb} the wheel base in m, θ_g the slope angle with respect to the horizon in degree.

2.1.2 *Traction force for moving condition (dynamic component)*

The major external forces acting on a vehicle are shown in figure 2 which include the aerodynamic resistance R_a, rolling resistance R_{roll}, grade resistance R_g ($W \sin\theta_g$), and tractive effort F_t. So that the traction force at moving condition could be found as:

$$F_{t(dynamic)} = \frac{Wa}{g} + R_{roll} + R_g + R_a \tag{3}$$

$$= \frac{Wa}{g} + f_r m_a g + m_a g \sin\theta_g + \frac{1}{2}\rho_{air} C_D A_f v^2 \tag{4}$$

where ρ_{air} is the air density in kg/m^3, C_D is the coefficient of aerodynamic resistance, A_f is the frontal area of the vehicle in m^2 and v is the travelling speed of the vehicle relative to the wind in m/s.

Vehicle traction on the maximum grade is due to rolling resistance and grade resistance. The aerodynamic resistance, R_a is considered zero as the vehicle moves on the maximum grades very slow ($v \cong 0$). Besides, the acceleration, a is also considered zero as the $v \cong 0$. So that the traction force at the maximum garde is

$$F_t = R_{roll} + R_a \tag{5}$$

$$= f_r W + mg \sin\theta_g \tag{6}$$

The maximum tractive force that the tire-ground contact can support is described by the product of the normal load and coefficient of road adhesion, μ. Any small amount over this maximum tractive force will cause the tire to spin on the ground.

$$F_{t,max} = \mu W \tag{7}$$

The propulsion power required to propel the vehicle at maximum speed, $P_{prop,max}$ in Watt or W is the product of the traction force and the maximum travelling speed which could be found from Equation 4, and based on maximum travelling speed (v_{max}):

$$P_{prop} = F_{t(dynamic)} \times V_{x,max} \tag{8}$$

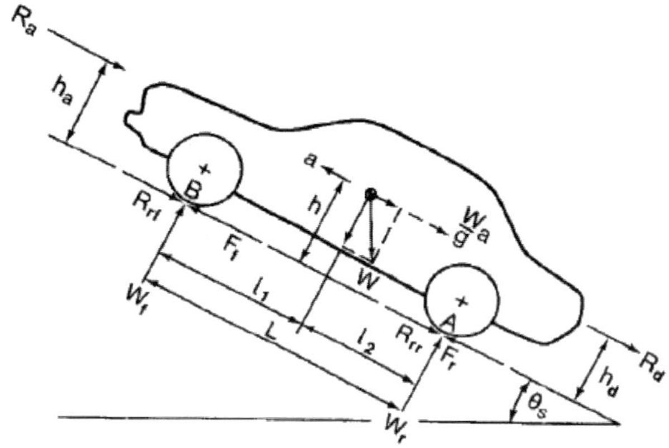

Figure 2. Forces acted on a car

By considering the car traction at the starting of 0.5 km/h, urban speed of 40 km/h, and maximum speed of 90 km/h (25 m/s), the traction force and propulsion power requirement are calculated and summarized in Table 1. Besides, it were considered that the adhesion coefficient value μ is 0.4, coefficient of rolling motion resistance, f_r is 0.02, Drag coefficient, C_D is 0.24, Frontal area, A_f of 2.86 m^2, and air density ρ_{air} equals to 1.2 kg/m^3.

The propulsion power for the vehicle can be estimated as following equation, where, P_{req} is the vehicle required propulsion power in kW, F_t is the traction force in N and V is the vehicle speed in m/s. Power required for the vehicle can be predicted as,

(i) Initial start-ups mode

$$P_{req(initial)} = F_{t(initial)} * \frac{V}{1000} \tag{9}$$

By using, $F_{t(initial)}$ is 2.29 kN and V is 2.78 m/s (10 km/h), the vehicle required propulsion power $P_{req(initial)}$ for the vehicle initial start-ups can be computed as 6.37 kN.

(ii) Dynamics mode

$$P_{req(dynamics)} = F_{t(dynamics)} * \frac{V_{crusing}}{1000} \tag{10}$$

By using, $F_{t(dynamics)}$ is 0.136 kN and V is 11.11 m/s (40 km/h), the vehicle required propulsion power, $P_{req\ (dynamics)}$ for the vehicle dynamics mode can be computed as 1.5 kW.

Based on table 1, a traction motor with rated power more than the calculated maximum power would be suitable to propel the car. Electrical motor with terminal voltage V_m of 100V and peak power of 18kW are considered in the design of the power supply. The battery size is estimated as 18 kWh to travel 160 km with speed of 100 km/h. It is noted that if the car is equipped with 20 kWh battery, it would get the traction power at initial start-ups, propel on 10% grade, and cruising speed (overdrive).

Table 1. Battery size required for 160 km travelling.

Traction Mode	Speed	Ft (kN)	Pprop (kW)	Ib (A)	Battery size (kWh)	Consideration
Initial start-ups (Reduction)	10 km/h	6.37	17.7	177	15	
Urban mode @ 10% gradiant	40 km/h	1.5	13.5	135	12	Motor rating voltage, V=100 volts
Crusing speed (Overdrive)	90 km/h	0.85	21.25	213	18	

Discharge current drawn from the battery I_b by the motor to meet the power demands of the car in different road conditions has been estimated as follows,

$$I_b = \frac{P_{prop}}{V_m} \tag{11}$$

In order to design the battery pack for the vehicle traction, battery cells with nominal voltage of 3.75V and rated capacity of 43Ah were considered. By connecting 6 cells in a 2-series, 3-parallel (2S3P) formation, 7.5V battery modules with 129Ah capacity are formed. The battery pack that have terminal voltage of 100V and 12.9kWh capacity are made by connecting 13 modules in series. The battery pack capacity is enough to supply the maximum power that required by the vehicle. The series/parallel configuration enables design flexibility to achieve the desired voltage and current ratings with a standard cell size. If higher currents are needed, it could be achieve by connecting two or more cells in parallel. On the other hand, if higher voltages needed two or more cells can be connected in series.

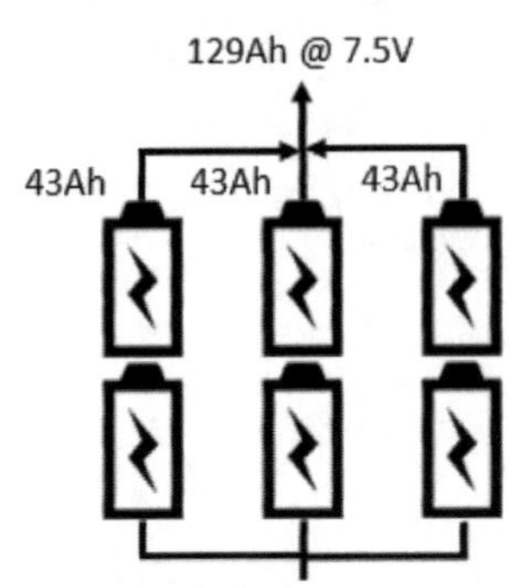

Figure 3. Battery module (2S3P) with 7.5V terminal voltage and 129Ah capacity

2.2 *Charging system simulation design*

The on-board charging system utilize a dc-to-dc step down or buck converter to supply a suitable power level to the battery as shown in figure 4. The current sensor captured the actual current of the buck converter output and fed them back to the controller. The controller work by using the difference, or error, between the actual and reference current in order to feed appropriate signal value to the pulse width modulator (PWM). The PWM signal then control the switching process of the buck converter in order to supply the desired current to the battery. The parameters value of the PID gains are obtain manually by try and error tuning method as discussed in [8, p. 483].

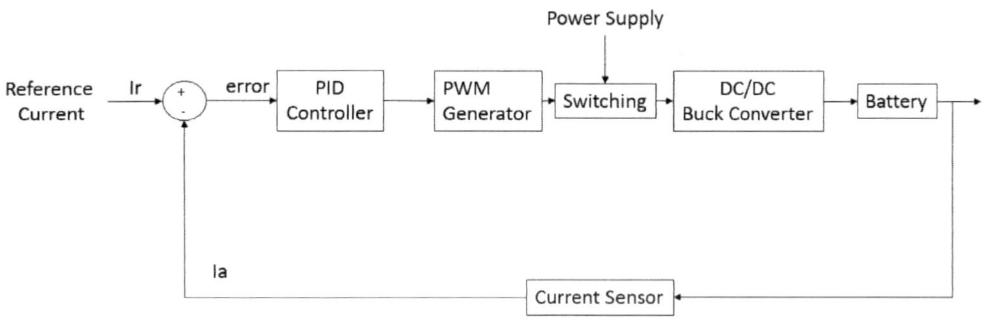

Figure 4. Buck converter (dc-to-dc) with closed-loop control system block diagram

Buck converter also known as step-down dc-to-dc converter consists of dc input voltage source V_S, controlled switch S, diode D, filter inductor L, filter capacitor C, and load resistance R as shown in Figure 5. Mathematical calculation on how to find suitable values for inductor and capacitor are shown here based on [9], [10] and [11].

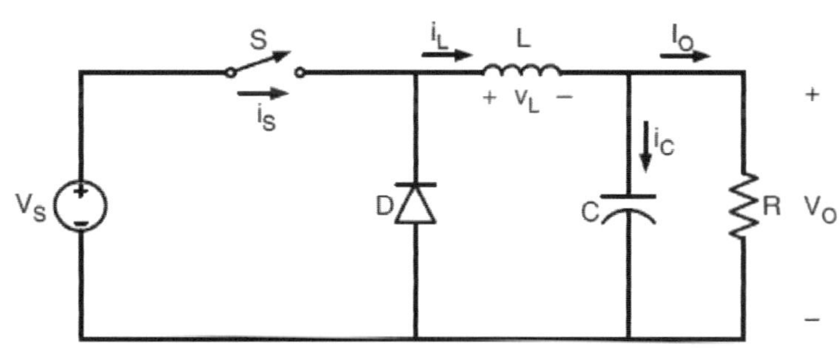

Figure 5. Basic diagram of a buck converter circuit

The duty cycle denotes the conducting time, t_1 of the switch as a proportion of the cycle period of T, where T = 1/f and f is switching frequency. The duty cycle, D is calculated as:

$$D = \frac{t_1}{T} \tag{12}$$

In a buck converter, the output current is the average value of inductor current I_L, and if the losses are neglected then the output current is estimated as below where it is depend on the duty cycle value [12]

$$I_{OUT} = \frac{I_{IN}}{D} \tag{13}$$

The minimum value for inductor L and capacitor C are found by the following equation:

$$L = \frac{V_{OUT} \times (V_{IN} - V_{OUT})}{\Delta I_L \times f_s \times V_{IN}} \tag{14}$$

$$C = \frac{\Delta I_L}{8 \cdot f_s \cdot \Delta V_{OUT}} \tag{15}$$

Where ΔI_L is the inductor current ripple, f_s is the switching frequency and ΔV_{out} is the output voltage ripple. The value for duty cycle, inductor and capacitor are calculated and summarized in table 2.

Table 2. Estimated values of converter parameters

Parameters	Calculated parameters for real size charger [a]	Calculated parameters for scale down charger [b]
Duty cycle, D (equation 12)	0.022	0.440
Inductor, L (μH) (equation 14)	685	94
Capacitor, C (μF) (equation 15)	30	120

[a] Consideration: V_i = 240 V, V_o = 5 V, f = 25 kHz, ΔI_L = 30%, ΔV_{out} = 10 mV
[b] Consideration: V_i = 240 V, V_o = 100 V, f = 25 kHz, ΔI_L= 30%, ΔV_{out} = 10mV

3. RESULT AND DISCUSSSION

Simualtion result of a real electric vehicles charging process is presented in this section. In order to validate these result, a small size battery with 1020 mAh rated capacity has been used for both the simulation and experimental of the charging process.

3.1 Simulation of charging process for Electric Vehicles

The battery charging process was then simulated by MATLAB Simulink tool. The charging currents are varies in order to observe the charging time for the 12.9 kWh battery of the vehicle accordingly. To standardize the charging process, the state of charge (SOC) of the battery is considered to be zero percent at the beginning of chaging process.

Table 3. Proposed EV battery charging parameters for simulation

Terminal voltage (V)	Battery capacity (Ah)	Initial state of charge (%)	Charging current / C-rate		
			Slow mode	Medium mode	Fast mode
100	129	0	0.1C – 0.2C	0.3C – 0.6C	0.7C – 1.0C

It is proposed that the C-rate for slow mode is from 0.1C to 0.2C, for medium mode is from 0.3C to 0.6C, and for fast mode is from 0.7C to 1.0C. The battery parameters that are considered for the charging system simulation in MATLAB Simulink are summarized in table 3. In the simulation, the constant current (CC) mode without constant voltage (CV) mode is considered for the battery charging process until the state-of-charge (SOC) reaches at 85%. There are two reasons for considering CC mode without CV during the charging process; extend the battery life span and shorten the charging time. Besides, the time taken to charge the battery to 100% of SOC is compared between simulated and calculated time. For calculated charging time, the following equation is considered:

$$t_{charging} = \frac{Q}{I_{charging}} \tag{16}$$

Where Q is the battery capacity in Ah and $I_{charging}$ is the charging current. For the simulated charging time, the data are extracted from the following simulation graphs.

In slow mode charging process, the charger sets the charging current to 0.1C and 0.2C which is equal to about 15 A and 30 A. The current level was maintains constantly as specified until the battery voltage measures 100 V and once the charger measures 85% state of charge, the charger then terminates the process. The result of the slow mode charging simulation is shown in figure 6.

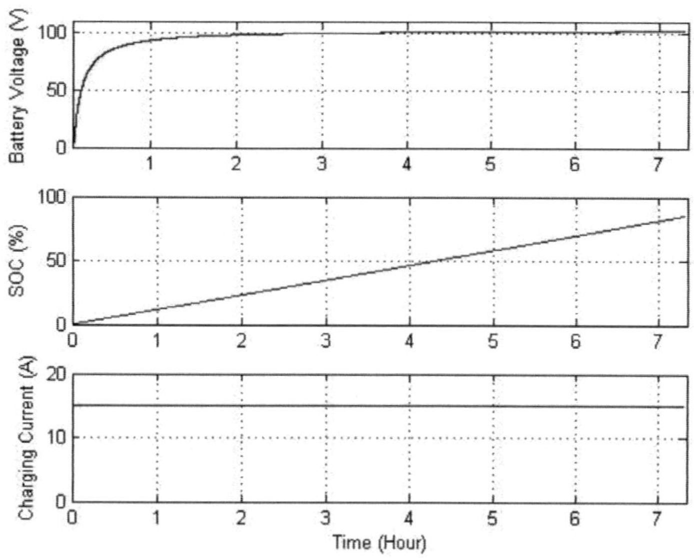

(a)

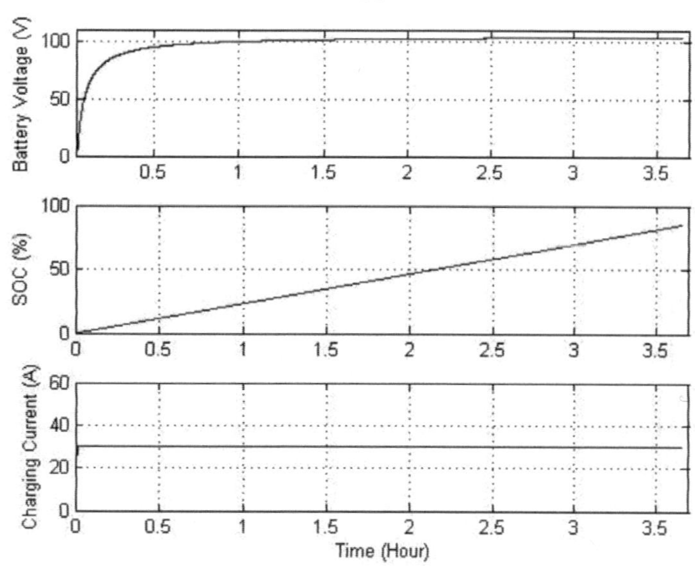

(b)

Figure 6. Simulation result for EV charging with slow mode for 85% of SOC, the charging current (a) 15 A and (b) 30A

The advantage of this mode is that it offers very less stress on the battery and this could prolong the battery life. Figure 6 show that the charging system needs about 7.31 hours for charging current of 15A abd 3.65 hours for charging current of 30A to get 85% SOC of 129 Ah capacity battery.

Medium mode charging was specified to have charging current from 0.3C to 0.6C and in this study they were set at 40 A, 50 A, 65 A, and 77 A. The current level was maintains constantly as specified until the charge voltage measures 100 V and the process was terminated when the state of charge reach 85%. The result of the medium mode charging simulation is shown in figure 7. The advantage of this mode is that it offers an intermediate trade-off between charge speed and stress on the battery.

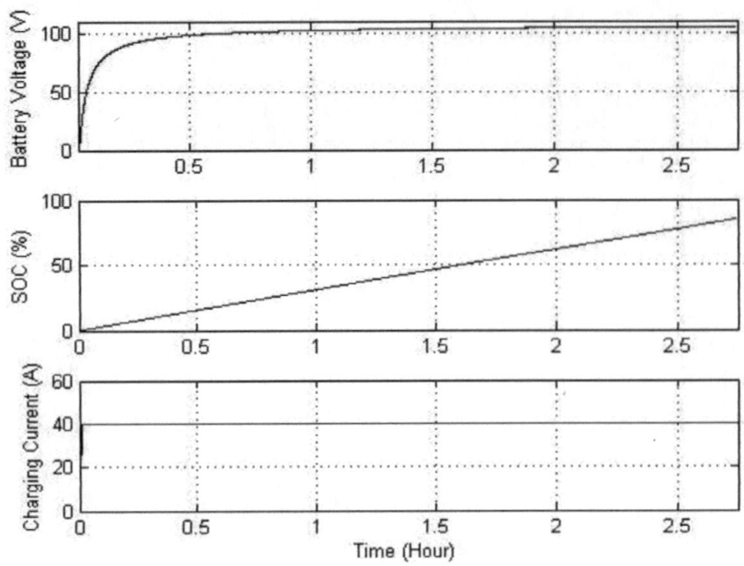

(a)

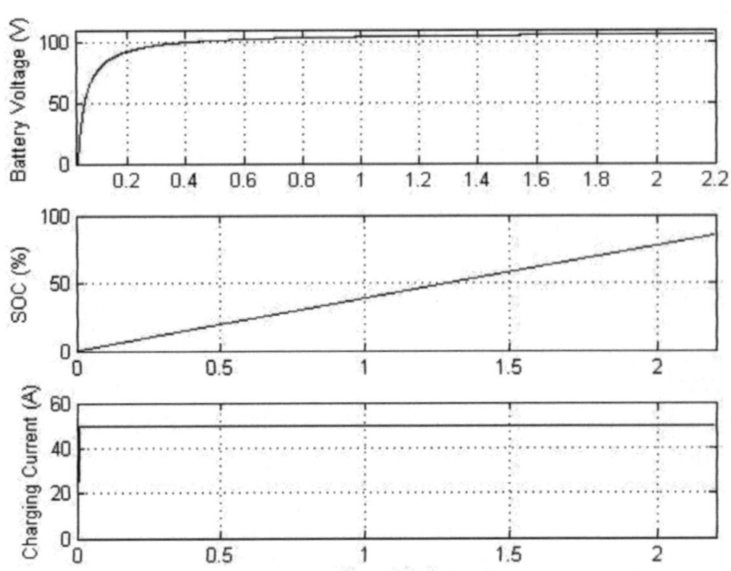

(b)

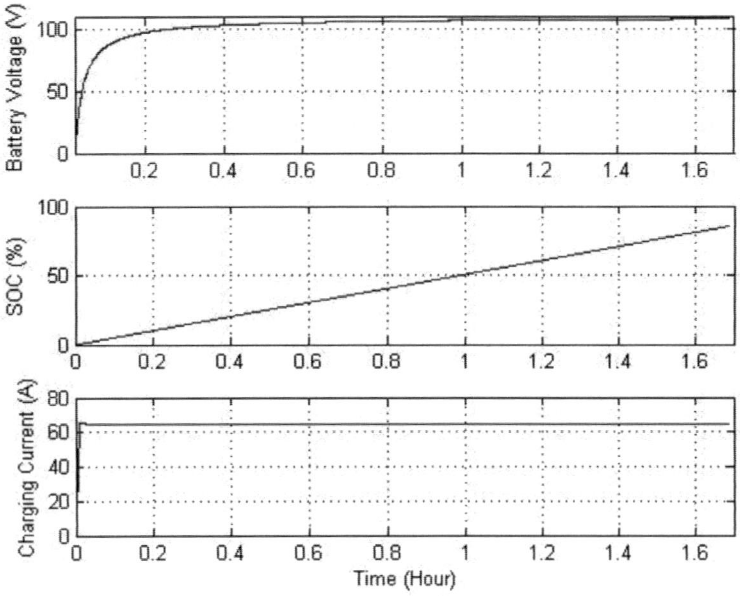

(c)

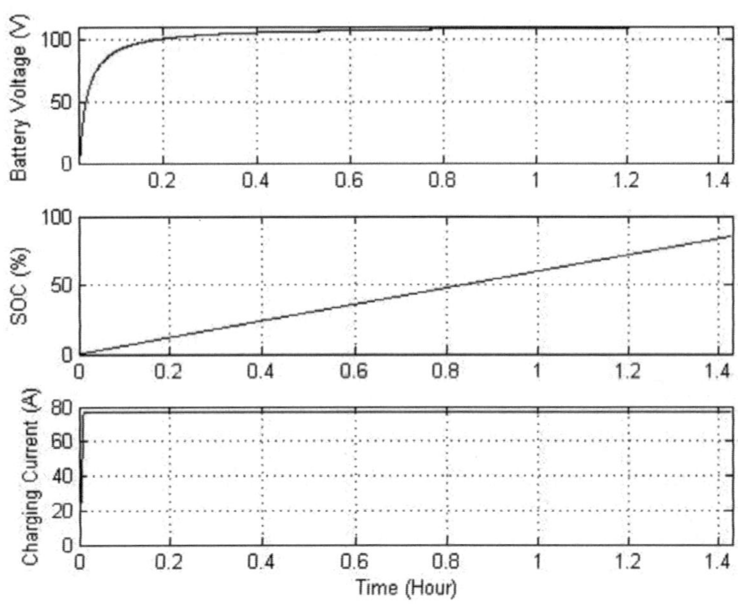

(d)

Figure 7. Simulation result for EV battery charging with medium mode for 85% of SOC, the charging current (a) 40A, (b) 50A, (c) 65A and (d) 77A

Figure 7 show that the charging system needs about 2.5 hours for charging current of 40A, 2.0 hours for charging current 50A, 1.6 hours for charging current 65A and 1.3 hours for charging current 77A.

In this study, the fast mode is considered to have the charging rate of 0.7C to 1.0C. The charger current was maintained constant at 90 A, 100 A, and 120 A and the charging process will be terminated when the state of charge reach 85%. The advantage of this charge mode is speed where the charging time less than one hour could be achieved and results in less stress on the battery. The following graph shows the result of the fast mode charging simulation.

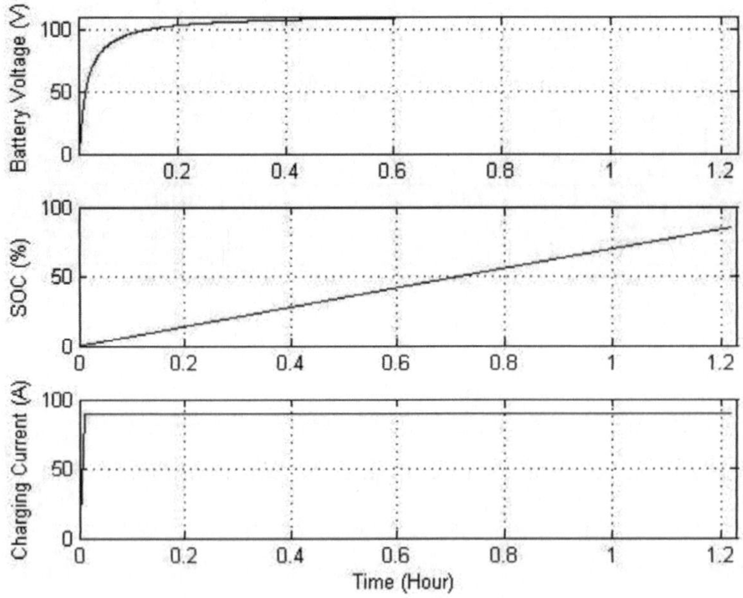

(a)

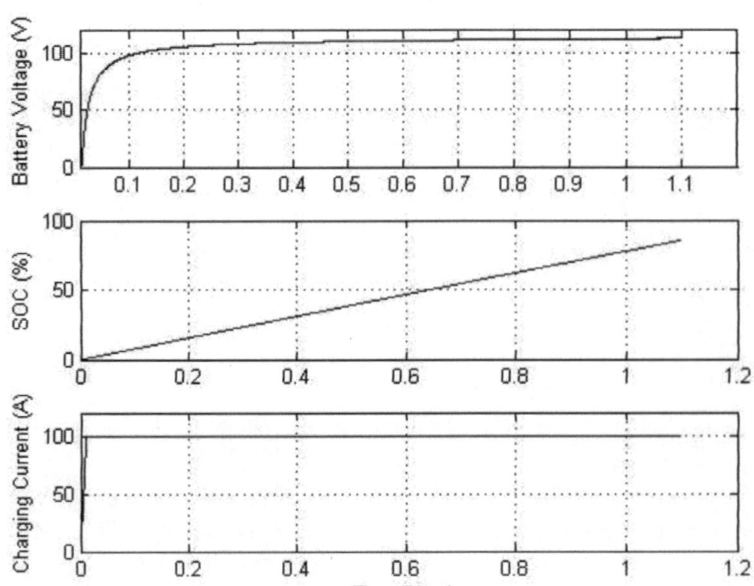

(b)

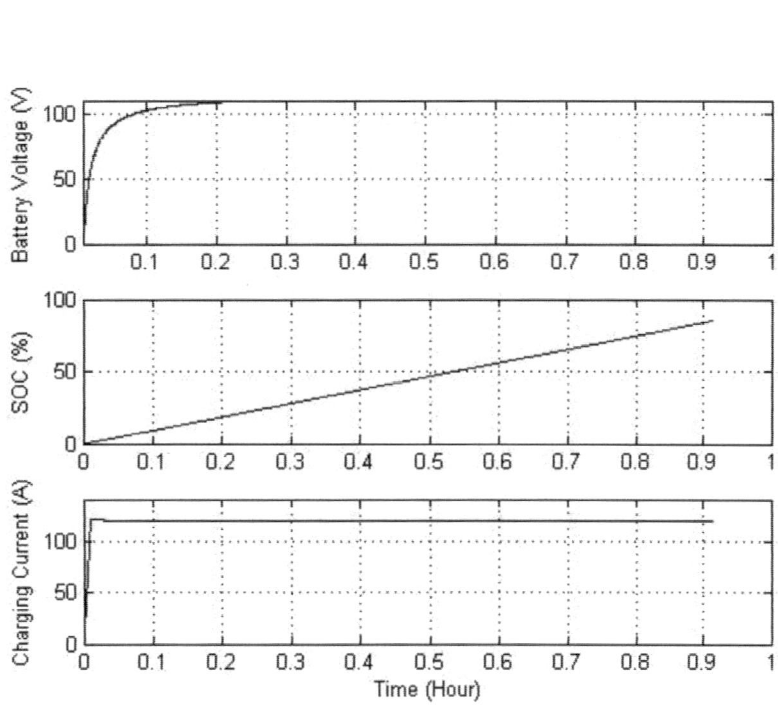

(c)

Figure 8. Simulation result for EV charging with fast mode for 85% SOC with the charging current (a) 90A, (b) 100 A and (c) 120 A

For the fast charging, the charging process needs about 1.22 hours for charging current of 90 A, 1.11 hours for 100 A and 0.9 hour for 120 A.

Table 4 summarized the simulation results for the slow, medium, and fast charging mode in terms of charging current, charging rate, and charging time. It is found that by charging the battery with constant current until it reach 85% SOC, and eliminate the saturation stage with constant voltage, about 13.9% to 16% of charging time was reduced.

Table 4. Result summary for EV charging simulation

Charging Mode	Charging Current (A)	Charging Rate (C)	Simulated Charging Time (Hours)		Charging Time Difference (%) (Between 100% and 85% SOC)
			100% SOC	85% SOC	
Slow	15	1/10	8.60	7.31	15.0
	30	1/5	4.30	3.65	15.1
Medium	40	1/3	3.23	2.71	16.0
	50	2/5	2.58	2.2	14.7
	65	1/2	1.98	1.68	15.2
	77	3/5	1.68	1.43	14.9
Fast	90	7/10	1.43	1.22	14.7
	100	4/5	1.29	1.11	13.9
	120	9/10	1.08	0.92	14.8

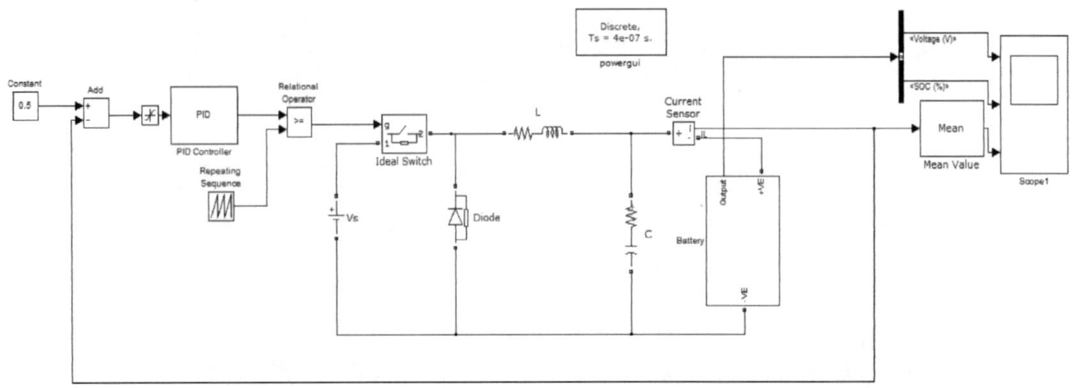

Figure 9. Diagram for charging process simulation in MATLAB Simulink tool

3.2 Simulation and experimental result of scaled-down battery charging process

Due to the availability and accessibility to a real electric vehicles, a small size battery is considered in order to mimic as well as to validate the charging process of a real vehicles. In the simulation a battery with 1020 mAh capacity has been charged with constant current of 500 mA, 700 mA, and 1000 mA until the state of charge reach 85%. The simulation results are shown in the following graphs.

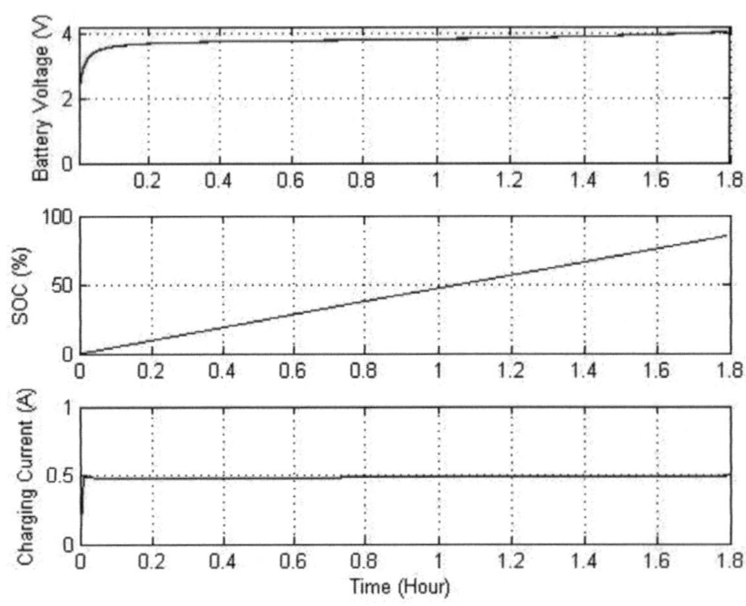

(a)

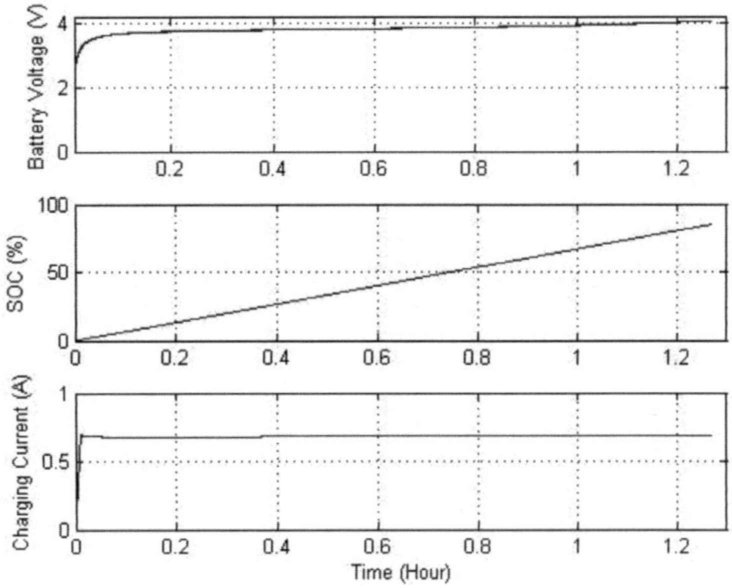

(b)

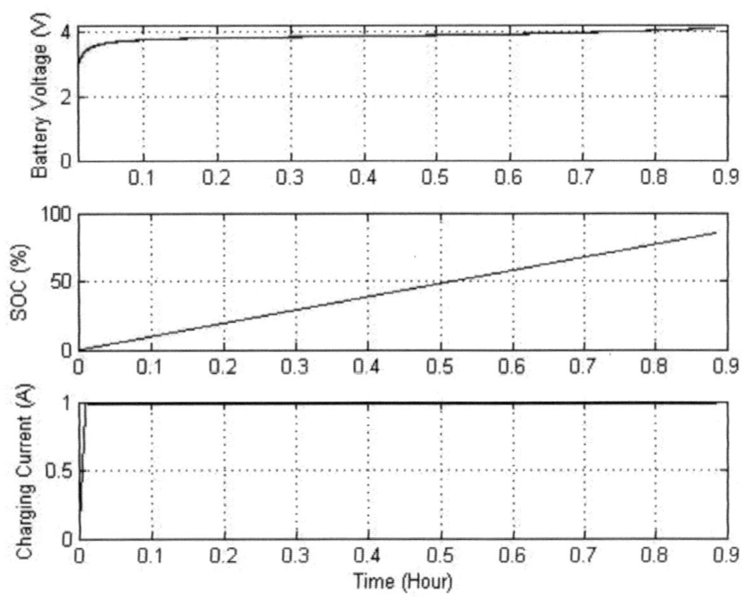

(c)

Figure 10. Simulation result for scale down battery charging with charging current (a) 500mA, (b) 700mA, (c) 1000mA

The simulation results are summarized in Table 5 in terms of charging current, charging rate, and charging time. It is found that charging time for calculated and simulation is almost the same with maximum different of 1.6 % due to fluctuation in simulation charging current. It was found that about 12.25% to 13.73% of the charging time could be reduced by implementing the constant current mode only and charging the battery up to 85% of SOC.

Figure 11. Developed charger prototype

Besides, the experimental has been conducted on a scale down battery pack to identify the charging time and efficiency and to validate the simulation result, which has made based on the developed model (figure 11). The charger was set to maintain the charging current that represent the three modes charger of slow, medium, and fast. It was assumed that the slow mode was 500 mA, medium mode was 700mA, and fast mode was 1000mA. The charger voltage was set to be at the maximum rated voltage of the cell which is 4.2 V. The battery was charged with constant current until the charger voltage and battery voltage were almost saturated. At this point, the charging current started to fall towards zero. The plotted battery voltage throughout the charging process corresponding with charging current and charging time are presented in the following graphs.

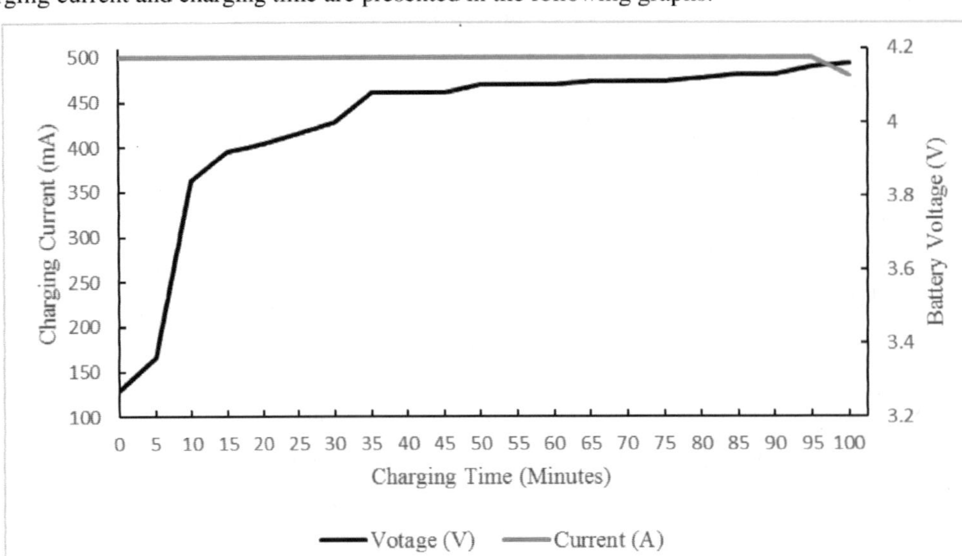

(a)

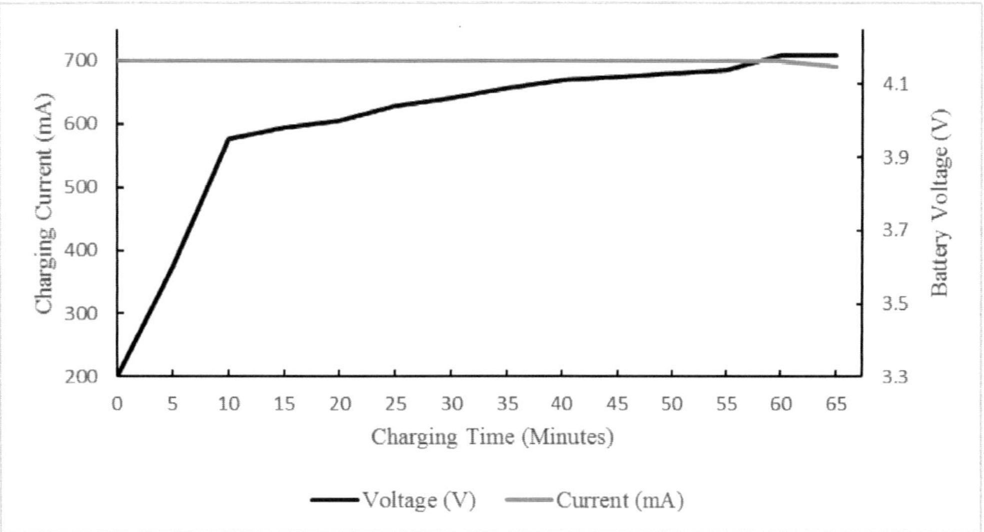

(b)

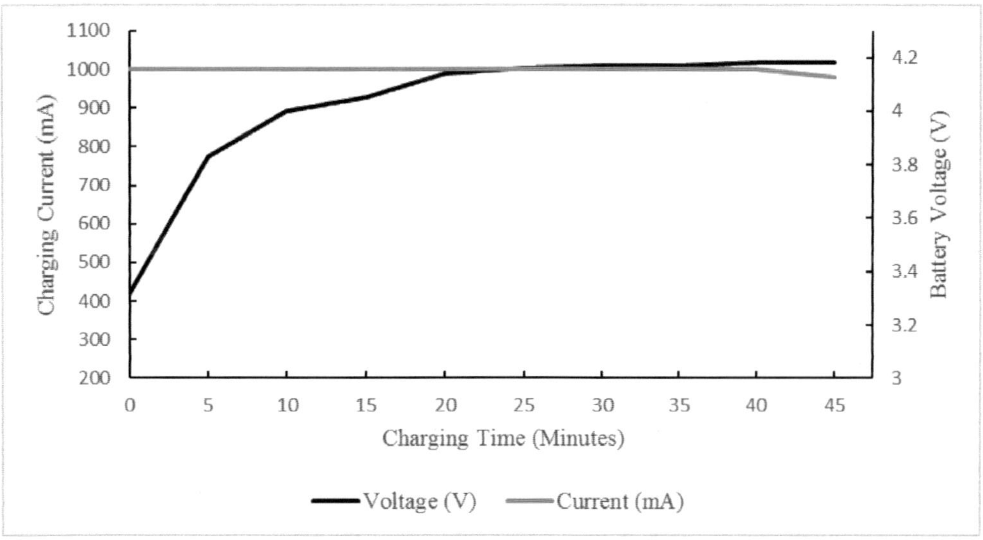

(c)

Figure 12. Developed charger performance for (a) 500 mA, (b) 700 mA, and (c) 1000 mA

Time taken to reach 4.2 V are about 91 minutes for slow mode, 60 minutes for medium modes, and 43 minutes for fast mode. Table 5 shows the comparisonon the charging time between simulation, conventional charger, and developed prototype charger. It was noted that for the simulation the initial battery voltage was considered at 0 V and battery SOC at 0%. While for the experiment, the initial battery was measured initially at around 3.3 V. It was found that the different charging time between the simulated and prototype charger is about 15%. The time taken to charge the battery is faster for the prototype one since the battery initially at 3.3 V and already contains the remaining charge.

Table 5. Charging time comparison table

Charging Mode	Charging time (h)		Charging time difference between simulation and prototype
	Simulation	Prototype	
Slow (500 mA)	1.79	1.52	0.27 hour (15.08%)
Medium (700mA)	1.27	1.079	0.19 hour (14.96%)
Fast (1000mA)	0.88	0.748	0.13 hour (14.77%)

4. CONCLUSION

The following conclusions have been made based on the contents of the paper:

1. The slow charging mode with charging current 15A is able to charge the battery overnight in residence or at office for 6-7 hrs, the medium charging mode with charging current 25-35 A is able to charge battery for 1.5 to 2 hrs, and the fast charging mode with charging current 100A is able to charge the battery in less than 1 hour. However, based on the power grid of Malaysia [13] the maximum 50A current can be used to charge the EV battey with the advanced charger. Therefore, the medium mode with charging current 50 A is the optimized passive mode for the EV's battery charging system.
2. Charging the battery with constant current until it reach 85% SOC, and eliminate the saturation stage which is continue with constant voltage, would reduce the charging time about 13.9% to 16%. This technique would also reduce the stress on the battery and thus prolong the battery life span.
3. The developed advanced charger prototype performance has been investigated and compared with the simulation model, the maximum variation has been found 15%, this closed agreement between the advanced charger prototype, simulation model and conventional charger valide the prototype

REFERENCES

[1] L Jie 2011 *Environmental Effects of Vehicle Exhausts, Global and Local Effects – A Comparison between Gasoline and Diesel* (Halmstad University)
[2] Nissan 2016 *Nissan Leaf* [Online]. Available: http://www.nissan.co.uk/GB/en/vehicle/electric-vehicles/leaf/main-features/main-features.html. [Accessed: 14-Feb-2016]
[3] N S Pearre, W Kempton, R L Guensler, and V V Elango 2011 Electric vehicles: How much range is required for a day's driving? *Transp. Res. Part C Emerg. Technol.* vol 19 n. 6 pp 1171–1184
[4] Mitsubishi 2016 *Mitsubishi Innovative Electric Car (MiEV)* [Online]. Available: http://www.mitsubishicars.com/imiev/specifications. [Accessed: 14-Feb-2016]
[5] SAE Hybrid Committee 2011 *SAE Charging Configurations and Ratings Terminology* vol 1
[6] J Y Wong 2001 *Theory of Ground Vehicles* 3rd ed vol 1 (Canada: John Wiley & Sons)
[7] A Rahman, A Hossain, A A H M Zahirul, and M Rashid 2012 Fuzzy knowledge-based model for prediction of traction force of an electric golf car *J. Terramechanics* vol. 49 no 1 pp 13–25
[8] R C Dorf and R H Bishop 2011 *Modern control systems* (Pearson) 12th ed.
[9] M H Rashid 2011 *Power Electronics Handbook: Devices, Circuits, and Application* (Florida: Elsevier Inc.) 3rd ed
[10] O Kilgenstein 1994 *Switch-Mode Power Supplies In Practice* (John Wiley and Son Ltd)
[11] Texas Instruments 2015 *Basic Calculation of a Buck Converter's Power Stage* (Texas)
[12] L M S Diogo and J M F D Costa 1997 A Low Cost Fuzzy Logic Controller for DC-DC Switching Converters *ISIE*
[13] Tenaga Nasional Berhad 2015 *TNB Electricity System* [Online]. Available: http://www.tnb.com.my. [Accessed: 16-Jun-2015].

Design Optimization of a Hybrid Electric Vehicle Powertrain

Firdause Mangun, Moumen Idres and Kassim Abdullah

Department of Mechanical Engineering, Kulliyyah of Engineering, International Islamic University Malaysia, P.O. Box 10, 50728 Kuala Lumpur, Malaysia

midres@iium.edu.my

Abstract. This paper presents an optimization work on hybrid electric vehicle (HEV) powertrain using Genetic Algorithm (GA) method. It focused on optimization of the parameters of powertrain components including supercapacitors to obtain maximum fuel economy. Vehicle modelling is based on Quasi-Static-Simulation (QSS) backward-facing approach. A combined city (FTP-75)-highway (HWFET) drive cycle is utilized for the design process. Seeking global optimum solution, GA was executed with different initial settings to obtain sets of optimal parameters. Starting from a benchmark HEV, optimization results in a smaller engine (2 *l* instead of 3 *l*) and a larger battery (15.66 kWh instead of 2.01 kWh). This leads to a reduction of 38.3% in fuel consumption and 30.5% in equivalent fuel consumption. Optimized parameters are also compared with actual values for HEV in the market.

1. Introduction

Over recent years, many researchers are working to improve HEV technology especially regarding fuel economy. Among these improvements are downsizing of internal combustion engine, implementation of energy recovery system, addition of supercapacitors, implementation of optimal energy management strategy, and various parameter optimizations.

Design optimization of powertrain has significant impact on fuel economy. The main components of HEV powertrain are internal combustion engine, electric motor-generator, battery pack, and supercapacitors. A proper sizing of these components will result in better fuel economy and ensures maximum benefit of supercapacitor addition. This will lead to higher fuel economy, improvement of vehicle performance, and increases of battery life.

In order to achieve such proper sizing, optimization offers a reliable solution. It is capable of determining a set of best values to meet design objective. In this study, Genetic Algorithms (GA) method was utilized. To achieve global optimization, different settings such as; different objectives and different GA options are being considered.

Ceforolini [1] conducted dynamic programming (DP) optimization for determining the best topology for HEV. Many topologies were considered in order to compute its best fuel economy achievable. Then, overall fuel consumption of different topologies were compared to find best topology. Ceforolini [1] concluded that parallel HEV shows the lowest fuel consumption. Thus, this work is based on parallel HEV due to its fuel economy benefits. This probably resulted from smaller component sizes, and it also requires only one electric motor-generator (EMG).

Multidisciplinary optimization of a series HEV was considered by Fan [2]. Optimization includes both components sizing and power management logic. The objective is to find minimal amount paid by consumers for fuel, electricity and battery. The design variables are number of battery banks, state of charge (SOC) threshold and engine torque and its speed. Optimizations were carried out using Genetic Algorithm (GA), Simulated Annealing (SA), Pattern Search and Nelder-Mead method. GA method was able to find optimal solution for this vehicle design. It shows improvement in fuel economy by about 4.3%, and demonstrated lower computational time than SA method.

Zhang and Mi [3] performed optimization for parallel HEV using DIRECT, SA, GA, and PSO method. Based on his results, PSO ranked at the bottom (least improvement in fuel economy), and SA ranked at the top (most improvement in fuel economy; 15.01%). GA optimization shows improvement about 7.12%. Although GA method is not as good as SA, it was mentioned earlier that SA requires significantly more time, about as twice as GA. This lower improvement shows by GA could had resulted from the fact that this optimization method uses random candidates to form initial population, this population that will yield final optimal results. Thus, to overcome this issue, different settings were considered to ensure best optimal can be achieved.

2. Methodology

In this section, optimization setup of powertrain design problem will be presented. Vehicle modeling and energy management strategy (EMS) are based on Mangun *et al.* [4]. A parallel HEV topology is utilized as shown in figure 1. Quasi-Static-Simulation (QSS) (Guzella and Amstutz [5]) using backward-facing approach is utilized instead of dynamic approach (forward-facing). The former computes fuel consumption from a known drive cycle. The latter computes fuel consumption from a known driver input. EMS is built using Fuzzy Logic to manage power flow between ICE and EMG, and to manage power flow between battery and supercapacitor.

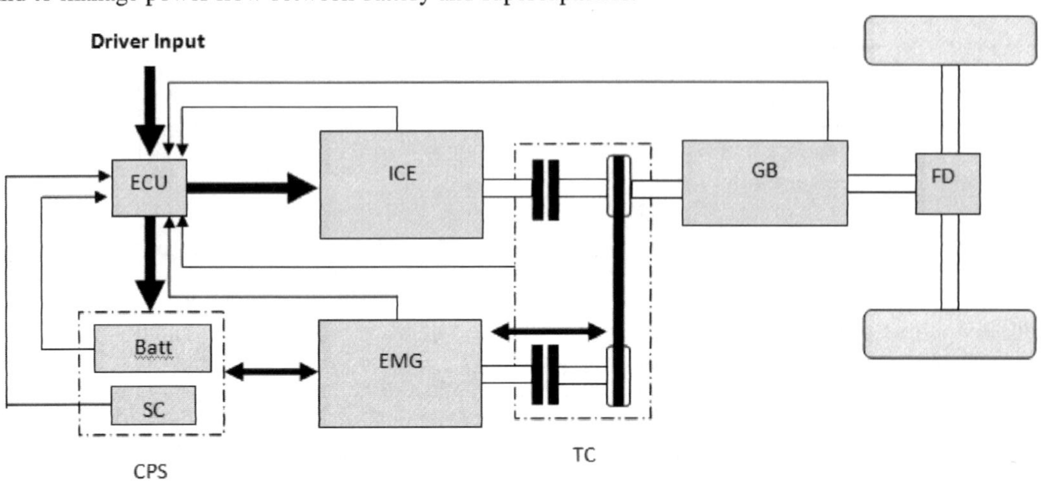

Figure 1. A parallel HEV with supercapacitor (ECU: electronic control unit, Batt: battery, SC: supercapacitor, ICE: internal combustion engine, EMG: electric motor-generator, TC: torque coupler, GB: gearbox, FD: final drive).

Table 1. Specifications of supercapacitor (per module).

Parameter	Symbol	Value	Unit
Rated capacitance	C_{SC}	165	F
Voltage	V_{SC}	48	V
Maximum current	I_{SC}	1900	A
Energy to weight ratio	SC_{ψ}	3.9	Wh/kg

Brief information on Fuzzy Logic-based EMS regarding supercapacitors is mentioned here. Despite presence of battery, supercapacitor usage is prioritized during regenerative braking and acceleration. This means, supercapacitor main energy source is regenerative braking, and it will only be used during acceleration. The specifications of supercapacitor are shown in table 1. The supercapacitor has a low self-discharge rate (as low as 20% per month [6]). Within the timeframe of operation of the vehicle (hours), the supercapacitor can be used for energy storage. Additionally, supercapacitor has adequate specific energy (energy per weight) to supply power during acceleration period which lasts for small time.

2.1. Mass Estimations

In order to ensure changes in vehicle mass are considered during optimization process, several sets of equations were utilized [7, 8, 9 & 10]. Equations for estimating the mass of ICE (M_{ICE}), EMG (M_{EMG}), battery (M_B), and supercapacitor ($M_{SC,total}$) are listed below (mass in kg):

$$M_{ICE} = 1.62\ P_{ICE} + 41.8 \tag{1}$$

$$M_{EMG} = 0.833\ P_{EMG} + 21.6 \tag{2}$$

$$M_B = \frac{B_{capa}}{B_\psi} \tag{3}$$

$$M_{SC,total} = N_{sc}\ M_{SC} \tag{4}$$

Vehicle curb mass (M_{vc}) is estimated using the ratio method:

$$M_{vc} = \left(\frac{M_n}{\gamma_n}\right) + (M_{SC,total})\ ; \qquad \gamma_n = \frac{M_{n,ref}}{M_{vc,ref}} \tag{5}$$

where P_{ICE} is rated power of ICE (kW), P_{EMG} is rated power of EMG, B_{capa} is battery capacity (kWh), B_ψ is battery specific energy (99 Wh/kg for Lithium-ion battery), N_{sc} is number of supercapacitors (in parallel), M_{SC} is mass of one unit of supercapacitor, γ_n is the ratio between reference vehicle nominal mass ($M_{n,ref}$) and vehicle curb mass ($M_{vc,ref}$), and M_n is vehicle nominal mass (mass of ICE, EMG and battery).

2.2. Genetic Algorithms method.

GA is based on natural evolutionary process that is occurring in nature, where it represents reproduction and evolutions of generations after generations, which selects the fittest or the best that survived. A built-in function in Matlab [11] used for GA optimizations. Objective function is

$$F_f(x) = m_f\ or\ m_{f,e} \tag{6}$$

Fuel consumption (m_f) can be directly obtained from simulation results. Equivalent fuel consumption $(m_{f,e})$ can be calculated as:

$$m_{f,e} = m_f + \frac{EE}{33.7 \times 1000 \times 3600 \times 0.264172\ \rho_f} \tag{7}$$

where EE (joule) is electrical energy consumed ($EE = Eb + Esc$, Eb is battery net discharge energy, Esc is supercapacitor net discharge energy), and ρ_f is gasoline density.

For design variables, lower and upper bounds were determined based on data for 5 HEV SUV (Mitsubishi Outlander, Toyota Highlander, Nissan Pathfinder, Volkswagen Tuoareg & Ford Escape), as presented in table 2. For example, ICE capacity range from 2.0 litres (Mitsubishi Outlander) to 3.6 litres (Toyota Highlander). EMG rated power and battery capacity range were determined by the same method. Since there are no data for the number of supercapacitors, its range is selected based on initial simulation runs.

Table 2. Design variables bounds.

Design Variables	Symbol	Lower Bound	Upper Bound	Units
ICE capacity (X_1)	l_{ICE}	2.0 (gasoline)	3.6 (gasoline)	Litres
EMG rated power (X_2)	P_{EMG}	20	130	kW
Battery capacity (X_3)	B_{capa}	2	20	kWh
No. of Supercapacitors (X_4)	N_{SC}	2	10	-

The curb mass of the vehicle is limited to 2000 kg (an average mass between lightest and heaviest HEV from dataset). This condition is implemented as a constraint for the optimization problem. This is represented by:

$$M_{v,c} < 2000 \tag{8}$$

Table 3. Summary of different optimization settings (sets).

Objective function	Set 1	Set 2	Set 3	Set 4	Set 5	Set 6
Minimize (m_f or $m_{f,e}$)	P=20	P=60	P=100	P=100 S=Tournament	P=100 C=Heuristic	P=100 MD= forward and backward

For best optimal search, different GA settings including population (P), selection (S), crossover (C), and migration direction (MD) were considered as in table 3. Number of optimization runs is 12, this means 6 cases for each objective function.

3. Results and Discussion

A combined drive cycle (CDC) that is closely representing real-world driving cycle is used for design optimization (figure 2). It consists of FTP-75 and HWFET from EPA city and highway drive cycle respectively.

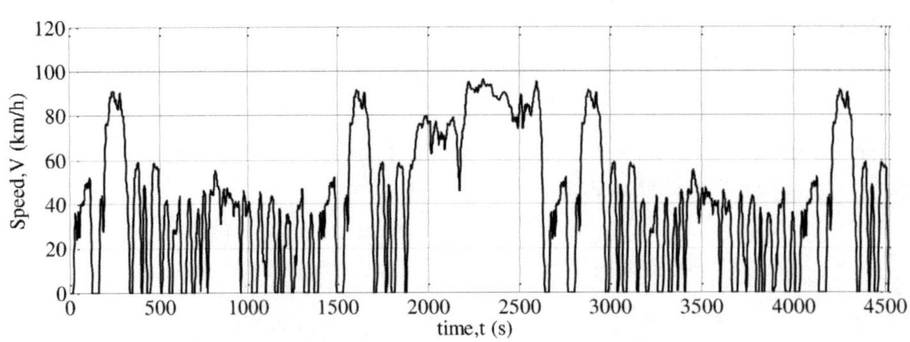

Figure 2. Combined FTP-75-HWFET drive cycle (CDC).

It is observed from table 4, that best fuel economy at set 3 and best equivalent fuel economy at set 5. The difference is 2%.

Table 4. Fuel consumption for different optimization settings.

Objective function		Set 1	Set 2	Set 3	Set 4	Set 5	Set 6
Minimize m_f	m_f (kg)	1.7362	1.6487	*1.4799*	1.6326	1.6728	1.5848
	$m_{f,e}$ (kg)	1.7489	1.7932	1.6671	1.7825	1.8157	1.7600
Minimize $m_{f,e}$	m_f (kg)	1.8053	1.6563	1.8734	1.7646	*1.4502*	1.9036
	$m_{f,e}$ (kg)	1.8581	1.7628	2.0027	1.8779	1.6447	1.91311

Table 5. Component sizes for both optimization objectives at optimal point

Objective function	l_{ICE} (l)	P_{EMG} (kW)	B_{capa} (kWh)	N_{SC} (unit)	$M_{v,c}$ (kg)	m_f (kg)	$m_{f,e}$ (kg)
Minimize m_f	2.0	24.7077	15.2434	5	1811	1.48	1.66
Minimize $m_{f,e}$	2.0	21.6430	15.6608	2	1778	1.45	1.65

As shown in table 5, the major difference between objective 1 and 2 (m_f and $m_{f,e}$ respectively) is that objective 2 (minimize $m_{f,e}$) shows lower number of supercapacitor. For both objectives, EMG rated power and battery capacity shows little difference. It was decided that setting optimization objective to minimize $m_{f,e}$ at option 5 (crossover by heuristic) yielded best optimal component size. This is because it has minimal fuel consumption and lower number of units of supercapacitors. The best optimal parameters are summarized in table 6.

Table 6. Optimal HEV powertrain parameters.

Parameter	Reference	Optimized	Market HEV	Market HEV
Name	Generic HEV	Generic HEV	Mitsubishi Outlander	Volkswagen Tuoareg
Configuration	Parallel	Parallel	Series-Parallel	ISAD system
Transmission	5-speed GB	5-speed GB	Single-speed GB	8-speed GB
ICE capacity, l_{ICE} (l)	3.0	2.0	2.0	3.0 Supercharged
EMG power, P_{EMG} (kW)	40	21.64	60 (front) 60 (rear)	35
Batt capacity, B_{capa} (kWh)	2.0093	15.66	12.4 (Li-Ion)	1.7 (Ni-MH)
No. of SC, N_{SC} (units)	2	2	-	-
Curb Mass, $M_{v,c}$ (kg)	1800	1778	1845	2329
Fuel consumption, m_f (kg)	2.3496	1.4502 (-38.3%)	-	-
Equivalent fuel consumption, $m_{f,e}$ (kg)	2.3674	1.6447 (-30.5%)	-	-

Table 6 lists the optimized HEV as compared to HEV available in the market. Compared to Mitsubishi Outlander, the optimized HEV has the exact same capacity of ICE (2 *l*) and slightly higher battery capacity (15.66 kWh compared to 12.4kWh). EMG size is significantly lower (21.64 kW compared to 60 kW). This is due to absence of multi-speed gearbox in Mitsubishi Outlander (for high torque output, gearbox is replaced by higher EMG rated power). Volkswagen Tuoareg has a multi-speed gearbox (for higher torque output), which leads to a relatively low EMG rated power (35 kW). This is compatible with the optimized HEV (21.64 kW), which has a multi-speed gearbox.

The improvement in equivalent fuel economy is a result of several design changes. Firstly, ICE was downsized from 3 litres to 2 litres (smaller ICE capacity consumes less fuel). Secondly, increment of battery capacity from 2.0093 kWh to 15.66 kWh, this increased electrical energy usage which has high efficiency. Moreover, larger battery capacity reduces 'on-board charging' by ICE. This leads to a less usage of engine only mode, and more electric and hybrid mode. Thirdly, vehicle curb mass was maintained around initial value (1800 kg). Despite the increment of battery size, the final vehicle curb mass is reduced to 1778 kg, mainly due to ICE downsizing. This leads to lower inertial and frictional forces acting on the vehicle, which improves fuel economy.

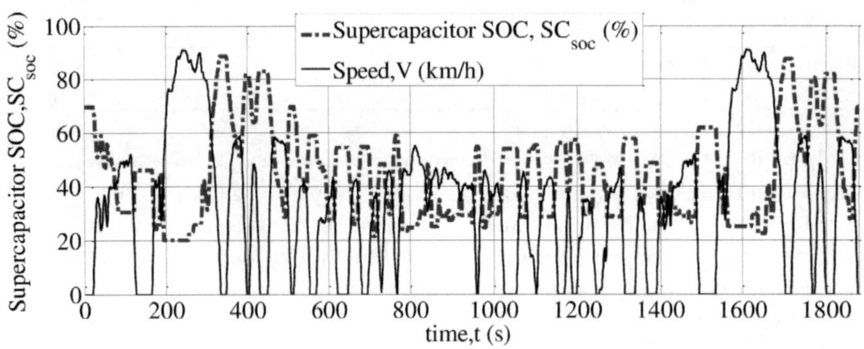

Figure 3. Supercapacitor state of charge (SOC) during initial period of combined cycle.

Figure 3 shows pattern of supercapacitor SOC for best optimal HEV, during combined drive cycle. During regenerative braking and acceleration periods, SOC has fast increase and decrease. This is compatible with supercapacitor characteristics since it has fast charging and discharging rates. For battery, its minimum SOC is limited to 55% to ensure longer life cycle. On the contrary, supercapacitor life cycle is significantly longer than battery life cycle. Hence, its minimum SOC is limited to a lower value (20% in this study). It was observed that supercapacitor voltage was fluctuating because it changes linearly with SOC during discharge operation (supplying power to EMG). However, this voltage fluctuation can be regulated by adding a DC-DC converter downstream of supercapacitor.

4. Summary

Parallel HEV powertrain optimization process was carried out using GA. It yielded improvement in fuel economy and equivalent fuel economy. Optimized HEV has smaller components sizes and higher fuel economy. It can be concluded that optimized HEV suits a plug-in type hybrid vehicle. This is because it has higher battery capacity than most non plug-in HEV. This work on plug-in parallel HEV is in progress. The moderate size of EMG had raised an important question about vehicle performance and gradeability. This will probably increase both ICE and EMG sizes and result in a reduction of electric energy storage size. Therefore, this issue will be investigated in future work by adding performance and gradeability constraints.

Acknowledgments
Authors gratefully thank Ministry of Higher Education Malaysia (MOHE) for financial support through MyMaster15 and FRGS15-231-0472.

References

[1] Cerofolini A 2014 *Optimal Supervisory Control of Hybrid Vehicles* (Bologna: University of Bologna) pp 153-155

[2] Fan B SM 2011 *Multidisciplinary Optimization of Hybrid Electric Vehicles: Component Sizing and Power Management Logic* PhD Thesis (Waterloo, Ontario, Canada: Univesity of Waterloo) pp 76-85

[3] Zhang X and Mi C 2011 *Vehicle Power Management: Modeling, Control and Optimization* (Berlin: Springer Verlag) pp 295-301

[4] Mangun M F, Idres M, Abdullah K 2015 Power management controller for hybrid electric vehicle using fuzzy logic *ARPN Journal of Engineering and Applied Sciences* **10** (Asian Research Publishing Network (ARPN)) pp 9682-9687

[5] Guzella L and Amstutz A 2005 *The QSS Toolbox Manual* (Zurich: Swiss Federal Institute of Technology (ETH))

[6] Diab Y, Venet P, Gualos H and Rojat G 2009 Self-discharge characterization and modeling of electrochemical capacitor used for power electronics applications *IEEE Transactions in Power Electronics* (IEEE) **24**(2) p 511

[7] Delucchi M, Burke A, Lipman T and Millar M 2000 *Electric and Gasoline Vehicle Lifecycle Cost and Energy-Use Model* Report for the California Air Resources Board, Institute of Transportation Studies, Davis, CA (University of California) pp 31-37

[8] Lui X, Diallo D and Marchand C 2011 Design methodology of hybrid electric vehicle energy sources: application to fuel cell vehicles *International Journal of Automotive Technology* (Berlin: Springer Verlag) p 435

[9] Malen D E, and Kundan R 2007 *Preliminary Vehicle Mass Estimation Using Empirical Subsystem Influence Coefficients* Mass Compounding Project Team report, Ann Arbor, Michigan, United States (InterProfessional Programs, University of Michigan) p 8

[10] Simpson A 2006 *Cost benefit analysis of plug-in hybrid electric vehicle technology* NREL/CP-540-40485 (Oak Ridge, TN,USA: National Renewable Energy Laboratory)

[11] MATLAB R2013a 2013 (Natick, Massachusetts, United States: The MathWorks, Inc)

IOP Conf. Series: Materials Science and Engineering **185** (2017) 012025 doi:10.1088/1757-899X/185/1/012025

A Novel Electro Conductive Graphene/Silicon-Dioxide Thermo-Electric Generator

*Ataur Rahman and Yusuf Abdi
Department of Mechanical Engineering
Kulliyyah of Engineering
International Islamic University Malaysia
50728 KL, Malaysia
*Corresponding Author: arat@iium.edu.my

Abstract. Thermoelectric generators are all solid-state devices that convert heat energy into electrical energy. The total energy (fuel) supplied to the engine, approximately 30 to 40% is converted into useful mechanical work; whereas the remaining is expelled to the environment as heat through exhaust gasses and cooling systems, resulting in serious green house gas (GHG) emission. By converting waste energy into electrical energy is the aim of this manuscript. The technologies reported on waste heat recovery from an exhaust gas of internal combustion engines (ICE) are thermoelectric generators (TEG) with finned type, Rankine cycle (RC) and Turbocharger. This paper has presented an electro-conductive graphene oxide/silicon-dioxide (GO-SiO2) composite sandwiched by phosphorus (P) and boron (B) doped silicon (Si) TEG to generate electricity from the IC engine exhaust heat. Air-cooling and liquid cooling techniques adopted conventional TEG module has been tested individually for the electricity generation from IC engine exhausts heat at the engine speed of 1000-3000rpm. For the engine speed of 7000 rpm, the maximum voltage was recorded as 1.12V and 4.00V for the air cooling and liquid cooling respectively. The GO-SiO2 simulated result shows that it's electrical energy generation is about 80% more than conventional TEG for the exhaust temperature of 500°C. The GO-SiO2 composite TEG develops 524W to 1600W at engine speed 1000 to 5000 rpm, which could contribute to reduce the 10-12% of engine total fuel consumption and improve emission level by 20%. .

1. Introduction

A thermoelectric produces electrical power from heat flow across a temperature gradient. As the heat flows from hot to cold, free charge carriers electrons in the material are also driven to the cold end. The resulting voltage (V) is proportional to the temperature difference (ΔT) via the Seebeck coefficient, α, ($V = \alpha \Delta T$). By connecting an electron conducting (n-type) and hole conducting (p-type) material in series, a net voltage is produced that can be driven through a load. A good thermoelectric material has a Seebeck coefficient between 100 μV/K and 300 μV/K; thus, in order to achieve a few volts at the load, many thermoelectric couples need to be connected in series to make the thermoelectric device [1].

A large amount of energy from the stream of exhaust gasses could potentially be used for waste heat recovery to increase the work output of the engine [2]. The authors [3,4,5, 6] stated that the waste heat produced by thermal combustion process generated gasoline could get as high as 30-40% is lost to the environment through an exhaust pipe. The engine energy distribution: shaft power 25-40%; heat rejection- coolant heat rejection 10-35%, exhaust enthalpy loss 20-45%, and engine external surface loss 2-10% [7]. The percentage of fuel energy converted into useful work only 10.4% and also found that 27.7% energy lost through exhaust pipe was investigated [8]. The authors (Dolz et al.2012) [9] reported that the value of exhaust gasses is 18.6% of total combustion energy. The authors (Wong et al.(2012) [10] stated that by installing heat exchanger exhaust energy can be saved up to 34%. The IC engine waste energy of IC engine can be recovered about 15% by using waste energy harvesting coolant based (weHSc) and exhaust based (weHSex) as an integrated unit [11,12]. Based on the above researches output it could be concluded that the engine waste energy can be turned into useful energy by using the thermal electric generator, turbocharging, exhaust re-circulating, heat recovered from coolant and etc.

An accurate estimate of the heat transfer between cylinder gasses and cylinder wall of a combustion engine is necessary for a precise calculation of power, efficiency, and emissions during engine development [13]. Several models exist for evaluating the heat transfer coefficient, of which the correlations [14, 15]. The authors (Gu et al. , 2009) [16] found that the engine cycle (Rankine Cycle) efficiency of several working fluids is very sensitive to evaporative pressure but insensitive to expander inlet temperature. The authors (Boretti, 2012) [17] stated that in a given temperature gradient for optimizing the work output, the working fluid's evaporation enthalpy should be as high possible. Thermoelectric materials can capture some of this heat, and produce electricity. The Thermocouple was first used by ([18] to measure the gun bore temperatures. The author (Stobart et al.2010) [19] explored the possibility of the thermoelectric generator (TEG) in vehicles in which they found that the 1.3 kW output of thermoelectric device could potentially replace the alternator of the small vehicle. By improving thermocouples, it would be possible to convert 3 to 5% of the waste heat into electricity, which would be efficient enough to recharge a vehicle's 12V battery. Therefore, The load on the engine is reduced thereby improving fuel efficiency by as much as 10% and require the temperature gradient at least 500°C [20, 1]. An increase of 20% of fuel efficiency can be easily achieved by converting about 10% of the engine waste heat into electricity [5, 21] . The authors (Yu and Chau, 2009) [6] has proposed and implemented an automotive thermoelectric waste heat recovery system by adopting a maximum power point tracker (MPPT) controller as tools for power conditioning and transfer.

The aim of this study to develop a graphene oxide (GO) and silicon dioxide (SiO2) composite sandwiched by N-type and P-type semi-conductive materials TEG to increase the temperature gradient of 500°C and increase the engine efficiency by 10% as stated [20].

2. Materials and Methodology

2.1 Mathematical modeling

This study has been conducted by studying the heat development of exhaust by using GT-suite software, the performance of conventional TEG with air and waste cooling techniques, and the development of GO/SiO_2 composites sandwiched by P-type and N-type semi-conductive materials. Energy is supplied to the engine in the form of chemical energy of the fuel and producing useful power and losses as heat through exhaust and coolant. First Law of thermodynamics states that energy is conserved. Therefore, the energy balance equation for the engine can be represented as,

$$\dot{m}_f h_f + \dot{m}_a h_a = P_{brake} + \dot{Q}_c + \dot{Q}_{exh} + (\dot{m}_a + \dot{m}_f) h_e + \dot{Q}_{rad} \quad (1)$$

with $\dot{Q}_C = \dot{m}_c c_{pc} \Delta T$ and $\dot{Q}_{exh} = \dot{m}_{exh} c_{exh} \Delta T$

where, $\dot{m}_a$ and $\dot{m}_f$ is the rate of flow of air and fuel to the engine as initial energy, $\dot{Q}_c$ and $\dot{Q}_{exh}$ is the heat transfer rate of the engine to the coolant and exhaust respectively, h_e is the enthalpy of unburned gas mixture.

$$p_{brake} = \dot{m}_f h_f + \dot{m}_a h_a - \dot{Q}_c - \dot{Q}_{exh} - (\dot{m}_a + \dot{m}_f) h_e - \dot{Q}_{rad} - \dot{Q}_{fric} \tag{2}$$

Equation (2) indicates that the P_{brake} will be increased if % of $\dot{Q}_c$ and $\dot{Q}_{exh}$ can be recovered and reduced the unburned gas mixture of $(\dot{m}_a + \dot{m}_f) h_e$. Recover of waste heat energy can increase the brake power, improve the fuel consumption and reduce the CO_2 emission. The flow of the exhaust at beginning of the exhaust manifold has high sonic velocity and kinetic energy. The kinetic energy of the exhaust is converted to additional enthalpy with an increase in temperature. The maximum temperature of the exhaust can be computed by applying the energy conservation law,

$$T_{max} = T_{ex} + \Delta T = T_{cy} \left(\frac{P_{ex}}{P_{cy}} \right)^{\left(k-1/k\right)} + \frac{V^2}{2gc_p} \tag{3}$$

where, V is the velocity of the exhaust in m/s which could be 694 m/s and c_p is the specific heat at constant pressure.

The simulation study on the effect of engine speed (rpm) to the heat transfer in cylinder wall and the volumetric efficiency has been conducted by using GT-suite software. At higher engine speeds, there is less time per cycle. Combustion occurs about the same engine rotation at all speeds, so the time of intake and combustion is less at higher speeds. The less time for intake and ignition and less time for heat transfer per cycle, causes the engine runs hotter as shown in Figures 1-2.Therefore, the wall of the cylinder becomes hotter which might affect the fuel lost. The author (Reitz, 2012) [22] reported that up to 30% of fuel energy is lost to the wall heat transfer. Therefore, the exhaust temperature will be high and it could be in the range of 300 – 2000°C. By using a thermoelectric generator waste heat energy of exhaust can be harvested. They reported that the technical feasibility of conversion of heat energy radiated from automobile engines produce power, using thermoelectric generators applying about 35 Ah chrage can be produced. The resulting voltage (V_v) is proportional to the temperature difference (ΔT) via the Seebeck coefficient, α, ($V_v = \alpha\Delta T$). A good thermoelectric material has a Seebeck coefficient between 100 μV/K and 300 μV/K; thus, in order to achieve a few volts at the load, many thermoelectric couples need to be connected in series to make the thermoelectric device. A thermoelectric generator converts heat (Q) into electrical power (P) with efficiency η.

$$P = \eta_{TEG} Q \tag{4}$$

The amount of heat, Q, that can be directed though the thermoelectric materials frequently depends on the size of the heat exchangers used to harvest the heat on the hot side and reject it on the cold side. The efficiency of a thermoelectric converter depends heavily on the temperature difference, $\Delta T = T_h - T_c$ across the device. This is because the thermoelectric generator, like all heat engines, cannot have efficiency greater than that of a Carnot cycle ($\Delta T / T_h$). The efficiency of a thermoelectric generator is typically defined as,

$$\eta = \frac{\Delta T}{T_h} \bullet \frac{\sqrt{1+ZT} - 1}{\sqrt{1+ZT} + T_c/T_h} \tag{5}$$

$$\text{with } zT = \frac{\alpha^2 T}{\rho k}$$

where, the first term is the Carnot efficiency and ZT is the merit for the device. Seebeck coefficient (α), electrical resistivity (ρ), and thermal conductivity (κ) are temperature (T) dependent material properties. Increasing the temperature gradient of TEG's module can increase the efficiency of thermoelectric generator. The basic physical phenomena is associated with the operation of thermoelectric generators (TEG), the Seebeck effect. Under steady state condition, the equation that governs the heat flow at the hot side, Q_N is:

$$Q_N = K_{TEG}\Delta T + S_{TEG}\left(T_H I\right) - 0.5I^2 R_{TEG} \quad (6)$$

where, K_{TEG} is the total thermal conductance of thermoelectric module, S_{TEG} is the total Seebeck coefficient of thermoelectric module, R_{TEG} is the total resistance of thermoelectric module, T_H is the temperature at hot side, ΔT is the temperature difference between hot and cold side and I is maximum current. The efficiency of thermoelectric generator can be determined,

$$\eta_{TEG} = \frac{S_{TEG}}{Q_H}\left(T_H - T_C\right) \quad (7)$$

where, η_{TEG} is the efficiency of thermoelectric generator.

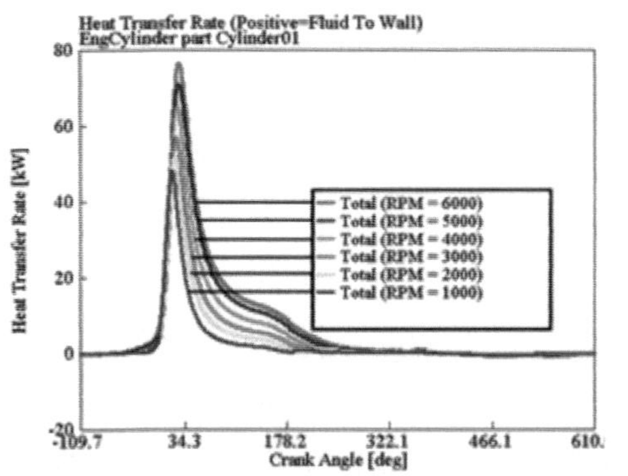

Figure 1. Heat transfer at different crank angle

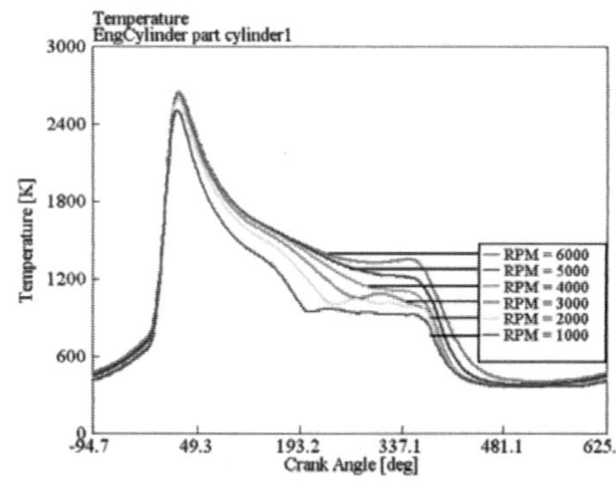

Figure 2. Temperature of the exhaust at different crank angle.

2.2 Development of TEG

There are two types of TEG has been developed from this study to identify the performance of GO/SiO_2 composite electrolyte semiconductive over the conventional semi-conductive TEG.

2.2.1 Conventional TEG

Many thermoelectric couples of N-type and P-type thermoelectric semiconductors (modules TECI-12708) are connected electrically in series and thermally in parallel to make a thermoelectric generator as shown in Figure 3,called as conventional TEG. The flow of heat drives the free electrons (e^-) and holes (h^+) producing electrical power due to the temperature gradient. The direct conversion of heat energy into electrical energy can be accomplished through the *Seebeck* effect. TEG module and exhaust gas pipe as shown in Figure 3, has been developed by using a TEG module TEC1-12708 with connecting 126 units of semiconductors (P-type and N-type) in series. The module specification: maximum voltage 15.4V, maximum current of 8A, temperature difference (gradient) ≤66°C and maximum allowable heat of 73.92W. Electricity generation in terms of Ah can be made more by making the more temperature difference even ≥500°C.

IOP Conf. Series: Materials Science and Engineering **185** (2017) 012025 doi:10.1088/1757-899X/185/1/012025

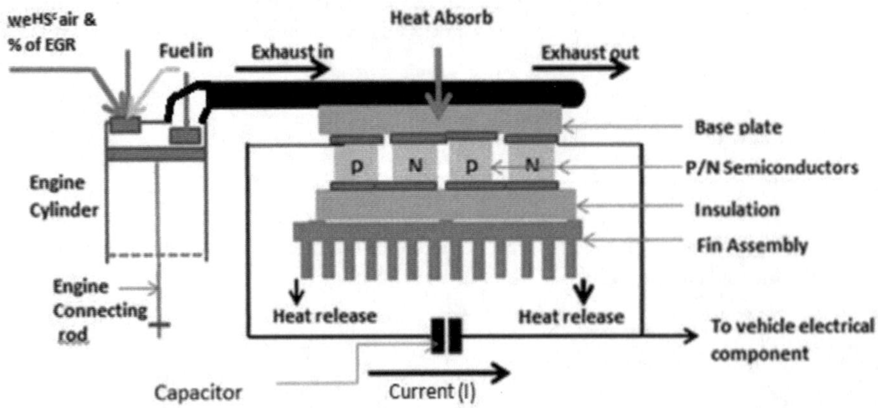

Figure 3. Model of TEG [12] [

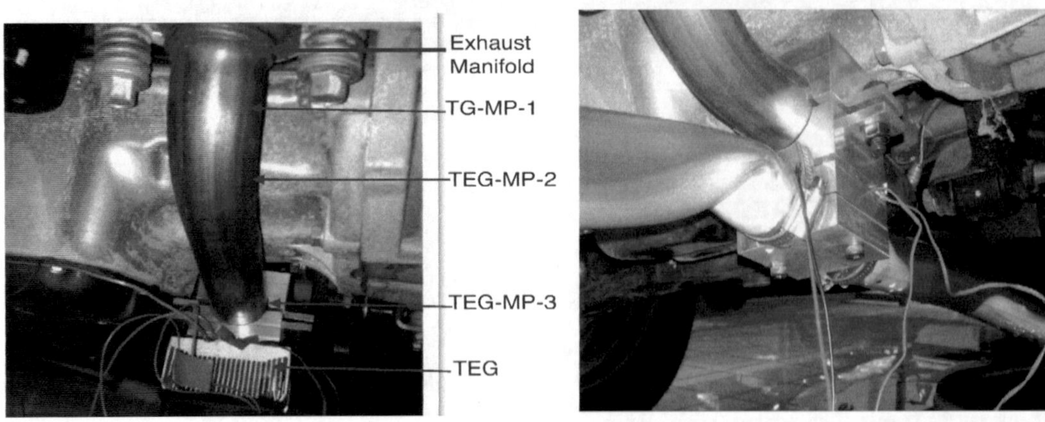

(a) (b)

Figure 4. Thermoelectric generator (a) air cooling and (b) liquid cooling.

The performance of TEG on electric power generation from exhaust has been investigated by developing two types of TEG cooling methods: fine-type air-cooling and liquid-type cooling as shown in Figure 4. In fin type air-cooling, a draft fan has been used to draw additional fresh air to the fin for increasing temperature gradient ($\Delta T = T_h - T_c$) with decreasing the temperature of cooling sides (T_c) of TEG. In liquid cooling, the liquid (H_2O) flow has maintained continuously to keep-down T_c. It is expected that the performance of liquid cooling TEG better than the air cooling at higher engine RPM as the heat rejection ($Q = hA\Delta T$) of liquid is greater than the air due to the values of heat transfer coefficient (h). The heat transfer coefficient is considered: h=100 W/m^{2}K for air and h=1000 W/m^{2}K for H_2O.

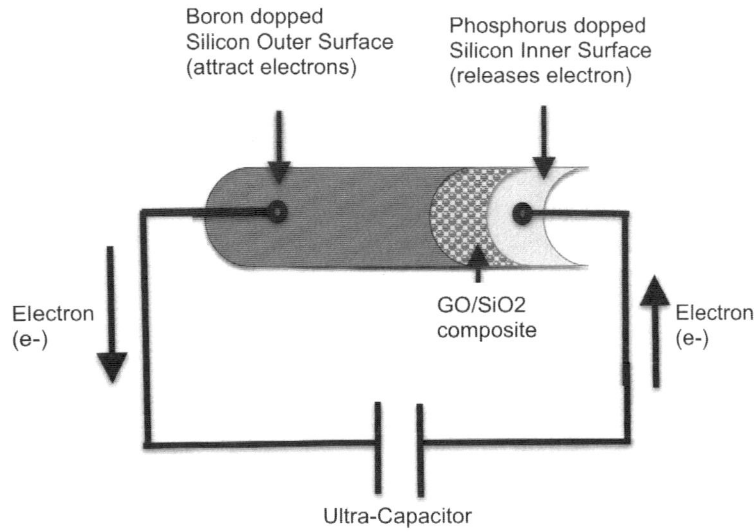

Figure 5. Schematic diagram of GO/SiO_2 composite semi-conductive TEG.

2.2.2 GO/SiO_2 composite electrolyte semi-conductive TEG

The GO/SiO2 composite TEG consists of a single cylindrical pellet of P-type and N-type semi-conductive material are connected electrically in series. The flow of heat through the GO/SiO2 composite drives the majority free electrons (e-) from N-type to P-type semiconducting surface and produces electrical power. The electron mobility is caused due to the temperature gradient. In GO/SiO2 composite TEG as shown in Fig. 5, the temperature gradient is expected to 200-1000°C. The GO/SiO2 composite has been sandwiched by phosphorous and boron doped silicon. The phosphorous-doped silicon (donor) releases electron while the boron-doped silicon (accepter) consumes the electrons. The GO/SiO2 transports the electrons towards the phosphorous dopped silicon due to the thermal diffusion. Therefore, the electricity generates. In the literature of this article, it is reported that if the temperature gradient reaches of the TEG to 500°C, the electric power can be generated about 1.5 to 2.0 kW, which would enhance the engine efficiency by 10%. The maximum temperature gradient is expected 1000°C for the engine rpm 7000rpm at the crank angle 49.3 degrees.

The GO/SiO2 composite allows traveling the electron from the inner surface to outer surface due to the high electric conductivity of the graphene oxide. The GO concentration with the SiO2 has considered based on the study [23]. While the inner and outer surface characteristics have optimized based on the exhaust temperature. The behavior of the inner and outer surface of the GO/SiO2 is same as conventional TEG (Fig.4). The characterization of GO/SiO2 composite has made based on the exhaust temperature and thermal conductivity of GO. The intrinsic mobility reaches the value as high as 2×105 cm2V-1S-1 at the corresponding resistivity of the graphene flakes 10-6 Ohm-cm, which is slightly less than the resistivity of silver (1.6x10-6ohm.cm). The magnitude of an electron mobility of graphene exceeds 2x105cm2V-1S-1 [24] while for carbon nano-tubes 100,000 cm2V-1S-1 [25]. Besides the outstanding electromagnetic and mechanical properties, the graphene shows unique thermal characteristics. The thermal conductivity of graphene at room temperature was discovered in the range of 3080-5300 Wm-1K-1. These values substantially exceed those of diamond (1000–2200 Wm-1K-1) which is considered to be the best bulk thermal conductor. Heat propagation in graphene occurs through conduction. The heat propagation in graphene is mostly due to acoustic phonons (atomic vibrations).

The law of heat conduction defines the time rate of heat transfer as, $\frac{\partial Q}{\partial t} = -k\oint \vec{\nabla}T.d\vec{s}$, where, dQ/dt is the heat transfer rate in W, T is the temperature, K, S is the surface through which the heat is flowing in m^2, and k is the thermal conductivity in $Wm^{-1}K^{-1}$. The equation indicates that the heat transfer rate is proportional to the temperature gradient and to the area of the heat flowing surface. The surface is normal to the temperature gradient and the heat conduction can be rewritten as, $\frac{\partial Q}{\partial t} = k\frac{dT}{dX}\int_0^w dy\int_0^k dz$, where w is the width of the flake in m, h is the thickness of the flake in m, and Wh is the cross-sectional area of the flake in m^2. The thermal contact resistance and the temperature rises at the contact due to the ohmic Joule heating ($Q=i^2R$) for higher current could occur in TEG. The thermal resistance for heat flow from the graphene oxide channel into the metal oxide through SiO_2 can be estimated as, $R_{C(T)} \approx \frac{1}{hL_TW_c} \cong \frac{1}{W_c}(hkt)^{-1/2}$, where, L_T thermal transfer length of the TEG in mm can be formulated as $L_T = (kt/h)^{1/2}$, h is the thermal heat transfer coefficient and W_c is the thickness of the GO/SiO_2 composite. Heat dissipation at the contacts consists of parallel paths to the underlying SiO_2 substrate, and side metal contact with $h_{g\text{-}ox} \approx 10^8 Wm^{-2}K^{-1}$ and $h_{c(sm)} \approx 10^7 Wm^{-2}K^{-1}$. Therefore, the total thermal resistance can be estimated as, $R_{c(T)} = (R_{c(GO/SiO_2)} + R_{c(sm)})$ .If the temperature develop at the material (N-type) due to the electron mobility, the total temperature develop at the N-type semi-conductive material can be defined as, $T_{N-type} = (T_{exh} + T_{electron\,mobility})$

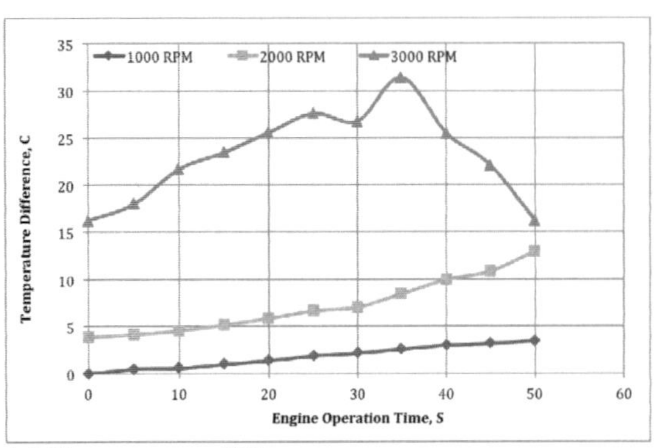

(a) Temperature diffrence

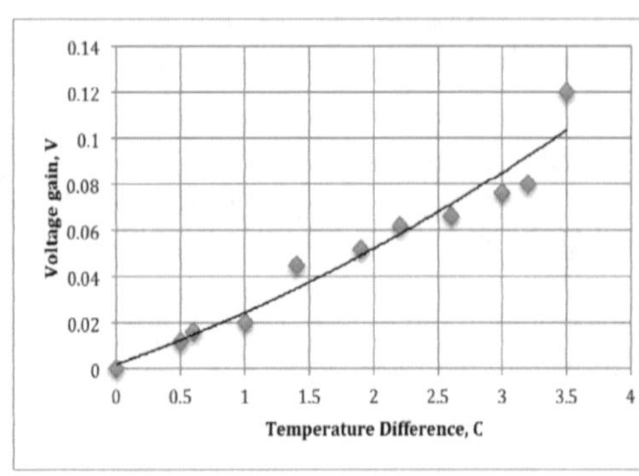

(b) 1000 RPM

The International Conference on Information Technology and Digital Applications IOP Publishing
IOP Conf. Series: Materials Science and Engineering **185** (2017) 012025 doi:10.1088/1757-899X/185/1/012025

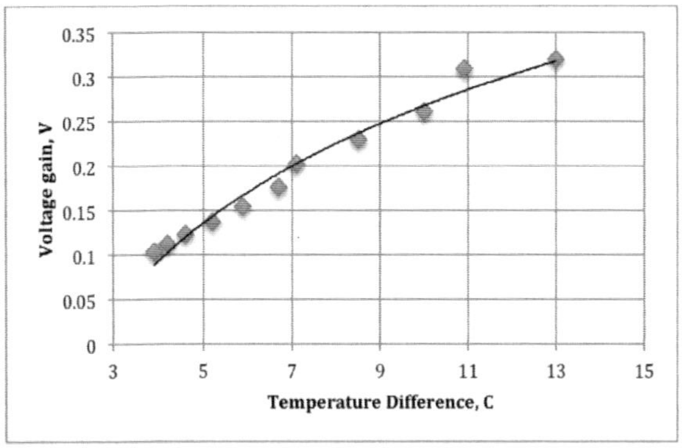

(c) 2000 RPM

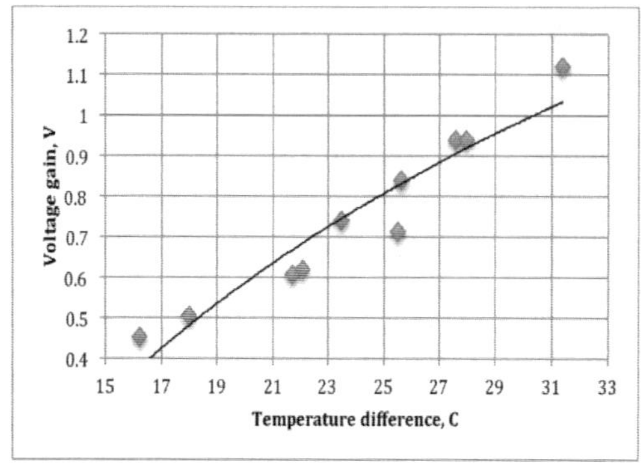

(d) 3000 RPM

Figure 6. Performance of fine type TEG (a) temperature difference vs engine operation times (b-c) Voltage gain for different engine RPM

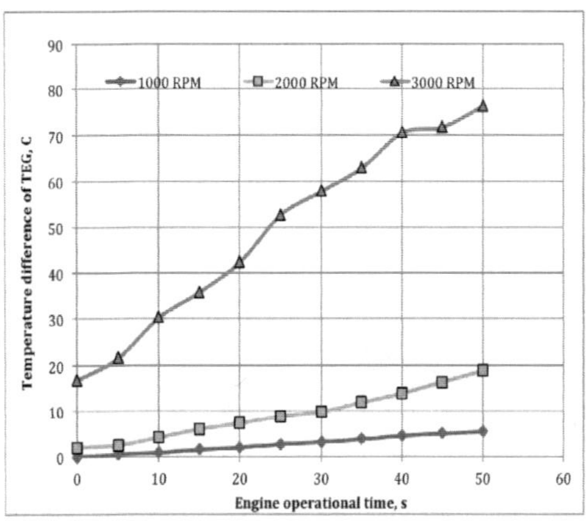

(a) Temperature difference

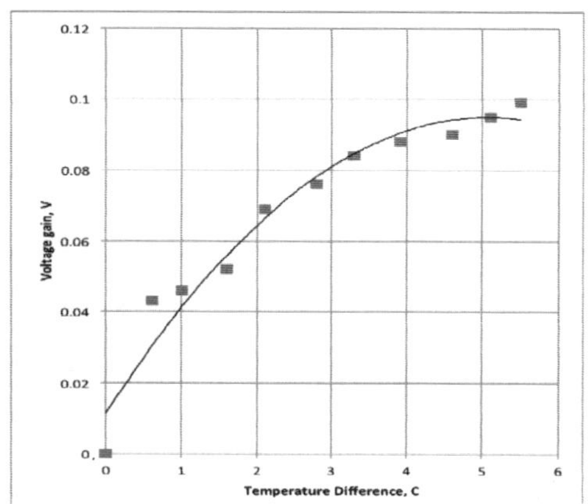

(b) 1000 rpm

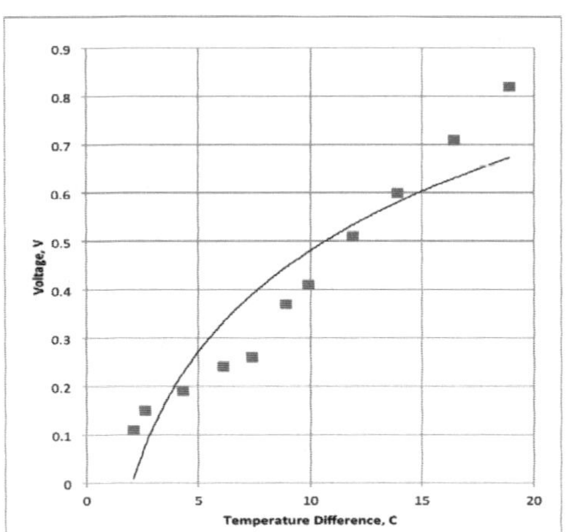

(c) 2000 rpm

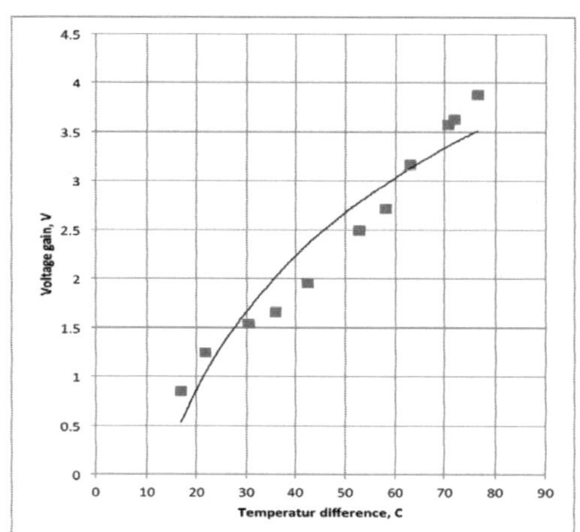

(d) 3000 rpm

Figure 7. Performance of liquid cooling type TEG (a) temperature difference for engine operation times (b-d) Voltage gain for different engine RPM

3. RESULT AND DISCUSSION

3.1 Performance investigation of conventional TEG

This thermoelectric generator is tested at Perodua Myvi (Malaysia Car). The installation part is attached to the closest exhaust manifold of internal combustion engine to gain the highest temperature to generate high voltage gain and prevent heat loss to the surroundings. Three mounting points of TEG model have been considered to investigate of TEG (Figure 4).

The TEG performance has been conducted with engine RPM from 2000 to 3000 rpm. Figures 6 and 7 show the voltage development by the TEG increases with increasing the RPM of the engine, which is due to the increment of fuel heating value. Figure 6(a) shows at engine 3000 RPM, the temperature difference (ΔT) decreases for the longer period and higher exhaust heat, which affects the TEG performance. This is could be due to the lower heat transfer co-efficient of air. The TEG performance for the engine rpm of 5000 is recorded very low. Figure 7(a) shows that the temperature differences of TEG have been increased with increasing the engine rpm from 1000 to 3000 rpm for the longer period and increases the TEG performance. This is because of higher heat transfer values of liquid. The test result also shows the TEG temperature differences has been increased until the engine reaches at 7000 rpm. This is normally not considered for the engine of the passenger car. The maximum engine rpm is considered for the engine for passenger car about 5000 rpm.

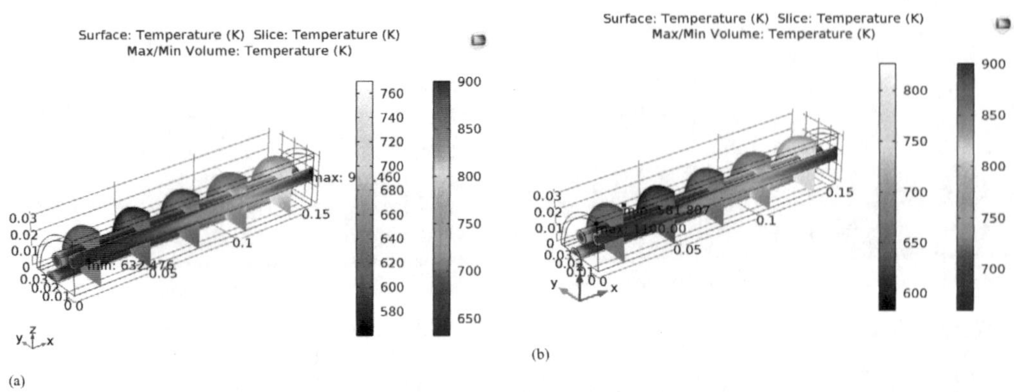

Figure 8. Temperature distributions in the GO/SiO_2 boundary (a) 900°C (b)1100°C.

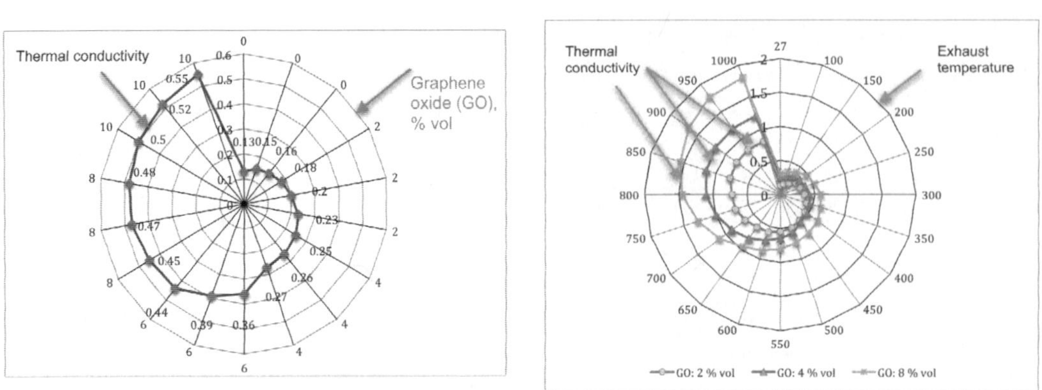

Figure 9. GO/SiO_2 composite silicon, phosphorus, and boron embedded TEG (a) Thermal conductivity for graphene oxide concentration (%vol.) at room temperature, (b) Thermal conductivity at exhaust temperature

3.2 GO/SiO_2 Composite TEG

The heat propagation in graphene has been conducted by using COMSOL Multiphysics software package. COMSOL package is a finite element analysis tool and it has modules for various physics phenomena: Acoustic Module, Structural Mechanics Module, Heat Transfer Module and others. The finite element analysis technique is a numerical method for finding approximate solutions of partial differential equations. The solutions are mostly based on dividing the simulated object into a large set of very small components and linearizing the differential equations within each of the components. During the model design by COMSOL software, the power of the heat source and the thermal conductivity of graphene has been specified. The thermal diffusion in the flake of GO causes the electron mobility and increases the electricity generation.

Figures 8(a) and 8(b), the result shows the energy exchange between the surface heating by exhaust and TEG surface. Figure 9(a) shows that the exhaust entering the heating pipes at 900 K and exit at 632°C to the surrounding. The temperature gradient 278°C, which causes the high electron mobility from an inner surface to the outer surface and generates electricity. Figure 8(b) shows the exhaust enters at 1100 K and exit at 581°C to the surrounding and develop the temperature gradient 529°C and causes the electricity generation.

Figures 9-10 show the GO/SiO2 thermal and electric conductivity. The result shows that both the thermal and electric conductivity has increased significantly with increasing the temperature, which causes the driving of electrons from N-type and P-type electro-conductive doped silicon. The liquid cooling TEG performance increases exponentially with increasing the engine speed. The experiment has been conducted for the maximum engine speed of 5000rpm. The power recovered from the exhaust of engine by using the module mentioned in earlier is about 200W if the exhaust is covered with four modules. Based on the increment of temperature difference, the liquid cooling TEG has able to recover 480W power. The 20 cm hollow cylindrical GO/SiO2 composite semi-conductive TEG is expecting to generate electricity about 1.6 to 2.0KW for the temperature differences of 360 to 500°C. Based on the increment of temperature difference, the liquid cooling TEG has able to recover 480W power from the exhaust heat.

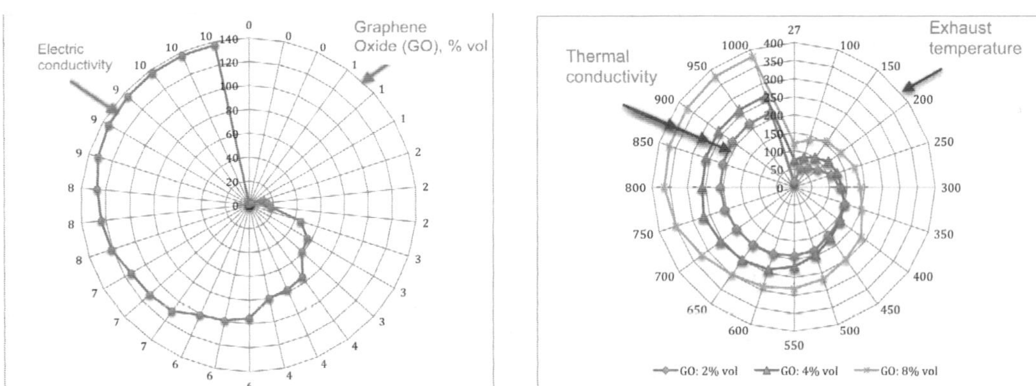

Figure 10. GO/SiO_2 composite with semi-conduction TEG (a) Electrical conductivity for different GO (vol%), (b) Elective conductivity at exhaust temperature

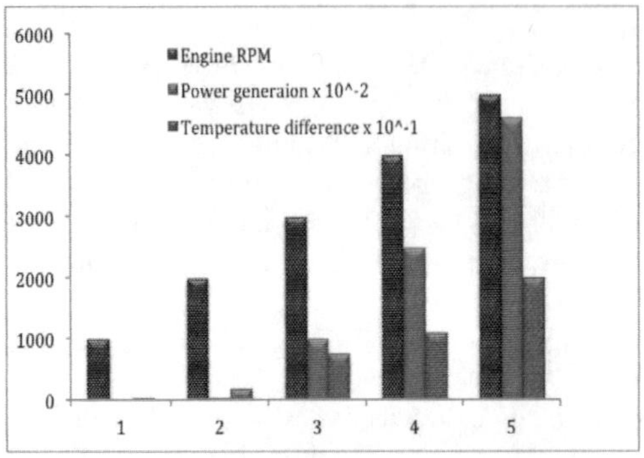

Figure 11. Power generation from engine exhaust by using single TEG module for air cooling.

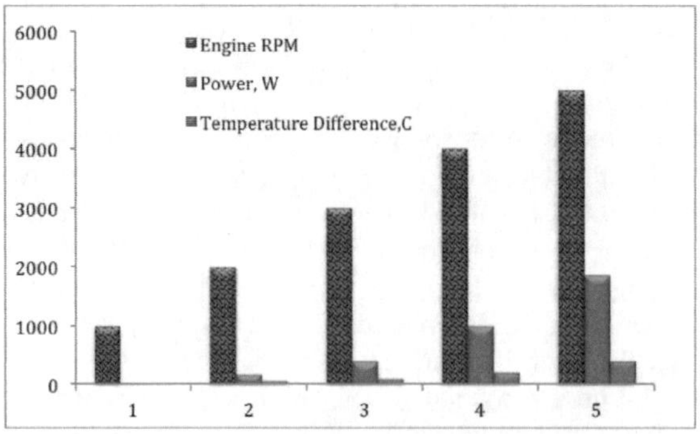

Figure 12. Power generation from engine exhaust by using single TEG module for liquid cooling.

Figures 11-12 shows the performance of conventional TEG. The conventional TEG with air cooling (Fig.11) shows the power generation at 5000 rpm is 450W while liquid cooling conventional TEG (Fig.12) is 1500W. The liquid cooling TEG performance for the 5000 rpm engine speed is 233% higher than the air cooling TEG. The basic reason for increasing power generation of liquid cooling over air-cooling is mainly for liquid higher heat transfer coefficient.

4. Conclusion

The graphene intrinsic mobility of 2×10^5 $cm^2V^{-1}S^{-1}$ at a carrier density of 10^{12} cm^{-2} and thermal conductivity at 300 $Wm^{-1}K^{-1}$ at room temperature (300K) shows that graphene is a promising material for the semi-conductive thermoelectric generator for the renewable of high electricity. The higher thermal diffusion for 500°C, the GO/SiO_2 composite semi-conductive hollow cylindrical TEG is expecting to generate electricity about 1.6 to 2.0KW. The higher electricity generation from the exhaust heat by using GO/SiO_2 composite could be useful to operate the air-conditioning system of the vehicle and meet the vehicle electrical load requirement.

REFERENCE

[1] G. J. Snyder. Thermoelectric Power Generation: Efficiency and Compatibility," in *Thermoelectrics Handbook Macro to Nano*, edited by D. M. Rowe (CRC, Boca Raton, 2006), Ch. 9.

[2] P.K Stobart et al. The potential for thermo-electric devices in passenger vehicle applications, SAE Paper nso. 2010-01-0833, Presented at SAE 2010 World Congress & Exhibition, April 2010, Detroit, MI, USA, Session: Advanced Hybrid Vehicle Powertrains (Part 2 of 3).

[3] M. Hatazawa et al. Performance of a thermoacoustic sound wave generator driven with waste heat of automobile gasoline engine. Transac- tions of the Japan Society of Mechanical Engineers, 70(689):292–9 (2004).

[4] F. Stabler. Automotive applications of high efficiency thermoelectrics, in DARPA/ONR program review and DOE high efficiency thermoelectric work- shop. San Diego, CA, 2002.

[5] J. Yang. Potential applications of thermoelectric waste heat recovery in the automotive industry, in International conference on thermoelectrics 2005: 155-159.

[6] C. Yu and K.T Chau.Thermoelectric automotive waste heat energy recovery using maximum power point tracking. Energy Conversion and Management 2009;50(6):1506–12.

[7] V. Sam. Shelton. www.old.me.gatech-edu/energy/ICEngines/11_HeatTansfer.pdf. Retrieving date: 13 May 2014.

[8] J.C Conklin and J.P Szybist JP. A highly efficient six-stroke internal combustion engine cycle with water injection for in-cylinder exhaust heat recovery. Journal of Energy, 35:1658–64 (2010).

[9] V. Dolz et al. HD Diesel engine equipped with a bottoming Rankine cycle as a waste heat recovery system. Part 1: Study and analysis of the waste heat energy. Applied Thermal Engineering. 269–78 (2012).

[10] T.Wang et al. Analysis of recoverable exhaust energy from a light-duty gasoline engine. Applied Thermal Engineering. http://dx.doi.org/10.1016/j.applthermaleng.2012.03.025.

[11] A. Rahman et al. Nonlinear modeling and simulation waste energy harvesting system of IC engine-Fuzzy Approach. Journal of Renewable and Sustainable Energy. Vol.5, Issue. 3, pp: 1-13 (2013).

[12] A. Rahman et al. Power generation from the waste of IC engine. Journal of Renewable and Sustainable Energy Review, Elsevier Publisher, Vol 51(2015), pp. 382-395.

[13] J.B., Haywood. *Internal Combustion Engine Fundamentals*. McGraw Hill, 1988.

[14] Woschni, G (1967) A universally applicable equation for the instantaneous heat transfer coefficient in the internal combustion engine. SAE paper 670931.

[15] W.J.D Annand, W.J.D. Instantaneous heat transfer rates to the cylinder head surface of a small compressionignition engine. Proc Inst. Mech. Engrs, 185(72):976–987, 1971.

[16] W. Gu et al. Theoretical and experimental investigation of an organic rankine cycle for a waste heat recovery system. Proceedings of the Institution of Mechanical Engineers, Part A: Journal of Power and Energy, 223:523–33, 2009.

[17] A. Boretti. Recovery of exhaust and coolant heat with R245fa organic Rankine cycles in a hybrid passenger car with a naturally aspirated gasoline engine. Applied Thermal Engineering 36(0):73–7, 2012.

[18] D. A Bendersky et al. A special thermocouple for measuring transient temperatures. Mechanical Engineering, 75:117–121, 1953.

[19] P.K Stobart RK. An availability approach to thermal energy recovery in vehicles. Proceedings of the Institution of Mechanical Engineers, Part D: Journal of Automobile Engineering, 221 (1), pp.70-79, 2007.

[20] J. Ahn et al. Ronney, and S. M. Haile, in The 5th International Fuel Cell Science, Engineering & Technology Conference (Fuel Cell 2007), 250832.

[21] R. Saidur et al. A review on compressed-air energyuse and energy savings. Renewable and Sustainable Energy Reviews, 14(4):1135–53, 2010.

[22] R.D Reitz. *Reciprocating internal combustion engines*. Engine research center, University of Wisconsin-Madison, 2012 Princeton-CEFRC.

[23] P. Wonjun et al. Electrical and thermal conductivities of reduced grapheme oxide/polystyrene composites. Applied Physics Letter, Vol. 104 (11), pp.1-13 (2014).

[24] T. Dürkop et al. Extraordinary Mobility in Semiconducting Carbon Nanotubes". Nano Letters **4**: 35, 2004.

[25] K. Bolotin et al. Ultrahigh electron mobility in suspended graphene". Solid State Communications **146**: 351, 2008.

IOP Conf. Series: Materials Science and Engineering **184** (2017) 012026 doi:10.1088/1757-899X/184/1/012026

HYBRID ENGINE POWERED CITY CAR: Fuzzy Controlled Approach

Ataur Rahman*, Mohiuddin AKM, Hawlader MNA, Sany Ihsan,
Department of Mechanical Engineering
Faculty of Engineering
International Islamic University Malaysia
50728 KL, Malaysia
*Corresponding author: arat@iium.edu.my

ABSTRACT. This study describes a fuzzy controlled hybrid engine powered car. The car is powered by the lithium ion battery capacity of 1000 Wh is charged by the 50 cc hybrid engine and power regenerative mode. The engine is operated with lean mixture at 3000 rpm to charge the battery. The regenerative mode that connects with the engine generates electrical power of 500-600 W for the deceleration of car from 90 km/h to 20 km/h. The regenerated electrical power has been used to power the air-conditioning system and to meet the other electrical power. The battery power only used to propel the car. The regenerative power also found charging the battery for longer operation about 40 minutes and more. The design flexibility of this vehicle starts with whole-vehicle integration based on radical light weighting, drag reduction, and accessory efficiency. The energy efficient hybrid engine cut carbon dioxide (CO_2) and nitrogen oxides (N_2O) emission about 70-80% as the loads on the crankshaft such as cam-follower and its associated rotating components are replaced by electromagnetic systems, and the flywheel, alternator and starter motor are replaced by a motor generator. The vehicle was tested and found that it was able to travel 70 km/litre with the power of hybrid engine.

1. Introduction

Transportation consumes more than 20% of the world's total primary energy and produces much of the world's air pollution. Transport service industry increasingly serves as a prerequisite in continuing and developing the productivity of other sectors in the society. If energy use trends for the worlds are projected to year 2100, transportation sector would then need to have twenty times more energy-efficient engine, which roughly equates to 169.312 km per litre. While, incomplete combustion of automobiles engine causes approximately 73.19% of green house gas (GHG) emissions [1]. The emission of CO has significant effects on fetal health like lung cancer, binds to hemoglobin in the blood stream and infant mortality based on the study of [2, 3, 4, 5, 6]. Furthermore, the emission of CO causes the increase of Global warming which lead to major climatic changes that affect the whole world, every single country and many other organisms. Thus, optimal hybrid electric vehicle (HEV) design is becoming a hot spot in recent years as a viable way to meet requirements on both vehicle performance and environment protection [7, 8, 9, 10, 11]. The authors [12] reported that it is difficult related to packaging, fuel storage, power-train cooling, and cost. This study presents a fuel efficient and eco-friendly hybrid engine powered DC motor propelled one seated Urban vehicle. The design flexibility of this vehicle started with whole-vehicle integration, which mainly based on light weighting, drag reduction, fuel-efficient and emission. The proposed vehicle chassis frame and body are made with aluminum except the chassis side bars and the cross members which leads the vehicle weight of 1.48 kN. Drive wheel's traction is digitally controlled by using DYNAMO, Electronic Control Module DC-DC converter is used to stabilize the voltage of DC motor for propel the car.

2. Methodology

The major target of the vehicle design is to make the vehicle fuel efficient and emission reduction with emphasizing light weighting and electric propulsion. Power train of this vehicle is consisted of hybrid engine, power controller and DC motors. The hybrid engine has been built with a 50 cc engine, a battery pack (lithium ion) of capacity1000 Wh, a 24 volts generator, a power regenerative mode. Power regenerative mode has been built connecting 4F capacitor with a 24 V motors by accelerator control switch. The car weight is 2.36 kN including driver weight of 0.6 kN. The total dynamic analysis of the car will be taken into accounts to estimate the desired output power of the proposed engine. It has been developed with downsizing the conventional engine by removing the moving parts such as starter, flywheel, alternator, numbers of pulleys and others. An electro-supercharger has been used to develop the pressurized additional air for creating massive turbulence effect and auto oxidation and atomization and enhancing burning efficiency of air-fuel mixture about 90% in shortly [11].

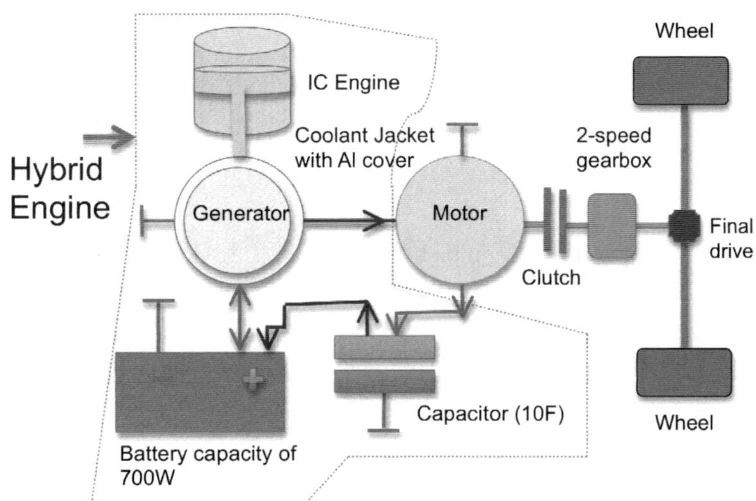

Figure 1. Vehicle driveline.

2.1 Drive Line Modeling

Figure 1 shows the drive ine of the developed vehicle. It consists of clutch, transmission shaft, propeller shaft, final drive, drive shafts and wheels. Using the generalized Newton's second law motion equation has derived the fundamental equations for the driveline. Relations between inputs and outputs are in the following be described as:

$$\left(J_w + \frac{W}{g} r_w^{\,2} + i_t^{\,2} i_f^{\,2} J_e\right)\ddot{\alpha}_w = i_t i_f \left(T_e - T_{fric.e}\right) - \frac{W}{g} c_r r_w^{\,2} \dot{\alpha}_w - \frac{1}{2} c_{air} A_f \rho_a v^2 - \frac{W}{g}\left(c_r + g \sin\phi\right) r_w \quad \textbf{(1)}$$

where, W is the vehicle weight in kN, r_w is the wheel radius, J_w and J_e are the polar moment of inertia of the wheel and engine crankshaft, i_t and i_f are the slippage of the transmission and final drive, respectively

The input torque of the generator ($T_{g(i)}$) can be considered as the torque of the clutch (T_c) when the clutch engaged and there is no internal friction. The input torque of the generator can be computed as:

$$T_t = T_e - T_{fric.e} - J_e \ddot{\alpha}_c = T_{g(i)} \quad (2)$$

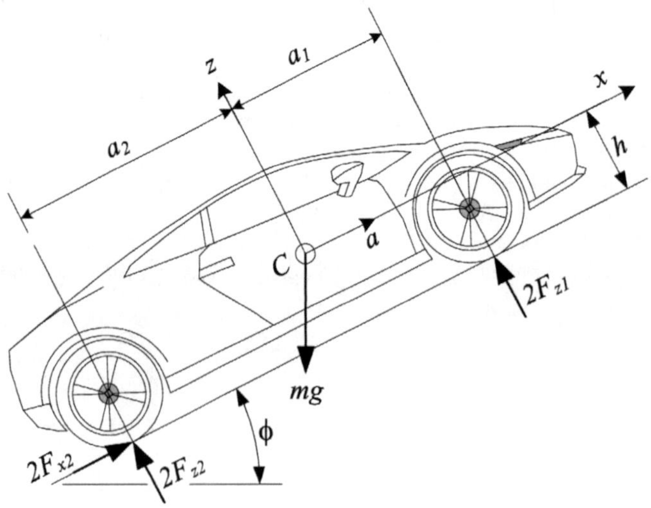

Figure 2. Vehicle on inclined road [13]

The torque of the drive DC motor can be calculated by considering the input power of the motor in terms of the electrical power, which is generated by the generator. The torque of the motor can be computed as:

$$T_m = \frac{\dot{\alpha}_g}{\dot{\alpha}_m}\left[T_e - T_{fric.e} - J_e\ddot{\alpha}_c\right]\frac{1}{\eta_g \eta_m} \tag{3}$$

where, T_m is the motor torque in Nm, $\ddot{\alpha}_c$ is the angular acceleration of the clutch, η_g is the gear box efficiency $\dot{\alpha}_m$ is the motor speed, rad/s, η_m is the efficiency of the motor. The dynamic analysis of the vehicle is conducted by considering the vehicle (Figure 2) at slope of 6.7% gradient ($\phi = \sin^{-1} 6.7\%$) The design parameters of the vehicle are: weight 2358 N including driver 0.59 kN, height centre-of- gravity of 0.75m, wheel base of 1.2 m, and the location of CG from front wheel 0.75m. The car is assumed to accelerate on inclined road with angle ϕ, the normal forces under the front and rear wheels, F_{zf}, F_{zr} are computed by using the modified equation [14] :

$$F_{z1} = \frac{1}{2}mg\left(\frac{a_2}{l}\cos\phi - \frac{h}{l}\sin\phi\right) - \frac{1}{2}ma\frac{h}{l} \tag{4}$$

$$F_{z1} = \frac{1}{2}mg\left(\frac{a_2}{l}\cos\phi + \frac{h}{l}\sin\phi\right) + \frac{1}{2}ma\frac{h}{l} \tag{5}$$

where, l is the wheel base in m, a_2 is the location of CG from the rear wheel in m, a is the acceleration of the vehicle in m/s^2 and h is the height of the vehicle in m.

Based on the characteristics of acceleration, the normal forces of the vehicle under the front and rear wheels can be classified as $\frac{1}{2}mg\left(\frac{a_2}{l}\cos\phi \pm \frac{h}{l}\sin\phi\right)$ is the static part and $\pm\frac{1}{2}mg\frac{h}{l}\frac{a}{g}$ is the dynamic part. The dynamic parts, $\pm\frac{1}{2}mg\frac{h}{l}\frac{a}{g}$, depend on acceleration a and height h of mass center CG, while the static parts, $\frac{1}{2}mg\left(\frac{a_2}{l}\cos\phi \pm \frac{h}{l}\sin\phi\right)$ are influenced by the slope angle φ and height h of mass center. Traction force of the car can be computed as, $F_x = F_{x1} + F_{x2} = \mu\left(F_{z1} + F_{z2}\right)$.

2.2 Hybrid Engine Model Development

An intelligent electromagnetic system has developed to control the fuel injection of the hybrid engine as shown in Figure 1. A high efficient generator has used with the 50 cc engine to generate power for the 24V motor to propel the car. A small battery pack has used just to start the engine and initially magnetized the rotor generator. A capacitor of 4F is connected with motor to recover the deceleration energy, which is 600 W when the vehicle decelerates from 90 km/h to 20 km/h and to charge the battery. The specification of hybrid engine has shown in Table 1. In continuous operation, the decelerated recovered energy by the capacitor has been used to operate the air-conditioning system of the ca

Table 1: Specification of hybrid engine powered car

Description	Unit	Values
VEHICLE		
Weight including driver weight	kN	2.36
Operating speed	kM/h	30-75
Traction torque (maximum)	Nm	25
HYBRID ENGINE		
Engine size	cc	50
Intake (air and fuel)	Ratio	25:1
Fuel intake	Method	Electromagnetic
Generator Capacity	V	24
Nominal voltage	V	24
Nominal current	A	86
Battery		
Nominal voltage	V	24
Capacity	Ah	86
Capacitor	F	4

The mass flow rate of air is related to the effective throttle area can be computed by using the equation [15]

$$\dot{m}_{thr} = \frac{C_D A_{thr} P_0}{\sqrt{RT_0}}\left(\frac{P_m}{P_0}\right)^{1/\gamma}\left[\frac{2\gamma}{\gamma-1}\left\{1-\left(\frac{P_m}{P_0}\right)^{(\gamma-1)/\gamma}\right\}\right]^{1/2} \quad \text{if} \quad \frac{P_m}{P_0} > \left[\frac{2}{\gamma+1}\right]^{\gamma/\gamma-1} \tag{7}$$

$$\dot{m}_{thr} = \frac{C_D A_{thr} P_0}{\sqrt{RT_0}} \gamma^{1/2} \left[\frac{2}{\gamma+1}\right]^{\frac{(\gamma+1)}{2(\gamma-1)}} \quad \text{if} \quad \frac{P_m}{P_0} \leq \left[\frac{2}{\gamma+1}\right]^{\gamma/\gamma-1} \tag{8}$$

with $$A_{thr} = -\frac{d^2}{2a}\left(1-a^2\right)^{1/2} + \frac{D^2}{2}\arcsin\left(1-a^2\right)^{1/2}$$

where, a= d/D, is the diameter ratio, D is the throttle bore diameter in m, d is the throttle pin diameter, γ is the ratio of the specific heat ($\gamma = c_p / c_v$, c_p and c_v are the heat coefficient at constant pressure and volume, respectively).

2.3 Control System of Hybrid Engine

Hybrid engine generates the power required to drive the vehicle via a combination of an internal combustion engine and an electric generator [16]. The reliable management of power flows or torque distributions may require the performance analysis of the engine such as engine sizing optimization. However, to assess the engine torque generation a number of indicators could be used. They are definable in dynamics with on-board information from the sensors: engine air fuel ratio *A/F*, vehicle travelling speed ω, and traction torque T_t. Improving the control strategy design for the on-board information is considered as one of the most promising concepts is management of power flow of a hybrid engine. Fuzzy logic controller (FLC) is used to determine proper engine air fuel ratio. Another fuzzy controller is used to control the generator output power.

The controlling variables of the hybrid engine powered car are: A/F ratio for optimum of 15:1 (minimum) and 25:1 (maximum), Vehicle speed of 0-35 km/h (optimum is 30 km/h at 0 degree slope road and 8 km/h at 6.7% gradient), vehicle traction torque: 0-25 Nm (optimum is 20 Nm based on wheel torque, which is equal to the motor shaft torque as the final drive ratio is considered as 1:1), motor output power is optimized as 400 W, which in turn gives 20 Nm shaft torque based on 30 km/h vehicle travelling speed. Engine torque: 0-35, Optimum is 6.45 Nm, optimum engine speed of 4000 rpm with 2.7 kW power to run the generator in order to produce the required current and voltage for powering the DC motors to propel the car. Battery power: 0-500 W, the optimum is 400 W based on 20 A current with 20 V. Generator current: 0-75 amp, generator produces maximum current 56 amps without cooling and 72 amps with cooling to meet the car maximum traction power.

Air-fuel ratio (AF), vehicle speed (VS) and engine traction torque (TW) are used as input parameters and vehicle torque generated from generator (TG) is used as output variable in FLC "Engine". For fuzzification of these factors the linguistic variables low (L), medium (M)and high (H) are used for the input parameters.Subsequently, the linguistic variables low (L), low medium (LM), medium (M), high (H) and very high (VH) are used for the output. The units of the input and output variables are: AF (%), VS (km/h), TW (Nm) and TG (Nm). For the three inputs and one output, a fuzzy associated memory or decision is formed as regulation rules. The creation of fuzzy rules is mainly depended on expert assessments, empirical analysis, system application, and so on. Subsequently, the rule bases can also be shaped from the neural network model. Total of 27 rules are formed in the cases under consideration. For example, Rule 5 and 15 can be interpreted as follows:

Rule 5: If AF = L and VS = M and TW = M, then TG = M, i.e. if the engine's AF is low and vehicle speed is medium and engine traction torque is medium, then torque generated from generator is medium.

Rule 15: If AF = M and VS = M and TW = H, then TG = H, i.e. if the engine's AF is medium and vehicle speed is medium and engine traction torque is high, then torque generated from generator is high.

Fuzzifications of the air-fuel ratio (AF), vehicle speed (VS),engine traction torque (TW) and torque generated from generator (TG)are made by aid follows functions. These formulas are determined by using measurement values.

$$AF(i_1) = \begin{cases} i_1; & 5 \le i_1 \le 20 \\ 0; & otherwise \end{cases} \tag{9}$$

$$VS(i_2) = \begin{cases} i_2; & 5 \le i_2 \le 30 \\ 0; & otherwise \end{cases} \tag{10}$$

$$TW(i_3) = \begin{cases} i_3; & 5 \le i_2 \le 20 \\ 0; & otherwise \end{cases} \tag{11}$$

$$TG(o_1) = \begin{cases} o_1; & 5 \le o_1 \le 25 \\ 0; & otherwise \end{cases} \tag{12}$$

In equations (9-12), i_1 is the first input variable (AF), i_2 is the second input variable (VS), i_3 is the third input variable (TW), and o_1 is the first output variable (TG). Prototype fuzzy sets for the fuzzy variables, namely, air-fuel ratio (AF), vehicle speed (VS),engine traction torque (TW) and torque generated from generator (TG)are set up using MATLAB FUZZY Toolbox.

3.3 Prediction of generator current

Results obtained from the engine torque generation of generator can be used thereafter in the fuzzy inference system (FIS) "Generator", which performs the prediction of generator current thus improving the management of power flow of a hybrid engine. In particular, the FIS "Generator" is intended for investigating the already obtained generator torque to assess the possible distribution in road-wheels due to the current transmission. As can be noticed from Figure 2, FIS "Generator" has the generator torque T_g, and battery power P_b as input variables for calculating the generator current. In accordance with the scheme from Fig. 2, in the second stage the vehicle seeks the value of a certain current generated from the generator (I) during engine operation. In particular, this part performs the generation of current on the basis of

- Engine torque generated from the generator (torque generation range is 0-35, optimum is 6.45 Nm at the optimum engine speed of 4000 rpm with 2.7 kW power to run the generator in order to produce the required current and voltage for powering the DC motors until zero level of the fuel).
- Battery power (battery power range is 0-500 W, optimum is 400 W based on 20 amp current with 20 V).

Implementation of fuzzy values into the system, torque generation from the generator (TG) and battery power (PB) are used as input parameters and generator current (I) is used as output variable in FIS "Generator". For fuzzification of these factors the linguistic variables low (L),low medium (LM),medium (M), high (H) andvery high (VH) are used for the input parameters.Subsequently, the linguistic variables low (L), low medium (LM), medium (M),high medium (HM), high (H) and very high (VH) are used for the output. Similar procedures have been employed for theFIS "Generator". The units of the input and output variables are: TG (Nm), PB (W) and I (amp). Total of 25 rules are formed in the cases under consideration. Fuzzifications of the torque generated from generator (TG), battery power (PB) and generator current (I)are made by aid follows functions. These formulas are determined by using measurement values.

$$TG(i_1) = \begin{cases} i_1; & 5 \le i_1 \le 25 \\ 0; & otherwise \end{cases} \tag{13}$$

$$PB(i_2) = \begin{cases} i_2; & 0 \le i_1 \le 100 \\ 0; & otherwise \end{cases} \tag{14}$$

$$I(o_1) = \begin{cases} o_1; & 0 \le o_1 \le 80 \\ 0; & otherwise \end{cases} \tag{15}$$

where, i and o indicate input and output of the fuzzy variables.

4.0 Result And Discussion

The city car (Figure 3) is developed with hybrid engine model. The car is tested both in laboratory as well as on the road surface at different slope. The engine tank was filled with 100 ml fuel in each test. The engine has been set at the optimum 4000 rpm to run the rotor generator in order to produce the required current and voltage to power the DC motors. In 1st test, the engine was stalled and unable to power the car to accelerate from start. The power deficiency of the generator during starting the car was overcome by using battery power, which, was connected in series with the output of the generator. It was found that the car was started to move from its first starting very smoothly with the power of generator and the additional discharge power of the battery. But the problem was incurred once the engine was at very high speed about 120 km/h. Figures 4-5 show the developed the fuzzy control surface to show the mesh plot of the example relationship betweentorque generated from generator (TG) and battery power (PB)on the input side andgenerator current (I)on the output side. The surface shown in the Figs 4-5 are the spatial interpretation of fuzzy rules "IF-THEN" for the chosen input and output variables. It shows that as the generator torque and battery powerincrease, there is concomitant increase in generator current steadily initially and further increase rapidly as expected. The current reaches the apex when the generator torque and battery power both reach their respective maximum level. Hence, all the components contribute significantly towards currentgeneration. Subsequently, it could be noted that the effect is highly significant at the higher level of battery power. In particular, current generation increases with the increase of battery power significantly, although the effect is less prominent at the lower level of generator torque. Thus, proper management of power flow or torque distribution is a critical issue at present in order to obtain energy efficient hybrid engine. The illustration shows that monitoring the output current and improving the control strategy design for the on-board information may lead the hybrid engine car (HEC) as efficient as possible. However, the reliable management of power flows or torque distribution may require the performance analysis of the engine such as engine sizing optimization.

Analytically, it is estimated that the DC motor need to get 360-400 watt to propel the car with speed of 35-45 km/h. Test result shows the hybrid engine able to produce the desired power for the car propelling at 4000 rpm. Figure.6(a) and 6(c) shows that the hybrid can generate the power 504 W at maximum 4000 rpm without the support of battery. Figures 6(b) and 6(d) shows that the hybrid engine was not in problem to generate the required power of the vehicle propelling at operating speed of maximum 4000 rpm. However, the engine was heaped once the speed reach to 5000 rpm and above. Hybrid engine produces 15-20 A to meet the power requirement of the DC motor to propel the car at 35-45 km/h constantly until the fuel supply stop (Figure 7). It is found that the car was propelled 14.3 km (4.6 km by 100 ml fuel, 7.6km by 700 W battery power and 2.1 km by deceleration power) with the hybrid engine powered. Equivalently for 1 liter of fuel, the car was propelling 57.4 km. The misalignment of the engine with the alternator due to the vibration, the engine was not run at high speed. The engine was heaped which was due to the safety features in the rectification circuit cut off the current supply when the current exceed the predetermine current, in this case 70A at 7000rpm. As the safety circuit activated, it cut the current supply and the voltage drop from 14.8V to less than 3 volt.

Figure 3. Hybrid engine powered city car

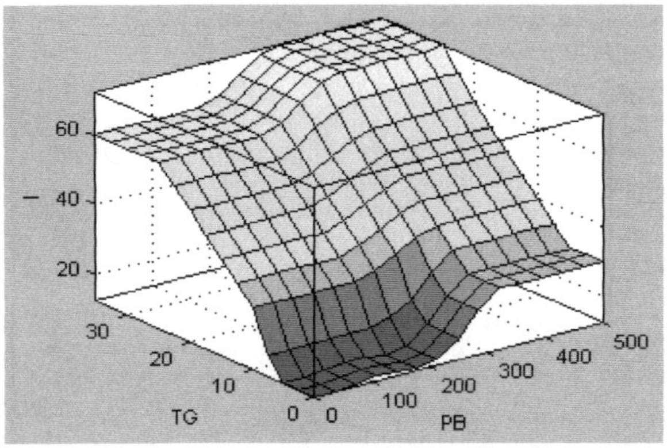

Figure 4. Control surfaces of the fuzzy inferring system for TG with VS and AF

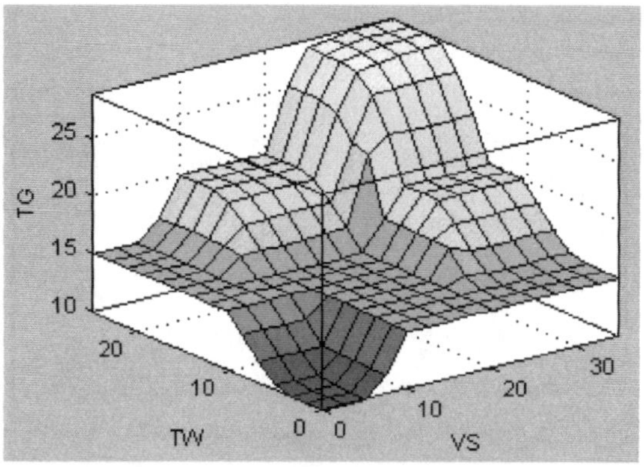

Figure 5. Control surfaces of the fuzzy inferring system for TG with TW and VS

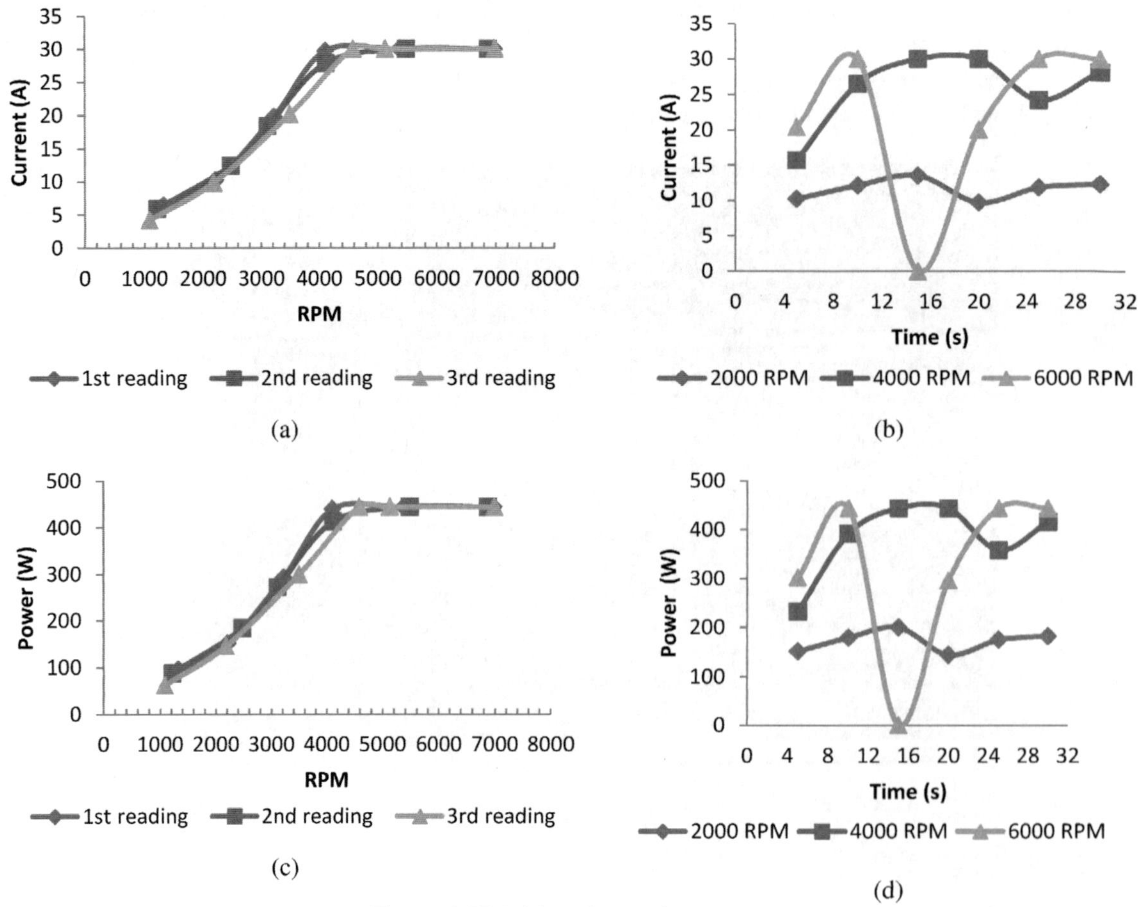

Figure 6. Hybrid engine performance

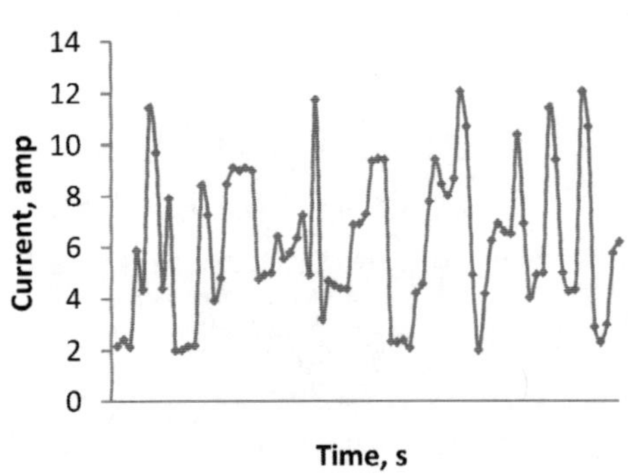

Engine rpm = 2000

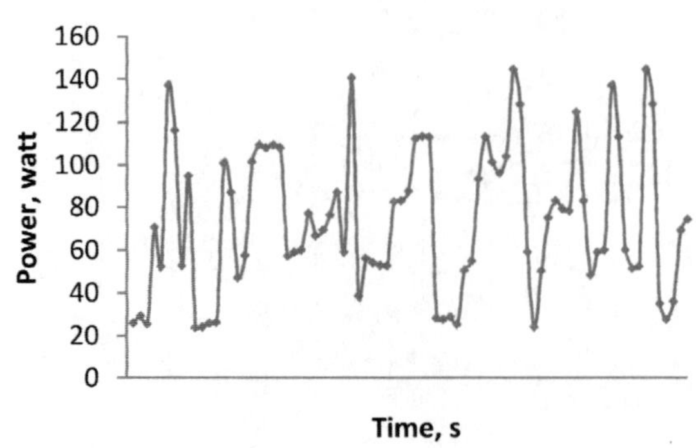

Engine rpm = 2000

IOP Conf. Series: Materials Science and Engineering **184** (2017) 012026 doi:10.1088/1757-899X/184/1/012026

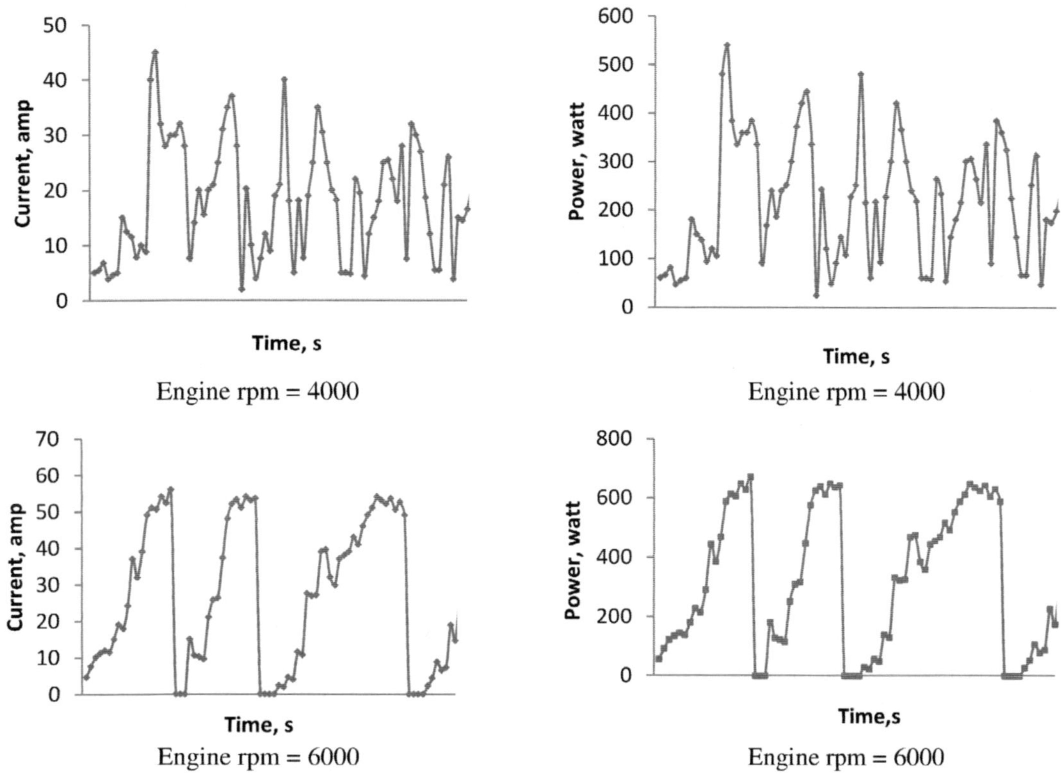

Figure 7. Hybrid engine field performance result.

Performance investigation of the has been investigated for the vehicle speed in the range of 30-80 km/h. The result as shows in Figure 7 shows that the vehicle was not in trouble when it was operated with speed of 30-50 km/h. While, the vehicle was in trouble when its accelerated for the speed 60-75 km/h. This is due to the control system of the hybrid engine generator and limited battery power capacity. The discharging capacity and thermal behavior of the battery has been investigate during operation of the vehicle. The result shows in Figures 8(a) and 8(b) shows that discharge current of the battery in the range 60-80 A for the vehicle speed in the range of 30 – 55 km/h. Figures 8(c) and 8(d) battery temperature raised to 39°C for the activation of cooling system and increased to 65°C without activation of cooling system. It could be mentioned that cooling system needs to activate all the way during vehicle operation to maintain the battery life as the lithium ion battery's recommended temperature would not be more than 50°C.

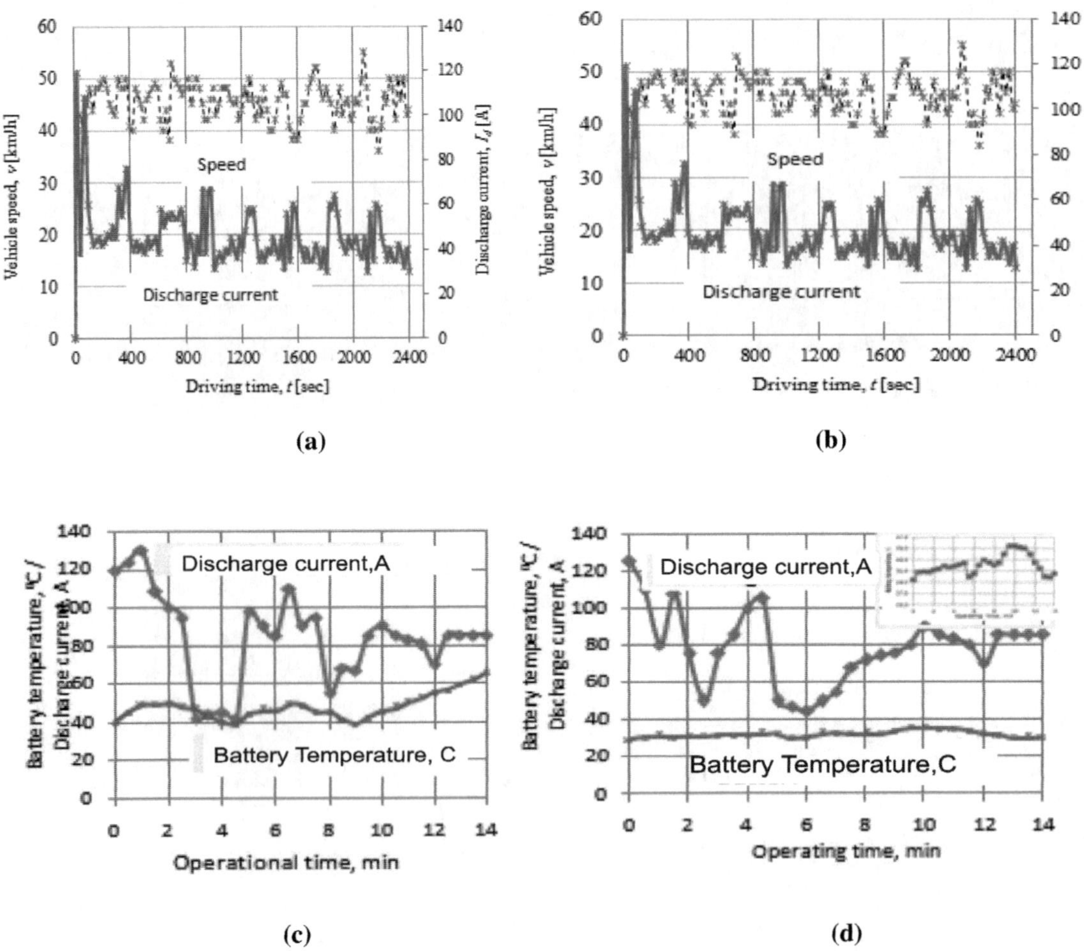

Figure 8. Battery discharge current and temperature.

5. CONCLUSION

Vehicle operation with the hybrid engine power is less efficient than its operated with the fully battery power as the efficiency of the generator is recorded as maximum 70%. The engine power can be increased by adding few extra field diodes and at the neutral point of the generator (Y type) to rectify the un-rectified power. The huge engine vibration causes lot of power transmission losses. With proper alignment of the engine and generator it could be eliminated. Using 100 ml fuel both in laboratory and road experiment tested the vehicle. It was found that the vehicle travelled 4.5 km with the power of 100 ml fuel.

REFERENCE

[1] EU report, 2001. EU Directive 80/1268/EEC Amended by 2004/3/EC". Retrieved from: http://ec. europa.eu/enterprise/sectors/automotive/documents /directives/directive-80-1268-eec_en.htm(Accessed on: September 13, 2010).

[2] J. Currie et al.Air pollution and infant health: Lesson from New Jersey. J. Health Econ., 28(3): 688-703, 2009.

[3] M.T.,Salam et al. Birth outcomes and prenatal exposure to ozone, Carbon MoNO2ide and particulate matter: Results from the children's health study. Env. Health Perspect., 113(11): 1638-1644.

[4] B. Ritz et al. Air pollution and infant death in Southern California, 1989-2000. Pediatrics, 118(2): 493-502, 2006.

[5] M.Wilhelm et al. Local variations in CO and particulate air pollution and adverse birth outcomes in los angeles county, California, USA. Env. Health Perspect., 113(9): 1212-1221, 2005.

[6] T.J Woodruff et al. Air pollution and postneonatal infant mortality in the united states, 1999-2002. Env. Health Perspect., 116(1): 110-115, 2008.

[7] M. Kim et al. Fuzzy control based engine sizing optimization for a fuel cell/battery hybrid mini-bus. J. Pow. Sour., 178: 706-710, 2008.

[8] A. Poursamad et al. Design of genetic-fuzzy control strategy for parallel hybrid electric vehicles. Cont. Eng. Pract., 16: 861-873, 2008.

[9] A. Rajagopalan et al. Development of fuzzy Logic and neural network control and advanced Emissions modeling for parallel hybrid vehicles. Center for Automotive Research, Intelligent Structures and Systems Laboratory, Ohio State University, USA, 2003.

[10] A. Rahman et al. Kinematics and nonlinear control of an electromagnetic actuated CVT system for passenger vehicle. J. Mech. Sci. Technol., 26(7): 2189-2196, 2012.

[11] A. Rahman et al. Power generation from the waste of IC engine. Journal of Renewable and Sustainable Energy Review, Elsevier Publisher, Vol 51. pp. 382-395, 2015.

[12] V.Ivanov et al. Fuzzy evaluation of tyre-Surface interaction parameters. J. Terramech., 47: 113-1130,

[13] Reza N. Jazar. *Vehicle Dynamics: Theory and Application*, Springer Science, LLC, 233 Spring Street, New York, 10013, USA, 2008.

[14] A. Rahman et al. Fuzzy knowledge-based model for prediction of traction force of an electric golf car. Journal of Terramechanics, Vol. 49(2012), pp. 13-25.

[15] B.H John. *Internal Combustion Engine Fundamental*. MaGraw-Hill, New York, pp: 930, ISBN: 007028637X, 1988.

[16] A. Rahman, Ataur. Development of a model of energy efficient hybrid engine towards green technology Int. J. of Electrical and Hybrid Vehicle. Vol.3, No.2, pp: 138-151, 2011.

IOP Conf. Series: Materials Science and Engineering **184** (2017) 012027 doi:10.1088/1757-899X/184/1/012027

An investigation into NVC characteristics of vehicle behaviour using modal analysis

Zahir Hanouf[1,4], Waleed F.Faris[2] and Kartini Ahmad[3]

[1, 2] Mechanical Engineering Department. International Islamic University Malaysia. 50728 Gombak Kuala Lumpur, Malaysia
[3] Science of Engineering Department. International Islamic University Malaysia. 50728 Gombak Kuala Lumpur, Malaysia.
E-mail: Zahir8@iium.edu.my

Abstract. NVC characterizations of vehicle behavior is one essential part of the development targets in automotive industries. Therefore understanding dynamic behavior of each structural part of the vehicle is a major requirement in improving the NVC characteristics of a vehicle. The main focus of this research is to investigate structural dynamic behavior of a passenger car using modal analysis part by part technique and apply this method to derive the interior noise sources. In the first part of this work computational modal analysis part by part tests were carried out to identify the dynamic parameters of the passenger car. Finite elements models of the different parts of the car are constructed using VPG 3.2 software. Ls-Dyna pre and post processing was used to identify and analyze the dynamic behavior of each car components panels. These tests had successfully produced natural frequencies and their associated mode shapes of such panels like trunk, hood, roof and door panels. In the second part of this research, experimental modal analysis part by part is performed on the selected car panels to extract modal parameters namely frequencies and mode shapes. The study establishes the step-by-step procedures to carry out experimental modal analysis on the car structures, using single input excitation and multi-output responses (SIMO) technique. To ensure the validity of the results obtained by the previous method an inverse method was done by fixing the response and moving the excitation and the results found were absolutely the same. Finally, comparison between results obtained from both analyses showed good similarity in both frequencies and mode shapes. Conclusion drawn from this part of study was that modal analysis part-by-part can be strongly used to establish the dynamic characteristics of the whole car. Furthermore, the developed method is also can be used to show the relationship between structural vibration of the car panels and the passengers' noise comfort inside the cabin.

1. Introduction

The size of requirement for industrial vehicles excelling in comfort and reliability has become very large for the last few decades. Due to the demands of high speed operation and the use of light weight and flexible structures in modern machinery, static measurements of stress/strain properties are not sufficient. One of the main areas that should be considered when investigating a vehicle is its NVC characteristics. This is leads to the improvement of its vibro-acoustic quality as we know that structural vibration has a strong relationship with the interior noise. Therefore, due to the important of NVC characteristics of vehicles it is become a competitive target for car selling. The more refined the car is, the more selling potential it will be. In improving noise, vibration and comfort (NVC) characteristics of a vehicle, it is a must to understand the dynamic behaviour and characteristics of each structural part of the vehicle. This issue has been lacking in the open scientific literature. The only available information is modal analysis of the body-in-white and chassis but the individual structural components for the whole car still not studied yet. It has been researched and studied for decades how computational and experimental modal analysis methods are used to solve noise and vibration problems in automotive and also the importance of the combination between them in order to provide a comprehensive understanding of the dynamic characteristics of vehicles prior and after manufacturing which is leading to improve their noise, vibration and comfort. This method has been used successfully for simple structure. However, for complex structure like cars still under investigation, because of no good linearity between the excitation force and the response force which for sure affect the quality of the results. Therefore, to solve this problem we have developed a new method. This method is based on modal analysis part by part and was validated by a comparison between experimental and computational modal analysis. It helps us to identify easily the individual structural component of the vehicle that can be radiate high vibration or contribute to the interior noise and then getting control of it. In this work, we are addressing this point by detailed study for a specific passenger car model as a case study using modal analysis part by part.

Knowing dynamic characteristics of a vehicle components such as natural frequencies, mode shapes and damping ratios are essential for vehicle comfort, structural safety evaluation, longer fatigue life, and structural health monitoring. By knowing the dynamic behaviour of a structure under its natural frequencies, we could determine the location of damaged parts of the structure before it starts causing losses of life, because damaged structure tends to exhibit excessive deformation under the same magnitude of excitation. Also by knowing the dynamic characteristics of a structure, engineers can design better structures.

Weight, stiffness and damping are the important factors that have a strong effect on the natural frequencies and mode shapes of a structure, which they can determine where these natural frequencies and mode shapes will exist. For that reason, we have to look into these frequencies and know how they affect the vibration behavior of the structure when a force excited it. This can help engineers to design good structures with improved vibration. Therefore, most of the developments of the structural vibration analysis tools in the new era of engineering came about due to the need of solving vibration problems. Modal analysis in both methods (experimental and computational) is one of the useful tools that provide an understanding of structural dynamic characteristics, operating condition and performance criteria that makes design of optimal dynamic behavior or solving structural dynamics problems in the existing designs.

2. Computational modal analysis in automotive

By the early 70's, Finite Element Analysis (FEA) was limited to expensive mainframe computers generally owned by the aeronautics, automotive, defence, and nuclear industries. Since the rapid decline in the cost of computers and the phenomenal increase in computing power, FEA has been developed to an incredible precision. Present day supercomputers are now able to produce accurate results for all kinds of parameters. There is a lot of researches conducted in the last decade provide a comprehensive source of information. This information can be used as a guideline to make this work more successful. The following are samples of researches performed on structural dynamic analysis of vehicles using computational modal analysis method.

A computational modal analysis was used to investigate the dynamic characteristic of a truck chassis design to determine the relationship between road excitations and the vibration and strength of the truck chassis. The responses of the truck chassis which include stress distribution and displacement under various loading condition were also investigated. There are two types of structural analysis carried out; normal mode analysis to determine the natural frequencies and mode shapes, and the linear static stress analysis to look into the stress distribution and deformation pattern of the chassis under static load. Different boundary conditions were used for each analysis. It was found that the road excitation is the main disturbance to the truck chassis as the chassis natural frequencies lie within the road excitation frequency range. Furthermore, the determination of the suitable mounting locations of components such as engine and suspension system was possible from the mode shape (Teo & Roslan 2007). Another work was also carried out by (Kassahun 2008) on static and dynamic analysis of a commercial vehicle structure (van body). The method used was finite element modelling and analysis for which the inputs were obtained from quarter car model analysis. The dynamic analysis includes only modal and random response analysis of the structures. In the dynamic analysis, damping properties of the structure were not considered, only the damping due to shock absorber of the vehicle was included in the analysis. The analysis was carried out for the main load bearing structure of the vehicle, i.e., chassis side frames and cross members, steel frames of the van body structure, and cab. The responses to static loads and random excitation caused by road roughness were determined. Components and, particularly areas which are much affected by the different loads were identified. A further method was reported and validated by (Abrams 2008) using an engine as a structural member of a formula of Society of Automotive Engineers (SAE) race car chassis. A modal analysis was used with a beam mesh of the chassis to provide relative comparison; with and without structural support from an engine. It was also used to validate the assumption that chassis stiffness would increase an increase of chassis stiffness. Results confirmed that providing structural support at engine mounting points of the chassis increased the stiffness and changed the first elastic mode of vibration from bending to torsion. Additionally, a static analysis was used to measure chassis deflection, and established a decrease in deflection when structural support was provided at the engine mounting points of the chassis. This report's conclusion indicated that using the engine as structural member significantly improves the stiffness of the chassis.

In a different study (Zhanwang & Chen 2000), finite element model for Mini-car Changan Star 6350 using SDRC/I-DEAS was developed. The model was imported to MSC. Patran to generate the mesh and the dynamic analysis of the model was accomplished using MSC.Nastran. The analysis included modal analysis of the load bearing structures, the chassis and the body structure frame, in which the first 100 natural frequencies were determined in torsion and bending modes, and transient dynamic analysis of the structures due to road loads that were introduced in the form of acceleration excitations. Results show good consistency with experimental results, therefore, ability of Mini car to meet dynamic design requirements was confirmed. The implementation of FE methods for dynamic analysis of heavy trucks was presented by (Johansson & Gustavsson 2000). The proposed approach involves the development of FE model of the vehicle and its subsystems, i.e. frame, superstructure, engine, cab, axles and chassis suspensions, and steering system. There was a focus on the analysis of the road, wheel, brake, and driveline induced vibrations, static load cases and fatigue life prediction. An example was presented which shows the lateral and vertical acceleration of the cab structure due to periodic excitation of axle due to wheel run out. It was concluded that, FE-based complete vehicle analysis has the advantage to cover broad range of analyses with one type of model.

An optimization of light weight bus side structures was reported (Lan et al. 2004). The optimization targeted the reduction of the number as well as weight of different parts in the whole body structure. For this purpose, two scenarios, with and without supporting structures between longitudinal waist beams of the side frame were considered. The FE-model for the structure without the supporting structure was developed using the computer package UG and mesh generated in MSC.Patran, while the FE Analysis was completed using ANSYS. All the results obtained were compared with experimental results. The study was stated that the structure with support enables significant alleviation of body weight and satisfactory performance requirements were achieved. Furthermore, a comparative study was conducted between analytical and numerical methods for modelling automotive dissipative silencers with mean flow (Kirby 2009). The results presented in this study demonstrate excellent agreement between analytic and numerical models provided a sufficient number of propagating acoustic modes were retained. However, the numerical mode matching method was shown to be the fastest method. Moreover, the hybrid finite element method was demonstrated to be as fast as the analytic technique. The possible causes of the failure of an exhaust system for heavy duty commercial trucks and the possible solutions were examined (Vascncellose et al. 2003). The researchers describe the use of computer simulation supported by experimental data on the design process of the exhaust system, with regard to structural behaviour. A finite element model was generated including the complete vehicle and the exhaust system. Static and dynamic analyses were performed in MSC.Nastran software simulating different loading conditions and system configurations. The results obtained assured the structural integrity of the modified exhaust system when implemented on the vehicle and also contribute to a better understanding of this system behaviour and its structural strength. Generally, numerical models validated with experimental data are a powerful tool during the development phases of a new vehicle, reducing project time and costs.

A numerical modelling approach for investigating the onset of squeal in a drum brake system was proposed few years back (Jinchun et al. 2006). The brake system model was based on the modal information extracted from finite element models for individual brake components. The component models of drum and shoes were coupled by the shoe lining material which was modelled as springs located at the centroids of discretized drum and shoe interface elements. The vibrational characteristics of the developed multi degree of freedom coupled brake system model were determined by a complex eigen-value analysis. The study showed that both the frequency separation between two system modes due to static coupling and their associated mode shapes played an important role in mode merging. In another research work (Jae-Yeol Park & Singh 2010). The effect of spectrally varying mount properties (including stiffness and damping) was examined analytically on the dynamics of power train motions under torque excitation. New analytical formulations were verified by comparing the frequency responses with numerical results obtained by the direct inversion method (based on Voigt type mount model and finally, an axiom for torque roll axis decoupling was provided by employing direct and adjoint eigenvalue problems.

Finally, an interesting approach for a Reduced Beam and Joint Modelling approach to optimize and analyse global vehicle body dynamics, specifically bending and torsion modes was developed currently (Donders et al. 2009). This is a concept of Computer-Aided Engineering (CAE) modelling approach based on predecessor FE models, which can be used to improve the fundamental body noise, vibration and harshness behaviour (global modes) by means of beam and joint modifications. Modifications in the body beam-like sections and in the joints were analysed using body reduced modal model. This small-sized model was used to quickly and accurately optimize the low-frequency vehicle performance. The work conclude that static Beam Force Analysis strategy can be used to provide complementary insight in the vehicle body design in an early design stage, which can be employed to improve the static body performance.

As can be seen the above researches focus on the finite element analysis of vehicles and their different components, i.e. engine, steering system, superstructure, cab and axles and chassis suspensions. Different computer packages such as ANSYS, ADAMS, DADS, MSC.patran/MSC.Natran and IDEAS were used to develop the virtual models of vehicles and their components in order to perform the required analysis the method still recognized as an acceptable tool for designing simple structural design. However, the confidence in the results for complex structures still questionable because of several simplifications in the model. Therefore, certainty in the computational results most often depends on the experience of the analyst who builds the finite element model.

3. Experimental Modal analysis in automotive

One common reason for experimental modal analysis is the verification or correction of the results of computational modal analysis approach. Below are some reviews on the application of experimental modal analysis in automotive.

In a case study work, (Verboven et al. 2006) presented a new developments for the nonparametric processing of modal test data. The application of the presented methods was studied for the case of a typical modal test of body-in white structure. An application of multi sine excitation technique permits improvement of the data quality, as well as the detection, qualification and quantification of nonlinear distortions during Frequency Response Function (FRF) measurements. Also, the study suggested that, to make the presented techniques available for multi-input modal testing, attention is paid to the design of optimal multi-input excitations by maximizing the Fisher information matrix as well as minimizing the crest factor of the applied excitation.

An investigation of the applicability of a non-parametric exponential time-window in the pre-processing of data acquired with arbitrary excitation to improve the modal parameters of an engine sub-frame of a car was done by (Verboven et al. 2005). This sub-frame structure was characterized by a low damping, thus Hanning window was applied to reduce possible effects of leakage. Also implemented a mixed nonparametric-parametric approach to minimize the effects of measurement errors introduced by both the effects of leakage and measurement noise. Engineers are often interested in obtaining the modal parameters for reasons of physical interpretation and applications of numerical model updating and modification prediction.

A successful implementation of modal control to arbitrary curved panels using experimentally evaluated mode shapes was demonstrated (Hurlebaus et al. 2008). The identification of the modal control parameters from a measured data base was of interest. The test object was a modern roaster car body. The floor panel has been selected for the experimental setup. Since, for such structures, the evaluation of modal parameters from numerical calculations of local modes is complicated because the results strongly depend on proper boundary conditions of the truncated structure. Thus, the modal data was identified using experimental modal analysis. The modal controller was implemented on a digital controller board, and experimental tests with the floor panel and centre panel of a car body were carried out to validate the proposed concept. The work ended to a reduction of structural vibrations, which also leads to a reduction in acoustic radiation.

According to the previous research studies, much work has been done on vehicle components such as engine, exhaust system, cabin structure, chassis suspension, bearing structure, power train system and drum brake system using either experimental modal analysis or computational modal analysis or both. In addition there is also some work has been done on the whole car but with using the experimental modal analysis only or computational modal analysis only. The researchers discussed how computational and experimental modal analysis methods are used to solve noise and vibration problems and came out with the importance of the combination between them in order to provide a comprehensive understanding of the dynamic characteristics of vehicles prior and after manufacturing which leading to improve the noise and vibration comfort (NVC). In this study we develop an accurate method to analyse the dynamic characteristics of whole car using modal analysis part by part in order to solve the noise and vibration problems of a passenger car in easy way and less errors. The confidence in the results for complex structures still questionable because of the several simplifications and assumptions applied on the structure. The main assumption involved in modal analysis is that the structural system is linear, i.e., structural displacements are directly proportional to applied loads.

4. Methodology

The main objective of this research is to implement an accurate and reliable method to investigate the dynamic characteristics of a complex structure, hence assures easier and faster implementation. In the first part of this work computational modal analysis part-by-part is utilized to determine and analyse the dynamic characteristics of a specific vehicle structure. The second part carried out experiment modal analysis part by part of the same vehicle. The method was validated by a comparison between experimental and computational modal analysis results. Modal analysis was utilized as a tool to find the dynamic properties of the passenger car namely natural frequency, mode shape and damping. In this paper only few vehicle panels are considered namely, trunk, hood, roof and door panels.

4.1. Computational Modal Analysis part by part

Computational modal analysis part by part was conducted using FEA technique. It is a computer simulation technique used in engineering analysis. In general, there are three phases in any computer-aided engineering task are pre-processing, analysis Solver and Post-Processing. The simulation results of any Finite element analysis depend on many variables. The most important ones are:

a) CAD Model.
b) Boundary Conditions and Initial Loading
c) Assumptions made

The Model used for this investigation is Dodge Neon car from NCAC (National Crash Analysis Centre).
The car model used for this study was designed for crash test and has several other parameters attached to it that is not related to modal analysis. Therefore we have to remove unwanted parts and components that pose a problem to the modal analysis and make the problem complicated. Two softwares have been used in this investigation namely VPG 3.2 and Ls-Dyna. VPG 3.2 software is very capable software with a lot of libraries that contain vehicle related car panels and Ls-Dyna has been used as its processor and solver. The car panels (trunk, hood, roof and door) were extracted from the original FEA model of the whole car then performed modal analysis test. Modal analysis is performed in two stages, firstly the modes are computed and then output to a binary database, secondly this database been used as input to supply modes for the transient dynamic analysis. The modes are computed using implicit analysis.

4.1.1. Results and discussion

The results of computational modal analysis are presented in table1, table 2 and table3. The first mode of vibration occurs at 48.63 Hz, the horizontal part appears to be not vibrating and the vertical part looks like flying freely with high amplitude at the bottom edge. The second mode of vibration occurs at 81.13Hz, the vertical part of the trunk appears to be oscillating in a second bending mode as we can see clearly that the panel exhibit two nodes and three antinodes, it also shows significant vibration at the two corners of the horizontal part of the trunk. At the third mode the panel shows third bending mode at both vertical and horizontal parts of the panel with high amplitude.

The fourth mode was at 105.43 Hz, both parts of the trunk showing the third bending mode with significant amplitude at the horizontal part. In the fifth mode both parts of the panel showing first bending mode. The sixth mode was at 133.81 shows significant vibration at the horizontal part.
Roof panel showed elastic bending for all modes. First and second modes are closed to each other (75.4 Hz and 76.4 Hz) are showing same second bending mode at the rear part of the roof. Most probably the rear beam is the main contributor to the second bending mode. The third and fourth modes have similar behaviour and also their frequencies are close (97.81 Hz and 100.5 Hz). They show first bending mode at the left and right sides of the roof. The fifth mode occurs at 102.9 Hz. At this mode the roof shows a third bending mode at the rear frame of the roof with high amplitude. The last mode was at natural frequency 104.5 Hz. It shows second bending mode at the front of the roof.

The hood panel showed the first mode at 58.83 Hz, it is a torsional mode, each two opposite corners are vibrating in phase. At the second mode the hood panel shows second bending mode, and occurs at 76.08 Hz. The third mode occurs at 86.8 Hz, it appears to be a second bending mode. There are two antinodes, one on the right side and the other one on the left side. The fourth mode of vibration occurs at 107.78 Hz. It shows high vibration on the left and right sides, and looks like flying mode. The fifth mode occurs at 117.6 Hz, and appears to be a third vertical bending mode. The last mode of vibration occurs at 125.7 Hz and only the left side of the hood has significant vibrating.

The fist mode of the door panel exhibits first bending mode at the bottom half of the door panel at 55.71 Hz. The second mode occur at 74.01 Hz. It exhibits second bending mode at the bottom part of the door panel with low amplitude. At the third mode the door panel does not show significant mode. The fourth mode occurs at 94.88 Hz. It shows first bending mode at the centre of the door. This mode is resulted from the vibration of horizontal crossing beam as shown in the figure. At the fifth mode, the door does not show significant bending mode. The last mode was at 105.45 Hz. At this mode the door panel shows second bending mode at the bottom half.

Table1: Comparison between computational and experimental mode shapes for Trunk and Roof panels.

Mode	Trunk panel		Roof Panel	
	Computational	Experimental	Computational	Experimental
1	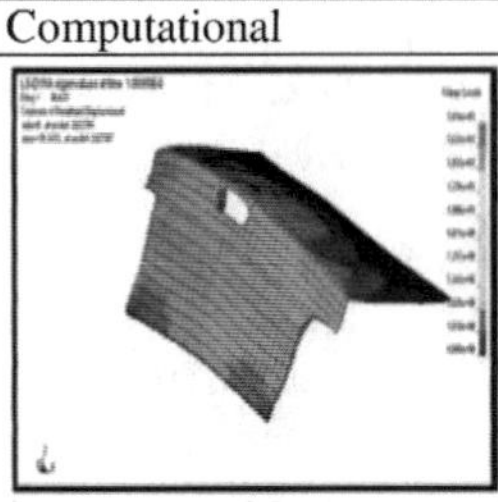	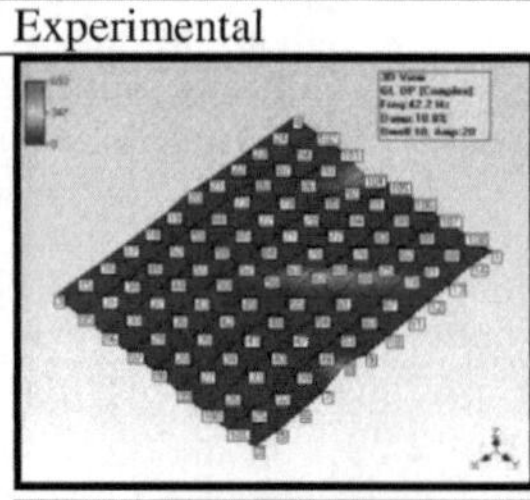	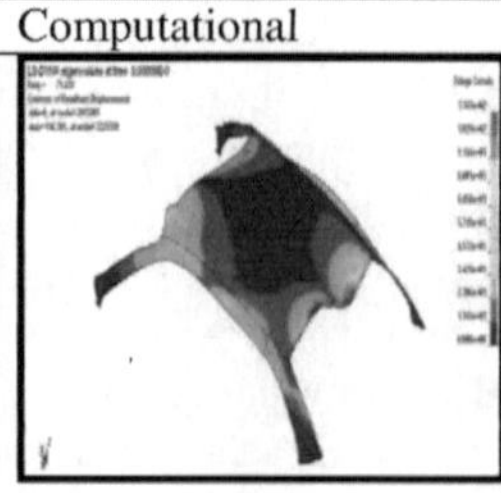	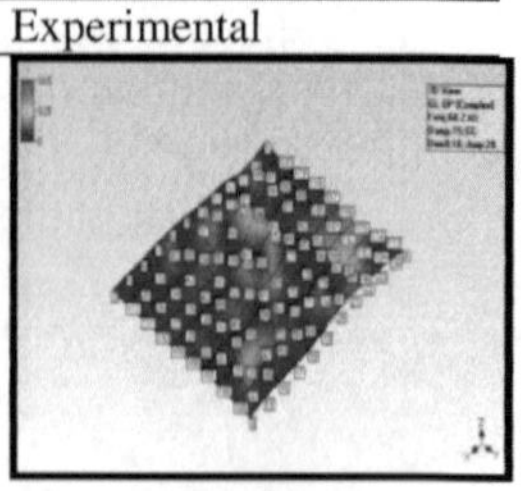
2	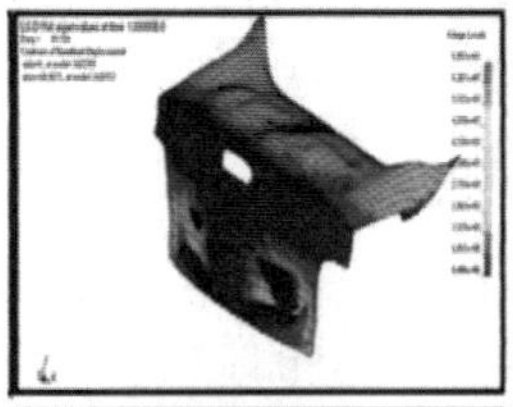	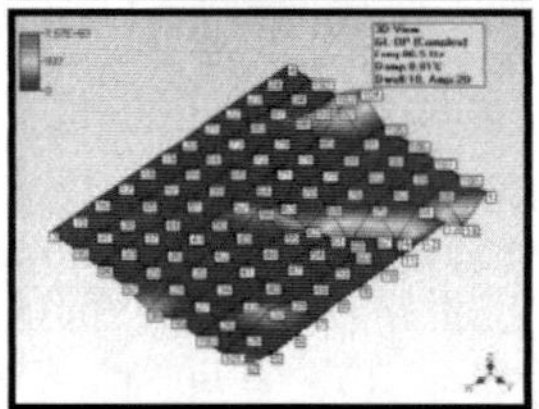	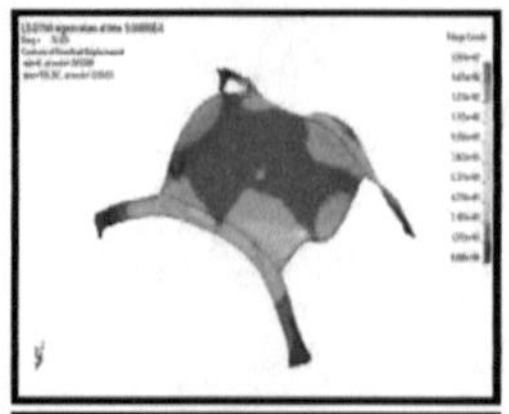	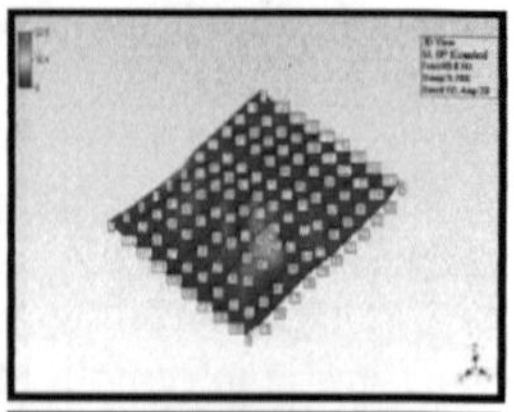
3	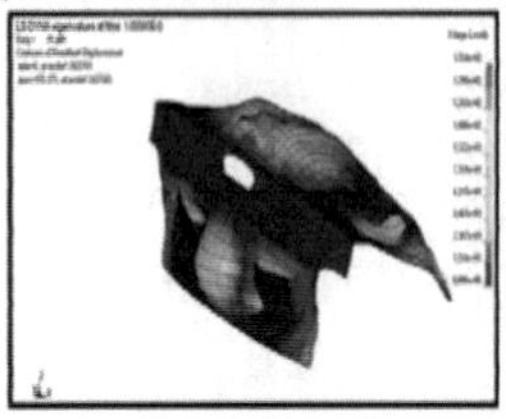	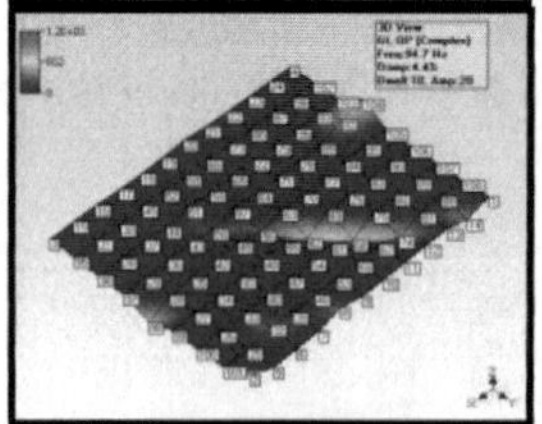	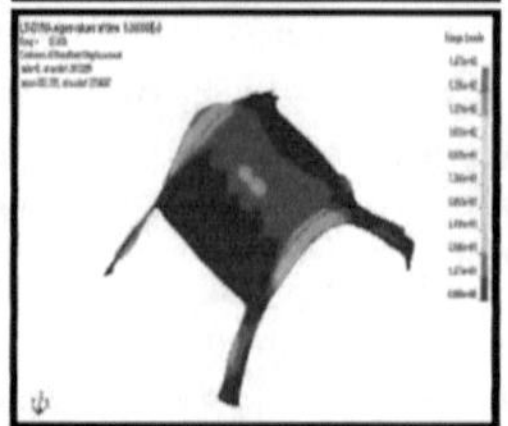	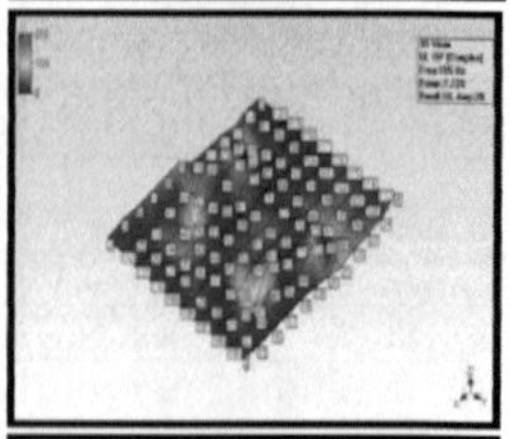
4	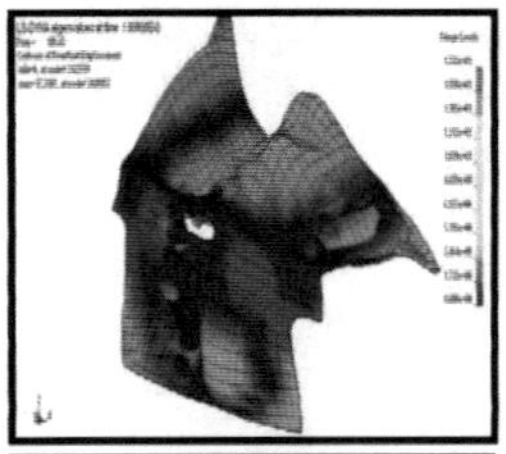	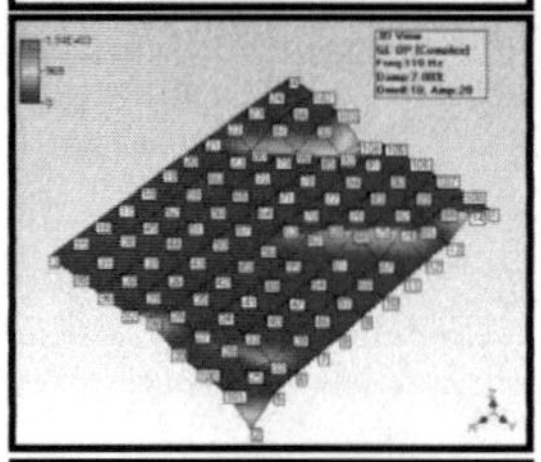	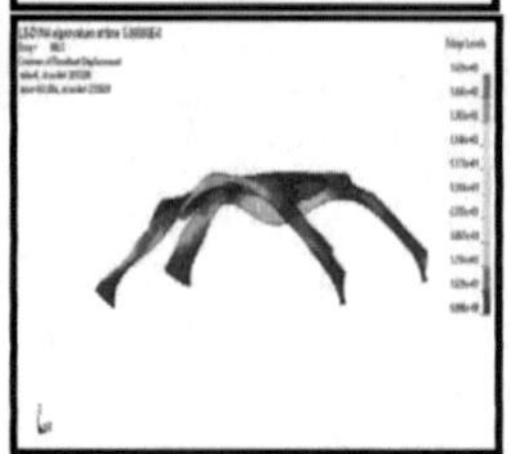	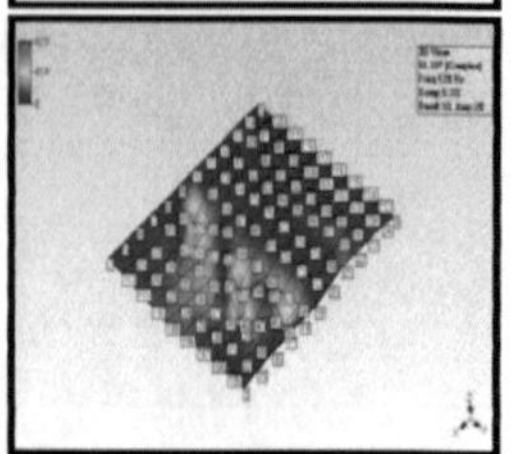
5	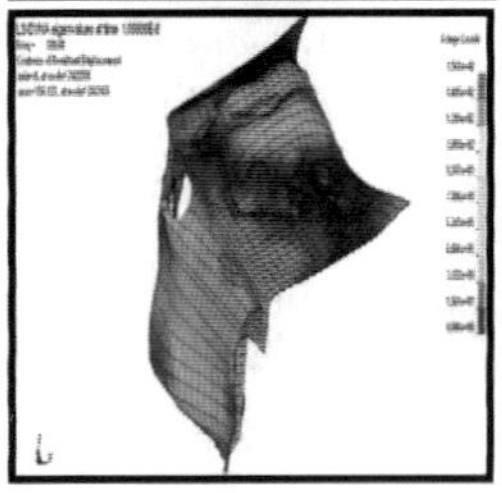	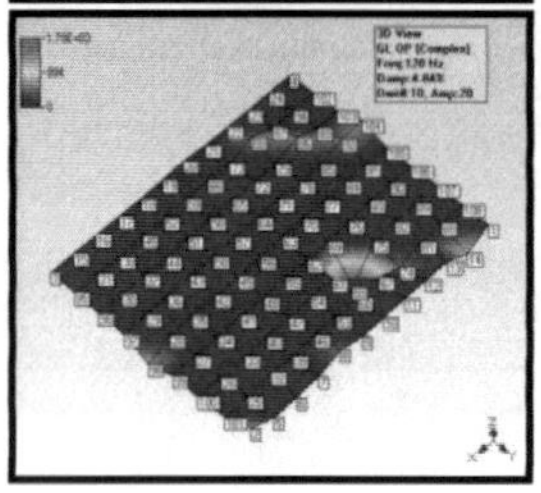		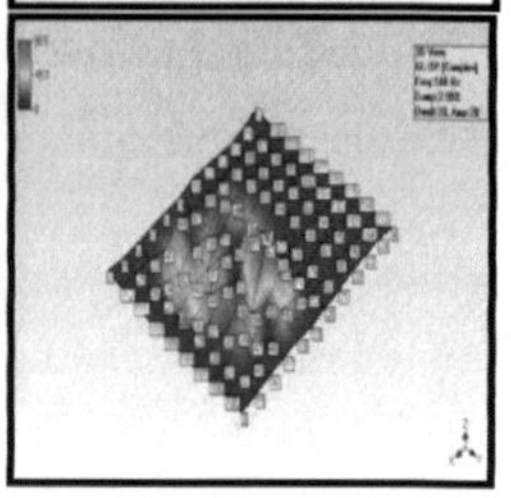

6	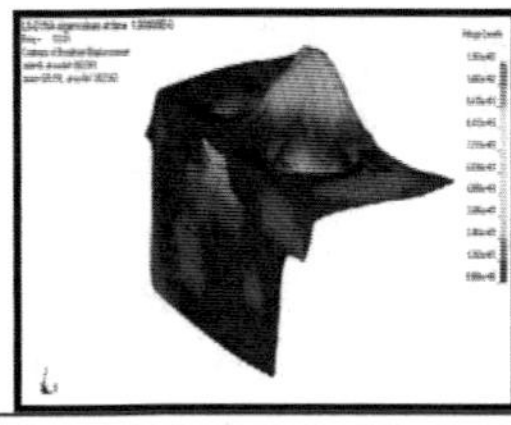	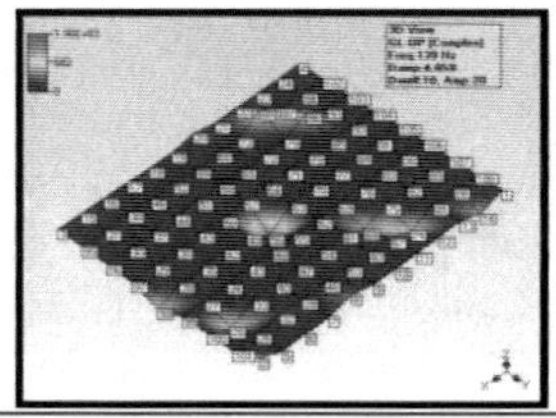	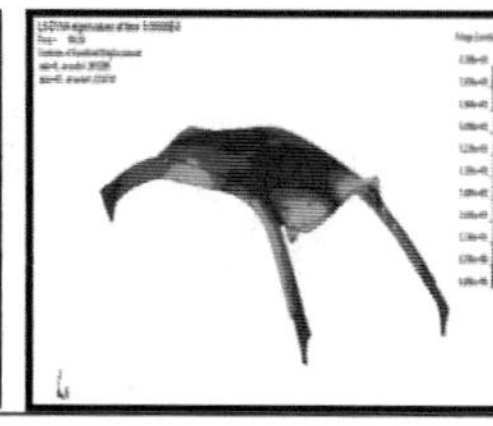	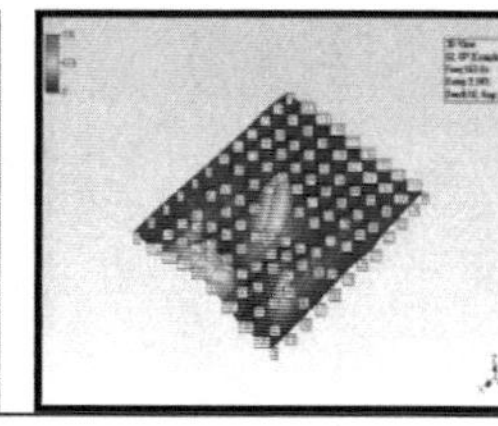

Table2:Comparison between computational and experimental mode shapes for Hood and Door panels.

Mode	Hood panel		Door Panel	
	Computational	Experimental	Computational	Experimental
1	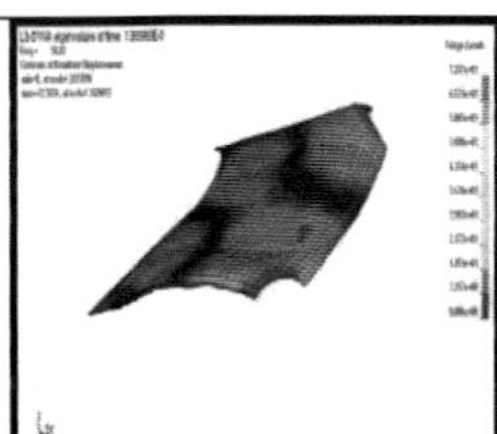	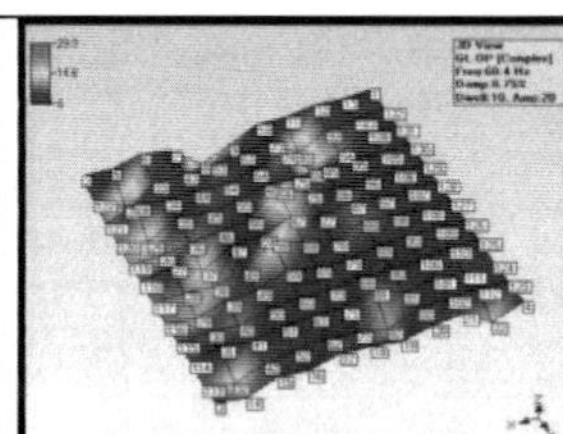	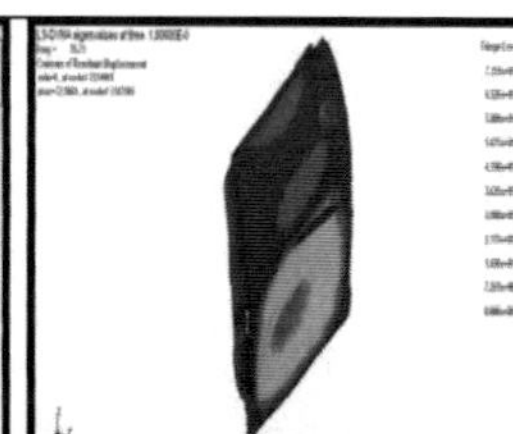	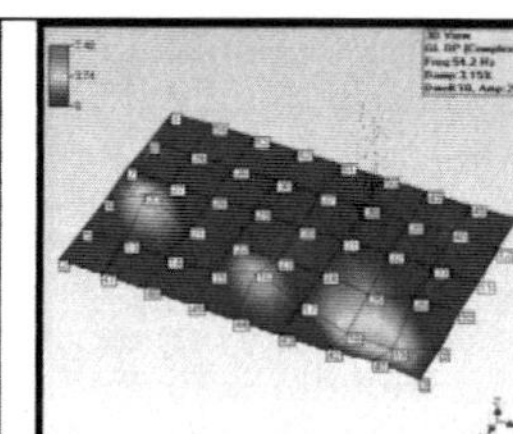
2	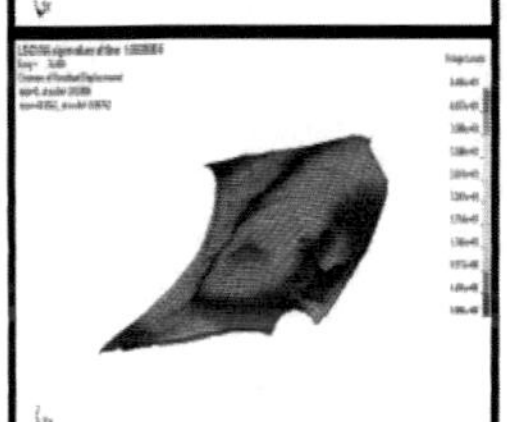		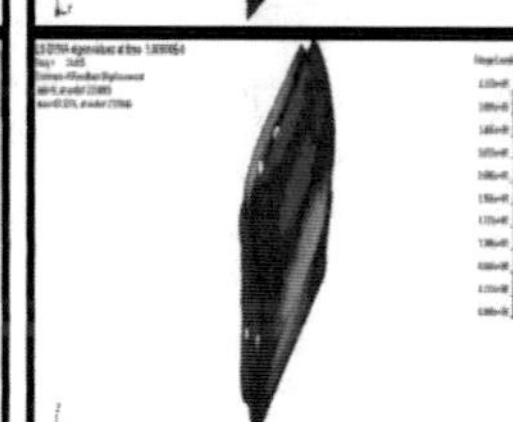	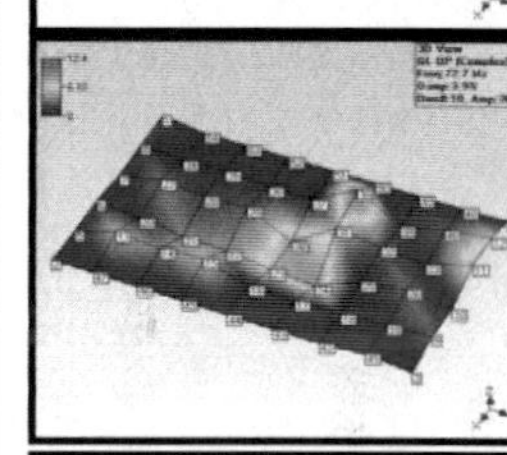
3	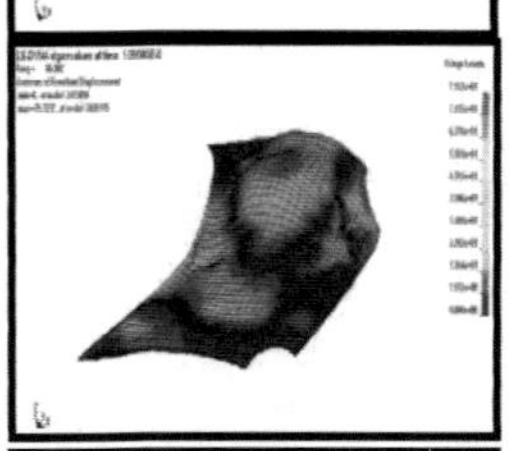		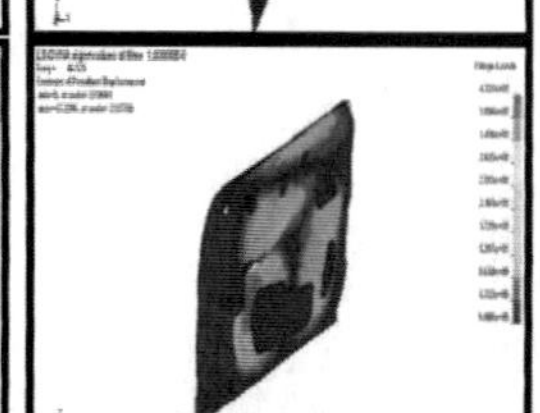	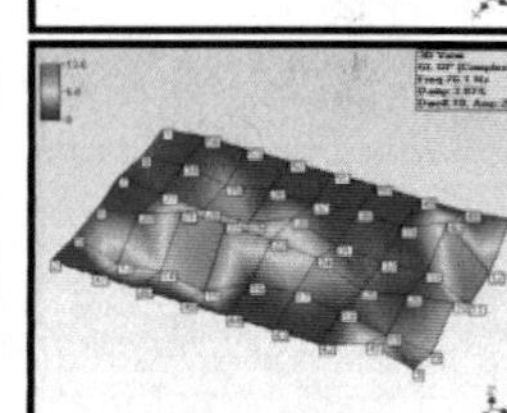
4	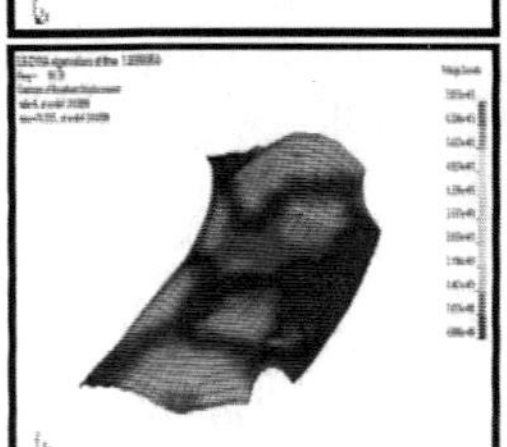	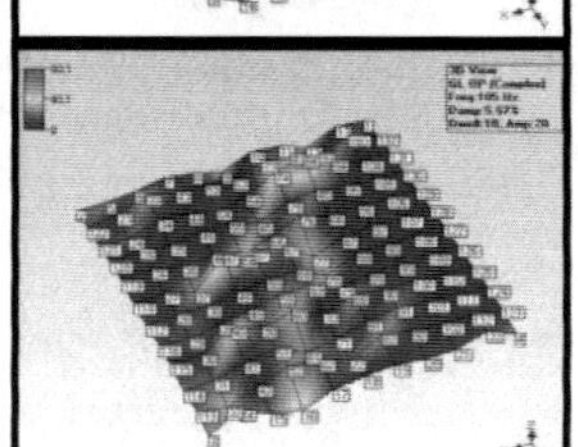	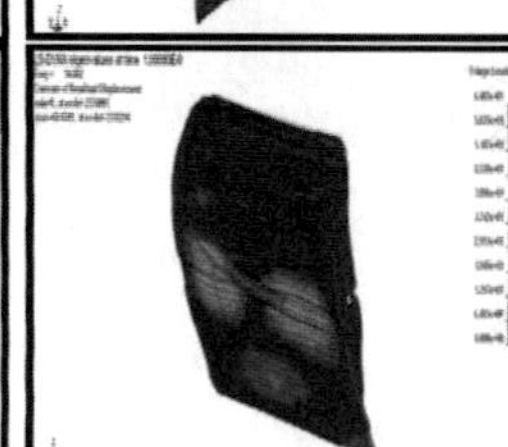	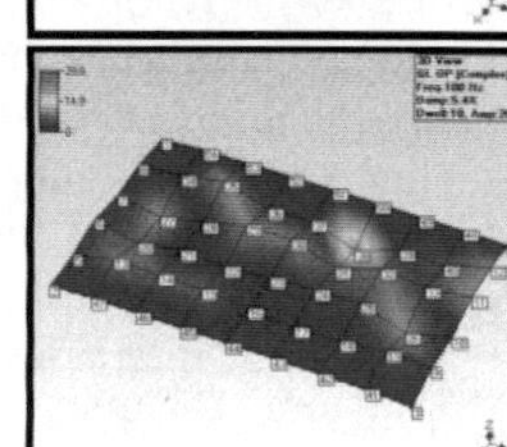

5
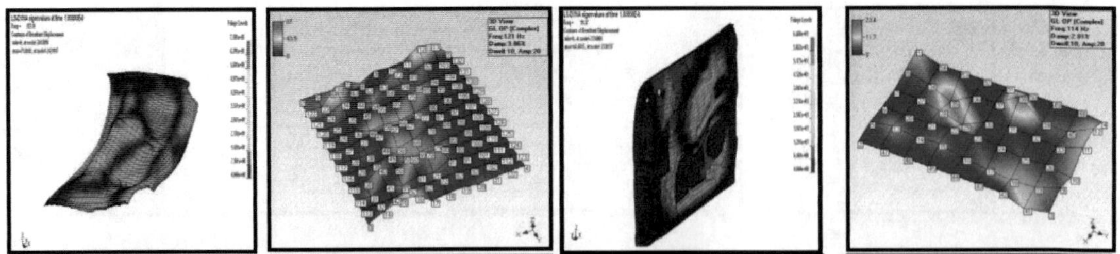

6
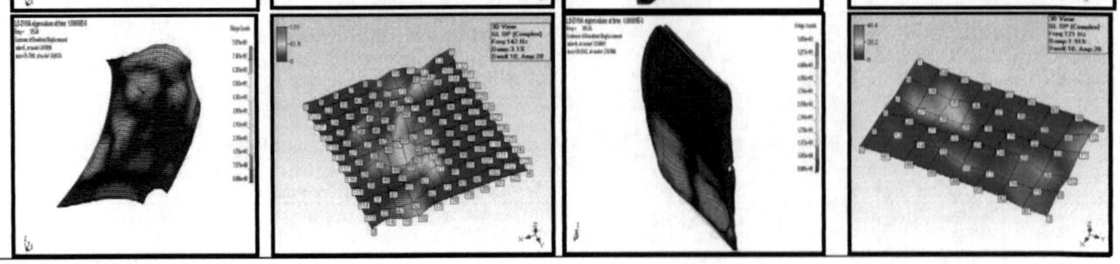

Table 3: Comparison between experimental and computational frequencies of panels (in Hz).

		MODE					
PANEL		1	2	3	4	5	6
Trunk	Exp. Freq	42.2	86.5	94.7	110	120	139
	Comp. Freq	48.63	81.13	95.4	105.43	120.68	133.81
Roof	Exp. Freq	60.2	89.8	105	129	148	163
	Comp. Freq	75.4	76.4	97.8	100.5	102.93	104.56
Hood	Exp. Freq	60.4	76.2	88.8	105	121	142
	Comp. Freq	58.83	76	86.8	107.78	117.59	125.68
Door	Exp. Freq	54.2	72.7	76.1	100	114	121
	Comp. Freq	55.7	74	86.5	94.88	99.37	105.45

4.2. Experimental Modal Analysis part by part and comparison with computational Modal Analysis

Experimental modal analysis was performed to determine a complete dynamic description of each car component (trunk, hood, roof and door). An impact hammer was used to excite the structure at a fixed point and six accelerometers were moved over each point on the structure. The force and accelerometers signals were fed into two three channels analysers to produce FRFs. Data was collected in B&K Pulse Labshop EMA software then transferred to ME'scope VES software for further analysis.

The quality assessment of the measurement of the frequency response functions is very important, because this leads to the quality of the modal parameters. Therefore some checks are important to ensure that the quality of the measured frequency response functions is sufficient to evaluate the dynamic parameters of the structure. The first check t is the coherence function, it is a frequency dependent indicator that shows which part of the output signal is coming from the real input signal and which part is coming from the additional measurement noise. If the value of the coherence function is near one then a strong linearity exists between the input and output and the influence of the noise is negligible. If the value of the coherence function is far away from one then no linearity exist between the input and output and the spectrum is dominated by the measurement noise. Figure1 shows the coherence measured in the experiment.

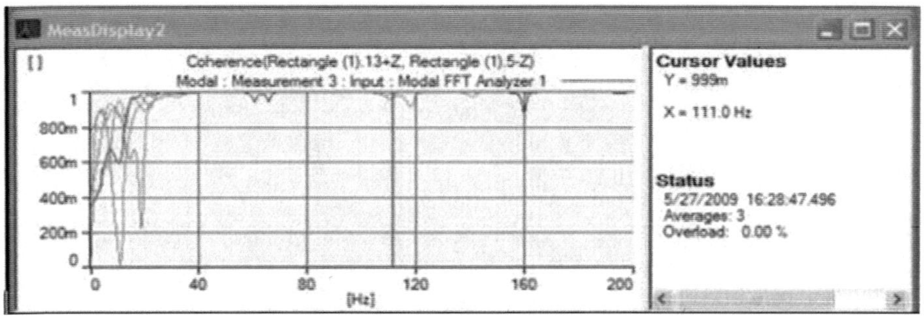

Figure 1. Coherence function

The second check was made by evaluating the reciprocity and reproducibility of several measured frequency response functions. Theoretically the reciprocity and reproducibility should be the same for all measured FRFs. The technique is to measure frequency response functions of a couple of points on the same structure for example front car door panel. Fist we excite point 5 and take the response at point 6 and vice versa, then excite point 3 and take the response at point 8 and vice versa. The results showed that the frequency response functions taken from the first couple of points are the same as shown in Figures 2c and 2d. The second couple of points also shows the same frequency response functions Figures 2a and 2b . Therefore, we conclude that the reciprocity and reproducibility of the measured frequency response functions is reasonably good.

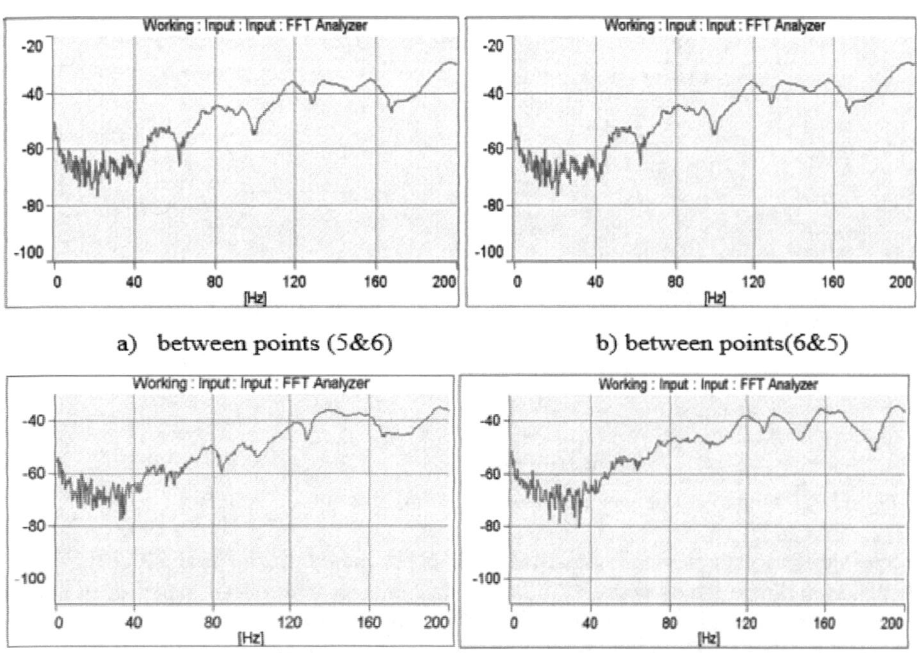

a) between points (5&6) b) between points(6&5)

c) between points (3&8) d) between points(8&3)

Figure 2. Frequency response function

5. General discussion

Comparison of computational frequencies and mode shapes with those obtained from the experimental modal analysis results indicated quite good agreement at majority modes for Trunk and Hood panels, whereas the Roof and Door panels showed lower agreement as can we see in table3. This is because of the complexity of Door and Roof panels structures compared to Trunk and Hood panels structures. Moreover the experimental panels and the computational panels indicated some components attached to the experimental panels which were not accounted in the computational panels due the difficulty to include them explicitly in the computational model. These masses affect the overall frequency response function.

Table3 shows the comparison between experimental and computational frequencies of all panels. Trunk panel shows fair agreement between the two methods except at the first mode as we can see in figure 3a. The roof panel does not show good agreement as the percentage different is quite high, it varies from 7% to 44%. We have also noticed that the last modes showed disagreement as seen in figure 3b. The third comparison was conducted on the Hood panel. Figure 3c shows good agreement at all modes as the percentage is less than 12%. Door panel does not show reasonable agreement especially at modes 3, 5 and 6 as the percentage different is 13%, 14% and 14% respectively. Table 3 and figure 3d.

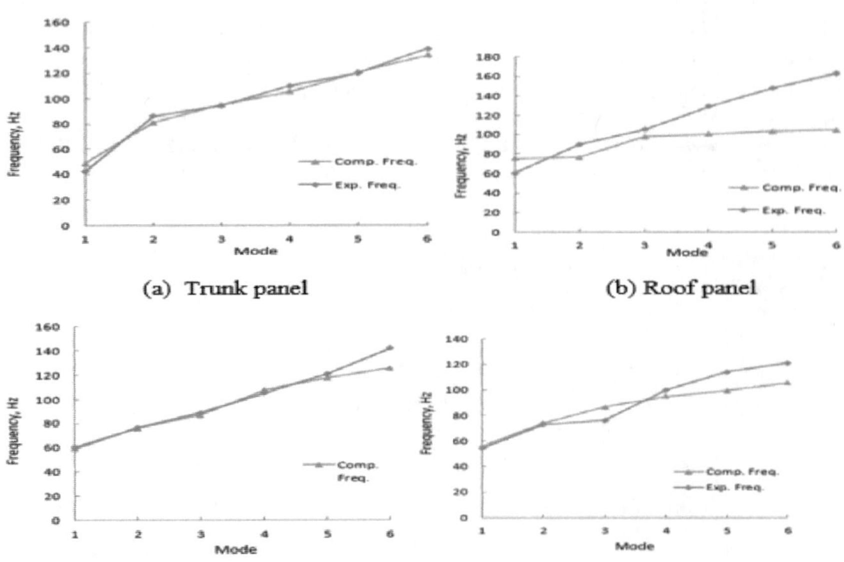

(a) Trunk panel (b) Roof panel

(C) Hood panel (d) Door panel

Figure 3: Comparison between experimental and computational panel frequencies

For a reasonable comparison, computational and experimental modal analysis require comparison mode shapes as well see table1 and 2. The first mode shape of the trunk panel does not show any vibration behaviour similarity. This is reasonable because their frequencies are far away from each other (42.2 Hz and 48.63). The second, third and fourth mode shapes are exhibit some similarities in their behaviour. All of them showing vibration in the horizontal and vertical part of the trunk panel. The fifth and sixth modes do not show any significant similarity. Mode shapes of roof panel are shown in table1. At the first mode both showing vibration at the front and right side of the panel. The third mode shape shows a fair similarity between the experimental and computational methods as both have significant vibration at both sides of the panel. The fifth mode also shows good similarity in the vibration behaviour at the rear part of the roof panel. The remaining modes do not show any interesting similarity. Hood panel shows reasonable similarity at modes 2, 4, and 5. At mode 2 both methods show significant vibration at the right sight of the panel. Mode 4 showing high vibration on left and right sides for both methods. The other modes do not show significant similarity. For door panel only two modes have similar behaviour, mode 1and 3, the others do not show significant similarity.

Results from the simulation model using modal analysis part by part are in quite good agreement with the experimental modal analysis results within the investigated panels. Slight discrepancies may be accounted for geometric and material approximations made during the simulation of the FE model. It has been noticed from the comparison results that the vibration amplitude of the experimental mode shapes is significantly less from those of computational modal analysis. This is obvious because of the extra damping treatment applied on the surfaces of the actual car panels to reduce the strength of the noise and vibration. The numerous damping mechanisms at work in a real structure cannot be easily modelled, so damping is not included in most FEA models. However, mode shapes can still be obtained from model without damping.

In conclusion the results obtained from all panels showed a good agreement between experimental and computational modal analysis at the frequency level. However the mode shapes failed to have good agreement. In addition, we have noticed that the computational mode shapes exhibit high amplitude, whereas the experimental mode shapes exhibit low amplitude. Probably this difference occurred because of the damping applied to the surface area of the panels to reduce the vibration amplitude which can affect the excitation and response functions.

6. Conclusion

Generally, we can conclude that modal analysis part by part is a valid method and can be implemented to determine the dynamic characteristics of any complex structure. The results showed fair agreement between experimental and computational modal analysis. However, still some discrepancies between the experimental and computational results. This could be because of: Firstly, boundary conditions: inaccurate boundary conditions can have influence on some or all vibration modes. Therefore, to overcome these problems we keep try other boundary conditions, which could be a further work for this research. Secondly, linearity between input and output noticed from the coherence signals obtained from experimental modal analysis tests for all panels. The results showed that the coherence at low frequencies was far away from one which resulted in large error at the first modes. Thirdly, the correspondence between computational model and the real model as the available FE model was not exactly the same as the real model. This can have great effect on the results and must be considered in future research.

The proposed approach can be considered as an effective tool to determine the dynamic characteristics of vehicle components that can be expected to be the source of the interior noise comfort.

References

[1] Teo H F and Roslan Abd. Rahman 2007 Statics and dynamics structural analysis of a 4.5 Truck Chassis. *J. Mekanikal*. **24**: 56-67.

[2] Kassahun M 2008 Static and dynamic analysis of a commercial vehicle with van body. University thesis. *Addis Ababa University*.

[3] Zhanwang Y and Chen Z 2000 Dynamic response analysis of minicar changan Star 6350. *Changan Automobile Groupe*.

[4] Johansson I and Gustavsson M 2000 FE-based vehicle analysis of heavy trucks, Part I: Methods and Models, *proc of 2nd MSC. Worldwide Automotive Conference*, MSC (Sweden).

[5] Lan J Chen and Lin J 2004 Comparative analysis for bus side structures and lightweight optimization *J. of Automobile Engineering* **218**(10):1067-75. Part D.

[6] Kirby R.A 2009 Comparison between analytic and numerical methods for modelling automotive silencers with mean flow *J. of Sound and Vibration*. **325**(3): 565-82.

[7] Vasconcellos N, dos Anjos and Argentino M 2003 Structural analysis of an exhaust system for heavy trucks. *Debis humaita IT Services* Latin America, Brazil.

[8] Jinchun H, Charles M K and Anil K.B 2006 Modelling of automotive drum brakes for squeal and parameter sensitivity analysis. *J. of Sound & Vibration*. **289**(1-2): 245-63.

[9] Jae-Yeol, Park and Rajaendra Singh 2010 Role of spectrally varying mount properties in influencing coupling between powertrain motions under torque excitation. *J. of Sound & Vibration*. **329** (14): 2895-2914.

[10] Donders S, Takahashi Y, Hadjit R, Van T, Brughmans T and Desmet W 2009 A reduced beam and joint concept modelling approach to optimize global vehicle body dynamics. *Finite Elements in Analysis & Design*. **45**: 439-55.

[11] Verboven P, Guillaume P, Vanlanduit S and Cauberghe B 2006 Assessment of nonlinear distortions in modal testing and analysis of vibrating automotive structures. *J. of Sound and Vibration*. **293** (1-2): 299-319.

[12] Verboven P, Parloo E, Cauberghe B and Guillaume P 2005 Improved modal parameter estimation for lowly damped systems using nonparametric exponential windowing techniques. *Proc Mech. Sys & Signal*. **19**(4): 675-99.

[13] Hurlebaus S, Stobener U & Gaul L 2008 Vibration reduction of curved panels by active modal control *J. of Computer & Structures*. **86** (3-4):251-257.

Effect of Stacking Layup on Spring-back Deformation of Symmetrical Flat Laminate Composites Manufactured through Autoclave Processing

M N M Nasir[1], M A Seman[2], L Mezeix[3], Y Aminanda[4], A Rivai[5] and K M Ali[6]

[1, 4]Department of Mechanical Engineering, International Islamic University Malaysia, 50728 Kuala Lumpur, Malaysia
[2, 5]Faculty of Mechanical Engineering, Universiti Teknikal Malaysia Melaka, 76100, Durian Tunggal, Malaysia
[3]Faculty of Engineering, Burapha University, 20131, Chonburi, Thailand
[6]Aerospace Malaysia Innovation Centre, 63000 Cyberjaya, Malaysia

E-mail: yulfian@iium.edu.my

Abstract. The residual stresses that develop within fibre-reinforced laminate composites during autoclave processing lead to dimensional warpage known as spring-back deformation. A number of experiments have been conducted on flat laminate composites with unidirectional fibre orientation to examine the effects of both the intrinsic and extrinsic parameters on the warpage. This paper extends the study on to the symmetrical layup effect on spring-back for flat laminate composites. Plies stacked at various symmetrical sequences were fabricated to observe the severity of the resulting warpage. Essentially, the experimental results demonstrated that the symmetrical layups reduce the laminate stiffness in its principal direction compared to the unidirectional laminate thus, raising the spring-back warpage with the exception of the $[45/-45]_S$ layup due to its quasi-isotropic property.

1. Introduction

The usage of composites in the aerospace industry has increased significantly in the past 30 years where leading manufacturers e.g. Airbus have been integrating composite-made structures into their latest airliners. The most recent has been with the A350 which saw more than 50% application of composite materials due to its higher specific stiffness compared to metals.

As with any other material, composites induce residual stress as a result of the manufacturing process which in this case involves curing at high temperatures inside an autoclave. The residual stresses will pre-stress the composite and decrease its overall strength. An observed consequence of this is a deviation of the final product from what was initially designed. This phenomenon is known as spring-back deformation. This issue will cause problems during the assembly stage because of poor fit-up between the mating structures which will compel the technicians to force fit the parts. Such practice will increase the internal stress levels of the structure and reduce its span life.

There are many factors to spring-back deformation. One is the change in mechanical properties of the laminate during the curing process. While fibre properties remain essentially constant, the matrix resin properties evolve as the resin polymerizes. The correlation between the development of residual stresses and the resulting warpage is more pronounced during the cool-down stage as observed in an experiment [1] when the thermal stresses that had been accumulated during the ramp and hold stages, were relieved. Another source of spring-back warpage is the difference of fibre orientation between individual plies i.e. anisotropic layups which results in in-plane stresses within the laminate. The severity of the warpage is more for an asymmetrical and unbalanced layup as discovered in another study [2] due to the multiple constraints that had been imposed as a result.

Nonetheless, tool-part interaction is seen as the most critical mechanism [3]-[6]. As illustrated in figure **1**, interfacial shear stresses will develop from the difference in stretching between both components during heat-up, generating a stress gradient through the part thickness that finally yields the spring-back warpage.

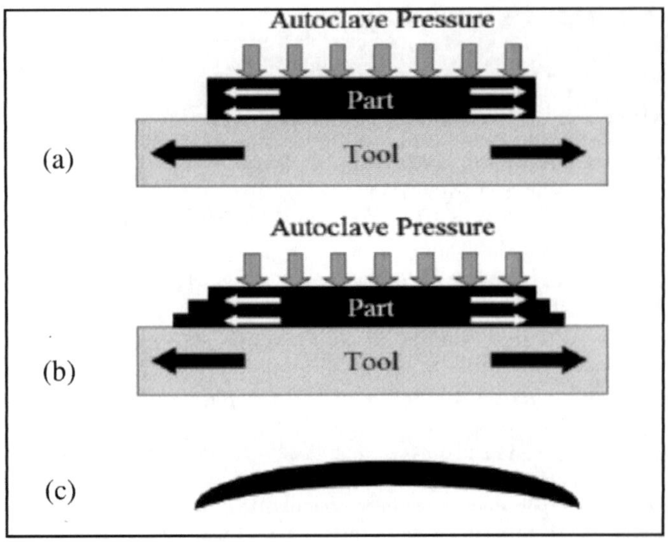

Figure 1. Part warpage due to tool-part interaction [6].

However, these are simply natural behaviors of laminates and are difficult to control. A study [7] categorized the controllable parameters of spring-back deformation into intrinsic and extrinsic parameters whereby intrinsic relates to part geometry and material properties whereas extrinsic parameters are facets of the manufacturing process. The current study investigates the effect of having a symmetrical ply layup which is an intrinsic parameter.

The varied stacking sequence of a laminate results in in-plane stresses within the plies due to the difference in the coefficients of thermal expansion in both the fibre and resin directions at the ply level [8]-[10]. With cross-ply composites [0/90] as shown in figure **2**, the 90° fibres impose a mechanical constraint on the 0° fibres during the processing phase as in where σ represents the residual stresses [11].

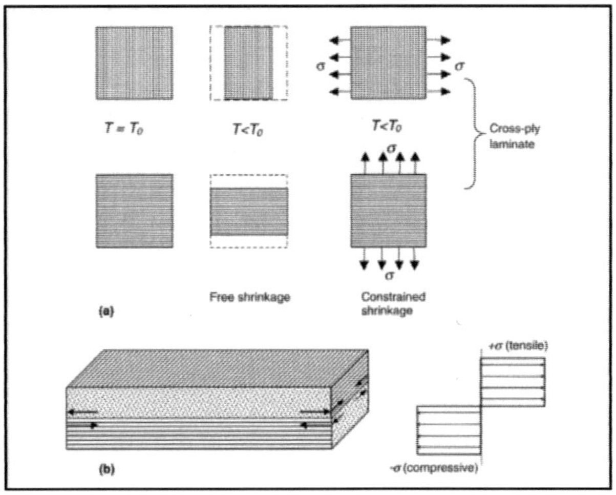

Figure 2. Schematic view of the (a) residual stress formation in an asymmetrical cross-ply laminate and the (b) residual thermal stresses when the laminate is constrained [11].

For the [0/90] layup, various stable geometries exist after curing as shown in figure 3 [2]. While laminates with a small in-plane to thickness ratio lead to saddle shapes i.e. (a) and (b), laminates with larger in-plane dimensions show cylindrical shapes i.e. (c) and (d). For angled laminates, angled cylindrical shapes are formed i.e. (e) and (f). These effects reduce with increasing laminate thickness and can be avoided using symmetrical laminates. This paper therefore describes and analyses the resulting warpage produced by symmetrically stacked laminates manufactured through autoclave processing.

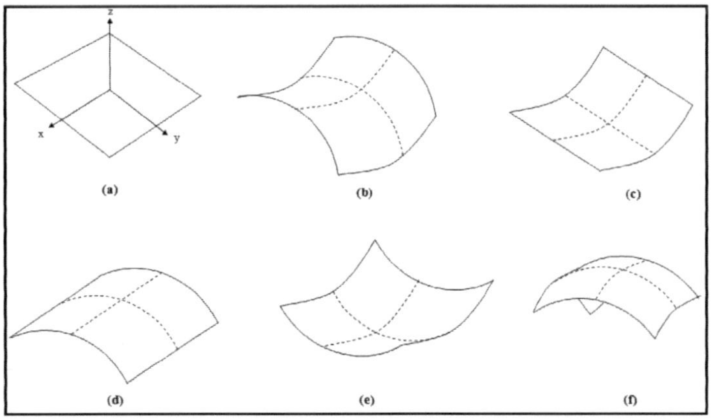

Figure 3. Part warpage due to asymmetrical layups; (a) original; (b) saddle-shaped; (c) and (d) cylindrical-shaped; (e) and (f) rotated cylindrical [2].

2. Experimental

2.1. Materials and parameter

The current study will employ tool made from S275JR carbon steel to manufacture laminate composites made from IMA/M21E of various symmetrical layups and observe the effect on spring-back deformation. Both components are fabricated by CTRM, the leading supplier of aerospace composite structures in Malaysia, based on the industry standards. The part samples were cut to dimensions of 300x300 mm^2 and laid up to a thickness of 4 plies with symmetrical sequences. The sequences are $[0/90]_S$, $[0/45]_S$, $[45/0]_S$ and $[45/-45]_S$ (see figure 4). The symmetrical configurations are for the purpose of keeping the research relevant to the aerospace manufacturing industry as it is recommended to manufacture aircraft structures with symmetrical and balanced stacking [12].

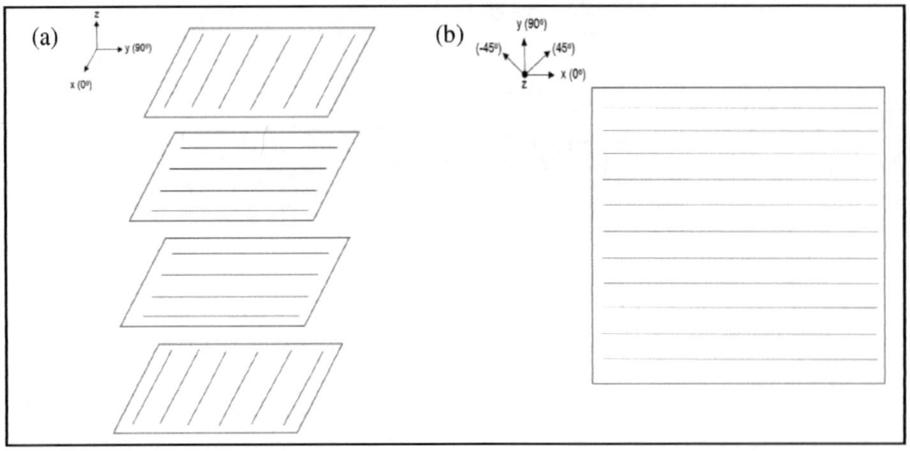

Figure 4. (a) Laminate stacking layup for $[0/90]_S$ and (b) fibre orientation of a ply.

2.2. Manufacturing process and measurement procedure
Firstly, polytetrafluoroethylene (PTFE) was placed on the surface of the tool to serve as release agent that allows easy separation of the laminate composite part from the tool after processing. Next, the prepreg was laid up on the tool according to the designed stacking and the release film was placed on top of the prepreg and consolidated for 5 minutes. Afterwards, the breather cloth was placed on top of the release film for final bagging. Once the vacuum test was performed during the final bagging stage, the system in
figure **5** was placed inside the autoclave and cured to the respective pressure and temperature profile.

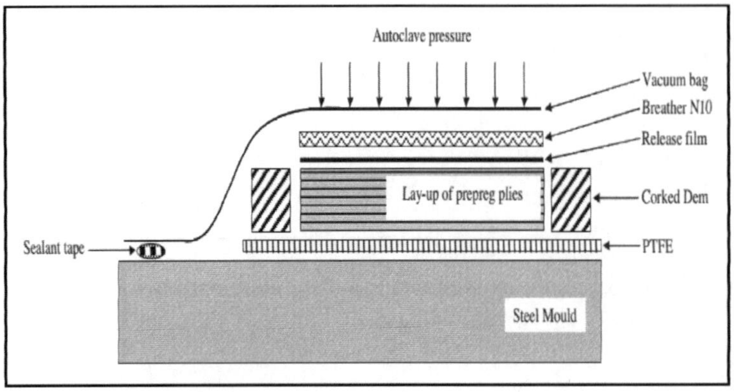

Figure 5. Final bagging system for autoclave curing.

IOP Conf. Series: Materials Science and Engineering **184** (2017) 012028 doi:10.1088/1757-899X/184/1/012028

The resulting warpage for a symmetrically-stacked laminate differs from a unidirectional laminate in that the warpage pattern is non-uniform (see figure 6). The spring-back warpage is measured as the maximum point of warpage relative to the reference point as shown in figure 6.

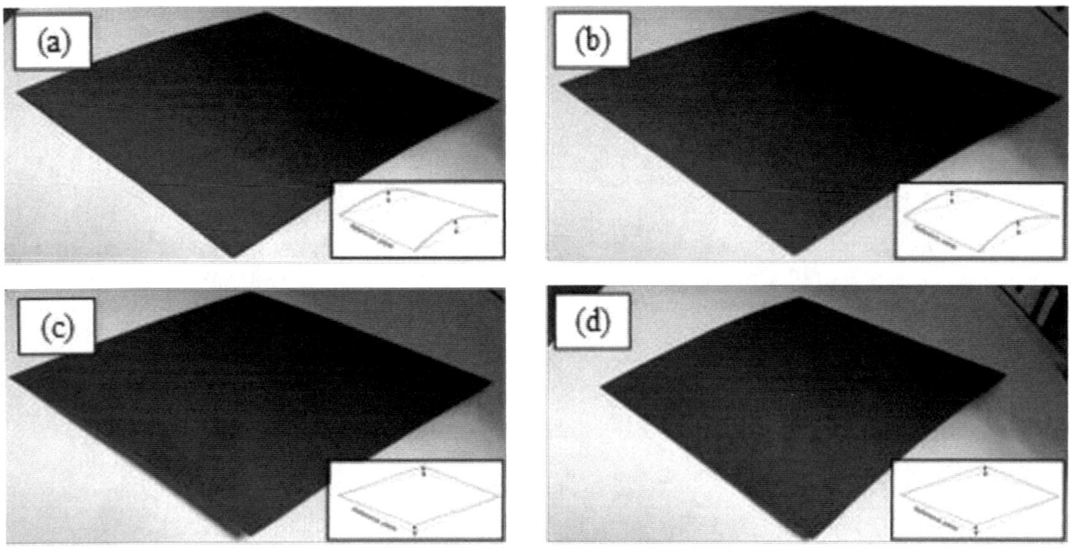

Figure 6. Laminate part warpage for (a) $[0/90]_S$; (b) $[0/45]_S$; (c) $[45/0]_S$ and (d) $[45/-45]_S$.

3. Results and discussion

A plot of the maximum warpage for all the tested samples is provided in figure 7. The author has also included the maximum warpage for the unidirectional laminate i.e. 0° from a previous accompanying study [13] as the baseline. The maximum warpage of all 5 configurations are tabulated in table 1 .

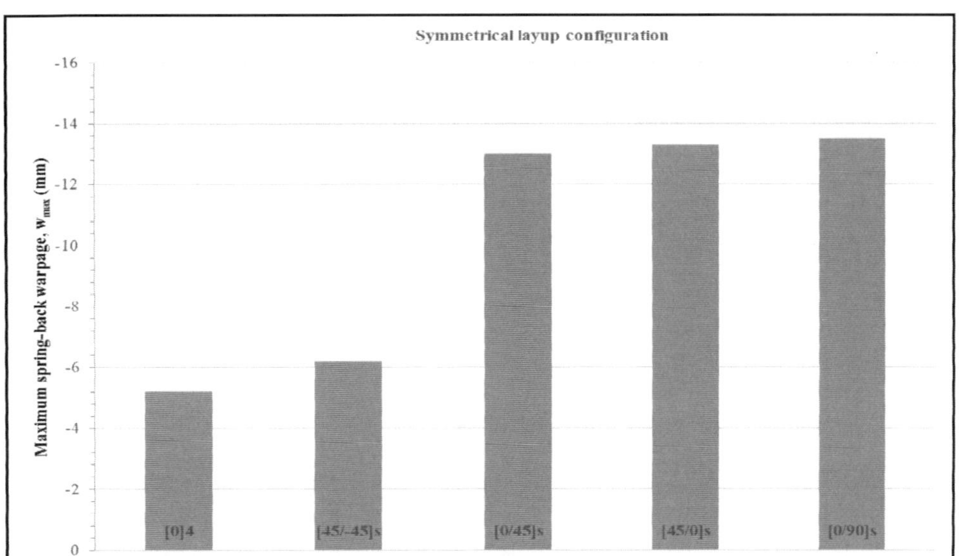

Figure 7. Maximum warpage for 300x300 mm^2 and 4 plies thickness with symmetrical configurations.

Table 1. Maximum warpage of the 300x300 mm^2 and 4 plies thickness laminate samples.

Laminate layup	Maximum warpage
$[0]_4$	-5.2 mm
$[45/-45]_S$	-6.2 mm
$[0/45]_S$	-13 mm
$[45/0]_S$	-13.3 mm
$[0/90]_S$	-13.5 mm

As mentioned earlier, it is standard practice to lay up the laminates symmetrically because it uncouples bending and the membrane response thereby preventing twisting of the laminates. Another reason is to supposedly prevent spring-back warpage during temperature cool-down but as shown in figure 6, figure 7 and table 1, there still exists warpage and in various forms depending on the ply layup configuration.

Table 2. Maximum warpage of the 300x300 mm^2 and 4 plies thickness laminate samples.

Laminate layup	Longitudinal stiffness	Transversal stiffness	Principal material direction
$[0]_4$	154 GPa	8.5 GPa	0°
$[45/-45]_S$	12.2 GPa	12.2 GPa	45°
$[0/45]_S$	83.1 GPa	10.3 GPa	10°
$[45/0]_S$	83.1 GPa	10.3 GPa	10°
$[0/90]_S$	81.3 GPa	81.3 GPa	10°

Table 2 shows the resultant stiffness and principal material direction of the laminate samples calculated using Classical Laminate Analysis (CLA). By relating the values with the experimental data in table 1, the layups that yield approximately the same warpage also possess close longitudinal stiffness and have the same principal material direction i.e. $[0/45]_S$, $[45/0]_S$ and $[0/90]_S$. The warpages of all 3 layups are also approximately 2.5 times greater than the $[0]_4$ layup which is a direct consequence of the longitudinal stiffness being almost 50% of the unidirectional laminate. However, the $[45/-45]_S$ layup possesses a lower longitudinal stiffness but still generated the same magnitude of warpage for a unidirectional laminate which might be due to the quasi-isotropic nature of $[45/-45]_S$ layup compared to the others which are significantly anisotropic. This is shown in figure 8 where the laminate stiffness is mapped at every angle between 0° to 90° for all the symmetrical layups.

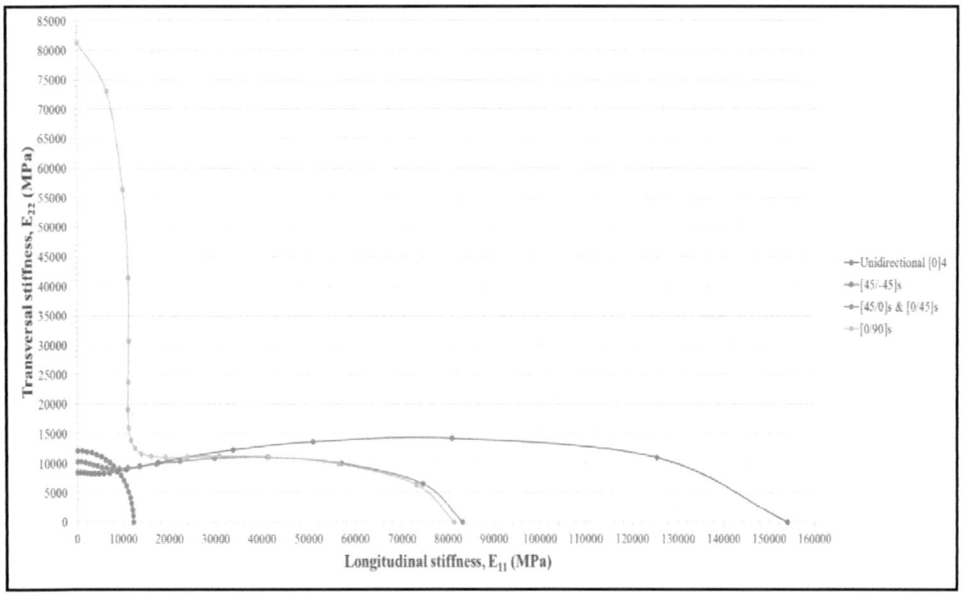

Figure 8. Laminate stiffness between 0° to 90°.

4. Conclusions and future work

The objective of this study was to investigate the effect of the stacking layup on the warpage of flat IMA/M21E laminates. The main conclusions are as follow:

- The stiffness reduction in laminate principle direction will significantly increase the severity of the warpage by more than 150% especially for orthotropic laminate.
- The severity of the warpage in a symmetrical layup could be greatly reduced if the laminate is quasi-isotropic in behaviour.

This study has increased the understanding on the part design parameters affecting spring-back warpage of laminate composites and demonstrates the importance of how warpages are defined and measured on symmetrical laminates. However, the conclusions drawn should be used with caution outside the scope of this study. There is also a need to investigate the effects of extrinsic i.e. process parameters e.g. cure cycle, tool material, tool surface conditions etc. as this is more practical in the aerospace manufacturing industry.

Finally, the detailed study on the mechanism triggered by the 45° fibre direction ply can be simulated using the FEA previously developed in the accompanying study of unidirectional laminates [13]. The analysis on the macromechanical stress distribution obtained from the FEA model would provide more understanding on the influence of the ply stacking sequence in the spring-back warpage especially the effect of the tool-part interaction.

5. Acknowledgements

Much appreciation goes to the Aerospace Malaysian Innovation Centre (AMIC) for the financial, operational and other related matters that ensure the progress of this research. Credit also goes to CTRM Sdn. Bhd for its support in the manufacturing and measurement of the laminate samples.

References

[1] Olivier P A 2006 A note upon the development of residual curing strains in carbon/epoxy laminates. Study by thermomechanical analysis *Compos. Part A* **37** 602-16

[2] Rapp S Shape memory polymers in fibre composite structures for shape adjustment [unpublished dissertation] Technical University of Munich 2011

[3] Ridgard C 1992 Accuracy and distortion of composite parts and tools: Causes and solutions *Tooling for Composites '93 (California)*

[4] Nelson R H and Cairns D S 1989 Prediction of dimensional changes in composite laminates during cure *34th Int. SAMPE Conf. on Symp. & Exhibition (Nevada)*

[5] Wisnom M R, Ersoy N, Potter K and Clegg M J 2005 An experimental method to study the frictional processes during composites manufacturing *Compos. Part A* **36** 1536-44

[6] Pagliuso S 1982 Progress in science and engineering of composites, Warpage, a nightmare for composite parts procedures *4th Int. Conf. on Composite Materials (Tokyo)*

[7] Albert C and Fernlund G 2002 Spring-in and warpage of angled composite laminates *Compos. Sci. Technol.* **62** 1895-1912

[8] Hyer M 1981 Some observations on the cured shape of thin unsymmmetric laminates *J. Compos. Mater.* **15** 175-94

[9] Kim K S and Hahn H T 1989 Residual stress development during processing of graphite/epoxy composites *Compos. Sci. Technol.* **36** 121-32

[10] Vo T, Mailen R and Jack D 2012 Laminate stiffness and curvature for laminated carbon fiber composites, experimental observation and model validation *Automotive Composites Conf. and Exhibition (Michigan)*

[11] Harris B 1999 *Engineering composite materials*, second ed. (London: IOM Communications)

[12] Federal Aviation Administration 2012 Advanced Composite Material *Aviation Maintenance Technician Handbook – Airframe* (Washington: FAA) chapter 7 pp. 1-58

[13] Mezeix L, Seman A, Nasir M N M, Aminanda Y, Rivai A, Castanie B, Olivier P, Ali K M 2015 Spring-back simulation of unidirectional carbon/epoxy flat laminate composite manufactured through autoclave process *Compos. Struct.* **124** 196-205

A study on the parameters affecting the properties of coated paper honeycomb

Nurdina Abd Kadir[1] , Yulfian Aminanda[1*], Mohd. Sultan Ibrahim Shaik Dawood[1], Hanan Mohktar[1]

[1]Mechanical Engineering Department of IIUM, Jln. Gombak, P.O. Box 10, 50728 Kuala Lumpur, Malaysia

*corresponding author : yulfian@iium.edu.my

Abstract. In this study, three parameters, namely the density of Kraft paper, the thickness of cell wall and the cell size of honeycomb were evaluated. A statistical analysis was performed to investigate the effect of the parameters on the specific compression strength(SCS) and specific energy absorption (SEA) of Kraft paper honeycomb. Based on the design of experiment, 27 configuration of Kraft paper honeycomb with different cell sizes (10mm,15mm,20mm), thicknesses of cell wall (1ply, 2ply, 3ply) and paper densities (80gsm,120gsm,175gsm) are subjected to quasi static compression loading. An analysis of variance (ANOVA) was used to examine the effect of each parameters and to obtain the optimum configuration of Kraft paper honeycomb. As a result, the thickness of cell wall shows as the most prominent parameter for specific energy absorption and specific compression strength. Meanwhile, the Kraft paper honeycomb with density 175gsm, 3 ply thickness of paper and 10mm cell size of honeycomb shows the optimum configuration with 724.80 J/kg of SEA and 9.35 MPa/kg of SCS.

1. Introduction

Sandwich panel containing commercial Kraft paper honeycomb core with resin pre-impregnated into the paper or no resin at all is widely used in furniture, doors, partitions, mobile homes, signs and similar construction. Meanwhile, special grade Kraft paper honeycomb core which dipped in a phenolic resin after expansion in order to increase its water resistance and strength is used in portable military shelter, aerospace and naval industry [1]. Some main advantages of these sandwich panel include reduced weight and production cost, lower materials usage and less damage during transportation [2]. Although honeycomb sandwich panels are widely used in engineering industries since many decades, there is limited information available concerning the properties and behavior of the lightweight paper honeycomb panel. Moreover, honeycomb core are the weakest part of sandwich panel and they may fail or collapse through cell material fracture or cell wall buckling, depending on the loading regime and the core configuration [3][4][5].

Up until today, many researchers investigated on honeycomb structure used for high technology application such as aerospace and naval industry like Nomex and metallic honeycomb[6][7][8]. Unfortunately, scarce studies can be found in the literature dealing with honeycomb sandwich panel used for low technology such as furniture or load bearing application. Recently, some manufacturers are aiming to use paper honeycomb panel or structure for load bearing application such as floor, decks, transportation pallet, load bearing wall and partition due to lower cost and material usage of paper honeycomb compared to solid wood based panel.

Investigation of the crushing phenomena of honeycomb structures under both quasi-static and dynamic loading conditions considering the effects of cell dimension and foil thickness have been performed earlier by Wu and Jiang [9] and Yamashita and Gotoh [10]. Wu and Jiang investigated the honeycomb specimens with cell size of 3.2 and 4.7 mm and Yamashita and Gotoh studied the quasi-static compression response of aluminium honeycomb in the thickness direction. Meanwhile, Dongmei wang [11] studied the impact behavior and energy absorption on the paper honeycomb sandwich panel where the effect of paper honeycomb structure factors on the impact behavior was analyzed.

In this research, honeycomb made from Kraft paper are manually manufactured with different parameters especially on the honeycomb cell sizes, thicknesses of cell wall and densities. Series of quasi static experimental tests were carried out to investigate the effect of the parameters on specific energy absorption and specific compression strength. The optimum configuration on specific energy absorption and specific compression strength were determined through statistical analysis of variance (ANOVA).

2.Specimen Fabrication and Methodology

Kraft Paper honeycomb coated with vanish were fabricated with different parameters by using honeycomb maker. The density of paper (80gsm, 120gsm, 175gsm), thickness of cell wall (1ply, 2ply, 3ply) and cell size of honeycomb (10mm, 15mm, 20mm) were taken as parameters (table 1).The height for all specimens was 45mm. According to full factorial design, 27 configuration are generated considering three parameters with three levels as shown in table 2. The paper honeycombs were subjected to quasi-static compression loading by using SHIMADZU Autograph AG-X 250 compression machine. As the bottom plate was fixed and the upper plate was subjected to vertical downward displacement of 22.5mm. The force acting on the specimens was measured by load cell, and the force-displacement curve was plotted. The rate of displacement was 0.5 mm/min. Then, statistical analysis of variance (ANOVA) was used to study the effect of parameters and to obtain the optimum configuration of Kraft paper honeycomb on the specific energy absorption (SEA) and specific compressive strength (SCS).

Table 1. Parameters and levels using in DOE.

Factor	Code	Unit	Level 1	Level 2	Level 3
Density of paper	A	gsm	80	120	175
Thickness of cell wall	B	ply	1	2	3
Cell size of honeycomb	C	mm	10	15	20

3. Results and Discussion of Compression Test

3.1. Force- Displacement curve

The sample of force-displacement curves of Kraft paper honeycombs tested in uniform compression are given in figure 1. The curves show that there is elastic behavior at the beginning of the indentation until a critical load is reached. After the peak load, a sharp drop is observed. Based on Aminanda et al [12], the sharp drop is corresponds to the beginning of vertical edge deformation. Then the force decreases to reach a plateau, which corresponds to the succession of fold forming and ends up by condensation of the honeycomb.

From the graph, the specific energy absorption (SEA) was determined by dividing energy absorption capability of specimen (EA) with mass of specimen (m). EA was estimated by the area under force-displacement curve and it was measured until 22.5mm of displacement. Meanwhile, specific compressive strength (SCS) was calculated by dividing maximum compressive stress (σ_{max}) with mass of specimen (m). Afterward, SEA and SCS will be analyze using ANOVA.

Table 2. The configuration of kraft paper honeycomb by general factorial.

Run	*A* *Density of paper* *(gsm)*	*B* *Thickness of cell wall* *(ply)*	*C* *Cell size of honeycomb* *(mm)*
1	175	2	10
2	80	2	20
3	80	2	10
4	120	3	15
5	175	3	20
6	80	3	15
7	120	2	10
8	120	1	20
9	120	2	15
10	120	3	10
11	80	1	20
12	80	3	20
13	175	2	20
14	80	1	10
15	175	1	20
16	175	2	15
17	120	1	10
18	120	1	15
19	80	2	15
20	80	3	10
21	175	1	10
22	120	3	20
23	175	3	20
24	175	1	15
25	120	2	20
26	175	3	10
27	80	1	15

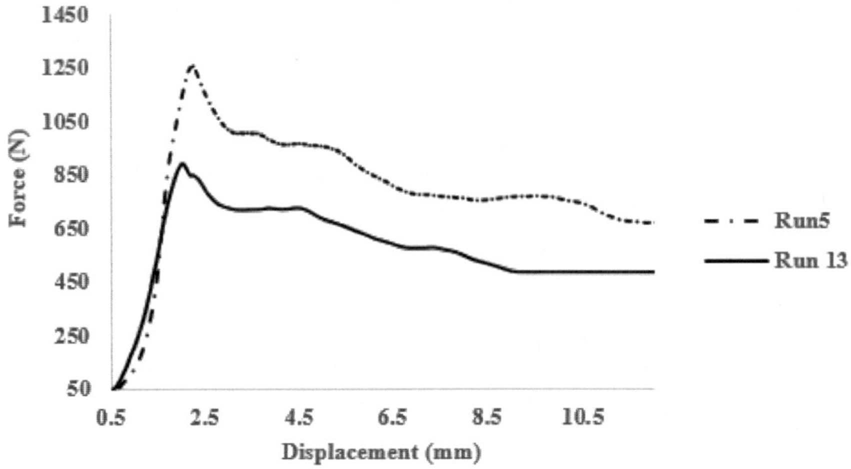

Figure 1. Force-displacement curves of selected kraft paper honeycomb.

3.2. Failure Mechanism

The final deformation of paper honeycomb after quasi-static compression can be observed in figure 2. From the figure, it can be seen that the failure mechanism of paper honeycomb consists of buckling of the cell walls and vertical edges, opening of the double thickness wall due to tearing, local wall buckling on the top and bottom surfaces, and tendency of unstable outer cells to buckle towards inward or outward direction. The mechanism is similar to Nomex honeycomb that reported by Aminanda et al. [1].

Figure 2. Condition of selected kraft paper honeycomb before (top) and after (bottom) quasi static compression test.

3.3. Analysis of Design of Experiment

3.3.1. Analysis of Specific Energy Absorption (SEA)

Table 3 shows the results of ANOVA associated with SEA obtained from Factorial method. The model for SEA was developed with 99% confident level and P-value less than 0.001 which indicated that this model is significant. *A*, *B* and *C* refers to density of paper, thickness of cell wall and cell size of honeycomb. From table 3, all parameters have significant effect on the SEA due to the P-value less than 0.05. The thickness of cell wall, *B* shows as most significant parameter with P-value less than 0.0001, following by density of paper, *A* and cell size of honeycomb, *C* with P-value of 0.006 and 0.0238, respectively. The P-value for interaction between density of paper and cell size of honeycomb is 0.1287 which indicates weak influence of the two parameters interaction to SEA value.

Figure 3 depicts the plot of main parameters to SEA. From the figure, It can be observed that SEA increases as the density of paper and thickness of cell wall increases. However, SEA decreases with decreasing of honeycomb cell size until 15mm. Then, SEA become constant as the bigger cell size. Eq. 1 showed the effect of density of paper, *A*, thickness of cell wall, *B* and cell size of honeycomb, *C* to the specific energy absorption.

$$SEA= 426.33 - 105.03A_1 + 27.55A_2 -145.94B_1 + 31.68B_2 + 62.86 C_1 - 8.35C_2 + 6.10A_1C_1 - 49.93A_2C_1 - 55.27A_1C_2 + 21.03A_2C_2 \quad (1)$$

Table 3. Analysis of variance (ANOVA) for SEA.

Source of data	Sum of square	Degree of freedom	Mean square	*F* value	Prob > *F*	Comment
Model	5.967E+005	10	59669	9.04	<0.0001	Significant
A	1.602E+005	2	80077.86	12.13	0.0006	
B	3.182E+005	2	1.591E+005	24.10	<0.0001	
C	62932.46	2	31466.23	4.77	0.0238	
AC	55395.17	4	13848.79	2.10	0.1287	

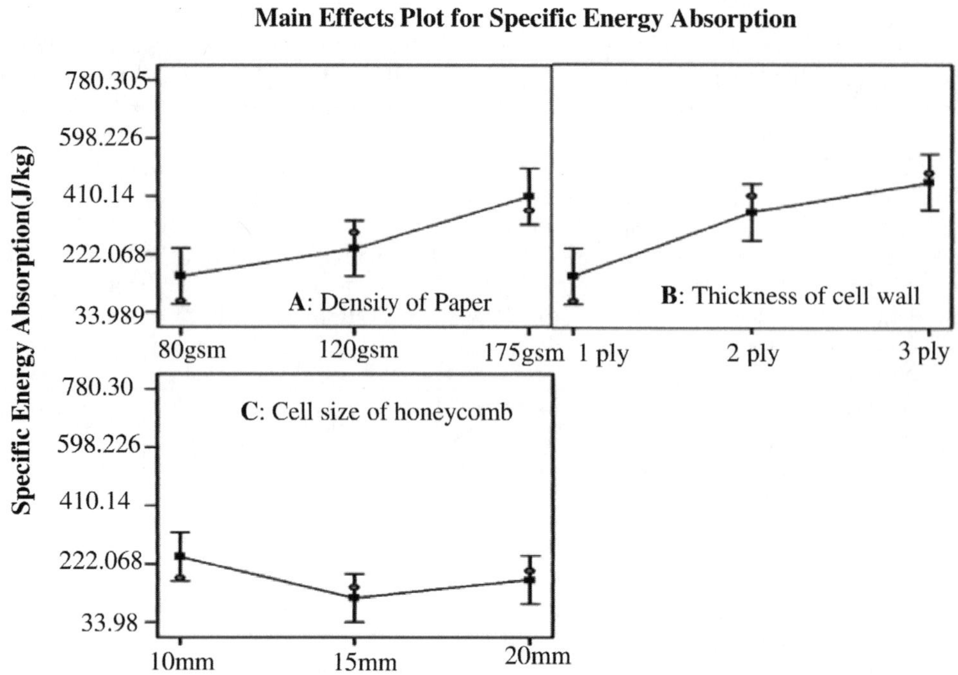

Figure 3. Main effect for specific energy absorption.

3.3.2. Analysis for Specific Compressive Strength (SCS)

The model for SCS was developed from ANOVA with 99% confident level and P-value less than 0.001 which indicated that this model is significant (table 4). The same parameters have been used for SEA analysis previously. From table 4, thickness of cell wall also displays as most significant parameter for SCS with P-value less than 0.0001 which similar to result of ANOVA for SEA. Meanwhile, both density of paper and honeycomb cell size showed the P-value of 0.003. There might have interaction between thickness of paper and cell size of honeycomb due to its P-value between 0.05 and 0.1. The P-value of 0.1348 for interaction between density of paper and thickness of paper implies no interaction between these two parameters. Main effect on the specific compression strength shows in figure 4 where it can be seen that the SCS increases with increasing of paper density and thickness. Meanwhile. SCS decreases with increasing the cell size of honeycomb. The effect of density of paper, *A*, thickness of cell wall, *B* and cell size of honeycomb, *C* on specific compression strength can be showed from Eq. 2.

$$SCS = 6.08 - 1.19A_1 + 0.49A_2 - 1.92B_1 + 0.40B_2 + 1.12C_1 - 0.23C_2 - 0.100A_1B_1 - 0.54A_2B_1 + 0.15A_1B_2 - 0.15A_2B_2 - 0.18B_1C_1 - 0.38B_2C_1 + 0.66B_1C_2 + 0.20B_2C_2 \quad (2)$$

Table 4. Analysis of variance (ANOVA) for SCS.

Source of data	Sum of square	Degree of freedom	Mean square	*F* value	Prob > *F*	Comment
Model	105.23	14	7.52	13.19	<0.0001	Significant
A	19.43	2	9.73	17.08	0.0003	
B	55.73	2	27.86	48.91	<0.0001	
C	18.98	2	9.49	16.66	0.0003	
AC	4.94	4	1.23	2.17	0.1348	
BC	6.12	4	1.53	2.69	0.0827	

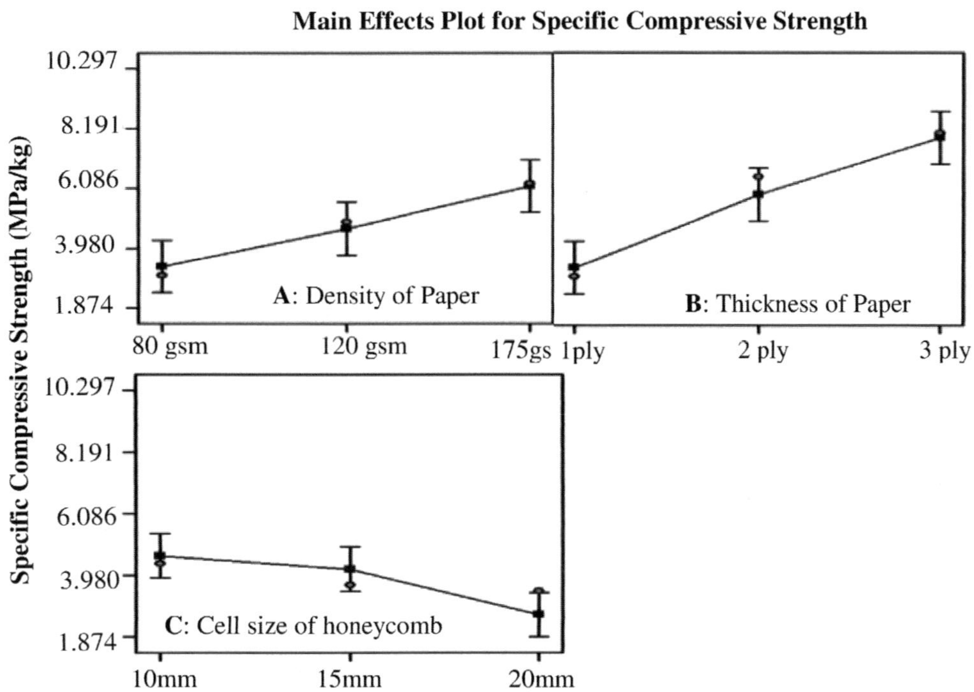

Figure 4. Main effect for specific compression strength.

3.3.3. Optimization Parameter of Paper Honeycomb

The optimum configuration parameters resulting higher SEA and SCS has been obtained through Design of experiments methods. As a result, the Kraft paper honeycomb with density 175gsm, 3 ply thickness of paper and 10mm cell size of honeycomb shows the optimum configuration with 724.80 J/kg of SEA and 9.35 MPa/kg of SCS. These finding was agreed with other researchers which found that the higher cell wall thickness and smaller cell size increase the energy absorption and crush strength [13][14].

4. Summary

In this study, the crushing behavior and effect of parameter under quasi -static compressive load was investigated. All three parameters are giving the significant effect on specific compression strength and specific energy absorption. The sample with highest density of paper, highest thickness of paper and smallest cell size of honeycomb shows the optimum performance for both specific compression strength and specific energy absorption.

References

[1] Bitzer T 1997 *Honeycomb Technology: Materials, Design, Manufacturing, Applications and Testing* (Springer Science & Business Media)

[2] Chen Z and Yan N 2012 Composites : Part B Investigation of elastic moduli of Kraft paper honeycomb core sandwich panels **43** 2107–14

[3] Asprone D, Auricchio F, Menna C, Morganti S, Prota A and Reali A 2013 Statistical finite element analysis of the buckling behavior of honeycomb structures *Compos. Struct.* **105** 240–55

[4] Castanié B, Bouvet C, Aminanda Y, Barrau J-J and Thevenet P 2008 Modelling of low-

energy/low-velocity impact on Nomex honeycomb sandwich structures with metallic skins *Int. J. Impact Eng.* **35** 620–34
[5] Castanié B, Aminanda Y, Bouvet C and Barrau J-J 2008 Core crush criterion to determine the strength of sandwich composite structures subjected to compression after impact *Compos. Struct.* **86** 243–50
[6] Wan Abdul Hamid W L H, Aminanda Y and Shaik Dawood M S I 2013 Experimental Investigation on the Energy Absorption Capability of Foam-Filled Nomex Honeycomb Structure *Appl. Mech. Mater.* **393** 460–6
[7] Zhu F, Zhao L, Lu G and Wang Z 2008 Deformation and failure of blast-loaded metallic sandwich panels—Experimental investigations *Int. J. Impact Eng.* **35** 937–51
[8] Aminanda Y, Castanié B, Barrau J-J and Thevenet P 2009 Experimental and numerical study of compression after impact of sandwich structures with metallic skins *Compos. Sci. Technol.* **69** 50–9
[9] Wu E and Jiang W-S 1997 Axial crush of metallic honeycombs *Int. J. Impact Eng.* **19** 439–56
[10] Yamashita M and Gotoh M 2005 Impact behavior of honeycomb structures with various cell specifications—numerical simulation and experiment *Int. J. Impact Eng.* **32** 618–30
[11] Wang D 2009 Impact behavior and energy absorption of paper honeycomb sandwich panels *Int. J. Impact Eng.* **36** 110–4
[12] Aminanda Y, Castanié B, Barrau J-J and Thevenet P 2005 Experimental Analysis and Modeling of the Crushing of Honeycomb Cores *Appl. Compos. Mater.* **12** 213–27
[13] Xu S, Beynon J H, Ruan D and Lu G 2012 Experimental study of the out-of-plane dynamic compression of hexagonal honeycombs *Compos. Struct.* **94** 2326–36
[14] Khoshravan M R and Najafi Pour M 2014 Numerical and experimental analyses of the effect of different geometrical modelings on predicting compressive strength of honeycomb core *Thin-Walled Struct.* **84** 423–31

Hardfacing of duplex stainless steel using melting and diffusion processes

H Lailatul[a*] **and M A Maleque**[b]

Department of Manufacturing and Materials Engineering
Kuliyyah of Engineering, International Islamic University Malaysia, P.O. Box 10, 50728 Gombak,
Kuala Lumpur, Malaysia.

[a*]Laila_7164@yahoo.com; [b]maleque@iium.edu.my

Abstract. Duplex stainless steel (DSS) is a material with high potential successes in many new applications such as rail car manufacturing, automotive and chemical industries. Although DSS is widely used in various industries, this material has faced wear and hardness problems which obstruct a wider capability of this material and causes problems in current application. Therefore, development of surface modification has been introduced to produce hard protective layer or coating on DSS. The main aim of this work is to brief review on hard surface layer formation on DSS using melting and diffusion processes. Melting technique using tungsten inert gas (TIG) torch and diffusion technique using gas nitriding are the effective process to meet this requirement. The processing route plays a significant role in developing the hard surface layer for any application with effective cost and environmental factors. The good understanding and careful selection of processing route to form products are very important factors to decide the suitable techniques for surface engineering treatment. In this paper, an attempt is also made to consolidate the important research works done on melting and diffusion techniques of DSS in the past. The advantages and disadvantages between melting and diffusion technique are presented for better understanding on the feasibility of hard surface formation on DSS. Finally, it can be concluded that this work will open an avenue for further research on the application of suitable process for hard surface formation on DSS.

1. Introduction

Duplex stainless steels have gained increasing interest in recent years for a number of applications in various industrial sectors of petrochemical process plant, marine engineering and automotive industries. These materials offer unique combination of high mechanical properties and corrosion resistance [1]. Although DSS is widely used in various industries, this material has faced wear and hardness problem in many applications [2]. Surface modification is a technique to improve the performance of engineering surface properties that has been developed for many years. This modification can be done by changing the surface microstructure without affecting the bulk properties of the material [3]. Previous works have approached many techniques to meet this requirement such as melting through TIG torch [4-8] and diffusion through gaseous thermochemical nitriding [9-14].

Surface melting by incorporation of ceramic particles can produce composite layer dispersed with ceramic particles which enhance the surface hardness [3]. In this process, the alloy powder in form of hard ceramic particles with desirable composition is homogeneously deposited onto the surface of the substrate material. The heat sources such as laser, electron beams or TIG torch were used to melt the alloy powder and substrate material [4]. A coated surface obtained by TIG torch has the potential to produce a dense coating with good metallurgical bonding to the base material. The commonly used hard carbide particles to produce ceramic composite coatings with this melting process are; WC [5], TiC [6] and SiC [7]. Dyuti et al. [15] have found that the composite layer can be produced on carbon steel by melting via preplacement of Ti and Al powder mixtures in nitrogen environment under a TIG torch with energy inputs between 540-675 J mm-1. The clad layer that formed with dendrite microstructure consists of Ti-Al nitrides which distributed homogeneously throughout the melt pool. The hardness of the composite layer depend on dendrite population and maximum hardness was 900 Hv.

In the case of gaseous thermochemical nitriding (also known as diffusion process), the hardening of the surface layer on duplex stainless steel is due to incorporation of nitrogen in the austenite lattice, forming expanded austenite. Expanded austenite without nitrides is obtained when high amounts of nitrogen dissolved in stainless steel at temperature below 450 °C [2]. The growth of expanded austenite has been studied by previous researchers using various surface treatments such as nitriding [9,10] carburizing [10] and hybrid process with integration of nitriding and carburizing [16]. Fewell et al. [17] had explored that the expanded austenite is the crystalline structure and not mixed phase with some grains having different lattice parameters.

Most of the findings have showed that surface modifications via melting and thermochemical diffusion techniques are proven methods to improve surface coating of many substrate materials in terms of surface hardness, tribological and corrosion resistance. Several research works found in literatures focusing on melting using different alloying powders and substrate materials via TIG torch [3-7] and laser cladding [18-21]. No attempt was made on the incorporation of SiC particulates hardfacing on DSS by TIG torch melting technique in the earlier works. Moreover, the comparative study between melting and diffusion processes also is not available in the literature. In this paper, an attempt has been made to review, consolidate and highlight the results obtained by various researchers on hard layer formation through melting and diffusion for understanding the importance of opening the research avenue in order to employ this techniques effectively and efficiently for DSS.

2. Processing route for hard surface layer

2.1. Melting Process

Tungsten inert gas welding (TIG) is one of the cleanest and most important melting technique for welding stainless steels. This technique is using a non-consumable tungsten electrode, shielded by inert gas, is used to strike an electric arc with the base metal, providing necessary heat to melt the workpiece [22]. Normally, the inert gases such as argon, helium or nitrogen are used as shielding gas for this welding. The common application of this process is rail car manufacturing, automotive and chemical industries. One of the most advantages of melting technique is the elimination of interfacial incompatibility of the material based on their nucleation and growth from the parent matrix phase. The disadvantage of this technique is the residual stress during welding may cause distortion to the material.

Many research works have been done on characterization of TIG welded mild steel and low alloy steel joints in the past. Buytoz [7] analyzed the microstructural properties of SiC particulates hardfacing on carbon steel with varying SiC between 30-45 μm. From the study, it can be observed that the microstructure of coating surface consists of primary dendrites and eutectic mixture of carbides and austenite. The microhardness of the surface layer was significantly improved to as high as 1135 Hv as compared to 220 Hv of the substrate material.

Maleque et al. [8] investigated the wear behaviour of TiC coated AISI 4340 (low alloy steel) produced by TIG surface melting. TiC coating layer was synthesized by preplacing a 1 mg/mm2 of 40 μm TiC powder on the surface of the steel. The maximum hardness achieved until 996 Hv and wear resistance increased two times higher than untreated alloy. The surface hardening and excellent wear resistance of treated alloy due to formation of dendrite precipitation. Another finding by Mridha et al. [23] has investigated the incorporation of TiC particulates on low alloy steel surfaces via TIG technique. In the study, the crack free composite layer of 0.5 mm to 1.0 mm thickness was developed. Figure 1(a) shows the cross sectional view of the track with heat input 1680 J/mm by preplacing a 1.0 mg/mm2 addition of TiC gave a hemispherical shape melt with the hardness of the resolidified layer improves between 750 to over 1100 Hv compared to base hardness of 300 Hv.

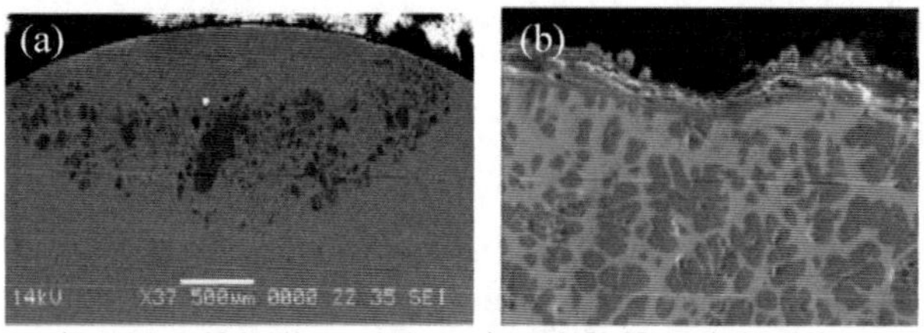

Figure 1. Melt cross section (a) 1680 J/mm, 1.0 mg/mm2 [23] and (b) Micrographs showing dendrite distribution at deeper melt depth [3].

Mridha et al. [3] investigated the influence of processing conditions powder content and energy input on the hardness and surface clad quality of plain carbon steel using a TIG torch with varying titanium powder addition and figure 1(b) shows the dendrites observed at the middle of the melt zone. The maximum surface hardness of around 2000 Hv was developed in most of the tracks and this corresponds to high concentration of TiN dendrites within the modified layer. In addition, Mridha et al. [15] explored the melting of multipass surface tracks in AISI 4340 low alloy steel incorporating titanium carbide powders. It was found that the first track, containing more unmelted and partially melted TiC particulates, developed a maximum hardness of 1000 Hv. It has been observed, that the overall hardness developed in the multipass tracks was 2.5-3.5 times that of the steel substrate hardness of 300 Hv.

Adeleke et al. [24] studied the surface alloying of CP-Ti using preplaced Fe-C-Si powder by TIG technique. The surface modified Cp-Ti alloy exhibited hardness values ranging from 600 Hv to 800 Hv which was 3 to 3.5 times higher than base metal hardness. The higher hardness observed at particular areas are attributed to high population and big sizes of dendritic microstructures of TiC.

2.2. Thermochemical diffusion process

This process is mainly applied to steels and consists of enriching the surface of the mateial to be treated with certain metalloids. The gasseous compound containing the carbon, nitrogen or boron are supplied and deposited to the surface of material via diffusion mechanism. Generally, the treatment is heated to a temperature ranging from 500 to 1000 °C to diffuse on the surface of material. Steels that have undergone this treatment are characterized by good resistance to wear, abrasion and fatigue [25]. The main advantages of this technique is the hardening of the entire surface of the treated specimen. However, the process must be controlled and ensure the process temperature is not too high to avoid the reduction of corosion resistance.

Several researchers have explored the opportunity to perform this technique in the past. Haruman et al. [26] investigated the low temperature fluidized bed nitriding of austenitic stainless steel. The nitriding performed in temperature range between 400 °C and 500 °C for various treatment times. The results showed the hard layer surface after diffusion process as shown in figure 2. The microhardness of the layer also increased up to 1200 Hv.

IOP Conf. Series: Materials Science and Engineering **184** (2017) 012030 doi:10.1088/1757-899X/184/1/012030

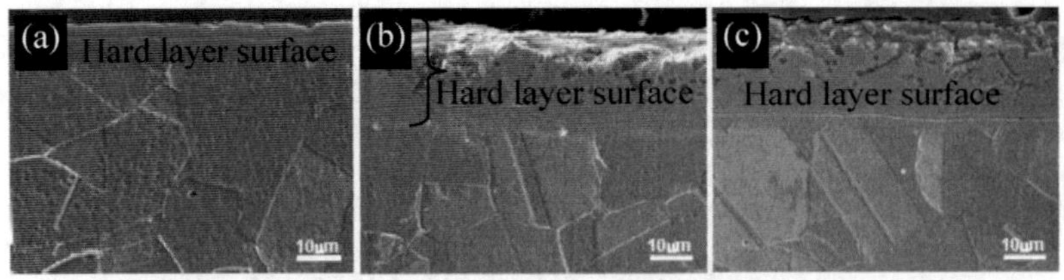

Figure 2. SEM micorgraphs of 500 °C of nitrided austenitic stainless steel: (a) 1 hour, (b) 3 hour (c) 6 hour [26].

Christiansen et al. [9,10] demonstrated that nitriding and carburizing of DSS are feasible to produce hard surface layer due to formation of expanded austenite. The nitriding was performed in a mixture of NH3 and H2. Expanded austenite without nitride/carbides was obtained when high amount of atomic nitrogen/carbon are dissolved in DSS at temperature below 450°C for nitrogen and 550°C for carbon. The hardness of hard layer increased until 1200 Hv compared to 400 Hv of substrate material. Nagatsuka et al. [27] investigated plasma nitriding treatment on DSS with temperature of 356 °C and 383 °C. Based on the result, the expanded austenite observed on treated specimens and produced hardened surface until 1000 Hv and good wear resistance.

2.3. *Laser cladding*

Laser cladding is the fusion of a different metal to a substrate surface, with a minimum of melting of the substrate. Surface coating by laser is a method that has been developed over the last two decades. The lasers minimal and easily controllable energy delivery makes it possible to alloy, impregnate, clad and harden components that are exposed to wear and corrosion. This process is employed which require a high productivity combined with flexibility without compromising on quality. A high and uniform quality with a low heat input makes this process suitable for a wide range of applications in which minimum distortion is desired. The disadvantages of this cladding technique is slow and expensive [28].

Laser surface alloying of Al with Fe is studied by Tomida et al. [21]. It is found that the hardness of laser alloyed layer increases until 1000 Hv with increasing of Fe Content. However, cracking occurred in the alloyed layer with higher hardness than 600 Hv because the brittle lump-like Fe2Al5 compound was produced on the surface. The wear resistance of the alloyed layer improved with increasing the hardness due to the formation of the fine Fe rich intermetallic compounds.

2.4. *Ion Implantation*

This treatment introduces energetic ions into the surface of austenitic stainless steel. Ion induces two types of change to the surface of the solid; structural alterations (creation of disorder) and chemical alterations (modification of the chemical composition of the surface). These changes provides significant result of physical, mechanical and chemical of the surface material of austenitic stainless steel [25]. The advantages of this process is can be easily and precise control of the dose and depth profile. The disadvantages of this process is damage recovery because ion implantation may causes damage to the crystal structure of the target which is often unwanted.

Fernandes et al. [29] investigated the mechanical properties of nitrogen rich surface layers on austenitic stainless teel treated by plasma immersion ion implantation. Based on the analysis, the harder surface formed due to the higher nitrogen concentration obtained by this technique. The wear rate of the treated samples also increased with nitrided layer with up to 500 nm of thickness.

3. Advantages and disadvantages of melting and diffusion techniques

Table 1 shows the matrix of the advantages and disadvantages of melting and diffusion processes on materials of the surface modification. It is important to understand of the process and to choose the best method in the application.

Table 1. Advantages and disadvantages matrix of melting and diffusion route of the process [30-34].

Melting route		Diffusion route	
Advantages	Disadvantages	Advantages	Disadvantages
Inert gas such as argon, helium and nitrogen	Expensive	Ammonia or methane gas	The usage of ammonia gas is toxic and harmful
Short processing time, flexibility in operation and processing precision	Operator must be very close to weld specimen during the process	Allow effective hardening of the entire surface	Long processing time to produce hard surface layer
Reduced distortion due to concentrated heat source	Heat source may contribute to residual stress	Relatively low processing temperature	Temperature control is important to avoid the reduction of corrosion resistance properties.
Ability to create composite hard surface layer	The process requires binders and additives	Hard nitride layer is easy to produce	Difficult to produce high purity material
Eliminate interfacial incompatibility based on their nucleation and growth from the parent matrix phase	The process require solid lubricant particles to withstand wear resistance at high temperature	The surface layer has excellent wear properties at room and high temperature	Require more time to develop hard layer
No spark or sputter	Care should be taken to protect skin and eyes due to arc rays is very bright	The treatment process is controlled properly in the tube furnace	Difficult to control the extraction of hazardous gas and require expensive setup for proper ventilation
The process and clad specimens are clean	Specimen cleaning must be taken care before cladding	The native oxide film on the surface is removed before treatment	The specimen needs to be soaked in concentrated hydrochloric acid

4. Concluding remarks

Hard facing development on duplex stainless steel using melting and diffusion processes are highlighted with the merits and demerits. It is clear that both processes are applicable to produce hard layer with better wear resistance properties. In summary, it can be concluded that many of the research works carried out on TIG torch melting technique and diffusion of thermochemical nitriding for other steels but very limited information is available in literature on DSS. In spite of considerable research on both treatments, no study has been conducted to compare and assess the characteristics features in terms of simplicity, microstucture and properties performance on DSS. Future study on the hard surface formation of DSS is expected to have the significant effect on surface hardness and wear resistance. Therefore, future work should be conducted to compare and assess the characterization of surface coated materials between these two processes.

References

[1] Charles J 2007 *Past, present and future of duplex stainless steel* (La Paine Saint Denis, France – Duplex conference) pp 1-10.

[2] Paijan L H, Berhan M N, Adenan M S, Yusof N F M and Haruman E 2012 Structural development of expanded austenite on duplex stainless steel by low temperature thermochemical nitriding process, *Advanced Materials Research* **576** pp 260-263.

[3] Mridha S and Dyuti S 2011 Effect of processing parameters on microstructure and properties TIG melted surface layer of steel, *Advanced Materials Research* **264-265** pp 1421-1426.

[4] Maleque M A, Bello K A, Idriss A N M and Mridha S 2013 Processing of TiC-CNT hybrid composite coating on low alloy steel using tig torch technique, *Applied Mechanics and Materials* **378** pp 259-264.

[5] Leunda J, Soriano C, Sanz and Navas V G 2006 Laser cladding of WC tool steels for die repair, *Physics Procedia* **261** pp 592-599.

[6] Mridha S, Ho S Y and Kum S T 1998 Surface addition of hard particles using TIG arc melting, *Processing and Fabrication of Advanced Materials* **6** pp 131-138.

[7] Buytoz S 2006 Microstructural properties of SiC based hardfacing on low alloy steel, *Surface & Coating Technology* **200** pp 3734-3742.

[8] Maleque M A, Ghazal B A, Ali M Y, Hayyan M and Ahmed A S 2015 Wear behaviour of TiC coated AISI 4340 steel produced by TIG surface melting, *Materials Science Forum* **819** pp 76-80.

[9] Christiansen T and Somers M A J *2006 Characterisation of low temperature surface hardened stainless steel*, sixth ed. (Ballerup, Denmark).

[10] Christiansen T and Somers M A J 2005 Low temperature gaseous nitriding and carburising of stainless steel, *Surface Engineering* **21** pp 5-6.

[11] Haruman E, Sun Y, Triwiyanto A, Manurung Y H P and Adesta E Y 2010 : submitted to Journal of Materials Engineering and Performance (2010).

[12] Baranowska J, Wysiecki M 2000 Influence of surface pretreatment on case formation during gaseous nitriding, *Surface and Coatings Technology* **25** pp 30-34.

[13] Baranowska J, Franklin E E, Pelletier C G N 2005 Tribological behaviour and mechanical properties of low temperature gas nitrided austenitic steel in relation to layer morphology, *Wear* **259** pp 432 -438.

[14] Baranowska J 2004 Characteristic of the nitride layers on the stainless steel at low temperature, *Surface and Coatings Technology* **180-181** pp 145-149.

[15] Mridha S, Idriss A N M, Maleque M A, Yaacob I I, Baker T N 2015 Melting of multipass surface tracks in steel incorporating titanium carbide powders, *Materials Science and Technology* **31** pp 1362-1369.

[16] Tsujikawa M, Noguchi S, Yamauchi N, Ueda N, Sone T and Nakata K 2007 Surface layer hardness of austenitic stainless steel via low temperature plasma carburizing and/or nitriding, *Materials Science Forum* **561-565** pp 2477-2480.

[17] Fewell M P, Mitchell D R G, Priest J M, Short K T and Collins G A 2000 The nature of expanded austenite, *Surface and Coatings Technology* **131** pp 300-306.

[18] Yan J, Gao M and Zeng X 2010 Study on microstructure and mechanical properties of 304 stainless steel joints by TIG, laser & laser TIG hybrid welding, *Optics and Lasers in Engineering* **48** pp 512-517.

[19] Murad A H I, Khourshid A and Sharef T 2012 Gas tungsten arc and laser beam welding processes effects on duplex stainless steel 2205 properties, *Materials Science and Engineering* **549** pp 105-113.

[20] Mridha S and Baker T N 2007 Incorporation of 3um SiC into Ti surface using a 2.8 kW laser beam of 186 & 373 MJ m^{-2}, *Journals of Materials Processing Technology* **185** pp 38-45.

[21] Tomida S and Nakata K 2003 Fe-Al composite layers on aluminum alloy formed by laser surface alloying with iron powder, *Surface and Coatings Technology* **174–175** pp 559-563.

[22] Sathiya P, Aravindan S, Soundararajan R and Noorul Haq A 2009 Effect of shielding gases on mechanical and metallurgical properties of duplex stainless steel welds, *Journal of Materials Science* **44** pp 114-121.

[23] Mridha S, Md Idriss A N, Baker T N 2012 Incorporation of TiC particulates on AISI 4340 low alloy steel surfaces via tungsten inert gas arc melting, *Advanced Materials Research* **45** pp 655-660.

[24] Adeleke S A and Maleque M A 2015 TIG melted surface modified titanium alloy for cylinder liner application, *International Journal of Automotive Engineering and Technologies* **4** pp 130-138.

[25] Takadoum J 2008 *Materials and Surface Engineering in Tribology* (France - Wiley-ISTE).

[26] Haruman E, Sun Y, Malik H, Sutjipto A G E, Mridha S and Widi K 2006 Low temperature fluidized bed nitriding of austenitic stainless steel, *Solid State Phenomena* **118** pp 125-130.

[27] Nagatsuka K, Nishimoto A and Akamatsu K 2010 Surface hardening of duplex stainless steel by low temperature active screen plasma nitriding, *Surface and Coatings Technology* **05** pp 5295-5299.

[28] Gedda H 2004 *Laser Cladding of Experimental and Theoretical Investigation* (Sweeden: Luleå University,

[29] Fernandes B B, Mandl S, Oliveira R M and Ueda M 2014 Mechanical properties of nitrogen-rich surface layers on SS304 treated by plasma immersion ion implantation, *Applied Surface Science* **310** pp 278-283.

[30] Solas and Smith F 2014 , *Thermal processes - Tungsten active gas shielded welding* , second rev. , Ireland.

[31] ASM International 2002 *Process selection guide - surface hardening of steels: understanding the basics* pp. 1-15.

[32] Mridha S, Idriss A N M, Maleque M A, Suryanto and Souad A 2012 Effect of voltage on the consolidation of TiC particulates on steel substrate fused by TIG welding arc, *International Journal of Mechanical and Materials Engineering (IJMME)* 7 pp 48-53.

[33] Maleque M A and Adeleke S A 2013 Surface alloying of CP-Ti using Fe-C-Si powder by tungsten inert gas torch technique, *International Conference on Mechanical, Industrial and Materials Engineering* pp 668-673.

[34] Bello K A, Maleque M A,Ahmad Z, Adebesi A A and Mridha S 2015 Preparation and characterization of TIG-alloyed hybrid composite coatings for high temperature solid lubrication, *Proceedings of Malaysian International Tribology Conference* pp 265-267.

Acknowledgement
Special thanks to the Ministry of Higher Education (MOHE) who has funded this research project under Fundamental Research Grant Scheme of FRGS15-203-0444.

Analysis of Fracture Mechanism for Al-Mg/SiC$_p$ Composite Materials

M A Maleque, A A Adebisi, and N Izzati

[1] Department of Manufacturing and Materials Engineering,
International Islamic University Malaysia
53100, Kuala Lumpur, Malaysia

E-mail: maleque@iium.edu.my

Abstract. The present study aims to examine the fracture mechnism of silicon carbide particle (SiC$_p$) reinforced aluminium matrix composite (AMC) material with 1 wt% addition of magnesium is fabricated using the stir casting process. The aluminium composite (Al-Mg/SiC$_p$) is investigated for fatigue life and impact strength considering reinforcement weight fraction and influence of temperature on fracture toughness. The fabricated composite was tested using fatigue testing machine and charpy impact tester. Fractographic observations were evaluated with the scanning electron microscopy (SEM) on the fracture surface. It was found that increasing the SiC$_p$ weight fraction increased the fatigue life of the composite. Moreover, the 20 wt% SiC$_p$ Al-Mg composite attained the highest number of cycle and fatigue life compared to other variations. The mechanism responsible for the phenomena includes load transfer from the Al matrix alloy phase to the high strength and stiffness of the incorporated SiC$_p$. The temperature variation influenced the impact strength of the composite and improved fracture toughness is achieved at 150 °C. It can be concluded from this study that reinforcement weight fraction and temperature affects the fracture behavior of the composites.

1. Introduction

Aluminium reinforced silicon carbide particle composite possess improved operational potential for critical structural components due to its attractive properties when compared to monolithic materials. These properties include improved strength and stiffness, high elastic modulus, hardness, wear resistance and low coefficient of thermal expansion. Moreover, the possible lightweight benefit of particle reinforced composite also makes it an alternative choice for researchers [1, 2]. Furthermore, the application of particle reinforced composite as component material for emerging structures involving cyclic loading and toughness strength requires sufficient amount of energy absorption and cyclic load resistance [3]. In order to achieve this, the influence of the particle reinforcement weight fraction and processing technique play a major role in determining the appropriate properties. Various studies [4, 5] have shown that, the stir casting technique is a simple and cost effective route to fabricate the particle reinforced composite. The influence of stir casting processing condition determines the intrinsic fatigue response of the composite. Therefore, by carefully choosing the reinforcement weight fraction of the particles with suitable casting conditions, it is possible to optimize the fatigue and impact properties of the aluminium matrix composite [6]. Moreover, the presence of reinforcement particle influences the toughness and fatigue life cycle of the composites under operational condition. Uygur [7] found that addition of hard ceramic particle in aluminium composite material reduces the low cycle fatigue behaviour under strain controlled cyclic loading conditions as compared to unreinforced aluminium alloy. This reduction is attributed the brittle nature of ceramic particles phase, high dislocation density at the interfaces, hydrostatic stress development and plastic flow constraint within the aluminium alloy. Moreover, the effect of temperature is also a vital factor in determining the energy absorbed and fracture properties of composites. However, studies have been conducted on the impact toughness behaviour of reinforced aluminium composite at room temperature [8]. In order to further understand the fracture mechanism of aluminium composite, this study aims to investigate the fatigue and impact strength of Al-Mg/SiC_p composite considering reinforcement fraction and influence of temperature on fracture toughness.

2. Methodology

2.1. Materials and method

Aluminium alloy 6061 is used as the matrix phase and silicon carbide particle (SiC_p) as the reinforcing phase material. The matrix phase contains magnesium and silicon as major alloying elements and are commonly used as components for automotive and aerospace applications due to their superior mechanical strength. The chemical composition is shown in Table 1.

Table 1. Chemical Composition of Al6061 by weight percent (wt %)

Composition	Si	Fe	Cu	Mn	Mg	Cr	Zn	Ti	Al
Al 6061	0.62	0.23	0.22	0.03	0.84	0.22	0.10	0.10	Bal

The SiC_p phase comprises of the mixture of three different particle sizes; 15 μm, 40 μm and 80 μm in the ratio 1:1:2 respectively. In order to obtain homogenous mixture of the SiCp, the Fritsch Pulverisette P5 planetary ball milling set-up is utilized with a powder to ball mixing ratio of 5:1 at a speed of 150 rpm for 15 min. [9]. The Al-Mg/SiC_p composite is produced using the stir casting technique and the process is explained as contained in [10]. The reinforcement weight fraction varies from 5, 10, 15, 20 and 25 wt% to the matrix alloy.

2.2. Fatigue testing

The fatigue test sample as shown in Figure 1 is produced to dimension from the cast billet using the ROMI M420 CNC lathe machine with cutting speed of 1500 rpm and feed rate of 0.2 mm/rev. Fatigue test was carried out using INSTRON 8874 Universal Testing Machine according to ASTM E466-15. Test sample of five different compositions 5, 10, 15, 20 and 25 wt% were tested. The constant amplitude sinusoidal loading at a stress ratio of 0.1 and a frequency of 50 Hz was applied. The test was conducted at a maximum stress levels of 200 MPa. Result was obtained by measuring the number of cycles each composite sample can withstand before fracture occurred using the S-N curve.

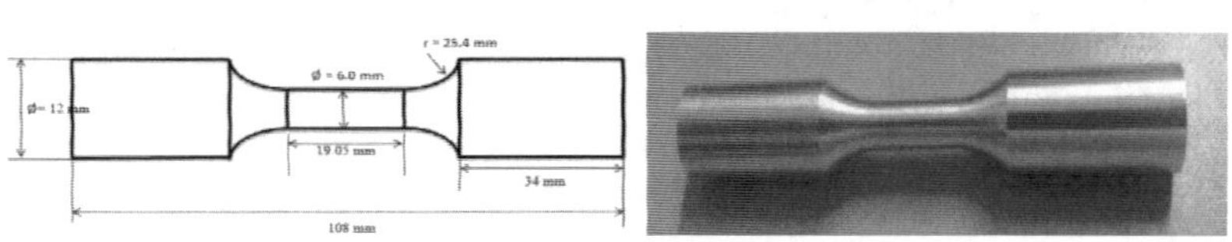

Figure 1. Standard fatigue test sample

2.3. Impact testing

The V-notched impact specimen was cut using EDM wire cut machine in order to obtain the desired rectangular shape of length of 55 mm with thickness and width of 10 mm and a V-notch angle of 45°. Figure 2 shows the actual V-notched impact specimen. The impact test is carried out using the INSTRON Charpy impact tester according to ASTM E23-12c. In this study, the specimens were tested at different temperature (-15, 25, 150, 200 °C). After all test has been conducted, the characterization of the fracture surface analysis is examined using JEOL, JSM 5600 field emission scanning electron microscope (FESEM).

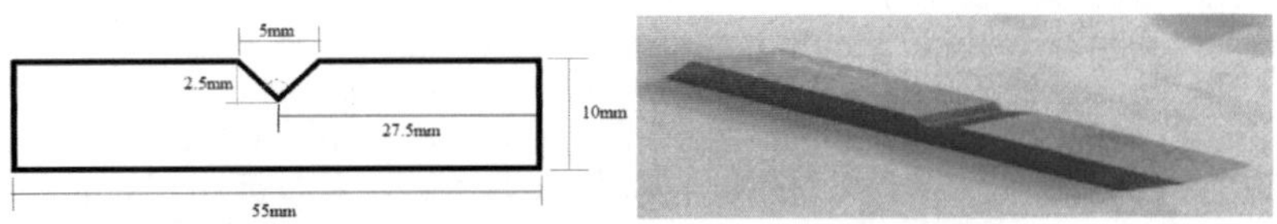

Figure 2. Standard V-notched impact test specimen

3. Failure Properties of Al-Mg/SiC$_p$ Composite

3.1. Fatigue life

The fatigue life of each composite material considering the reinforcement weight fraction was investigated by the number of cycles it can withstand before fracture failure. A maximum stress of 200 MPa was applied to the composite samples and results obtained is presented by S-N curve in Figure 3 The result shows that fatigue life of SiC$_p$ reinforced Al-Mg composite increases as the reinforcement weight fraction increases resulting higher load bearing capacity of the composite materials. Moreover, increasing the amount of reinforcement fraction initiates increase in elastic modulus, yield strength and ultimate strength with a corresponding decrease in ductility.

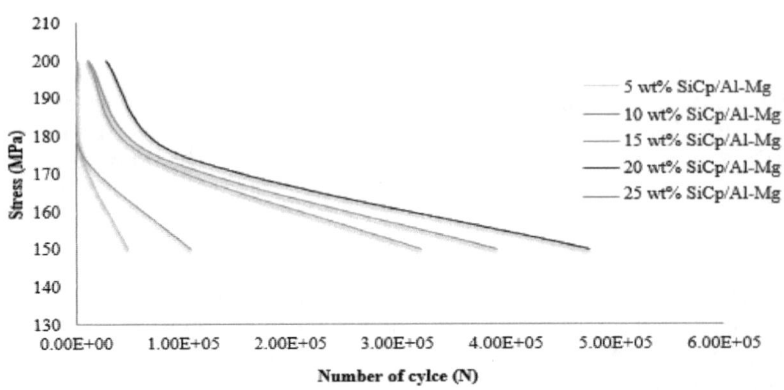

Figure 3. S-N curve profile of fatigue life strength with different reinforcement weight fraction

From Figure 3, Al composite with 5 wt% SiC$_p$ has the shortest fatigue life strenght at maximum stress of 200 MPa followed by 10 wt% SiC$_p$ and 15 wt% of SiC$_p$ composite respectively. SiC$_p$ reinforcement with < 10 wt% SiC$_p$ shows that it is insufficient to withstand the cyclic stress applied to the composite which means reinforcement is insufficient in the composite formulation to transfer the load to the matrix. Higher weight fraction of SiC$_p$ reinforcement increases the elastic modulus of the composite materials, thus, the number of cycle is expected to increase. It is observed from the S-N curve profile that 20 wt% SiC$_p$ Al-Mg composite has the highest fatigue life. Though, 25 wt% SiC$_p$ shows a slightly shorter fatigue life than 20 wt% SiC$_p$ reinforced aluminium composite at maximum stress level of 200MPa. This is as a result of decrease in cyclic ductility which is due to the presence of substantial amount of SiC$_p$ reinforcement which induces brittleness in the fatigue properties [11].

3.2. Impact toughness

The toughness is analyzed considering the influence of temperature on 20 wt% SiC$_p$ Al-Mg composite. Figure 4 shows the impact toughness of the composite considering temperature variaitons. However, the unreinforced aluminium alloy attained impact toughness energy of 14 J. The ductile nature of Al alloy makes it to undergo plastic deformation which is responsible for the high energy impact especially at room temperature. The impact toughness Al-Mg alloy decreased with the presence of reinforced SiC_p. Considering the temperature effect, the toughness is increased as the temperature rises. However, this is not the case for temperature above 150°C as indicted in the Figure 4. At higher temperature beyond 150°C, the microstructure of the Al alloy experiences overaging which increases diffusion rate and accelerate the dissolution and growth rates, therefore enhances the particle coarsening. The maximum solute solubility in the matrix alloy also increases and precipitates dissolve to release solute in solid solution. These mechanisms increases the interparticle spacing which in turn soften the Al alloy. Due to this condition, energy absorbed before plastic deformation is lower as the failure occurs in the matrix alloy phase or near the matrix-reinforcement interface at temperature between 200 to 250°C. Thus, for temperature higher than 150°C, the impact toughness starts to drop.

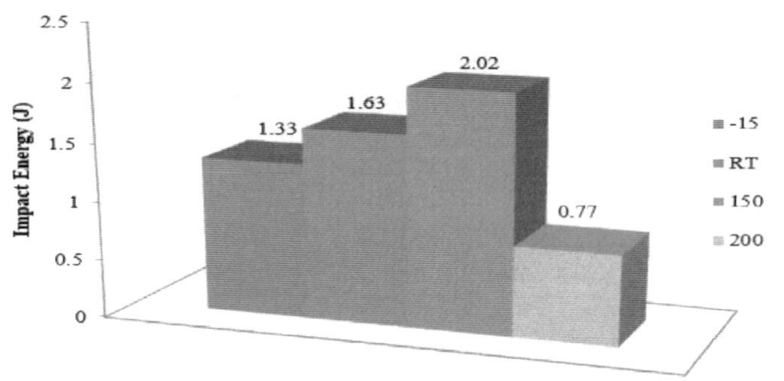

Figure 4. Impact toughness of 20 wt% SiC_p Al-Mg composite considering temperature variation

4. Fracture surface analysis of Al-Mg/SiC_p Composite

4.1. Fatigue fracture

The fatigue resistance of the Al-Mg/SiC_p composite increases with increasing the weight fraction of SiC_p reinforcement which inturn increases the fatigue life of the composite material. Basically, fracture occur in composite material by crack initiation which subseqently proceeds at the debonded interface between Al-Mg alloy and SiC_p. The study shows that presence of SiC_p either assisted or resisted the crack propagation of the composite. Due to the presence of brittle ceramic SiC_p with different weight fraction (5, 10, 15, 20, 25 wt%), the composite materials exhibit tendencies of brittle behavior. Al-Mg composite reinforced with 5 wt% and 10 wt% SiC_p shows some ductile behavior before fracture as observed in Figure 5(a). The ductile properties is characterized by the dimple structures observed on the fractured surface. These dimple structures are produced by the decohesion of SiC_p from Al alloy matrix. Figure 5(b) also exhibited the ductile tendencies resulting in necking and plastic deformation of the matrix alloy before failure.

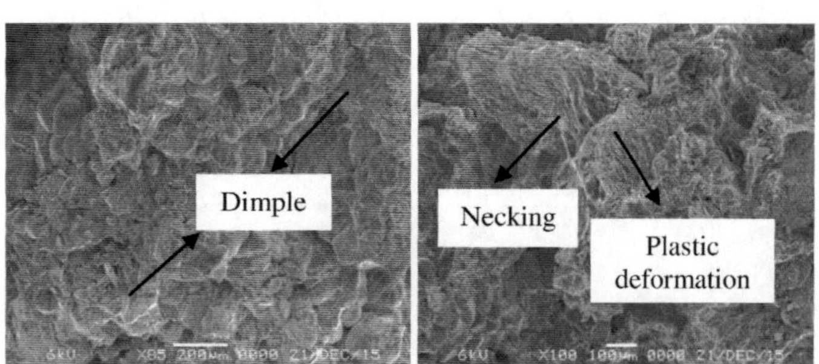

(a) (b)

Figure 5. Fractgraphy analysis of (a) 5 wt% SiC_p and (b) 10 wt% SiC_p Al-Mg composite

Figure 6(a) shows the fractured surface of Al-15 wt% SiC_p composite. The surface analysis shows the combination of dimple ductile behaviour of the matrix alloy phase and the brittle failure of the reinforcement. Some of the SiC_p were observed to crack due to high stress intensity. Moreover, the larger particles are prone to crack while the fine particles decohere from matrix alloy phase to form void nucleation.

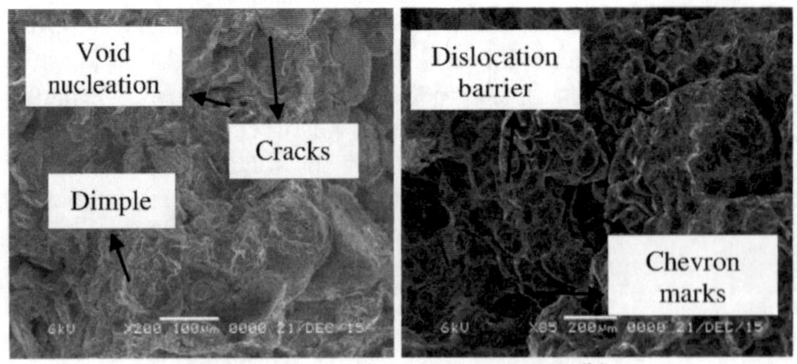

(a) (b)

Figure 6. Fracture surface analysis of (a) 15 wt% SiC_p and (b) 20 wt% SiC_p Al-Mg composite

Al-Mg composite reinforced with 20 wt% of SiC has longest fatigue life because the crack growth is hindered by the effect of SiC_p as shown in Figure 6(b). The presence of SiC_p provides a tough barrier to the dislocation motion. The SiC_p act as the load bearer and also resist the crack propagation which is evidence with the higher number of cycles before failure. With the dispersion of 20 wt% SiC_p in Al-Mg alloy, the load is transferred more effectively. From Figure6(b), the microstructure shows that there is no void nucleation around the SiC_p as it acts as stress reliever and also the presence of the chevron marks on the fractured surface proves that Al-20wt%SiC experienced brittle fracture.

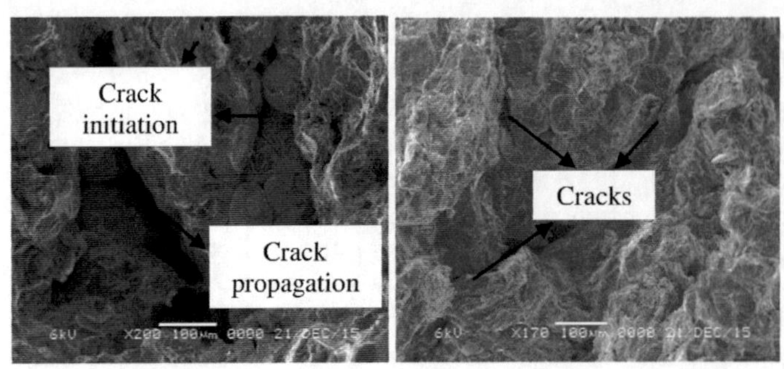

Figure 7. Crack initiation and propagation for 25 wt% SiC_p/Al-Mg composite

The 25 wt% reinforced SiC_p Al composite shows shorter fatigue life strength compared to 20 wt% SiC_p composite due to excessive crack initiation spotted around the particles as a result of the weight fraction of reinforcement. These regions serves high stress concentration, which in turn develops to crack initiation sites as shown in Figure 7. As the cracks approaches the interphase area, SiC_p tend to absorb the crack energy, which eventually fracture with rapid failure. In this condition, SiC_p no longer act as stress relievers but behaves in a brittle manner as the crack propagates through the particles.

4.2 Impact fracture

The impact toughness fracture surface of Al-Mg alloy tends to exhibit higher ductility properties compared to the SiC_p reinforced Al-Mg composite. Figure 8(a) shows the cup and cone shape of fracture surface which indicates necking before major failure occurred. The rough surface indicates that the matrix alloy shows a gross plastic deformation during fracture. It proved that the material needed high energy for the fracture to occur. The effect of temperature is observed on failure analysis using surface fracture micrograph. At room temperature, the Al-Mg alloy experienced a transgranular fracture as the crack propagations occurs across the grain boundaries as shown in Figure 8(b). When the Al-Mg alloy is reinforced with SiC_p, a transformation from ductile to brittle phase takes place depending on the wt % of reinforcement. The fracture process propagates with little or no plastic deformation before the composite material fails. The presence of hard ceramic SiC_p act as stress concentration areas which leads to formation of unstable crack resulting in low energy absorption and impact energy in resisting the load.

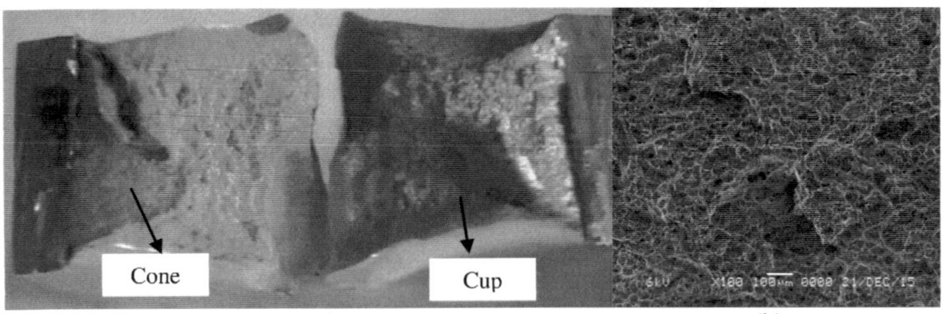

(a) (b)

Figure 8. Failure surface of Al-Mg alloy (a) Cup and cone; and (b) transgranular fracture at room temperature

At temperature of -15 °C, the 20 wt% SiC_p/Al composite exhibits flat cleavages with small areas of fracture surfaces, hence, low energy is required for the fracture to occur. Further analysis shows that the composite material exhibited brittle fracture consisting transgranular cleavage facets as shown in Figure 9(a). At room temperature the composite experiences fast propagation of transgranular crack across the grains. These are represented with river lines as shown in Figure 9(b) between the cleavages which converged in the direction of local crack propagation. As the propagation of the crack occurs rapidly, chevron marks are spotted at the crack area instead of dimple microstructure.

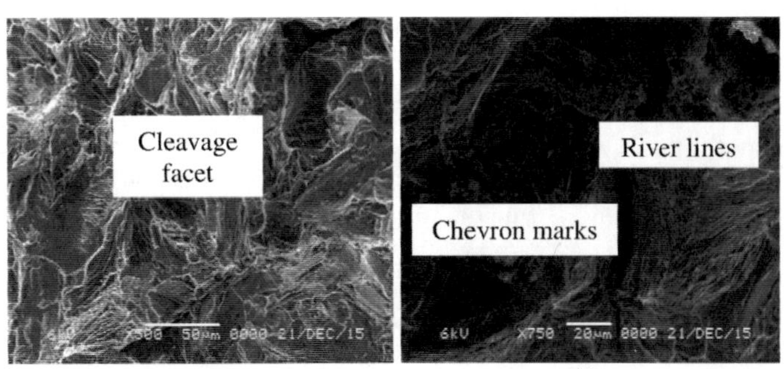

(a) (b)

Figure 9. Impact fracture surface of 20 wt% SiC_p/Al-Mg at (a) -15 °C and (b) room temperature

At 150 °C, the 20 wt% SiC_p/Al composite was found to exhibit some external necking prior to fracture. In addition, high elastic loads are built up in the reinforcement, thus induces matrix yield and plastic deformation before interfacial debonding as shown in Figure 10(a). However, as the temperature increases, the bonding between matrix and reinforcement become weaker, where the particles served as void nucleation sites. At temperature of 200 °C, the composite exhibited ductile characteristics with the presence of dimples observed on the fractured surface. Though, the crack propagates catastrophically with high speed through the grains and dominantly at the interface between matrix and particulate reinforcement, resulting in lower energy absorption compared to the composite material at temperature of 150 °C. Void nucleation can be seen around the SiC_p, providing crack propagation across the matrix-reinforcement interface which serves as preferential sites for cavity initiation as shown in Figure 10(b).

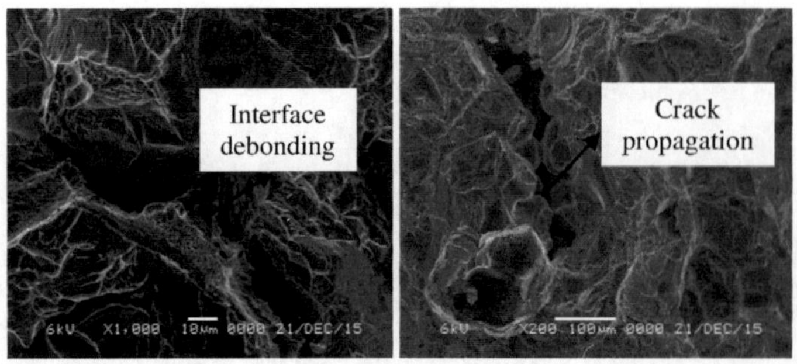

(a) (b)

Figure 10. Impact toughness feature of 20 wt% SiC_p/Al-Mg at (a) 150 °C and (b) 200 °C

5. Conclusion

The following results can be concluded from the study;

1. The stir casting process was successfully utilized in fabricating Al-Mg/SiC_p composite with different weight fraction of the reinforcement phase.
2. The reinforcement weight fraction of 20 wt% SiC_p recorded the optimum and longest fatigue life strength due to the highest number of cycle attained compared to other reinforcement variations which in turn influences the elastic modulus. On the other hand, the 25 wt% SiC_p shows a slightly shorter fatigue life than 20 wt% SiC_p composite at maximum stress level of 200MPa. This can be ascribed to decrease in cyclic ductility which is due to the presence of substantial amount of SiC_p reinforcement which induces brittleness in the fatigue properties.
3. With 20 wt% reinforcement the crack growth is hindered by the effect of SiC_p which provides a tough barrier to the dislocation motion and also act as load bearer in order to resist the crack propagation.
4. The impact toughness of the composites decreased compared to the unreinforced Al-Mg alloy. However, the influence of temperature play a role in the toughness value of the composite.

Acknowledgement

The authors acknowledge the support of the Ministry of Education and International Islamic University Malaysia for financial support to conduct this research work under Project No.: PRGS12 002-0002.

References

[1] Maleque M A, Adebisi A A and Shah Q H 2012 Energy and cost analysis of weight reduction using composite brake rotor, *Int. J. Veh. Struc. Syst.* **4**(2) 69-73.

[2] Adebisi A A, Maleque M A and Shah Q H 2014 Performance assessment of aluminium composite material for automotive brake rotor, *Int. J. Veh. Syst. Model. Test.* **9**(3-4) 207-217.

[3] García-Romero A M, Egizabal P and Irisarri A M 2010 Fracture and fatigue behaviour of aluminium matrix composite automotive pistons, *Appl. Compo. Mater.* **17**(1) 15-30.

[4] Hashim J, Looney L and Hashmi M S J 1999 Metal matrix composites: production by stir casting method, *J. Mater. Process. Tech.* **92-93** 1-7.

[5] Balaji V, Sateesh N and Hussain M M, 2015 Manufacture of aluminium metal matrix composite (Al7075-SiC) by stir casting technique, *Mater. Today: Proc.* **2**: 4-5 3403-3408.

[6] Arivukkarasan S, Dhanalakshmi V, Suresh A and Aruna M 2013 Performance study on fatigue behaviour in aluminium alloy and alumina silicate particulate composites, *J. Appl. Sci. Eng.* **16**(2)127-134.

[7] Uygur I, 2002 Low cycle fatigue properties of 2124/SiC_p Al-alloy composites, *Turkish J. Eng. Env. Sci.* **26** 265-274.

[8] Ozden S, Ekici R and Nair F 2007 Investigation of impact behaviour of aluminium based SiC particle reinforced metal matrix composites, *Compo. Part A: Appl. Sci. Manufac.* **38**(2) 484-494.

[9] Zhao N, Nash P and Yang X 2005 The effect of mechanical alloying on SiC distribution and the properties of 6061 aluminium composite, *J. Mater. Process. Technol.* **170** 586–592.

[10] Adebisi A A, Maleque M A, Ali M Y and Bello K A 2016 Effect of variable particle size reinforcement on mechanical and wear properties of 6061Al–SiC_p composite, *Compos. Interfaces*, **23**(6) 533-547.

Micro Wire Electro Discharge Grinding: Optimization of Material Removal Rate and Surface Roughness

Mohammad Yeakub Ali *, **Mohamed Abd Rahman, and Rosmarina Nordin**

Department of Manufacturing and Materials Engineering
International Islamic University Malaysia
PO Box 10, 50728 Kuala Lumpur, Malaysia
* E-mail: mmyali@iium.edu.my

Abstract. This paper presents the analysis and modelling of material removal rate (MRR) and surface roughness (R_a) by micro wire electro discharge grinding (micro-WEDG) with control parameter of gap voltage, feed rate, and spindle speed. The data were analyzed and empirical models are developed. The optimized values of MRR and R_a are 0.051 mm^3/min and 0.25 μm respectively with 110 V gap voltage, 38 μm/s feed rate, and 1315 rpm spindle speed. The analysis showed that gap voltage has significant effect on material removal rate while spindle speed has significant effect on surface roughness.

1. Introduction

Micro manufacturing field continuously improved from few decades before due to encouragement by increase in demand on micro scale products. Thus, scaling down the traditional macro-manufacturing process to micro- manufacturing is continuously developing in order to improve its efficiency as well as accuracy [1].There are various techniques in order to produce multiple types of micro- products. Micro- electromechanicals systems (MEMS) particularly comprises of techniques such as photolithography, LIGA, laser fabrications, etc. while non-MEMS based manufacturing techniques involves mechanical machining, EDM, laser cutting, and etc. Micro wire electro discharge grinding (micro-WEDG) is one of the non-MEMS based manufacturing [1]. In micro-WEDG the uses of grinding wheel is replaced with micro wire which acts as an electrode. When fine rod need to be ground, it acts as a workpiece and allowed to be rotated while the electrode (micro wire) is held stationary. This process utilizes the electrical sparks or thermal energy produced to remove the unwanted materials from the workpiece in order to make the desired shape. The electrode is fixed on the table, and the table is controlled to give the feed. This process is broadly used for forming very thin rods with high-aspect ratio, which can be used as tool electrodes for micro-EDM drilling or milling (micro-tools) as well as micro- parts [2]. Comparing the usage of grinding wheel with WEDM, WEDM is more flexible.

In order to enhance the performance of WEDG particularly in accuracy and repeatability, special wire guide mechanism need to be applied. Usage of wire guide was introduce where machining accuracy is not sufficient enough for aiming size of rod around Ø50 μm. This due to wire vibration amplitude which range 10 μm and more also due to wire deviation produced by original curl or by machining force [2].The concept of wire guide is there is no vibration and wire deviation at point where the wire contacts with the wire guide. Besides that, suitable tension of wire is required in order to keep contact with the wire guide. However, excessive high tension applied is not necessary as long as the middle part of the wire is fix by the tension applied. High material removal rate (MRR) is achieved at maximum during rough-cut stage. This is due to increasing in energy in each discharge produce, which depends on the energy stored by the capacitor, this theory is based on Eq.1. Meanwhile, during finish cut, lower voltage is used in order to produce good surface finish [3].

$$E = \frac{1}{2} CV^2 \qquad (1)$$

where, E = diacharge energy, C = capacitance, and V = voltage

Mefuz et al. (2008) investigated in the influence of feed rate, capacitance, and voltage on five response variables, which are surface roughness, maximum peak-to valley roughness height, tool wear ratio, and material removal rate. Based on the research, capacitance and voltage are the most influential factor on surface roughness; meanwhile, for MRR it was found that feed rate, capacitance, and voltage carries significant effect in order to increase MRR [4]. The same results also obtain by Periyanan et al. (2011), by using Taguchi method as design of experiment found that feed rate, capacitance, and voltage have significant effects on MRR in micro-WEDG [5]. Thus, the objective of this paper is to investigate the MRR and R_a with three varying parameters, gap voltage, feed rate, and spindle speed.

2. Experiment

Taguchi's L9 orthogonal array statistical approach was used to design the experiment with a total of 9 experiments. The experiments were conducted using micro-WEDG machine, DT-110 Mikrotools (Mikrotools Inc., Singapore) to complete the analysis of MRR and R_a. The experimental parameters are shown in Table 1 with controlled parameters; gap voltage, feed rate, and spindle speed. The workpiece used was tungsten rod with 1.00 mm diameter and copper wire as the electrode wire. Before an experiment is started, tungsten rod is cut into length of approximately 20 mm. After that, the weight of the workpiece is measured using electronic balance in order to obtain its initial weight. Then, air gun was used to dry the workpiece after the experiment is done to obtain its final weight. These steps were repeated for all the remaining experiments. Alicona Infinite Focus (Alicona Imaging, Austria) was used to determine the R_a value of the machined area. The coaxial white light produce by light source which then delivered through a beam splitter to a series of selectable, infinity-corrected objectives contained in a six-place turret. Fine light produced is then used to focus on workpiece. After that, the workpiece reflects the light back and projected through the beam splitter on colour digital sensor, thus resulting image which similar to conventional light microscope. All the data are tabulated in Table 2. The results were analyzed using analysis of variance (ANOVA) approach. Analysis of variance (ANOVA) approach was used to check the adequacy of the model. The analysis mainly intended to shows the interaction effects of the process variables on the responses. Main effect was the direct effect of independent variables while interaction effect was the join effect of two independent variables of the responses.

3. Analysis and Discussions

3.1 Material removal rate (MRR)

The developed model is shown in Eq. 2. Based on ANOVA, the model was developed based on 95% level of confidence and the model shows F-value of 8.77 indicates the model is significant. There is small influence, 2.93% of noise on the model developed. Furthermore, F-value and Prob> F proves that factor v and interaction effects of fn are the most influential factors on MRR by indicating the Prob>F value less than 1%. The Prob> F value of interaction effects of fn shows the above 95% level of confidence which express this interaction brings great impact on MRR. For MRR, inverse transformation was used for developing the model as shown in Eq. 2.

$$1/MRR = 365.33 - 4.61v + 6.31f + 0.15n - 5.57 \times 10^{-3}fn \quad (2)$$

where, MRR = material removal rate (mm^3/min), v = voltage (V) and f = feed rate (μm/s)
Furthermore, as mentioned above, voltage have great influence on MRR because when higher energy spark strikes, larger amount of material was removed by melting and evaporation [6]. Figure 1a shows the sugar colour graph of machined area obtain by using Alicona surface roughness machine when 110 V gap voltage, 30 μm/s feed rate, and 500 rpm spindle speed were used. The graph helps to shows the levels of the peaks presence on the surface of the workpiece by different colour labels. Figure 1b shows the contour graph of 1/MRR against the feed rate and spindle speed. With the increase in feed

IOP Conf. Series: Materials Science and Engineering **184** (2017) 012032 doi:10.1088/1757-899X/184/1/012032

rate from 20 μm to 30 μm and increase in spindle speed 500 rpm to 1000 rpm causes reduction in MRR value, but as the feed rate further increase to 40 μm and spindle speed increase to 1500 rpm, the value of MRR increase. This is because as increase in speed of rotating workpiece, the amount of eroded volume per crater also increases [7]. Besides that, when more speed is applied, the discharge energy is better and melts more material.

Table1. Experimental parameters

			Level		
Control Parameters	Factor	Unit	I	II	III
Gap voltage	v	V	100	105	110
Feed rate	f	μm/sec	20	30	40
Spindle speed	n	rpm	500	1000	1500
Fixed Parameters:					
Workpiece material		Tungsten (Ø 1.00 mm)			
Wire electrode		Copper wire			
Capacitance (μF)		0.4			
Polarity		Workpiece positive			
Dielectric fluid		EDM-3			
Threshold (%)		25			

Table 2. Experimental results

Expt	Parameters			Response					
	Gap voltage, v (V)	Feed rate, f (μm/s)	Spindle speed, n (rpm)	Time taken , t (min)	Δ Weight, w (g)	MRR (mm^3/ min)	R_a (μm)	MRR S/N value	R_a S/N value
1	110	20	1500	8.70	0.004	0.0230	0.4107	-32.7654	9.7144
2	110	30	500	27.5	0.0134	0.0257	1.285	-31.8013	-2.1794
3	110	40	1000	21.9	0.009	0.0222	0.3268	-33.0729	7.7295
4	100	20	1000	23.11	0.0106	0.0238	0.5638	-32.4685	-5.0885
5	100	30	1500	22.02	0.0103	0.0242	0.3933	-32.3237	8.1055
6	100	40	500	51.72	0.0112	0.0111	1.7965	-39.0935	4.9775
7	105	20	500	27.86	0.0112	0.0207	1.1877	-33.6806	11.6108
8	105	30	1000	51.97	0.0110	0.0110	0.2957	-39.1721	10.5830
9	105	40	1500	22.68	0.0103	0.0232	0.2627	-32.6902	11.6108

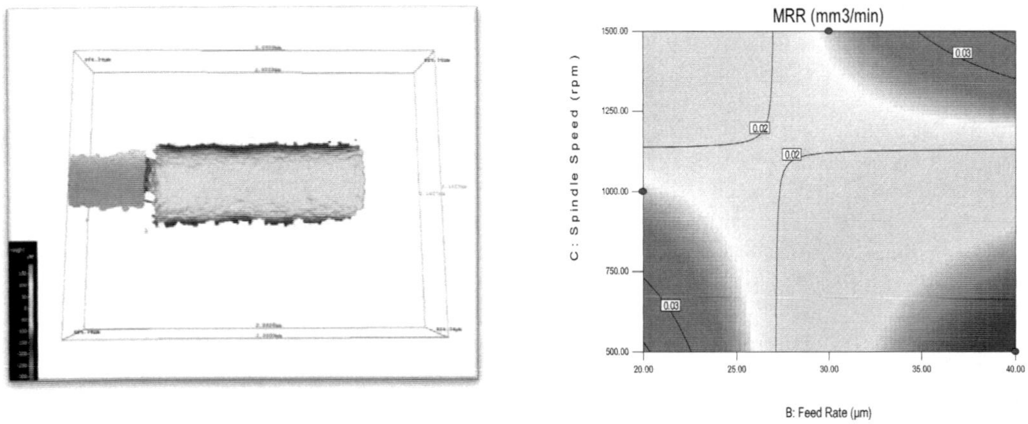

Figure 1. (a) Sugar colour graph of workpiece produce by Alicona when 110 V gap voltage, 30 µm/s feed rate, and 500 rpm spindle speed (b) countour graph of 1/MRR vs. feed rate and spindle speed.

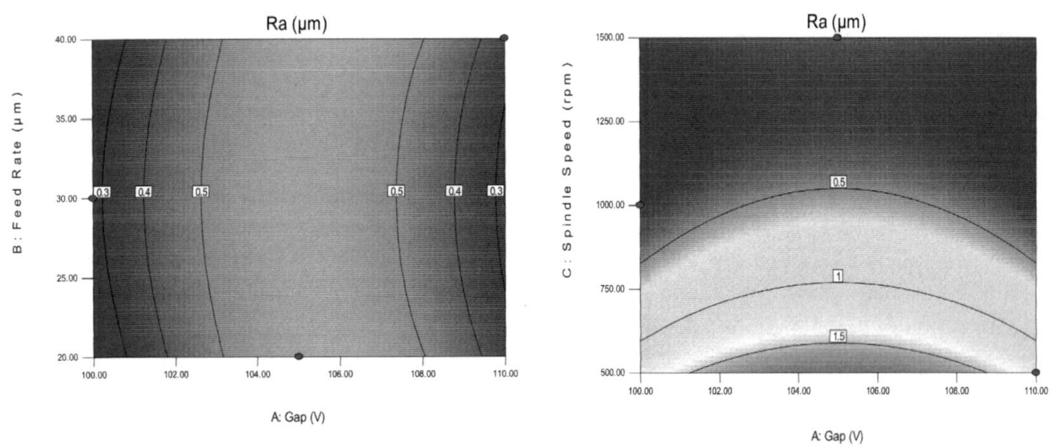

Figure 2. Countour graph of (a) R_a vs. gap voltage and feed rate, (b) R_a vs. gap voltage and spindle speed.

3.2 Surface roughness (R_a)

The developed model is shown in Eq. 3. Based on ANOVA, the model was developed based on 95% level of confidence and the model shows F-value of 1266.95 indicates the model is significant. There is small influence, 99.00% of noise on the model developed. Furthermore, F-value and Prob> F proves that factor n, v^2, n^2, and interaction effects of v^2n and vf^2 are influential factors on R_a by indicating the Prob>F value less than 1% which shows very influential influence.No power transmission was used for modelling R_a.

$$R_a = 4.23 - 6.33 \times 10^{-3}n - 8.13 \times 10^{-5}v^2 + 1.98 \times 10^{-6}n^2 + 1.18 \times 10^{-7}v^2n + 1.09 \times 10^{-6}\,vf^2 \quad (3)$$

where, R_a = material removal rate (mm^3/min), v = voltage (V) and f = feed rate (μm/s)

Figure 2a shows the contour graph of R_a vs. gap voltage and feed rate. Gap voltage has a strong influence on R_a. R_a increase as voltage increase from 100 V to 103 V, then R_a value reduce as the voltage increase to 110 V. In the first phase, higher discharge energy reduces the ability to produce good surface roughness [8]. High discharge energy produce stronger spark, stronger spark erodes high amount of debris from workpiece causes uneven crater to be produce. When debris is produced, it will be trap in between plasma channel and yield unwanted spark. Consequently, high amount of energy discharge is engaged to spark with debris, whereas only small portion of discharge energy is available to removed material from the workpiece. Meanwhile, in second phase, further increase in gap voltage, reduces the R_a, this may be due to different theories which stated that discharge energy strongly effect the discharge crater geometry. It must be assumed that the plasma channel able to breaks down with increasing circumferential speed. This affects the discharge crater becomes smaller in relation to crater area and volume, thus lower R_a is obtained. Besides that, Figure 2b shows the contour graph of R_a vs. gap voltage and spindle speed. R_a decreased as spindle speed increase. This condition happens due to the effect of the rotational movement on the position and expansion of the plasma channel of a single discharge [9].

3.3 Signal to noise (S/N) ratio for MRR

Signal to noise is vital in Taguchi experimental design and proposed to maximize the performance of the objectives. Table 2 shows the S/N value for MRR and Ra. Based on Taguchi method, S/N ratio is defined as signal to ratio where signal signifies the desirable value and noise represents the undesirable value. S/N value calculated for MRR as shown on Table 2 above is achieved by calculation using Eq. 4. Since the objective is to maximize as much as possible the value of MRR, equation of 'Larger-the–better' is used for S/N ratio calculation. Figure 3 shows the effect of three control parameters which are gap voltage, feed rate and spindle speed graphically. The analysis was done with the helps from MINITAB 14 software. The software facilitate in producing S/N ratio response table, which used to further analyze the results. Based on Figure 3, factors at 110 V, 20 μm/s, and 1500 rpm gives the maximum MRR. Gap voltage and spindle speed is observed which provide significant effect on MRR. Even though, feed rate do not show significant influence but it still cannot be neglected.

$$\text{Larger-the-better:} \qquad S/N = -10 \log \frac{1}{n}\left(\sum \frac{1}{y^2}\right) \quad (4)$$

3.4 Signal to noise (S/N) ratio for R_a

Taguchi method helps in reducing fluctuation causes by noise and aim to achieve consistent performance, thus for R_a which target to smallest value as possible. Due to that signal to noise value for R_a is calculated by using Eq. 5 and show in Table 2, which used for 'Smaller-the-better' [12]. Factors level at 105 V, 30 μm/s, and 1500 rpm are recommended for minimization of R_a value as shown in Figure 4. Feed rate provide least influence on R_a, while spindle speed give highest influence followed by gap voltage.

$$\text{Smaller-the-better:} \qquad S/N = -10 \log \frac{1}{n} \sum y^2 \quad (5)$$

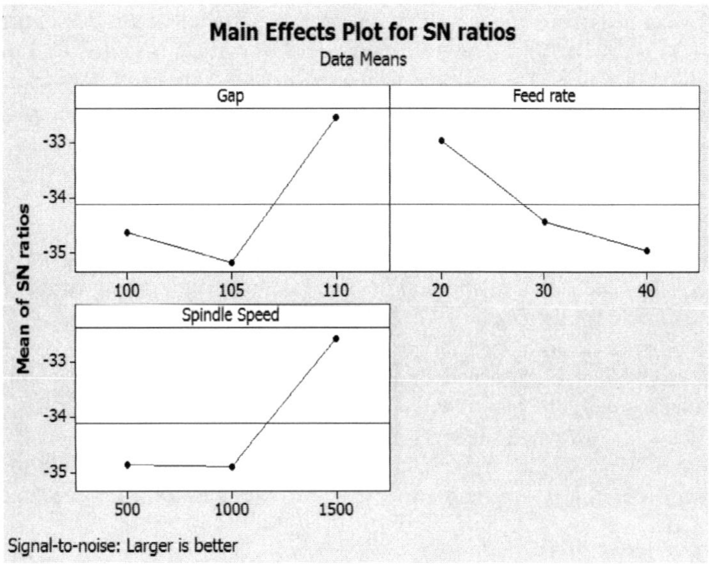

Figure 3. S/N ratio curves of gap voltage, feed rate, and spindle speed for (a) MRR and (b) R_a

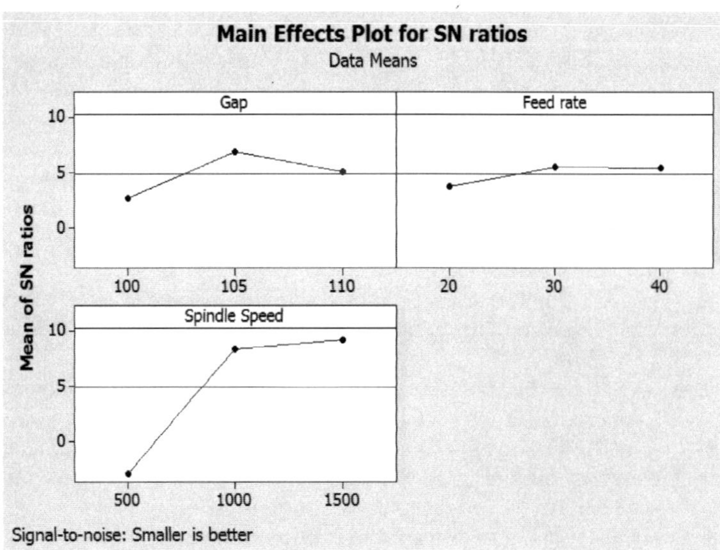

Figure 4. S/N ratio curves of gap voltage, feed rate, and spindle speed for (a) MRR and (b) R_a

4. Optimization and Verification

Multi-responses optimization is done to obtain optimum set of parameters for micro-WEDG process. The optimization was done by selecting the gap voltage, feed rate, and spindle speed to be in range with maximum MRR and minimum R_a. The optimized result for maximum MRR was 0.05863 mm^3/min and minimum R_a was 0.255 µm with 110 V gap voltage,38 µm/s feed rate, and 1315 rpm spindle speed. Experiment was conducted for verification using results obtained from the optimization.

The actual values obtained from the experiments were compared with the optimize results. The percentage error for MRR is 14.76 %. The percentage error for MRR is quite big. Electronic weighing balance might be the factor, which contribute to the error since due to its low resolution. To reduce this error, increase the amount of material remove and using higher resolution electronic weighing balance may give positive impact on the results.

5. Conclusions

The purpose of this paper is to study the effect of machining parameters, which are gap voltage, feed rate, and spindle speed on MRR and R_a of micro-WEDG. In order to analyse the results of the experiment, ANOVA has been used as method of analysis and empirical equations for both response were developed. This study shows that:

1. Based on ANOVA, it is observed that gap voltage provides significant influence on MRR where further increase in gap voltage, MRR increase. Meanwhile, spindle speed provides highest influence on R_a where further increase in spindle speed, reduce the R_a.
2. The maximum MRR is found to be 0.05863 mm^3/min meanwhile, Ra is found to be 0.255 μm. The optimum parameters obtained are 110 V gap voltage, 38 μm/s feed rate, and 1315 rpm spindle speed.

Acknowledgements

The authors would like to thank the Ministry of Science, Technology and Innovation (MOSTI), Malaysia for financial support under Science Fund Research Project SF 15-016-0066.

References

[1] Qin Y, Brockett A, Ma Y, Razali A, Zhao J, Harrison C, and Loziak D 2010 Micro manufacturing: research, technology outcomes, and development issues *Int. J. Adv. Manuf. Technol.* vol 47 pp 821-837

[2] Masuzawa T, Fujino M, Kobayashi K, Suzuki T, and Kinoshita N 1985 Wire electro-discharge grinding for micro-machining *CIRP Ann. Manuf. Tech.* vol 34 pp 431-434

[3] Weingärtner E, Jaumann S, Kuster F, and Boccadoro M 2010 Special wire guide for on-machine wire electrical discharge dressing of metal bonded grinding wheels *CIRP Ann. Manuf. Technol.* vol 59 pp 227-230

[4] Mehfuz R and Ali M Y 2009 Investigation of machining parameters for the multiple-response optimization of micro electrodischarge milling *Int. J. Adv. Manuf. Technol.* vol 43 pp 264-275

[5] Periyanan P R, Natarajan U, and Elango A 2011 Optimization of parameters on material removal rate in micro-WEDG process *Int. J. Eng. Sci. Technol.* vol 3 pp 66-76

[6] Marigoudar R N and Sadashivappa K 2013 Effect of machining parameters on MRR and surface roughness in machining of ZA43/SiCp composite by WEDM *Int. J. Applied Sci. Eng.* vol 11 pp 317-330

[7] Weingärtner E, Wegener K, and Kuster F 2012 Wire electrical discharge machining applied to high-speed rotating workpieces *J. Mater. Process. Technol.* vol 212 pp 1298-1304

[8] Rees A, Brousseau E, Dimov S S, Gruber H, and Paganetti I 2008 Wire electro discharge grinding: surface finish optimisation *Proc. Int. Conf. Multi Material Micro Manufacture (4M), Cardiff*

[9] Uhlmann E, Piltz S, and Jerzembeck S 2005 Micro-machining of cylindrical parts by electrical discharge grinding *J. Mater. Process. Technol.* vol 160 pp 15-23

Investigation of Kerf in Micro Wire Electro Discharge Machining

Mohammad Yeakub Ali*, Siti Nur Zalikha Khamarruzaman and Asfana Banu

Department of Manufacturing and Materials Engineering
International Islamic University Malaysia
PO Box 10, 50728 Kuala Lumpur, Malaysia
* E-mail: mmyali@iium.edu.my

Abstract. Precision and accuracy are among those of the most important factors in the field of machining processes. Micro wire electro discharge machining (micro-WEDM) is used widely in the machining process to fabricate micro parts with high precision and accuracy. However, in micro-WEDM, minimal kerf remains as a critical issue. This paper investigates the kerf in micro-WEDM of stainless steel (S304). Experimental investigation is carried out with 70 μm diameter tungsten electrode wire with two varying parameters voltage and capacitance. The kerf data are analyzed and an empirical model is developed using Design Expert software. The optimum parameters for minimum kerf are found to be 10nF capacitance and 90V voltage.

1. Introduction

Wire electrical discharge machining (WEDM) is a non-traditional machining process that is widely used for machining delicate components. WEDM involves a series of process that is cooling and heating where the machining efficiency, accuracy and surface quality are influenced by the electrical discharge energy. The main challenges in this machining process are to control the kerf, accuracy and surface finishing. This is because WEDM relies heavily on the operators' experiences and the tables of parameters provided by the builder of the machines. Furthermore, the discharges of the electrified current in a controlled manner by means of a thin wire, is guided alongside the desired cutting path. This heightened precision allows for complicated, three dimensional cuts, and produces highly accurate punches, dies, and stripper plates. The common machining parameters such as pulse-on and off, feed rate, current, polarity, voltage and vibration should be choose properly as it can affect the overall performance [1-4].

Kerf is a width of the machined slots which is one of the most vital characteristics of WEDM [5, 6]. The corner errors and kerf variation are usually caused by the wire tool deflection and vibration in the discharge gap. These are the main factors that affect the WEDM machining accuracy. However, the kerf variations have higher influences on dimensional accuracy in micro-WEDM compared to the conventional WEDM. This is because, the relative error found in miniature parts produce by the micro-WEDM are bigger than the corresponding values in conventional WEDM [7]. Besides that, a stable machining performance in micro-WEDM is related to the debris free machined kerf. It is evident from the debris tracking analysis that the most debris are left out from the kerf section under any constant fluid flow rate. More effectively, debris can be excluded and high micro-WEDM performance is obtainable with the improvement of jet flushing conditions of the working fluid from the nozzles [8]. In

IOP Conf. Series: Materials Science and Engineering **184** (2017) 012033 doi:10.1088/1757-899X/184/1/012033

machining micro structures, precision control of kerf is very important and it is related to machining processes, electrode wire diameter and also machining vibration. As such, it is difficult to control the minimum kerf. Therefore, the objective of this paper is to study the optimal parameters of voltage and capacitance for minimum kerf.

2. Experiment

Taguchi's L9 orthogonal array statistical approach was used to design the experiment. Taguchi method was chosen because it requires minimum experimental cost thus minimizes the effect of the sources of variation [9]. Besides, it reduces the number of runs needed for the experiment compared to the full factorial approach [10].A total of 9 experiments were conducted using micro-WEDM machine, DT-110 Mikrotools (Mikrotools Inc., Singapore) to complete the analysis of kerf. The experimental parameters are shown in Table 1 with controlled parameters; voltage and capacitance. The workpiece used was stainless steel (S304) with 70 μm diameter tungsten electrode wire. The wire-workpiece gap is thoroughly maintained by a computer controlled positioning system and the gap ranges around from 0.025mm to 0.075mm. The measurement of kerf is taken using scanning electron microscope (SEM) machine. Kerf will be measured by the width of the gap present. Then, the difference in gap will be compared to the tool width used that is 70μm. The example of the measured kerf at five different positions is shown in Figure 1. The results of average kerf are tabulated in Table 2. The results were analyzed using analysis of variance (ANOVA) approach.

3. Analysis and Discussions

Based on the results tabulated in Table 2, run 8 shows the smallest kerf with 90.00 μm when the capacitance is 10 nF and voltage is 90V. This is probably because when the capacitance decreases, small amount of energy is dissipated. Thus, it emits a weak sparks and leads to a smaller material erosion [11]. Therefore, with a smaller value of capacitance and voltage, the cutting width is to be reduced significantly.

Analysis of variance (ANOVA) of average kerf is shown in Table 3. The model was developed with 95% confidence level. The F-Value of 58.42 implies that the model is significant. There is only 0.08% chance that the F-Value could occur due to the noise. Values of Prob>F less than 0.05 indicates that the model terms, v(voltage), c (capacitance), v^2 and c^2 are found to be significant. Looking at the Prob>F, the factor c (capacitance) and c^2 is the most influential on kerf. The Prob>F for c and c^2 indicates that they have more than 99.00% confidence level. The Prob>F for v and v^2 has a confidence level of at least 95.00% thus showing that these factors have a very good influence on kerf itself. The analysis shows the effects of the relation of the process variables with each other either directly or indirectly. Statistically, F-Value providing confidence level as to whether the values are significantly different. For a large F-Value, it indicates that the performance characteristics are affected by the variation in process parameters [12].

Therefore, the final equation for this statistical model in terms of the actual factors is expressed by Eq. 1. This equation can be used to make predictions about the response for each given levels for each factors. However, the equation should not be used to get the relative impacts of each factor. This is because the coefficients are scaled to accommodate the units of each factors and the intercept is not at the centre of the design space. Furthermore, regression model has helped generating an equation to trace the statistical connections between the process's parameters and the response variable [13].

$$Kerf = 294.62 - 4.28v + 0.13c + 0.02v^2 - 2.4 \times 10^{-4}c^2 \quad (1)$$

where, v = voltage (V) and c = capacitance (nF).

Table1. Experimental parameters

Control Parameters	Factor	Unit	Level I	Level II	Level III
Voltage	*v*	V	90	100	110
Capacitance	*c*	nF	400	100	10
Fixed Parameters:					
Workpiece material		Stainless steel (S304)			
Wire electrode		Tungsten (Ø 70 µm)			

Table 2. Taguchi's L9 orthogonal array experimental design and measured kerf values

Expt.	Voltage, *v* (V)	Capacitance, *c* (nF)	Kerf at different point (µm) A	B	C	D	E	Average Kerf (µm)
1	90	400	107.00	103.00	109.00	102.00	96.00	103.40
2	100	100	92.70	98.70	98.70	99.30	101.00	98.08
3	90	100	95.30	95.30	97.30	107.00	98.70	98.72
4	100	400	105.00	102.00	100.00	103.00	98.00	101.60
5	110	100	109.00	103.00	103.00	100.00	103.00	103.60
6	100	10	90.70	88.70	90.00	94.70	86.70	90.16
7	110	400	109.00	103.00	107.00	101.00	104.00	104.80
8	90	10	90.70	88.70	87.30	91.30	92.00	90.00
9	110	10	96.00	94.40	92.00	91.30	88.40	92.42

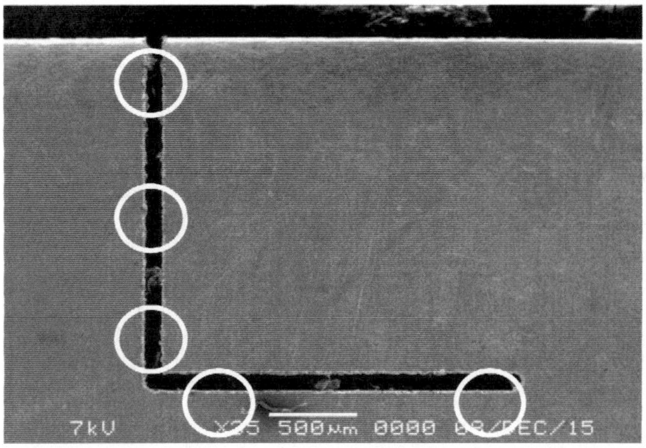

Figure 1. Five different points taken for calculating average kerf (From top - point A, B, C, D and E)

Table 3. ANOVA of average kerf

Source	Sum of Squares	Degree of Freedom (DF)	Mean Square	F-Value	Prob>F
Model	272.12	4	68.03	58.42	0.0008
v	12.61	1	12.61	10.83	0.0302
c	230.89	1	230.89	198.27	0.0001
v^2	9.77	1	9.77	8.39	0.0443
c^2	74.94	1	74.94	64.35	0.0013
Residual	4.66	4	1.16		
Cor Total	276.78	8			

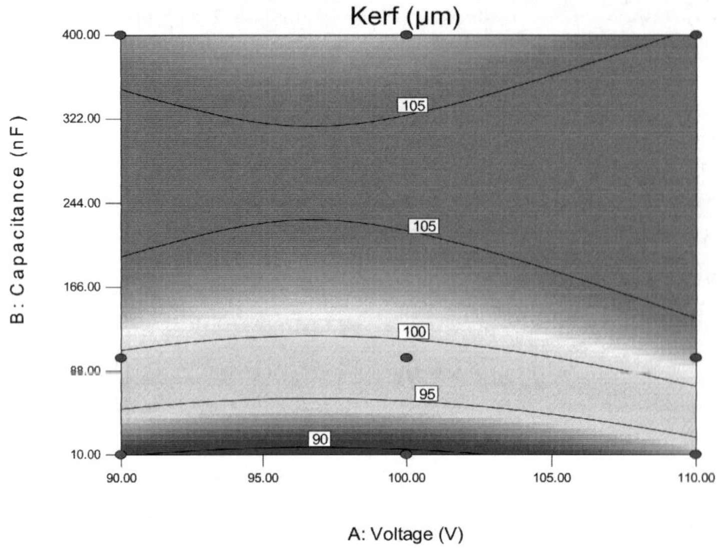

Figure 2. Contour plot of average kerf vs. capacitance and voltage

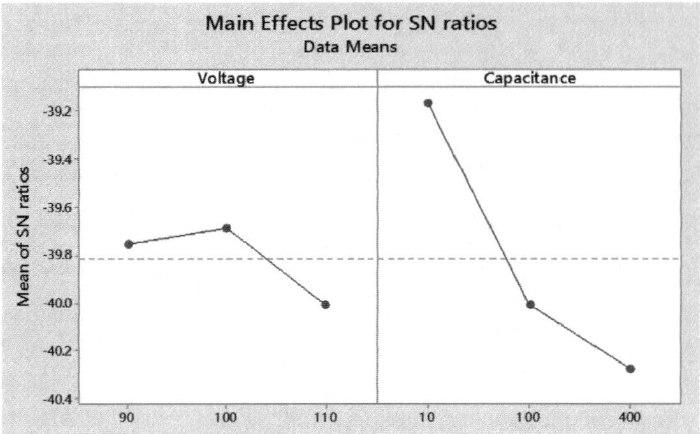

Figure 3. S/N ratio curves of voltage and capacitance for average kerf

Figure 2 shows the contour plot capacitance and voltage on kerf at constant wire tension and speed. Capacitance has a strong influence on kerf. The kerf increases with the increase of capacitance and voltage. Along with the increase of capacitance, a stronger sparks was emitted thus dissipating a large energy [11]. Therefore, this condition leads to a very large kerf to be produced. With increasing capacitance during the process, the cutting width get larger as it was affected by a high spark energy.

According to Taguchi method, the S/N ratio is the ratio of signal to noise where signal can be regarded as the preferable value while noise can be regarded as the value that was not preferred. The result of kerf was used to calculate the Signal to Noise ratio (S/N) using Eq. 2. Kerf is desired to be at minimum ratio, therefore the lower the better characteristic is used for S/N ratio calculation. The most favourable setting will be the one which could achieve the lowest S/N ratio [10].

$$S/N_{LB} = -10 \times \log\left(\frac{1}{r}\sum y^2\right) \tag{2}$$

where S/N LB = Signal to Noise ratio for the Lower the Better, y = output characteristic (Kerf) and r = number of trials

The S/N ratio for the two control factors for the experiment which is voltage and capacitance are shown respectively in Figure 3. The analysis was done using Minitab 17 software as a predictor to measures the performances and the interaction in between the factors. Analysis of the results leads to the decision that voltage at level 2 and capacitance at level 1 gives minimum kerf. Although the other factors at other level did not show any significant effect, but they still cannot be neglected. From Figure 3, capacitance seems the most important factor that affects the response as a whole.

4. Optimization and Verification

The optimized result for minimum average kerf was 89.61 µm with 93 V voltage and 10 nF capacitance. Experiment was conducted for verification using results obtained from the optimization. The actual values obtained from the experiments were compared with the optimize results. Average kerf had been increased from 89.48 µm to 89.61 µm with an error of 0.145%. The percentage error of average kerf is relatively small shows that the emperical model is significant.

5. Conclusion

The purpose of this paper is to investigate minimum kerf of micro-WEDM on stainless steel voltage and capacitance as the controlled parameters. ANOVA approach was used for analysis of minimum kerf and an emperical equation was developed. This study shows that:

1. Based on the experiments, the minimum average of kerf is found to be 90.00 µm with 90 V voltage and 10 nF capacitance.
2. Based on ANOVA analysis, kerf is strongly influenced by the capacitance. At a lower capacitance value, a low energy is dissipated when a weak sparks was emitted. Thus causes a smaller kerf.
3. The model predicts minimum kerf (89.61 µm) when 93 V voltage and 10 nF capacitance are used. The predicted value and experimental kerf value is within 0.145%. As such, the result developed an emperical model that is significant to represent the relationship of the parameters. Therefore, it is also useful to use it to predict the performance characteristics of machining stainless steel by micro-WEDM process.

Acknowledgements

The authors would like to thank the Ministry of Science, Technology and Innovation (MOSTI), Malaysia for financial support under Science Fund Research Project SF 15-016-0066.

References

[1] Liao Y S, Huang J T, and Su H C 1997 A study on the machining-parameters optimization of wire electrical discharge machining *J. Mater. Process. Tech.* vol 71 pp 487-493

[2] Patil P A and Waghmare C A 2014 A review on advances in wire electrical discharge machining *P. Int. Conf. Research and Innovation in Mechanical Engineering. Springer*, India, pp 179-189

[3] Tosun N and Cogun C 2003 An investigation on wire wear in WEDM *J. Mater. Process. Tech.* vol 134 pp 273-278

[4] Debroy A and Chakraborty S 2013 Non-conventional optimization techniques in optimizing non-traditional machining processes: a review *Manage. Sci. Letters*. vol 3 pp 23-38

[5] Hoang K T and Yang S H 2015 Kerf analysis and control in dry micro-wire electrical discharge machining *Int. J. Adv. Manuf. Technol.* vol 78 pp 1803-1812

[6] Ghodsiyeh D, Golshan A, and Shirvanehdeh J A 2013 Review on current research trends in wire electrical discharge machining (WEDM) *Indian J. Sci. Technol.* vol 6 pp 4128-4140

[7] Di S, Chu X, Wei D, Wang Z, Chi G, and Liu Y 2009 Analysis of kerf width in micro-WEDM *Int. J. Mach. Tool. Manu.* vol 49 pp 788-792

[8] Okada A, Uno Y, Onoda S, and Habib S 2009 Computational fluid dynamics analysis of working fluid flow and debris movement in wire EDMed kerf *CIRP Ann. Manuf. Techn.* vol 58 pp 209-212

[9] Mahapatra S S and Patnaik A 2006 Parametric optimization of wire electrical discharge machining (WEDM) process using Taguchi method *J. Barz. Soc. Mech. Sci. & Eng.* vol 28 pp 422-429

[10] Durairaj M, Sudharsun D, and Swamynathan N 2013 Analysis of process parameters in wire EDM with stainless steel using single objective Taguchi method and multi objective grey relational grade *Procedia Eng.* vol 64 pp 868-877

[11] Somashekhar K P, Ramachandran N, and Mathew J 2010 Material removal characteristics of microslot (kerf) geometry in μ-WEDM on aluminum *Int. J. Adv. Manuf. Tech.*vol 51 pp 611-626

[12] Tosun N, Cogun C, and Tosun G 2004 A study on kerf and material removal rate in wire electrical discharge machining based on Taguchi method *J. Mater. Process. Tech.* vol 152 pp 316-322

[13] Pandaa A K and Singh R K 2013 Optimization of process parameters by Taguchi method: catalytic degradation of polypropylene to liquid fuel *Int. J. of Multidisciplinary and Current Research* pp 50-54.

Dimensional Accuracy of Micro-Electro Discharge Milling

Mohammad Yeakub Ali [*], **Mohamed Abd Rahman, Siti Nur Zuhaida Zunairi and Asfana Banu**

Department of Manufacturing and Materials Engineering
International Islamic University Malaysia
PO Box 10, 50728 Kuala Lumpur, Malaysia
* E-mail: mmyali@iium.edu.my

Abstract. This paper investigates the dimensional accuracy and tool wear rate (TWR) in micro electro discharge (micro-ED) milling of stainless steel (S304) workpiece and tungsten as the tool electrode with control parameter; voltage and feed rate. Dimensional accuracy is represented by the difference of micro-channel width between the two ends of the channel. The TWR and the micro-channel width data are analyzed and empirical model are developed. The optimum values for minimum TWR and minimum micro-channel width are found to be 91 V voltage and 10 μm/s feed rate.

1. Introduction

Electro discharge machining (EDM) or spark machining removes materials that are electrically conductive with rapid and repetitive spark discharges from electric pulse generators in the presents of the dielectric fluid between the tool and the workpiece. Dielectric fluid is used to dip the workpiece and tool electrode and is supplied continuaously to the machining area to create the spark as well as washes away the debris.There is no physical cutting forces exist between the tool and the workpiece [1]. In micro field, micro-EDM is one of the branch of the EDM. Micro-EDM and EDM has very similar material removal mechanism except the tool size, discharge energy, and axes movement resolutions are in micron level in micro-EDM [2, 3, 4, 5]. Micro-ED milling has the similar feed system as the conventional milling machine which the workpiece is fed to the direction of tool electrode while the tool is revolving. Micro-ED milling is the real competitor based on three main characteristics which it has high machining aspect ratio, ability to machine any strong metal that conducts electric and low machining expense [4]. Micro-ED milling is also able to machine a very complicated shape which makes it way more competitive than other process [6].

Among all the common parameters in micro-EDM process, some play important role and have significant effects on to micro-ED milling. The parameters include feed rate, voltage, capacitance, spindle speed, and polarity. These parameters affects different output of the process including surface roughness, dimensional accuracy, material removal rate (MRR), and tool wear rate (TWR) in certain ways.Micro-ED milling is a precise process when compared to the micro milling and micro laser milling. The dimensional stability of this machining process is highly depends on the discharge energy applied [4]. This is because the discharge strongly affects the TWR. Electrode wear is not desirable in this process since it reduces the diameter of the tool electrode relative to machining time. This will reduce the width of the channel generated as it approaches the end. Hence, the accuracy is lower when TWR is high. Besides that, when the voltage and capacitance are high, the discharge energy is high and resulted in high TWR and low dimensional accurracy. Discharge energy is shown in Eq. (1) [4].

Therefore the objective of this paper is to investigate the TWR and micro-channel width with two varying parameter, voltage and feed rate.

$$E = \frac{1}{2}CV^2 \tag{1}$$

where, E = discharge energy, C = capacitance, and V = voltage

2. Experiment

Micro-ED milling process is conducted by using a multi-purpose miniature commercial machine tool, DT-110 Mikrotools (Mikrotools, Singapore). Three level general factorial statistical approach was used to designed the experiment for micro-channel width and TWR. A total of 9 experiments were conducted using stainless steel (S304) as the workpiece with dimension of 25 mm × 15 mm ×0.5 mm and tungsten as the electrode. The experimental parameters are shown in Table 1. Capacitance was kept constant throughout the experiments. Micro-channel width was measured using scanning electron microscope,SEM (JEOL JSM-5600) at the both end of the channel. The differences of micro-channel width between the two ends of the channels were calculated. The difference of width for each micro-channel was calculated in order to observe and analyze the dimensional accuracy of the process and denoted as δ. The width difference represents the level of accuracy of the micro-ED milling. The smaller the value of δ obtained, the higher the accuracy of the process. Figure 1 shows the schematic diagram for the micro-channel as well as the SEM images of both ends of the micro-channel width where D and d are measured. Note that $\delta = D - d$.

On the other hand, TWR was calculated by using Eq. (2) [4]. All the experimental and calculated results are tabulated in Table 2.

$$TWR = \frac{V_T}{t} \tag{2}$$

where, TWR = tool wear rate, VT = the volume of the material removed from the electrode tool, t = the machining time.

Table 1. Experimental parameters

Control Parameters	Factor	Unit	Level *I*	Level *II*	Level *III*
Voltage	*v*	V	90	100	110
Feed rate	*f*	μm/s	10	20	30

Fixed Parameters:	
Workpiece material	Stainless steel (S304)
Electrode	Tungsten (Ø 800 μm)
Spindle speed (rpm)	1000
Polarity	Workpiece positive
Threshold	24
Machining length (mm)	3.0
Depth of cut (mm)	0.30
Capacitance (μF)	0.4

Table 2. Experimental results

Expt.	Parameters		Response			
			Micro-channel width (μm)			TWR
	Voltage, v (V)	Feed rate, f (μm/s)	D	d	$\delta = D - d$	($\times 10^{-3}$ mm^3/min)
1	90	10	809	441	368	1.866
2	90	20	797	345	452	2.185
3	90	30	810	357	453	2.189
4	100	10	814	460	354	2.306
5	100	20	803	408	395	3.047
6	100	30	822	452	370	3.302
7	110	10	817	454	363	2.518
8	110	20	853	383	470	3.101
9	110	30	865	353	512	3.988

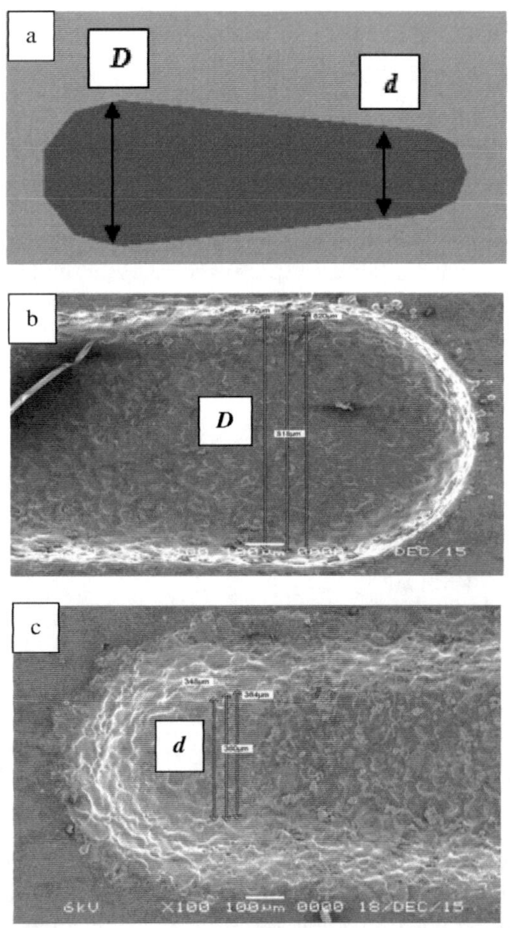

Figure 1. a) Schematic diagram of micro channel generated; SEM image for micro-channel width measurement for experiment 3 with 90 V voltage and 30 μm/s feed rate (b) D and (c) d

3. Analysis and Discussions

3.1 Micro-channel width

The developed model is shown in Eq. 3. Based on ANOVA, model F-Value of 256.72 indicates the model is significant. This is due to less than 0.39 % chance of noise could occur in the model F-Value. Based on F-Value, factor v^2 have the most influence on micro-channel width with highest value of 470.36, followed by factor f^2 and v. The interaction effect, v^2f gives F-Values of 199.37 followed by the vf. In coloumn Prob>F, the values less than 0.05 indicates the factors are significant. Therefore, it is concluded that all the factors are significantly affects the micro-channel width difference except for the f factor which is 0.1114.

$$\text{Width difference} = -3328.56 + 73.33v + 504.07f - 0.377v^2 - 0.357f^2 - 9.94vf + 0.0505v^2f \qquad (3)$$

where, v = voltage (V) and f = feed rate (μm/s)

Figure 2a shows the contour plot for micro-channel width difference against voltage and feed rate. From the figure, it is obvious that the voltage and the feed rate strongly influenced the micro-channel width difference thus influenced the dimensional accuracy of micro-ED milling. It can be observed that the accuracy increases with the increases of voltage and feed rate. However, it is also observed that, when the two factors are continuously increased then the accuracy decreases. In the earlier phase, at a specific voltage (100 V), when the feed rate decreases, the micro-channel width difference also decreases. This is because low feed rate lowers the cutting speed and slows down the erosion of the material from the electrode which results in low tool wear [4]. The reduction in electrode diameter is small hence the width difference of the micro-channel is not so big. This consequently increases the accuracy of micro-ED milling. As the voltage and feed rate increase further, the micro-channel width difference begins to rise again. This is because, spark energy are increased causing stronger spark resulting in greater amount of material to be eroded from the electrode [5]. This reduces the diameter size of the electrode and the difference of the micro-channel width becomes larger which results in lower accuracy.

3.2 Tool Wear Rate (TWR)

The developed model is shown in Eq. 4. Based on ANOVA, model F-Value of 42.36 indicates the model is significant. This is due to less than 2.32 % chance of noise could occur in the model F-Value. Based on F-Value, factor f have the most influence on micro-channel width with highest value of 89.55, followed by factor v and vf. In coloumn Prob>F, the values less than 0.05 indicates the factors are significant. Thus, it is concluded that all the factors are significantly affects the TWR except for the v^2, f^2, and vf^2.

$$\text{TWR} = WRnd\ oloumn\ v + 0.412f - 0.00244v^2 - 0.0163f - 0.00332vf + 0.000155vf^2 \qquad (4)$$

where, TWR = tool wear rate, v = voltage (V) and f = feed rate (μm/s)

Figure 2b shows the contour plot for TWR against voltage and feed rate. From the figure, it is obvious that TWR increases with the increasing of voltage and feed rate. When the voltage is at the highest value, the TWR is concurrently at the highest value too. The higher the voltage, the spark energy also has the tendency to increase. Spark energy will determine the magnitude of the crater generated during the machining process [4]. Hence, the higher the spark energy, the bigger the magnitude of crater generated during the machining which results in high TWR and the condition is vice versa for the low voltage. TWR have the highest value when the feed rate is high. When the feed rate is high, the tool cutting speed will increase and increases the quantity of the material removed from the electrode[4]. Thus, larger amount of material removed from the electrode causing higher TWR and vice versa for the condition of low feed rate value.

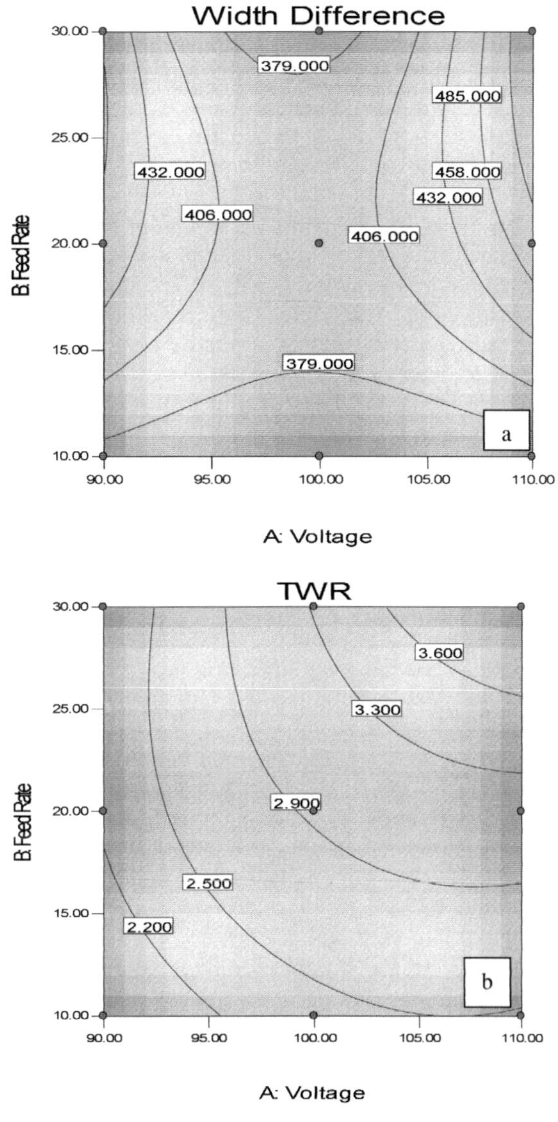

Figure 2. Countor graph of (a) micro-channel width difference vs. feed rate and voltage and (b) TWR vs. feed rate and voltage

4. Optimization and Verification

Multi-responses optimization is done to obtain optimum set of parameters for micro-ED milling process. The optimization was done by selecting the feed rate and voltage to be in range and TWR and micro-channel width difference to be minimum. The optimized result for minimum TWR was 1.866 × 10^3 mm^3/min and minimum micro-channel width difference was 368.378 µm with 91 V voltage and 10 µm/s feed rate. Experiment was conducted for verification using results obtained from the optimization. The actual values obtained from the experiments were compared with the optimize results. TWR and micro-channel width difference had been increased to 1.904 × 10^3 mm^3/min and 399 µm respectively. The percentage error for TWR and micro-channel width difference are 2.036 % and 8.313 %

respectively. The percentage error for TWR is quite small while the percentage error for micro-channel width is a bit high. This might be caused by the error during estimating process of the width of the micro-channel using SEM. Since SEM is used for estimating the width, the most accurate measurement of the micro-channel width could not be obtained. However, these two percentage errors are considered quite acceptable since they are still in an acceptable range.

5. Conclusions

The purpose of this paper is to determine the optimum machining parameters in order to obtain the high dimensional accuracy of micro-ED milling. Two factors that have been chosen for this study are voltage and feed rate while as for the responses, tool wear rate (TWR) and micro-channel width difference were taken into consideration to determine the accuracy of the machining process. ANOVA approach was used for analysis and emperical equation were developed. This study shows that:

1. The dimensional accuracy of micro-ED milling is represented by the micro-channel width difference. The smaller the difference of the width between the beginning and the ending of the micro-channel, the process is said to have higher dimensional accuracy.
2. Voltage have strong influence on micro-channel width difference. The higher the energy, the greater the amount of material removed from the electrode which causes the reduction of electrode diameter. Hence, the accuracy is lower when the voltage is higher.
3. As for the TWR, the model shows that both feed rate and voltage individually have strong effects along with the interaction effect of feed rate and voltage on the electrode wear. High magnitude of debris are generated causing large amount of undesired flash released, resulting in the high TWR [5].
4. Multiple-response optimization shows the minimum values of width difference of micro-channel and TWR that can be achieved are 368.378 μm and 1.866 mm^3/min respectively when the voltage is 91V and feed rate is 10 μm/s.
5. Based on the verification experiment, it can be observed that the value of percentage error is 8.313 % and 2.306 % for micro-channel width and TWR respectively. This shows good prediction accuracy. Hence, it can be concluded that gap voltage of 91 V and feed rate of 10 μm/s are verified to be the optimum parameters to obtain high dimensional accuracy and low TWR.

Acknowledgements

The authors would like to thank the Ministry of Science, Technology and Innovation (MOSTI), Malaysia for financial support under Science Fund Research Project SF 15-016-0066.

References

[1] Jahan M P, Wong Y S, and Rahman M 2009 A study on the quality micro-hole machining of tungsten carbide by micro-EDM process using transistor and RC-type pulse generator *J. Mater. Process. Tech.* vol 209 pp 1706-1716

[2] Masuzawa T 2000 State of the art of micromachining, *CIRP Ann. Manuf. Techn.* vol 49 pp 473-488

[3] Mikesic I J, Fleisig R V, and Koshy P 2009 Electrical discharge milling with oblong tools *Int. J. Mach. Tool. Manu.* vol 49 pp 149-155

[4] Karthikeyan G, Ramkumar J, Dhamodaran S, and Aravindan S 2010 Micro electric discharge milling process performance: An experimental investigation *Int. J. Mach. Tool. Manu.* vol 50 pp 718-727

[5] Mehfuz R and Ali M Y 2009 Investigation of machining parameters for the multiple-response optimization of micro electrodischarge milling *Int. J. Adv. Manuf. Technol.* vol 43 pp 264-275.

[6] Ali M Y, Hamad M H, and Karim A I 2009 Form characterization of microhole produced by microelectrical discharge drilling. *Mater. Manuf. Process.* vol 24 pp 683-687.

Optimization of tribological performance of SiC embedded composite coating via Taguchi analysis approach

M A Maleque[1], K A Bello[1, 2], A A Adebisi[1] and N Akma[1]
[1] Department of Manufacturing and Materials Engineering, International Islamic University Malaysia.Darul Salam, 53100 Gombak, Selangor, Malaysia.
[2] Department of Metallurgical and Materials Engineering, Ahmadu Bello University Zaria, Nigeria

E-mail: maleque@iium.edu.my

Abstract. Tungsten inert gas (TIG) torch is one of the most recently used heat source for surface modification of engineering parts, giving similar results to the more expensive high power laser technique. In this study, ceramic-based embedded composite coating has been produced by precoated silicon carbide (SiC) powders on the AISI 4340 low alloy steel substrate using TIG welding torch process. A design of experiment based on Taguchi approach has been adopted to optimize the TIG cladding process parameters. The L9 orthogonal array and the signal-to-noise was used to study the effect of TIG welding parameters such as arc current, travelling speed, welding voltage and argon flow rate on tribological response behaviour (wear rate, surface roughness and wear track width). The objective of the study was to identify optimal design parameter that significantly minimizes each of the surface quality characteristics. The analysis of the experimental results revealed that the argon flow rate was found to be the most influential factor contributing to the minimum wear and surface roughness of the modified coating surface. On the other hand, the key factor in reducing wear scar is the welding voltage. Finally, a convenient and economical Taguchi approach used in this study was efficient to find out optimal factor settings for obtaining minimum wear rate, wear scar and surface roughness responses in TIG-coated surfaces.

1. Introduction

Metal components frequently lose functionality through severe environmental condition such as wear, corrosion, high temperature and high loading. In order to prevent such damages, advances in modification of critical engineering surfaces are frequently undertaken by researchers to extend the service life of engineering components [1, 2]. Over the years, this surface modification approach has been adopted by various industries, such as aerospace, petrochemical, automotive and general machinery [3]. The unique advantage of surface modification method is that the base material of the component can be selected for strength and price, while the surface properties can be altered independently for specific tribological conditions to which the critical sections of the part is subjected to in service [4-6].

Ceramic materials such as SiC particles are usually used in surface modification for improving tribological performances of steels because of their excellent physical and chemical properties. The incorporation of ceramic particles into molten metal surfaces to produce hard composite layers is popular because this method can tailor the surface to suite the requirements of specific applications

[7]. High power laser and electron beam melting techniques have been used extensively for processing such composite layers, which are reported to increase wear and corrosion resistance significantly. However, these techniques are expensive and difficult and hence have a limited application. An alternative, more economical and novel method was developed using conventional TIG torch melting for surface modification work. This TIG surface melting process is simpler, cheaper to establish and flexible in operation, economical in time, energy and manufacturing procedure, compared to laser and electron beam processing [2, 8, 9].

Many authors have analyzed composite layer properties and characteristics under varying TIG welding conditions to [5, 6, 10]. It is evident from the studies that optimization of material surface properties largely depends on the understanding and control of the fusion process parameters that can influence both the mechanical and tribological performances of the reinforced composite layer. The TIG surface modification method is governed by many important parameters such as arc current, traveling speed, arc gap length, electrode size, argon gas flow rate and precoated powder content and composition [11]. Usually, the desired process parameters are determined based on experimental data and handbook values, which does not guarantee the chosen parameters will yield optimal performance characteristics. To obtain the best functional response of TIG processed coatings, a suitable design of experiment tool can be used to for precise identification of prominent factors for process optimizations. Recently, a robust empirical technique called Taguchi design has become widely known for parametric optimization in order to compensate for the shortcomings of conventional experimental methods. Unlike full factorial design, Taguchi requires minimal number of experiments to determine optimal process settings with least variability. Precisely, Taguchi's is a simple and efficient approach to optimize designs for quality and performance at a minimum cost and time [8, 10-11].

In the present study, Taguchi experimental design based on L9 orthogonal array was used in investigating the effect of essential TIG control parameters on wear rate, wear track widths and surface roughness of the TIG-processed SiC composite coatings. Moreover, the optimal parameters for processing SiC-embedded coating on AISI 4340 steels were established.

2. Experimental details

2.1. Materials

AISI 4340 low alloy steel with the dimension of 100 mm × 45 mm × 14 mm was used as the substrate material in this work. The surface of the base material were ground using silicon emery paper and thoroughly cleaned in acetone to remove all the contaminants in form of oxide layers and grease. The chemical composition of the steel substrate can be found elsewhere [4]. Silicon carbide powder (99.5% purity, 40-50 μm) supplied by Sigma Aldrich was used for surface alloying. The SiC particulates weighed at the proportion of 0.5 mg/mm^2 was mixed with organic binder (PVA) and preplaced onto the surface of the dry substrate samples. Organic binder was primarily used to prevent the powders from blowing away under the flow of shielding gas during TIG glazing operation. The pre-coated substrates were then dried in an oven to remove moisture before surface irradiation under TIG arc torch. Further experimental detail on this process can be found elsewhere [2, 5].

2.2. Planning experiment

The design of experiments involves the use of statistical tools to analyze the effects of controllable factors on the response performances. Taguchi design approach was used for planning of experimental runs in this study. Taguchi method is well known to be simple and efficient for optimizing and improving performance by using standard orthogonal arrays (OA) with minimal number of experiments. In this study, the effect of TIG process variables such as welding current, travel speed, arc voltage and argon gas flow rate are considered to be studied using standard Taguchi L9 (3^4) design. The L9 OA chosen in this study is a modified Taguchi array that can accommodate up to 4 factors at three levels. The individual effects of all the factors are studied on the response variable on order to obtain optimal TIG welding parameters that would yield best tribological behavior. The tribological response variable in this case are wear rate, wear tracks width and surface roughness. Nine experiments are required by Taguchi L9 array to optimize the TIG welding parameters. The factor levels and their corresponding response are assigned to the columns of the modified array as shown in Table 1.

Table 1: The design matrix for experimental runs

Experimental Runs	Current (A)	Speed (mm/s)	Voltage (V)	Argon flow rate (L/min)	Wear rate μg/cm	Wear tracks width (μm)	Surface roughness (μm)
1	70	1.0	20	15			
2	70	1.5	25	20			
3	70	2.0	30	25			
4	80	1.0	25	25			
5	80	1.5	30	15			
6	80	2.0	20	20			
7	90	1.0	30	20			
8	90	1.5	20	25			
9	90	2.0	25	15			

2.3. Developing surface coating

Surface melting was performed using TIG process. A thoriated tungsten electrode (3.6Ø mm) was used to generate a stable arc for the glazing action at a fixed height of 2 mm from the sample. The energy density of the TIG torch was controlled by the supply of current and voltage to produce series of multiple overlap tracks. Details of the processing conditions used in this work are presented in columns 2, 3, 4 and 5 of Table 1. During the melting process, a streamed of pure argon gas was supplied to avoid excessive oxidation of the molten pool.

2.4. Tribological tests

The wear behavior of the SiC embedded steel was assessed using CSM ball-on-disk tribometer for 30 minutes duration at 30.6 N load, constant speed (100 rpm) under dry non lubricated condition and at ambient temperature (30°C). The embedded steel samples have the dimension of 10 mm × 15 mm × 11 mm were mechanically fixed chamber and alumina ball was used as a counterpart material. The track radius was fixed for all trial of experiments at 3 mm. The wear rate was calculated based on the relationship in Equation 1:

$$\text{Wear rate, } W_r = \frac{\Delta w}{2\pi rnt} \quad (1)$$

where *Wr* is the wear rate in terms of volume loss, *Δw* is weight loss, *n* is revolution per minute, r is the radius of ball distance from the centre of the steel disc and *t* is the sliding time.

3. Results and discussion

3.1. Wear rate analysis

The results of experiments obtained for wear rate of SiC embedded low alloy steel under selected conditions of TIG melting processes are presented in the second column of Table 3. Based on Taguchi approach, the results of the quality characteristic are transformed into signal-to-noise (S/N) ratio for the optimum parameter analysis. The S/N ratio for minimum wear rate can be determined based on the smaller-the-better (STB) criterion, which is calculated as logarithmic transformation of loss function as shown below:

$$\frac{S}{N} = -10 \log \frac{1}{n}\left(\Sigma Y^2\right) \quad (2)$$

Where n is the number of experiment, Y is the experimental data and S/N is the signal-to-noise ratios for wear rates. However, the concept ("smaller-the-better") was equally applicable to surface roughness and wear track widths performance index assessment purposes. The signal-to-noise ratio is measured in decibel scale (dB) and the results calculated using Eq. (2) is given in Table 2. The analysis of S/N ratio is done by computing the mean of the S/N ratio for each level of TIG welding parameters. The separation of each process parameter by S/N ratio is done based on the experimental design array is orthogonal. The S/N ratio single-response Table for wear rate is presented in Table 3.

Table 2: Results of wear rate with S/N ratio

Expt. no	Wear rate (µg/cm	S/N ratio (dB)
1	1.39703	-2.90409
2	1.46776	-3.33310
3	1.27320	-2.09793
4	1.75070	-4.86425
5	1.30861	-2.33619
6	2.19280	-6.81999
7	1.89217	-5.53923
8	1.27324	-2.09820
9	1.39703	-2.90409

Table 3: S/N ratio response table for wear rate

Levels	Current (A)	Speed (mm/s)	Voltage (V)	Argon flow rate (L/min)
1	-2.778	-4.436	-3.941	-2.715
2	-4.673	-2.589	-3.700	-5.231
3	-3.514	-3.941	-3.324	-3.020
Delta	1.895	1.849	0.616	2.516
Rank	2	3	4	1

It can be observed from Table 3 that the effect of argon flow rate and current are ranked higher and are considered to be very significant whereas welding voltage and speed are found to have slightly effects on the performance characteristic. However, the aim of the analysis is to identify the levels of the combining factors which present minimum wear rate. Figure 1 shows the influence of process variables on the wear rate. As far as the minimization of wear rate is concerned, the results of S/N response analysis indicate that the optimal combination of parameters occur at factor levels corresponds to the current of 70 A, speed of 1.5 mm/s, voltage of 30 V and argon flow rate of 15 L/min.

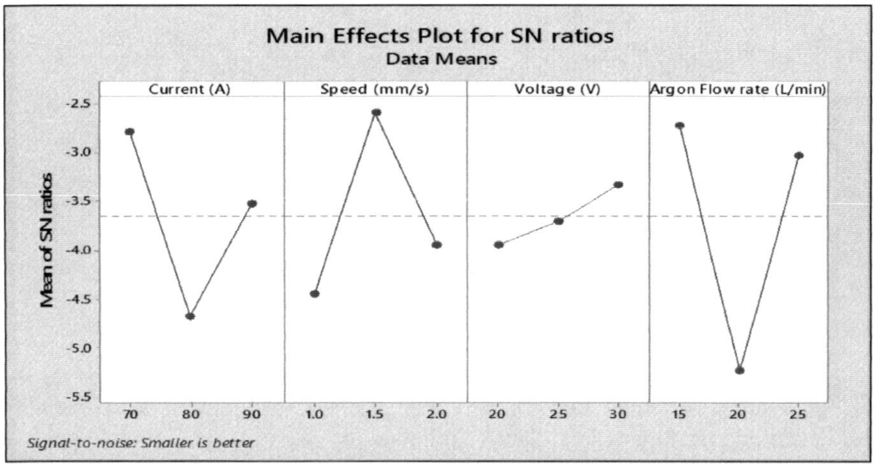

Figure 1. Main effects plot for wear rate of SiC embedded LAS under TIG torch melting.

3.2. Wear tracks width Analysis

The second response which is wear tracks width for each TIG torch process parameters at different levels were calculated. The normal method of calculating the desirable factors levels is to look at simple averages of the results. However, the variability of results within a trial condition cannot be judged by this method. Thus, signal-to-noise ratio analysis is done considering wear tracks width as the performance index. The analysis is carried out using the smaller-the-better (STB) criterion and is expressed as Equation 2. Table 4 shows the experimental results for wear tracks width tests and the corresponding S/N ratio for each experiment. Since the experimental design is orthogonal, it is possible to separate out the effect of each control factor at different levels. All the calculations are performed using Minitab software. The S/N response table shown in Table 5 includes ranks based on Delta value (the highest average of each factor minus the lowest average of the same); rank 1 is assigned to the parameter with highest Delta value, rank 2 to second highest Delta value and so on. In this case voltage has the highest Delta value thus rank 1 is assigned to voltage (V).The corresponding main effects plot for S/N ratio is shown in figure 2. The significance of each parameter can be judged by the inclination of plot. The parameter with highest inclination line has greater significance than the rest on the wear behaviour of the material. From the main effects plot, it is seen that the parameter voltage V is the most significant parameter while other parameters, argon flow rate (L/min) and speed (mm/s) are also significant parameters in controlling the wear behavior of the coated steel samples. The optimal process parameter combination is the one that yields minimum S/N ratio and thus the same for minimum wear tracks width is found to occur at a current of 70 A, speed of 1.0 mm/s, voltage of 25 V and argon flow rate of 15 L/min.

Table 4: Results of wear tracks width with S/N ratio

Expt. no	Wear tracks width (μm)	S/N ratio (dB)
1	212.156	-46.5331
2	250.251	-47.9675
3	560.991	-54.9791
4	293.118	-49.3408
5	356.213	-51.0342
6	346.486	-50.7937
7	480.862	-53.6404
8	352.109	-50.9335
9	264.624	-48.4526

From the main effect plots in Figure 2, the effects of individual process parameters on the wear tracks width of the composite coated steel can be clearly seen. Thus, from figure 2, it is observed that wear tracks width decreases with highest of voltage, followed by argon flow rate and speed. However, current has almost no effect on wear tracks width.

Table 5: S/N ratio response table for wear tracks width

Levels	Current (A)	Speed (mm/s)	Voltage (V)	Argon flow rate (L/min)
1	-49.83	-49.84	-49.42	-48.67
2	-50.39	-49.98	-48.59	-50.80
3	-51.01	-51.41	-53.22	-51.75
Delta	1.18	1.57	4.63	3.08
Rank	4	3	1	2

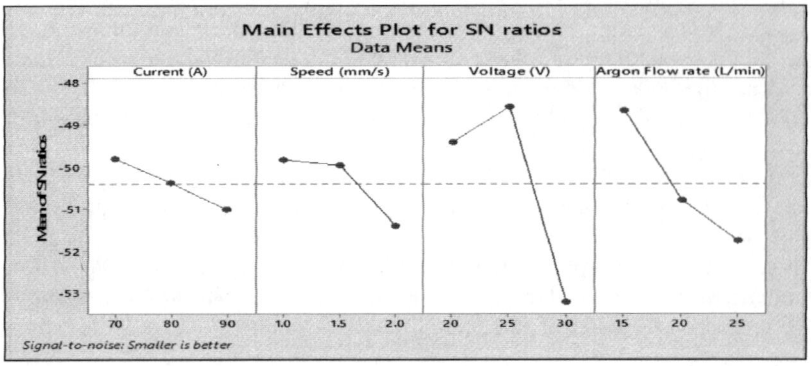

Figure 2. Main effects plot for wear tracks width of SiC embedded LAS under TIG torch melting.

3.3. Surface roughness analysis

Table 6 shows the effect of various factors i.e. current, speed, voltage and argon flow rate on the surface roughness of SiC embedded alloy steel after wear test. The response analysis of S/N ratio is done by computing the mean of the S/N ratio for each factor level. The separation of each process parameter by the S/N ratio was possible since the experimental design array was orthogonal. The S/N ratio single-response table for surface roughness is presented in Table 7. Since the surface roughness is a 'smaller the better' type of quality characteristic and because the objective is to minimize the surface roughness, therefore, the S/N ratio for 'smaller the better' type of response was used as given in Equation (2) for wear rate.

Table 6: Results of surface roughness with S/N ratio

Expt. no	Surface roughness (μm)	S/N ratio (dB)
1	7.2096	-17.1582
2	5.8929	-15.4066
3	6.6398	-16.4431
4	9.1722	-19.2495
5	13.7000	-22.7344
6	3.2907	-10.3458
7	4.7235	-13.4853
8	9.3172	-19.3857
9	9.0207	-19.1048

From the results of Taguchi analysis shown in Table 7, it is clear that argon flow rate (AFR) is the most significant factor on the surface roughness, followed by travel speed, welding voltage and current. The delta contribution of argon flow rate to the surface roughness is 6.59. Figure 3 shows that the higher S/N ratio corresponds to higher performance characteristics. Hence, the optimal combination of parameters for minimum surface roughness can be seen in Figure 3 as the current of 70 A, speed of 2.0 mm/s, voltage of 20 V and argon flow rate of 20 L/min.

Table 7: S/N ratio single-response for surface roughness

Levels	Current (A)	Speed (mm/s)	Voltage (V)	Argon flow rate (L/min)
1	-16.34	-16.63	-15.63	-19.67
2	-17.44	-19.18	-17.92	-13.08
3	-17.33	-15.30	-17.55	-18.36
Delta	1.11	3.88	2.29	6.59
Rank	4	2	3	1

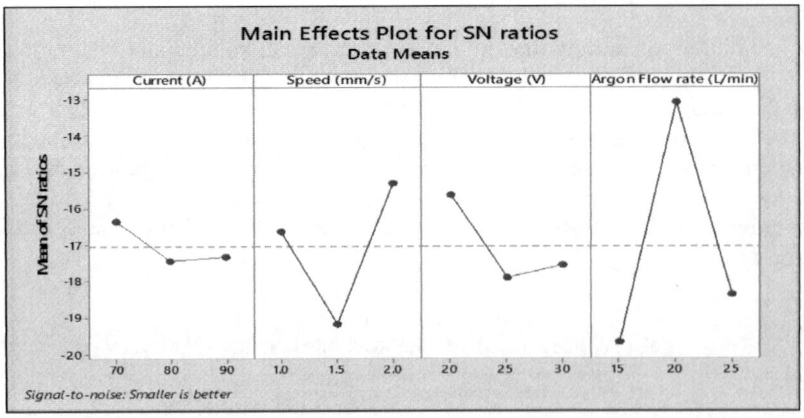

Figure 3. Main effects plot for surface roughness of SiC embedded LAS under TIG torch melting

4. Conclusion

Based on the analysis of the experimental result using Taguchi method, the following conclusions can be draw from the present studies on fry sliding wear test for SiC composite coatings developed by TIG torch surface melting technique:

1) The optimal combination of process parameters for minimum wear rate is found to be a current of 70 A, speed of 1.5 mm/s, voltage of 30 V and argon flow rate of 15 L/min. Meanwhile for wear tracks width, the optimal processing parameters occurs at current of 70 A, speed of 1.0 mm/s, voltage of 25 V and argon flow rate of 15 L/min and the analysis of surface roughness resulted in optimal parameter levels of 70 A current, speed of 2.0 mm/s, voltage of 20 V and argon flow rate of 20 L/min.
2) Argon gas flow rate has the highest influence on the general tribological performance of SiC embedded coatings.
3) The application of Taguchi approach was successful for optimizing tribological behavior of SiC embedded coating on LAS in this study and can also be applied in optimizing other coating performance characteristics.

Acknowledgement

The authors acknowledge the financial assistance from Research Management Center of International Islamic University Malaysia under the grant no RMGS-12-007-0020.

References

[1] Bell T, Morton P H, Bloyce A 1994 *Mater Sci Eng,* **184** 73-86.
[2] Adeleke S A and Maleque M A 2014 *Adv. Mater. Res.* **1024** 207-210.
[3] Kirchgabner K, Badisch E and Franek F 2008 *Wear* **265** 772-777.
[4] Maleque M A, Bello K A, Idriss A N M and Mirdha S 2013 *Appl. Mech. Mater.* **378** 259-264.
[5] Mirdha S, Idriss A N M, Maleque M A, Suryanto S and Souad A 2012 *Intl. J. of Mech Mater Eng.* 7 48-53.
[6] Ulutan M, Yildirim M, Buytoz S and Celik O N 2010 *Tribol T* **54** 67–69.
[7] Buytoz S 2006 *Surf Coat Technol.* **200** 3734-42.
[8] Bello K A, Maleque M A, Zuraida A and Mirdha S 2015 *Adv Mater Res.* **115** 238-242.
[9] Maleque M A, Ghazal B A, Ali M Y and Ahmed A S 2015 *Mater Sci Forum,* **819** 76-80.
[10] Pasupathy J and Ravisankar V 2013 *Int J Sci Eng Res.* **4** 25-28.
[11] Peng D X 2012 *Ind. Lubr. Tribol.* **64** 303-311.

Process Capability of High Speed Micro End-Milling of Inconel 718 with Minimum Quantity Lubrication

Mohamed Abd Rahman*, Mohammad Yeakub Ali, Abdul Rahman Shah Rosli, and Asfana Banu

Department of Manufacturing and Materials Engineering, International Islamic University Malaysia, P.O. Box 10, 50728 Kuala Lumpur, Malaysia
*E-mail: mrahman@iium.edu.my

Abstract. The demand for micro-parts is expected to grow and micro-machining has been shown to be a viable manufacturing process to produce these products. These micro-products may be produced from hard-to-machine materials such as superalloys under little or no metal cutting fluids to reduce machining cost or drawbacks associated with health and environment. This project aims to investigate the capability of micro end-milling process of Inconel 718 with minimum quantity lubrication (MQL). Microtools DT-110 multi-process micro machine was used to machine 10 micro-channels with MQL and 10 more under dry condition while maintaining the same machining parameters. The width of the micro-channels was measured using digital microscope and used to determine the process capability indices, Cp and Cpk. QI Macros SPC for Excel was used to analyze the resultant machining data. The results indicated that micro end-milling process of Inconel 718 was not capable under both MQL and dry cutting conditions as indicated by the Cp values of less than 1.0. However, the use of MQL helped the process to be more stable and capable. Results obtained showed that the process variation was greatly reduced by using MQL in micro end-milling of Inconel 718.

1. Introduction

Micro-machining such as micro-milling machining is increasing applied in the production of micro-channels with dimensions below 1 mm as demands for 3-dimensional (3D) miniature components increase [1]. Micro-channel production is considered as an advanced technology and the market demand is very high considering the uses of these micro-products in electronic devices, medical devices and also in many miniature components and micro-parts of aerospace sector. The demand for parts and components made from so called super-alloys is expected to increase. Benefits of these materials may be taken advantage of in the manufacture of turbine engine, engines in energy sector, air-industry, cosmonautics, mining and for medical equipment. A synergy between the use of this class of materials and micro-machining may well be expected in the production of quality micro-products sought for now and more so in the future. One of these is Inconel 718 which is a high-strength, corrosion-resistive, precipitation hardened alloy which may be used in the temperature range from -252°C to +700°C in many applications [2].

In metal cutting processes both at conventional and micro levels, metal cutting fluids are required to serve as lubricants, coolants and chips flusher throughout a machining process [1,3]. High production machining typically generates high cutting zone temperature that might cause premature tool failure and wear [4]. Therefore the usage of cutting fluid is vital in order to keep the machining process running smoothly and ensuring a good work piece surface quality while prolonging the cutting tool life. Minimum quantity lubrication or MQL is the recent method used to deliver metal cutting

IOP Conf. Series: Materials Science and Engineering **184** (2017) 012036 doi:10.1088/1757-899X/184/1/012036

fluid to the tool-work interface. It was introduced to help reduce high machining cost as well as possibly eliminate health and environmental problems associated with the use of metal cutting fluids. On the extreme, dry machining may be carried out without using coolant or metal cutting fluid. In a review [3], MQL appears to be studied largely with the conventional machining processes and less with micro-machining processes.

While many studies were done on the use of MQL in machining process, smaller numbers were carried out on its use in micro-machining [1,3]. Those that did focused mainly on issues related to tool wear and surface finish [5,6]. In order to produce quality micro-products, stable and capable micro-machining processes are critical. There is a need to investigate characteristics of micro-machining processes such as the micro end-milling in terms of its stability and capability. To improve our understanding in this aspect, the process capability study was carried out on micro end-milling of Inconel 718 under MQL and dry cutting conditions.

2. Experiment

In this study, straight micro-channels were machined on Inconel 718 using micro end-milling available on Microtools DT-110 Integrated Multi-Process machine (see Figure 1). Ten (10) micro-channels were machined with a minimum quantity of Accu-Lube LB-2000 (LB-1) lubricant supplied by a Bluebe FK MQL system. Another set of 10 micro-channels was machined under dry cutting condition. Two-flute tungsten carbide end-mill cutters with diameter of 0.8 mm and nose radius of 0.1 mm were used. Table 1 provides the machining parameters used for both MQL and dry cutting conditions.

The width of each micro-channel was measured using a digital microscope with 400X magnification. Measurements were carried out at three (3) places along the length of each micro-channel and the average was calculated. QI Macros for Excel was used to analyze the results and determine the process capability.

3. Results and Discussion

The mean and standard deviation of the micro-channels for MQL and dry cutting conditions were 795.38 μm and 4.22 μm and 793.12 μm and 9.10 μm respectively. The mean width of the micro-channels machined under MQL cutting condition was closer to the targeted value of 800 μm that that under dry condition. Using the process capability of ± 5 μm during fabrication of micro-channels using a chemical etching method [7], the results clearly indicated that using MQL condition improved the capability of micro end-milling process of Inconel 718 in terms of accuracy and precision.

Figure 2 and 3 below showed the control charts developed for the machining process under dry machining and MQL conditions respectively. Machining under dry condition seems to be relatively less stable than under MQL condition as shown by the general downward trend in Figure 2. Both charts indicated the progressive wear of the cutters. However, the wear of the cutter used in dry machining of the workpiece appeared to be higher.

Dispersion in the width values of micro-channels machined under dry condition was found to be higher as indicated by the broader distance between the control limits in Figure 2. This serves as an evidence of a more consistent micro end-milling process when carried out using MQL.

The frequency distribution in Figure 4 obtained for dry cutting machining process indicated that 5 micro-channels were within the specification limits while the remaining 5 were out of tolerance. In the case of MQL, 7 out of 10 micro-channels were within the specification limits as indicated in Figure 5. The values of Cp for dry cutting and MQL conditions are 0.17 and 0.36 respectively. Values below 1.0 are indicative of incapable processes [8]. There are numerous factors that may contribute to the capability of a micro-machining process. However, based on apparently equal machining parameters, micro end-milling process of Inconel 718 using MQL was shown to be more capable. This may be explained by the lubricating effect presence hence resulting in less cutter wear.

Table 1. Experimental parameters

Parameters	Level
Depth of Cut (mm)	0.15
Feed Rate (mm/min)	2
Spindle Speed (rpm)	28,000

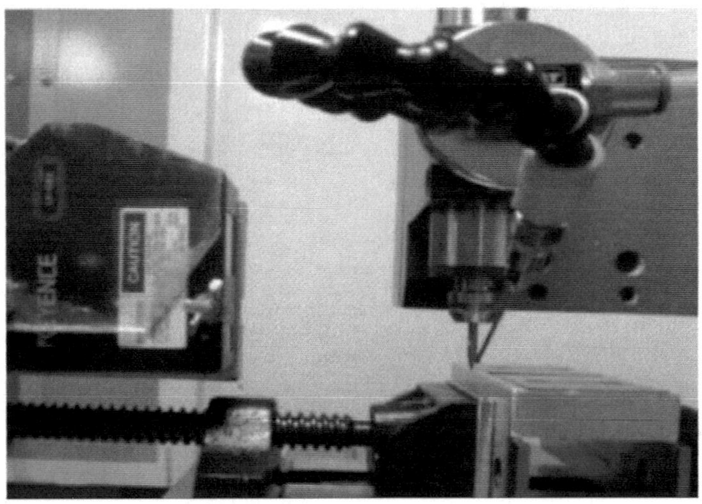

Figure 1. Micro Milling Setup with MQL

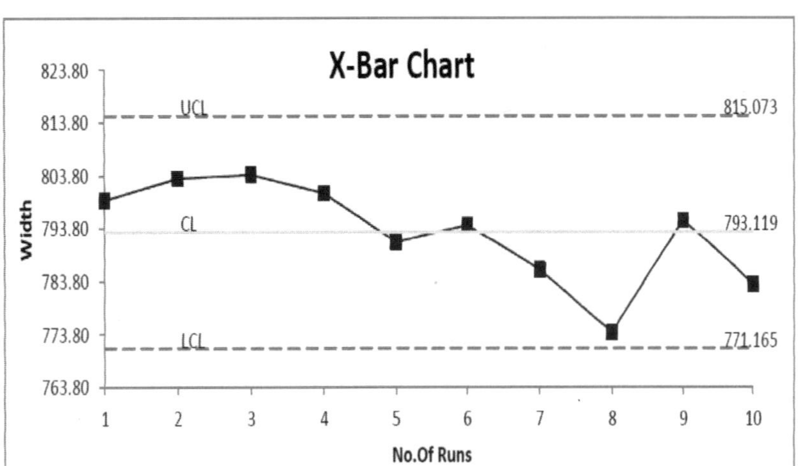

Figure 2. X-Bar chart for dry machining

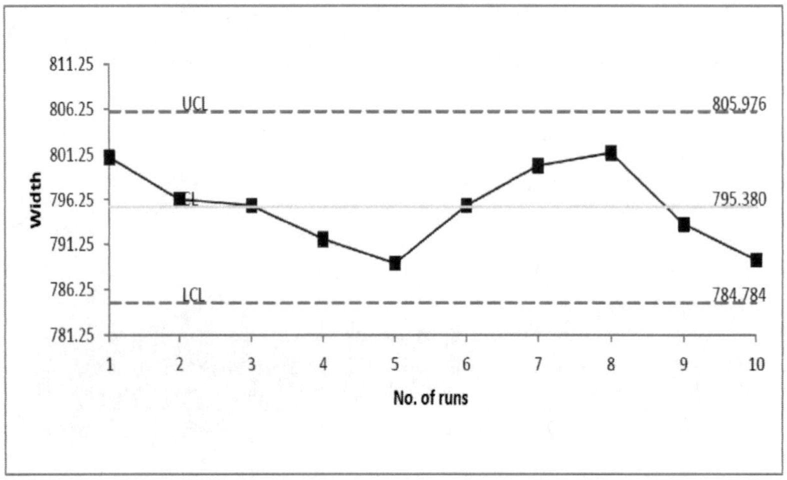

Figure 3. X-Bar chart for MQL machining

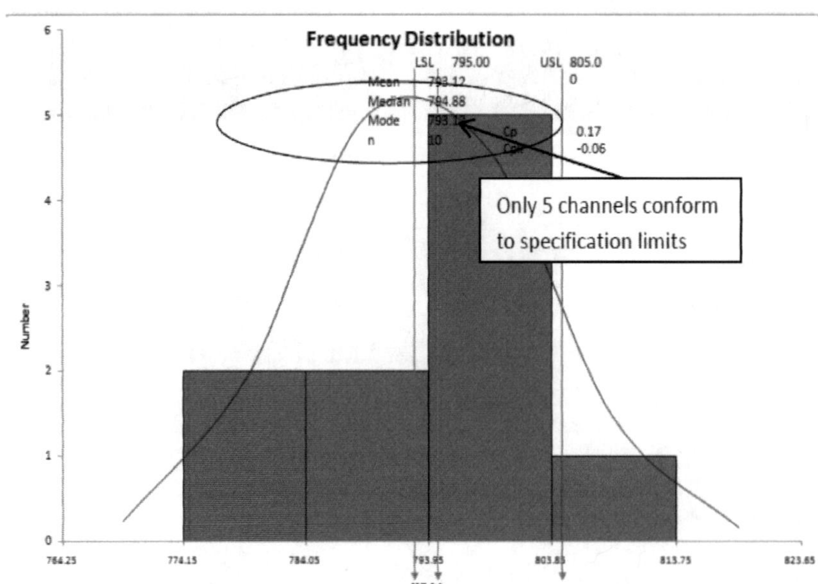

Figure 4. Frequency distribution for dry machining

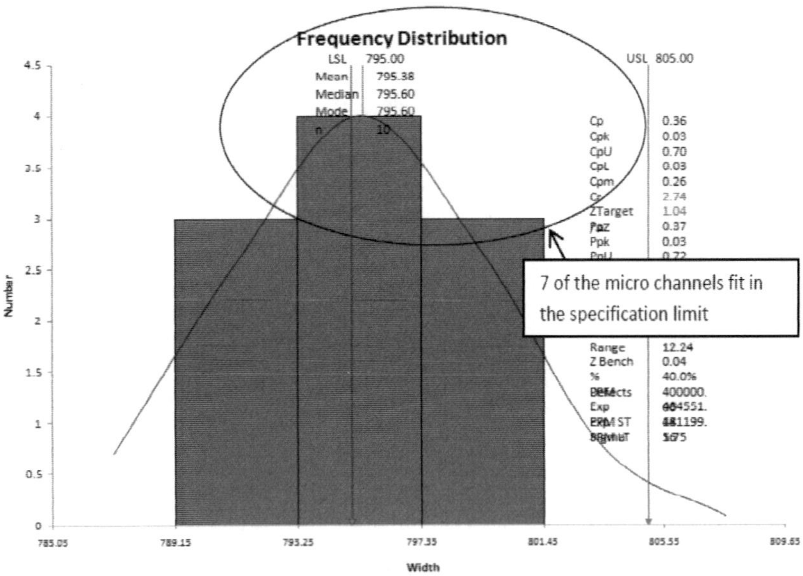

Figure 5. Frequency distribution for MQL machining

The centering of the machining process as indicated by the values of Cpk needs to be investigated further. Many factors may affect the stability of cutter during the machining process which may include vibration and machine dynamics. Further studies need to be carried out to focus on the factors affecting the capability hence quality of this increasingly important class of manufacturing process.

4. Summary

Micro-channels on Inconel 718 have been successfully machined using high-speed micro end-milling and the width of the micro-channels has been measured and analyzed using statistical process control tools. From the study the following may be concluded:

- Micro end-milling of Inconel 718 using MQL resulted in a more stable and consistent process as indicated by much smaller process dispersion compared to machining under dry cutting condition.
- Micro end-milling of Inconel 718 using MQL and under dry cutting condition were both shown to be incapable which may be attributed to numerous factors. However, the use of MQL clearly helped to make the process to be more capable.
- More studies need to be carried out in order to improve the understanding on significant factors affecting the capability hence the quality of the micro-machining processes.

Acknowledgement

The authors would like to express their thanks to the Ministry of Education, Malaysia for financial support under Research Grant FRGS 14-156-0397.

References

[1] Zhang Y and Jun M B G 2014 Atomized water and oil sprays as a single jet for cutting fluid delivery in micro-milling *9th Int. Workshop on Microfactories*

[2] Miroslav J and Mrkvica I 2012 Milling possibilities of material Inconel 718 *Manuf. and Ind. Eng.*, vol 11 ISSN 1338-6549

[3] Boubekri N and Shaikh V 2012 Machining using minimum quantity lubrication: a technology for sustainability *Int. J. of Applied Science and Technology* vol 2 pp 111-114

[4] Dhar N R, Kamruzzaman M, and Mahiuddin A 2005 Effect of minimum quantity lubrication (MQL) on tool wear and surface roughness in turning AISI-4340 steel *J. Mater. Process. Technol.* pp 299–304

[5] Mohamed A R, Atiqah N, Ali M Y, and Chowdhury M S H 2013 Tool vibration due to high speed micro end milling parameters *Applied Mechanics and Materials* vol 372 pp 364-368

[6] Ali M Y, Mohamed A R, Asfana B, Lutfi M, and Fahmi M I 2012 Investigation of vibration and surface roughness in micro milling of PMMA *Applied Mechanics and Materials* vol 217 pp 2187 - 2193

[7] Burger J, Holland H, Berenschot E, Seppenwoolde J H, ter Brake M, Gardeniers H, and Elwenspoek M 2001 169 Kelvin cryogenic microcooler employing a condenser, evaporator,flow restriction and counterflow heat exchangers *MicroElectro Mechanical Systems 2001 MEMS 2001. The 14th IEEE International Conference* pp 418 – 421

[8] Besterfield D H 2012 *Quality Improvement* 9th Edition Prentice Hall

Effects on Vibration and Surface Roughness in High Speed Micro End-Milling of Inconel 718 with Minimum Quantity Lubrication

Mohamed Abd Rahman*, Mohammad Yeakub Ali, Amir Saddam Khairuddin[c]

Department of Manufacturing and Materials Engineering, International Islamic University Malaysia, P.O. Box 10, 50728 Kuala Lumpur, Malaysia
*E-mail: mabdrahman2@gmail.com

Abstract. This paper presents the study on vibration and surface roughness of Inconel 718 workpiece produced by micro end-milling using Mikrotools Integrated Multi-Process machine tool DT-110 with control parameters; spindle speed (15000 rpm and 30000 rpm), feed rate (2 mm/min and 4 mm/min) and depth of cut (0.10 mm and 0.15mm). The vibration was measured using DYTRAN accelerometer instrument and the average surface roughness Ra was measured using Wyko NT1100. The analysis of variance (ANOVA) by using Design Expert software revealed that feed rate and depth of cut are the most significant factors on vibration meanwhile for average surface roughness, Ra, spindle speed is the most significant factor.

1. Introduction

Recently, micro-milling process offers many benefits to manufacturing industries through numerous products which can be produced using this method. In order to meet the demand of miniature parts and components with increasing number of functions, micro-machining has an important role in today's manufacturing technology especially in Micro Electro-Mechanical systems (MEMS) [1]. In machining operation, the aim is to have a good surface finish. Most machining processes do not produce good surface finish due to vibration that occurs during these processes resulting in low dimensional accuracy, poor surface finish and bad stability to the system as well as tool damage and machine tool damage [2].

Machining parameters especially feed rate, depth of cut and spindle speed if not properly controlled can increase the vibration in the machining process. Too high vibration is very dangerous and may lead to serious injury. Therefore, vibration must be minimized to stabilize the cutting process, improve surface roughness and prevent tool from breaking. Machining parameters and properties of the workpiece need to be fully considered before machining process [3]. Forced and self-excited vibration limits the stability of cutting and lead to tool and machine damage. It also limits the productivity of machining operation.

Cutting fluids are used in machining process to reduce temperature and contact between tool-work piece, to reduce thermal deformation of work piece and to wash away chips formation between tool and work piece zone. As a result, dimensional accuracy is achieved and surface finish and tool life are improved [4]. However there are many bad effects of using cutting fluids which include danger to health of user and environment due to the airborne mist [5]. Not only that, it contributed to major expenses in machining operation [6].

Dry cutting is very useful to minimize costs of using cutting fluids in addition to being safer and eco-friendly. However, dry cutting causes faster tool wear because of higher temperature due to greater friction and adhesion between chip and tool [7]. Therefore, minimum quantity lubrication (MQL) was introduced. Instead of using large amount of cutting fluids in flood cooling, only small amount of cutting fluids is used in MQL system at the cutting zone. By minimizing the use of cutting fluids in MQL, costs related to cutting fluids in machining process can be reduced [8].

Since less amount of fluids is used in MQL, some economical advantage results. MQL is also efficient in improving surface roughness of machined part. However, due to vibration, most of machining operations do not often produce good surface finish. Many studies on surface finish for micro-milling with dry cutting and flood cooling have been done. There are however few studies done on the impact of MQL in alleviating the effects of vibration and improving surface roughness of machined surface in high speed micro-milling. Therefore, in this study the effects of cutting tool vibration as well as the surface roughness in micro end-milling of Inconel 718 with MQL were investigated.

2. Experiment

The experiments were conducted using Mikrotools Integrated Multi-Purpose Machine Tool DT 110 suitable for high, middle and low spindle speeds for various processes such as micro-milling. MQL Bluebe FK type with Accu-Lube LB-200 (LB-1) was used for lubrication and cooling. DYTRAN Instrument accelerometer was used to detect the vibration on the machine. Figure 1a shows the accelerometer attached to the outer spindle during the machining process. The Multi-Channel Orchestra System was used to analyze the vibration detected by the accelerometer. For a good operating of the Orchestra system, the 01dB-Metravib software displayed the result of vibration data (see Figure 1b). Wyko NT1100 was used to measure the surface roughness of the micro-channels machined.

The workpiece used was Inconel 718 with dimensions of 50 mm x 40 mm x 7 mm. For cutting tool, tungsten carbide end mill with two flutes with Ø 0.5 mm diameter was used. Two-level factorial design was used as the experimental design to conduct 8 experiments. The experimental parameters and the results are shown in Table 1 and Table 2 respectively.

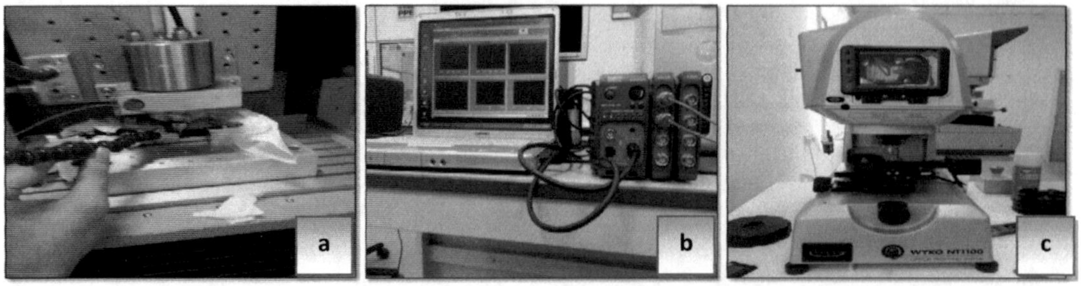

Figure 1. (a) position of accelerometer, (b) vibration monitor using Metravib software, and (c) Wyko NT1100 Optical Profiler

Table 1. Experimental Parameters

Control Parameters	Factor	Level	
		I	II
Depth of Cut, DOC (mm)	A	0.10	0.15
Feed Rate, v (mm/min)	B	2	4
Spindle Speed, N (rpm)	C	15000	30000

Table 2. Experimental results

No	DOC (mm)	v (mm/min)	N (rpm)	Avg. Tool Vibration (Hz)	Ra (μm)
1	0.10	2	15000	98.3	0.12
2	0.15	2	15000	98.5	0.17
3	0.10	4	15000	98.7	0.15
4	0.15	4	15000	99.7	0.18
5	0.10	2	30000	98.1	0.22
6	0.15	2	30000	99.6	0.20
7	0.10	4	30000	99.3	0.21
8	0.15	4	30000	99.8	0.24

3. Results and Discussion

The results of vibration generated during micro end-milling with different spindle speed, feed rate and depth of cut were analyzed. Analysis of variance results showed the main and interaction effects of the process variables on the responses (see Table 3). The model was developed with 95% confidence interval. The Model F-value of 7.34 implies the model is significant. There is only a 4.20% chance that a "Model F-Value" this large could occur due to noise. Values of "Prob > F" less than 0.0500 indicate model terms are significant. In this case A and B are significant model terms.

Table 3. ANOVA for response surface 2FI Model of average vibration

Source	Sum of Squares	DF	Mean Square	F Value	Prob > F
Model	2.72	3	0.91	7.34	0.0420 (significant)
A	1.28	1	1.28	10.34	0.0324
B	1.12	1	1.12	9.09	0.0394
C	0.32	1	0.32	2.59	0.1831
Residual	0.50	4	0.12		
Corr. Total	3.22	7			

Figure 2a shows the model graph of contour for spindle speed versus feed rate with constant depth of cut of 0.13 mm. From the graph, the average vibration is at maximum when it is at highest spindle speed and at highest feed rate. The average vibration starts decreasing slowly for further decrease in spindle speed and feed rate. However, Figure 2b shows the model graph of contour for spindle speed versus depth of cut with constant feed rate 3 mm/min. From the graph, average vibration gradually increases when the spindle speed and depth of cut increases. The tool vibration starts decreasing for further decrease in depth of cut and spindle speed. Meanwhile, Figure 2c shows the model graph of contour for feed rate versus depth of cut with constant spindle speed of 22500 rpm. From the graph, average vibration is at maximum value when the depth of cut increases with highest feed rate. The average vibration decreases when both feed rate and depth of cut decreases.

IOP Conf. Series: Materials Science and Engineering **184** (2017) 012037 doi:10.1088/1757-899X/184/1/012037

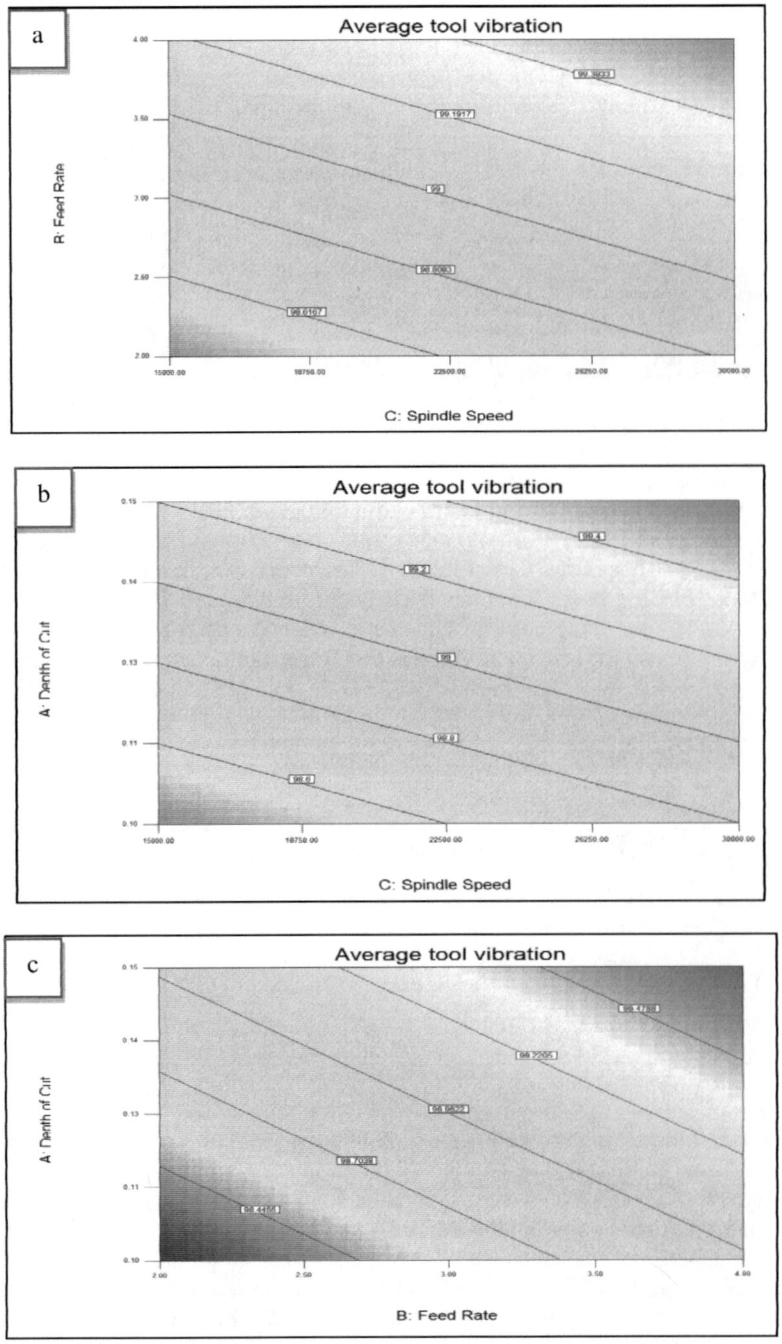

Figure 2. Contour plot for average vibration (a) spindle speed vs. feed rate with depth of cut 0.13 μm, (b) spindle speed vs. depth of cut with feed rate 3 mm/min, and (c) feed rate vs. depth of cut with spindle speed 22500rpm

Analysis of variance (ANOVA) approach was used to check the adequacy of the model for average surface roughness Ra. ANOVA results for response surface 2FI Model is shown in Table 4. The model was developed with 95% confidence interval. The Model F-value of 9.32 implies the model is significant. There is only a 2.81% chance that a "Model F-Value" this large could occur due to noise. Values of "Prob > F" less than 0.0500 indicate model terms are significant. In this case C are significant model term.

Table 4. ANOVA for response surface 2FI Model of average surface roughness

Source	Sum of Squares	DF	Mean Square	F Value	Prob > F
Model	9.437E-003	3	3.146E-003	9.32	0.0281(significant)
A	1.013E-003	1	1.013E-003	3.00	0.1583
B	6.125E-004	1	6.125E-004	1.81	0.2492
C	7.813E-003	1	7.813E-003	23.15	0.0086
Residual	1.350E-003	4	3.375E-004		
Corr. Total	0.011	7			

Figure 3a shows the model graph of contour for spindle speed versus feed rate with constant depth of cut of 0.15 mm. From the graph, average surface roughness is at maximum when feed rate and spindle speed are at highest value. The average surface roughness starts decreasing for further decrease in feed rate and spindle speed. However, Figure 3b shows the model graph of contour for spindle speed vs. depth of cut with constant feed rate of 3 mm/min. The graph indicates that average surface roughness decreases when the spindle speed and depth of cut decrease. Meanwhile, Figure 3c shows the model graph of contour for feed rate vs. depth of cut with constant spindle speed of 22500 rpm. From the graph, average surface roughness is at maximum value when the depth of cut increases with highest feed rate. The average surface roughness decreases when both feed rate and depth of cut decrease.

4. Conclusions

The purpose of this paper is to investigate vibration and average surface roughness, Ra of Inconel 718 in micro end-milling and the results were analyzed using ANOVA. So, from this investigation, it can be concluded that:

1. The highest value for average tool vibration is 99.8 Hz and the lowest value for average tool vibration is 98.1 Hz.
2. The highest value for average surface roughness is 0.24 □m and the lowest value for average surface roughness is 0.12 □m.
3. For average tool vibration, depth of cut and feed rate are significant factors while for average surface roughness, spindle speed is the significant factor in this experiment.

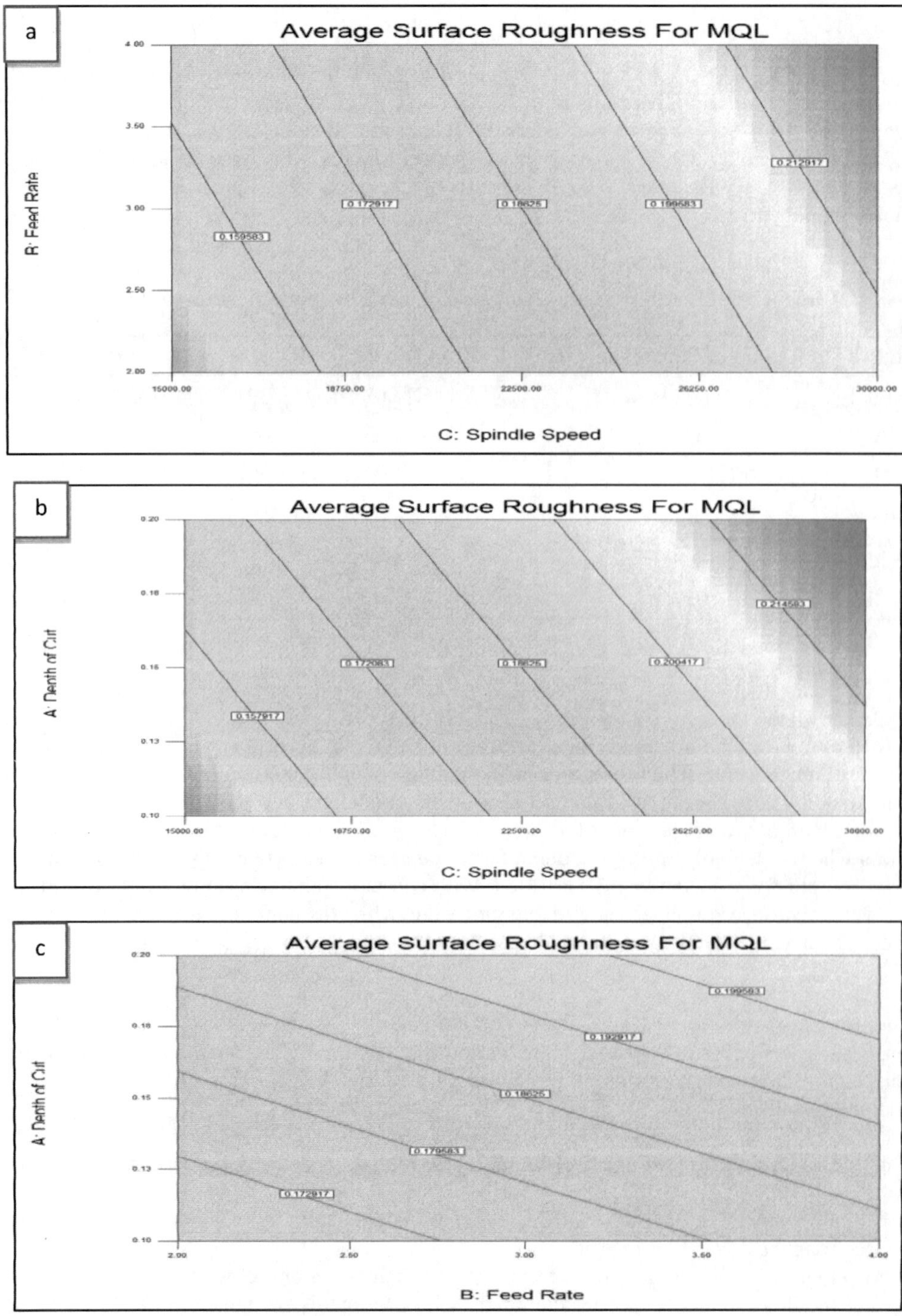

Figure 3. Contour plots for average surface roughness (a) spindle speed vs. feed rate with depth of cut 0.15 μm, (b) spindle speed vs. depth of cut with feed rate 3 mm/min, and (c) feed rate vs. depth of cut with spindle speed 22500 rpm

Acknowledgement
The authors would like to express their thanks to the Ministry of Education, Malaysia for financial support under Research Grant FRGS 14-156-0397.

References
[1] Gupita A and Ahmad A 2007 Microsensors based on MEMS technology *Defense Science Journal* vol 57 pp 225-232.
[2] Al-Regib E, Ni J, and Lee S H 2003 Programming spindle speed variation for machine tool chatter suppression *Int. J. Mach. Tool Manu.* vol 43 pp 1229-1240.
[3] Kassab S Y and Khoshnaw Y K 2007 The effect of cutting tool vibration on surface roughness of workpiece in dry turning operation *Eng. & Technology* vol 25 pp 879 – 889
[4] Cakīr O, Yardimeden A, Ozben T, and Kilickap E 2007 Selection of cutting fluids in machining processes *J. Achievements in Materials and Manufacturing Engineering* vol.25 pp 99 – 102
[5] Li K M and Chou S Y 2010 Experimental evaluation of minimum quantity lubrication in near micro-milling *J. Mater. Process. Tech.* pp 2163 – 2170
[6] Klocke F and Eisenblätter G 1997 Dry cutting. *CIRP Annals: Manufacturing Technology* pp 519-526
[7] Davoodi B and Tazehkandi A 2014 Experimental investigation and optimization of cutting parameters in dry and wet machining of aluminum alloy 5083 *J. Clean. Prod.* vol 68 pp 234-242
[8] Abhang L B and Hameedullah M 2010 Experimental investigation of minimum quantity lubricantsin alloy steel turning *Int. J. Eng. Sci. Tech.* vol. 2

Hexagonal Hollow Tube Based Energy Absorbing Crash Buffers for Roadside Fixed Objects

M. S. Uddin[1,*] Nurul Amirah Shafie[1,2] and Grad Zivkovic[2]
[1]School of Engineering, University of South Australia, Mawson Lakes, SA 5095, Australia
[2]Automotive Safety Engineering Pty Ltd, Lonsdale, SA 5160, Australia

Email: * Mohammad.Uddin@unisa.edu.au

Abstract. The purpose of this study was to investigate the deformation of the energy absorbing hexagonal hollow tubes in a lateral compression. The aim is to design cost effective and high energy-absorbing buffer systems, which are capable of controlling out-of-control vehicles in high-speed zones. A nonlinear quasi-static finite element analysis was applied to determine the deformation and energy absorption capacity. The main parameters in the design were diameter and wall thickness of the tubes. Experimental test simulating the lateral compressive loading on a single tube was performed. Results show that as the diameter and the thickness increase, the deformation strength increases. Hexagonal tube with diameter of 219 mm and thickness of 4 mm is shown to have the highest energy absorption capability. Compared to existing cylindrical and octagonal shapes, the hexagonal tubes show the highest energy absorption capacity. Hexagonal tubes therefore can be regarded as a potential candidate for buffer designs in high speed zones. In addition, they would be compact, cost effective and facilitate ease of installation.
Keywords: Hexagonal hollow tubes, energy absorbing buffer, finite element simulation, deformation strength.

1. Introduction

Enormous amount of people around the world are involved in road accidents every year. One of the unforgiving types of crashes is vehicle's impacts with roadside fixed objects, e.g. utility poles or trees. More importantly, such collisions result in fatal injuries to occupants, in addition to the cost of repairing damaged objects or structures involved [1]. For instance, between 2000 and 2011 alone in South Australia, 36% of fatal road crashes are the result of vehicle's impact with fixed objects [2]. While improved vehicle, road design and safety have decreased the fatality rate to some extent, such events have still been remained as a serious concern across the world.

In order to minimize the risk of collisions with fixed objects, passive safety measures such as roadside barriers are employed. Commercial buffers, often known as 'impact attenuators' such as QuadGuard [3] and Absorb360 [4] are used as the breakaway barriers and they are often found to be large in size and highly expensive. Cylindrical hollow tube based metallic energy absorbing buffers are recently applied for roadside fixed objects [5]. They seem to have been quite successful; however, they suffer from issues such as assembly of tubes in buffers is not strong, causing tubes dislodged under the high impact collisions. The tube's energy absorption capacity is even low due to its geometric shape and hence they may not suitable to be employed in high speed roadways. In other words, the tube/s or buffers will be deformed or broken too quickly, leaving the vehicles/occupants to be impacted seriously, instead of taking the impacts up by buffers, during collisions. Larger tubes may be considered to solve the issue. However, they incur further cost and large space for installation. Thus, a new design of energy absorbing buffer is sought, which will have high energy absorption capability, be compact and cost effective. The current paper aims to investigate the deformation and energy absorption ability of a new hollow hexagonal tube based crash buffers for roadside fixed objects. Using FE simulation and experiment, a parametric study is performed to evaluate the effect of tube's geometric parameters, e.g. diameter and thickness, on the deformation. The benefit of hexagonal tube over existing tubes (e.g. cylindrical and octagonal) and its potential use for buffer designs are discussed and addressed.

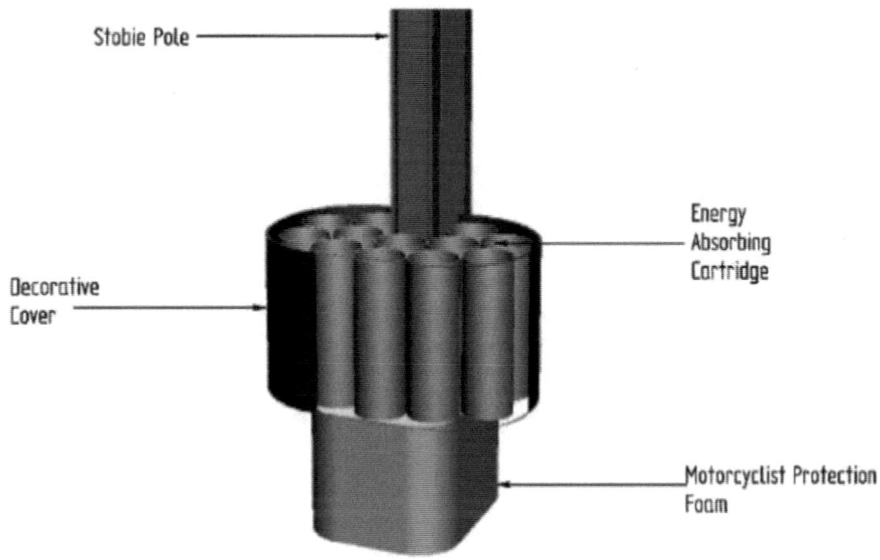

Figure 1. Existing energy absorbing buffer system

2. Issues with current buffer designs

Hollow tube based impact attenuators are widely used to prevent damages and injuries in many applications such as automotive, rail. The kinetic energy of the out-of-control vehicles are absorbed through the progressive deformation of the tubes, hence minimizing impacts on vehicles and occupants. Such deformable bodies are generally installed within the vehicle structure, which often don't consider impacts with roadside fixed objects. Automotive Safety Engineering (ASE) Pty Ltd first introduced cylindrical hollow tube based crash buffers which are used around the fixed objects such as Stobie poles to protect the pole and more importantly, vehicle occupants in the event of impacts with fixed objects. Such design and technology have been tested successfully and preliminary well received by the state governments for its use in roadways [6].

Figure 1 shows an example of cylindrical tube based crash buffer installed around the pole. As can been seen in figure 1, due to their nearly round shape, the buffers can take impacts at both frontal and sideways. The problem with the current cylindrical tube design is however that it is quite difficult to assemble tubes together due to its round faces. Under the impact, the tubes are often displaced and/or dislodged from the original locations, and hence, the buffer loses its deformation or energy absorption capacity. Because of its geometry, the deformation rate of the cylindrical tube is relatively high, making it too soft and flexible. As a result, the tubes cannot withstand large impact, i.e. it deforms and fails too quickly at even a small impact force. To tackle this, often large diameter and thick tubes are needed to be used, causing an increase of cost and installation space. In place of cylindrical geometry, octagonal shaped tubes are another option often considered as an energy absorbing structure [7]. It has flat faces which can be used to assemble with nearby tubes to form a crash buffer. Recent analysis showed that they may not be a complete solution as they either behave like a cylindrical tube or often pose fabrication complexity because of their inherent structure with having more geometric transitions. Therefore, it is important to seek for a new geometric shape which can be strong, flexible/deformable, cost effective and capable of absorbing large impact energy.

3. Materials and method

3.1. *Proposed hexagonal tube.*

This study proposes a hollow hexagonal tube as energy absorbing structure. The new shape is a modification of octagonal tube previously designed and studied by the same authors. The new hexagonal shape has six flat faces for easy and secured assembly of tubes together. Because it has a more rigid structure the deformation and energy absorbing capacity would be expected to be large, as opposed to cylindrical and octagonal tubes. It is a simple shape and requires less processing steps, thus reducing associated costs. Figure 2 shows an example of proposed hexagonal tube and its geometry (outer diameter = 195 mm and wall thickness = 1.5 mm). The tubes possess vertical and horizontal slots to increase the flexibility to deform. Horizontal slots also allow fastening tubes together by bolt-nut fasteners. In order to find an optimum geometry which will offer the best deformation ability, we have done a parametric study of the proposed shape by varying its outer diameter (of 165, 195 and 219 mm) and wall thickness (of 1.5, 2, 2.5, 3 and 4 mm) while the length or height (of 510 mm) is kept constant. Geometry of tubes are grouped by '*Tube A*' (diameter = 165 mm), '*Tube B*' (diameter = 195 mm) and '*Tube C*' (diameter = 219 mm) according to the size order. The values are chosen based on their availability at the local suppliers.

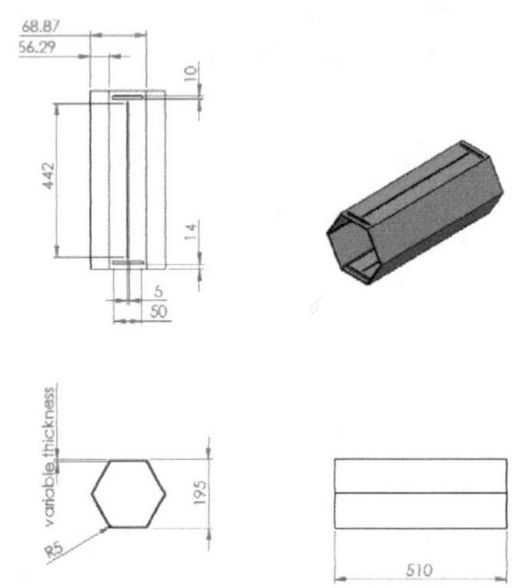

Figure 2. Definition of hexagonal tube

3.2. *Simulation.*

A nonlinear quasi-static simulation using finite element (FE) method was conducted in order to gain a greater understanding of how the individual tubes. We performed FE analysis on single tubes with geometric parameters mentioned in the previous section. The simulation setup was designed in such a way that it could be replicated by a real experiment for the purpose of comparison. As can be seen in Fig. 3, the FE model consists of a tube placed between a top and a bottom plate, where, the bottom plate was fixed while the top plate was allowed to move only in the vertical direction. The top plate was assigned a vertical displacement that would force the tube to be compressed laterally. A reaction force probe was applied to the bottom surface of the top plate to estimate the force against the displacement.

Finite element code built in Solidworks software tool was applied to solve the contact problem. 8-node solid elements were used in the FE model. The material of the tubes was mild steel. The multi-linear isotropic hardening stress-strain data of the material was obtained from experimental study [8]. Frictional contact between the tube and the plates was studied and showed a negligible effect. Hence, the contact was assigned as frictionless throughout FE study. This was found to reduce computational effort in simulation.

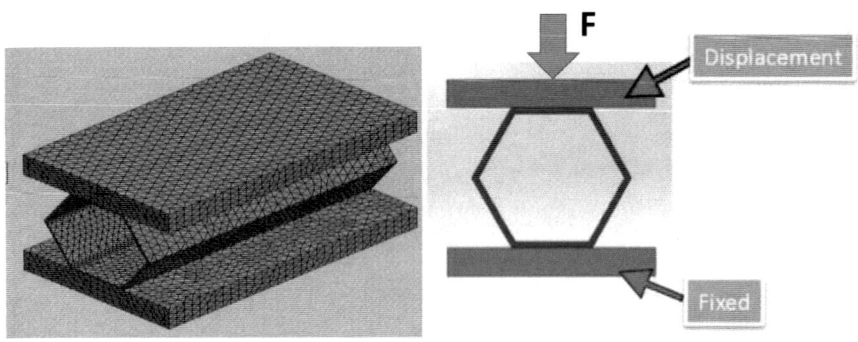

Figure 3. FE model of hexagonal tube

3.3. Experiment.

Experiments involving lateral compression were conducted in order to investigate the deformation of the individual tubes. The aim was to validate the quasi-static FE simulation. The tests involved the deformation of a tube located between two plates at top and bottom. The bottom plate was fixed and the top plate (push plate) was attached to a 20-tone hydraulic cylinder so that the tube was deformed in a controlled way that would enable the reaction force acting on the top plate to be recorded. Figure 4 shows the experimental setup. At a certain rate of push plate movement, reaction load and displacement were measured by a hydraulic pressure gauge and a linear encoder, respectively. In order to verify the efficacy of FE modelling, as a representative, *Tube C* (d_o = 219 mm) with t = 2mm was used as the specimen. The tube was deformed up to 213 mm until it becomes approximately flattened (75% of its total deformation).

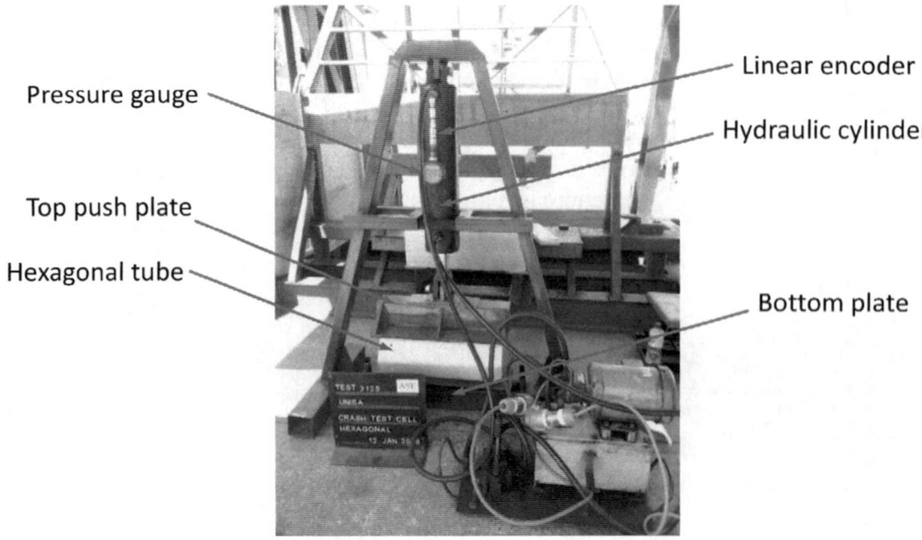

Figure 4. Experiment setup

4. Results and discussion

4.1. Comparison between simulation and experiment.

Figure 5 shows a comparison of tube deformation pattern of *Tube C* obtained from experiment and simulation. Both experiment and simulation display almost the same trend of the deformation until the tube reaches its maximum deformation level. The tube undergoes both elastic and plastic deformations. Figure 6 illustrates a comparison of load-displacement data obtained from experiment and simulation for *Tube C*. Results show that both simulation and experimental plots follow approximately the same trend of deformation as the loading increases. At displacement up to about 45 mm, the tube undergoes elastic deformation till the yield state and at this zone, simulation result matches with experiment. After that, the plastic deformation is more dominant till 200 mm. Experiment shows about 20% lower strength in the plastic zone than simulation. This would be due to the slight difference in material properties used in simulation and experiments. It is to be noted that mild steel tube specimen used in experiments might have pores, cracks and internal stress concentration, thus lowering the strength of the material. Overall, the results for a single tube justify the accuracy of the current FE model, thus indicating that the model would be able to reasonably estimate the accurate deformation of other tubes and potential buffer system consisting of a series of individual tubes.

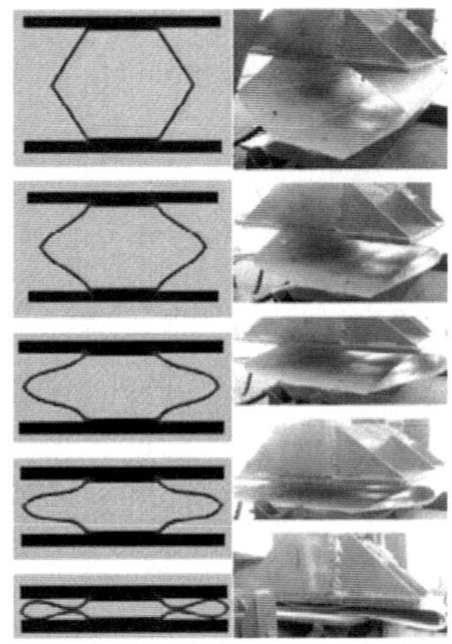

Figure 5. Comparison of hexagonal tube deformation between simulation and experiment

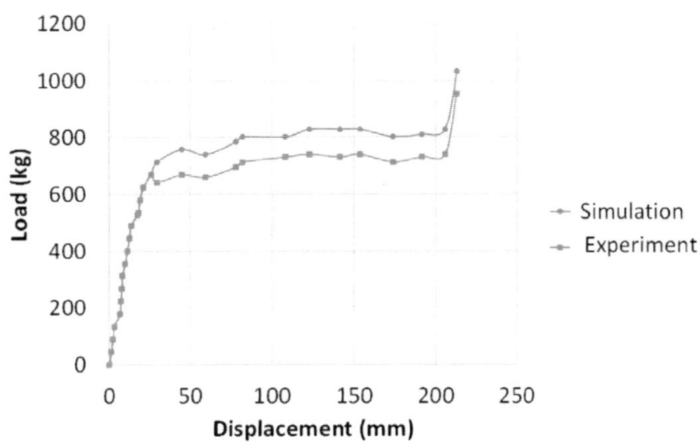

Figure 6. Comparison of load-displacement of *Tube C* between simulation and experiment

4.2. Effect of hexagonal tube geometry.

Using the same FE simulation procedure, the effect of various geometry of the hexagonal tube on the deformation profile was investigated, in which, the diameter and the thickness were varied. Figures 7a, b and c shows load-displacement with respect to different thicknesses (t) for *Tube A* (d_o = 165 mm), *Tube B* (d_o = 195 mm) and *Tube C* (d_o = 219 mm), respectively. It can be seen that for a given tube diameter, the tube shows higher deformation strength with the increase of thickness. This is quite obvious that higher thickness means more rigidness of the tube, which requires larger force to deform. Also, as the tube diameter increases, the amount of force requires deforming the tube increases. This will eventually increase the energy absorption capacity of the tube. It can be noticed from figure 7 that *Tube C* shows the highest deformation strength among all. Further, it appears that *Tube C* with thickness of 4 mm can absorb the highest energy and can be chosen as a potential candidate for buffer design. In particular, as is seen from figure 7c, when thickness is increased from 3mm to 4 mm, the deformation strength of *Tube C* increases significantly, i.e. the large force is needed for a small amount of deformation. It is often recommended that larger diameters may be used to take up high impact loading in the event of vehicles hitting objects at high speeds, e.g. 80-100 km/h. However, it must be noted that other thicknesses may be considered and the final decision in selecting an appropriate geometry would be based on the analysis of feasibility in terms of the number of tubes and the associated costs incurred to the buffer designs. In general, for a given buffer size, tubes with less energy absorbing capability will increase the number of tubes required for the buffer, which will definitely increase fabrication and handling costs.

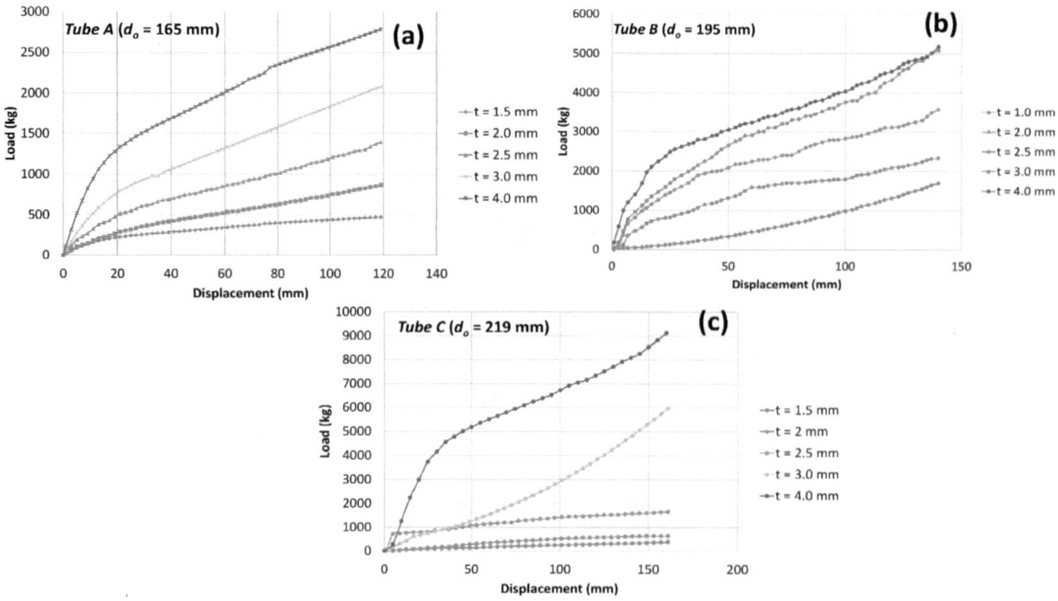

Figure 7. Load-displacement plot with respect to thickness for (a) *Tube A,* (b) *Tube B* and (c) *Tube C*

4.3. Comparison between hexagonal and existing tube designs.

In order to validate the performance of hexagonal shape, we further performed a comparison of the deformation of hexagonal *Tube C* (d_o = 219 mm and t = 4 mm) with existing tubes –cylindrical and octagonal. In this regard, following the same procedure, FE simulation was performed to estimate load-displacement results of tubes. During simulation, we have considered tubes with the same diameter and thickness for the sake of a balanced performance comparison. Energy absorption capacity is another indicator to measure the deformation ability of the tubes. We have estimated energy absorbed by the tubes, where, the energy is estimated as the area under the load-displacement curve. Figure 8 shows comparison of energy absorption capability between hexagonal, octagonal and cylindrical tubes. Hexagonal tube exhibits the highest energy absorption capacity among all. Octagonal shape shows the second highest deformation ability. The underlying reasons for difference in deformation capacity would be simply the inherent geometric variation due to their shapes. The results clearly show that hexagonal tubes can be used for buffer designs for high speed impacts. As can further be seen from figure 8, the hexagonal tube (*Tube C,* d_o= 219 mm, t =4 mm) has the energy absorption capacity of about 23.5 kJ at its 75% deformation. For instance, a vehicle with mass of 2000 kg travelling at 80 km/h possesses the kinetic energy of 94 kJ. It is expected that this energy will be equally absorbed by the buffer and the vehicle structure. Therefore, the buffer to be designed will absorb energy of 247 kJ. Thus, the number of tubes required for the buffer to absorb energy can be estimated by dividing the total energy by the energy absorption capacity of an individual tube (= 247 kJ/23.5 kJ), which is about 11 tubes. Accordingly, it can be easily predicted that buffers made of octagonal and cylindrical tubes will require 14 and 18 tubes, respectively. The results imply that hexagonal tube based buffers will be much more compact and cost-effective. However, to make a comprehensive conclusion of this, a detailed analysis in terms of arrangement of tubes within the buffer, fabrication, handling and maintenance cost of the tubes and buffers must be conducted. This has been left for our future work.

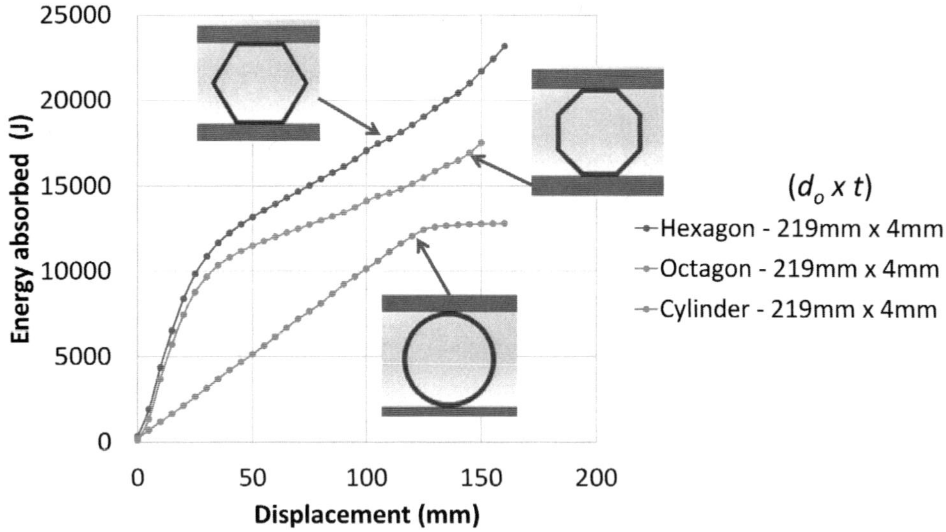

Figure 8. Comparison of energy absorption capacity among hexagonal, octagonal and cylindrical tubes

5. Conclusion

This paper studied new hexagonal hollow tube based energy absorbing crash buffers for roadside fixed objects. A FE simulation was performed to investigate the effect of different geometry on deformation and energy absorbing capacity of individual tubes. Experiments were conducted to verify FE simulation. The key conclusions drawn out of the study are as follows.

- FE simulation matched with experimental results, indicating that the FE model proposed is accurate enough to estimate deformation of the tubes.
- Parametric study showed that larger diameter and larger thickness increases the deformation strength of the tubes.
- For a given diameter and thickness, the hexagonal tube was found to show the highest energy absorption capacity and hence can be regarded as potential candidate for buffer designs.

Acknowledgements

The authors would like to acknowledge the Automotive Safety Engineering Pty Ltd for providing the support of carrying out experimental tests.

References

[1] Wang Y G, Chen K M, Ci Y S, and Hu L W (2011) *Safety Performance Audit for Roadside and Median Barriers Using Freeway Crash Records*: Case Study in Jiangxi, China. *Scientia Iranica*, **18**, 1222-30. http://dx.doi.org/10.1016/j.scient.2011.11.020
[2] BITRE (2011) *Cost of road crashes in Australia,* Bureau of Infrastructure, Transport and Regional Economics, Canberra, Australia
[3] Bolyan (2015) *Quadguard Cushion System*, Bolyah Group, Australia.
[4] RMS (2015) *Absorb360 crash cushion*, Road management solutions, Australia.
[5] Zivkovic G and Winfield M (2010) *A Study into the Lateral Compression of a 115mm Diameter Steel Tube* (Adelaide, Automotive Safety Engineering Pty Ltd)
[6] Zivkovic G and Day J (2012) *Safety Barrier Product information Report* (Adelaide,

Automotive Safety Engineering Pty Ltd)

[7] Singh H (2014) *Design and optimization of octagonal tube based energy absorbing crash buffers for high speed roadways*, Undergraduate Thesis (Adelaide, University of South Australia)

[8] Uddin M S, and Quintel J (2014) *Design and analysis of energy absorbing crash buffers for fixed objects in high speed roadways*, 8th Australasian Conference on Applied Mechanics (ACAM8), 23-26 November, Melbourne, Australia

IOP Conf. Series: Materials Science and Engineering **184** (2017) 012039 doi:10.1088/1757-899X/184/1/012039

Improvement of surface quality using silicon carbide powder added dielectric fluid during EDM

Ahsan Ali Khan[1*], Nurul Nasuha Binti Mohamad Royani[2], Muataz Hazza Faizi Al-Hazza[3], Mohd Radzi Haji Che Daud[4] and AKM Mohiuddin [5]

1,2,3,4 Department of Manufacturing and Materials Engineering, International Islamic University Malaysia, Jalan Gombak, 53100 Kuala Lumpur, Malaysia

5 Department of Mechanical Engineering, International Islamic University Malaysia, Jalan Gombak, 53100 Kuala Lumpur, Malaysia

* aakhan@iium.edu.my

Abstract. In recent years there is an increasing interest of Electrical Discharge Machining (EDM) process, with particular emphasis on the potential of this process for surface modification. In this study, the effect of various EDM parameters on the material removal rate (MRR), tool wear rate (TWR) and surface roughness (SR) using powder-mixed electrical discharge machining (PMEDM) was carried out by using Silicon Carbide (SiC) powder. The workpiece material used was high carbon steel AISI 1050 and the electrode tool material was copper (Cu). Experiments have been designed using full factorial design of experiment (DOE) and the 3 factors were designed for 2 levels. The input process parameters used were the concentration of powder from 0.0 to 5.0g/L, peak current ranging from 2.5A to 4.5A and pulse-on time ranging from 8μs to 10μs. The result shows that when silicon carbide powder is mixed into dielectric fluid during EDM process, it gives minor effect on material removal rate (MRR), but improves tool wear rate (TWR) and surface roughness (Ra). However, as peak current and pulse-on time increase gradually, all responses follow the same trend.

1. Introduction

Electrical Discharge Machining (EDM) is one of the most used nonconventional material removal process applications due to its ability to produce geometrically complex shapes and its ability to machine materials irrespective to their hardness and toughness. A spark is produced at the point of smallest inter-electrode gap generating temperature in the range of 8,000-12,000°C, causing erosion and vaporization of material. Based on Pecas and Henriques [1], die sinking version of electrical discharge machining (EDM) is a current manufacturing technology widely disseminated in several industrial sectors. Recent studies show that one type EDM, powder mixed electrical discharge machining (PMEDM) is being investigated to improve machining efficiency and surface finish. PMEDM is one of the advanced techniques that have been used to enhance the capability of EDM. Using PMEDM a mirror like finish can be obtained on the already machined components at relatively high machining rate [2]. PMEDM can enhance EDM process by controlling the process parameters setting. Bhattacharya, [3], M.Y Ali et al. [4] mentioned in their studies that under the same particle concentration, the smallest suspended particle powder led to the best surface finish. Kumar et al. [5] reported that when the silicon carbide particles are suspended in the dielectric fluid, it makes uniform in discharging energy dispersion; it displays multiple discharging effects within a single input pulse in the PMEDM. This phenomenon results in reducing TWR and improving Ra. In the present study the work material, the electrode material and additive powder were AISI 1050 steel, copper and silicon carbide respectively.

2. Experimental Procedure.

The main composition of the workpiece material AISI 1050 steel is: Mn: 0.6-0.9%, C: 0.47-0.55%, S: 0.05% and P: 0.04%. Its general properties include: density 7.85 g/cm^3, melting point 1400-1450°C, modulus of elasticity 210 GPa, electrical resistivity 0.163 μΩ.m, thermal conductivity 49.8 W/mK and thermal expansion11.9 μm/mK. The main characteristics of copper electrode were: density 8.96 g·cm−3; melting point 1083 °C, boiling point 2925 °C and electrical resistivity 1.96 μ Ω/cm. Properties of silicon powder are: powder size 0.1-1.0 micron, density 3.21 g/cm³, thermal conductivity 1.0-5.0 W/cm K, melting point 2987 °C and electrical resistivity 1x10^9 μΩ cm. Kerosene was used as dielectric fluid. Rectangular holes of 15mm x 10 mm were machined on the workpiece. Experiments were conducted on MITSUBISHI EX 22 die-sinking EDM machine. Work surface roughness was measured using Mitutoyo SURFTEST SV.500. Weight of the workpiece and the electrode before and after each machining was measured by an electronic balance. Experiments were conducted according to DoE. Two level factorial design has been used to plan and analyze the experimental results.

3. Results and Discussions.

In the present study, two level factorial design has been used to plan and analyze the experiments. The detailed plan of the experiments is shown in Table 1. Peak current, spark on time (T-on) and powder concentration are taken as the input parameters. MRR, TWR and Ra were the responses.

Table 1: Experimental plan

Std	Run	Block	Factor 1 A:Ip A	Factor 2 B:Ton microseconds	Factor 3 C:Concentratio g/L	Response 1 MRR g/min	Response 2 TWR g/min	Response 3 Ra micrometer
3	1	Block 1	4.5	10	0			
7	2	Block 1	4.5	10	5			
1	3	Block 1	4.5	8	0			
4	4	Block 1	2.5	10	0			
8	5	Block 1	2.5	10	5			
5	6	Block 1	4.5	8	5			
2	7	Block 1	2.5	8	0			
6	8	Block 1	2.5	8	5			

3.1 Material Removal Rate (MRR).

Fig.1 shows the relationship between MRR with design parameters of peak current (Ip) and pulse on time (Ton). It can be seen that as the peak current and pulse on time increase, MRR tends to increase too. This is due to the fact that sparks with more energy are generated during machining with higher current that would assist to remove more material from the workpiece. So, the relationship between current and MRR is directly proportional. If Ton is increased, more time is available for the workpiece to absorb heat. As a result, MRR increases.

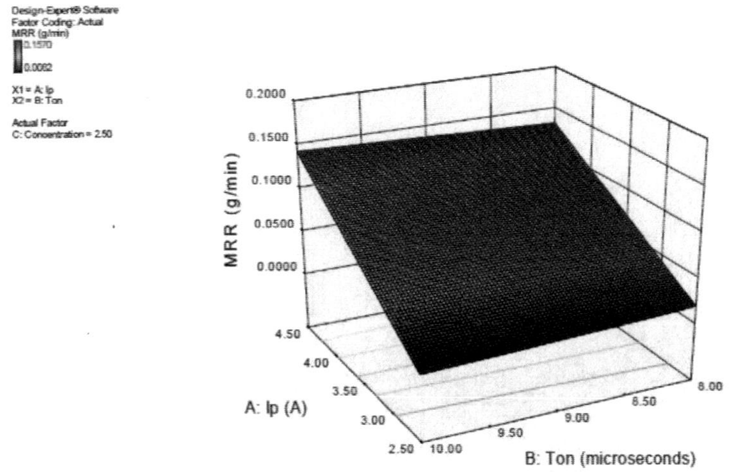

Fig.1: 3D surface plot of MRR as a function of current and Ton

Fig.1: 3D surface plot of MRR as a function of current and Ton

Fig.2 displays the effect of peak current and concentration of SiC carbide powder on MRR. It can be seen from the figure, the value of MRR increases with increase in peak current. However, the value of MRR decreases as powder concentration is increased to 5.0g/L. This is because silicon carbide is nonconductive and it decreases the effective current. Silicon carbide reduced the intensity of the sparks. Due to this phenomenon, MRR slightly decreases.

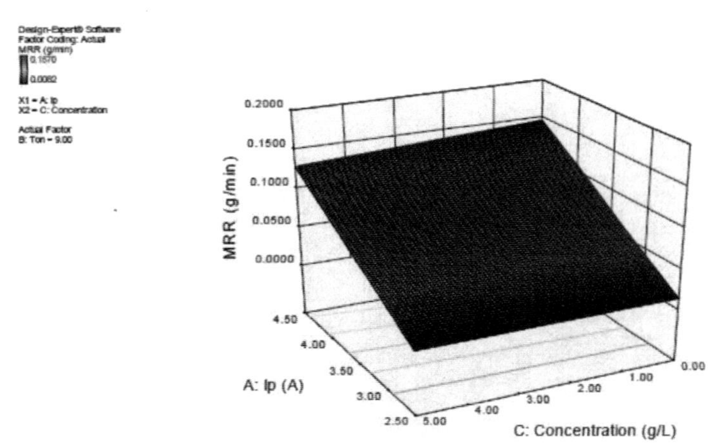

Fig.2: Interaction effect of current and SiC concentration on MRR

3.2. Tool Wear Rate (TWR).

The interaction between peak current and pulse on time for TWR is demonstrated in Fig.3. As can be noticed from this figure, high peak current gives remarkable effect in increasing the TWR. Higher current intensifies the sparks and more material is eroded from the electrode. Ton also results increase in TWR, but its effect is not as strong as that of current. As Ton is increased, electrode gets more time to absorb heat and erodes more. This reason causes the TWR to increase.

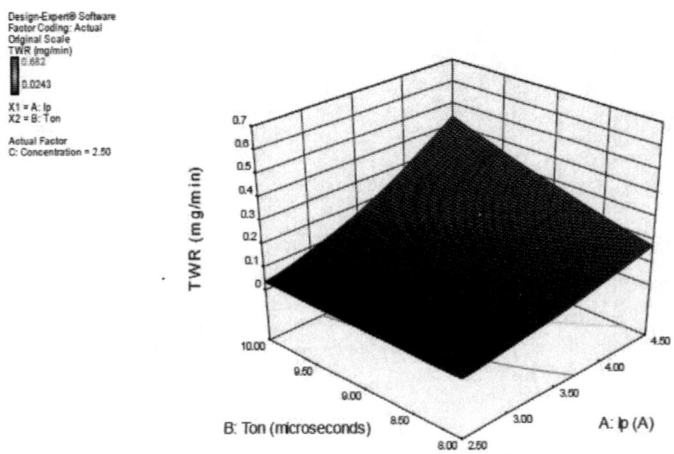

Fig.3: Interaction effect of current and Ton on EWR

Fig.4 illustrates the relationship of current and SiC powder concentration on EWR. It shows that though with increase in current EWR increases sharply, EWR decreases. SiC is an electrically nonconductive material. Its presence in dielectric reduces the effectiveness of the spark and so TWR reduces.

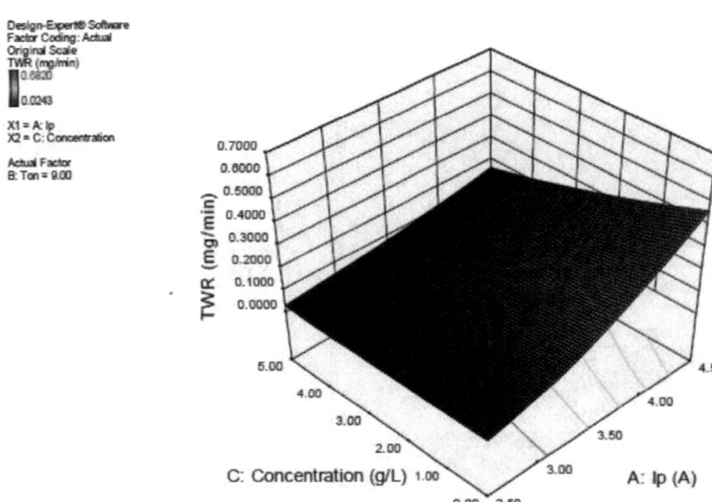

Fig.4: Interaction effect of current and SiC concentration on EWR

3.3. Surface Roughness (Ra).

Current of higher value results sparks of higher energy and deeper craters are formed on the work surface. As a result work surface becomes rough (Fig.5). It can be observed that current has a very strong effect on work surface roughness Ra. However the effect of Ton on Ra is not significant. With increase in Ton, work surface receive more heat and deeper crates are produced on the surface. But this effect is not prominent.

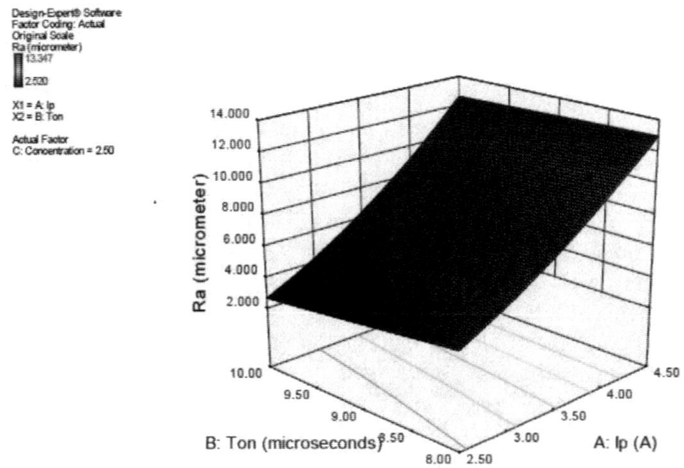

Fig.5: Interaction effect of current and Ton on Ra

Interaction effect of peak current and SiC concentration on Ra is illustrated in Fig. 6. It is obvious from the figure that Ra increases with increase in current. This effect is very significant. But as powder concentration is increased, it gives smoother surface. SiC carbide is electrically nonconductive. Its presence in dielectric reduces the effective current of the plasma channel. Therefore, craters on the work surface are not deeper, which results a smoother surface.

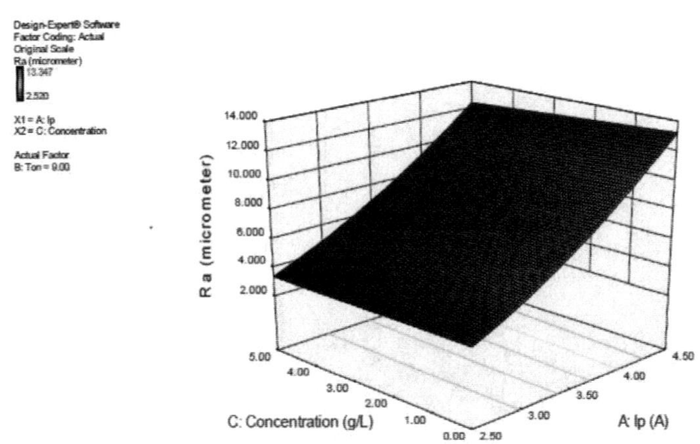

Fig.6: Interaction effect of current and SiC concentration on Ra

4. Conclusion

High carbon steel AISI 1050 was machined using EDM. SiC powder was added to the dielectric during EDM. Peak current, Ton and SiC powder concentrations were the variables for the study. It was found that MRR, Ra and TWR increase with increase in current. If Ton is increased, MRR, TWR are increased and at the same time work surface becomes rougher. Addition of SiC to dielectric fluid reduces MRR and TWR, but work surface becomes smoother.

Acknowledgments
Authors acknowledge the contribution and encouragement from colleagues and technical staff of the department.

References

[1] Pecas P and Henriques E 2009 Intrinsic innovations of die sinking electrical discharge machining technology: estimation of its impact. *The International Journal of Advanced Manufacturing Technology*, vol 44 n10 pp 880-889.
[2] Zhang Y, Liu Y, Yang S, Ji R Cai, B Li H and Wang F 2012 A review of the current understanding and technology of powder mixed electrical discharge machining (PMEDM) *International Conference on Mechatronics and Automation* pp 2240-2246.
[3] Bhattacharya A, Batish A and Singh G 2012 Optimal parameter settings for rough and finish machining of die steels in powder-mixed EDM. *Int J Adv Manuf Techno,* pp 537–548.
[4] MY Ali, N Atiqah and Erniyati 2011 Silicon carbide powder mixed micro electro discharge milling of titanium alloy. *International Journal of Mechanical and Materials Engineering (IJMME)* vol 6 n3 pp 338-342.
[5] Kumar S and Singh R 2010 Investigating surface properties of OHNS die steel after electrical discharge machining with manganese powder mixed in the dielectric *Int J Adv Manuf Technol* pp 625-633.

Improving machining performance using alumina powder-added dielectric during electrical discharge machining (EDM)

Ahsan Ali Khan[1*], Marina Binti Abu Bakar[2], Muataz Hazza Faizi Al-Hazza[3], Mohd Radzi Haji Che Daud[4] and AKM Mohiuddin[5]

1,2,3,4 Department of Manufacturing and Materials Engineering, International Islamic University Malaysia, Jalan Gombak, 53100 Kuala Lumpur, Malaysia

5 Department of Mechanical Engineering, International Islamic University Malaysia, Jalan Gombak, 53100 Kuala Lumpur, Malaysia

* aakhan@iium.edu.my

Abstract. Alumina powder was added to dielectric fluid during Electrical discharge machining (EDM) of AISI 304 steel. The main machining parameters of current (A), tool diameter (mm) and powder concentration (g/L) were chosen to determine the EDM machining characteristics of material removal rate (MRR), tool wear rate (TWR) and surface roughness (Ra). This work adopted two-level full factorial experiments. The significant factors that affect the MRR and tool TWR are current, tool diameter and powder concentration. The MRR increases when the current and tool diameter are higher. The presence of powder concentration also increases the MRR. The higher current used and the presence of powder concentration increase the TWR. Meanwhile, a smaller tool diameter gives higher value of TWR. Current and powder concentration are the significant factors to determine the surface roughness of work surfaces. Higher current applied gives high surface roughness, but higher powder concentration gives improved surface finish.

1. Introduction

Electrical discharge machining (EDM) is a unique technique of machining electrically conductive materials with using a series of electrical sparks between the electrode and workpiece. During electric sparks temperature rises from 8,000 to 20,000 °C sufficient to melt and vaporize any work material. One of the recent developments in EDM is powder mixed EDM (PMEDM). PMEDM can distinctly improve the surface finish and surface quality to obtain near mirror-like surfaces at relatively high machining rate. Ming and He [1] stated that when kerosene is mixed with some additive powder during EDM, the material removal rate (MRR) will increase. Addictive can increase the discharge gap, lower the discharge voltage and make the discharging current more even. It was concluded that PMEDM will affect the gap distance between electrodes. It was reported that higher conductivity of dielectric will easily collapse the insulating strength thus extends the gap distance. Pecas and Henriques [2] observed an improvement in the performance of conventional EDM when used with a powder-mixed dielectric. The additives with different thermo-physical properties, including aluminum (Al), chromium (Cr), copper (Cu), and silicon carbide (SiC) were studied. The particle size of additives in the dielectric oil influences the surface quality of EDM work. The smallest particles (70-80 nm) give better surface finish of machined work, the bigger the particle size leads to poor improvement of surface roughness. However, particle size has opposite impact on the recast layer. Kumar et al. [3] reported that when the aluminum particle suspended in the dielectric fluid, it makes uniform of the discharging energy dispersion; it displays multiple discharging effects within a single input pulse in the PMEDM. In the present study the work material, the electrode material and additive powder were AISI 304 steel, copper and alumina respectively.

2. Experimental Procedure.

The main composition of the workpiece material AISI 304 steel is: Cr: 18-20%; Ni: 8-10.5%; Mn: 2.0%; Si: 1.0%; C: 0.08%; P: 0.04% and S: 0.03%. Its physical properties include hardness 187 HB; density 7.9×10^3 kg/m³; melting point 1400-1455 °C; thermal conductivity 48 $Wm^{-1}K^{-1}$ and yield strength 207 MPa. The main characteristics of copper electrode were: density 8.96 $g \cdot cm^3$; melting point 1084.62 °C, boiling point 2562 °C and electrical resistivity 1.96 μΩ cm. Properties of alumina were: sizes 30-45 microns; density 3.95-4.1g/ cm^3, melting point 2072 °C and boiling point 2977 °C. Kerosene was used as dielectric fluid. Experiments were conducted on MITSUBISHI EX 22 die-sinking EDM machine. Work surface roughness was measured using Mitutoyo SURFTEST SV.500. Weight of the workpiece and the electrode before and after each machining was measured by an electronic balance. Experiments were conducted using two-level full factorial design.

3. Results and Discussions

In this study, two-level full factorial design was used to plan and analyze the experiments. The detailed plan of the experiments is shown in Table 1. Current, tool diameter and powder concentration were taken as the input parameters while MRR, TWR and Ra were the responses.

Table 1: Experimental plan for the investigation

Std	Run	Block	Factor 1 A:A current	Factor 2 B:tool diameter mm	Factor 3 C:powder con g/L	Response 1 MRR g/min	Response 2 TWR g/min	Response 3 Ra micrometer
6	1	Block 1	6.5	14	3			
2	2	Block 1	6.5	14	0			
3	3	Block 1	2.5	20	0			
1	4	Block 1	2.5	14	0			
5	5	Block 1	2.5	14	3			
4	6	Block 1	6.5	20	0			
7	7	Block 1	2.5	20	3			
8	8	Block 1	6.5	20	3			

3.1 Material Removal Rate (MRR).

Influence of current and electrode diameter on MRR are shown in Fig.1. The higher current applied will increase the value of MRR. This is due to the fact that sparks with more energy are generated during machining that would assist to remove more material from the workpiece. So, the relationship between current and MRR is directly proportional. This finding agrees with the research done by Abbas et al. [4]. When the tool diameter area is large, the heat loss to surrounding is less compared to small area. This results the MRR value to be higher when large tool diameter is used.

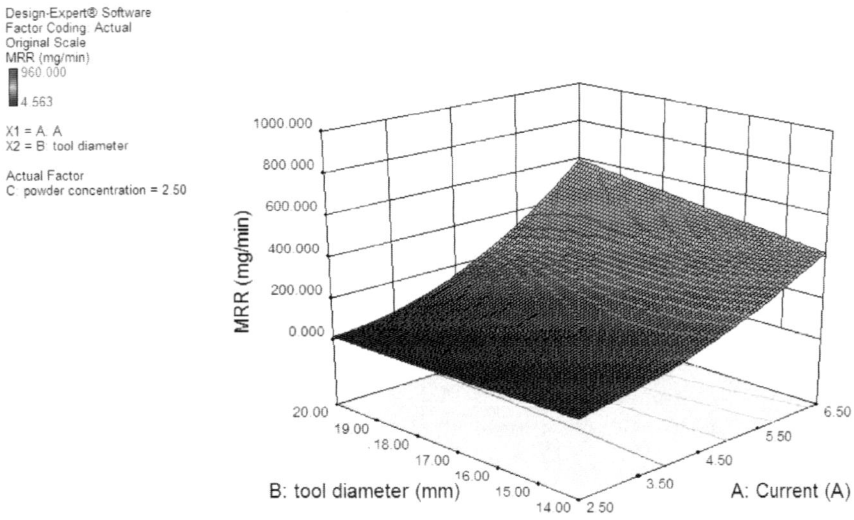

Fig.1: 3D surface plot for MRR as a function of the tool diameter and current

The relationship of current and powder concentration on MRR is presented in Fig.2. It is evident from the Fig.2 that increase in powder concentration facilitates MRR. Increase in powder concentration breaks down the main plasma channel into small sparks. As a result spark is distributed over the machining zone more evenly and thus increasing MRR.

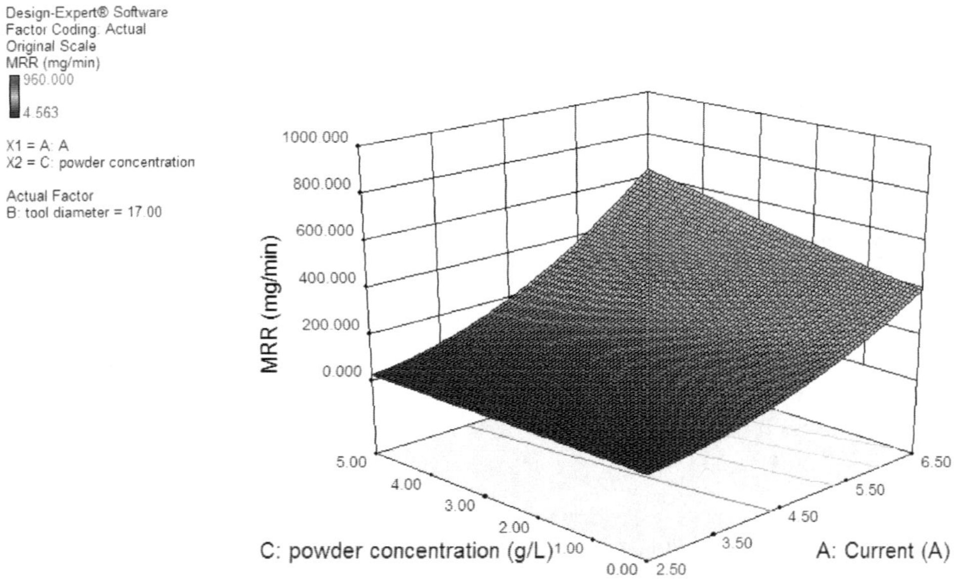

Fig.2: 3D surface plot for MRR as a function of the powder concentration and current

3.2. Tool Wear Rate (TWR).

The higher current applied will result in higher value of TWR. This was due to the fact that more heat was generated with a higher value of current that produced high intensity of sparks which in turn caused the tool material to be removed from the surface. The relationship between current and tool diameter on TWR is illustrated in Fig.3. Current and TWR are almost directly proportional. However, the relationship between tool diameter and TWR was inversely proportional at high level of current. The lower tool diameter, higher will be the higher value of TWR. When the diameter is smaller, energy concentration is more and more material is removed from the tool material.

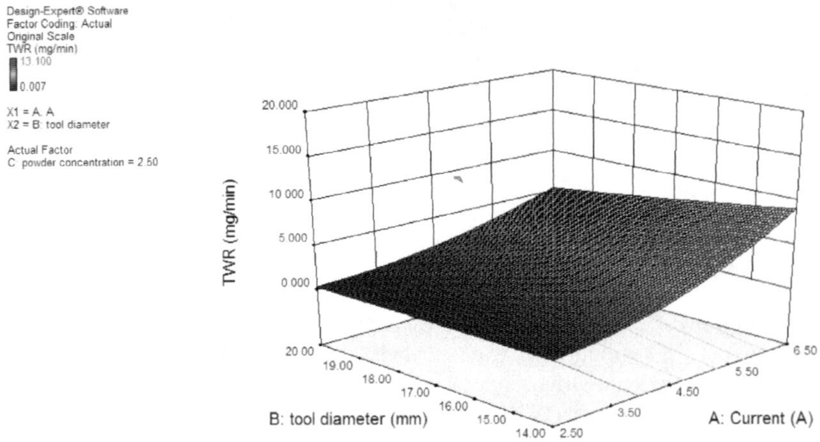

Fig.3: 3D surface plot for TWR as a function of tool diameter and current

However, with increase in powder concentration TWR reduces (Fig.4). Powder concentration increases the spark gap and improves flashing away of the machined debris from the machining zone and facilitates cooling of the electrode. As a result TWR decreases.

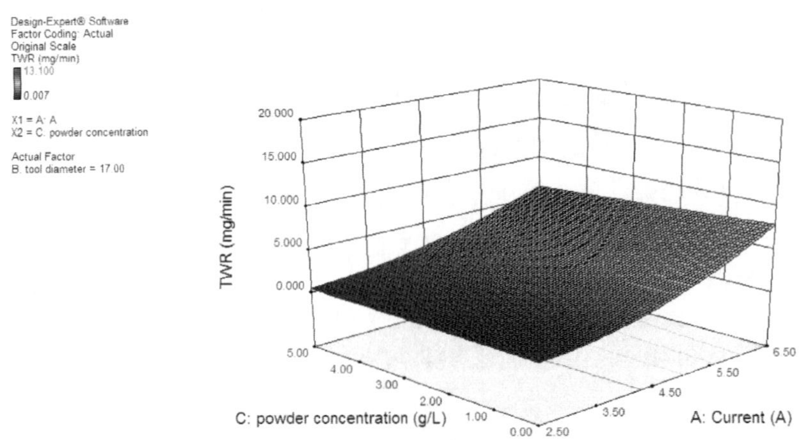

Fig.4: 3D surface plot for TWR as a function of powder concentration and current

3.3. Surface Roughness (Ra).

The higher current applied increases the value of surface roughness which means poorer surface condition. The intensity of sparks increases due to high current which produces deeper craters and makes the work surface rougher (Fig.5). Similarly, a higher diameter of the electrode shows poor surface finish. With increase in diameter of the tool, flashing of the dielectric becomes poorer and more debris is deposited on the surface and the surface becomes rougher. From Fig.6 it can be observed that higher concentration of powder improves work surface finish. The higher the concentration of powder, the more scattered will the spark discharge be and the smaller will the craters be on the surface.

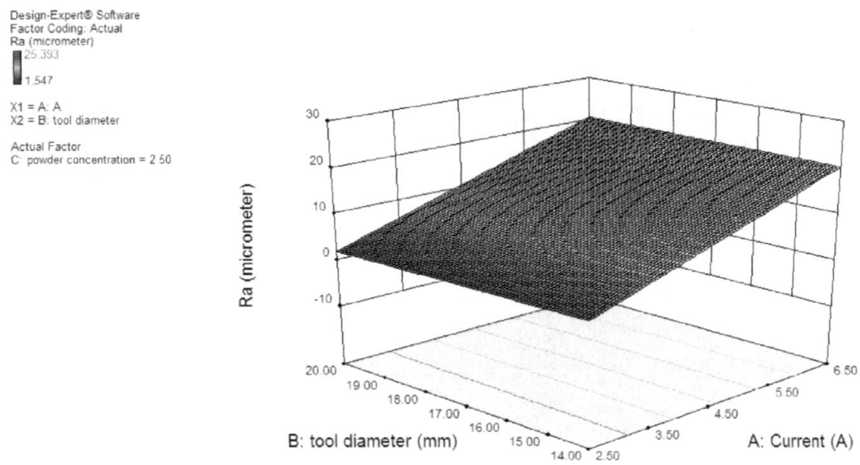

Fig.5: 3D surface plot for Ra as a function of tool diameter and current

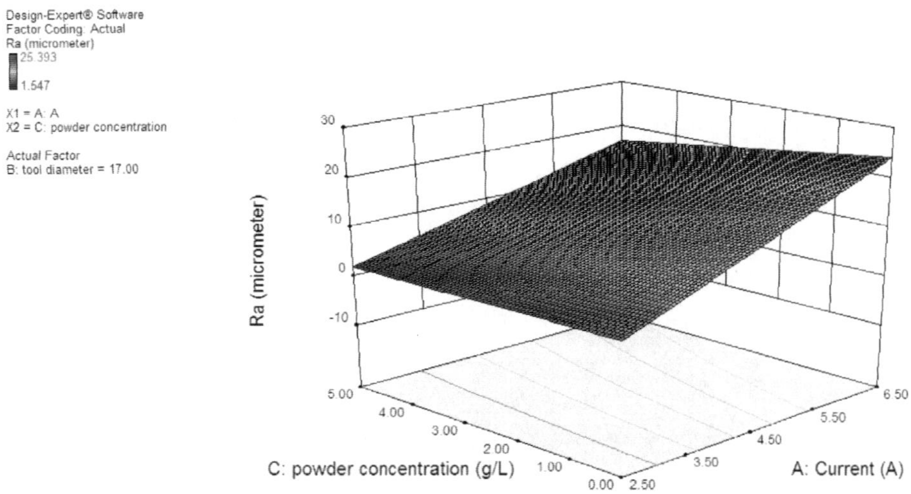

Fig.6: 3D surface plot for Ra as a function of powder concentration and current

4. Conclusion

EDM was done on AISI 304 steel with current, electrode diameter and alumina concentration in the dielectric as variables. It was found that MRR, Ra and TWR increase with increase in current. With increase in electrode diameter MRR and Ra increase, but TWR reduces. Alumina powder addition to dielectric shows significant improvement in EDM, since it causes increase in MRR, but reduces TWR and Ra.

Acknowledgments

Authors appreciate the help from colleagues and encouragement of technical staff of the Department of Manufacturing and Materials Engineering, International Islamic University Malaysia.

References

[1] Ming QY, He, LY 1995 Powder-suspension dielectric fluid for EDM *Journal of Materials Processing Technology* 52 pp 44-54.
[2] Pecas P and Henriques E 2003 Influence of silicon powder-mixed dielectric on conventional electrical discharge machining *International Journal of Machine Tools and Manufacturing vol* 43 pp1465-1471.
[3] Singh P, Kumar A and Beri N 2010 Some experimental investigation on aluminium powder mixed EDM on machining performance of hastelloy steel *International Journal of Advanced Engineering Technology* vol1 pp 28-45
[4] Abbas NM, Solomon DG and Bahari M F 2007 A review on current research trends in electrical discharge machining (EDM) *International Journal of Machine Tools & Manufacture* vol 47 1214-1228.

Mechanical Impedance Modeling of Human Arm: *A survey*

A Ahmad Puzi[1,a], S N Sidek[1,b] and F Sado[1,c]

[1] Department of Mechatronics Engineering, International Islamic University Malaysia, Kuala Lumpur, Malaysia

[a]asmarani8487@gmail.com
[b]snaim@iium.edu.my
cabdfsado1@gmail.com

Abstract. Human arm mechanical impedance plays a vital role in describing motion ability of the upper limb. One of the impedance parameters is stiffness which is defined as the ratio of an applied force to the measured deformation of the muscle. The arm mechanical impedance modeling is useful in order to develop a better controller for system that interacts with human as such an automated robot-assisted platform for automated rehabilitation training. The aim of the survey is to summarize the existing mechanical impedance models of human upper limb so to justify the need to have an improved version of the arm model in order to facilitate the development of better controller of such systems with ever increase in complexity. In particular, the paper will address the following issue: Human motor control and motor learning, constant and variable impedance models, methods for measuring mechanical impedance and mechanical impedance modeling techniques.

1. Introduction

Motor control is important for effective human movement. Human motor control is the process by which human use their brain and cognition ability to activate and coordinate the muscles and limbs involved in improving the performance of a dedicated motor skill [1]. Most activities that are performed with the hands involve interaction with the environment. This interaction imposes forces on the hand and can also destabilize motion. However, human have excellent capabilities to manipulate objects. This means that the central nervous system (CNS) is able to adapt to various task dynamics. For instance, one may have difficulty in opening a door for the first time due to unknown friction. However, after many trials the appropriate force to be exerted will be learned, and one will open the door without difficulty and even without thinking about it. This situation may be regarded as impedance control [2] which can been described as an effective strategy of the nervous system to deal with kinematic variability such as due to neuromuscular noise and environmental perturbations.

In early works, impedance was assumed to be constant which allows the use of linear models [3]. In constant impedance model, they ignored environmental perturbations whereas in active impedance the perturbations and other noise elements are included.

Lance [4] was among the earliest who defined spasticity as a motor disorder characterized by a velocity dependent increase in tonic stretch reflexes (muscle tone) with exaggerated tendon jerks, resulting from hyper excitability of the stretch reflex, as one component of the upper motor neuron syndrome. High value in spasticity's assessment reflects higher restriction in the functional range and speed of motion of the limb. Spasticity assessment has close relation to the mechanical impedance of the upper limb. Hogan et.al [1] reported that muscle co-activation invokes mechanical impedance of the limb and it is significant to perform tasks during specific therapy session. In addition, Burdet et.al [5] found that human learns how to stabilize the unstable movement by varying the mechanical impedance of the affected muscles

Generally, mechanical impedance consist of a set of mass, spring and damper. These elements are important to measure the ability of the muscle in performing specific task [6]. The parameters are also used to describe the dynamic motion of a person. Mechanical impedance which is defined as the ratio of the force applied at a point to the resulting velocity at that point is strongly related to muscle activation [7]. Mass, spring and damper elements in mechanical impedance have been represented as inertia, stiffness and damping of the limb respectively. Therefore, mechanical impedance modeling is useful as it has direct relation with the level of assessment made on the muscle quality. However there is less discussion on the variability in the parameter of the mechanical impedance. In this paper, a review on human motor control and motor learning, measuring of mechanical impedance and different mechanical impedance modeling techniques is reported.

2. Literature Review

The paper is composed of four sections where in section 2.1 motor control and learning function for human will be elaborated. The subsequent section explains on the different between constant and variable impedance. Then, the method of measuring mechanical impedance was discussed in section 2.3. Section 2.4 summarizes the findings from other works that use different modeling to represent mechanical impedance of human arm.

2.1. Human motor control and Motor Learning

In order to develop model of human limb mechanical impedance, the understanding of human motor control in motor learning process is required. Human have the ability to quickly adapt to the movement changes according to internal and external disturbances. Therefore, when the human arm movement is interrupted, the arm tends to automatically return to the desired trajectory [8]. From that, human learns to control the muscle. Central Nervous Systems (CNS) determines the right muscles to be activated with correct amount of force and timing in order to track the desired trajectory. Furthermore, movement needs feedforward mechanism to strategize the force for the task to be done successfully. Shadmehr et.al [9] explained how the feedforward mechanism took place during task performance. The results showed that the feedforward control managed to reduce the movement error-trial and compensate the disturbances.

2.2. Constant and Variable Impedance

Each parameter of mechanical impedance is important to measure the ability of the muscle in performing a specific task. The value of the parameters varies at every instant in the trajectory and also affected by the variation in environment. In early works, impedance parameter was assumed to be constant and linear.

Hondori et.al [10] described mechanical impedance in terms of spring, mass and damper. Definition of impedance has been refer to average stiffness, defined as ratio between generalized force (force or torque) to generalized displacement (linear or angular) for the system [1]. The system that we focus is upper limb as it is important for human daily activities such as grasp, flexion and extension. Mass, spring damper is represented of inertia, stiffness and viscosity respectively, whereby participate in order to perform the specific task smoothly and effectively.

In a view by Nikooyan et.al [11] muscle can be represented as a spring-damper lumped system, and it can be divided into two type of system models namely passive and active models In passive model, the spring and damper was treated as constant parameters with additional nonlinearities term. In early modeling, constant impedance introduced whereby elbow to wrist joints represented by two-DOF. The work explained the dynamic response to vibration input to the arm and acknowledge that the arm system may be used as a low-pass mechanical filter that attenuates high frequencies. While Holmlund revealed that human impedance varies with direction, gender, level and posture by experiment done in measuring mechanical impedance during different condition and factor. Kistemaker et.al [14] claimed that when using constant coefficient for stiffness and damping factor, it leads to large error in the estimated parameter. They suggested that parameter estimation should use the moving-time-window algorithm. The use of time-window method provides mean to determine the time-dependent changes in joint stiffness and damping in human upper limb based on the data of the movement. It is crucial to model the upper limb using variable impedance parameter for stiffness and damping factor so as they can be adaptable in the application that requires interaction between human and systems. The model falls under the second model type that is active model.

2.3. Measuring Mechanical Impedance

In more recent rehabilitation robotic systems mechanical impedance plays vital role in making the system more human like [15]. It poses a challenge to measure the mechanical impedance of the limb thus researchers came out with different measuring devices with different techniques to measure the parameters.

Hondori et.al [16] proposed an experiment to measure dynamic mechanical impedance. He proved that human mechanical impedance varied when the limb moved. However the relation between the change in mechanical impedance and the learning process was not established. The system allowed the measurement of real and imaginary parts of mechanical impedance parameter separately where the real part was identified as mechanical resistance, which was also independent of frequency and the imaginary part was identified as mechanical reactance which varied with frequency. In his paper, he reported that postures affects mechanical impedance during reaching motion. In general the force and velocity are needed to measure mechanical impedance. Hence, the measurement system were developed to be able to measure and record the given force and velocity [17].

In the manual ways of conducting upper limb rehabilitation training for stroke patient, the therapist measures the muscle stiffness subjectively.. The Modified Ashworth Scale (MAS) is the common assessment tool used by the therapist to measure the affected muscle quality or muscle tone of stroke patient. The scale is widely used to grade muscle spasticity. The grade is based on the relationship between the range of motion and resistance force during extension/flexion in the joint. According to Bohannon & Smith, there is high interrater reliability when using the MAS to assess the spasticity. Thus, introducing a quantitative measurement based on the MAS may help to overcome the lack of the traditional method. Uchiyama [18] had done the quantitative evaluation of spasticity in upper limb using mathematical model. He assumed that elastic coefficients were proportional to elbow angle and muscle activities. The elbow joint angle, torque, and electromyograms (EMGs) of the biceps brachii, triceps brachii, and brachioradialis muscles were measured. Furthermore, the limitation of this research only take account for elasticity response without depend on viscosity.

Characterization of average elasticity based on MAS assessment were established here. Likewise, Nima [19] conducted experiment using correlation method (Pearson's correlation test) to analyse

IOP Conf. Series: Materials Science and Engineering **184** (2017) 012041 doi:10.1088/1757-899X/184/1/012041

relationship for both elasticity and viscosity with the MAS. Meanwhile, Ayuni [20] developed spasticity mathematical model based on the MAS +1 scored. Hence it shows strong positive linear correlation coefficient. The elbow joint system was developed as a one degree of freedom system to represent part-task trainer. Table 1 summarized the measuring method of mechanical impedance and spasticity.

Table 1: Measuring method of mechanical impedance and spasticity

Author	**Measuring method**	**Remarks**
Noor Ayuni, 2015	Spasticity of grade 1+ on MAS based mathematical model	Muscle tone is proportional to the velocity of the motion.Thedata collected focus only for motion with grade of 1+ in MAS. There is no muscle model used to measure the mechanical impedance and it involve two dynamics equations to represent the following case; 1) Muscle tone catch during slow and fast motion stretching 2) Resistance through range of motion
Nima, 2011	Quantitatively evaluation of spasticity during passive flexion and passive extension	Analyzed relationship between viscous, elastic and MAS using correlation method (Pearson's correlation test) and torque and position was the output.
Hondori, 2011	Smart mug to measure geometrical impedance	Vibration is applied to make perturbation in any direction. The result is impedance highly depend on wrist's posture and hypothesis is bigger diameter gives most stable condition and vice versa. Input is centrifugal force and output is velocity
Hondori, 2010	On Line learning process	Mechanical impedance changes with expertise (learning function). Experiment done by using electrical motor (electrical impedance) as sensor instead of force sensor and transduction matrix method. Force (torque) is the input and velocity (angular velocity) is the output.
Uchiyama, 2005	Quantitatively evaluation of spasticity during passive flexion and passive extension	Characterize average elasticity based on MAS assessment based on muscle model without damper while include EMG signal. The inputs are elbow joint angle, torque and EMG (biceps brachii, triceps brachii and brachioradialis). Least square method(experimental) and Range-kutta method (estimation) used.

2.4. Mechanical Impedance Modeling

Biologically, muscle comes with two sections which are thick (myosin) filaments and thin (actin). This part shown in figure 1. Myosin filaments slide against actin which tend to shorten the activated muscle. Neural activation signals is receive when muscle is activated. That signal consist of several spikes. The amount of force produce is depends on the frequency and magnitude of spikes.

In addition, muscle tension is count on both length and velocity. Experiment did by Burdet [21] to measure stiffness (K) and damping (B) for a cat's muscle. As a result, when the length is equal to half of the initial length, muscle cannot generate force and same goes with the velocity. However, the force increases as length or velocity increases. Hence, the impedance of a single muscle changes with the force it generates.

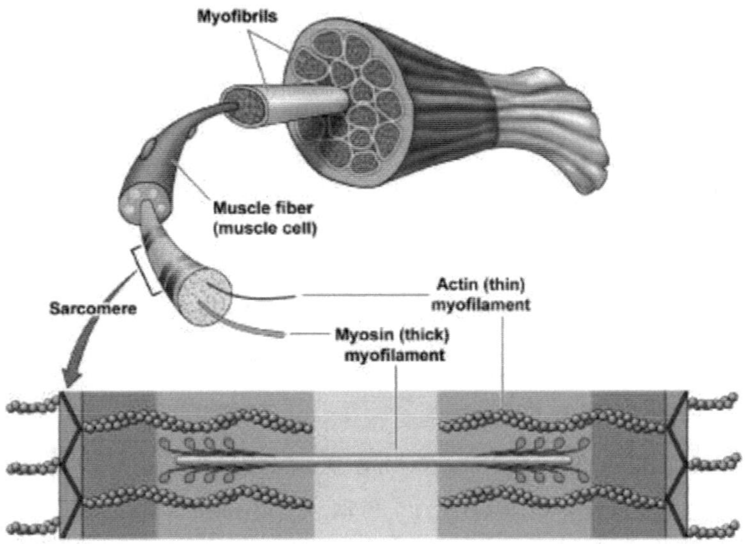

Figure 1: Myosin and actin filaments in a muscle; figure adopted from http://sites.duke.edu

From the aspect of biomechanics systems, previous study has been introduces two type muscle models which are Maxwell model and Voight model. As it can be seen from figure 2 Maxwell model consists of a spring in series with a damper while Voight model has the spring in parallel with a damper. From a prospective of the input it shows that force step input and displacement step input tests from Voight's model is more realistic than Maxwell model [22]. Even though the Voight's model is more realistic, the limitation of these two model is none of them are capable of modeling the active contractile property of a muscle. After that many researchers come out with new modeling based on Voight model in order to predict the mechanical impedance of human's upper limb.

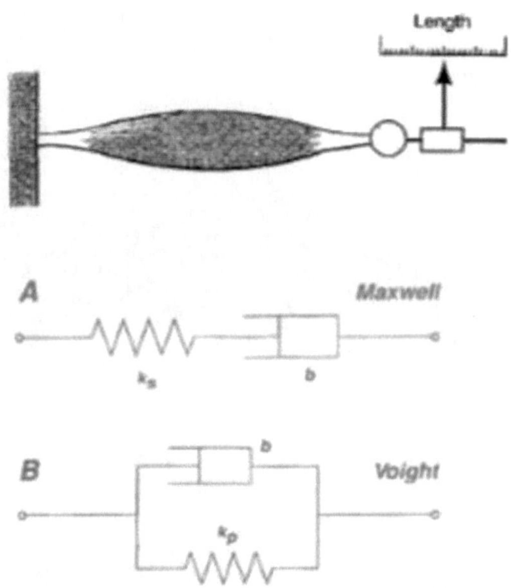

Figure 2: Maxwell and Voight models; figure adopted from (McMahon, 1984)

Mechanical impedance model was illustrated by one or more mechanical elements into one system as described its function. Each element represents the function of the real systems. Mechanical impedance modeling is an important stage in order to determine quantitative assessment of the system which in this case is human arm. In this section, different modalities will be elaborated. It can be represents in two ways which are structure model and mathematical model. Previous studies had greatly used the mass-spring-damper systems (MSD) in representing mechanical impedance of human arm [23]-[25]. Equation 1 corresponds to the mass-spring-damper model shown in Figure 3. This is second order dynamic equation where $m_e(t)$, $b_e(t)$, and $k_e(t)$ are the impedance parameters which denote the mass, damping factor, and stiffness of the arm, respectively; and $f_e(t)$ representing the force exerted to arm.

Researchers have been done a quite number of experiments on mechanical impedance of human arm. Several disparate mathematical models are proposed for the representation of human arm movement. Recent years, MSD model has been improved to investigate mechanical impedance during movement of the arm [23]. They have included muscle activation as dependent parameter [26]. However this model was developed assuming that the system has simple joints and does not consider complex muscle mechanics and geometry because when dealing with muscle it is not easy as the shape is very irregular . As reported by Speich et al. [24] and Rahman et al. [27] was developed a model with five parameters with additional spring and damper to better approximate the dynamics systems. Then, Shuaijie et al. [28] was studied the mechanical impedance during maintained posture and reaching movements in order to analyzed human impedance changes depending on the situation. The similar study by Tasnuava et al. [29] introduced Rayleigh's dissipation factor for the damping action. Lagrangian approach is applied to develop the mathematical model of human arm during movement.

David et al. [30] distinguish two models which are viscoelastic model and Hill's model. Viscoelastic model only consider spring and damper while Hill's model consist of three elements which are contractile element (CE), and constant value of spring. These models preferred use when it interacts with musculoskeletal systems and mass neglected. Table 2 summarizes different modeling of human arm mechanical impedance.

$$m_e(t)\ddot{x}(t) + b_e(t)\dot{x}(t) + k_e(t)x(t) = f_e(t) \quad (1)$$

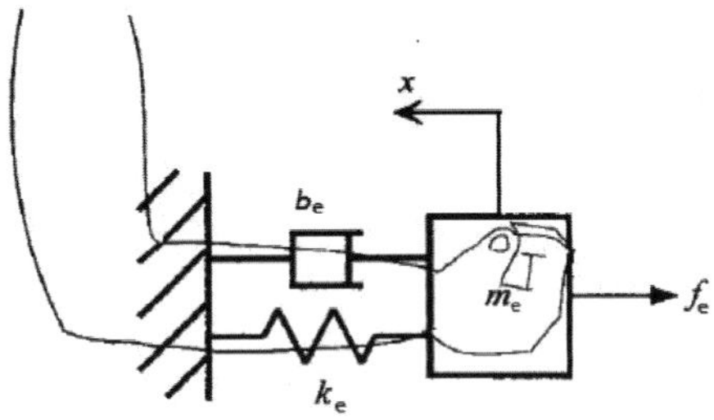

Figure 3: Mass-spring-damper model for human arm's mechanical impedance

Table 2: Different modeling of human arm's mechanical impedance

Author	Model of Human Arm	Remarks
David 2013	Based on human neuromuscular system impedance.	Using Hill muscle model and viscoelastic model acts as machine. Experimentally for subjects to differentiate between 'machine' and 'human'.
Choudry 2013	Lagrangian based model during dynamics planner motion	Lagrangian method reduces the mathematical complexity and consider active movement only.
Wang 2011	Considered that hand inertia, viscocity and stiffnes constant during disturbance because of low degree and short duration of the disturbance.	Noisy and unstable force uncontrolled.
Artemiadis 2010	End-point dynamic based on second order system in 3D space.	
Speich 2005	Five parameters model by additional spring and damper.	A comparison of the five parameter model with a two parameter spring-damper model suggests that in some cases the use of additional model may not significant improvement.

Rahman 2002	Spring and damper systems representation of biceps and triceps muscle. Mass refers to the arm. A second order equation was considered as the model for arm dynamics.	For calculations, ARX (Auto Regressive eXogenous) model presented and minimum jerks applied.

3. Conclusion

In this paper, different structure and mathematical models of mechanical impedance in modeling human arm is discussed. In particular, the impedance values have direct relation to human motor control and motor learning. It can also be represented as constant or dynamics parameters where the value varies upon changes in force and velocity as well as posture and the environment.

Acknowledgement
The work presented was carried out in the Biomechatronics Research Laboratory of International Islamic University Malaysia. The authors wish to gratefully acknowledge the grant funding from the Ministry of Science, Technology and Innovation (MOSTI) Malaysia (SF15-015-0065).

References

[1] Rosenbaum and David A 2009 *Human motor control* Academic press.
[2] Hogan and Neville 1984 Adaptive control of mechanical impedance by coactivation of antagonist muscles *IEEE Transactions on Automatic Control* **29** 681-690.
[3] Tsuji T, Morasso P G, Goto K and Ito K 1995 Human hand impedance characteristics during maintained posture *Biological cybernetics* **72** 475-485.
[4] Lance J W 1980 The control of muscle tone, reflexes, and movement Robert Wartenbeg Lecture *Neurology* **30** 1303-1303.
[5] Burdet E, Osu R, Franklin D W, Milner T E and Kawato M 2001 The central nervous system stabilizes unstable dynamics by learning optimal impedance *Nature* **414** 446-449.
[6] Hondori H M and Shih-Fu L 2010 Perturbation-based measurement of real and imaginary parts of human arm's mechanical impedance In *2010 Annual International Conference of the IEEE Engineering in Medicine and Biology* 5911-5914.
[7] Tsumugiwa T, Yokogawa R and Hara K 2002 Variable impedance control based on estimation of human arm stiffness for human-robot cooperative calligraphic task *Robotics and Automation Proceedings ICRA'02 IEEE International Conference* **1** 644-650
[8] Won J and Hogan N 1995 Stability properties of human reaching movements *Experimental Brain Research* **107** 125-136
[9] Shadmehr R and Mussa-Ivaldi F A 1994 Adaptive representation of dynamics during learning of a motor task *The Journal of Neuroscience* **14** 3208-3224.
[10] Hondori H M and Shih-Fu L 2009 A method for measuring human arm's mechanical impedance for assessment of motor rehabilitation *Proceedings of the 3rd International Convention on Rehabilitation Engineering & Assistive Technology* ACM 17.
[11] Nikooyan A A and Zadpoor A A 2011 Mass–spring–damper modelling of the human body to study running and hopping-an overview *Proceedings of the Institution of Mechanical Engineers, Part H: Journal of Engineering in Medicine* 0954411911424210.
[12] Qassem W 1996 Model prediction of vibration effects on human subject seated on various cushions *Medical engineering & physics* **18** 350-358.
[13] Holmlund P and Lundström R 1998 Mechanical impedance of the human body in the horizontal direction *Journal of sound and vibration* **215** 801-812.

[14] Kistemaker D A, Van Soest A J K and Bobbert M F 2007 A model of open-loop control of equilibrium position and stiffness of the human elbow joint *Biological cybernetics* **96** 341-350.

[15] Vanderborght B, Albu-Schäffer A, Bicchi A, Burdet E, Caldwell D G, Carloni R, ... and Garabini, M 2013 Variable impedance actuators: A review Robotics and autonomous systems **61** 1601-1614.

[16] Hondori H M and Shih-Fu L 2010 A Novel Device for Measuring Mechanical Impedance during Dynamic Tasks *BIODEVICES 2010* 64-68.

[17] Palazzolo J J, Ferraro M, Krebs H I, Lynch D, Volpe B T and Hogan N 2007 Stochastic estimation of arm mechanical impedance during robotic stroke rehabilitation *Neural Systems and Rehabilitation Engineering, IEEE Transactions* **15** 94-103.

[18] Uchiyama T, Kato R, Obata S and Uchida R 2006 Quantitative Evaluation of Spasticity in Upper Limbs in Hemiplegic Subject Using a Mathmatical Model In *2005 IEEE Engineering in Medicine and Biology 27th Annual Conference* IEEE 6612-6614.

[19] Ardabili N S, Abdollahi I, Khorramymehr S, Shirzad H and Bahadorany H 2011 Quantitative evaluation of spasticity at the elbow of stroke patients *Biomedical Engineering (ICBME), 2011 18th Iranian Conference* IEEE 131-136.

[20] Zakaria N A C, Komeda T, Low C Y, Hanapiah F A and Inoue K 2015 Spasticity mathematical modelling in compliance with modified ashworth scale and modified tardieu scales *Control, Automation and Systems (ICCAS), 2015 15th International Conference* IEEE1893-1897.

[21] Burdet E, Franklin D W, Osu R, Tee K P, Kawato M and Milner T E 2004 How are internal models of unstable tasks formed? *Engineering in Medicine and Biology Society, 2004. IEMBS'04. 26th Annual International Conference of the IEEE* **2** 4491-4494.

[22] McMahon T A 1984 *Muscles, reflexes, and locomotion*. Princeton University Pre

[24] Speich J E, Shao L and Goldfarb M 2005 Modeling the human hand as it interacts with a telemanipulation system *Mechatronics* **15** 1127-1142.

[23] Choudhury T T, Rahman M M, Khorshidtalab A and Khan M R 2013 Modeling of Human Arm Movement: A Study on Daily Movement *Computational Intelligence, Modelling and Simulation (CIMSim), 2013 Fifth International Conference* IEEE 63-68.

[25] Artemiadis P K, Katsiaris P T, Liarokapis M V and Kyriakopoulos K J 2010 Human arm impedance: Characterization and modeling in 3D space *Intelligent Robots and Systems (IROS), 2010 IEEE/RSJ International Conference on* IEEE 3103-3108.

[26] Tee K P, Burdet E, Chew C M and Milner T E 2004 A model of force and impedance in human arm movements *Biological cybernetics* **90** 368-375.

[27] Rahman M, Ikeura R and Mitzutani K 2002 Impedance characteristics of human arm for cooperative robot *International Conference on Control, Automation and Systems* 1455-1460.

[28] Wang S, Zuo G, Xu J and Zheng H 2011 Human hand impedance characteristics during reaching movements *Biomedical Engineering and Informatics (BMEI), 2011 4th International Conference* IEEE **3** 1282-1286.

[29] Lin D C, Godbout D and Vasavada A N 2013 Assessing the Perception of Human-Like Mechanical Impedance for Robotic Systems.*Human-Machine Systems, IEEE Transactions* **43** 479-486.

[30] Palladino J L 2014 Modeling muscle's nonlinear viscoelastic dynamics *Engineering in Medicine and Biology Society (EMBC), 2014 36th Annual International Conference of the IEEE* 828-831.

Human Location Detection System Using Micro-Electromechanical Sensor for Intelligent Fan

S. Parnin[1] and M. M. Rahman[2]
Department of Mechatronics Engineering, Kulliyah of Engineering, International Islamic University Malaysia, 53100 Kuala Lumpur, Malaysia.

[1]sullehaparnin@gmail.com, [2]mozasser@iium.edu.my

Abstract. This paper presented the development of sensory system for detection of both the presence and the location of human in a room spaces using MEMS Thermal sensor. The system is able to detect the surface temperature of occupants by a non-contact detection at the maximum of 6 meters far. It can be integrated to any swing type of electrical appliances such as standing fan or a similar devices. Differentiating human from other moving and or static object by heat variable is nearly impossible since human, animals and electrical appliances produce heat. The uncontrollable heat properties which can change and transfer will add to the detection issue. Integrating the low cost MEMS based thermal sensor can solve the first of human sensing problem by its ability to detect human in stationary. Further discrimination and analysis must therefore be made to the measured temperature data to distinguish human from other objects. In this project, the fan is properly designed and program in such a way that it can adapt to different events starting from the human sensing stage to its dynamic and mechanical moving parts. Up to this stage initial testing to the Omron D6T microelectromechanical thermal sensor is currently under several experimental stages. Experimental result of the sensor tested on stationary and motion state of human are behaviorally differentiable and successfully locate the human position by detecting the maximum temperature of each sensor reading.

1. Introduction

Human life is getting more comfortable and less burden with many intelligent inventions by scientist and engineers. Since then, the intervention of human to the appliances becoming less as the automated system is rising. The key to the automatic devices is the use of a good sensor, transducer, actuator and excellent control system. Ability in making the decision with respect to the event that is driven by the changing environment is what describes the intelligent of the devices. Creating an environment that is able to respond to the human that inhabits it is becoming the central of this research. Smart devices is the device that are not ignorant about their environment, surrounding, situation, task and location where it is located [8]. Smart or intelligent device therefore start from the selection of the sensors.

Aside from home automation, this project build the standing fan especially for old citizen and disable people. In a situation that they are unable to monitor the fan at all time, the fan is expected to understand their needs in an intelligent way by estimating the extreme angle swing. The power consumed for one fan is depends on its operating time and its actuator ratings. This means that to reduce power consumed, the fan must only operated when it is in need where human presence in the operating area. Therefore the need to sense people will become ever more pressing as the accurate information from the sensor is required. Energy wastage, inefficient angle of fan, non-autonomous is the effect that must be faced by the user of standing fan. This is because the operating principle of normal standing fan is the discrete system that works based on powered on the device. Therefore there is no other way to save the energy consumed unless the user wisely used the fan when in needed and turn it off while finished. The underlying reason to those mentioned problem are because of non-automated system, no controller used. Only involving current supply to turn on and rotates the devices, also, there is no sensor for interacting with the environment condition. Standing fan which retain all of the mentioned properties make it fail to differentiate and analyze the environment it served.

All of those factors will cause the fan fail to differentiate between human and static things. Human motion cannot be expected and can be said dynamic. Things such as furniture and appliances can be group as static things. In a place such as indoor room, human and things can be differentiated by using suitable sensors even if human is in static. Together with the suitable sensor, approximate distance is possible to calculate. And more, fan also unable to estimate the range of human presence. Human presence can be further described by its position with respect to the sensing devices. The range of human to the sensing devices may be in terms of distance, angle, left or right side, height and etc.

The ultimate goal is for the sensor system able to differentiate the presence of human and non-human, either is in static or a dynamic. Therefore, the selection of the suitable sensors for human sensing such as the presence, and location was focused to the MEMS-IR and distance measuring sensors. A compatible and yet, rational human sensing method must be developed to ensure a stable standing fan performance. As such it is expected that to analyze the temperature data from the MEMS sensor, and the decision of approximate fan extreme swing angle. At the end of this project, we will compare the fan performance using two different sensor system i.e Ultrasonic (US) and Thermal Sensor. However, this paper will only cover the sensory system by using thermal sensor.

2. Comparable study

Related researches were selected to be analyzed further in terms of their objectives, technique, and application to determine the research gaps. Research gap increases the chances of one research to be continued and developed more. It differentiates one research to another even it is in the same field and can be derived from one of the elements in the tabulated data or it can also be derived from any of the elements of interest. There exist a clear gap among the research in terms of research methodology and application.

In this research, the important parameters are the distance of human from the sensor and human body temperature. The performance of the standing fan will be compared between the US and MEMS. At the presence, US is used widely in home automation and rehabilitation such as indoor lightning system, mobile rescue robot, and robotic lamp, service type robot, security system, , animation, walking support system for blind people, shoes obstacle detection system, home security system and patient body movement for radiotherapy emission. Variety of techniques used for human presence and detection such as Kalman Filter, Neural Network, learning machine and any other preprocessing methods have been used by the past researchers.

This research is the continuation of [1], though it has the same application, further the technique of localization detection and fan actuating system is different. Pyroelectric or motion sensor have been implemented in [1] which only can detect the temperature variation which cause a signal loses when human stays longer or in 'static' at the same spot. Therefore the standing fan suffer one major drawback which in turn cause the inefficient on and off during fan operation as it fail to sense the present of occupants in a room. Therefore, MEMS IR thermal sensor is used to replace pyroelectric motion sensor which solve the mentioned drawback.

Other MEMS types of sensor similar to IR thermal sensor is MEMS accelerometer which can be applied in rehabilitation for visually challenge [6], home automation [2], human assistance [3, 4], and machine navigation [5]. Meanwhile the application for human sensing using US are also found in detection of human motion [9, 11], presence and localization [9, 10], rehabilitation [14], tracking and identification for smart home [12]. Those are some of the related research that have common implementation of US and MEMS.

3. Heat based human sensing method

The most common human sensing method is image processing using the camera. Though there are other method by sensors such as radar, motion, pressure, and etc. The detection accuracy of human at the early stage would determine the other stages of sensing detection. The stages of human sensing can be grouped in the Spatio Temporal Properties (STPs) which consists the sensing of human presence, location, counting, tracking, and identity [13]. Standing fan is a type of a swing electrical appliances. Therefore, Identity sensing is not necessary. Meanwhile, counting and tracking may also be added afterwards. Figure 1 shows the stages of the human sensing covered in the research scope.

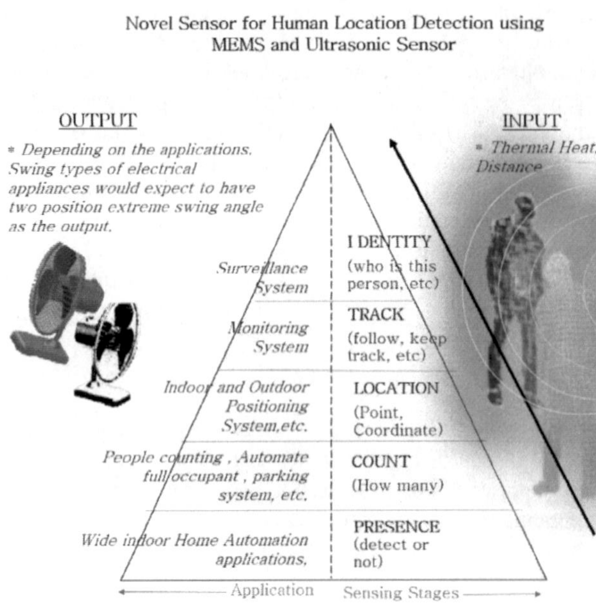

Figure 1. Human-sensing from presence detection to the identifying identity. Only two sensing scope will be needed in this project which are presence and location

To this date, research for applying MEMS based sensor for human sensing is not as much as the other sensor such as camera and radar. Some of the possible application of this human detection heat-based sensory system is home automation, security, and other indoor types of application. The methods of human sensing in this research rather use heat as input variable. For an indoor workspaces, the possible movable occupants would be human and pets. While, other things such as appliances remain stationary. However, those three (human, pets, and electrical appliances) share one common property which is heat. The challenge in this research is to eliminate the temperature data from those that is categorized as non-interest subject (i.e pets and appliances).

Table 1 shows the temperature range of human and pets at the skin surface. Human temperature is slightly higher than animals. Temperature range of cat and dog is overlap at the value of 38.3 to 39.1 °C. Even if they did have the range of heat value, other factors such as type and thickness of the clothes would affect the readings of the thermal sensor. The challenges in this research is heat is changing or transferring. Body heat reflect value change depends on the clothes. Clothes such different thickness, color, and fabric will reflect different temperature. Body heat also changes in an

Table 1. Range of temperature for human and pets

Object	Temp range (^{o}C)
Human- Armpit	36.1 to 37.2
dog	38.3 to 39.2
Cats	37.7 to 39.1

indoor and can be loosed to indoor surrounding. As over the time a person temperature in an indoor will be approximately same as the surrounding. Body heat fluctuates about 0.5^{o}C everyday depends on time and activities. Also, body heat is varying at different part of the body. Therefore, a proper height of the sensor location should be arranged to at least target the sensor to the upper part of the human body.

4. MEMS-IR thermal sensor

The D6T series sensors are made up of a cap with silicon lens, MEMS thermopile sensor chips, and dedicated analog circuit and a logic circuit for converting to a digital temperature value on a single board through one connector. This project uses Omron D6T-8L-06 series. It works by I2C communication protocol. It is as small as (14x18) mm size and a very sensitive infrared temperature sensor. Instead of measuring temperature on a single point, it can detect it in an array of pixels. The view angle of sensor are 62.8^o and 6° for X and Y direction respectively. We could estimate the (Field of View) FOV for the different distance of detection as in table 2. It is the theoretical value for the width and height of each pixel. The width and height of FOV is increasing linearly as the increase of the detection distance, X (m).

Table 2. Theoretical FOV ranging.

Detection Distance, X(m)	Width of FOV, Wpn(m)	Width of FOV per Pixel, W(m)	Height of FOV, H(m)	Detection power
Setting by default	*Wpn = 2X * Sin (62.8/2)*	*wpn = Wpn / n , where n=8*	*H= 2X*Sin (6/2)*	*Decreases over distance X*
1	1.04202	0.13025	0.10467	
2	2.08404	0.26051	0.20934	
3	3.12606	0.39076	0.31402	
4	4.16808	0.52101	0.41869	
5	5.21010	0.65126	0.52336	

[n = number of pixel, X = detection distance, Wpn = total width of FOV, wpn = width of FOV per pixel, H = height of FOV]

Thermal sensor used can detect the non-contact temperature up to 5 meters and nevertheless the detected temperature falls with distance. Given the same amount of power supply, the intensity should be decrease as the object move further from the FOV of thermal sensor. Even though the FOV area is getting bigger as it is further than the sensor, the data collected can be unstable and noisy. Detection temperature gets higher in response to the closeness of object because the sensor detects an average temperature over the targeted area. Temperature of the object detected would depends on the pixel background and objective temperature. Depending on the area distance, the temperature value [7] can be described as in equation 1.

$$T_X = \frac{T_B + T_O}{A} \tag{1}$$

Where,

$$T_X = Detected\ temperature$$
$$T_B = Background\ temperature$$
$$T_O = Objective\ temperature$$
$$A = Pixel\ area$$

Detected temperature is therefore the summation of background and objective temperature over the pixel area. Figure 2 shows the set up for the experiment and the enlarged view of the thermal sensor. The FOV of thermal sensor is illustrated by the eight pixel (i.e. Pixel 0 to Pixel 7). The total width and height of the FOV is denoted by W and H respectively.

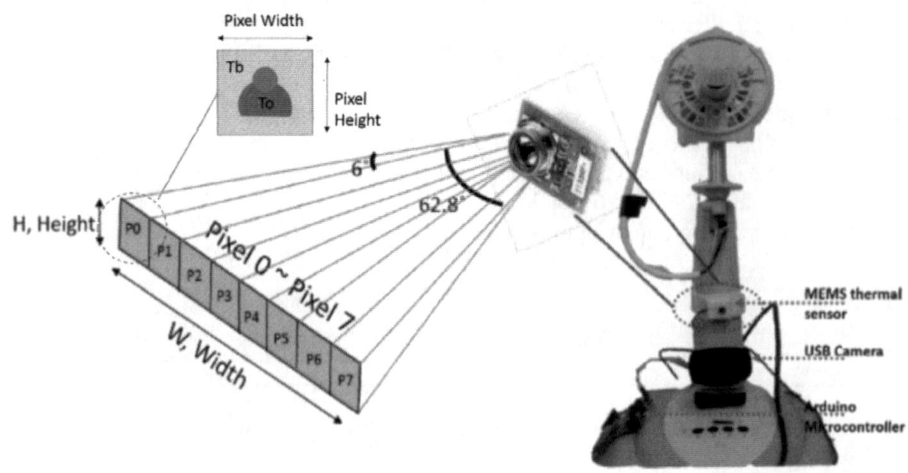

Figure 2. Standing fan with MEMS IR thermal sensor and enlarged view of the sensor FOV

5. Behavior of the measured data

In the non-contact temperature module, only the surface temperature measurement data can be acquired. Any objects of the same temperature will read the same. If so, it is impossible to differentiate between human, animals and electrical appliances as those three are the common occupants in the indoor room. One solution to differentiate it is by analyzing the data measured by the thermal sensor in terms of pattern and behavior using any available software. Formula or set of conditional ranging may be developed and improved the accuracy of the human detection.

Experiment for collecting the data was done using Arduino, an open source software microcontroller based and a Processing as a visual development tool. USB camera is used and aligned with the MEMS thermal sensor to record the experimental data and time frame images, though it is only be used for the purpose of describing the FOV of the sensor during experiment. A graphical user interface program using the Processing development language is developed for displaying the current image and current temperature from the thermal sensor. This preliminary experiment is done in two modes, mainly 'Stationary' and 'Dynamic motion'.

5.1. *Human in stationary*

Here, we define 'stationary' as in the application as such as a person sitting in front of the fan FOV and did not move the main joint such it cause the body to move from one point to another. A hand and head movement may be acceptable but not the main body. Two sets of experiment was conducted concluded that MEMS thermal sensor is able to read twice of temperature data in one second. Table 3 summarized the number of temperature array read from MEMS thermal sensor.

Table 3. MEMS thermal sensor reading per second

Experiment	Time Tested (Seconds)	Number of Data Array read	Data Temperature reading per second
1	370	792	2.14
2	439	1067	2.43

A four seconds temperature and FOV images with a distance of 1.5 meters away from sensor are recorded as in figure 3.a. The occupants sit at the chair allowing the sensor to target and sense the upper part of the body. Main body of the subject remain stationary while only the small movement of head, neck, shoulder, arm, wrist and hands are made. Experiment was conducted in an indoor room of approximately 26°C (i.e. T_B=26^0C). While the temperature on the sensor surface is recorded between 26.4 to 26.5 °C.

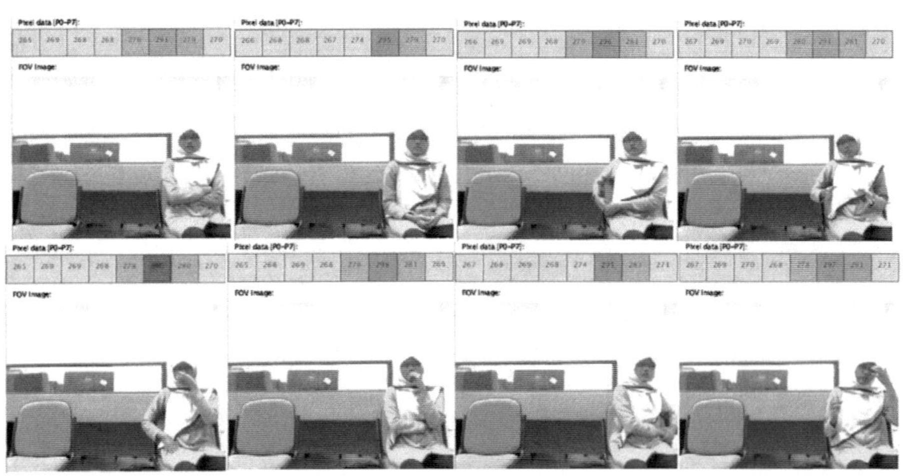

(a)

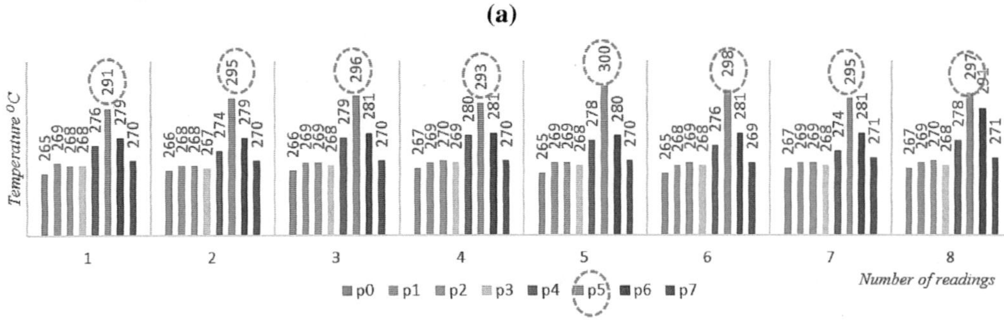

(b)

Figure 3 (a). The 4 seconds image of the occupant position in stationary (b). The 4 seconds temperature data at Pixel 0 to Pixel 7 for occupant in stationary

The corresponding temperature reading for figure 3(a) is shown in figure 3(b). All the temperature in figure 3(b) is the detected temperature, T_X. As long as the main body did not make the obvious move, the behavior of each pixel remain in the same pattern for all 8 readings. As marked in the red circles, the region in which the main body located is at pixel 5 as it always give the highest temperature value. Pixel 5 temperature value ranging from 29.1 to 30.0°C to give the changes in just 0.9°C. While the other Pixel's data fluctuates between 5°C. The pattern remains similar until the end of both experiments stated in table 3.

5.2. *Human in motion*

Motion in this research context is the continuous movement that result in the changing of the current position. Human could be in dynamic state if he is walking, running, and moving the main body to another pixel of FOV. A person walk 1.5m away across the sensor with normal speed are recorded as in figure 4. Result shows that the peak or maximum temperature changes over pixel, from P0-P7 or from P7-P0 while the temperature of other pixels fluctuates within only 5°C.

The shifting of the maximum temperature value (i.e red circles) followed the direction of the occupant's motion. Therefore, this result will not only give the current location but can be used to check the occupant's path. This will only works when the occupant is walking by maintaining the speed to be constant. In a condition that the occupant is running across the sensor FOV faster than the sensor reading per second as mentioned in table 3, the sensor will not be able to record the similar pattern of maximum pixel's shift shown in figure 4. Different occupant's behavior will result in different pattern of the measured data.

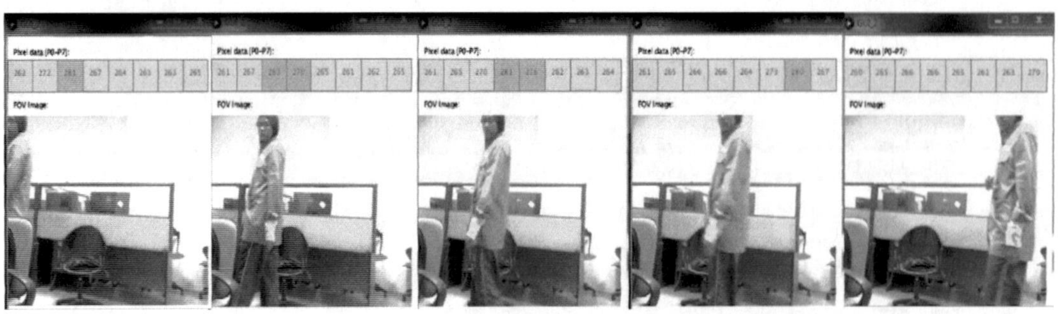

(a)

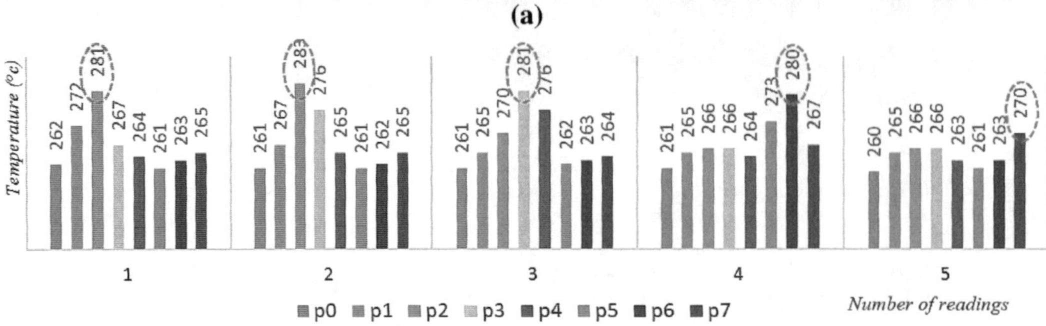

(b)

Figure 4. (a). The images of the moving occupant. (b). The corresponding temperature in duration of 2.5 seconds.

6. Conclusion and future works

By looking further ahead to the future, there would be many promising application to the sensory system for human location detection. As the preliminary experiment resulted in the comparable behavior for both static and dynamic, further analyze should be done to find the suitable swing angle of the fan by using either mathematical computational or preprocessing. It is clearly indicated from the result that the location of occupants can be determine by the highest and rare temperature value. Swinging angle of the fan head is supposed to swing at the suitable range of with respect to the human location.

An additional MEMS thermal sensor will be added in parallel and connected to the (Serial Data Line) SDL and (Serial Clock Line) SCL pins of the microcontroller via PCA9545 I^2C-bus switch. Additional sensors will widen the sensor FOV's width depending on the sensor-to-sensor placement. Finally, the design of the gear system of the fan blade motor and fan swing angle motor should be come afterwards.

Acknowledgement

We acknowledge the financial support of the Research Management Centre, IIUM, under Research Endowment Grant (EDW B14-161-1046).

References

[1] M. M. Rahman and F. Radzi, 2015 Sensory System to locate Human for Smart Fan. *1st International Workshop on Mechatronics Education*

[2] D. V. Babu, D. K. Yadav, N. K. Yadav, and V. K. Verma 2015 MEMS Based Smart & Secure Home Automation System with Multi-Way Control & Monitoring Facility using Smart Phone *J. Chem. Pharm. Sci.* **9** 357–359

[3] S. Kaur and H. C. Vashist 2013 Automation of Wheel Chair Using MEMS **3** 227–232

[4] S. V Patel and M. E. Ashalatha 2014 Human Fall Detection Using MEMS Accelerometer," *Int. J. Emerg. Technol. Adv. Eng.***4** 370–373

[5] N. Pranathi and S. M. Ahmed, 2013 Tri-Axis Motion Detection using MEMS for Unwired Mouse Navigation System in the Future Generation Machines *Int. J. Adv. Res. Comput. Commun. Eng.* **2** 3672–3675

[6] T. Ayesha. S. Smitha, and M, Rumana, 2015 Hand Gesture Based Home Automation for Visually Challenged *Int. J. Innov. Eng. Research Technol.* **2** 1–7

[7] OMRON Corporation, 2013 Infrared MEMS Thermal Sensor D6T Product Series : Infrared MEMS Thermal Sensor

[8] A. Schmidt and K. Van Laerhoven, 2001 How to build smart appliances? *IEEE Pers. Commun.*, **8** 66–71

[9] D. Espès, A. M. Pistea, C. Canaff, I. Iordache, P. Le Parc, and E. Radoi, 2013 New Method for Localization and Human Being Detection using UWB Technology: Helpful Solution for Rescue Robots

[10] D. Caicedo and A. Pandharipande 2012 Ultrasonic arrays for localized presence sensing *IEEE Sens. J.* **12** 849–858

[11] J. M. Sabatier and A. E. Ekimov 2006 Ultrasonic Methods for Human Motion Detection in Battlefield Acoustic Sensing for ISR Application **9** 9–12.

[12] M. Fitzpatrick and N. Matthiopoulos 2013 Real Time Person Tracking and Identification using the Kinect Sensor

[13] T. Teixeira, G. Dublon, and A. Savvides 2010 A Survey of Human-Sensing: Methods for Detecting Presence, Count, Location, Track, and Identity *ACM Comput. Surv.* **5** 1–35

[14] S. Senthilkumar and R. Vinothraj 2012 Design and study of ultrasound-based automatic patient movement monitoring device for quantifying the infraction motion during teletheraphy treatment. *Journal of Applied Clinical Medical Physics* **13** 6

Design and development of hybrid energy generator (photovoltaics) with solar tracker

Mohiuddin A K M[1,a], Mohamad Syabil Bin Sabarudin[1], Ahsan Ali Khan[2], Sany Izan Ihsan[1]

[1]Department of Mechanical Engineering, [2]Department of Manufacturing & Materials Engineering,
Kulliyyah of Engineering, International Islamic University, Malaysia
P.O.Box 10, 50728 Kuala Lumpur, Malaysia

E-mail: [a]mohiuddin@iium.edu.my

Abstract. This paper is the outcome of a small scale hybrid energy generator (hydro and photovoltaic) project. It contains the photovoltaics part of the project. The demand of energy resources is increasing day by day. That is why people nowadays tend to move on and changes their energy usage from using fossil fuels to a cleaner and green energy like hydro energy, solar energy etc. Nevertheless, energy is hard to come by for people who live in remote areas and also campsites in the remote areas which need continuous energy sources to power the facilities. Thus, the purpose of this project is to design and develop a small scale hybrid energy generator to help people that are in need of power. This main objective of this project is to develop and analyze the effectiveness of solar trackers in order to increase the electricity generation from solar energy. Software like Solidworks and Arduino is used to sketch and construct the design and also to program the microcontroller respectively. Experimental results show the effectiveness of the designed solar tracker sytem.

1. Introduction

Global trends in energy supply and consumption are unsustainable. The major energy carriers, the fossil fuels, besides being depleted, it also causes severe damage to environment and health. Thus, finding a better energy resource nowadays is a very challenging task for mankind. The demand of energy resources is increasing rapidly since the industrial revolution. Fossil fuel energy resources like petroleum are finite and will be depleted sooner or later. The renewable energy like wave energy, wind energy, and solar energy have become the main energy resources to replace the fossil fuels. These renewable energies have been extensively explored and developed in order to fulfill the human needs.

Among the renewable energy sources in this world, solar energy is the most abundant and most reliable compared to others. This is because our Earth receives a very large amount of sunshine throughout the year. Solar energy system has widely been used in the tropical weather countries especially in the South Asian countries.

Taking into account, the difficulties of supplying electricity to the people living in the remote areas, this project aims to design and develop a small scale hybrid energy generator that basically made up of two environmental friendly energy sources which are solar energy and hydro energy. This paper will cover the solar energy component of the project.

The main objective of this paper is to show the effectiveness of the developed solar photovoltaic component for a small scale hybrid energy generator (hydro and photovoltaic) including the solar tracker system. For a photovoltaic hydro hybrid system, a low operating cost photovoltaic array and a hydro plant with a generator were used. The array usually provides 70 percent to 90 percent of the annual energy and the hydro plant with the generator provides the remainder.

Solar energy is a type of non-conventional energy. It is the energy emitted by the sun. The energy emitted by the sun is generated by nuclear fusion in the sun's core that converts approximately 650 000 000 tons of hydrogen to helium every second. The products of the fusion are mainly heat and electromagnetic radiation. The heat stays in the sun as it is essential in preserving the thermonuclear reaction. Whereas, the electromagnetic radiation including the visible light, ultra-violet radiation as well as infra-red light radiates out into space and propagates in all direction. Only a small fraction of the total radiation produced reaches the earth. To be exact, approximately two-thirds of this radiation is reflected back into space, but the remaining energy is greater than one hundred times the amount of power presently available on the Earth. Moan and Smith [1] state that "Most of the energy sources on the planet are indirectly derived from the Sun". It is estimated that approximately 170 000 terawatts (TW) of solar radiation is constantly impacting the surface of the Earth. Thus, the solar energy is considered as a renewable source of energy.

The two main components required to have a functional solar energy generator are collector and power pack as storage unit. The collector basically traps and collects the solar energy from the sun that falls on it and converts it to electricity, heat or even both. The storage unit is needed to store some of the energy generated in order to be used when needed such as in emergency or during night or during cloudy days. It also can hold the excess energy produced during where energy produces is in maximum productivity.

Photovoltaics (PVs) or solar panels are one of the ways of capturing solar energy. This is the systems that produce direct current electricity that were converted from the sun based on the principle of photovoltaic effect. When sunlight falls on certain metals, like silicon, the electrons get excited and escape from the metal, and then these are collected by another metal and passed through wires in a steady stream. Thus the electron flow set up constitutes the electric current [2]. Some materials, like selenium exhibit electric properties when exposed to light. Photovoltaic cells convert sunlight directly to electricity using solid–state, crystalline materials [3]. When these materials are crystallized with semiconducting elements like silicon, a PV cell is formed and electricity can be conducted. The basic unit of solar photovoltaic (SPV) is a solar cell which is a wafer of electron emitting metal.

Solar photovoltaic system or solar array is an energy power system which is designed to supply usable solar power from sun by using photovoltaic cells. This system consists of several crucial components such as solar panels, solar inverter, charge controller, cables, battery bank, generator and other electrical accessories needed to set up this system. Solar panels is used to absorb and directly converting the sunlight into electricity while solar inverter will convert the electrical current from alternating current (AC) to direct current (DC).

The system's efficiency can be improved by integrating a solar tracking component with the system. The solar tracking is a system by which a solar panel is used to detect the source of sunlight in order to ensure that the photovoltaic solar panels receive a great amount of sunlight. In order to collect and convert the solar energy, photovoltaic solar panel is the best choice. Solar panel is basically a set of solar photovoltaic (PV) modules that connected electrically and built on an assisting mechanism. The solar panel contains PV cells that collect and convert the sunlight that enters the panel into electric current by means of PV effect. Since solar panel produces direct current, the current will be converted into alternating current by using the inverter. Furthermore, a typical PV system generally includes a solar panel, sometimes a battery and an inverter and/or solar tracker and some wiring.

Moreover, there are three essential materials that always used to make solar cell which are semiconductor compound, silicon and other conception (dye sensitized). The principle works of the solar modules based on the movement of the electrons in the N-type silicon and P-type silicon that will generate electricity [4].

Apparently, most of the photovoltaic cells on today's market operate at efficiency less than 15 percent approx. The photovoltaic cells production and its installation have increased throughout the globe. The available data shows an increase in areas of Far East and in Europe, although only a significant increase in North America and the rest of the world [5].

2. Solar tracker

In order to increase the solar yield and electricity generation, solar trackers and concentrators were usually used. This type of combination can increase the maximum input from the sunlight. Generally, the Photovoltaic system must be facing the sun throughout the daylight in order to collect the maximum sunlight as possible. Thus, a maximum power output can be harnessed from the Photovoltaic system. Besides, this combination also is the best possibility and can be practically used at small or at large scale Photovoltaic systems.

Recent development of solar panel which is known as concentrated solar power (CSR) is a technology of a solar tracking system which is focusing on a large area of sunlight. In this system, the actuator will rotate and move the solar panel to facing parallel to the sun in order to get the maximum sunlight. To do that, a proportional-integral-derivative (PID) controller is used to stabilize the actuator and to maximize the collection of sunlight. There are two types of trackers that are available in the market. One is standard photovoltaic (PV) and the other is concentrated photovoltaic (CPV) trackers. Their characteristics are different from each other and depend on specific conditions such as their actuation architecture, actuator type, their intended applications, number and orientation of the axes and also their vertical supports and foundation.

In terms of the mechanical design of the trackers, they are classified into two types which are single axis trackers and two axis trackers. In single axis trackers, the system has only one degree of freedom which is the rotation axis, whereas in two axis trackers, the system has two degree of freedom as shown in Fig. 1.

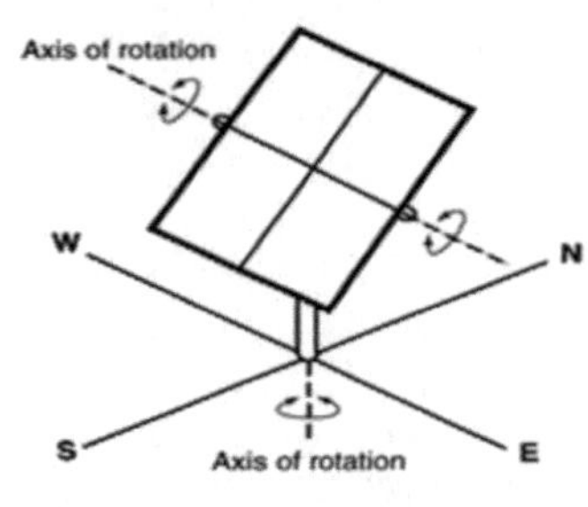

Figure 1. Two axis tracking system

In order to model the solar trackers, one may select from five types of single axis tracker and two types of two axis trackers. The single axis trackers are horizontal single axis trackers (HSAT), horizontal single axis tracker with tilted modules (HTSAT), tilted angle axis trackers (TSAT), vertical single axis trackers (VSAT), and polar aligned angle axis trackers (PSAT). Two common uses for two axis trackers are tip-tilt dual axis trackers (TTDAT) and azimuth-altitude dual axis trackers (AADAT). The actuator and sensor specifications are the most vital components for designing the tracker.

3. Design and experiment

The system is equipped with sensors and actuator which are used for the solar tracking purposes. The data collected by the sensor will be transmitted to the microcontroller and the actuator will move based on the data from the microcontroller. The amount of energy that has been collected will be received by microcontroller, and the information will be displayed at the LCD screen. The information will help us to choose the perfect spot in order to get the maximum energy input. The sunlight energy collected will be compared between the light-dependent resistors (LDR) and the Arduino microcontroller will interpret which one received the highest amount of sunlight and sends response to the servo motor to face the solar panel towards the direction with highest amount of sunlight. The flow chart of solar tracker's activity is shown in Fig. 2.

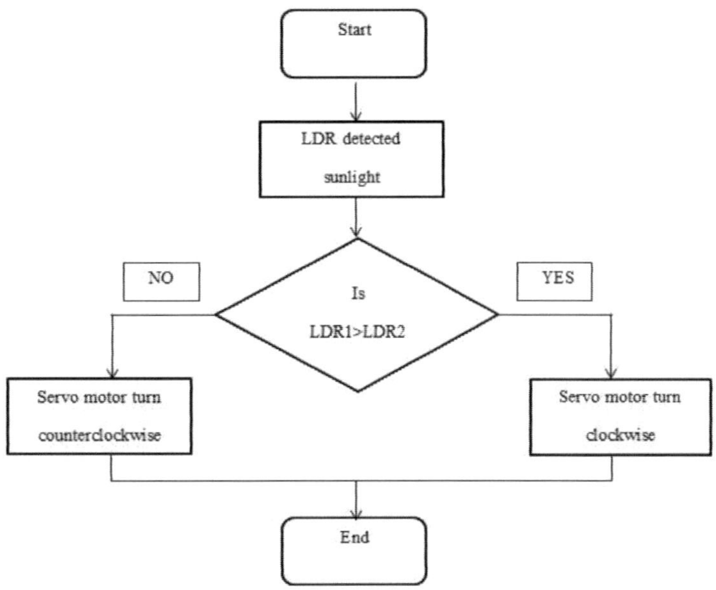

Figure 2. Solar tracker's flowchart

Monocrystalline solar panel is used in this project which can produce 50 Watts and the voltage is 12 volts. Digital servo motor was used as an actuator for the solar tracker system. It was selected since it is of light weight and high torque. The function of the servo (like angle of rotation and speed) can be controlled by Arduino programming.

Light dependent resistor (LDR) or simply a photoresistor is basically a light controlled resistor. This resistor operates as when the light intensity increases, the resistor value decreases which makes it suitable to be used as the sensor for the solar tracking system. Two sets of LDR were used in this system, which sense and receive the intensity of the sunlight. The data from the sensors are then compared between them and the microcontroller will choose the LDR which receives the highest sunlight intensity. Hence, the servo will rotate according to the data from the microcontroller, which eventually make the solar panel facing towards the highest light intensity in order to maximize the energy collected by solar panel.

In order to program the servo motor, a micro controller is needed and the suitable microcontroller used is Arduino UNO. The specification of the microcontroller is given in Table 1. Arduino is basically a microcontroller which manipulates the data whether in analog form or digital form. This open-source physical computing has a simple microcontroller board and a development environment for writing the codes of the system.

Table 1. Arduino specification

IOP Conf. Series: Materials Science and Engineering **184** (2017) 012043 doi:10.1088/1757-899X/184/1/012043

Specification	Details
Microcontroller	ATmega 328
Operating voltage	5 V
Input voltage (recommended)	7 – 12 V
Input voltage (limit)	6 – 20 V
Digital I/O pins	14 (6 pin for PWM output)
Analog pin	6
Direct current per I/O pin	40 mA
Flash memory	32 kB (ATmega 328), 0.5 kB used by boot loader
SRAM	2 kB (ATmega 328)
EEPROM	1 kB (ATmega 328)
Clock speed	16 MHz

Generally, it is used to form an interactive object, collecting inputs from sensors, actuators actuating and output is displayed. The voltage sensor circuit can be used to measure the voltage output of the solar panel, hydro generator and also power of the battery. The digital pin from the Arduino UNO will be connected in between the resistor 100k and 20k.

Liquid-crystal display (LCD) is a display panel which uses the light modulating properties of liquid crystal. Generally, it is used to display fixed images or arbitrary images just like one in the computer display. It is made up of a series of segments filled with liquid crystals and arrayed in front of backlight so that images will produce in monochrome or colour. The LCD is used in this project to show the data of the microcontroller and the voltage data from the solar panel and generator.

An inverter is used to convert the DC current produced by the solar panel to AC current. Battery selection for the power pack is chosen based on the storage capacity and also the voltage it can supply. The suitable battery that was chosen is lithium-ion battery for its high life cycle and affordability. The battery can store about 2500 mAh and the voltage is 3.7 volt. Thus, the system needs a total of 4 batteries 2 for generator and 2 for solar. Since battery's voltage is 3.7 volt, a voltage regulator (7805) is needed for 5 volt voltage output. In addition, a diode also is needed so that the solar panel and hydro generator are supplying voltage to the battery and not the other way.

4. Results and discussions

The charge controller box is needed in order to monitor and control the rate of charge of the battery. The controller will control the battery voltage by comparing the battery voltage and the battery voltage whether the system should continue charging the battery or not.

The experiment was conducted on the rooftop of Block E2, Kulliyyah of Engineering, International Islamic University Malaysia (IIUM). The experiment was done from 7 am to 7 pm. The intensity of sunlight also was measured using the solarimeter. The result of the output voltage and current was recorded separately for the system with the aid of solar tracker and without the aid of solar tracker in Table 2 and Table 3 respectively.

Table 2. Output voltage and current collected by solar panel using solar tracker

Time	Sunlight intensity (W/m^2)	Voltage (V)	Current (A)	Power (W)
7 am	357	10.01	0.12	1.20
8 am	415	10.12	0.18	1.82
9 am	624	11.23	0.23	2.58
10 am	760	11.56	0.28	3.24

11 am	872	12.50	0.34	4.25
12 noon	940	12.53	0.35	4.39
1 pm	1022	12.59	0.41	5.16
2 pm	976	12.56	0.39	4.90
3 pm	805	12.53	0.35	4.39
4 pm	788	12.23	0.33	4.04
5 pm	612	11.98	0.30	3.59
6 pm	550	11.65	0.24	2.80
7 pm	414	11.01	0.22	2.42

Table 3. Output voltage and current collected by solar panel without using solar tracker

Time	Sunlight intensity (W/m^2)	Voltage (V)	Current (A)	Power (W)
7 am	342	9.78	0.11	1.08
8 am	476	10.22	0.15	1.53
9 am	543	11.11	0.21	2.33
10 am	643	11.54	0.27	3.12
11 am	840	12.35	0.30	3.71
12 noon	915	12.43	0.33	4.10
1 pm	972	12.47	0.35	4.36
2 pm	892	12.12	0.30	3.63
3 pm	765	11.97	0.30	3.60
4 pm	700	11.12	0.26	2.89
5 pm	634	11.00	0.23	2.53
6 pm	556	10.57	0.19	2.01
7 pm	441	9.23	0.13	1.20

Both results show the same trend even though different sunlight intensity was received during the experiment. The system with the aid of solar tracker shows more voltage and power produced than the system without the aid of the solar tracker. This shows that the solar tracker system eventually helped the solar panel to harvest more sunlight and produce more output voltage and power. Figs. 3 and 4 show the graphs of the voltage versus time for the system with the aid of solar tracker and without the aid of the solar tracker respectively. Both of the graphs show the same trend of increase of the output voltage as the time approaching the midday.

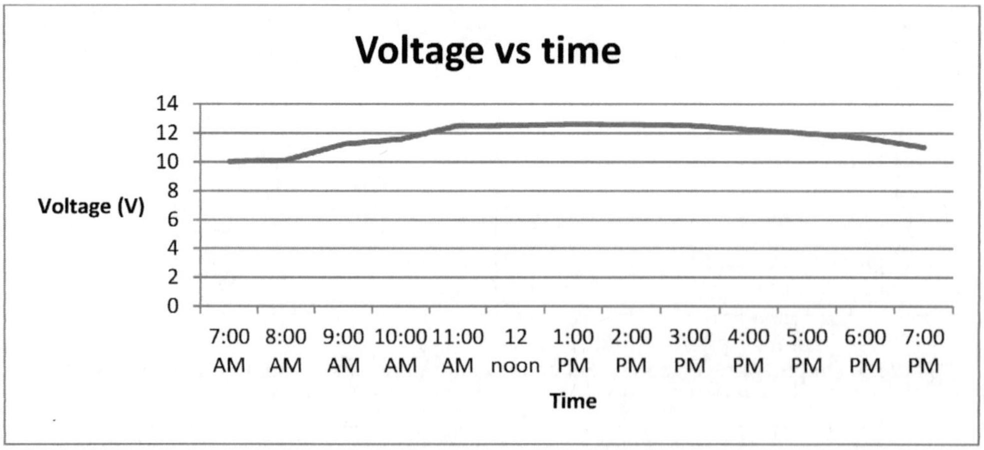

Figure 3. Voltage versus time for the system with the aid of solar tracker

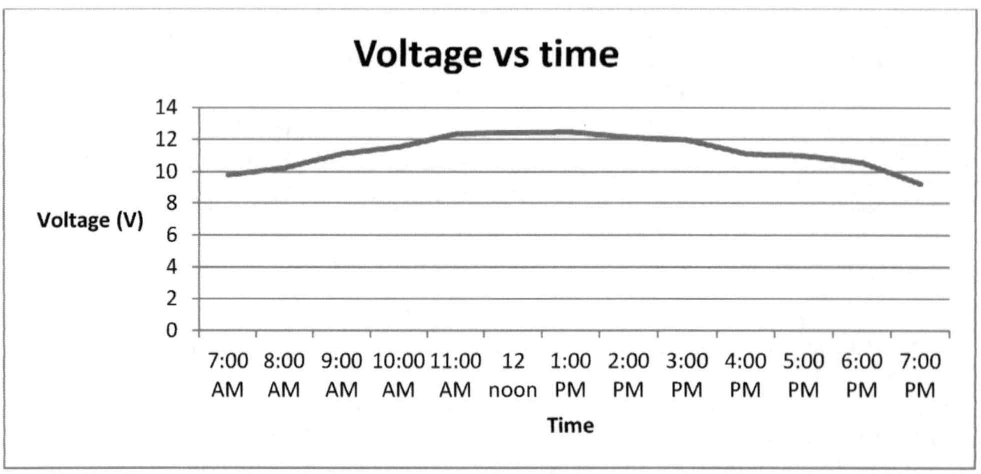

Figure 4. Voltage versus time for the system without the aid of solar tracker

Fig. 3 shows the voltage output for the system with the aid of solar tracker and Fig. 4 shows the same for the system without the aid of solar tracker. It is clear from the figures that the voltage output in Fig. 3 is much higher than that of Fig. 4. First visible changes in the graphs start from the third hour of the experimentation. It was observed that the voltage output continues to increase steadily for the system with the aid of solar tracker afterwards which is not the case for the system without the aid of solar tracker. Even though both graphs show the same trend, but we can see the difference at the end hours of the experiment. Graph in Fig. 3, has almost steady trend at the end hours of the experiment. This shows that the solar tracker system helps solar panel to maximize the process of harvesting the solar energy.

4. Conclusion

The following conclusions can be drawn from the present study:

1. The sunlight intensity affects the voltage output of the solar panel.
2. Maximum voltage output of 12.59 V was successfully collected by the solar panel with the help of solar tracker.
3. Solar tracker system was successfully integrated with the solar panel to maximize the power output from the solar panel system.

Acknowledgements
The authors would like to acknowledge the RIGS of International Islamic University Malaysia for the support to carry out this work.

References

[1] Moan J L and Smith, Z A 2007 *Energy Used Worldwide*. Oxford, England: ABC-CLIO,Inc.
[2] Chaulan, D S and Srivastava S K 2004. *Non-Conventional Energy Resources.* Delhi, India: New Age International (P) Ltd.
[3] Boyle G, ed. 1996 *Renewable Energy: Power for a Sustainable Future*, UK: Oxford University Press.
[4] Dobrzanski L A, Drygala A, Giedroc E and Macek M 2012 Monocrystalline silicon solar cells applied in photovoltaic system, *Journal of Achievements in Materials and Manufacturing Engineering*, **53**(1) 7-12.
[5] Shinwari, W K, Ali F and Nayyar A H 2004 Electric Power Generation from Solar Photovoltaic Technology: Is It Marketable in Pakistan? *Pakistan development review* **43**(3) 267-294.

Improvement Design of Parabolic Trough

S. I. Ihsan[1,a*], M. A. I. M. Safian[1], M. A. M. Taufek[1] and A.K.M. Mohiuddin[1]
[1]Department of Mechanical Engineering,
Faculty of Engineering, International Islamic University Malaysia,
P.O.Box 10, 50728 Kuala Lumpur, Malaysia

E-mail: [a]sihsan@iium.edu.my

Abstract. The performance of parabolic trough solar collector (PTSC) has been evaluated using different heat transfer working fluids; namely water and SAE20 W50 engine oil. New and slightly improved PTSC was developed to run the experimental study. Under the meteorological conditions of Malaysia, authors found that PTSC can operate at a higher temperature than water collector but the performance efficiency of collector using engine oil is much lower than the water collector.

1. Introduction

Energy production in Malaysia is based on fuels and natural gas. By using fuels or natural gas, it is easy to generate power or electricity for many purposes but it will cause big consequences to the environment such as air pollutants due to the production of carbon dioxide and carbon monoxide. These cases will lead to human made natural disaster such as greenhouse effect, thinning of the ozone layer, acid rain and many other bad effects. In order to conserve the environments, other alternatives must be taking into consideration such as renewable energy. For example, the main source of energy on earth which is from the sun can be converted to electricity by using solar thermal collector which is known as Parabolic Trough Solar Collector. Sun will produce the ultraviolet rays to earth. Despite of polluting the environment with many harmful gases, the energy from the sun can be used to generate electricity. This research will be focused on the usage of Parabolic Trough Solar Collector in these modern days in order to conserve the environments.

In understanding the working mechanisms of a parabolic trough collector, solar radiation coming from the will hit the parabolic trough's mirror. In two dimensional (2-D), the incident rays will be reflected to one focus point due to the shape of a parabola. In three dimensional (3-D) views, the reflected rays from the solar radiation will produce one line which is called the focal line. The position where the focal line exists will be occupied with the heat collection elements consisting of receiver tube, heat transfer fluid (HTF) inside the receiver tube. The solar energy will be absorbed by the receiver tube and the heat collected is transferred to the HTF. This process is called the conversion of solar energy to thermal energy. Lastly, the HTF can be used for many application purposes such as generating electricity, production of hot water and many other applications.

According to Anam [1], "Parabolic trough is one of the most popular and demanding concentrated solar thermal technology available nowadays. It is efficient and environment friendly. But, the present day parabolic trough holds some deficiencies which if solved can increase the efficiency more". The parabolic trough consists of parabolic shaped trough frame, base that holds the trough, reflective coating or mirrors, heat collection tube, and fluids.

2. Parabolic trough design

A modified design of parabolic trough collector from the one developed by the previous student [2] is used for comparison in this work. Specifically, a tempered glass cover is added to the trough in order to trap solar radiation reflected from the surface to reduce heat losses. The designs are shown in Figure 1. The HTF is filled in the collector tube and the temperature and duration of exposure are measured. The process will be repeated for several times for several days. Two types of HTF are used in the experiment, which are water and engine oil. Their required properties are shown in Table 1. In the previous design, the instantaneous thermal efficiencies obtained were around 24% for water and 18-19% for engine oil [2]. With the introduction of the glass cover, the efficiency is expected to increase.

(a) Previous design [2] (b) Modified design

Figure 1. Parabolic trough solar collector design

Table 1. Heat Transfer Fluid Properties [3]

Heat Transfer Fluid (HTF)	Water	SAE20 W50 engine oil
Specific Heat (kJ/kg °C)	4.167	2.34
Density (kg/m^3)	1000	824

3. Results and discussion

For each working fluid, the experiments were repeated five times on different days to ensure consistency of data. The experiments were conducted on the roof top of E3, Kulliyyah of Engineering building since the area is flat and unshaded. Solar irradiation is measured using pyranometer available in the lab. Figures 2 and 3 shows the temperature increase in the collector and the solar irradiation, for water and engine oil as the HTF, respectively.

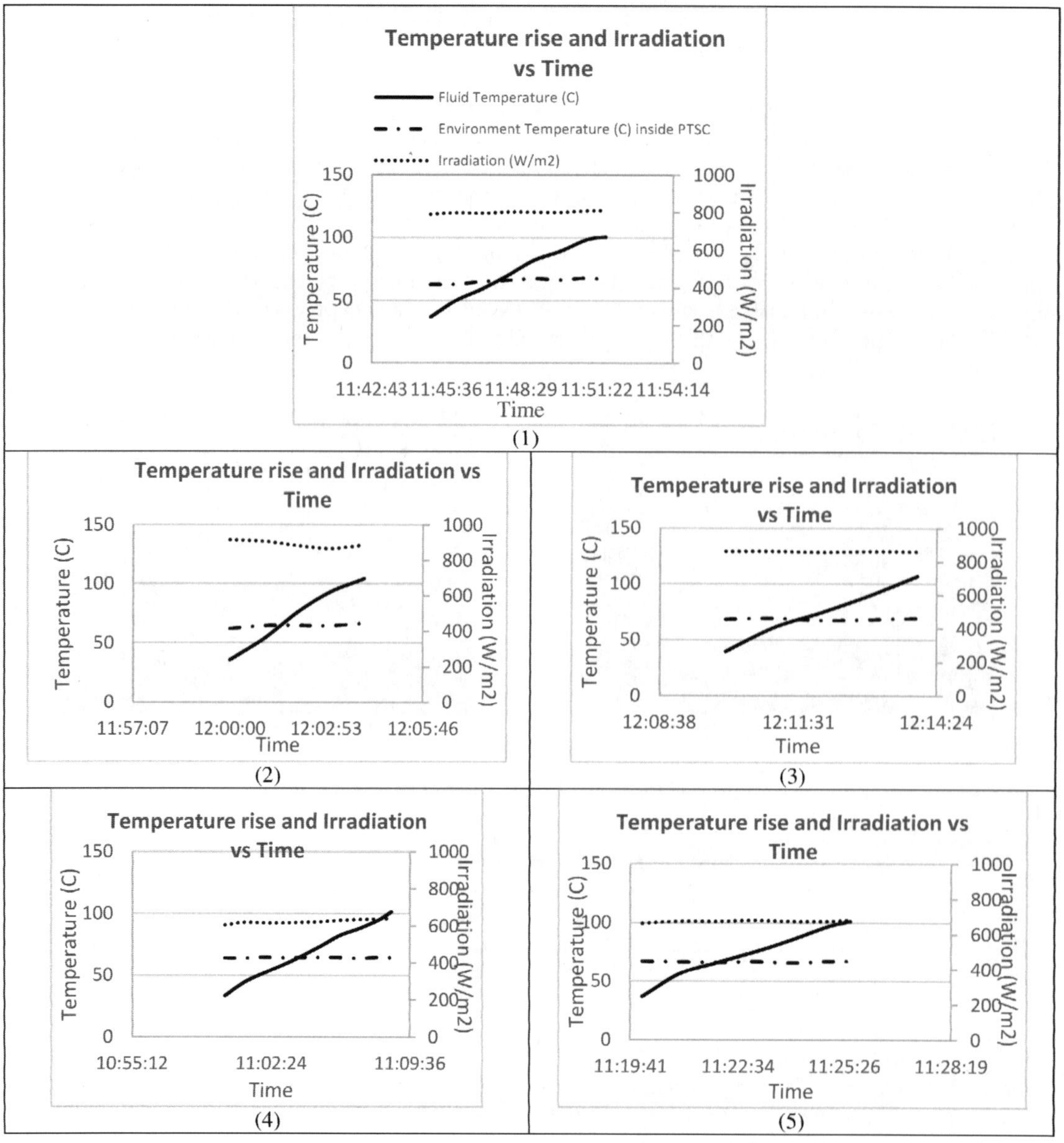

Figure 2. Temperature rise and irradiation vs time for water based collector

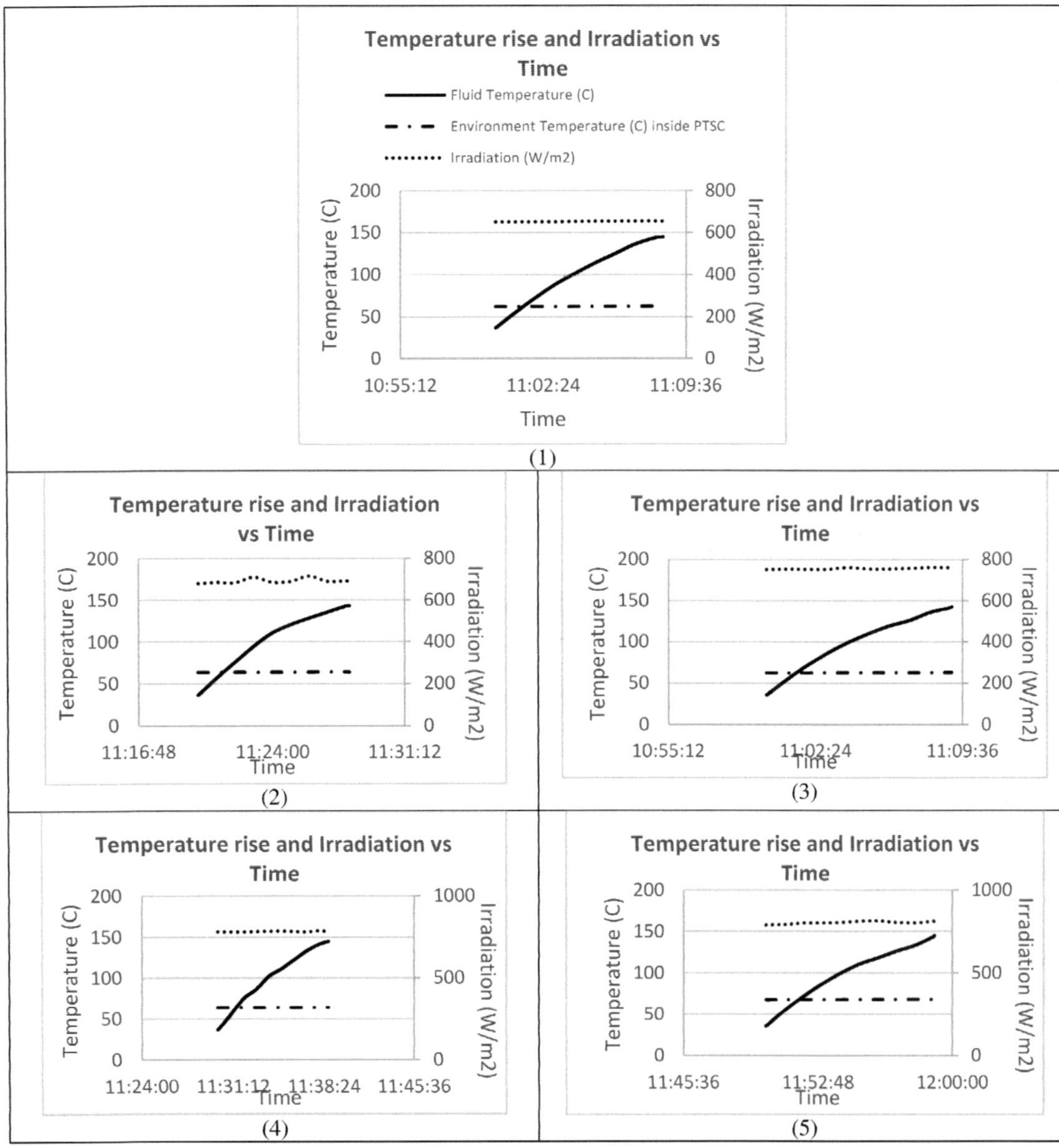

Figure 3. Temperature rise and irradiation vs time for engine oil based collector

As expected, the temperature in the working fluid increases as the parabolic trough is exposed to the sun. Complete data collected are summarized in Table 2 and 3 for both water and engine oil working fluid, respectively. The performance of parabolic trough solar concentrator was determine by obtaining values of instantaneous thermal efficiency, useful energy and the temperature of heat transfer fluid after being exposed to irradiation. The useful energy Q_u is calculated from the measurement of initial and final temperatures and the fluid mass, m_{fluid} as follows [4].

The useful energy, $Q_u = m_{fluid} C_P (T_f - T_i)$

The instantaneous thermal efficiency, $\eta_{th} = \frac{Q_u}{A_c I_b}$

Table 2. Summary of experimental results using water as the HTF

Experiment no	1	2	3	4	5
Duration (s)	402	240	240	517	336
Average Solar Irradiation (W/m^2)	801.2	891.7	860.2	624.5	675.2
Initial Temperature (°C)	36.7	35.9	39.4	33.3	36.9
Final Temperature (°C)	100	100	100	100	100
Temperature inside PTSC (°C)	65.7	64.5	68.3	64.4	66.7
Useful Energy Collected (kJ)	126.3	127.9	120.9	133.1	125.9
Total Solar Energy (kJ)	322.1	214.0	206.4	322.9	226.9
Instantaneous Thermal Efficiency (%)	39.2	59.8	58.6	41.2	55.5

Table 3. Summary of experimental results using SAE20 W50 Engine oil as the HTF

Experiment no	1	2	3	4	5
Duration (s)	505	492	545	525	543
Average Solar Irradiation (W/m^2)	652.4	692.7	755.2	782.2	883.8
Initial Temperature (°C)	36.9	36.5	35.8	36.3	35.4
Final Temperature (°C)	144.6	143.1	142.3	144.3	145.3
Temperature inside PTSC (°C)	62.4	63.6	62.5	63.7	67.9
Useful Energy Collected (kJ)	120.7	119.4	119.3	121.0	123.1
Total Solar Energy (kJ)	329.5	340.8	411.6	410.7	479.9
Instantaneous Thermal Efficiency (%)	36.6	35.0	29.0	29.5	25.7

It can be observed from the tables that, even though the value varies from one experiment to another, trough with water as the HTF has consistently higher instantaneous efficiency compared to engine oil indicating that water is more suitable for low temperature thermal recovery application. Comparing to the performance obtained by the previous student [2], results also indicate significant thermal efficiency improvement is obtained by using the modified parabolic trough. This is mainly attributed to the fact that heat losses to the surrounding through convection is prevented by using the glass cover.

4. Conclusions

The study found that for low temperature thermal recovery application, water is more suitable HTF compared to SAE20 W50. Significant increase in the instantaneous efficiency from the previous results by Mohd Taufek [2] was obtained through glass cover by reducing as much as possible heat losses to the surroundings.

Acknowledgements

The work is funded by the International Islamic University Malaysia (IIUM) Research Initiative Grant Scheme (RIGS). The authors would like to acknowledge the support from the Department of Mechanical Engineering, IIUM for the support to carry out this work.

References

[1] Anam, K. (2012). An Improved Model for Solar Parabolic Trough. *International Conference on Informatics, Electronics & Vision (ICIEV),* 2012, 420-423.

[2] Mohd Taufek, M. (2015). *Design and Analysis of Performance Trough Thermal Collector Performance.* Final Year Project Report.

[3] *Thermophysical Properties: Engine Oil, Unused.* (2011, November 13). Retrieved from Thermal - Fluids Central: https://www.thermalfluidscentral.org/encyclopedia/index.php/Thermophysical_Properties:_Engine_Oil,_Unused

[4] Cengel, Y. A., Ghajar, A. J., & Kanoglu, M. (2011). Fundamental of Thermal Radiation. Heat and Mass Transfer Fundamentals and Applications. New York: Mc Graw Hill.

[5] Iqbal, S., Iqbal, U., Khan, M. U., Saeed, M., & Waqas, A. (2014). Design, Fabrication and Analysis of Solar Parabolic Trough Collector for Steam Engine. *Robotics and Emerging Allied Technologies in Engineering (iCREATE), 2014 International Conference on*, 296-299.

[6] *The Basics of Synthetic Oil Technology*. (n.d.). Retrieved from Machinery Lubrication: http://www.machinerylubrication.com/Read/28671/basics-of-syntic-oil-technology

The Investigation of a Ride Quality of Nonlinear Half-Car Model

Fadly J. Darsivan

Department of Mechanical Engineering, Faculty of Engineering, International Islamic University Malaysia

fjd@iium.edu.my

Abstract. The paper discusses the ride quality response of a half car model when a bilinear model and a power law model were implemented in the suspension of half car vehicle. The models were subjected to a step and double pothole disturbances. The nonlinearity of the model is contributed by the power law damper models which are located both at the front and at the rear of the vehicle. Passenger car dampers were initially tested on the damper test bench where the dynamic characteristics were obtained. From the experimental results it clearly showed that the automotive damper behaves nonlinearly. From the force-velocity graphs a backbone model was then developed and plot. Based on this graph two types of non-parametric model namely the bilinear model and power law model were then obtained through the Lavenberg-Marquardt curve fitting algorithm (LMA) method. From the multibody system simulations they clearly showed that the bounce and pitch responses of the bilinear damper model illustrates an overestimations of responses compared to the nonlinear damper model which may lead to a less accurate representation of a realistic damper.

1. Introduction

The damper or shock absorber is a part of the component in the suspension assembly that is responsible in determining the ride quality and stability of the vehicle. Dampers are tuned differently for diverse types of vehicles and the different tuning would depend on the vehicles' capacity and utility.

While some researchers assumed the linear behavior of the shock absorber others have observed and concluded that the automotive damper is a nonlinear element and thus obtaining the nonlinear mathematical model is essential [1,2]. The representation of a single damping coefficient to a shock absorber may not be accurate since its physical characteristic would require it to have a varying damping coefficient at different level of disturbance frequencies [3,4]. When implemented onto a vehicle dynamics simulation the vehicle with a linear shock absorber model tends to overestimate its responses.

The purpose of this paper is to compare the variation of responses between a bilinear damper model with a power law damper model. The dampers were employed to a half car vehicle representation travelling on irregular road surfaces where it was subjected to bounce i.e. vertical motion and pitch i.e. rotational motion about its lateral axis.

2. Damper Models

The damper models implemented in this paper can be classified into two types i.e. bilinear model and power law model.

An experiment was conducted initially in order to obtain the characteristic of a typical passenger car damper. The data was analysed and the Lavenberg-Marquardt Algorithm (LMA) to attain the empirical models of both the bilinear and the power law models. Figure 1 shows the backbone data of the damper. In this case it shows the two regions of the damper i.e. rebound (moving up) and compression (moving down) [5]. To be more precise the rebound damping force is much higher compared to the compression region [6].

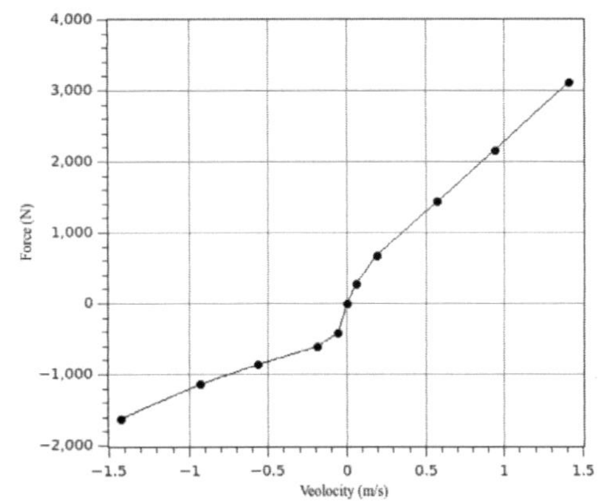

Figure 1. Backbone data of a passenger car damper.

2.1. *Lavenberg-Marquardt Curve Fitting*

The curve fitting process using the LMA is one of the steps in determining the accuracy of a realistic damper model. By using an opensource data analysis software the experimental results of the passenger car damper were plot as in figure 1. The LMA were implemented for two different models i.e. the bilinear model and the power law model with maximum iterations of 1000 and an error of 0.0001 for each respective model.

In statistics the adjusted R^2 can be used as an indicator of a good fit in which the value of the index can be between 0 and 1 where the value of closer to 1 indicating the better fit. The following table shows the results of the adjusted R^2 for both the bilinear model and the power law model after the implementation of the LMA. Table 1 clearly indicates that the power law model gives a better representation of a realistic damper. The accuracy of the model is critical since it will give the designer a good and correct sense of the ride quality of a vehicle.

Table 1. Adjusted R^2 Values for the models.

	Adjusted R^2	
	Rebound	Compression
Bilinear Model	0.981	0.770
Power Law Model	0.997	0.958

2.2. Bilinear model

Since the rebound and compression damping forces are asymmetric thus the bilinear model was used. The term bilinear indicates that the damper would have the following characteristics:-

$$F_d(v) = \begin{cases} d_1 v & ; v \geq 0 \\ d_2 v & ; v < 0 \end{cases} \tag{1}$$

Where $F_d(v)$ is the damping force that changes with respect to velocity, d_1 and d_2 are the slopes of the model during rebound and compression respectively and v is the velocity of the piston rod of the damper. The bilinear model depicted in (1) can be illustrated as in figure 2. By implementing the LMA it was found that the values of d_1 and d_2 are 2287 N.s/m and 1220.5 N.s/m respectively for the front damper and for the rear damper the values of d_1 and d_2 are 1488.2 N.s/m and 943.55 N.s/m respectively.

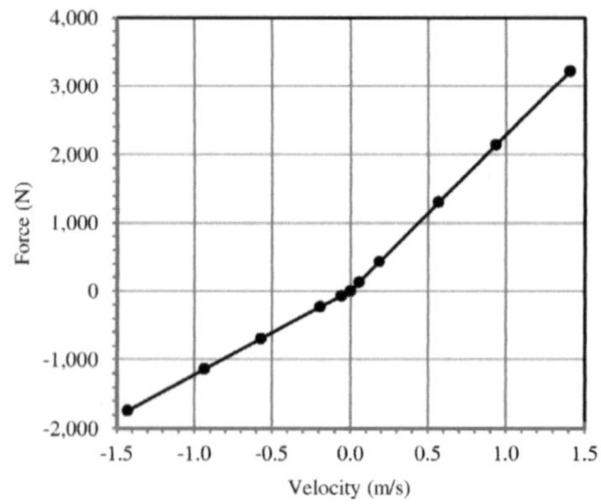

Figure 2. Bilinear damper model.

2.3. Power Law model

The power law model can be defined in the following form:-

$$F_d(v) = \begin{cases} c_1 v^{e_1} & ; v > 0 \\ c_2 v^{e_2} & ; v < 0 \end{cases} \tag{2}$$

Where c_1 and e_1 are the constants of the model for rebound and c_2 and e_2 are the constants for compression and can be illustrated as in figure 3 . From the figure it clearly indicates that the damping forces during rebound and compression are not equal and this is in agreement with the experimental result shown in figure 1.

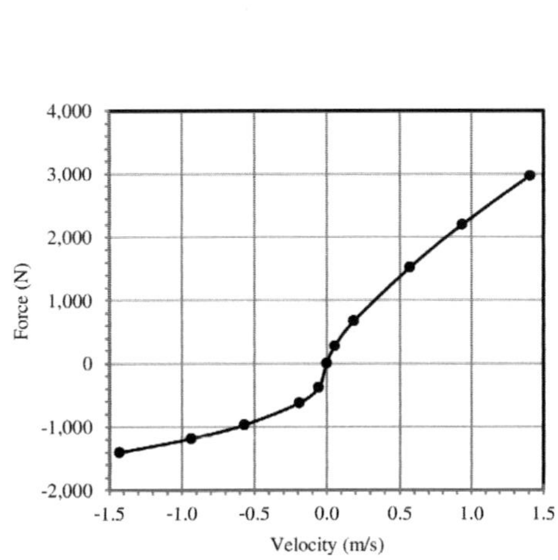

Figure 3. Power law damper model.

The values of the constants are obtained from the LMA and presented in table 2.

Table 2. Power Law Model Constant Values

Symbol	Values	
	Front	Rear
c_1	2302.1	1573.12
c_2	1217.8	946.34
e_1	0.74	1.03
e_2	0.41	0.4

3. Theoretical Background

A half car model is a representation of a vehicle that consists of a body, a front wheel and a rear wheel. Between the wheels are the assembly of the suspension that includes the suspension coil spring and the damper. Figure 4 represents the schematic diagram of half car model.

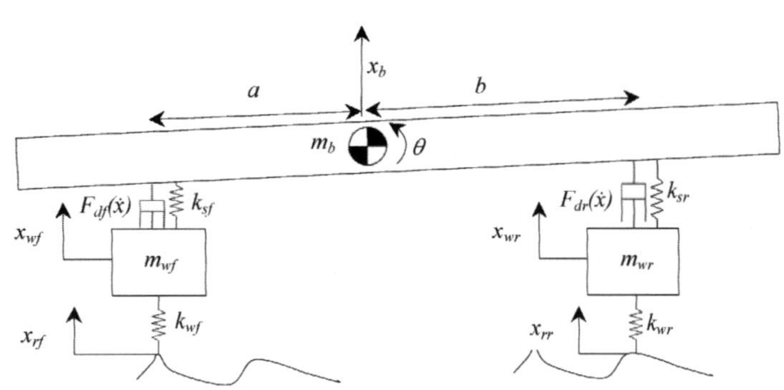

Figure 4. Schematic diagram of a half car model.

The symbols denoted in figure 4 can be summarized and some of their values are presented in table 2.

IOP Conf. Series: Materials Science and Engineering **184** (2017) 012045 doi:10.1088/1757-899X/184/1/012045

Table 3. Half Car Model Parameters.

Symbol	Connotation	Value	Unit
m_b	Mass of body	720	kg
J	Body moment of inertia	737.4	kg.m^2
m_{wf}	Mass of front wheel	40	kg
m_{wr}	Mass of rear wheel	40	kg
a	Front to CG distance	1.08	m
b	Rear to CG distance	1.48	m
x_b	Body displacement at CG		m
θ	Body rotation (pitch angle)		rad
x_{wf}	Front wheel displacement		m
x_{wr}	Rear wheel displacement		m
x_{rf}	Road displacement at the front wheel		m
x_{rr}	Road displacement at the rear wheel		m
$F_{df}(\dot{x})$	Front suspension damping force		N.s/m
$F_{dr}(\dot{x})$	Rear suspension damping force		N.s/m
k_{sf}	Front suspension spring constant	1.996e+004	N/m
k_{sr}	Rear suspension spring constant	1.996e+004	N/m
k_{wf}	Front wheel stiffness	1.755e+005	N/m
k_{wr}	Rear wheel stiffness	1.755e+005	N/m

The equations of motion of the half car model in the figure above can then be derived as in the following relationship:-

$$\begin{gathered} m_s a_b + F_{df}(\dot{x}) + F_{dr}(\dot{x}) + k_{sf}(x_b - a\theta - x_{tf}) + k_{sr}(x_b + b\theta - x_{tr}) = 0 \\ J\alpha - F_{df}(\dot{x})a + F_{dr}(\dot{x})b - k_{sf}(x_b - a\theta - x_{tf})a + k_{sr}(x_b + b\theta - x_{tr})b = 0 \\ m_{wf}a_{wf} - F_{df}(\dot{x}) - k_{sf}(x_b - a\theta - x_{tf}) + k_{wf}(x_{wf} - x_{rf}) = 0 \\ m_{wr}a_{wr} - F_{dr}(\dot{x}) - k_{sr}(x_b - b\theta - x_{tr}) + k_{wr}(x_{wr} - x_{rr}) = 0 \end{gathered} \tag{3}$$

In (3) the term $\dot{x}$ is the relative velocity between the body and the wheels a_b is the linear acceleration of the body and α is the angular acceleration of the body. The relationship of (3) was solved by implementing the *Runge-Kutta* method through a multibody dynamic software.

4. Half-Car Model Simulation

The simulation of the half car model was done by using a multibody dynamics simulation software and the parameters in table 3 are used. To compare the ride responses between the bilinear damper and the power law damper model the damping forces $F_d(\dot{x})$ in (3) were replaced with (1) and (2) respectively. The purpose of the simulation is to provide further proofs that the response of the half car model equipped with the bilinear damper model gives results which are different from the more accurate power law damper model.

As a disturbance to the vehicle two types of road profiles were implemented namely the step input and the double pothole disturbance input. The disturbances are illustrated as in the following figure.

IOP Conf. Series: Materials Science and Engineering **184** (2017) 012045 doi:10.1088/1757-899X/184/1/012045

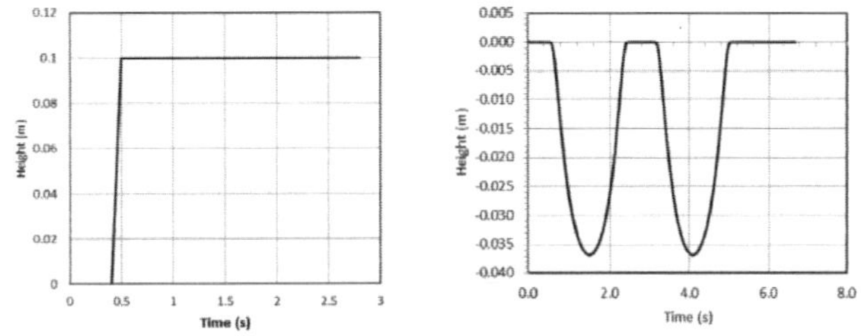

(a) (b)

Figure 5. Road disturbances (a) step input (b) double pothole.

As a measurement to ride responses the vertical and angular accelerations are used and from the data obtained the root mean square (rms) values will be calculated and the settling time of the body be taken.

5. Results

The results from the simulation can be divided into two groups based on the road disturbances. The main objective of this simulation is to compare the percentage variation between the bilinear model and the power law model. For each of the road disturbance the time history of the vertical position and acceleration and the pitch angle and angular acceleration of the body are presented. The vertical position and pitch angle indicate how fast the body move to its stable position after the disturbances have passed while the vertical acceleration and angular acceleration indicates the quality of the ride in which the smaller magnitudes of accelerations provide a better ride quality

Figure 6 below shows the response of the body's vertical position and its pitch angle when subjected to a step road disturbance. From the figure it can be seen that there is a variation with respect to the overshoot response for both magnitudes. The most obvious being the response of the pitch angle where the overshoot of the power law damper model is almost doubled. Another obvious observation is the oscillatory response of the bilinear damper model in which the settling time of the aforementioned damper is larger i.e. at approximately 3 seconds compared to the power law damper which is approximately 1 second.

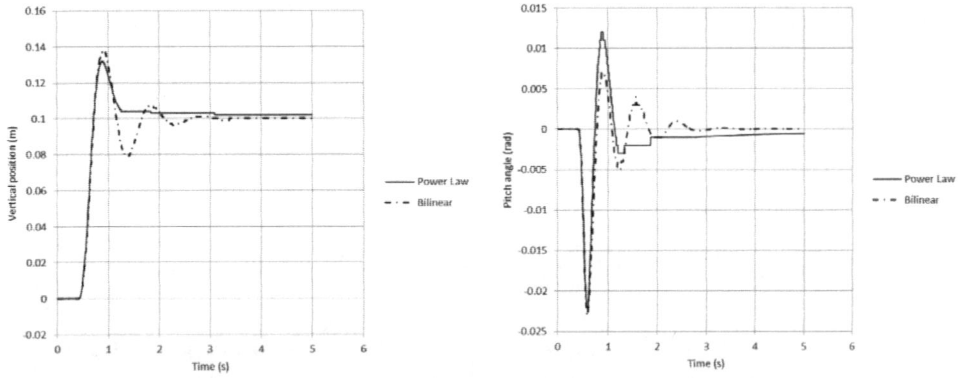

Figure 6. Vertical position and pitch angle response of the vehicle subjected to step disturbance.

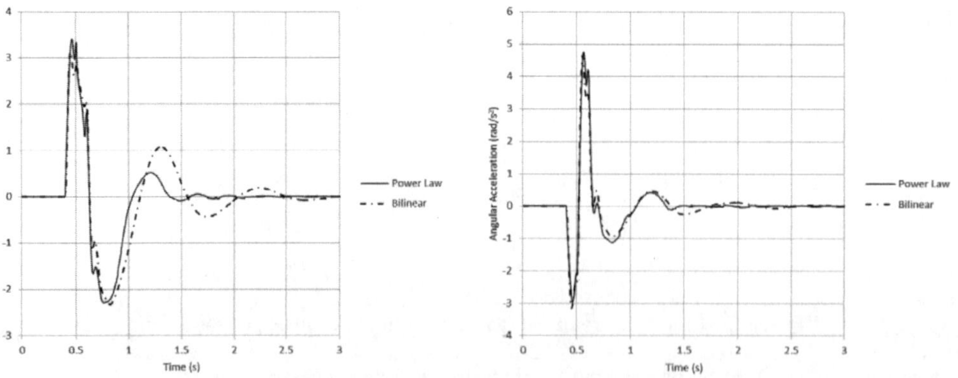

Figure 7. Vertical acceleration and angular acceleration response of the vehicle subjected to step disturbance.

Figure 7 shows the vertical acceleration measured at the centre of gravity of the body and its angular acceleration when the vehicle is subjected to the same step disturbance. From the vertical acceleration response it clear shows that a larger oscillation and larger settling time for the bilinear damper. However, an almost similar response can be seen for the angular acceleration result with some variation in the settling time of the vehicle.

Figure 8 is the vertical displacement and pitch angle response of the vehicle when it travelled to a double pothole at a speed of 10 m/s. The variations of results are again visible in the displacement response. The bilinear damper has shown an overestimation of displacement of approximately 20% more than the power law damper. However, a very close agreement of pitch angle response can be seen for both dampers with small magnitude of oscillations for the bilinear damper before reaching stable condition.

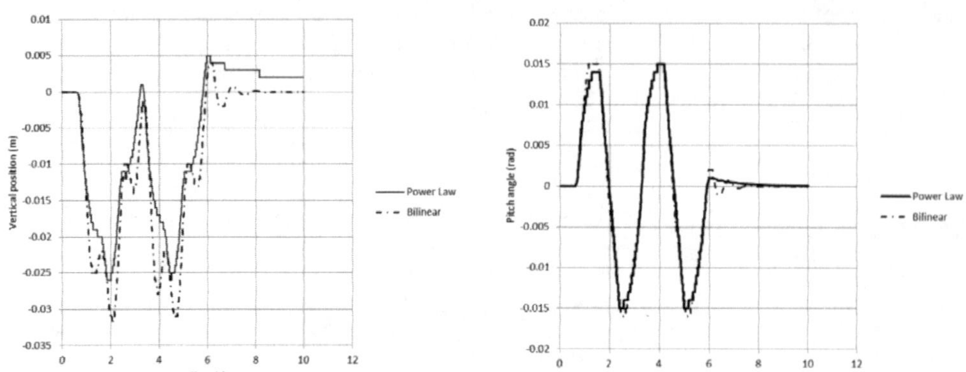

Figure 8. Vertical position and pitch angle response of the vehicle subjected to double pothole disturbance.

The following figure indicates the ride quality for both dampers when subjected to the double pothole road input. The most obvious observations from the figures are the larger oscillations of the power law damper and the overestimations of response for the bilinear damper especially in the positive direction. The insert shows a close-up view of the vertical acceleration between 3 and 4 seconds.

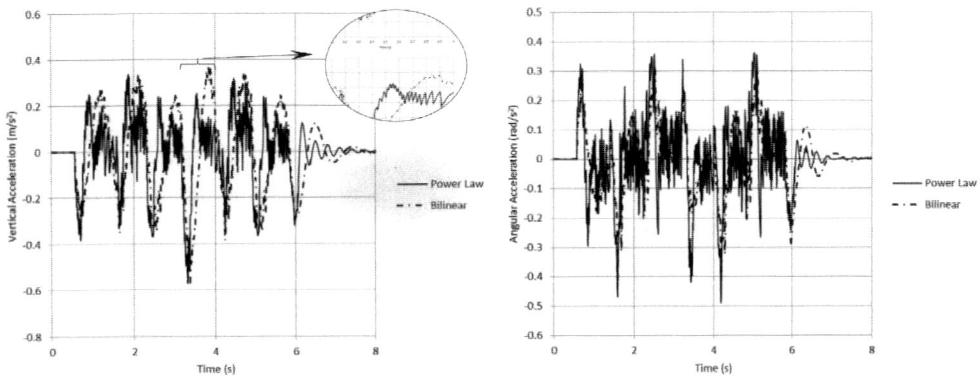

Figure 9. Vertical acceleration and angular acceleration response of the vehicle subjected to double pothole disturbance.

To summarise the variations of response for all the quantities measured the following table presents the percentage difference of their root mean square (RMS) values. From the table it shows that in some responses the bilinear damper model could overestimate as much as 50% than the power law model damper. Since the power law model depicts a closer representation of a realistic damper therefore these percentage variations of the bilinear damper may not provide a good sense of simulated ride performances and stabilities.

Table 4. RMS values of measured quantities.

Road Disturbance	Quantity	RMS Value Nonlinear	Bilinear	%Difference
Step	θ (pitch angle)	0.003	0.004	36.1
	α (angular acceleration)	0.480	0.658	37.1
	x (vertical displacement)	0.069	0.094	37.1
	a (vertical acceleration)	0.491	0.744	51.5
Double Pothole	θ (pitch angle)	0.007	0.008	3.6
	α (angular acceleration)	0.111	0.095	-13.9
	x (vertical displacement)	0.011	0.014	23.8
	a (vertical acceleration)	0.127	0.163	27.8

6. Conclusions

An experiment was conducted to obtain the characteristic of a typical passenger car damper. From these data two types of damper models i.e. bilinear model and power law model were mathematically generated through the Lavenberg-Marquardt curve fitting algorithm. From the curve fitting algorithm it can be concluded that the power law model gives a better representation of a more realistic damper compared to the bilinear damper. By employing both dampers into a half car vehicle model the results of the simulations shows some degree of variations with respect to the body vertical position and pitch as well as the vehicle's quality of ride. From the result of the simulations both dampers have shown some degrees of variations with respect to its overshoot, stability and settling time. The RMS values of the responses indicate that the bilinear model may not be an accurate damper representation to be implemented in the half car vehicle model.

References

[1] Rao, M. D., Gruenberg, S., & Torab, H., 1999. Measurement of dynamic properties of automotive shock absorbers for NVH. *Society of Automotive Engineers*.

[2] Duym, S., Steins, R. and Reybrouock, K., 1997. Evaluation of shock absorber models. *Vehicle System Dynamics*, **27**(2), pp. 109-127.

[3] Dixon, J. (2008). *The shock absorber handbook*. John Wiley & Sons.

[4] Lee, K. (1997). Numerical modelling for the hydraulic performance prediction of automotive monotube dampers. *Vehicle System Dynamics*, **28**(1), 25-39.

[5] Kasteel, R. V., Cheng-Guo, W., Lixin, Q., Jin-Zhao, L., & Guo-Hong, Y. (2005). A new shock absorber model for use in vehicle dynamics studies. *Vehicle system dynamics*, **43**(9), 613 631.

[6] Gillespie, T.D., 1992. *Fundamentals of Vehicle Dynamics*. Pensylvania: Society of Automotive Engineers, Inc.

Tensile Mechanical Property of Oil Palm Empty Fruit Bunch Fiber Reinforced Epoxy Composites

A L Ahmad Ghazilan[1,a] *, H Mokhtar[1,b], M S I Shaik Dawood[1,c], Y Aminanda[1,d] and J S Mohamed Ali[1,e]
[1]Department of Mechanical Engineering
Kulliyyah of Engineering, International Islamic University Malaysia
Jalan Gombak, 53100 Kuala Lumpur, Malaysia

E-mail: [a]ahmadluqman141@gmail.com, [b]hananm@iium.edu.my, [c]sultan@iium.edu.my, [d]yulfian@iium.edu.my, [e]jaffar@iium.edu.my

Abstract. Natural, short, untreated and randomly oriented oil palm empty fruit bunch fiber reinforced epoxy composites were manufactured using vacuum bagging technique with 20% fiber volume composition. The performance of the composite was evaluated as an alternative to synthetic or conventional reinforced composites. Tensile properties such as tensile strength, modulus of elasticity and Poisson's ratio were compared to the tensile properties of pure epoxy obtained via tensile tests as per ASTM D 638 specifications using Universal Testing Machine INSTRON 5582. The tensile properties of oil palm empty fruit bunch fiber reinforced epoxy composites were lower compared to plain epoxy structure with the decrement in performances of 38% for modulus of elasticity and 61% for tensile strength.

1. Introduction

Natural fiber from various biodegradable natural resources have offered potential improvement in reinforcing material compared to synthetic fiber reinforced matrix composites. Advantageous properties of low cost, low density, biodegradability, and better insulating and thermal resistance have served for improvement in natural fiber reinforcement. Through a remarkable development locally and globally, these materials have become preferable choices which undergo extensive research, application and transformation. With non-renewable, non-degradable and expensive resources of synthetic fiber, natural-recycled resources of natural fiber is becoming an important engineering material with a wide range of properties. From the engineering, technological and structural perspective, natural fiber is accepted due to its properties of stiffness, flexibility, impact resistance and modulus of elasticity. Several applications of industrial natural fiber reinforcements have been realized lately, such as the use of flax-sisal fiber mat reinforced epoxy matrix to produce door panels for Mercedes Benz E-Class, flax fiber reinforced polypropylene composite for rear shelf trim panels for the year 2000 model Chevrolet Impala, and natural fiber reinforced composites in structural applications such as ceiling and partition boards [1].

Depending on the utilization of natural fiber, the classification of the plants is divided into two distinguished parties which are the primary and secondary. Primary plants are classified so as the plants are grown only for their fiber content, i.e. jute, hemp, kenaf and sisal plant. The secondary classification consists of the fibers obtained as a secondary product, i.e. pineapple, coir and oil palm plants. Six common types of natural fiber which can be extracted from plants are the bast, leaf, seed, core, grass and reeds [2]. The growing interest in investing towards the development of natural fiber, especially oil palm fiber from empty fruit bunch, is mainly due to its vast availability from biodegradable waste, environmental advantages, economical production, and low specific weight with high specific strength when compared to conventional synthetic fibers. However, its natural property of hydrophilicity makes the natural fiber high in moisture absorption, and its weak adhesion ability to hydrophobic matrices needs special treatment to enhance the performance of reinforced natural fiber [3]. The mechanical performance of lower ultimate strength, lower elongation properties, large variability of performance prediction and poor weather resistance have restricted natural fibers in handling loading applications. Therefore it is essential to improve the mechanical performance of oil palm fiber in natural fiber reinforced matrix composites.

Oil palm (Elaies Gueneensis) is planted in 42 countries. Malaysia, Indonesia, West Africa, Latin America and India are the leading countries that produce oil palm products [4] [5]. 73% of fibrous sources can be extracted through the separation of fibrous mass of its full fruit bunch [6] by using extraction methods of physical or chemical retting process. The exceptional performance possessed by oil palm empty fruit bunch reinforced composite for structural loading application have been reported to have a similar performance to that of coir fiber which has high specific strength [7]. In fact, the high cellulose content and toughness value of oil palm empty fruit bunch fiber makes it suitable for composite application. However, the overall limitation of its mechanical performance needs specific modification to drop the hydrophilicity factor of oil palm fiber by physical and chemical treatment such as corona, plasma, silane, alkaline, acetylation, coupling agent and enzyme treatment method. High interlocking and adhesive capability with matrix material has made alkaline treatment the most effective method in treating most natural fibers [8]. Vacuum bagging techniques for fabricating natural fiber reinforced composite will provide good pysical and mechanical performance compared to wet lay-up process due to its process control, improved equipment and cost effective method [9].

The major performance of natural fiber reinforced composite depends on the composition properties of natural fiber and polymer synthetic matrix material. Mostly, the strength and stiffness of the structural load is carried by its fiber whereas the shape, surface and environmental resistance is carried out by its matrix. Thermoplastic and thermosets are the two type of dominating matrix materials that offer various potential in natural fiber reinforced composite applications. Epoxy resin as matrix material is suitable for natural fiber reinforced composite due to its good adhesion capability, good heat and chemical resistance, exemplified electrical insulator properties and superior mechanical performance [10]. It has been reported that mechanical properties of tensile strength for oil palm fiber-epoxy resin decreases from 47.78 MPa to 46.10 MPa with the increasing amount of oil palm fiber from 35% to 55% respectively [11]. Another study on the mechanical properties of oil palm fiber epoxy composite of 10% fiber, reported that a 7m T-Beam is able to support 200kg of load with a maximum deflection of 0.2mm, thus suggesting the potential usage of oil palm fiber-epoxy resin composite for intermediate load bearing bridge application [12].

2. Material and Method

2.1 Material

Untreated oil palm empty fruit bunch fiber is obtained from Malaysia Palm Oil Board in Bangi, Selangor, Malaysia. Zeepoxy HL002 TA/B epoxy resin, epoxy hardener, vacuum bagging consumables and materials was supplied by Sky Tech Enterprise.

2.2 Fabrication of Oil Palm Composites

Tangled and intermixed oil palm empty fruit bunch fiber was carded and combed using an electrical motor carding machine to produce untangled and organized fiber layer. For the fabrication of oil palm empty fruit bunch fiber composite, a 400 mm x 400 mm bottom stainless steel plate and 300 mm x 300 mm upper stainless steel plate, and four 200 mm x 5 mm flexible resin stopper were used as a mold. A layer of mylar sheet was placed on each side of the plate as a releasing agent, and the epoxy resin and hardener were mixed thoroughly according to its specific fiber-matrix volume fraction ratio. Layers of oil palm empty fruit bunch fiber were stacked accordingly on the bottom plate and the epoxy resin and hardener mixture was impregnated through the layers of fiber. Vacuum bagging materials were then placed on top of the wet fibers and then covered by the upper plate mold. Vacuum bagging film encloses the lay-up to provide a confined space which is sealed around the edges of the mold by using sealant tape. The vacuum pump will draw out the air inside the confined space to produce an airtight vacuum bag until the curing process is done.

2.3 Tensile Characterization of Composites

Tensile test specimen was prepared according to the standard size and specification of ASTM D 638 [13] using CNC milling. Tensile strength and modulus of elasticity for plain epoxy matrix as well as for empty fruit bunch fiber reinforced composite were tested by using INSTRON 5582 Ultimate Testing Machine, quarter bridge BX 120-3AA strain gauges and D4 Data Acquisition conditioner. This measurement was conducted according to ASTM D 638 procedures at a test speed of 5mm/min. Five specimens were tested for each case and its average value is tabulated accordingly.

3. Results and discussion

Measurement of load and strain data were made simultaneously during tensile testing process. Tensile strength is calculated by dividing the maximum load by the average cross-sectional gauge length area of the specimen. Modulus of elasticity is calculated by extending the initial linear portion of stress-strain curve then by dividing the difference in stress corresponding to the section on this straight line by the corresponding difference in strain value. Cross-plot graphs of stress and strain data of plain epoxy and EFB palm fiber composite specimen from tensile testing are plotted on Figure 1. Results of tensile tests of plain epoxy and empty fruit bunch fiber reinforced composite at 20% fiber volume are shown in Table 1.

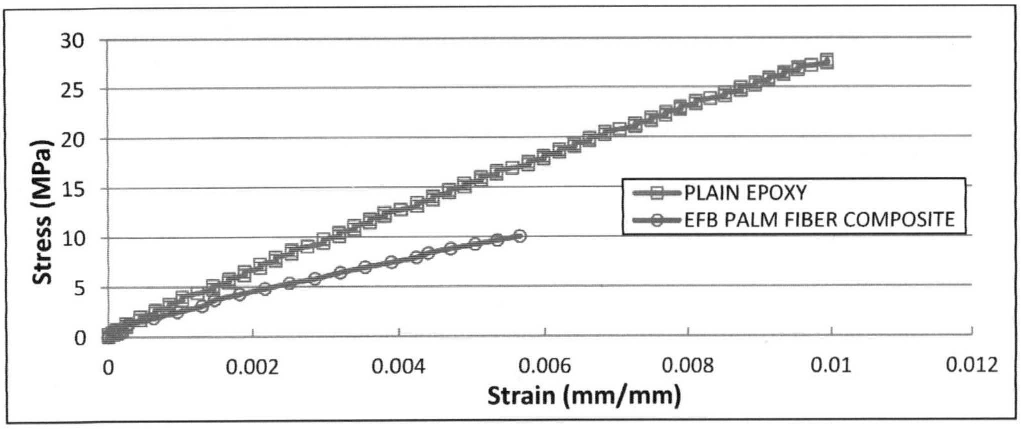

Figure 1. Tensile Stress vs Strain Curve.

Table 1. Tensile test results.

Composites	Tensile Strength (MPa)	Tensile Modulus (MPa)
Pure Epoxy	25.889	2805.09
EFB Composite	10.064	1712.13

It can be observed that tensile strength decreases by 61% and modulus elasticity decreases by 38% when oil palm fiber composite is compared to plain epoxy which is similar to findings of previous studies which reported that the ultimate tensile strength decreased by 24% when comparing plain epoxy resin to 35% fiber content of oil palm fiber epoxy composite [11]. Previous studies on short and randomly oriented fiber composite such as oil palm fiber reinforced composite, found that the axial load applied on the composite structure are first exerted on the matrix material and then transferred to the fiber throughout the fibers end and cylindrical surface. Therefore the load carrying capability and capacity is mainly carried out by the matrix rather than the fiber, which is completely different to the load carrying capability of synthetic or conventional long and unidirectional fiber composites.

The hydrophilic factor or poor adhesion capability between oil palm fiber and epoxy matrix will lead to inefficient stress transfer from epoxy matrix during loading conditions. In another research, it was also reported that the value of composite tensile strength decreased from 111 MPa to 24 MPa when glass fiber is replaced by oil palm fiber during tensile loading conditions [14]. This reduction of strength is due to the strong pectin interface bonding of oil palm individual fibers which cause fiber to not be loaded uniformly to each other, which then structurally failed to support stress transferred from the epoxy matrix. Hence, the incorporation of oil palm fiber in reinforced composite fails to act as a major reinforcement compared to synthetic fiber composite such as carbon or glass fiber composite. Thus, more advanced method of fiber treatment or hybrid fiber combination need to be taken up in order to improve the surface fiber adhesion capability and strength properties oil palm properties.

4. Conclusion

The performance of oil palm empty fruit bunch reinforced composites in this research was evaluated as a preliminary research on the tensile properties of EFB oil palm fiber reinforced composite. In this study the tensile properties of plain epoxy and oil palm fiber composites clearly indicates that empty fruit bunch oil palm fiber failed to act as major reinforcement for composite structure under tensile loading conditions whereby plain epoxy provide better tensile properties as stress is concentrated more in the epoxy matrix rather than fiber. Less surface interfacial adhesion capability with short and random fiber orientation also restrict the empty fruit bunch oil palm fiber to act as a strong alternative on reaching the desired specific properties under composite structure loading condition. Although empty fruit bunch oil palm fiber as a reinforcement fiber corresponds to a weaker tensile performance in composite structures, it has been reported that it thrives in impact strength when compared to pure epoxy due to its high fracture toughness [7]. With targets to provide superior mechanical properties and cost-effective fabrication method, more challenges need to be overcome in the development of natural fiber reinforced composite as an alternative to replace synthetic fiber reinforced composite.

Acknowledgements

The authors would like to acknowledge the Department of Mechanical Engineering of International Islamic University Malaysia for the support to carry out this work as well as the Ministry of Higher Education (MOHE) kindly provided funding for the research through the "Fundamental Research Grant Scheme" (FRGS) project no FRGS13-082-0323.

References

[1] John M J and Thomas S Biofibres and biocomposites 2008 *Carbohydrate Polymers* **71** 343-364

[2] Faruk O, Bledzki A K, Fink H -P, and Sain M Biocomposites reinforced with natural fibers: 2000–2010 2012 *Progress in Polymer Science* **37** 1552-1596

[3] Sgriccia N, Hawley M C and Misra M Characterization of natural fiber surfaces and natural fiber composites 2008 *Composites Part A: Applied Science and Manufacturing* **39** 1632-1637

[4] Abdul Khalil H P S, Siti Alwani M, Ridzuan R, Kamarudin H and Khairul A Chemical composition, morphological characteristics and cell wall structure of malaysian oil palm fibers 2008 *Polymer-Plastics Technology and Engineering* **47** 273-280

[5] Joseph S, Joseph K and Thomas S Green composites from natural rubber and oil palm fiber: physical and mechanical properties 2006 *International Journal of Polymeric Materials and Polymeric Biomaterials* **55** 925-945

[6] Wirjosentono B, Guritno P and Ismail H Oil palm empty fruit bunch filled polypropolene composites 2004 *International Journal of Polymeric Materials and Polymeric Biomaterials* **53** 295-306

[7] Jawaid M, Abdul Khalil H P S and Abu Bakar A Mechanical performance of oil palm empty fruit bunches/jute fibres reinforced epoxy hybrid composites 2010 *Materials Science and Engineering: A* **527** 7944-7949

[8] John M J and Anandjiwala R D Recent developments in chemical modification and characterization of natural fiber-reinforced composites 2008 *Polymer Composites* **29** 187-207

[9] *SP Systems Guide to Composites.*

[10] Shinoj S, Visvanathan R, Panigrahi S and Kochubabu M Oil palm fiber (OPF) and its composites: a review 2011 *Industrial Crops and Products* **33** 7-22

[11] Kalam A, Sahari B B, Khalid Y A and Wong S V Fatigue behaviour of oil palm fruit bunch fibre/epoxy and carbon fibre/epoxy composites 2005 *Composite Structures* **71** 34-44

[12] Bakar M, Natarajan V, Kalam A and Kudiran N Mechanical properties of oil palm fibre reinforced epoxy for building short span bridge 2007 *Experimental Analysis of Nano and Engineering Materials and Structures* 97-98

[13] International A Standard test method for tensile properties of plastics 2014 *ASTM D638-14*

[14] Hariharan A B A and Khalil H A Lignocellulose-based hybrid bilayer laminate composite: part I-studies on tensile and impact behavior of oil palm fiber-glass fiber-reinforced epoxy resin 2005 *Journal of Composite Materials* **39** 663-684

Surface Roughness Optimization Using Taguchi Method of High Speed End Milling For Hardened Steel D2

Muataz Hazza Faizi Al-Hazza[1,a], Nur Asmawiyah bt Ibrahim[2], Erry T. Y. Adesta[3] Ahsan Ali Khan[4] and Atiah Bt. Abdullah Sidek[5]
[1,2,3,4,5] Department of Manufacturing and Materials Engineering, Faculty of Engineering, International Islamic University Malaysia, 53100 Kuala Lumpur, Malaysia

E-mail: [a]muataz@iium.edu.my

Abstract. The main challenge for any manufacturer is to achieve higher quality of their final products with maintains minimum machining time. In this research final surface roughness analysed and optimized with maximum 0.3 mm flank wear length. The experiment was investigated the effect of cutting speed, feed rate and depth of cut on the final surface roughness using D2 as a work piece hardened to 52-56 HRC, and coated carbide as cutting tool with higher cutting speed 120-240 mm/min. The experiment has been conducted using L9 design of Taguchi collection. The results have been analysed using JMP software.

1. Introduction

High speed end milling for hardened steel is one of the new and advanced machining processes. This advanced process merge three advanced processes: high speed milling, hard milling and dry milling. Increasing speed of machining will increase the material removal rate and then increases the machining productivity. However, maintain the surface roughness should be considered with increasing productivity. The main challenge in high speed end milling of hardened steel is the increasing of flank wear rate that may affect the final surface roughness. Pare et al., [1], mentioned that the factors which influence the final result of the surface roughness can be divided into two categories which are controllable and uncontrollable factors. The controllable factors are cutting speed, feed rate and depth of cut. While the uncontrollable factors are tool geometry and material properties for both work pieces and cutting tool.

There are several researcher had done a researched on the consequence of cutting parameters to the surface roughness, flank wear and tool life. One of the researcher, Ezuwanizam [2] had find the optimal cutting parameters for cutting speed, feed rate and depth of cut in optimize the tool life of carbide insert coated with TiN in milling aluminum 6061. The result shows that tool wear in end milling decreases as the cutting parameters increase. However, axial depth of cut does not give significant effect compared to cutting speed and feed rate. Xuan-Truong and Minh-Duc [3]had explained the effect of cutting parameters on flank wear and surface roughness through the transformation of natural logarithm. They found that cutting speed has given major significant effect on flank wear and surface roughness is mostly influenced by feed rate. While for depth of cut, they conclude that it only gives a minor effect for both surface roughness and flank wear especially at low cutting speed and feed rate. However, determining the best cutting parameters to achieve the optimum value that can give higher productivity with maintain minimum surface roughness is the aim for any manufacturer.

2. Taguchi method

Taguchi method is a tool of finding the best combination of an input (cutting parameters, cutting condition, work piece and cutting tool material properties) for producing a high quality of product and services. Malvade and Nipanikar [4] mentioned that Taguchi methods more focus on developing the design for manufacturing process for creating high quality product compared to statistical process control which tries to control the factors that affect the product quality. Based on Moshat et al. researched, they stated that Taguchi method is one of the efficient tools in designing an experiment for creating a high quality product manufactured and was developed by Genichi Taguchi. Taguchi's orthogonal array (OA) is a step where provides less number of experiment run with high balanced combination of inputs. While Taguchi signal-to-noise ratios (S/N) is statistical measurement for the performance of the experiment for desired output. S/N ratios can be divided into three categories which is Nominal-is-Best (NB), Lower-the-Better (LB) and Higher-the-better (HB) and tits depending on the quality of the product or process being optimize. Highest S/N means the optimal parameters combination (Mahapatra and Chaturvedi, 2009).

3. Research Methodology

The experiment was conducted using the high-speed end milling modeled NEXUS 410A-II with vertical centre. The carbide inserts are attached to a 30 mm diameter of end milling cutter holder, While for the work piece, hard steel material which is D2 are chosen as the work material of this experiment as shown in the figure 13. Work material used is hardened steel D2 with dimension of 200 mm x 100 mm x 50 mm hardened to 52-56 HRC. For this experiment, a fresh carbide inserts will be used for each run of the experiment. The result and analysis of using Taguchi method using JMP statistical analysis software in analyzing and identify the optimizing the machining parameters in high speed end milling. Table 1 shows the experiment design using L9.

Table 1. Ranges of cutting parameters of the Experiment.

Cutting parameters	Min.	Max.
Cutting speed(m/min)	120	240
Feed rate (mm/tooth)	0.05	0.15
Depth of cut (mm)	0.10	0.20

4. Statistical analysis

Tables 2 and 3 show the statistical details necessary for determining whether a significant relation exist between the variables.

IOP Conf. Series: Materials Science and Engineering **184** (2017) 012047 doi:10.1088/1757-899X/184/1/012047

Table 2. Analysis of variance table.

Source	DF	Sum of Squares	Mean Square	F Ratio
Model	4	0.06781083	0.016953	16.5830
Error	4	0.00408917	0.001022	Prob > F
C. Total	8	0.07190000		0.0093*

Table 3. Summary of fit table.

RSquare	0.943127
RSquare Adj	0.886254
Observations (or Sum Wgts)	9

Based on the results in tables 2 and 3, the RSquare of 0.943127 is in reasonable agreement with the RSquare Adj of 0.886254 and Values of 'Prob > F' less than 0.05 and significant for both models. Therefore, the value of RSquare shows an adequate signal. The S/N ratio were developed using JMP statistical analysis. Table 4 summarizes the results of surface roughness and Figure 1 shows the relationship between the cutting speed, feed rate and depth of cut with the signal to noise ratio. Signal-to-noise ratio is a method used to measure the performance of the experiment. Signal means the desirable effect to the response or mean while noise is an undesirable effect or standard deviation to the response characteristics.

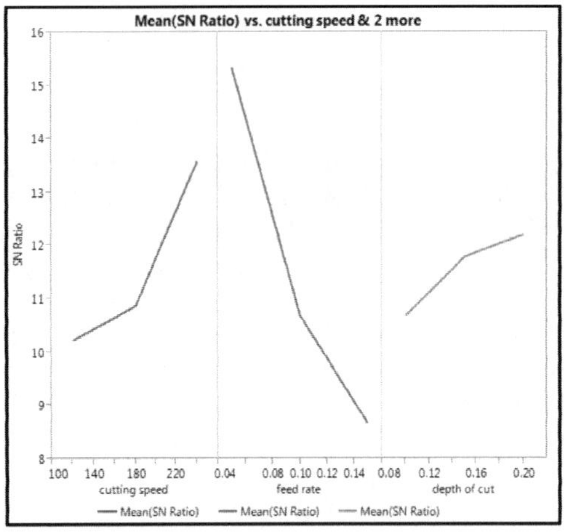

Figure 1. S/N ratio for surface roughness.

Table 4. Summarize result for analyzing process.

V_c (m/min)	F (mm/tooth)	DOC (mm)	Roughness Average, Ra (mm)
120	0.05	0.1	0.24
120	0.1	0.15	0.29
120	0.15	0.2	0.42
180	0.05	0.15	0.19
180	0.1	0.2	0.32
180	0.15	0.1	0.385
240	0.05	0.2	0.11
240	0.1	0.1	0.27
240	0.15	0.15	0.31

At the high cutting speed, the surface roughness is improved as the feed rate decrease as shows in the figure 4a. In the other hand, at low cutting speed, surface roughness is increased as feed rate increase. This result was in line with the research done by [7] which also observed similar result that, surface roughness can be minimized by utilizing lower values of feed rate with higher cutting speed. While depth of cut is relatively affect the depth of cut according to figure 4b and 4c which the color of contour was only blue. In order to investigate the effect of cutting parameters combination of high speed end milling on flank wear rate of insert carbide, 3D contour plot were generated using JMP software. Figure 2a, 2b and 2c below shows the interaction between cutting parameters and surface roughness.

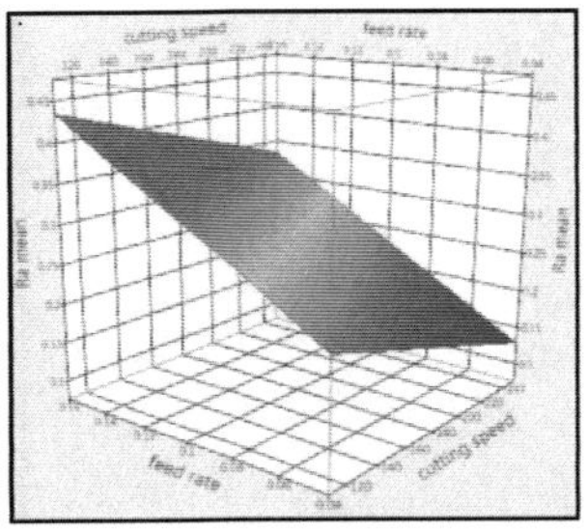

Figure 2a. Effect of cutting speed and feed rate on surface roughness.

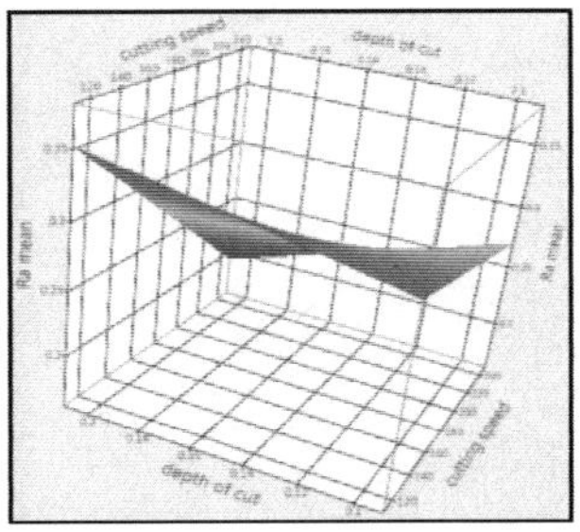

Figure 2b. Effect of cutting speed and DOC on surface roughness.

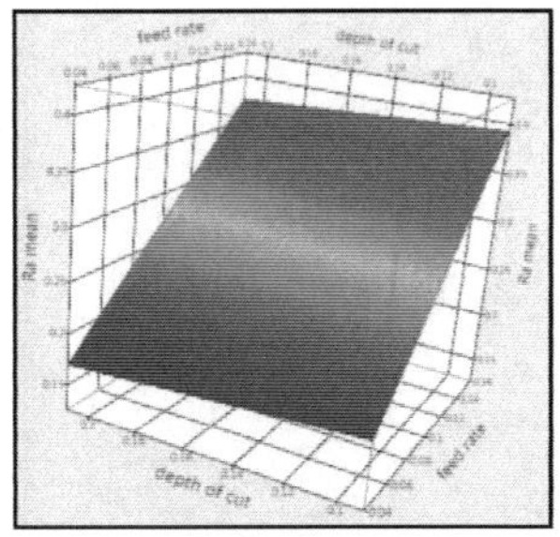

Figure 2c. Effect of feed rate and DOC on surface roughness.

The scatter diagram have been created using JMP software. Figures 3a, 3b and 3c show the scatter diagram between the cutting speed, feed rate and depth of cut on the surface roughness and signal to noise respectively.

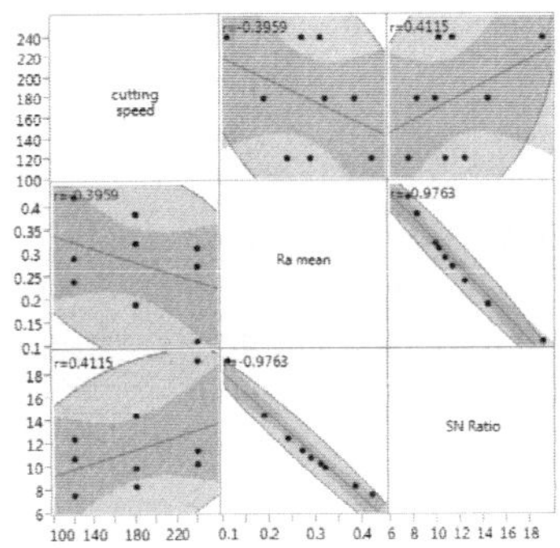

Figures 3a. Scatter diagram of cutting speed with the surface roughness and signal to noise ratio.

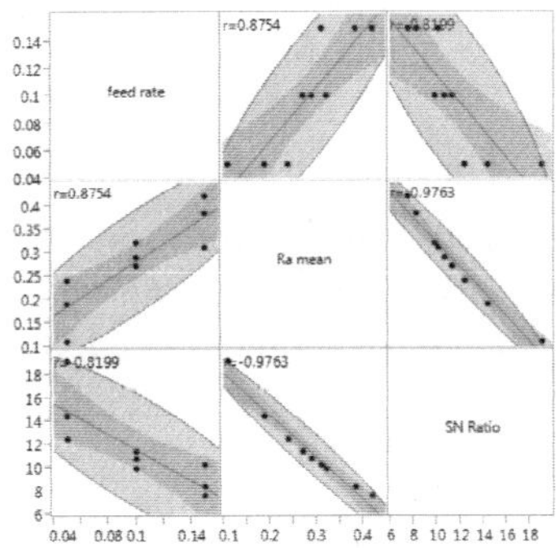

Figures 3b. Scatter diagram of feed rate with the surface roughness and signal to noise ratio.

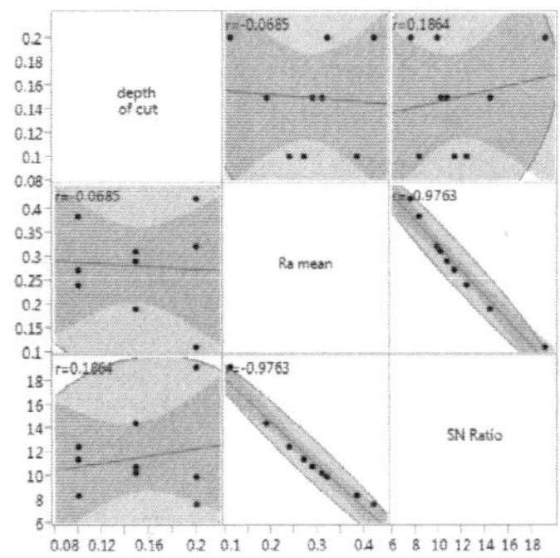

Figures 3c. Scatter diagram of depth of cut with the surface roughness and signal to noise ratio.

From the above figures 3a, 3b and 3c, it was found that the most effect factor on surface roughness and signal to noise ratio is the feed rate with correlation equal to 88% and 82% on mean surface roughness and signal to noise ratio respectively.

4.1 Desirability function

Desirability function method has been used to determine the best cutting levels to achieve the minimum surface roughness with reducing the signal to noise ratio. Figure 4 show the desirability function created by JMP software.

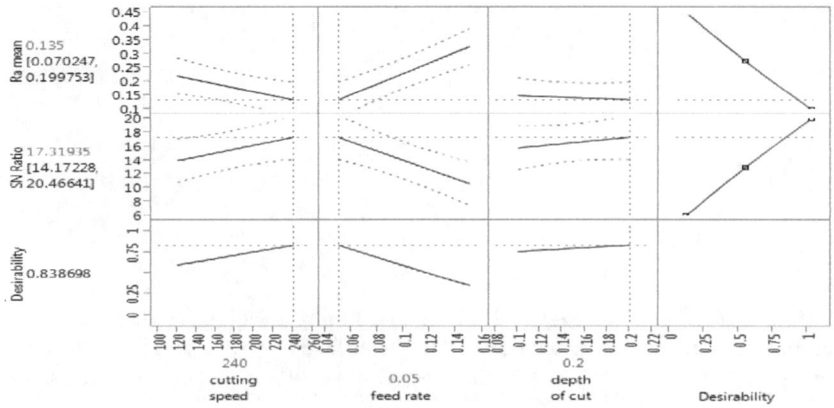

Figure 4. Desirability function graph.

The graph show that the maximum desirability was 83% can give Ra equal to 0.135 and S/n of 17.3. These values can achieved with cutting speed of 240mm/min, feed rate of 0.05 mm/tooth and depth of cut of 0.2 mm.

5. Summary

The experimental process was successfully done and the data was collected through High Speed End Milling process. It was found that the main effective factor on surface roughness was the feed rate. Increasing the feed rate has a negative effect on the roughness in contrast the cutting speed has a positive effect on the surface roughness while depth of cut is relatively affect.

References

[1] Pare, V., Agnihotri, G. and Krishna, C. M. 2011 Optimization of Cutting Conditions in End Milling Process with the Approach of Particle Swarm Optimization. International Journal of Mechanical and Industrial Engineering 1(2) 21-25.

[2] Mohd Ezuwanizam, M. 2010 Optimization of tool life in miling (Doctoral dissertation, Universiti Malaysia Pahang).

[3] Xuan-Truong, D. and Minh-Duc, T. 2013 Effect of Cutting Condition on Tool Wear and Surface Roughness during Machining of Inconel 718 Int J Adv Engg Tech/IV/III/July-Sept 108, 112.

[4] Malvade, N. V. and Nipanikar, S. R. 2014 Optimization of Cutting Parameters of End Milling On Vmc Using Taguchi Method Journal of Engineering Research and Studies.

[5] Moshat, S., Datta, S., Bandyopadhyay, A. and Pal, P. 2010 Optimization of CNC end milling process parameters using PCA-based Taguchi method International Journal of Engineering, Science and Technology 2(1) 95-102.

[6] Mahapatra, S. S. and Chaturvedi, V. 2009 Modelling and analysis of abrasive wear performance of composites using Taguchi approach International Journal of Engineering, Science and Technology 1(1) 123-135.

[7] Vorasri, M., Jirapattarasilp, K., and Kaewkuekool, S. 2011 The effect of high-speed milling on surface roughness of hardened tool steel World Academy of Science, Engineering and Technology 59 469-472.

[8] Kasim, M. S., Sulaiman, M. A., Hassan, C., Haron, C., AG, J., AI, G. and Zaid, M. 2013 Tool Life of TiAlN PVD Coated Carbide Tool in High-speed End Milling of Untreated Inconel 718 under Minimum Quantity Lubrication Condition Sains Malaysiana 42(12) 1721-1726

Flank wear analysing of high speed end milling for hardened steel D2 using Taguchi Method

Muataz Hazza Faizi Al-Hazza[1,a], **Nur Asmawiyah bt Ibrahim**[2], **Erry T. Y. Adesta**[3] **Ahsan Ali Khan**[4] **and Atiah Bt. Abdullah Sidek**[5]

[1,2,3,4,5] Department of Manufacturing and Materials Engineering, Faculty of Engineering, International Islamic University Malaysia, 53100 Kuala Lumpur, Malaysia

E-mail: [a] muataz@iium.edu.my

Abstract. One of the main challenges for any manufacturer is how to decrease the machining cost without affecting the final quality of the product. One of the new advanced machining processes in industry is the high speed hard end milling process that merges three advanced machining processes: high speed milling, hard milling and dry milling. However, one of the most important challenges in this process is to control the flank wear rate. Therefore a analyzing the flank wear rate during machining should be investigated in order to determine the best cutting levels that will not affect the final quality of the product. In this research Taguchi method has been used to investigate the effect of cutting speed, feed rate and depth of cut and determine the best level s to minimize the flank wear rate up to total length of 0.3mm based on the ISO standard to maintain the finishing requirements.

1. Introduction

High speed hard end milling is an advanced technology that merging three advanced machining processes: high speed milling, hard milling and dry milling. This advanced process increases the productivity and better final quality compare with the conventional machining. There are several researcher had done a researched on the consequence of cutting parameters on flank wear rate and tool life [1-4]. Ezuwanizam [5] claimed the optimal cutting parameters for cutting speed, feed rate and depth of cut in optimize the tool life of carbide insert coated with TiN in milling aluminum 6061. Xuan-Truong and Minh-Duc [6] had explained the effect of cutting parameters on flank wear through the transformation of natural logarithm. They found that cutting speed has given major significant effect on flank wear and surface roughness is mostly influenced by feed rate. While for depth of cut, they conclude that it only gives a minor effect for both surface roughness and flank wear especially at low cutting speed and feed rate. However, in this research taguchi method will be used to analyse and investigate the three independent factors: cutting speed (Vc), feed rate (f) and depth of cut on the flank wear rate ($Vb_{r)}$ up to flank wear length equal to 0.3 mm.

2. Taguchi method

Taguchi method is a tool of finding the best combination of an input (cutting parameters, cutting condition, work piece and cutting tool material properties) for producing a high quality of product and services. Malvade and Nipanikar [7] mentioned that Taguchi methods more focus on developing the design for manufacturing process for creating high quality product compared to statistical process control which tries to control the factors that affect the product quality. Taguchi's orthogonal array (OA) is a step where provides less number of experiment run with high balanced combination of inputs. While Taguchi signal-to-noise ratios (S/N) is statistical measurement for the performance of the experiment for desired output. Signal is a mean ratio while noise is a standard deviation. S/N ratios can be divided into three categories which is Nominal-is-Best (NB), Lower-the-Better (LB) and Higher-the-better (HB) and tits depending on the quality of the product or process being optimize. Highest S/N means the optimal parameters combination [8].

3. Experiment procedure

The experiment was conducted using the high-speed end milling modeled NEXUS 410A-II with vertical centre. The carbide inserts are attached to a 30 mm diameter of end milling cutter holder, While for the work piece, hard steel material which is D2 are chosen as the work material of this experiment as shown in the figure 13. Work material used is hardened steel D2 with dimension of 200 mm x 100 mm x 50 mm hardened to 52-56 HRC. For this experiment, a fresh carbide inserts will be used for each run of the experiment. The result and analysis of using Taguchi method using JMP statistical analysis software in analyzing and identify the optimizing the machining parameters in high speed end milling. Table 1 shows the experiment design using L9.

Table 1. Range of cutting parameters of the experiment.

Cutting parameters	Min.	Max.
Cutting speed(m/min)	120	240
Feed rate (mm/tooth)	0.05	0.15
Depth of cut (mm)	0.10	0.20

4. Statistical analysis

The ANOVA and S/N ratio were developed using JMP statistical analysis. Finally, JMP also used for finding the optimum cutting parameter combination in order to get the lower surface roughness and flank wear. Signal-to-noise ratio is a method used to measure the performance of the experiment. Signal means the desirable effect to the response or mean while noise is an undesirable effect or standard deviation to the response characteristics. The smaller-the-better output used in this analysis since the lower flank wear and surface roughness are desirable. The equation for calculating the smaller-the-better output as follow:

$$\mathrm{S/N} = -10\log_{10}\left[\frac{i}{n}\left(\sum_{i=1}^{a} y_i^2\right)\right]$$

y_i = measured experiment result in ith
n = the number of measurement in each run.

Table 2 summarizes the results of flank wear.

Table 2. Summarize result for analyzing process.

V_c (m/min)	f (mm/tooth)	DoC (mm)	Flank Wear Rate, (mm/min)
120	0.05	0.1	0.0027
120	0.1	0.15	0.0027
120	0.15	0.2	0.0031
180	0.05	0.15	0.0047
180	0.1	0.2	0.0056
180	0.15	0.1	0.0061
240	0.05	0.2	0.0077
240	0.1	0.1	0.0039
240	0.15	0.15	0.0095

Tables 3 and 4 show the statistical details necessary for determining whether a significant relation exist between the variables.

Table 3. Analysis of variance table.

Source	DF	Sum of Squares	Mean Square	F Ratio
Model	4	113.01009	28.2525	12.6611
Error	4	8.92579	2.2314	Prob > F
C. Total	8	121.93587		0.0153*

Table 4. Summary of fit model.

RSquare	0.926799
RSquare Adj	0.853599

Based on the results in tables 3-4, the Rsquare are in reasonable agreement with the RSquare Adjusted and Values of 'Prob > F' less than 0.05 and significant for both models. The signal to noise ratio has been developed as shown in Figure 1.

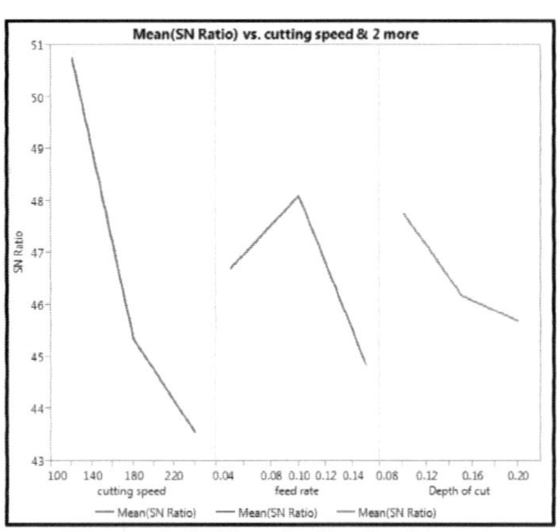

Figure 1. S/N ratio for flank wear.

The best cutting parameters for minimizing the noise factors were cutting speed of 120 m/min, feed rate of 0.10 mm/tooth and depth of cut of 0.10 mm. According to the slope of the graph, it shows that cutting speed gives most significant effect to the flank wear mean rate. This situation may happen because of the machining environment. As the flank wear increase due to the noise effect of vibration and temperature. In order to investigate the effect of cutting parameters combination of high speed end milling on flank wear rate of insert carbide, 3D contour plot were generated using JMP software. Figure 2a, 2b and 2c below shows the interaction between cutting parameters and flank wear mean rate of the insert carbide.

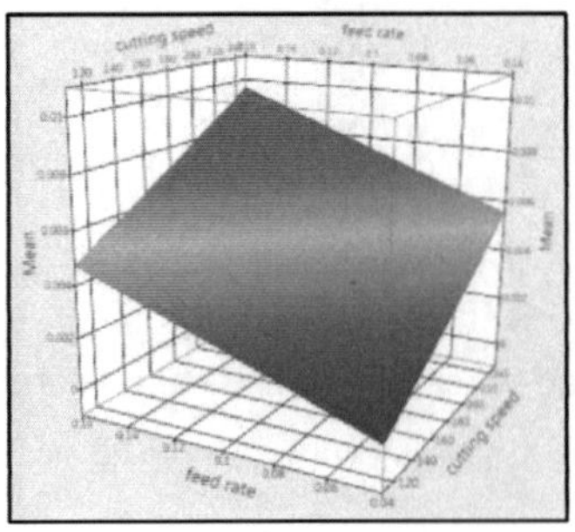

Figure 2a. Effect of cutting speed and feed rate on flank wear mean.

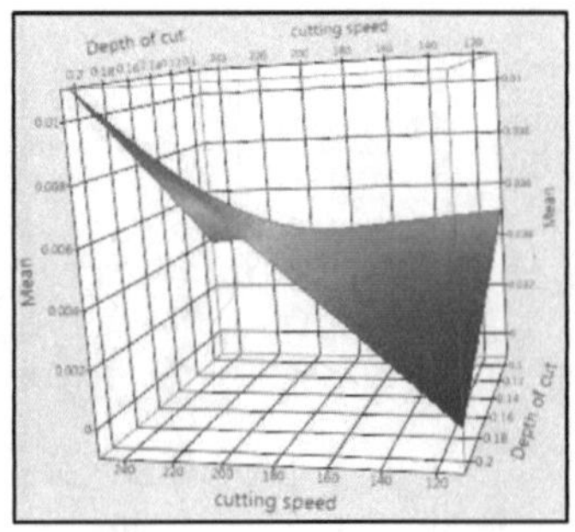

Figure 2b. Effect of cutting speed and DOC on flank wear rate.

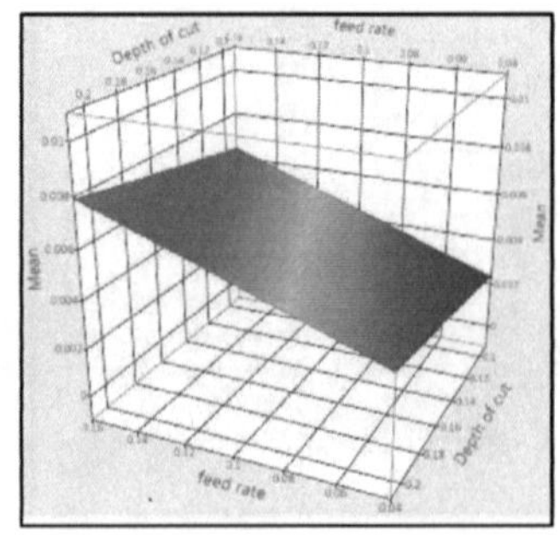

Figure 2c. Effect of feed rate and DOC on flank wear rate.

4.1. Scanning Electron Microscope (SEM) Image

Scanning Electron Microscope (SEM) is one of the electron microscope used for capturing the image of the small sample for example the flank wear of the insert. The insert was placed inside the SEM and the flank wear surface of the insert was analyzed as shows in the figure below. The image in figure 3a was taken using 40x magnification and figure 3b is the close up view for taking the measurement of the flank wear. The SEM image was captured in order to explain on how the flank wear measurement being taken using Hisomet II tool maker microscope. The flank wear reading was taken by measuring the average width of the wear on the insert. Figures 3a and 3b show the measurement for the flank wear of the insert by using SEM machine.

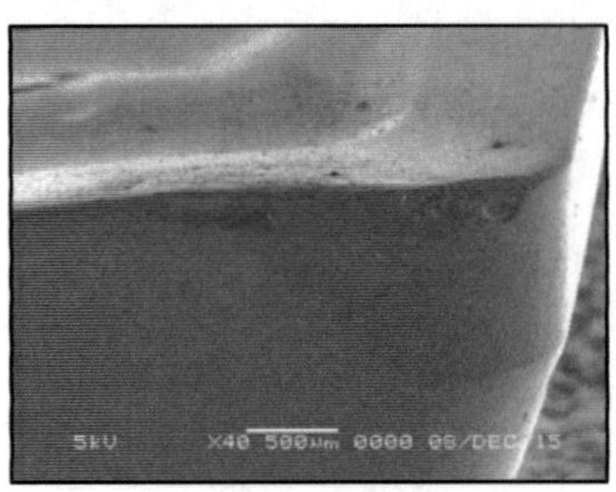

Figure 3a. Flank wear image captured by SEM.

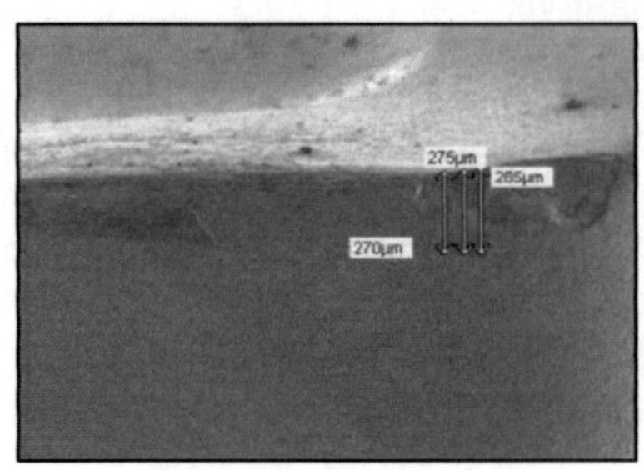

Figure 3b. Closed up of flank wear image for measurement.

Figure 4 show the graph relationship between flank wear of the insert and the machining time was constructed in order to determine the tool life of the insert as the machining time gradually increase. For determining the best tool life, the feed rate and depth of cut need to be optimized and the cutting speed need to be reduce. The relationship between the flank wear length and the machining time for different runs has been concluded on figure. Finally the desirability function graph is shown in Figure 5.

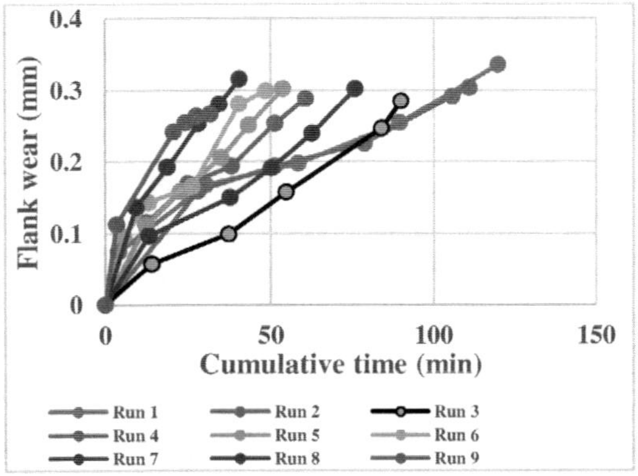

Figure 4. The relationship between flank wear and machining time.

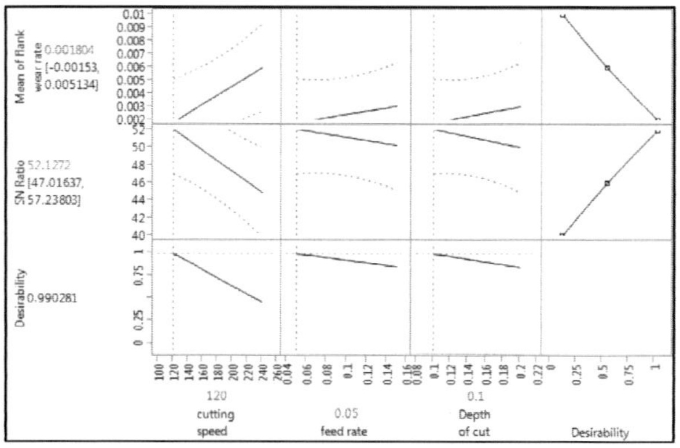

Figure 5. Optimization result for minimizing flank wear and maximizing S/N ratio.

5. Summary

The experimental process was successfully done and the data was collected through High Speed End Milling process. The data were analyzed using JMP statistical software. The statistical analysis shows a high correlation between the cutting parameters combination and the output responses. The new model for flank wear rate have been developed in terms of cutting speed, Vc (m/min), feed rate, f (mm/tooth) and depth of cut, DOC (mm). The desirability function was implemented to minimize the flank wear rate and maximize the signal noise ratio. It was found that with desirability 0f 98%, the cutting speed 120 mm/min, feed rate of 0.05 mm/tooth and 0.01 mm of depth of cut as shown in the Figure 5.

References

[1] Wang, B. and Liu, Z. 2016 Cutting performance of solid ceramic end milling tools in machining hardened AISI H13 steel International Journal of Refractory Metals and Hard Materials 55 24-32.

[2] Wang, C. Y., Xie, Y. X., Qin, Z., Lin, H. S., Yuan, Y. H. and Wang, Q. M. 2015 Wear and breakage of TiAlN-and TiSiN-coated carbide tools during high-speed milling of hardened steel Wear 336 29-42.

[3] Jozić, S., Bajić, D. and Stoić, A. 2015 Flank wear and surface roughness in end milling of hardened steel Metalurgija, 54(2) 343-346.

[4] Zhang, S., Li, J. F. and Wang, Y. W. 2012 Tool life and cutting forces in end milling Inconel 718 under dry and minimum quantity cooling lubrication cutting conditions Journal of Cleaner Production 32 81-87.

[5] Mohd Ezuwanizam, M. 2010 Optimization of tool life in miling (Doctoral dissertation, Universiti Malaysia Pahang).

[6] Xuan-Truong, Duong, and Tran Minh-Duc July-Sept 2013 Effect of cutting condition on tool wear and surface roughness during machining of Inconel 718 Int J Adv Engg Tech/IV/III 108 112.

[7] Malvade N. V., and Nipanikar S. R. 2014 Optimization of Cutting Parameters of End Milling On Vmc Using Taguchi Method Journal of Engineering Research and Studies.

[8] Mahapatra S. S. and C.Vedansh 2009 Modelling and analysis of abrasive wear performance of composites using Taguchi approach International Journal of Engineering, Science and Technology 1 no. 1 123-135.

International Conference on Mechanical, Automotive and Aerospace Engineering 2016 IOP Publishing
IOP Conf. Series: Materials Science and Engineering **184** (2017) 012049 doi:10.1088/1757-899X/184/1/012049

Modelling and control of Base Plate Loading subsystem for The Motorized Adjustable Vertical Platform

N M H Norsahperi[1,a*], S Ahmad[2,b], A F M Fuad[3,c] , I A Mahmood[4,d], S F Toha[5,e] , R Akmeliawati[6,f] and F J Darsivan[7,g]

[1-6] Department of Mechatronics Engineering, Kulliyyah of Engineering, International Islamic University Malaysia, Gombak, 53100, Malaysia
[7] Department of Mechanical Engineering, Kulliyyah of Engineering, International Islamic University Malaysia, Gombak, 53100, Malaysia

[a]nmheartziq@gmail.com , [b]salmiah@iium.edu.my , [c]amirfuhairah@gmail.com, [d]am.iskandar@iium.edu.my , [e]tsfauziah@iium.edu.my, [f]rakmelia@iium.edu.my and [g]fjd@iium.edu.my

Abstract. Malaysia National Space Agency, ANGKASA is an organization that intensively undergoes many researches especially on space. On 2011, ANGKASA had built Satellite Assembly, Integration and Test Centre (AITC) for spacecraft development and test. Satellite will undergo numerous tests and one of it is Thermal test in Thermal Vacuum Chamber (TVC). In fact, TVC is located in cleanroom and on a platform. The only available facilities for loading and unloading the satellite is overhead crane. By utilizing the overhead crane can jeopardize the safety of the satellite. Therefore, Motorized vertical platform (MAVeP) for transferring the satellite into the TVC with capability to operate under cleanroom condition and limited space is proposed to facilitate the test. MAVeP is the combination of several mechanisms to produce horizontal and vertical motions with the ability to transfer the satellite from loading bay into TVC. The integration of both motions to elevate and transfer heavy loads with high precision capability will deliver major contributions in various industries such as aerospace and automotive. Base plate subsystem is capable to translate the horizontal motion by converting the angular motion from motor to linear motion by using rack and pinion mechanism. Generally a system can be modelled by performing physical modelling from schematic diagram or through system identification techniques. Both techniques are time consuming and required comprehensive understanding about the system, which may expose to error prone especially for complex mechanism. Therefore, a 3D virtual modelling technique has been implemented to represent the system in real world environment i.e. gravity to simulate control performance. The main purpose of this technique is to provide better model to analyse the system performance and capable to evaluate the dynamic behaviour of the system with visualization of the system performance, where a 3D prototype was designed and assembled in Solidworks. From the Solidwork, the model was translated to Simmechanics with the system coordinate and specification i.e mass and inertia and actuator model was designed by using Simpower for simulating the system. Then, the system was integrated with controller by using conventional Proportional-Derivative (PD) controller with 0% steady state error, (ess) and 22.4% overshoot, (P.O) as the results.

1. Introduction

Satellite is an artificial body placed in an orbit round the earth or another planet used for collecting information and communication according to Oxford Dictionaries. Satellite will be orbiting around the outer space of the world to provide high efficiency on communication system and data interchange from one place to even the rural area. Malaysia has launched two satellites since 2000 named TiungSAT-1 (2001) and RazakSAT (2009), [1]. ANGKASA is the responsible agency in the space research in Malaysia, has taken another step by building up the Satellite Assembly, Integration and Test Centre (AITC). All operations such as assembly, integration works and launch and environmental testing now can be done in AITC.

Many types of chambers with specific function have been installed in the AITC. Thermal Vacuum Chamber (TVC) is the one of the chamber that is used for thermal vacuum test, which is critical test for satellite. It is located in the one of the ANGKASA's clean room, [2]. The MAVeP was proposed to be one of the facilities at the TVC for transferring satellite from the loading bay into the TVC. As the chamber is located in clean room that has a platform with limitation in space, the precision and accuracy of all mechanisms is important to maximize the safety of the satellite. The Figure 1 shows the overview of the TVC and the space available for the system to operate.

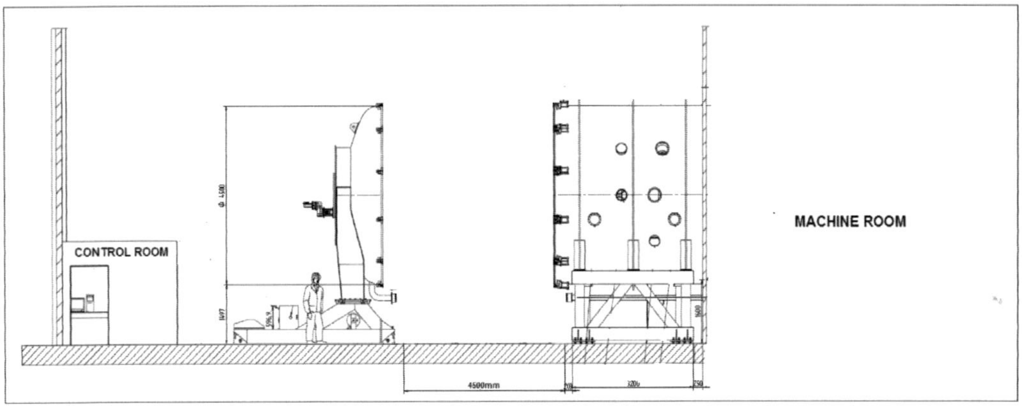

Figure 1. The TVC dimension and overview, [1].

Many subsystems are involved to transfer the satellite such as Mobility, Elevation, Locking and Base plate horizontal loading subsystems. After the satellite is fabricated, the satellite will be placed at the loading bay. The satellite will be transferred by MAVeP from the loading bay into the clean room by Mobility subsystem. After it has safely parked in front of TVC, the platform, which is at stowed position will elevate the satellite to the height where the rails of the TVC is to be reached and the beam will be extended to join the existing rail in the TVC for load transfer. By utilizing the Base plate Horizontal loading Mechanism, the satellite will be transferred from the MAVeP into the TVC to undergo the extreme thermal and pressure tests. Figure 2 visualizes the overview of MAVeP in 3D software.

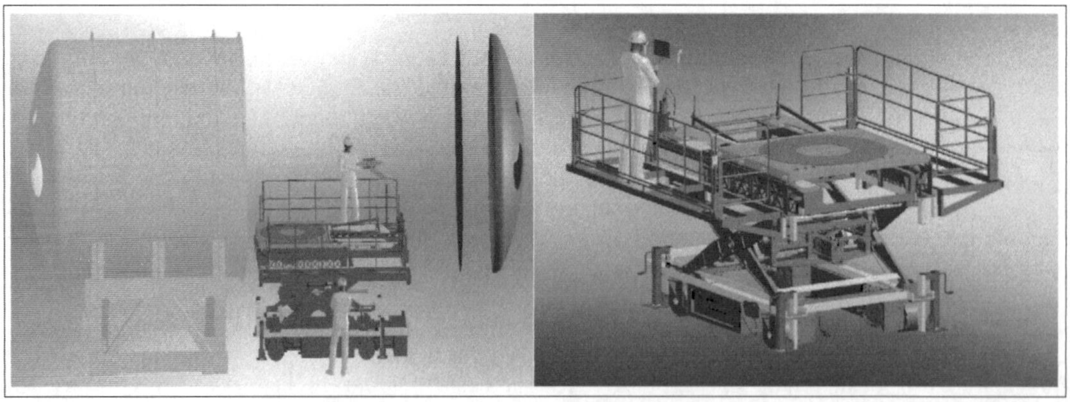

Figure 2. Visual overview of MAVeP in 3D software.

The Base plate subsystem needs to produce horizontal motion. The additional mechanical attachment is required to convert the angular motion produce by motor to the horizontal motion. Table 1 summarizes the common mechanical attachments that have been considered to demonstrate the motion. The Base plate Subsystem motion will have 0.25 to 0.5 metre per minute (m/min) velocity with maximum 3 meters travel distance as its specification requirements. For the motorized platform for satellite test, low velocity is preferred as the operators will be on the platform during the process (loading and unloading). In [3, 4] the researchers stated that, high velocity can cause vibration and long settling time. In order to prevent vibration, velocity profile is used to control the acceleration and reduce the vibration. The Based on the table, the rack and pinion is the most suitable and meets the system specification.

Table 1. Summary of the Common Mechanical Attachment.

	Rigid Chain, [5, 6].	Ball Screw, [7].	Rack Pinion, [8].
Price	High	Medium	Low
Availability	Mostly at United State	Local & Oversea	Local & Oversea
Accuracy	Medium	0.03 mm	0.05 mm (Total travel)
Advantages	Used for lifting the stage	Used for Z-axis motion	Used for horizontal motion
	Suitable for short distance	Suitable for short distance	Suitable for long distance (>3m)
		Low speed travel	Less backlash
			Low maintenance cost
			High speed travel
Disadvantages	Product availability	Fast motion can produce vibration	Travel depends on the rail guide
	High maintenance cost	Easy to have high backlash	

System identification needs the actual plant to obtain the system input and output data, which currently is not available. By implementing the physical modelling technique, the extensive knowledge about the dynamic behaviour of plant is crucial, [9]. The technique will represent the system with mathematical modelling by utilizing the theory such as Kirchhoff's Laws (KCL and KVL) and Newton's Laws, [9, 10]. In order to avoid un-modelled dynamics behaviour and uncertainty of the system by using physical model and reduce the time consumption to fabricate the system, 3D virtual model is the best method as it is very flexible and easy to construct, [10, 11]. Besides, the Propotional-Integration-Derivative (PID) control was implemented in the plant to enhance the elevation performance of the Base plate Subsystem. Figure 3 shows the control system architecture to reduce the positional error. A PD controller is commonly adopted in various types of system to produce an accurate positional system i.e. Robot Arm and actuator, [12,13]. The controller is capable to reduce the overshoot and manipulate the rise time. Yet, it will increase the steady state error in the system, [14]. The problem can be solved by adding an additional controller.

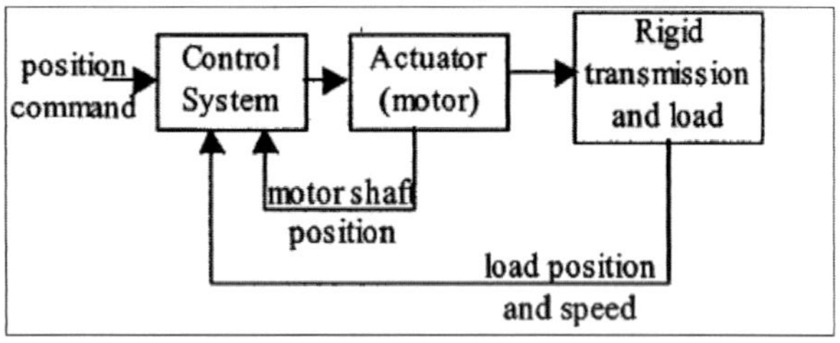

Figure 3. Control system architecture for precision position, [15].

2. Horizontal Base plate Loading Mechanism

2.1. Modelling of Base plate Subsystem

The common techniques to represent the system for control are system identification and physical modelling. Both techniques are time consuming and required extensive understanding about the dynamical behaviour of the system. By using 3D virtual, the system can be represented as close as to the actual system with visualization. Figure 4 shows the system model with dimension of 2.0 m width and 3.5 m length together with load and base plate mass approximately 1000 kg by using Solidworks and SimMechanics inspired from [16].

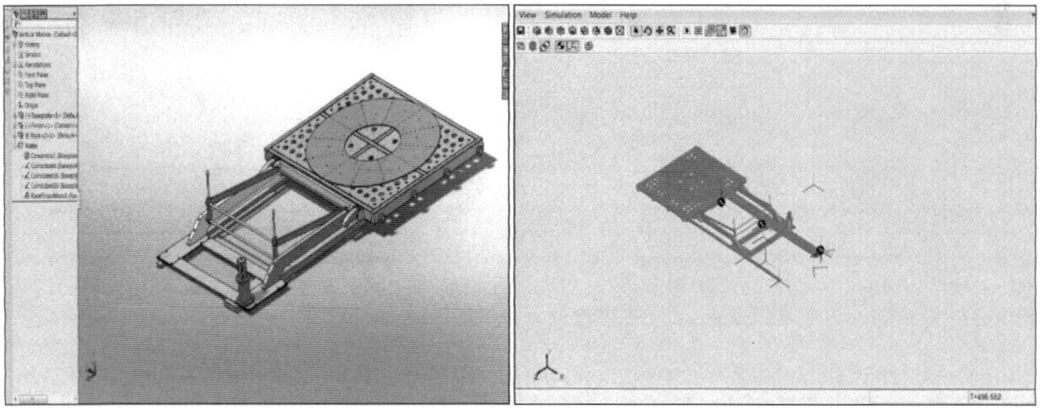

(a) (b)

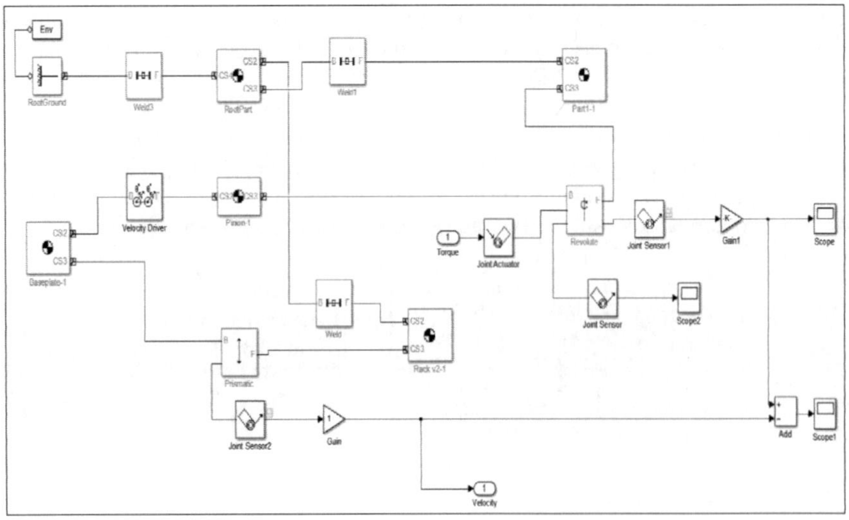

(c)

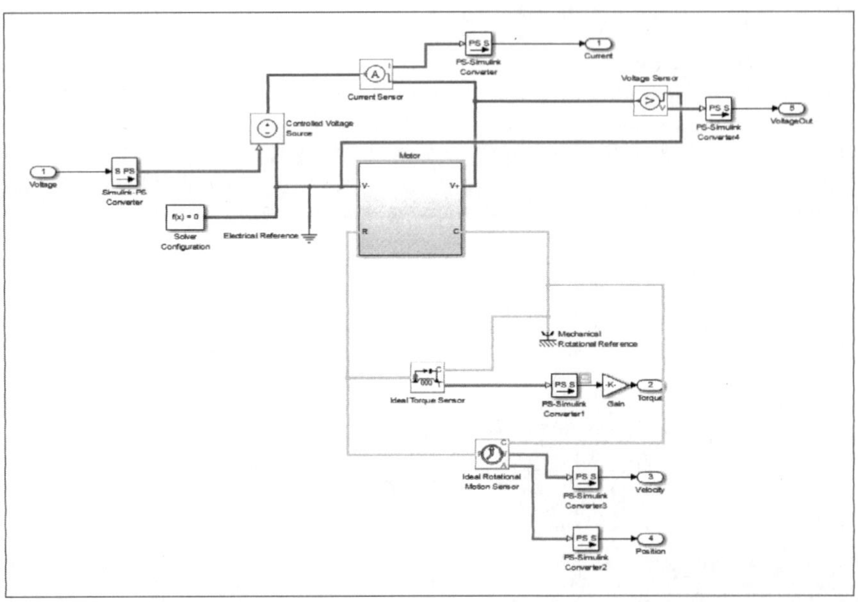

(d)

Figure 4. (a) Modelling the system by Solidworks; (b) Converted model from Solidwork to Sim Mechanics; (c) Translated model in SimMechanics; (d) Motor Model in SimPower.

The most crucial step is to set up both softwares where the Solidwork is to be linked with the Matlab software by command prompt in Matlab workspace. Secondly, all elements; base plate, rack, pinion and slider were assembled in Solidwork software and all *mates* in Solidwork should be compatible with Simmechanics as the software will translate the system based on configuration in Solidwork. The *mate* configuration is available on Matlab website. After the assembly process has been done, the system should be saved as .xml. Some of advanced *mates* such as rack and pinion and *cam* are incapable to be translated directly from Solidwork. Therefore, such *mates* can be set-up in

Simmechanics. Then, open the file by command prompt (*mech_import*) in Matlab workspace. All system configurations such as joints, actuators and sensors can be checked and modified in Simmechanics i.e system coordinate and components' mass. Based on coordinate, Simmechanics will run the model by using equation of motion by referring the coordination motion. Motor model was designed in Simulink and all motor parameters such as rated torque and rated speed for motor are based on Sanyo Denki motor R2AA08020F specifications.

2.2. *Control of Horizontal Motion of Baseplate Subsystem*

The conventional Proportional-Derivative (PD) control was first designed to simulate the model developed in SimMechanics. The PID equations can be represented as in equation (1).

$$u(t) = K_p e(t) + K_i \int_0^t e(\tau)d\tau + K_d \frac{de(t)}{dt} \quad (1)$$

Table 2 describes the gain parameters used for the controller. The gains that were tuned heuristically by using Ziegler-Nichols (ZN) method and followed by fine-tuning (FT), where critical gain value, K_{cr} and critical time, P_{cr} are 20 and 1200 seconds respectively, have produced satisfactorily system response for position of base plate horizontal loading. ZN technique is only feasible to be conducted if experimental data is available. In this paper, ZN technique is used to provide a good educated guess the starting point for tuning the controller parameter rather than trial and error.

Table 2. Gains Parameters.

Gain	K_p	K_i	K_d
ZN	12	0	1800
FT	8	0	1800

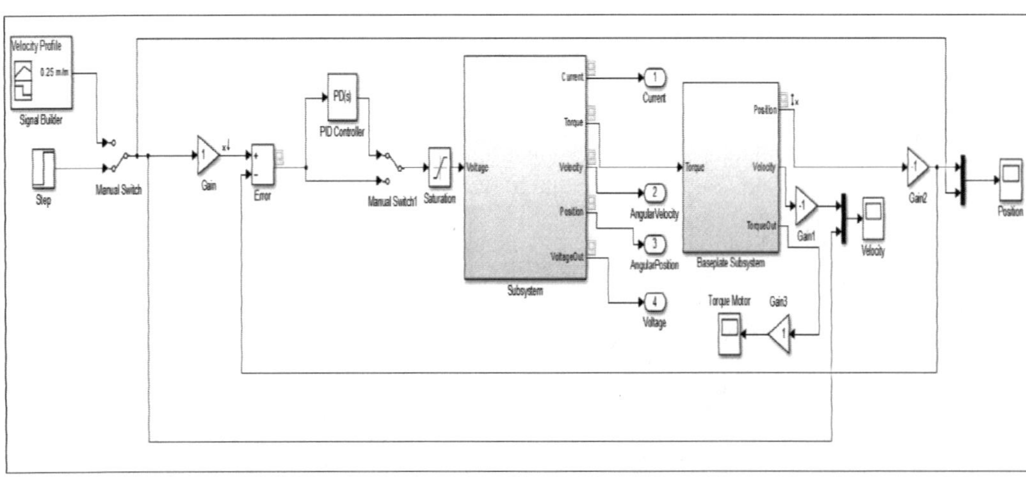

Figure 5. The Proportional-Derivative Control for Base plate Subsystem.

Based on Figure 5, the control was designed to improve the accuracy of the horizontal loading system. The control will compare the actual position with the desired position and compute the error. The error occurs will be manipulated by controller to force the system to perform as the desired performance. As the satellite has countless of sensitive electrical and electronics board, any vibration or jerk occurs must be minimized. Therefore, the control should have less steady state error than 1.0 mm, percentage of overshoot should be less than 50% and appropriate rise time.

3. Results and Discussion

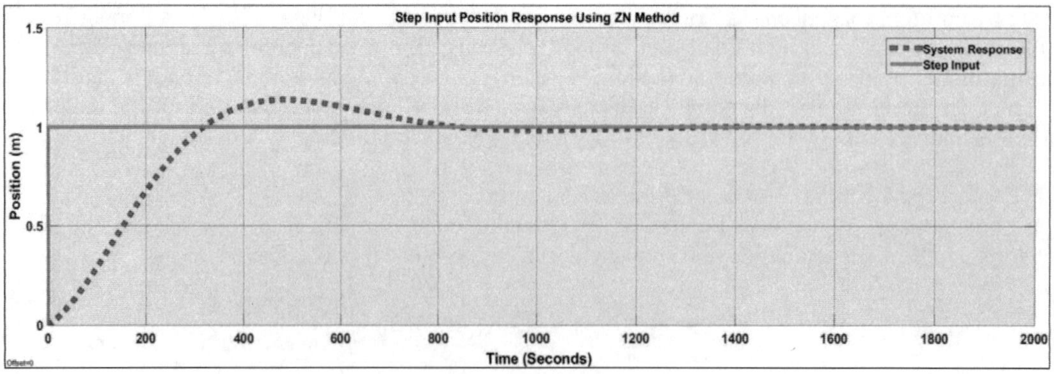

(a)

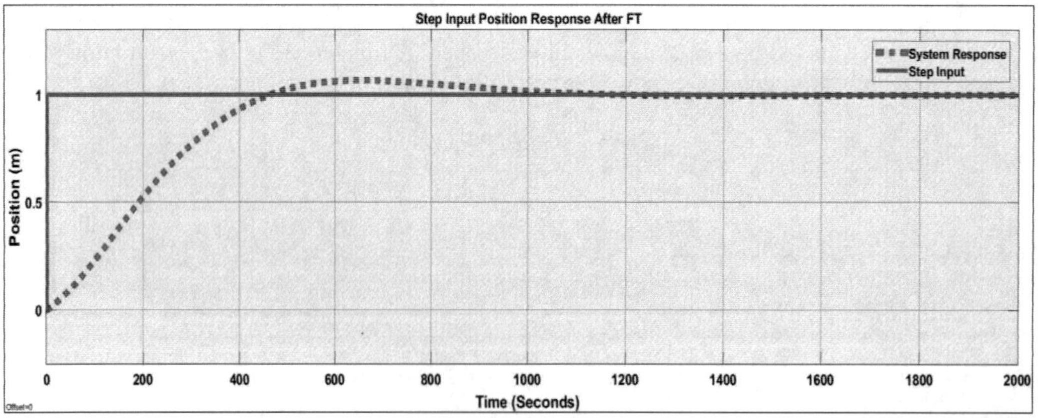

(b)

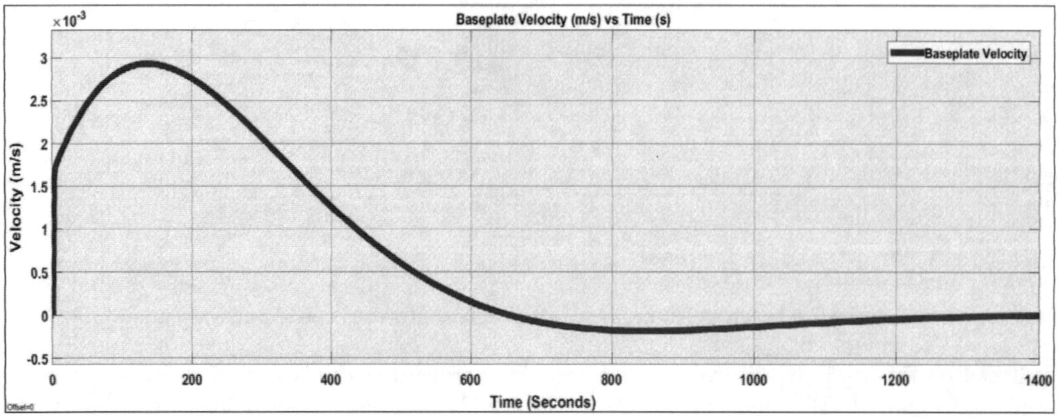

(c)

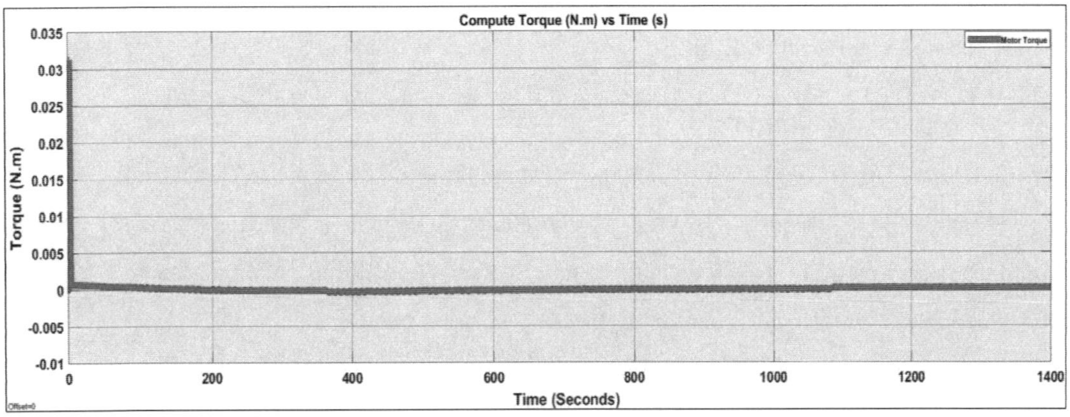

(d)

Figure 6. (a) Step Input Position Response Using ZN Gain; (b) Step Input Position Response After Fine-Tuning; (c) Baseplate Velocity (m/s); (d) Motor Torque (N.m).

Figure 6 (a) shows that the time travel to achieve the desired destination is approximately 1400 seconds for 1 meter travel distance with 27.8% of overshoot and 0 % steady state error. The system rise and settling time are 128 seconds and 1020 seconds respectively after the system was tuned by ZN method. After Fine-Tuning, the system overshoot, rise time and settling time are reduced to 22.4%, 147 seconds and 820 seconds respectively while steady state error remains constant. The integral gain is neglected as the gain has increased the system overshoot during Fine-Tuning process and PD controller has computed 0% steady state error even without integral gain. Meanwhile the Figure 6 (c) and (d), the motor needs 0.031 N.m torque to produce 0.0029 m/s linear velocity of the baseplate. The initial torque is high and then the torque is approaching to zero.

4. Conclusion

In summary, the model of the MAVeP system that represents the Base plate horizontal loading subsystem has been successfully modelled with 3D model technique using Solidworks and Simscape capability. The main challenge is to model the system in the software as close as possible to the actual system and to determine the system validity and parameters. The modelled system has been tested with PD control to evaluate the system performance. Based on the result, the system has 0% steady state error with 22.4% overshoot. The control system performance can be improved by computational method such as Genetic Algorithm to optimize control parameter. In future, the control will be integrated into the real system to evaluate the actual control performance. As the initial torque is high and step input can produce vibration, therefore, the velocity profile will be included to reduce initial torque and vibration in the system.

Acknowledgements

I would like to show my gratitude to the Government of Malaysia in Ministry of Science, Technology and Innovation as the funder of the project and also to ANGKASA and Universiti Putra Malaysia for their cooperation to realizing the mission of this project.

References

[1] Agensi Angkasa Negara 2015 *Design and Development of Motorised Adjustable Vertical Platform for Satellite Assembly, Integration and Test Facility* (Banting: ANGKASA)

[2] Woo L L, M Ismail and M D Subari 2009 *Setting-up the Assembly, Integration and Test Centre in Malaysia* (Istanbul:IEEE)

[3] H Li, M D Le and W Lin 2009 *Motion Profile Design to Reduce Vibration of High-Positioning Stages* (Singapore:IEEE)

[4] X Li 2010 *Optimizing S-Curve Velocity for Motion Control* (Alabama:ResearchGate)

[5] G Hengeveld, N Idsardi and B Witjes 2015 *Airstacking Machine* (Twente:University of Twente)

[6] Information on http://www.serapid.com/en/rigid-chains-can-lift-push-and-pull-loads.

[7] Enkon Pro 2011 *Hercules* (Canada: Enkon Information Systems)

[8] Information on http://www.camaster.com/wp-content/uploads/2013/09/Rack-and-Pinion%20or%20Screw%20Driven%20System.pdf.

[9] S F Toha 2014 *Basic Concepts of Physical Modelling and System Identification* (Gombak:IIUM)

[10] M T Islam, C Yin, S Jian and L. Rolland 2014 *Dynamic Analysis of Scissor Lift Mechanism Trough Bond Graph Modelling* (Canada: IEEE)

[11] S Shamsulkamar 2014 *Modelling and Control of 6-DOF of Industrial Robot by Using Neuro-Fuzzy Controller* (Johor: Universiti Tun Hussein Onn Malaysia)

[12] M A Azman, A A M Faudzi, C C Kai and etc. al. 2013 *P-Adaptive Nuero-Fuzzy and PD-Fuzzy Controller Design for Positional Control of a Modified Single Acting Pneumatic Cylinder* (NSW: IEEE)

[13] C Murrugarra, G Fernandez and O De Castro 2008 *Design of a PD Position Control Based on the Lyapunov Theory for a Robot Manipulator Flexible-Link* (Kunming: IEEE)

[14] Y Y Tan, A Irwan & M M Alam 2015 *PD-FLC with Admittance Control for Hexapod Robot's Lef Positioning in Seabed* (Kota Kinabalu:IEEE)

[15] A S Kulkarni and M A El-Sharkawi 2002 *Intelligent Precision Position Control of Elastic Drive Systems* (Canada:IEEE).

[16] Information on http://www.mathworks.com/company/newsletters/articles/modeling-machine-constraints-in-simmechanics-and-simulink.html#rackpinion

IOP Conf. Series: Materials Science and Engineering **184** (2017) 012050 doi:10.1088/1757-899X/184/1/012050

Modeling and simulation for heavy-duty mecanum wheel platform using model predictive control

A. F. M. Fuad [1,a*], I. A, Mahmood[2,b], S. Ahmad[3,c], N. M. H. Norsahperi[4,d], S. F. Toha[5,e], R. Akmeliawati[6,f] and F. J. Darsivan[7,g]

[1-6] Department of Mechatronics Engineering, Kulliyyah of Engineering, International Islamic University Malaysia

[7] Department of Mechanical Engineering, Kulliyyah of Engineering, International Islamic University Malaysia, Gombak, 53100, Malaysia

E-mail: [a]amirfuhairah@gmail.com, [b]am.iskandar@iium.edu.my, [c]salmiah@iium.edu.my, [d]nmheartziq@gmail.com, [e]tsfauziah@iium.edu.my, [f]rakmelia@iium.edu.my and [g]fjd@iium.edu.my

Abstract. This paper presents a study on a control system for a heavy-duty four Mecanum wheel platform. A mathematical model for the system is synthesized for the purpose of examining system behavior, including Mecanum wheel kinematics, AC servo motor, gearbox, and heavy duty load. The system is tested for velocity control, using model predictive control (MPC), and compared with a traditional PID setup. The parameters for the controllers are determined by manual tuning. Model predictive control was found to be more effective with reference to a linear velocity.

1. Introduction

The Mecanum wheel is an omni-directional wheel type, invented in 1973 by BengtIlon with the Swedish company, Mecanum A.B. [1] The Mecanum wheel design incorporates passive rollers mounted at 45 degrees to the wheel plane around a solid axis. The roller design allows for free movement in any direction without changing orientation. In addition, the Mecanum wheel design also allows for free rotation around its central axis with a relatively low driving torque, as well as minimal friction.

Both of these features allow for good maneuverability in areas that are very constrained and do not allow traditional wheels to navigate. [2] They also have the same torque requirements as a traditional wheel setup. However, they do come at a tradeoff, which is less traction with the floor surface, causing them to slip at lower torque levels compared to traditional wheels. [3] They are also very susceptible to rough floors, which may affect their ability to travel in the desired direction.

Mecanum wheels have been used in a variety of applications. Recent uses have seen it in space exploration, where NASA has explored hazardous terrain exploration options with Mecanum robots. Driven by brushless servomotors, NASA's OmniBot allowed agile movements in 2 degrees of freedom. In areas where human personnel are unable to be deployed, the OmniBot's ability to navigate areas freely without being limited by conventional wheel limitations is very powerful. [4,5]

Commercial applications towards the usage of Mecanum wheels also exist. Airtrax Sidewinder ATX-3000 forklifts are a good example, where the forklift's usage of Mecanum omnidirection wheels allows it to travel freely in tight spaces such as warehouses, though it has since halted production in 2013. [6]

The Mecanum wheels' usage in tight spaces allow for very specific uses in research environments as well. Malaysia's space agency, ANGKASA, has been exploring its usage in developing a satellite testing lifter platform. With total loads of up to 5 tons, the lifter platform is able to transport heavy satellite testing loads in a sterile clean room environment. With the application of Mecanum wheels, they were also able to achieve this in a very small area while minimizing shock with respect to the satellite's sensitive nature, which would have been normally very difficult with regular wheels. [13]

There have been many control schemes that have been presented for the Mecanum wheel platform. First, kinematics and dynamics are derived for a four-wheel setup, as in Tlale and de Villiers [7], and from there, we can use several methods. On the basic level, Viboonchaicheep et al. [8] have presented a position rectification control using Symptomatic Rectification (correction after inputs) and Preventive Rectification (indirect control parameter correction). More complete methods, which increase the effective maximum speed, include Lin and Shih, with nonlinear adaptive control with Lyapunov stability theory [9], as well as Tsai and Wu, with fuzzy control, through kinematic and joint-space dynamic modeling. [10]

This work focuses on proposing a simpler method, which will use a single Model Predictive Controller that reads a single velocity input and drives four servo outputs. The input is derived from the angular velocity of each of the four outputs. From the derivation of kinematics from the Mecanum wheel system as well as a simulation of an AC servo system, we can obtain a Mecanum wheel platform with linear and rotational velocity control using Model Predictive Control.

2. Modeling of Mobility System

2.1. AC Servomotor Equations.

From [11] and [12], we obtain the AC servomotor system, which consists of an AC motor coupled to a gearbox, as well as a load rigidly fixed to the output shaft.

Consider a motor, gearbox, and load with moments of inertia J_m, J_g and J_l. The motor is driven by a input control field voltage in its stator, denoted by V_s, and has two electrical constants K_1 and K_2. K_1 is calculated from the motor's stall torque, in N.m., divided by its rated voltage, and K_2 is calculated by its stall torque, in N.m., divided by its no-load speed at rated voltage in radians. The gearbox has a reduction ratio of N between the motor and the load. The load has a damping coefficient of c, in N.m.s/rad, and its angular displacement output is denoted as θ_m.

The motor's rotational motion can then be described by the following nonlinear differential equation:

$$(J_m + J_g + \frac{1}{N^2} J_l)\frac{d^2\theta_m}{dt^2} + c\frac{d\theta_m}{dt} = K_1 V_s - K_2 \frac{d\theta_m}{dt} \quad (1)$$

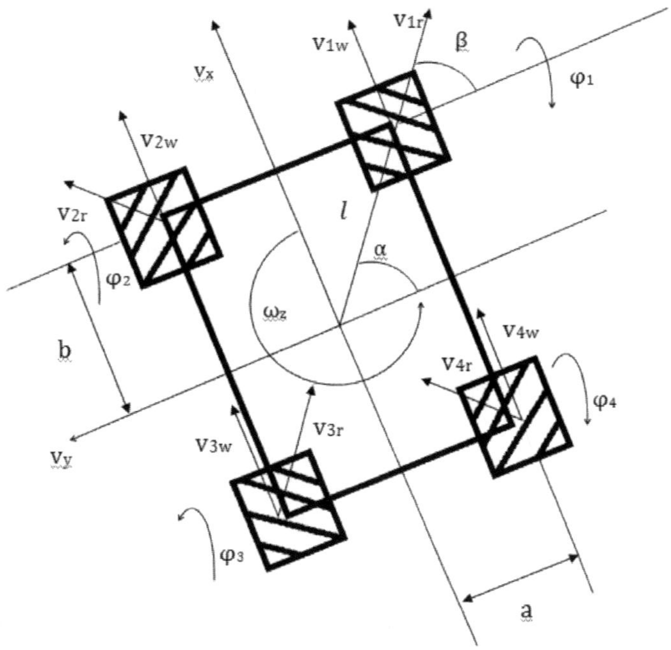

Figure 1. Configuration of Mecanum platform.

2.2. *Mecanum Platform Equation.*

Taking from [9], let the velocity in Cartesian coordinates be described as $v_c = [v_x\ v_y\ \omega_z]^T \in R^{3x1}$ where v_x and v_y are the linear velocities and ω_z is rotation about z-axis. The wheel angular velocities are described as $\omega_w = [\varphi_1\ \varphi_2\ \varphi_3\ \varphi_4]^T \in R^{4x1}$, with rollers angled at $\gamma = 45^o$ to the wheel axis, taken as clockwise looking outwards from the mounted frame as shown in Figure 1.

The wheels' linear velocities can be computed as $v_w = r\omega_w \cos\gamma$. Taking into consideration the Cartesian velocity vector, v_w can also be written as the following:

$$v_{wi} = r\omega_w \cos\gamma = [\sin(\alpha+\beta+\gamma) \quad -\cos(\alpha+\beta+\gamma) \quad -l\cos(\beta+\gamma)][v_x \quad v_y \quad \omega_z]^T \tag{2}$$

Where α is the angle between the geometric center of the platform and the wheel center, l is the distance between these positions, and β is the angle between the wheel axis and the line denoted by l. Main wheel radius is r. Let $v_{ci} = [v_{xi}\ v_{yi}\ \omega_{zi}]^T \in R^{3x1}$, where v_{ci} is the velocity in terms of inertial frame. The rotation matrix of the robot frame in terms of the inertia frame, $^R R_i(\theta)$, is as follows:

$$^R R_i(\theta) = \begin{bmatrix} \cos\theta & \sin\theta & 0 \\ -\sin\theta & \cos\theta & 0 \\ 0 & 0 & 1 \end{bmatrix}, \tag{3}$$

Where θ is the angle between the robot frame and inertial frame. $v_c = {}^{R}R_i(\theta)\, v_{ci}$. Assuming l and r are equal across all wheels, we can substitute the v_c term in v_w with ${}^{R}R_i(\theta)\, v_{ci}$. The α, β and γ values can be substituted for the following. Note that $\tan^{-1}(\frac{b}{a}) = \alpha$. Note that the derivation from [9] of the values in Table 1 is slightly different from this derivation, due to the modified configuration of the wheels, where this paper uses a diamond-like shape formation, and Lin and Shih use an X-shaped formation. All other derivations are the same.

Table 1. Wheel angular parameters.

Wheel	**α**	**β**	**γ**
1	$\tan^{-1}(\frac{b}{a})$	$-\tan^{-1}(\frac{b}{a})$	$-(\pi/2 + \pi/4)$
2	$\pi - \tan^{-1}(\frac{b}{a})$	$\tan^{-1}(\frac{b}{a})$	$(\pi/2 + \pi/4)$
3	$\pi + \tan^{-1}(\frac{b}{a})$	$-\tan^{-1}(\frac{b}{a})$	$-(\pi/2 + \pi/4)$
4	$2\pi - \tan^{-1}(\frac{b}{a})$	$\tan^{-1}(\frac{b}{a})$	$(\pi/2 + \pi/4)$

The substitution allows us to find the inverse kinematics equation,

$$\begin{bmatrix} \varphi_1 \\ \varphi_2 \\ \varphi_3 \\ \varphi_4 \end{bmatrix} = -(\sqrt{2}/r) \begin{bmatrix} -\sqrt{2}/2 & \sqrt{2}/2 & l\sin(\pi/4+\alpha) \\ -\sqrt{2}/2 & -\sqrt{2}/2 & l\sin(\pi/4+\alpha) \\ \sqrt{2}/2 & -\sqrt{2}/2 & l\sin(\pi/4+\alpha) \\ \sqrt{2}/2 & \sqrt{2}/2 & l\sin(\pi/4+\alpha) \end{bmatrix} \begin{bmatrix} \cos\theta & \sin\theta & 0 \\ -\sin\theta & \cos\theta & 0 \\ 0 & 0 & 1 \end{bmatrix} \begin{bmatrix} v_x \\ v_y \\ \omega_z \end{bmatrix}, \quad (4)$$

with the Jacobian defined as

$$J_0 = \begin{bmatrix} -\sqrt{2}/2 & \sqrt{2}/2 & l\sin(\pi/4+\alpha) \\ -\sqrt{2}/2 & -\sqrt{2}/2 & l\sin(\pi/4+\alpha) \\ \sqrt{2}/2 & -\sqrt{2}/2 & l\sin(\pi/4+\alpha) \\ \sqrt{2}/2 & \sqrt{2}/2 & l\sin(\pi/4+\alpha) \end{bmatrix} \begin{bmatrix} \cos\theta & \sin\theta & 0 \\ -\sin\theta & \cos\theta & 0 \\ 0 & 0 & 1 \end{bmatrix}. \quad (5)$$

Obtaining the pseudoinverse of J_0, which is J_0^+, we can find the forward kinematics equation to put into the final model.

$$J_0^+ = (J_0^T J_0)^{-1} J_0^T. \quad (6)$$

$$\begin{bmatrix} v_x \\ v_y \\ \omega_z \end{bmatrix} = -(\sqrt{2}/2)(r) J_0^+ \begin{bmatrix} \varphi_1 \\ \varphi_2 \\ \varphi_3 \\ \varphi_4 \end{bmatrix}. \quad (7)$$

3. Simulation Performance Analysis between Model Predictive Control (MPC) and Proportional-Integral-Derivative (PID)

Model predictive control (MPC) is an advanced control method that uses a set of plant outputs, state variables, model functions and process variable references and restrictions in order to form a "current model" of the system. It uses the current model to form future predictions of a finite prediction horizon. Because of this, MPC is also termed as "receding horizon control".

With this model, we have used Mathworks' MPC Controller Simulink block in order to form a linearized model of the plant. From there, the operating points of the system are extracted and the plant is sampled discretely with a period of 0.1s, prediction horizon of 1s, and a total run length of 10s. In addition, the MPC is compared against a PID model using PI controller. Both control methods were tuned to have almost zero overshoot and as low of a response time as possible. The reference value was a set linear velocity of 0.083m/s, or 5m/min.

Final value of PI controller are P = 5.28 and I = 19.5013. Table 2 shows the reference values used in simulating the model, taken from actual servos and figures from a proposed heavy-duty Mecanum wheel platform. The values are taken from servomotor SANYODENKI R2AAB8100FXP11, and gearbox SEW-EURODRIVE K87AQA115/1.

Table 2. Reference values in simulation.

Servo Rotor Moment ($kg.m^2$)	$6.7 \cdot 10^{-4}$	**Gearbox Ratio**	197.37
Gearbox Moment ($kg.m^2$)	$1.4 \cdot 10^{-4}$	**Stall Torque over Rated Voltage (Nm/V)**	$1.736 \cdot 10^{-2}$
Wheel Moment ($kg.m^2$)	3.987	**Stall Torque over No-Load Speed (Nm/rad)**	$1.216 \cdot 10^{-2}$
Platform Moment ($kg.m^2$)	61.935	**Wheel Radius (m)**	0.2032
Mass of Platform without Wheels (kg)	6000	**Mass of Wheel (kg)**	165.283
Wheel Axle Width (m)	1.5	**Wheel Axle Spacing (m)**	3.15

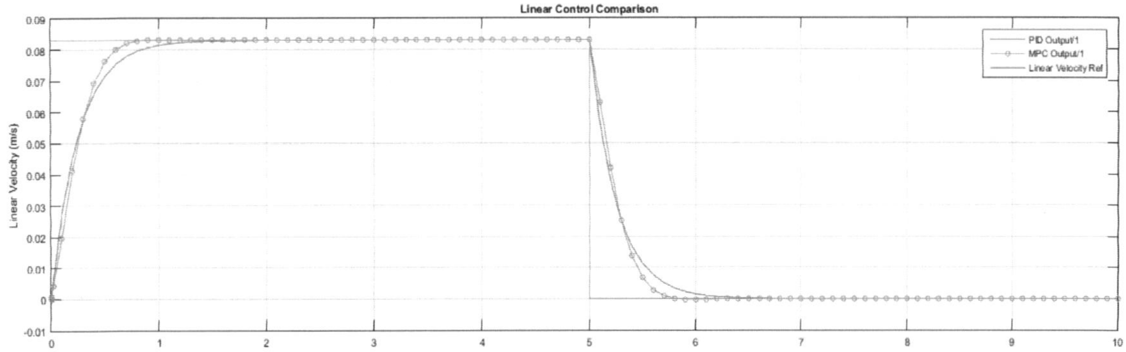

Figure 2. Linear velocity comparison between MPC and PID controller.

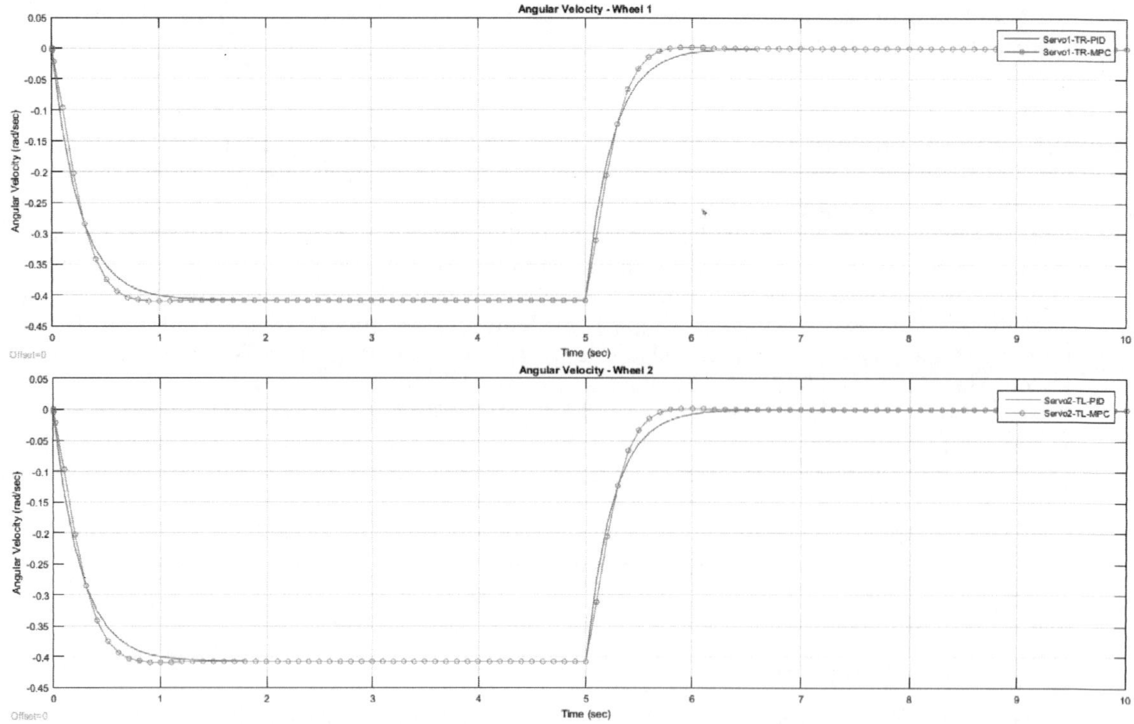

Figure 3(a) and (b). Angular velocity comparison between MPC and PID controller for front wheels on the right and left respectively.

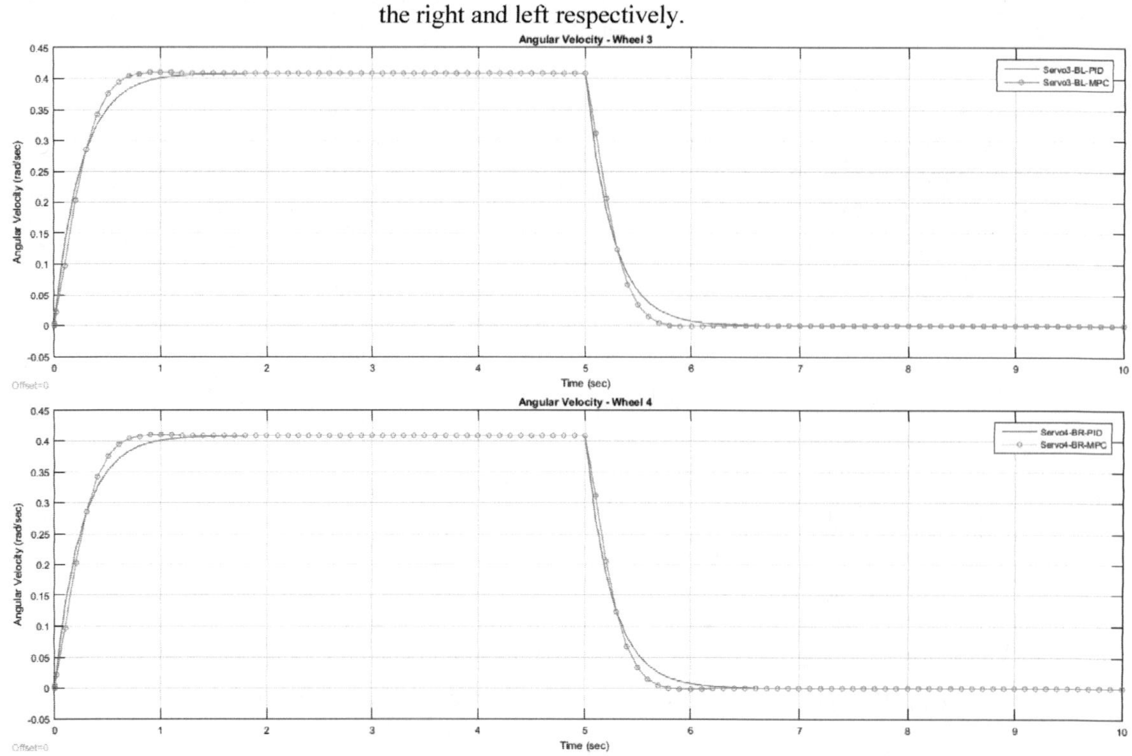

Figure 3(c) and (d). Angular velocity comparison between MPC and PID controller for rear wheels on the left and right respectively.

Figure 2 and Figure 3 shows the method in which the simulation was tested, where the controller simulated a desired velocity in the X-direction for the reference value of 0.083m/s for 5 seconds, and 0m/s for 5 seconds thereafter. As seen in the figures, the MPC control reaches the target velocity faster than the PID control by roughly 1 second. The limitation of this study is that there is no dynamics analysis, due to the very low speed tested for the platform. Factors such as frictional force reducing the speed were not taken into account.

4. Conclusion

In this paper, a mathematical model was derived for the kinematics of a Mecanum wheel platform driven by AC servomotors coupled to a gearbox and load. Two controller methods are tested: Model Predictive Control (MPC) and PID control.

Tuning was done automatically using computer simulation, and a test with velocity tracking was used to illustrate the difference between MPC and PID control. MPC controller was shown to have a more robust response compared to PID controller. In coming studies, more controller types can be tested, including Non-Linear Model Predictive Control (NMPC).

Finally, we would like to thank Mr. Dan Fisher, University of St. Thomas, Minnesota, USA, for helping us with the basic Mecanum wheel structure simulations.

References

[1] Han KL, Kim H, Lee JS. The sources of position errors of omni-directional mobile robot with Mecanum wheel. In *Systems Man and Cybernetics (SMC), 2010 IEEE International Conference* on 2010 Oct 10 (pp. 581-586). IEEE.

[2] Borenstein J, Feng L, Everett HR. *Navigating mobile robots: Systems and techniques*. AK Peters, Ltd.; 1996 Oct 1.

[3] Mecanum Force Vector Analysis [Internet]. 2016 [cited 17 December 2015]. Available from: http://www.chiefdelphi.com/media/papers/2390

[4] Lippit TC, Jones WC. "OmniBot mobile base". *KSC Re-search and Technology Report*, NASA, USA. 1998.

[5] Kennedy Space Center [Internet]. 2016 [cited 19 October 2015]. Available from: https://spinoff.nasa.gov/spinoff2000/ard8.htm.

[6] Omni Directional, Airtrax home [Internet]. 2016 [cited 17 December 2015]. Available from: http://www.airtrax.com

[7] Viboonchaicheep P, Shimada A, Kosaka Y. Position rectification control for Mecanum wheeled omni-directional vehicles. In *Industrial Electronics Society, 2003. IECON'03. The 29th Annual Conference of the IEEE* 2003 Nov 2 (Vol. 1, pp. 854-859). IEEE.

[8] Lin LC, Shih HY. Modeling and adaptive control of an omni-mecanum-wheeled robot. *Intelligent Control and Automation*. 2013 May 24;**4(02)**:166

[9] Tsai CC, Wu HL. Nonsingular terminal sliding control using fuzzy wavelet networks for Mecanum wheeled omni-directional vehicles. In *Fuzzy Systems (FUZZ), 2010 IEEE International Conference* on 2010 Jul 18 (pp. 1-6). IEEE.

[10] Morimoto S, Hatanaka K, Tong Y, Takeda Y, Hirasa T. Servo drive system and control characteristics of salient pole permanent magnet synchronous motor. *IEEE Transactions on Industry Applications*. 1993 Mar;**29(2)**:338-43.

[11] Dülger LC, Kireçci A. Motion control and implementation for an AC servomotor system. *Modelling and Simulation In Engineering*. 2007 Mar 19;2007.

[12] Leng EW, Salim H, Sabri SF, Ismail M. Design and development of Motorized Adjustable Vertical Platform (MAVeP) for satellite test facility. In2015 International Conference on Space Science and Communication (IconSpace) 2015 Aug 10 (pp. 424-427). IEEE.

Effect of Stitching on the Tensile Mechanical Property of Empty Fruit Bunch Oil Palm Fiber Reinforced Epoxy Composites

A L Ahmad Ghazilan[1,a] *, H Mokhtar[1,b], M S I Shaik Dawood[1,c], Y Aminanda[1,d] and J S Mohamed Ali[1,e]

[1]Department of Mechanical Engineering
Kulliyyah of Engineering, International Islamic University Malaysia
Jalan Gombak, 53100 Kuala Lumpur, Malaysia

E-mail: [a]ahmadluqman141@gmail.com, [b]hananm@iium.edu.my, [c]sultan@iium.edu.my, [d]yulfian@iium.edu.my, [e]jaffar@iium.edu.my

Abstract. Stitching has proven to be an effective reinforcement for improving through-thickness (out-of-plane) mechanical properties with the incorporation of z-direction reinforcements of synthestic fiber. There are drawbacks to stitching as it may cause focal stress concentration, and fiber misalignment and breakage, but at the same time, it can also improve the tensile performance of the composite. Tensile properties, such as tensile strength and modulus of elasticity, were obtained from tensile tests performed on both stitched and unstitched oil palm fiber composites as per ASTM D638 specifications using Universal Testing Machine INSTRON 5582. The test results indicated that stitching natural, short, untreated and random empty fruit bunch oil palm fiber reinforced epoxy composites improve the tensile strength and elastic modulus due to the increase in matrix volume percentage and additional tensional resistance .

1. Introduction

The interest in synthetic fibers have moved on to natural fibers as environment is one of the key concerns. Natural fibers are not only environmentally friendly, but also offer potential improvement in reinforcing materials when compared to synthetic fibers. The advantages of natural fibers are mainly the low cost, low density, biodegradability, and better insulating and thermal resistance. Synthetic fibers are becoming unpopular due to it being non-renewable, non-degradable and expensive whereas natural fibers are becoming more accepted due to its properties of stiffness, flexibility, impact resistance and modulus of elasticity. Natural fiber reinforcements being used in the industry are the use of flax-sisal fiber mat reinforced epoxy matrix to produce door panels for Mercedes Benz E-Class, flax fiber reinforced polypropylene composite for rear shelf trim panels for the year 2000 model Chevrolet Impala, and natural fiber reinforced composites in structural applications such as ceiling and partition boards [1].

The performance of natural fiber reinforced composite depends on the composition properties of natural fiber and polymer synthetic matrix material. Mostly, the strength and stiffness of the structural load is carried by its fiber whereas the shape, surface and environmental resistance is carried out by its matrix. Thermoplastic and thermosets are the two type of dominating matrix materials that offer various potential in natural fiber reinforced composite applications. Epoxy resin as matrix material is suitable for natural fiber reinforced composite due to its good adhesion capability, good heat and chemical resistance, exemplified electrical insulator properties and superior mechanical performance [2]. Vacuum bagging techniques for fabricating natural fiber reinforced composite provide good physical and mechanical performance compared to wet lay-up process due to its process control, improve equipment and cost effective method [3]. It has been reported that tensile strength for oil palm fiber-epoxy resin decreases from 47.78 MPa to 46.10 MPa with increasing value of oil palm fiber content from 35% to 55% respectively [4]. Another study on the mechanical properties of 10% oil palm fiber content composite with epoxy resin polymer matrix for a 7m T-Beam configuration is reported to be able to support 200kg of load with a maximum deflection of 0.2mm, thus suggesting the potential usage of oil palm fiber-epoxy resin composite for intermediate load bearing bridge application [5].

Weaving, braiding, knitting and stitching have become major methods of 3D fiber structure to improve the performance of fiber reinforced composites. The driving factors of 3D fiber structural development to undergo such technological improvement are the demand in reducing fabrication cost, increasing through thickness mechanical properties of laminate composites, and improving the impact resistance of fiber reinforced composites. This technological innovation have undergone major transformation within the aircraft, marine and automotive industry since the urge of weight-reduction factor upon manufacturing better physical as well as mechanical properties of composite structural components [6]. Despite having various advantages of weight reduction factor and cost effective fabrication factor, the primary drawback of fiber reinforced composites as structural components are the poor through thickness mechanical properties or poor interlaminar strength. The stitched fiber reinforced composite was initially investigated in the 1980s by the aircraft industry for an improvement towards through thickness mechanical properties, and it was found to have advantageous improved properties which are the cost effective method for joining stacked plies compared to other 3D fiber structure material [7].

Stitched fiber reinforced composite as an advanced fiber reinforced composite structure offer various advantages such as improved through-thickness mechanical properties, reduction of damage area and enhanced energy absorption capability despite adding extra steps in its composite fabrication process. On the other hand, the major drawbacks and limitations exerted for stitched fiber reinforced composites are the localized damage of in-plane fiber resulting from needle penetration of the prepreg layers of fibers which then cause lower flexural strength of stitched composites compared to unstitched composites. Material discontinuity can also occur at the stitching point where it can lead to high stress concentration during load bearing condition thus reducing the performance of the composite mechanical properties. In addition, one of the major drawbacks of stitched or other through thickness reinforcing methods is the detrition of the in-plane properties such as compressive or tensile properties. Apparently, these significant drops are due to the misalignment of fiber distribution which creates resin rich pockets surrounding the stitch point. These resin rich pockets reduce fiber content percentage becoming a high stress concentration point thus reducing in-plane mechanical properties [8] [9].

2. Material and Method

2.1 Material

Untreated oil palm empty fruit bunch fiber is obtained from Malaysia Palm Oil Board in Bangi, Selangor, Malaysia. Zeepoxy HL002 TA/B epoxy resin, epoxy hardener, vacuum bagging consumables and materials was supplied by Sky Tech Enterprise. Nylon thread of linear density of 210 denier-15 ply was supplied by HLF FISHING SUPPLIES Sdn. Bhd.

2.2 Stitch Preparation of Oil Palm Composites
Tangled and intermixed oil palm empty fruit bunch fiber was carded and combed using an electrical motor carding machine to produce untangled and organized fiber layer. Individual layers of dry palm fibers are stitched manually with 210 denier 15 ply thread (0.0158 mm^2 cross sectional area, 4.395 MPa tensile strength, and 43.603 MPa tensile modulus) using modified lock stitch method with 10 mm stitching length or pitch as shown in Figure 1.a. The stitches are straight rows and evenly spaced as shown in Figure 1.b. For the fabrication of stitched empty fruit bunch oil palm fiber composite, a 400 mm x 400 mm bottom stainless steel plate and 300 mm x 300 mm upper stainless steel plate, and four 200 mm x 5 mm flexible resin stopper were used as a mold. A layer of mylar sheet was placed on each side of the plate as a releasing agent. Epoxy resin and epoxy hardener were mixed thoroughly according to its specific fiber-matrix volume fraction ratio. A layer of stitched empty fruit bunch oil palm fiber is stacked accordingly on the bottom plate. The stirred epoxy resin and hardener were impregnated through the layer of stitched fibers. Vacuum bagging materials were then placed on top of the wet fibers and covered by the upper plate mold. Vacuum bagging film enclosed the lay-up to provide a confined space which is sealed around the edges of the mold by using sealant tape. The vacuum pump draws out the air inside the confined space to produce airtight vacuum bag until curing process is complete.

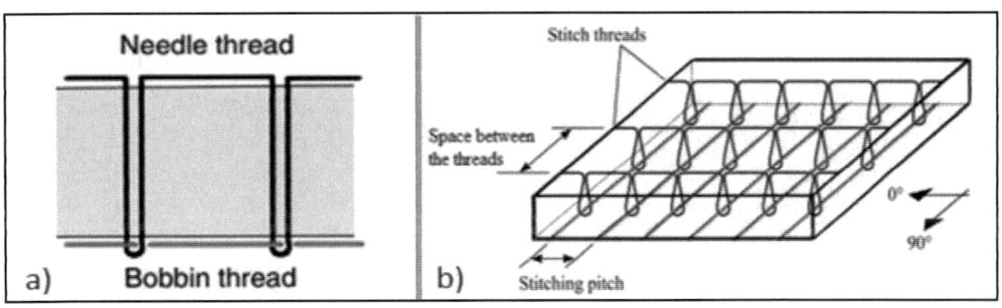

Figure 1. a) Modified lock stitch method and b) Schematic view of stitching.

2.3 Tensile Characterization of Composites
Tensile test specimen was prepared according to the standard size and specification of ASTM D 638 [10] using CNC milling. Tensile strength and modulus of elasticity for stitched and unstitched empty fruit bunch epoxy composite were tested by using INSTRON 5582 Ultimate Testing Machine, quarter bridge BX 120-3AA strain gauges and D4 Data Acquisition conditioner. This measurement was conducted according to ASTM D 638 procedures at a test speed of 5mm/min. Five specimens were tested for each case and its average value is tabulated accordingly.

3. Results and discussion
Measurement of load and strain data were made simultaneously during tensile testing process. Tensile strength is calculated by dividing the maximum load by the average cross-sectional gauge length area of the specimen. Modulus of elasticity is calculated by extending the initial linear portion of stress-strain curve then by dividing the difference in stress corresponding to the section on this straight line by the corresponding difference in strain value. Cross-plot graphs of stress and strain data of plain epoxy and EFB palm fiber composite specimen from tensile testing are plotted on Figure 1. Results of tensile tests of stitched and unstitched empty fruit bunch fiber reinforced composite at 20% fiber volume are shown in Table 1.

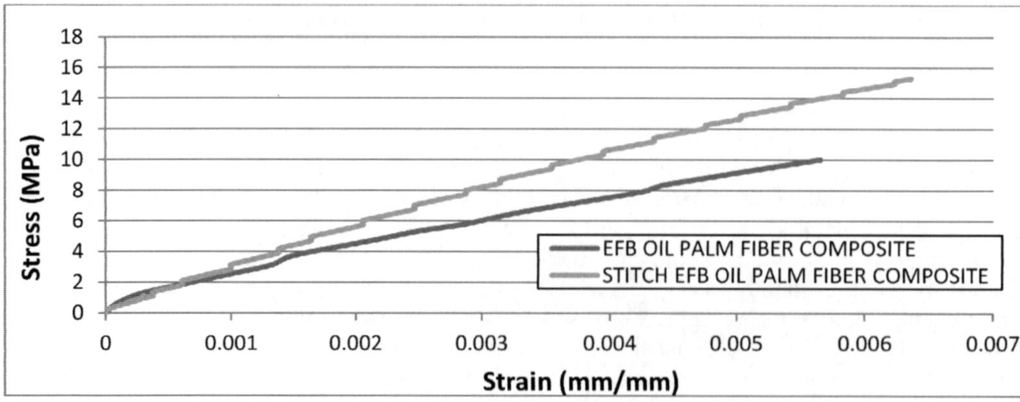

Figure 2. Tensile Stress vs Strain Curve.

Table 1. Tensile test results.

Composites	Tensile Strength (MPa)	Tensile Modulus (MPa)
Unstitched EFB Composite	10.064	1712.13
Stitched EFB Composite	16.717	2504.18

Based on Table 1, both tensile strength and tensile modulus of elasticity for the stitched composite correspond to a greater tensile performance when compared to the unstitched composite; i.e. 16.717 MPa and 2504.18 MPa respectively. Most reported studies have concluded that in-plane strength for stitched composite will induce a degradational performance due to decremental laminate integrity caused by undulation or resin pockets created around the stitches, as well as fiber misalignment, fiber crimping and fiber breakage during stitching process. Only one study reported an inconsistent result, in which both tensile strength and tensile modulus of elasticity for stitched composite have better performance compared to unstitched composite, and for this study it was believed that fiber breakage after stitching was neglected since the stitching process was done on dry preform fiber layers [11].

Importantly, empty fruit bunch fiber in this study was considered as short and randomly oriented fiber. The axial load applied to the composite structure is not exerted directly on the fiber material, but first, applied to the matrix material and later transferred to the fiber throughout the fibers end and cylindrical surface. This condition induces the fact that for short and randomly oriented fiber composites, the load carrying capability and capacity is mainly being carried out by the matrix, which is completely different to the load carrying capability of synthetic or conventional long and unidirectional fiber composites. Stitching introduces extra resin rich pockets surrounding the thread which increases the percentage value of resin to fiber material volume which then reduces the tensile performance under axial or in-plane load.

However, with natural, short and randomly oriented fiber composite, the load carrying capability mainly falls upon the matrix, thus by introducing Z-direction stitches, it increases the resin rich pockets surrounding the thread areas which will then increase the matrix volume percentage. The connected, continuous and axially aligned stitched thread will also serve as an additional tension resistance for axial load. Combining both factors of increased volume percentage of matrix material and extra tensional resistance from continuous axially aligned thread, the tensile performance, i.e., tensile strength and modulus of elasticity of stitched palm fiber composite was improved by 40% and 32% when compared to unstitched composite. It has been reported in one study that the slight increment of stitched composite tensile strength is due to the increased fiber volume fraction from the additional stitching thread. The benefit from increased fiber volume fraction have compensated the drawback factor of a fiber breakage and stress concentration at resin pocket around stitches [12].

4. Conclusion

The tensile performance of z-directional stitching for empty fruit bunch oil palm reinforced epoxy composite is evaluated to accommodate an improvement to conventional natural fiber reinforced composite. There are drawbacks with the incorporation of stitches for tensile performance due to focal stress concentration, fiber misalignment and fiber breakage. However, introducing through thickness reinforcement into preform natural fiber by stitching provides a possibility of improving the tensile performance of natural fiber reinforced matrix composites. In this study the tensile properties of stitched empty fruit bunch oil palm reinforced epoxy composite have shown better performance in both tensile strength and tensile modulus compared to unstitched empty fruit bunch oil palm fiber reinforced composite. This clearly indicates that by introducing stitches in through thickness direction of natural, short, untreated and randomly oriented empty fruit bunch oil palm fiber reinforced epoxy composite improves the tensile strength and modulus of elasticity due to the increased matrix volume percentage and additional tensional resistance from the stitched thread. Although stitched empty fruit bunch oil palm fiber as a means of reinforcement has corresponded to a weak tensile performance when compared to inorganic synthetic fiber composites, it is believed that empty fruit bunch oil palm fiber reinforced composite thrives in impact strength due to its high fracture toughness [13].

Acknowledgements

The authors would like to acknowledge the Department of Mechanical Engineering of International Islamic University Malaysia for the support to carry out this work as well as the Ministry of Higher Education (MOHE) kindly provided funding for the research through the "Fundamental Research Grant Scheme" (FRGS) project no FRGS13-082-0323.

References

[1] John M J and Thomas S Biofibres and biocomposites 2008 *Carbohydrate Polymers* **71** 343-364

[2] Shinoj S, Visvanathan R, Panigrahi S and Kochubabu M Oil palm fiber (OPF) and its composites: a review 2011 *Industrial Crops and Products* **33** 7-22

[3] *SP Systems Guide to Composites.*

[4] Kalam A, Sahari B B, Khalid Y A and Wong S V Fatigue behaviour of oil palm fruit bunch fibre/epoxy and carbon fibre/epoxy composites 2005 *Composite Structures* **71** 34-44

[5] Bakar M, Natarajan V, Kalam A and Kudiran N Mechanical properties of oil palm fibre reinforced epoxy for building short span bridge 2007 *Experimental Analysis of Nano and Engineering Materials and Structures* 97-98

[6] Vacuum Infusion Processing (VIP), RTM Light (LRTM, Resin Infusion, VARTM, SCRIMP) [Online]. Available: http://www.moldedfiberglass.com/processes/vacuum-infusion-molding

[7] L. Tong, A. P. Mouritz, and M. Bannister, 3D fibre reinforced polymer composites: Elsevier, 2002. L. Tong, A. P. Mouritz, and M. Bannister, 3D fibre reinforced polymer composites: Elsevier, 2002.

[8] Mouritz A P, Leong K H and Herszberg I A review of the effect of stitching on the in-plane mechanical properties of fibre-reinforced polymer composites 1997 *Composites Part A: Applied Science and Manufacturing* **28** 979-991

[9] Dransfield K, Baillie C and Mai Y-W Improving the delamination resistance of CFRP by stitching—a review 1994 *Composites Science and Technology* **50** 305-317

[10] International A Standard test method for tensile properties of plastics 2014 *ASTM D638-14*

[11] Kang T J and Lee S H Effect of stitching on the mechanical and impact properties of woven laminate composite 1994 *Journal Composite Material* **28** 1574-1587

[12] Mouritz A P Fracture and tensile fatigue properties of stitched fiberglass composites 2004 *Journal of Material: Design and Application* **218** 87-93

[13] Jawaid M, Abdul Khalil H P S and Abu Bakar A Mechanical performance of oil palm empty fruit bunches/jute fibres reinforced epoxy hybrid composites 2010 *Materials Science and Engineering: A* **527** 7944-7949

IOP Conf. Series: Materials Science and Engineering **184** (2017) 012052 doi:10.1088/1757-899X/184/1/012052

3D Point Smoothening Using Modified Local Regression for Reverse Engineering Process

Nur Ilham Aminullah Abdulqawi[1] and Mohd Salman Abu Mansor[1,*]

[1]School of Mechanical Engineering, Engineering Campus, Universiti Sains Malaysia, 14300 Nibong Tebal, Pulau Pinang, Malaysia

[*]Corresponding author's e-mail: mesalman@usm.my

Abstract. This research will introduce an existing statistical curve fitting tool i.e. local regression into reverse engineering process. Instead of using the local regression on statistical data, it will be applied on 3D point clouds obtained from 3D scanners. As most 3D scanners capture the smallest surface details such as textured and non-smooth surfaces, it makes the 3D point cloud to surface conversion process problematic. By modifying this existing statistical tool of local regression with the proposed algorithm, it helps to smooth out rough surfaces obtained from 3D point cloud, minimising the error when generating the virtual surface from smoothened point cloud rather than from raw point cloud. The related method of performing the smoothening process on the 3D point will also be included in this paper. At the end, comparison of the smoothened point cloud and the unsmoothened point cloud will be visualised and discussed.

1. Introduction

Reverse engineering process always starts with using 3D scanners to obtain 3D point cloud data, but the 3D scanners also detect every single detail on the object's surface, including irregular surfaces. In actual situation, objects do not have perfectly smooth surfaces, moreover some objects have textured surfaces, and hence the data produced by 3D scanner are usually considered as non smooth data or known as noisy data. To overcome this problem, some of the conventional reverse engineering processes need to be modified for improvement.

Computer programming has contributed to industries over the recent years ranging from basic computer applications to complex data processing tools. Computer programming has also become one of the common computing tool for scientists and engineers to perform various technical and research works, making computer programming more popular in research institutions and universities recently.

This paper will target on computer programming and its application on smoothening 3D points and its effect on the CAD model's surface. A modified local regression as a tool will also be adapted to this 3D point smoothening process. With this modified local regression smoothening introduced in reverse engineering process, it can benefit the designers by reducing errors on surface generation process.

The objective of this research is to minimise error on the surface generation by including point smoothening in the beginning of the process flow.

2. Related Works

There are a few similar research works that have been done on setting up methods on creating 3D CAD model by reverse engineering. Bénière et al. [1] had introduced a template that comes with a standard three steps of process from mesh convert to geometric primitives, geometric primitives to wireframe, then only from wireframe to 3D solid model, however the examples showed had very distinctive geometrical outlook and does not come out raw from a 3D scanner. Another work of Schnabel et al. [2] focused on point cloud conversion to geometrical primitives, and the primitives are sub-classified into basic shapes. This approach is useful for point cloud that resembles regular form objects, free-form objects will have more difficulties in order to achieve the same result, unless the point cloud data are reduced to approximation of basic shapes.

Even for mesh to solid conversion process, it is not always a smooth process because meshes generated from 3D point cloud are not always flawless. That is why a lot of methods had been introduced to repair or overcome those flaws on meshes. Marchandise et al. [3] had introduced mesh repair algorithm for mainly hole filling purposes. Some other works are more specific on mesh process only like Müller and Schwab [4] had targeted mesh refinement for finite element models. Another mesh processing work of Wei et al. [5] proposed a two step approach to smoothen mesh, the method is seen to be successful but required more computation time compared to smoothening points.

To overcome the mesh to solid conversion error, one must target the real root cause which originates from the irregularity of the point cloud data. Khanna and Rajpal [6] had a slightly different approach which directly uses point cloud to convert to curves and lines, at the same time the point cloud are being curve fitted relatively so that the final produced curves are smooth, however the approach does not aim to proceed to produce any surface or solid. There are also works on detecting and removing unwanted point cloud data in order to generate smooth final results, Nurunnabi et al. [7] also [8] Wang and Feng, both used their own methods to detect outlier points. The research methods of either reducing the unwanted points or perfecting point cloud is important because the refined point clouds are more ready to be processed for mesh and proceed to final CAD model.

3. Methodology

Conventionally, converting from 3D point cloud to 3D CAD model starts with mesh generation from 3D point cloud first, followed by smoothening the mesh, and then the subsequent processes. The conventional process flow found in most commercial CAD softwares is to generate the mesh first and then smoothen the mesh later. The methodology proposed here is different where point smoothening is introduced before mesh generation. Figure 1 is the illustration of proposed process flow versus the common CAD software process flow concept which includes mesh smoothening process.

This data smoothening is seen as the most critical process, as successful point data smoothening can lead to successful subsequence processes such as mesh generation, surface generation and solid generation.

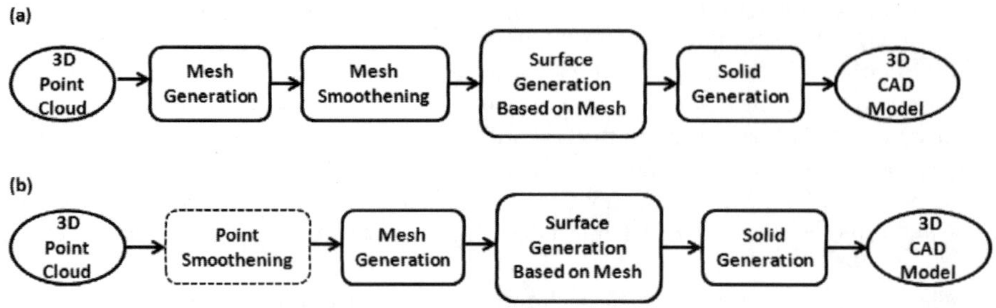

Figure 1. (a) Typical reverse engineering process for 3D point cloud to 3D CAD model conversion versus (b) Proposed reverse process with point smoothening for 3D point cloud to 3D CAD model conversion.

IOP Conf. Series: Materials Science and Engineering **184** (2017) 012052 doi:10.1088/1757-899X/184/1/012052

For 3D point smoothening, the local regression method is modified to suit the three dimensional applications specifically for reverse engineering, which is a local regression of using weighted linear least squares and a second degree polynomial model on the points. Furthermore, by doing iterative weighting on a specific range using the tricube function of this weight function can be simplified as

$$w_i = \left(1 - \left(\frac{|z-zi|}{r \times n}\right)^3\right)^3 \tag{1}$$

where, n is the total number points and r is the user defined constant

After specifying the range and input it to constant r, the specific i_{th} data point to be smoothed will have the largest weight influence on the fitting and data points outside the span have zero weight and no influence to the fitting. The smoothening will be based on weighted least square regression of second degree polynomial and the fitted x,y,z value is shown as

$$\hat{z}_i \equiv \sum_{i=1}^{N} w_i \left[a + b(x,y)_i + c(x,y)_i^2\right. \tag{2}$$

The polynomial constants a, b and c are derived from the least square method based on [9] and applied for second degree polynomial properties. To obtain the constants a, b and c, the least square method equation can be rearranged as matrix for solution. The solution for a, b and c constant is shown as

$$\begin{bmatrix} a \\ b \\ c \end{bmatrix} = \begin{bmatrix} \sum_{i=1}^{N} 1 & \sum_{i=1}^{N}(x,y)_i & \sum_{i=1}^{N}(x,y)_i^2 \\ \sum_{i=1}^{N}(x,y)_i & \sum_{i=1}^{N}(x,y)_i^2 & \sum_{i=1}^{N}(x,y)_i^3 \\ \sum_{i=1}^{N}(x,y)_i^2 & \sum_{i=1}^{N}(x,y)_i^3 & \sum_{i=1}^{N}(x,y)_i^4 \end{bmatrix}^{-1} \begin{bmatrix} \sum_{i=1}^{N} z_i \\ \sum_{i=1}^{N} z_i\,(x,y)_i \\ \sum_{i=1}^{N} z_i^2\,(x,y)_i \end{bmatrix} \tag{3}$$

The smoothening process is done based on the i_{th} point's distance to the next $i_{th}+1$ as a condition to fit this 3D point cloud regenerated from 3D scanner. The condition set for smoothening is that the furthest neighbour point has to be less than 0.05 of the overall sizing of the point cloud. Therefore the smoothened point cloud can be represented by

$$PC_{smoothen} = \begin{cases} [x, y, \hat{z}\ ] & , d \leq 0.05D \\ [x, y, z\ ] & , d > 0.05D \end{cases} \tag{4}$$

where, d is the distance between points of the consecutive order
D is overall sizing of the point cloud

4. Implementation and Discussion

The point smoothening effect may not be significant to the average human eye, but when it comes to the processes after mesh generation, smoother points will help prevent errors in subsequence processes, that is why smoother mesh are preferred over the raw point cloud when comes to reverse engineering processes. Figure 2 shows the comparison of the noisy data from an object, named as A, scanned with 3D scanner and the after smoothened point cloud.

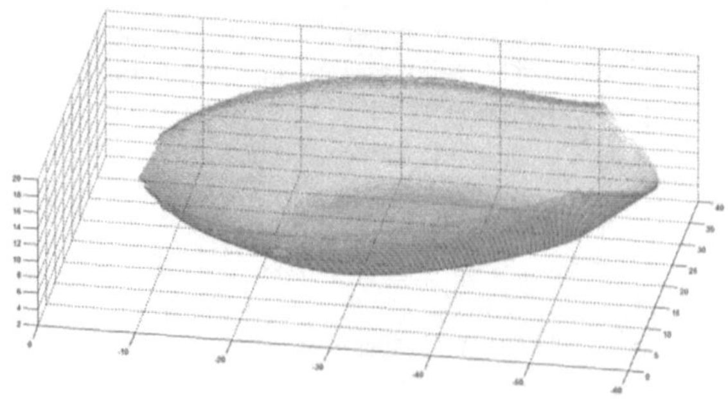

Figure 2. Illustration compare on raw 3D point cloud (magenta) and smoothened 3D point cloud (cyan) in the same scatter plot.

For fair comparison, both mesh sizes are set at 0.41mm distance from each point. Figure 3 shows the mesh comparison of mesh resulted from raw 3D point cloud versus the mesh resulted from smoothen 3D point cloud.

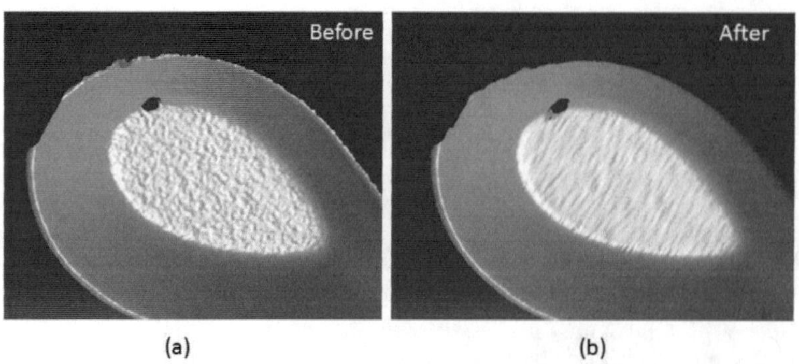

Figure 3. (a) 3D mesh from raw 3D point cloud using 0.41 mm mesh size (b) 3D mesh from smoothened 3D point cloud using 0.41 mm mesh size

The resulted smoothening effect from this proposed method should be discussed further in order to conclude that it has its own advantages. A small portion of point cloud data is extracted as per Figure 4 for further analyses. The points selected are labeled as A to K.

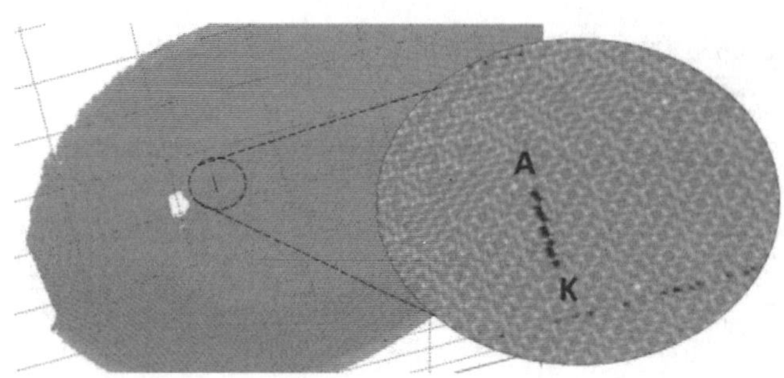

Figure 4. The selected portion of point cloud of A-K

IOP Conf. Series: Materials Science and Engineering **184** (2017) 012052 doi:10.1088/1757-899X/184/1/012052

The smoothening effect of point A-K is plotted in Table 1 and is visualised in Figure 5. The table also further computes on the distance of each points with the mean value of the selected portion. This is done to find out the surface roughness of each meshes, so that the smoothening effect can be numerically justified based on the surface roughness value.

Table 1. Extracted point data of A-K and difference of before and after smoothened together with calculated smoothening effect based on surface roughness difference

Point	X	Y	Z_{raw}	Z_{smooth}	Distance from mean (raw)	Distance from mean (smooth)	Squared distance (raw)	Squared distance (smooth)	Surface Roughness	
									R_{RMS} (raw)	R_{RMS} (smooth)
A	17.360	-7.405	11.589	11.674	-0.068	-0.041	0.005	0.002		
B	17.489	-7.426	11.592	11.683	-0.065	-0.032	0.004	0.001		
C	17.618	-7.448	11.597	11.692	-0.060	-0.023	0.004	0.001		
D	17.750	-7.448	11.603	11.700	-0.054	-0.015	0.003	0.000		
E	17.880	-7.460	11.610	11.709	-0.047	-0.006	0.002	0.000		
F	18.009	-7.472	11.626	11.717	-0.031	0.002	0.001	0.000	0.072	0.025
G	18.141	-7.471	11.648	11.724	-0.009	0.009	0.000	0.000		
H	18.271	-7.480	11.679	11.732	0.022	0.017	0.000	0.000		
I	18.401	-7.489	11.717	11.739	0.060	0.024	0.004	0.001		
J	18.532	-7.496	11.762	11.746	0.105	0.031	0.011	0.001		
K	18.663	-7.494	11.808	11.753	0.151	0.038	0.023	0.001		
Smoothening Effect = | R_{RMS} (smooth)- R_{RMS}(raw)| / R_{RMS}(raw) × 100%									65.19 %	

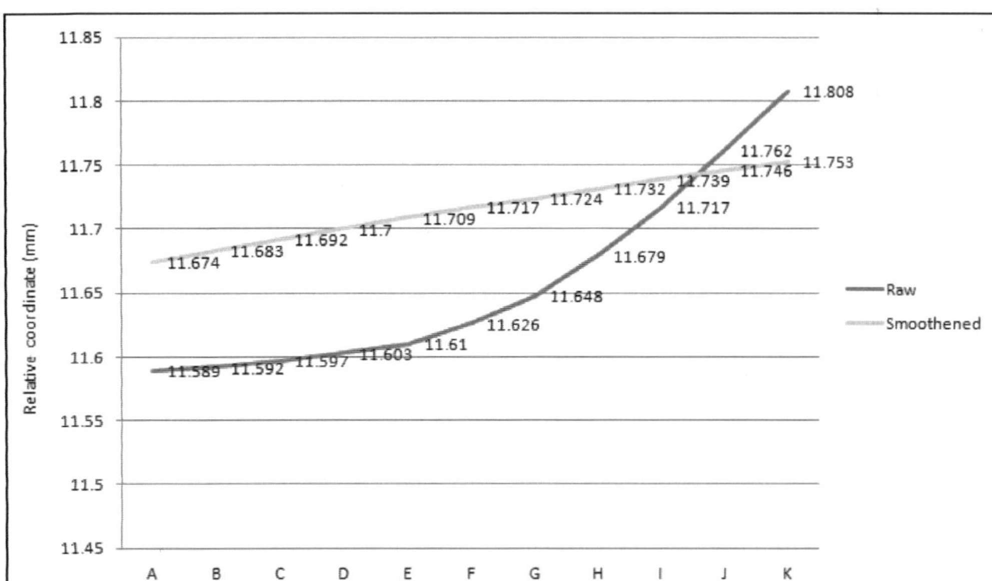

Figure 5. The effect of selected portion of point cloud of A-K before and after smoothened.

The tabulated data and plot showed smoothening effect numerically and graphically. The smoothening effect is recorded at 65.19% by based on comparing the surface roughness by root mean square, $R_{RMS.}$

Another specimen, named as B, is tested for the same smoothening effect. The smoothened effect is showed in Figure 6 as mesh.

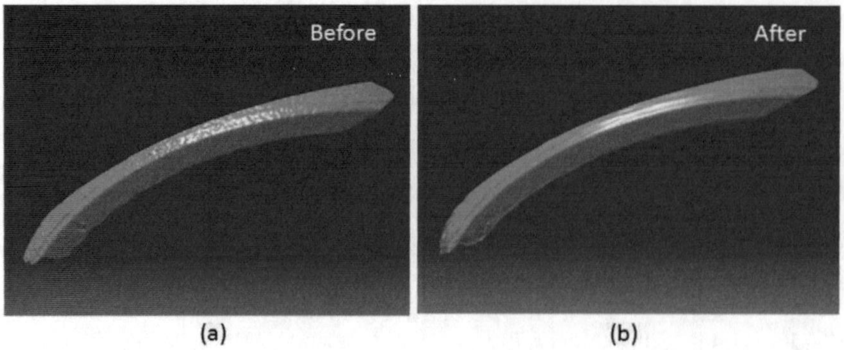

Figure 6. (a) 3D mesh from raw 3D point cloud using 1.3 mm mesh size (b) 3D mesh from smoothened 3D point cloud using 1.3 mm mesh size

The smoothening effect of this second specimen, B is also being quantified to another set of selected points, labelled as L to V as per Figure 7.

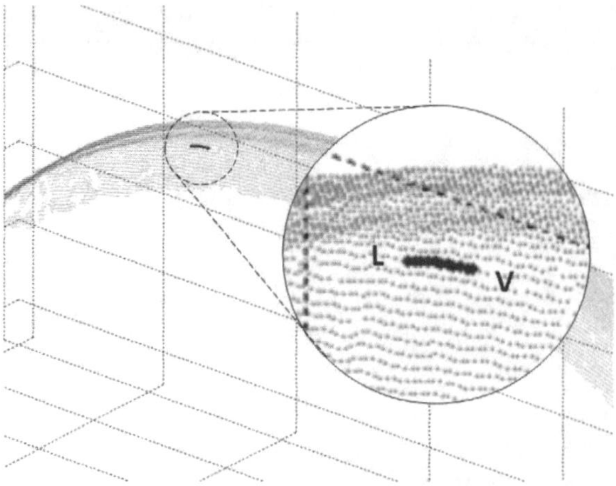

Figure 7. The selected portion of point cloud of L-V

The smoothening effect of point L-V is plotted in Table 2 and is visualised in Figure 8.

Table 2. Extracted point data of L-V and difference between before and after smoothened together with calculated smoothening effect based on surface roughness difference

Point	X	Y	Z_{raw}	Z_{smooth}	Distance from mean (raw)	Distance from mean (smooth)	Squared distance (raw)	Squared distance (smooth)	Surface Roughness R_{RMS} (raw)	Surface Roughness R_{RMS} (smooth)
L	27.691	-100.681	10.405	10.407	0.137	0.140	0.019	0.020		
M	27.826	-100.686	10.382	10.379	0.114	0.112	0.013	0.013		
N	27.962	-100.695	10.346	10.351	0.078	0.084	0.006	0.007		
O	28.097	-100.701	10.322	10.323	0.054	0.056	0.003	0.003		
P	28.231	-100.705	10.305	10.295	0.037	0.028	0.001	0.001		
Q	28.366	-100.709	10.289	10.267	0.021	0.000	0.000	0.000	0.095	0.089
R	28.501	-100.715	10.262	10.239	-0.006	-0.028	0.000	0.001		
S	28.638	-100.725	10.222	10.211	-0.047	-0.056	0.002	0.003		
T	28.775	-100.735	10.179	10.183	-0.089	-0.084	0.008	0.007		
U	28.912	-100.746	10.136	10.155	-0.132	-0.112	0.018	0.013		
V	29.049	-100.756	10.096	10.127	-0.172	-0.140	0.030	0.020		
Smoothening Effect = $\lvert R_{RMS}(\text{smooth}) - R_{RMS}(\text{raw})\rvert / R_{RMS}(\text{raw}) \times 100\%$									6.91 %	

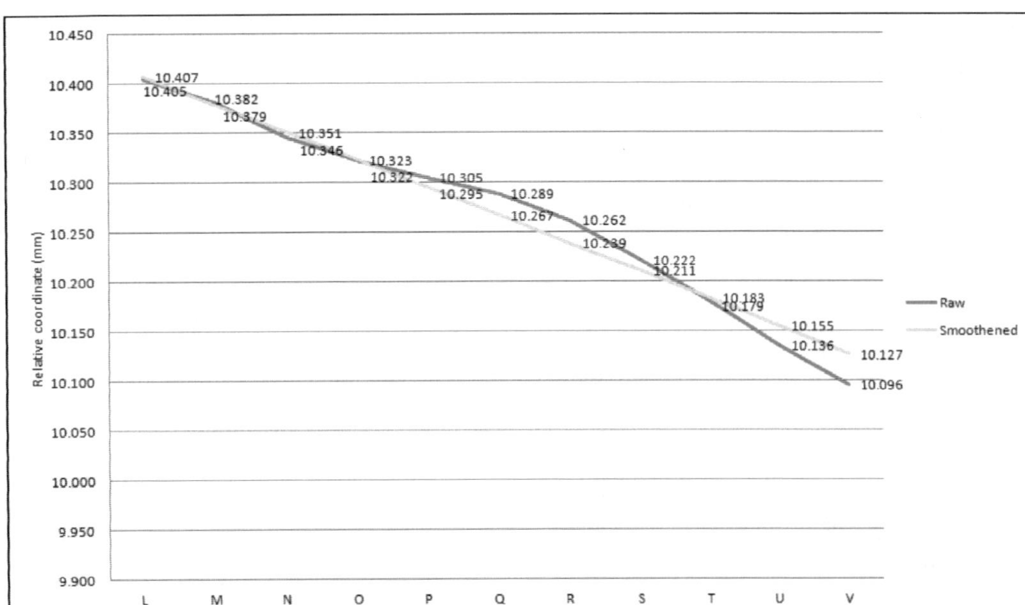

Figure 8. The effect of selected portion of point cloud of L-V before and after smoothening.

The tabulated data and plot showed smoothening effect numerically and graphically. The smoothening effect is recorded at 6.91% by based on comparing the surface roughness by root mean square, R_{RMS}.

For the second specimen object B, the smoothening effect of selected area is 10 times less significant compared to first specimen object A, based on the R_{RMS} percentage value. This is due to object A's point cloud has larger number of points compare to object B's. The equation used has a weight factor that is influenced by the number of points. The weight factor is inversely proportional to the number of points, making point cloud with higher number points less influenced by the weight of nearer points during smoothening. Hence this explained the point smoothening effect is significant for object A compare to object B.

Using computer to run the smoothening prior mesh generation can reduce subsequent error compared to using conventional reverse engineering process by CAD software. It is because most software does not have any point smoothening feature but mesh smoothening features instead, then the smoothened mesh has to go through various modification in order to come out as a surface.

Even though some CAD software has direct mesh to surface conversion option, most of the time errors will occur if the geometry of the object is not recognized by the CAD software. Instead to overcome this problem, users have to always create the drafting lines first, then users have to fill the surface according to the line boundaries one by one in order to create the complete surface model. Compared to this proposed methodology, after 3D point is smoothened, the subsequent tasks of creating surface model from point cloud will be easier with less error prompted by the CAD software since the smoothened points has better geometry representation than irregular raw points.

4.1. Benchmark with Mesh Smoothing Toolbox

Even though point cloud smoothening process is different from mesh smoothing process, but both results are comparable due to the point cloud can be easily triangulated into mesh representation. Mesh generation from smoothened point cloud using modified local regression and smoothened mesh from raw mesh using smoothing toolbox of CATIA software are supposed to have similar effect of removing textures found on raw 3D scanned virtual surfaces. Figure 9 is showing the effect of mesh generated from point cloud prior went through with modified local regression point smoothening, and also the mesh smoothened with CATIA's mesh smoothening toolbox with different iterative numbers.

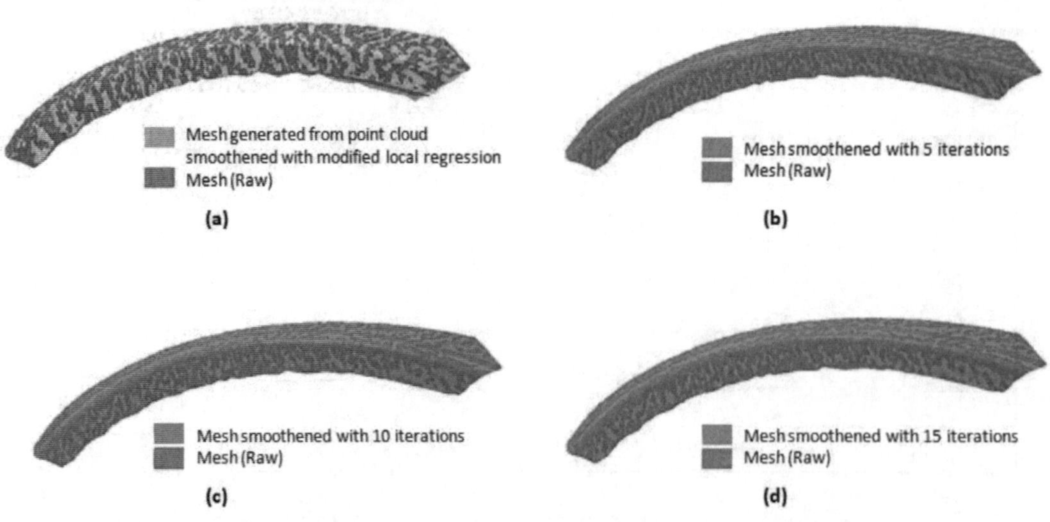

Figure 9. (a) Mesh smoothed with modified local regression point smoothening followed by mesh generation and mesh smoothed with mesh smoothing toolbox from CATIA with (b) 5 iterations, (c) 10 iterations and (d) 15 iterations.

Meanwhile the meshes are then mapped together for deviation analysis in order to verify smoothening performance of the modified local regression technique. Figure 10 is showing the different combination of meshed mapped together for deviation analysis.

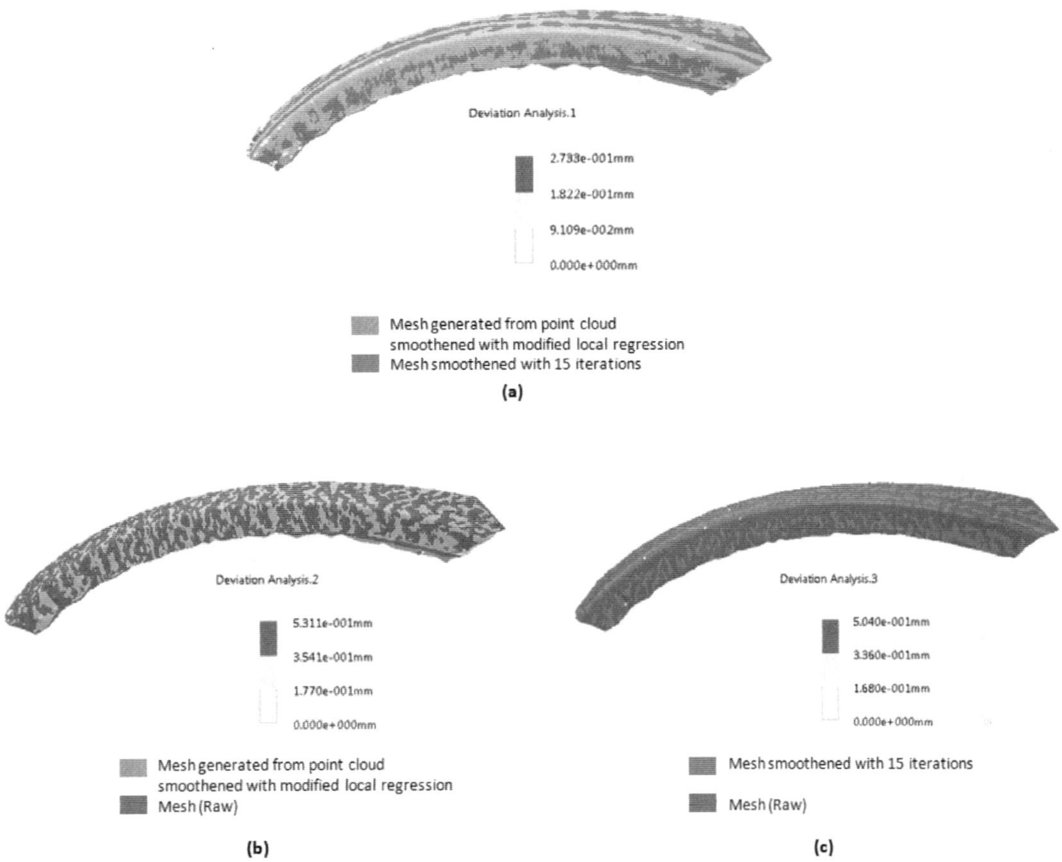

Figure 10. The deviation analysis of (a) deviation analysis 1, (b) deviation analysis 2 and (c) deviation analysis 3

The analysis based on Figure 10 shows that mesh generated from point smoothening using modified local regression does not deviate much from the mesh smoothened with CATIA's mesh smoothing toolbox. By referring to the figure's result of comparison for both mesh generated by point smoothening using modified local regression versus the mesh smoothened with 15 iterations CATIA's mesh smoothing toolbox, majority of the deviation are at level of 9.109×10^{-2} mm.

Meanwhile the figure also shows that both methods of point cloud smoothening and mesh smoothening are comparable when mapped with the raw mesh. The deviation level is almost similar to each other except on the deviation locations.

5. Conclusion

This proposed modified local regression as a tool to smoothen 3D point clouds can lead to smoothened mesh. The smoothening effect of 3D points based on the tested specimens is recorded at 65.19% and 6.91% for both tested objects. This point smoothening technique will help reduce at least some certain percentage amount of error in subsequence process namely mesh generation. This means that smoothening of 3D points will reduce the error in creating the virtual surfaces in reverse engineering process, hence easing the users by reducing manual tasks of using 3D CAD software in reverse engineering.

Besides, this 3D point smoothening also introduces another process sequence for reverse engineering, where the point data are smoothened at the beginning rather than the mesh later.

Acknowledgement

This research is supported by Universiti Sains Malaysia under the Research University Grant (No: 814247). The first author also would like to thank the tuition fee sponsorship of MyMaster by the Ministry of Higher Education Malaysia.

References

[1] Bénière R, Subsol G, Gesquière G, Le Breton F and Puech W 2013 A comprehensive process of reverse engineering from 3D meshes to CAD models *Computer-Aided Design* **45** 1382–93

[2] Schnabel R, Wahl R and Klein R 2007 Efficient RANSAC for Point-Cloud Shape Detection *Computer Graphics Forum* **26** 214–26

[3] Marchandise E, Piret C and Remacle J-F 2012 CAD and mesh repair with Radial Basis Functions *Journal of Computational Physics* **231** 2376–87

[4] Müller F L and Schwab C 2015 Finite Elements with mesh refinement for wave equations in polygons *Journal of Computational and Applied Mathematics* **283** 163–81

[5] Wei M, Shen W, Qin J, Wu J, Wong T-T and Heng P-A 2013 Feature-preserving optimization for noisy mesh using joint bilateral filter and constrained Laplacian smoothing *Optics and Lasers in Engineering* **51** 1223–34

[6] Khanna K and Rajpal N 2015 Reconstruction of curves from point clouds using fuzzy logic and ant colony optimization *Neurocomputing* **161** 72–80

[7] Nurunnabi A, West G and Belton D 2015 Outlier detection and robust normal-curvature estimation in mobile laser scanning 3D point cloud data *Pattern Recognition* **48** 1404–19

[8] Wang Y and Feng H-Y 2015 Outlier detection for scanned point clouds using majority voting *Computer-Aided Design* **62** 31–43

[9] Weisstein E Least Squares Fitting From MathWorld- A Wolfram Web Resource *http://mathworld.wolfram.com/LeastSquaresFitting.html* n.p.n.d. Web. 16 Jan 2016.

Influence of Ti addition on fracture behaviour of HSLA steel using TIG melting technique

Musa M. H. A.[1,a] *, **M. A. Maleque**[1,b], **M. Y. Ali**[1,c],
[1]Department of Manufacturing and Material Engineering
Kulliyyah of Engineering, International Islamic University, Malaysia
P.O.Box 10, 50728 Kuala Lumpur, Malaysia

E-mail: [a]mosahossin@yahoo.com, [b]maleque@iium.edu.my, [c]mmyali@iium.edu.my

Abstract. The welding process is a critical stage in the production of structural parts and the microstructure and mechanical properties of the welded joints must be appropriate in order to guarantee the reliability and durability of the components. The fracture toughness behaviour, which accounts for the residual strength of the component in the presence of flaws or cracks, is one of the most important properties to be evaluated in terms of microstructure and mechanical properties. In this present study, the surface of high strength low alloy (HSLA) steel was surface modified with the preplacement of pure Titanium (Ti) powder using a tungsten inert gas (TIG) arc heat source, at 100 ampere current with a voltage 30 V and a constant traversing speed of 1.0 mm/s using Argon shielded gas. The effect of preplaced Ti powder on the strength and toughness properties of the modified HSLA steel surface was investigated. The results indicated that the tensile and yield strength of HSLA steel decreased by ~12% and ~14%, respectively. While the impact toughness increased by ~33% and the ductility decreased by ~50%. The fractography analysis results by scanning electron microscopy (SEM) were also presented in this paper.

1. Introduction

High-strength, low-alloy (HSLA) steels are aimed to provide better mechanical properties and/or greater resistance to atmospheric corrosion than conventional carbon steels. They are designed to meet specific mechanical properties, such as yield strength greater than 275 MPa [1, 2]. Due to the high strength-to-weight ratio, HSLA steels are widely employed as a suitable material for oil and gas pipelines, chemical, petrochemical and marine applications [3]. However, owing to their low toughness and low strength due to welding process, their potential application is restricted. The welding process is a critical stage in the production of structural parts and the microstructure and mechanical properties of the welded joints must be appropriate in order to guarantee the reliability and durability of the components. The fracture toughness, which accounts for the residual strength of the component in the presence of flaws or cracks, is one of the most important properties to be evaluated [4]. One of the ways to overcome the problem is to modify the welding area on the substrate material by melting microalloying elements (Ti, V, Nb) via high energy input. Since then, research works showed improvement in surface hardness under this processing technique [5, 6]. On the other hand, the modified surfaces may affect the strength and toughness of particular material properties. Mechanical properties of HSLA steels are strongly connected to their microstructure obtained after welding that are generally performed in order to achieve a good hardness and/or tensile strength with sufficient ductility [7].

The development of surface alloying by adding a Ti powder into a molten alloy pool created by surface melting using TIG torch has been actively researched to obtain superior wear and corrosion resistance of low alloy steel [7-9]. The TIG torch with adequate power density is scanned over the material which melts a thin layer of the substrate surface by absorbing heat energy from the torch in a short time interval; the melt matrix interface moves toward the substrate at a rate which depends on the scanning speed and creates a melt pool. Upon solidification, the solidified layer creates a metallurgical bond with the substrate material [10]. Very stable Ti-rich carbonitrides can constrict austenite grain coarsening during austenitization. Small Ti additions may increase toughness and it was found to have significant influence on refinement of grains and increase of ferrite/pearlite ratio. Ti is the only micro-alloy Carbo-nitride that is stable at the high temperature attained in the HAZ during welding [11]. Increasing the titanium level in the range 0.01%-0.03% improving the toughness, particularly when the nitrogen and oxygen levels approach stoichiometry. However, if the titanium and nitrogen levels were sufficiently high to permit formation of TiN during solidification, then the resultant particles could act as crack nuclei which increase fracture transition temperature. It is well known that Ti, Nb, and V carbonitrides dissolve during the heating cycle of fusion welding, and precipitate if the cooling rate is slow enough [12]. It follows that the effect of any microalloying addition on HAZ toughness properties depends markedly on the heat input used (on the thermal cycle enabling particle solution, possible re-precipitation and subsequent hardening), and on the effect of soluble titanium on the phase transformations occurring during cooling after welding [13]. In another study the better combination of microstructure and impact properties was obtained in the range of 0.02–0.05% titanium by increasing the formation of acicular ferrite in the microstructure [14].

There is a little publication focusing on the effect of the surface modification technique by melting using TIG torch on the mechanical behavior of HSLA steel. Therefore, in the present work, HSLA steel was evaluated with Ti addition using TIG torch to modify the surface. Samples were taken from the chosen steel in order to provide information on the fracture toughness and mechanical behavior of the welding zone. The mechanical properties characterization was performed by means of tensile test and impact toughness. The obtained results provide a support for understanding on how the surface modification process affect the materials properties. The fractography analysis results by scanning electron microscopy (SEM) were also presented in this paper.

2. Experimental

A 10 mm thick HSLA steel plate with chemical composition containing 0.08% C was used as substrate material for this investigation and the chemical composition of the used HSLA steel is presented in Table 1. The specimen with the dimensions of 250 mm × 200 mm were cut out from the plate by using CNC machine. Tensile and impact samples were prepared according to ASTM E8-M-04 and ASTM E 23-11, respectively. The sample surface was abraded with SiC paper and degreased using acetone. The 45 μm size titanium powder of 99.97% purity was used as a preplacement and the amount was 1 mg per mm^2 area of the specimen surface. The Ti powder was first made into a pasty mass by mixing with PVA (polyvinyl acetate) binder and then placed on the substrate surfaces. The use of binder was mainly to prevent the powder from blowing away during melting operation under the flow of shielding gas [15]. The Ti powder preplaced specimen was then dried in an oven at 60 °C for 1h to remove moisture.

Table 1. Chemical Composition of the HSLA steel (wt. %)

C	Si	Mn	P	S	Cr	Ni	Cu	Mo	V	Al	Ti	Nb	Fe
0.08	0.17	1.41	0.014	0.0017	0.024	0.016	0.011	0.003	0.022	0.032	0.015	0.019	Bal.

A Miller TIG torch source attached with a semi-automatic traversing arm was used for melting purpose. The direct current electrode negative (DCEN) mode was used and A 3.2 mm diameter tungsten electrode was used to strike an arc between the electrode and the workpiece in pure Argon atmosphere with a gas flow rate of 20 L/min to prevent excessive oxidation. The arc was produced using a current of 100 A and voltage of 30 V. The glazing of the preplaced powder layer was done under the torch at a constant electrode traverse speed of 1.5 mm/s. The optimum parameters of TIG torch technique was chosen from elsewhere [6].

The Ti preplaced HSLA steel was tested using tensile testing machine in order to determine the tensile strength (σ_{uts}), yield strength (σ_y), as well as percent of elongation (E %). The tensile test is done at room temperature on Instor-5582 universal testing machine at a crosshead speed 2.5mm/min (three specimens were tested and the average value was presented). Also, the 10mm X10mm X 55 mm Charpy impact specimens were prepared to evaluate the impact toughness using INSTRON Charpy testing machine with 400J capacity at 25 °C. Impact energies were recorded and fracture surface was studied by using SEM.

3. Results and discussion

The results are presented and discussed in the following sections in terms of tensile properties, impact toughness, and fractography.

3.1 Tensile properties

The tensile test results of HSLA steel with and without Ti addition is presented in Fig. 1. Fig 2 and Table 2. Modified HSLA steel showed that the tensile strength was decreased by~12%, the yield strength decreased by ~14%, and also ductility reduced compared to as-received HSLA steel evaluated in this work by ~50. It shows a significant variation in the tensile properties due to the influence of Ti addition using TIG melting technique that change the chemical composition and microstructural morphology [16].

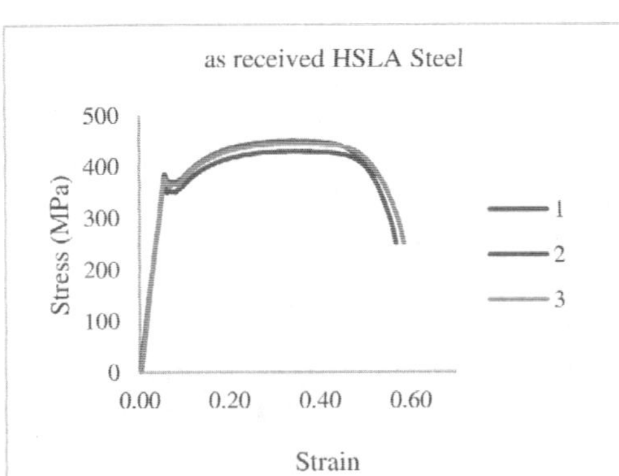

Fig. 1. Stress – Strain curve for as received HSLA steel

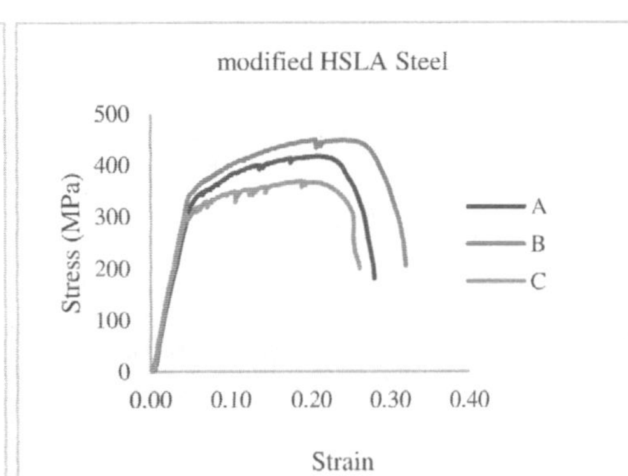

Fig. 2. Stress – Strain curve for modified HSLA steel with Ti addition.

IOP Conf. Series: Materials Science and Engineering **184** (2017) 012053 doi:10.1088/1757-899X/184/1/012053

Table 2. Tensile test results.

Steel	σ_y (MPa)	σ_{uts} (MPa)	E (%)
As received HSLA steel	370	445	42
Modified HSLA steel with Ti addition	320	390	21

3.2 Impact toughness

The Charpy impact test result is shown in Table 3. The general behavior is that the absorbed impact toughness corresponding to modified HSLA steel with Ti addition is higher compared to as-received HSLA steel. A significant increase in the impact toughness occurred for the test conducted which showed ~33%. It is also remarkable that though decrease in the tensile properties of the modified HSLA steel, but toughness properties of this steel was increased significantly which is required for pipeline applications. Improvement in HSLA steel toughness was obtained by the addition of Ti on the surface which produce new microstructural feature through the formation of intermetallic compound. Similar observation was done by other researchers [17].

Table 3. Charpy impact toughness test results

Sample No.	1	2	3	Average
Impact toughness of received HSAL steel (J)	88	85	83	85
Impact toughness of Ti addition HSAL steel (J)	100	115	125	113

3.3 Fractography

Further SEM examination on the fracture surfaces was conducted closely to the notch area and at the middle area especially in the TIG torch melting zone and base metal fractured surfaces. These impact tested specimens of base metal and welded zone were analyzed using a scanning electron microscope and EDX. It shows that the morphology of fracture surface of Charpy impact notch samples mainly consists of quasi-cleavage facets in the region of TIG torch melting zone. In addition, some river-like lines on quasi-cleavage facets and some cracks are also observed as shown in Fig. 4. However, the base metal shows that numerous dimples, which indicating a better deformation with higher ductility as indicated in Fig. 3.

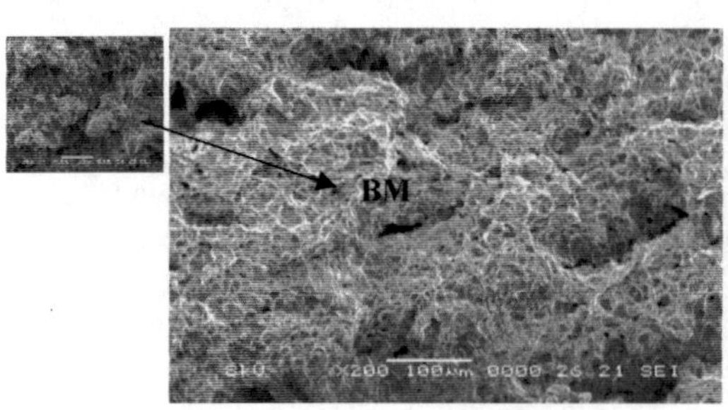

Fig. 3. SEM image of fractured surfaces of impact samples of base metal.

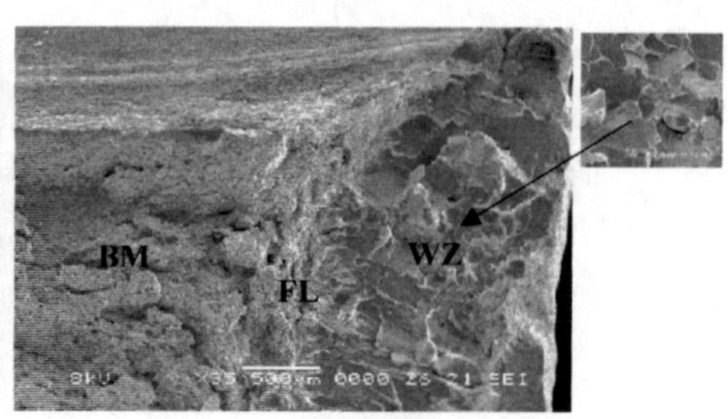

Fig. 4. SEM image of fractured surfaces of impact samples of TIG melting zone

The EDX elemental analysis of the three zones (base metal, fusion line and TIG torch welding zone) is shown in Fig. 5. It shows dissolution of preplaced Ti powder in HSLA steel whereby the melting of Ti depends on the energy input and the Ti content. In HSLA steels, crack initiation and propagation are dependent on the type and distribution of the brittle phase, as well as the characteristics of the matrix microstructure, inclusion particles, particularly the grain size. Figure 6 indicate the location of cleavage fracture mechanism was transgranular leading to brittle fracture, and the location of ductile mechanism was intergranular decohesion for both samples, impact and tensile fractures.

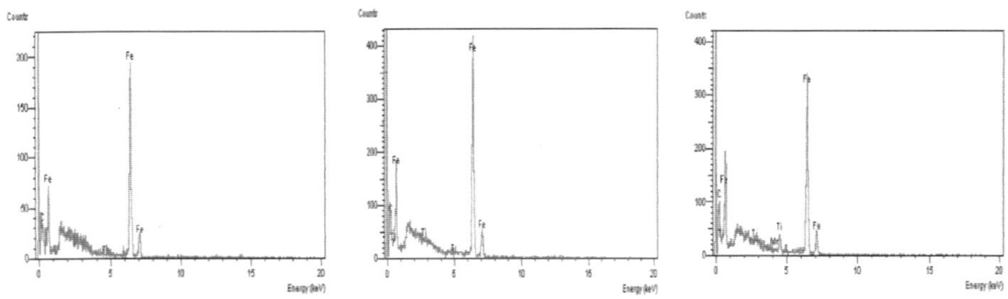

(a) (b) (c)

Fig.5. EDX analysis traces of Ti added HSLA steel taken from the fracture surfaces; a) Base metal, b) Fusion line and c) Weld zone.

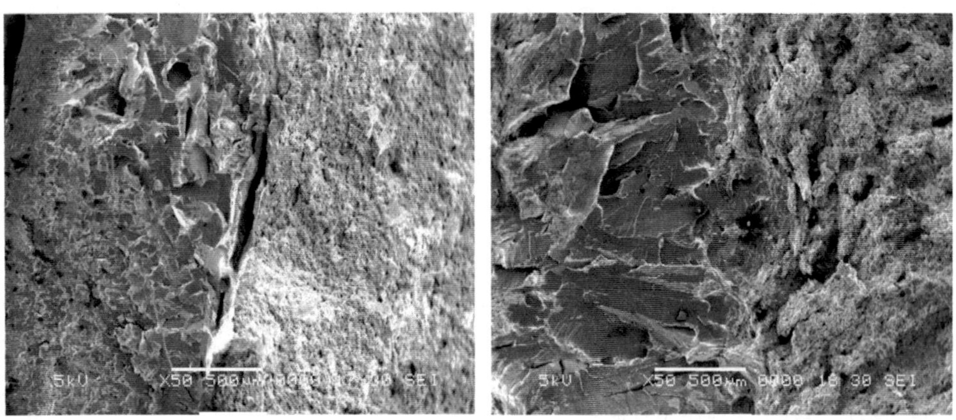

(a) (b)

Fig. 6. SEM fractographs of Ti added HSLA steel; a) Tensile fracture, and b) Impact facture

4. Conclusion

The following conclusion can be drawn from the present study:

1. The additional of 1mg/mm^2 Ti powder in HSLA steel has a significant influence on the mechanical properties and impact toughness representing lower strength and higher toughness.
2. The impact toughness was improved by 33%, however, ultimate tensile strength was decreased by 12% with the addition of Ti under TIG torch melting technique.
3. Improvement in toughness was obtained due to new microstructural feature through the formation of intermetallic compound.
4. Fractography observations revealed that the fracture characteristics were most likely brittle fracture in welding area, base metal exhibited ductile fracture resulting higher impact toughness for the modified HSLA steel.

Acknowledgements

The authors would like to acknowledge the Department of Manufacturing & Materials Engineering of International Islamic University Malaysia for the support to carry out this work.

References

[1] Paules J R 1990 Practical considerations in microalloying with vanadium, niobium, or titanium *Proc. of the Inter. Symp. on Micro-alloyed Vanadium Steels Cracow* 19–32
[2] Honeycombe R K W and Bhadeshia H K D H 2006 Steels: Microstructure and Properties *Elsevier Ltd 3ʳ*.
[3] Nathan S Ragu and et al 2015 Effect of welding processes on mechanical and microstructural characteristics of high strength low alloy naval grade steel joints *Defense Technology* **11.3** 308-317.
[4] Anderson T L 2005 *Fracture Mechanics: Fundamentals and Applications* (Boca Raton, FL: Taylor and Francis) p 621.
[5] Adeleke S A and Maleque M A 2015 Thin surface layers of iron-based alloys deposited by TIG hard facing *Tribology Online* **10(6)** 434-440.
[6] Maleque M A, Ghazal B A, Ali M Y and Ahmed A S 2015 Wear Behaviour of TiC Coated AISI 4340 Steel Produced by TIG Surface Melting *Materials Science Forum*, **819** 76-80.
[7] Mebarki N and et al 2004 Relationship between microstructure and mechanical properties of a 5% Cr tempered martensitic tool steel *Materials Science and Engineering* ***A*** **387** 171-175.
[8] Maleque M A and et al 2013 Processing of TiC-CNT hybrid composite coating on low alloy steel using TIG torch technique *Applied Mechanics and Materials* **378** Trans Tech Publications.
[9] Mridha, Shahjahan and et al 2015 Melting of multipass surface tracks in steel incorporating titanium carbide powders *Materials Science and Technology* **31.11** 1362-1369.
[10] Mridha, Shahjahan and et al 2012 Effect of voltage on the consolidation of TiC particulates on steel substrate fused by TIG welding arc *International Journal of Mechanical and Materials Engineering* **7.1**.
[11] El-Faramawy, Hoda S, Saeed N, Ghali, and Mamdouh M Eissa 2012 Effect of Titanium Addition on Behavior of Medium Carbon Steel *Journal of Minerals and Materials Characterization and Engineering* **11.11** 1108.
[12] Suzuki S and Weatherly G C 1985 Characterization of Precipitates and Grain Growth in Simulated HAZ Thermal Cycles of Ti--Nb Bearing Steel Weldments *HSLA Steels: Metallurgy and Applications* 675-682.
[13] Graville B A and Rothwell A B 1978 Effect of niobium on HAZ toughness using the instrumented Charpy test *Weld. Res. Abroad ;(United States)* **24.9**.
[14] Beidokhti, Behrooz, Koukabi A H and Dolati A 2009 Effect of titanium addition on the microstructure and inclusion formation in submerged arc welded HSLA pipeline steel *Journal of Materials Processing Technology* **209.8** 4027-4035.
[15] Bello K A, Maleque M, Ahmad A Z and Mirdha S 2015 Optimization of Hardness Behaviour of TIG Modified Ceramic Coating Using the Taguchi Approach *Advanced Materials Research* **1115** 238-242.
[16] Moon, Joonoh, and Changhee Lee 2009 Behavior of (Ti, Nb)(C, N) complex particle during thermomechanical cycling in the weld CGHAZ of a microalloyed steel *Acta materialia* **57.7** 2311-2320.
[17] Yan W, Shan Y Y and Yang K 2006 Effect of TiN inclusions on the impact toughness of low-carbon microalloyed steels *Metallurgical and Materials Transactions* **A 37.7** 2147-2158.

IOP Conf. Series: Materials Science and Engineering **184** (2017) 012054 doi:10.1088/1757-899X/184/1/012054

Sustainable Manufacturing framework from Islamic Perspective

S Zain[1,a] *, E Y T Adesta[1,b], A F Ismail[2,c] and Z Ahmad[1,d]

[1] Department of Manufacturing and Materials Engineering, Kulliyyah of Engineering, International Islamic University Malaysia, 53100 Kuala Lumpur, Malaysia

[2] Department of Mechanical Engineering, Kulliyyah of Engineering, International Islamic University Malaysia, 53100 Kuala Lumpur, Malaysia

Email : [a] sz1396@gmail.com, [b] faris@iium.edu.my, [c] eadesta@iium.edu.my, [d] zuraidaa@iium.edu.my

Abstract. Islam is a complete way of life and offers holistic and comprehensive solutions to the human life. Manufacturing is one of the fields in which the voice of Islam is hardly heard. The efforts to relate and integrate manufacturing practices with Islamic principles and to eventually create a holistic solution is still lacking. This study is an attempt to develop a framework of sustainable manufacturing practices from the Islamic perspective. This framework may offer optional solutions to the current problems faced by the world such as global warming, climate change, pollutions, diseases etc. due to human ignorance. The framework was developed based on a preliminary research on the recent study of Maqasid Shari'ah, which is used as the main foundation. The developed framework comprises of three main domains, namely social, economic and environment. The framework is hoped to offer alternative solutions in producing high quality products, whilst taking into consideration the society, economy and environment, in line with the teachings of Islam and in accordance to our subservience to Allah SWT.

1. Introduction

The global manufacturing sector continues to grow and remains critically important to most developed and developing countries, socially and economically. Manufacturing activities, on the other hand, have enormously contributed to many adverse environmental problems such as the depletion of natural resources, the increase in global warming, global climatic changes and pollutions due to an inherent increase in the consumption of resources, high levels of consumption in energy and materials, inefficient factory operations as well as high volumes of waste generation [1]. Sustainable development and sustainable manufacturing have been introduced to respond to the problem in a holistic and global manner.

Sustainable manufacturing can be defined as "the creation of manufactured products that use processes that minimize negative environmental impacts, conserve energy and natural resources, are safe for employees, communities, and consumers and are economically sound" [2]. Since a decade ago, several sustainable manufacturing frameworks have been developed to respond to the alarming global concern and in line with that, some manufacturers have embraced sustainability in their corporate strategy [3]. In this study, an alternative framework of sustainable manufacturing has been established. The framework is developed as an attempt to highlight Islam as a holistic solution provider to the current social, economic and environmental problems faced globally. Even though there is a serious lack in efforts to relate Islam and manufacturing in the literature as well as in manufacturing practices, the motivation of this study is the belief that Islam provides complete guidance for all aspects of life, be it individual, social, political, environmental, economic and international. A preliminary research on the recent studies of the Maqasid Shari'ah has been used as the main foundation in developing the sustainable manufacturing framework.

2. Literature Review

2.1 Sustainable Manufacturing Framework

In this research, several recent sustainable manufacturing models and frameworks are discussed and referred to in developing a new framework. In 2009, Kibira D et al. proposed a framework that encompasses four interacting complex domains which include the three common domains of environmental, financial and social, as well as the manufacturing domain [4]. The proposed framework also highlights a set of influential factors within each domain which has been identified in an acceptable common taxonomy and was designed to be pertinent to the modelling of sustainable manufacturing from the global level to the community level [4]. In the same year, the Organisation for Economic Co-operation and Development (OECD) which was based in Europe, took an advanced step by developing a sustainable manufacturing toolkit in the form of comprehensive universal indicator sets [5]. The main function of the toolkit is to support the end users to benchmark the environmental performance of manufactured products and processes in relation to financial viability [5].

Meanwhile, in 2012, Dombrowski U et al. proposed a more practical framework by analysing the existing Lean Production System and attempting to incorporate the social and environmental sustainability elements in production activities. [6]. The focus of the framework, however, was restricted in the design and production activities only. One of the recent proposals of sustainable manufacturing framework was developed by Ocampo L and Ocampo C O in 2015. Their proposed framework was an attempt to formulate a sustainable manufacturing system that is capable of maintaining the competitiveness of the manufacturers and at the same time fulfilling the demands for sustainability [1]. A common characteristic of the four proposed frameworks above, basically, is to retain or increase the sustainability of social and economic elements and at the same time enhance the sustainability in environmental element effectively and comprehensively.

2.2 Islam and development

Islam represents more than a belief system, it is a well-defined set of values providing specific guidance for essentially every aspect of life, applicable in all times and places. Within its firmly established principles and values, Islam has all the essentials that correspond to development to adopt modern ideas and technologies that can improve life. In the framework of the religion, Islam offers inspiration which can be translated into policies, regulation and law for development in order to achieve a sustainable relationship between mankind and the natural world created by Allah SWT. Islam offers great advantage for environmental conservation, protection and sustainable development alongside with the sustainability in society and economy. The recent global environment crisis has brought Islam into the limelight. The crisis offers an opportunity for Islam to become a significant option as a solution provider for the global manufacturing industries.

2.3 The Framework of Maqasid Shari'ah

Maqasid Shari'ah is translated as the objectives behind the Islamic rulings (shari'ah). In a more specific manner, Maqaṣid Shari'ah is the aims or fundamental purposes of Shari'ah upon which it was established [7,8]. In general, Maqasid Shari'ah is considered to be the grand framework that provides the basic philosophical foundations that give guidelines and directions for ensuring the realisation of maslahah (benefit) and the prevention of mafsadah (harm) in any element [7,9]. It is important to understand the meaning of the shari'ah itself before discussing further the concept of Maqasid Shari'ah. In this regard, al-Qayyim I (1993) stated that "The foundation of the Shari'ah is wisdom and the safeguarding of people's interests in this world and the next. In its entirely it is justice, mercy and wisdom. Every rule which transcended justice to tyranny, mercy of its opposite, the good to the evil, and wisdom to triviality does not belong to the Shari'ah although it might have been introduced therein by implication. The Shari'ah is God's justice and mercy amongst His people. Life, nutrition, medicine, light, recuperation and virtue are made possible by it. Every good that exists is derived from it, and every deficiency in being results from its loss and dissipation for the Shari'ah, which God entrusted His prophet to transmit, is the pillar of the world and the key to success and happiness in this world and the next" [10].

From the above definition, in a broad sense, Shari'ah can be regarded as a significance mechanism to offer solution to the modern and global crisis. Maqasid shari'ah, therefore, is not only deals with the modes of worship for the individual, but also proposes directives for a collective life containing various aspects such as socio-economic affairs, politics, health and protection of diseases, global warming, climate change as well as international relations [11].

Maqasid Shari'ah has evolved since its introduction as a branch of knowledge in Islamic studies. Imam al-Shāṭibī (14th century), a great scholar from Andalusia, who is the renowned master architect of Maqasid Shari'ah, divided the general objectives of shari'ah into three sub-categories: the Necessities (*ḍarūriyyah*), the Needs (*ḥājiyyah*) and the Embellishments (*taḥsīniyyah*) [12]. The categories are briefly discussed below:

(1) The Necessities *(Ḍarūriyyah)* can be explained as the interests of lives which people essentially depend upon, and whose neglect leads to total disruption and chaos in the society [7,8]. Since the introduction of Maqasid Shari'ah, this primary component, basically, comprises of the five objectives of the shari'ah: protection of faith (*dīn*), life (*nafs*), intellect (*'aql*), posterity (*nasl*) and wealth (*māl*) [7, 8]. Recently, in line with the flexibility of the shari'ah, two more essential elements of Necessities have been introduced, i.e. protection of environment *(bi`ah)* and health *(sihhah)* [11]. These seven elements of necessities are essential as the foundation for the establishment of peace and well-being in this world and the hereafter.

(2) The Needs *(Ḥājiyyah)* can be defined as "the interests that supplement the essential interests" [7]. This component is required to free the society from distress and difficulty by lightening hardships in applying the Necessities.

(3) The Embellishments (Taḥsīniyyah). This component refers to the interests whose "realization leads to improvement and the attainment of that which is desirable" [12]. Examples of this component are the avoidance of lavishness in consumption, cultivation of fortitude and observance of cleanliness in personal appearance.

3. Result and Discussion

3.1 Integrated framework of Sustainable Manufacturing

In the last decade, the concept of Triple Bottom Line [13,14] which comprises of the three domains of sustainable development; social, economic and ecological/environment has emerged [15]. Nowadays, the established consensus acknowledged globally is an equal consideration of the three domains [16].

In manufacturing, nevertheless, the actualisation of the concept from macro to intermediate level is still far from perfect [6]. The challenge is to create a healthy and long term competitiveness of manufacturing players by minimizing the use of natural resources and producing eco-friendly waste generation, and at the same time enhance the quality of life of the stakeholders as well as society [13].

Meanwhile, frameworks based on the Maqasid Shari'ah have been developed and successfully implemented in various sectors especially in the financial (banking, takaful etc.) and political (policies, acts, etc.) sectors [8,9]. The big question that may be asked in this study is how to integrate between the Triple Bottom Line of sustainability and the principles of Islam, in this case, the Maqasid Shari'ah. In another word, how to develop a framework of Sustainable Manufacturing based on the principles in the Maqasid Shari'ah. The first approach applied in this study was integrating the five + two elements of the Necessities *(Ḍarūriyyah)* in Maqasid Shari'ah with the three domains in the Triple Bottom Line framework.

From the authors' perspectives, the seven elements of the Necessities in the Maqasid Shari'ah can be integrated into the three domains of sustainability in the Triple Bottom Line. The Protection of wealth (*māl*) and the protection of environment *(bi`ah)* can be directly matched to the economical and environmental domains respectively. Meanwhile, the remaining elements which are the protection of faith (*dīn*), life (*nafs*), intellect (*'aql*), posterity (*nasl*) and health *(sihhah)* can be integrated into the social domain as the elements are clearly related to the establishment of social well-being. The structure for this integrated framework can be displayed by a three-sided pyramid as shown in figure 1, where each side represents a domain of sustainable manufacturing.

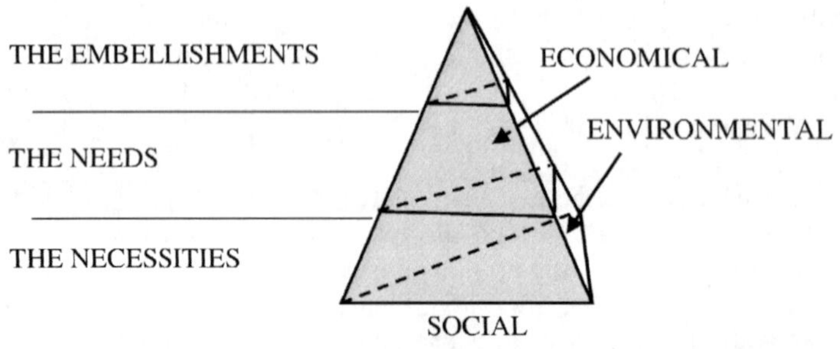

Figure 1. The structure of the integrated framework of Sustainable Manufacturing.

3.2 Sustainable Manufacturing Framework based on the Maqasid Shari'ah

The next phase of the framework development is more significant. At this stage, the elements of Maqasid Shari'ah are embedded and infused into the manufacturing activities, based on the structure of integrated framework which has been developed beforehand. The framework of sustainable manufacturing proposed in this study is shown in figure 2. The framework incorporates manufacturing activities which are divided into four main components, i.e. Natural Resources Consumption, Production, Consumerism and Disposal.

Manufacturing activities are normally initiated with the extraction of natural resources from eco-systems and mines to provide the raw materials as well as to generate energy to support the production operation. Manufacturers and industrialists as the vital players in manufacturing industry, they consume the raw materials for the purpose of developing and producing products, whether existing or new, and sell the products to fulfil market demands [4,5]. End users, subsequently, use the products in accordance to their needs until eventually disposing the products after they are considered to have no further use. In each component in the proposed framework, all of the three domains of sustainability are incorporated, as an approach to optimize the interaction between the domains in gaining sustainability. Furthermore, in each domain, two crucial elements of Maqasid Shari'ah; maslahah (benefit) and prevention of mafsadah (harm) are embedded and outlined more specifically.

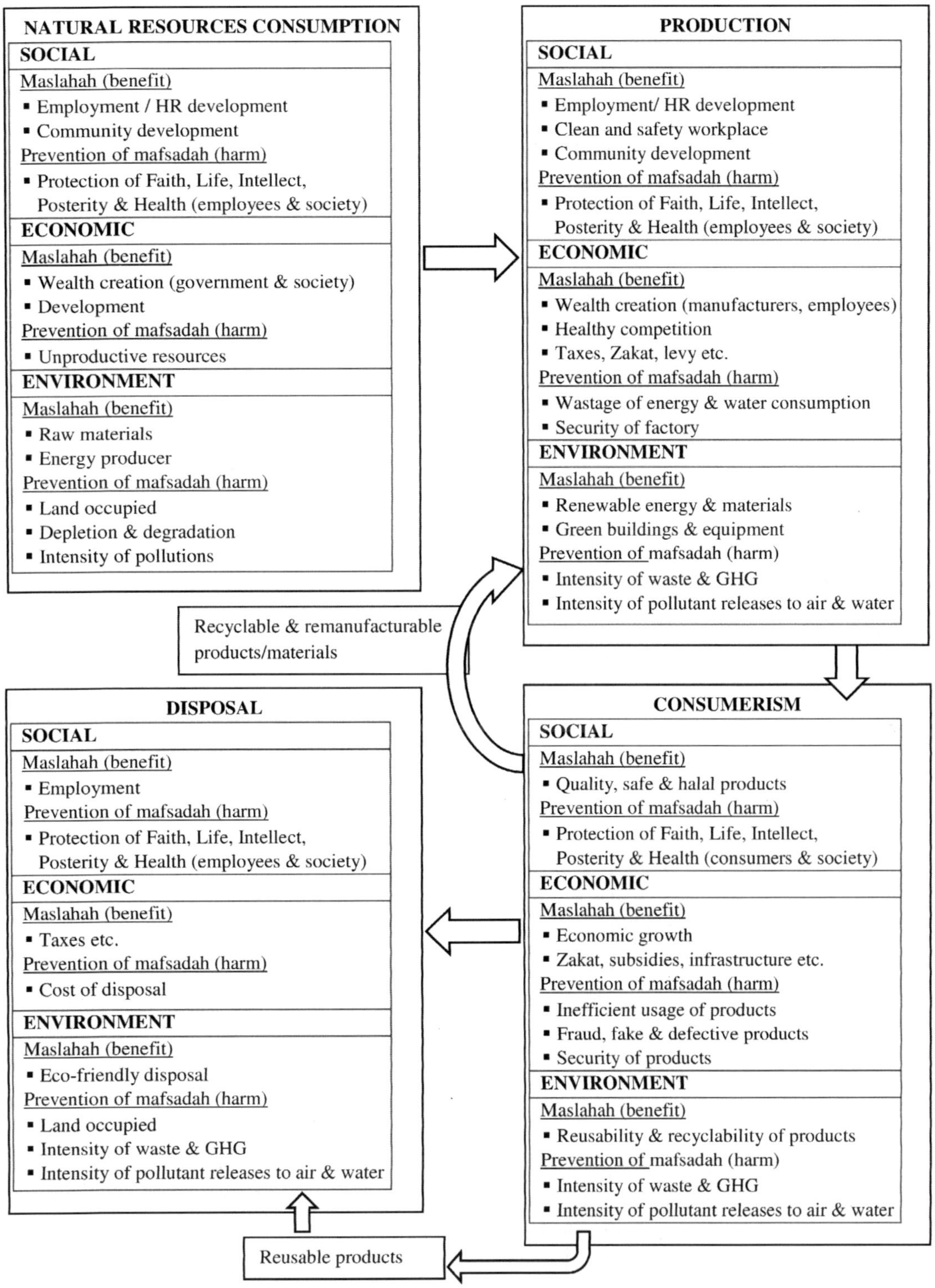

Figure 2. A proposed framework of Sustainable Manufacturing based on Islamic perspective.

Basically, the focal point of the framework is the social domain. This statement may look biased, but the fact is to achieve a decent standard of living, the society definitely needs economic stability and a healthy environment. In manufacturing, society can be categorised into three groups; business resources (employers, employees, suppliers), clients/consumers and communities who are not involved directly in any of the manufacturing activities. The employers, employees and supplies involve in many activities at the workplace that may lead to accidents, diseases and psychological effect such as stress. The risk of this group being exposed to accidents and diseases is higher than other groups. Meanwhile, the clients and consumers may be exposed to the risks when they are using the products. The community, while not being involved in any manufacturing activities directly, can also be exposed to some risks but the level of risks are considered the lowest among the groups.

As mentioned earlier, there are five elements of necessities in Maqasid Shar'iah which have been integrated in the social domain in the framework. All of the groups need protection of faith (e.g. for Muslim, from products that may lead to polytheism), protection of life (e.g. from death and injuries), protection of intellect (e.g. from materials/products/activities that weaken the mind), protection of posterity (e.g. from materials/products/activities that harm the reproductive system) and protection of health (from diseases). The holistic and strict protections can be executed by introducing, implementing and enforcing suitable acts, regulations and rules as well as inculcating positive cultures among the society. The consumers, specifically, should enjoy consuming quality, safe and value-for-money product(s) produced by the industry in the most efficient manner. This group should be protected from fake and defective products and also non-halal products (for Muslim). Moreover, manufacturing activities offer a great deal of employment opportunity to the community. As a result, the authority needs to plan an excellent human resources development programme by providing better education and training system as well as developing sufficient infrastructure to support the manufacturing activities. This will significantly boost the development in the area and consequently enhance the quality and standard of living of the society.

Wealth accumulation and management are very crucial at the organisation level such as the government, manufacturers, and retailers as well as at the family/individual level. In the proposed framework, all of the parties should enjoy financial benefits *(maslahah)* from the manufacturing activities, and simultaneously avoid any financial problems and crises *(mafsadah)*. In general, manufacturing activities may boost business activities, hence generate income to the government, manufacturers, retailers, stakeholders as well as employees. The consumers and society, meanwhile should obtain indirect financial benefits such as subsidy, zakat, scholarship etc. On the other hand, the framework suggests prevention of financial problems such as security of factory/workplace (from fire, flood etc.), cost-effectiveness of operations, restriction of frauds and fakes, avoidance of defective products, and minimised costs of operations and disposals. The ultimate objective of this economic domain is to create wealth for the stakeholders, increase economic growth and finally benefit the society.

Due to the increasing global environment problems, the environmental domain in the framework may be given an immense amount of attention. In the first place, the framework suggests efficient planning of extraction and consumption of natural resources by optimizing the land consumption, increasing the use of renewable resources (energy and raw materials) as an alternative and minimising the consumption of non-renewable resources (metals, oil, gas, coal, etc.). At the production sites, the framework suggests efficient usage of energy and water for operation, increase in the consumption of renewable raw materials, increase in the development of products that can be recycled and remanufactured, and equipping the factory with green technology. The manufacturing players are also suggested to eliminate or at least minimise the generation of emission and waste to the air, water or soil as well as the emission of noise as much as possible.

Consumers hold a huge responsibility in contributing to environmental sustainability. In the first place, they should choose to purchase products that has environmentally friendly characteristics such as reusable, recyclable, remanufacturable and can be disposed in the eco-friendliest manner. Then, they must use the product efficiently and minimise waste as much as possible. This habits will significantly help reduce the air and water pollutions and create healthier environment in a long run. Finally, in the disposal components, the framework suggests more protection of land for landfill and introducing eco-friendlier solutions of waste management.

4. Conclusion

Sustainable manufacturing is a response to the increasing deterioration of the global environment due to uncontrolled and unsustainable manufacturing activities in the ever globalizing market. This strategy is in line with the principles of Islam where Islam encourages interaction, optimization and balance of the three main domains in sustainability; social, economic and environment. This paper has proposed a framework of sustainable manufacturing based on the preliminary research of recent studies on the Maqasid Shari'ah. The framework was developed by embedding and infusing the elements of Maqasid Shari'ah into the manufacturing activities, based on the initial structure of integrated framework. The proposed framework was further established by outlining the elements of the maslahah (benefit) and prevention of maslahah (harm) in each domain of every component. This framework is an initial effort to expand the application of Maqasid Shar'iah as a framework in the manufacturing sector. The proposed framework, hopefully, will be improved continuously and expanded in more specific function in each manufacturing activity.

References

[1] Ocampo L and Ocampo C O 2015 A proposed sustainable manufacturing strategy framework, *Business Systems and Economics* Vol. 5 (1)

[2] http://www.trade.gov/competitiveness/sustainablemanufacturing/how_doc_defines_SM.asp [accessed 1 April 2016]

[3] Dubey R, Gunasekaran A and Chakrabarty A 2015 World class sustainable manufacturing : framework and performance measurement system *International Journal of Production Research* 53(17) : 5207-5223

[4] Kibira D, Jain S and McLean C R 2009 A System Dynamics Modeling Framework for Sustainable Manufacturing, *27th Annual System Dynamics Society Conference*

[5] Nasr N Z, Hilton B and German R 2011 A Framework for Sustainable Production and a Strategic Approach to a Key Enabler: Remanufacturing, Proceedings of the 8th Global Conference on Sustainable Manufacturing: *Advances in Sustainable Manufacturing*, Abu Dhabi, Springer pp 191-196

[6] Dombrowski U, Mielke T and Schulze S 2012 Lean Production Systems as a Framework for Sustainable Manufacturing, in : G. Seliger (Eds.), *Sustainable Manufacturing*, Springer Berlin Heidelberg, pp. 17-22

[7] Auda J 2008 Maqasid al-Shari'ah - a beginner's guide, *The International Institute of Islamic Thought*, Cromwell Press, London

[8] Dusuki A W and Bouheraoua S 2011 The framework of Maqasid al-Shari'ah and its implication for Islamic finance, *Islam and Civilisational Renewal*, Vol. 2 (2) pp. 316-336

[9] Laldin M A and Furqani H 2013 Developing Islamic finance in the framework of Maqasid al-Shari'ah: Understanding the ends (maqasid) and the means (wasa'il), *International Journal of Islamic and Middle Eastern Finance and Management*, Emerald Insight, Vol. 6 Iss: 4, pp.278 - 289

[10] al-Qayyim I 1993 *Medicine of Prophet*, Pearl Publishing House, USA

[11] Deuraseh N 2012 New essential values of daruriyyah (necessities) of the objectives of Islamic Law (Maqasid al-Shari`ah), *Journal Hadhari* 4 (2) pp. 107 - 116

[12] Kamali M H 1989 Sources, Nature and Objectives of Shari'ah *Islamic Quarterly* 33 pp. 215-35

[13] Jovane F, Yoshikawa H, Alting L, Boër C R, Westkämper E, Williams D, Tseng M, Seliger G and Paci A M 2008 The incoming global technological and industrial revolution towards competitive sustainable manufacturing, *CIRP Annals-Manufacturing Technology*, 57 pp. 641-659

[14] Elkington J 1998 Cannibals with forks: the triple bottom line of 21st century business, New Society Publishers, Gabriola Island

[15] Spear S and Bowen H K 1999 Decoding the DNA of the Toyota Production System *Harvard Business Review* n 9/10: 97-106

[16] Jovane F, Westkämper E and Williams D 2009 The Manufuture Road-Towards Competitive and Sustainable High-Adding-Value Manufacturing, Springer-Verlag, Berlin

Hierarchical extreme learning machine based reinforcement learning for goal localization

Nouar AlDahoul [1, a], Zaw Zaw Htike [1], Rini Akmeliawati [1]

[1] Department of Mechatronics Engineering

International Islamic University Malaysia, Kuala Lumpur, Malaysia

[a] nouar.aldahoul@live.iium.edu.my

Abstract. The objective of goal localization is to find the location of goals in noisy environments. Simple actions are performed to move the agent towards the goal. The goal detector should be capable of minimizing the error between the predicted locations and the true ones. Few regions need to be processed by the agent to reduce the computational effort and increase the speed of convergence. In this paper, reinforcement learning (RL) method was utilized to find optimal series of actions to localize the goal region. The visual data, a set of images, is high dimensional unstructured data and needs to be represented efficiently to get a robust detector. Different deep Reinforcement models have already been used to localize a goal but most of them take long time to learn the model. This long learning time results from the weights fine tuning stage that is applied iteratively to find an accurate model. Hierarchical Extreme Learning Machine (H-ELM) was used as a fast deep model that doesn't fine tune the weights. In other words, hidden weights are generated randomly and output weights are calculated analytically. H-ELM algorithm was used in this work to find good features for effective representation. This paper proposes a combination of Hierarchical Extreme learning machine and Reinforcement learning to find an optimal policy directly from visual input. This combination outperforms other methods in terms of accuracy and learning speed. The simulations and results were analysed by using MATLAB.

1. Introduction

Goal localization is one of the important tasks in indoor navigation systems. Visual data from a camera can be used, instead of traditional sensors such as GPS, to get information about the position of the agent and the goal. This observed data may be noisy and needs to be processed before being applied to a path planning algorithm.

The reinforcement learning suffers from the curse of dimensionality problem [1]. This problem appears when the agent needs to process and analyse high dimensional data (hundreds or thousands of dimensions). This problem can be solved by using one of the dimension reduction techniques. Traditionally dimensionality reduction techniques, such as Principle Component Analysis (PCA), Independent Component Analysis (ICA), and local linear embedding, are used to transform raw data to a lower dimensional manifold. Recently, feature learning techniques have been exploited to learn good representation automatically.

The autonomy of a learning system is achieved by its ability to adapt to the changes in the environment by finding suitable representations automatically. It is so important to stop depending on manual engineering that uses hand-crafted pre-processing stages and replace it by recently proposed deep models.

Unsupervised learning of deep auto encoder network was integrated into batch-reinforcement learning in [2, 3]. The near-optimal policy was demonstrated automatically by learned feature spaces in grid-world like task. Deep fitted q-iteration was proposed for this object.

A class-based active detection model was proposed in [4]. It learns to localize object known by the system. The agent analyses the content of region to select the next best action. DeepQNetwork is used to learn the localization policy. Pre-trained convolutional neural network is utilized to extract features from the current region.

Another system that uses raw visual input data is pole balancing controller, it was proposed to learn a control policy by using reinforcement learning [5]. The deep encoder neural network is utilized for dimensionality reduction of the raw images.

A real slot car controller is able to learn the optimal actions autonomously [6]. It is based on cluster-reinforcement learning which was added to a Fitted-Q batch reinforcement learning to approximate the Q function.

An artificial deep Q-network agent was demonstrated to compete human in game playing [7]. The inputs are the pixels and the game scores. The objective is to produce actions that maximize accumulated scores.

This paper proposes a deep model that is based on Hierarchical Extreme learning machine [8] in a visual reinforcement learning task. The combination of deep learning and Reinforcement learning is very fruitful to have a complete system that gets visual data as input and gives optimal policy as output. Fig. 1. Shows the block diagram of the proposed system. The deep learning is used to learn suitable features from high dimensional data. The model free reinforcement learning is utilized to learn the optimal policy.

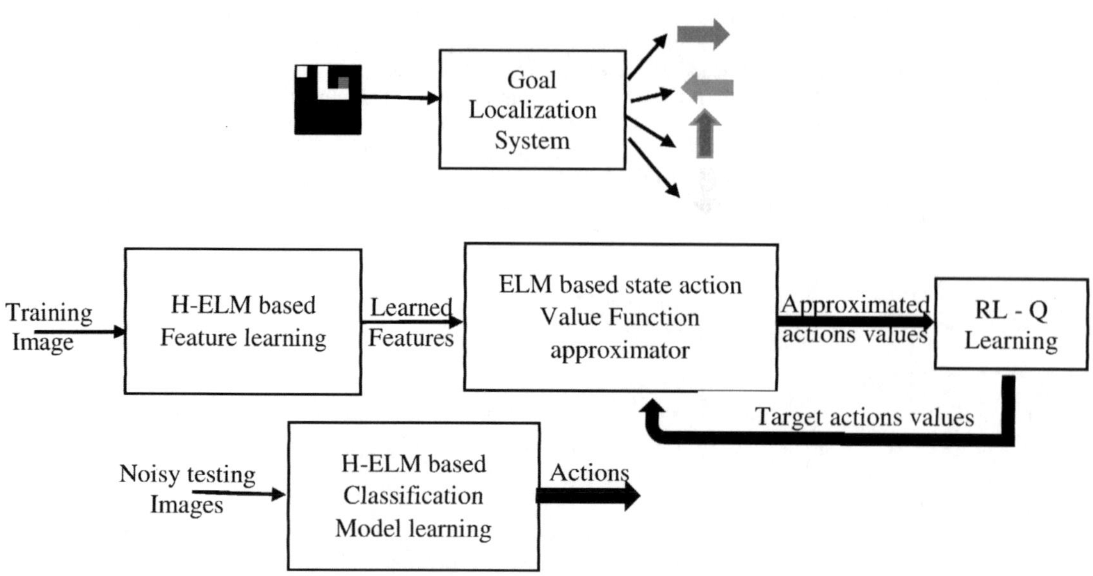

Figure 1. Block diagram of the proposed system

2. Methodology

2.1 Extreme Learning Machine for classification and regression

Extreme learning machine (ELM) is a feedforward neural network that has one hidden layer. The high generalization and high speed learning are behind the success of this learning method [9]. The biases and weights of the hidden layers are set randomly but the weights of output are calculated analytically.

$$f(x) = \sum_{i=1}^{L} F_i(x, W_i, b_i).\ \beta_i \quad , \quad W_i \in R^d \quad , \quad b_i, \beta_i \in R \tag{1}$$

Where $F_i(\cdot)$ is the activation function of i th hidden node, W_i is the input weight, b_i is the bias, and β_i is the weight of output, L is the number of nodes in the hidden layer.

$$\beta = H^{\dagger}T \quad , \quad \beta = H^T\left(\frac{1}{\lambda} + H.H^T\right)^{-1}.T \tag{2}$$

Where H is the output matrix of the hidden layer, $H^{\dagger}$ is the Moore–Penrose generalized inverse of H, T is the matrix of target, and λ is the regulation coefficient.

2.2 Hierarchical ELM for feature learning

When dealing with visual data such as images, the deep architecture of extreme learning machine is required [8]. This architecture can achieve self-taught feature learning by unsupervised elm-based sparse encoder. H-ELM gives better generalization and less learning time. The elm-based sparse encoder is built by using fast iterative shrinkage-thresholding algorithm (FISTA). This encoder is used as a basic component of the H-ELM. Deep architecture can be achieved by stacking multiple encoders. It guarantees better data recovery and reduces the testing time by reducing the number of neural nodes. Return to H-ELM paper [8] for more details.

2.3 Reinforcement Learning

Reinforcement learning is an important learning method that focuses on how agents should take optimal actions in the environment to maximize the discounted cumulative reward [1]. See Equation 3.

$$R = \sum_{t=0}^{\infty} \gamma^t\ r_{t+1} \tag{3}$$

Where $0 < \gamma < 1$ is the discounted factor.

RL framework is formulated as a Markov decision process (MDP). The main difference between the traditional learning and the model free reinforcement learning is that there is no need to have a prior knowledge about the model of the environment.

The main components of the reinforcement learning model are:

1. Environment states S.
2. Environment observations O.
3. Agent actions A.
4. Relation between states (S_t, a_t, S_{t+1}).
5. Reward R.

One of the most famous and used RL algorithms is Q-learning. It is a model free RL algorithm. Its core is a value iteration update. The value function is found by Equation. 4. The optimal policy can be found by Equation. 5.

$$Q(s,a) = Q(s,a) + \alpha\left(R + \gamma\ Q(s',a') - Q(s,a)\right) \tag{4}$$

$$\pi(s) = arg_a \max\left(Q(s,a)\right) \tag{5}$$

Where Q is the value function, α is the learning rate.

2.4 The proposed architecture

The proposed architecture is a combination of H-ELM and Q-learning RL. The H-ELM is used to learn good features so as to find representations for input images. The Q-learning is utilized to use the learned features as states in MDP model to produce optimal actions.

The main steps of the proposed method are as follows:

1) **Initialization** Initialize the weights W_0 of the encoder's hidden layers randomly. Initialize episode counter c and sample counter s. Initialize value function Q (s, a) = 0.
2) **Exploration** explore the environment by using ϵ-greedy policy to collect training data (observations O_t). These observations are used as inputs to H-ELM based deep model. Increment s with each new observation.
3) **Encoding** transfer all training data from observation space to feature space by applying auto encoder on the observations to get feature vectors Z_t , Z_t = Encode (O_t).
4) **Reinforcement Learning** Find the state action value for each feature vector by using Equation4 and by utilizing supervised ELM based value function approximator. If the convergence is achieved, return the approximated value function, greedy policy and the encoder. Else, repeat the steps from 2 to 4.
5) **H-ELM based classification** after mapping between the observations and the values of actions, the H-ELM based classifier is used to map each noisy observation with the best action results from the previous steps.

3. Experimental Results

3.1 The environment description and simulation

In this goal localization task with noisy image, the agent observes an image with 30*30 = 900 pixels. Gaussian noise is added to the images with zero mean and variance $\sigma = 0.1$. The agent should localize the goal region with a few number of moves (shortest-path problem).

The MDP (Markov Decision Process) model of the environment contains: a reward of 0 is given for any movement not leading to the goal area or obstacles. A reward of -1 is given to any movement colliding with obstacles. A reward of 1 is given to a goal ending movement. The rules of transition between different states are also given. Four actions are performed to move the agent to 4 directions: right, left, up and down.

The number of raw observations (images) is determined by the number of agent's positions in the image. In our case, there are 36 observations. Fig. 2. Shows the 36 observations without (a) and with (b) Gaussian noise. The training and testing data consists of 3600 noisy images for training and 3600 for testing. For each agent position, there are 100 images.

3.2 Accuracy analysis

The proposed method gives high performance with 100 % accuracy. Fig. 3. Shows the confusion matrix of the testing data. After finding efficient features, reinforcement learning is applied to find optimal actions. Fig. 4. Shows the average predicted state value. The convergence is achieved after only few steps.

The accuracy of different architectures of H-ELM (different number of nodes in the hidden layers) is shown in Table1. In Table 2, the comparison between H-ELM and PCA features is demonstrated. H-ELM seems to produce better accuracy with the same number of features. The Fig. 5. Displays the curve of PCA Eigen values.

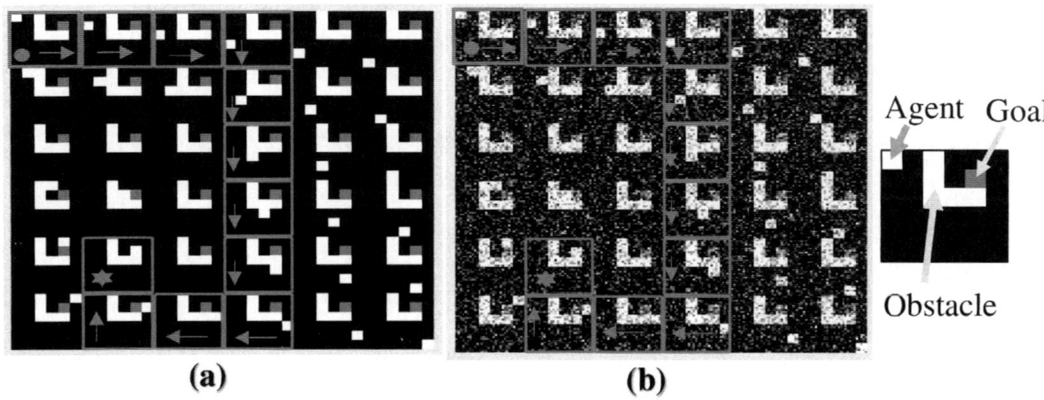

Figure 2. Training and testing images

a) Images without noise: red boxes refer to optimal actions from start (closed circle) to the goal (*), grey square represents the goal region.
b) Noisy images used to test the robustness of the system.

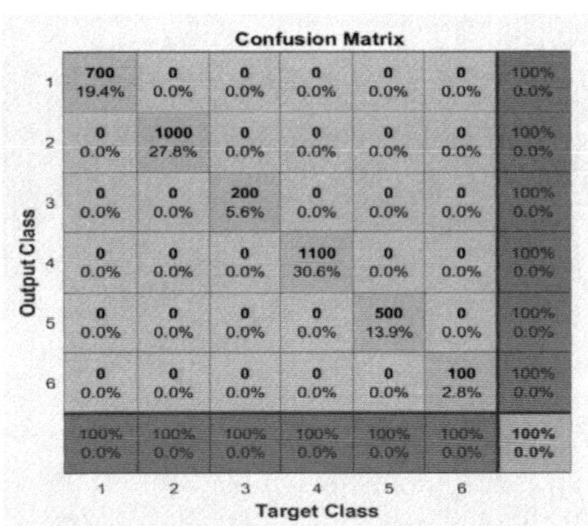

Figure 3. Confusion matrix of testing data

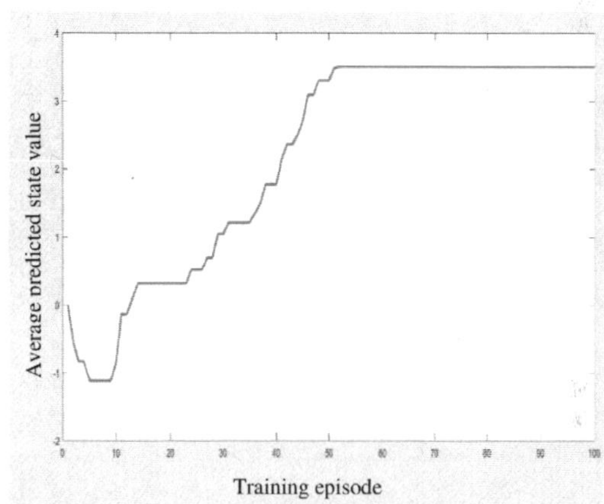

Figure 4. Average predicted state value curve

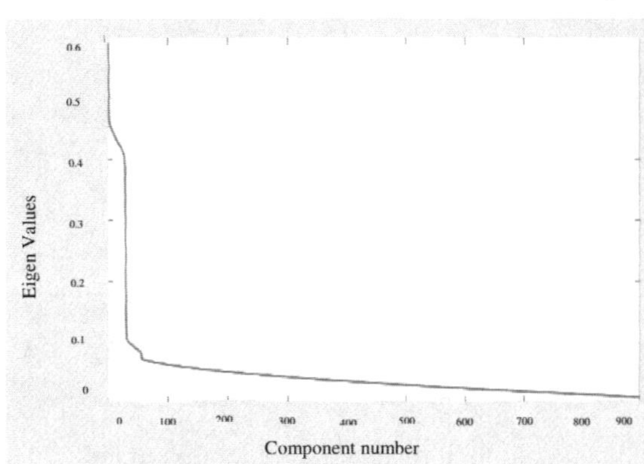

Figure 5. PCA Eigen values curve.

Table 1. Accuracy of testing noisy images with different model architectures

Number of nodes in first and second hidden layers	Accuracy %
500-200	99.97
500-100	99.94
500-50	99.97
500-20	97.63
500-10	97.47
500-7	96.52
500-5	87.91
500-3	66.11
200-3	69.88
400-3	76.52
400-5	90.83
400-7	95.72
400-10	97.02
400-100	100
400-200	100

Table 2. Accuracy comparison between PCA and H-ELM with different number of features

Number of features	PCA Accuracy %	H-ELM Accuracy %
50	100	**100**
20	97.17	**98.83**
10	97.08	**97.94**
5	93.61	**93.63**
3	71.33	**76.52**

3.3 Speed analysis

The combination of Hierarchical Extreme learning machine and Reinforcement learning outperforms other methods in terms of learning speed. Table 3 shows the training time of H-ELM and traditional stacked auto encoder [10]. The RL needs many episodes before achieving the convergence. From Fig.4, it is clear that the number of required episodes is almost 50. In each episode, the data samples are trained. Therefore the total time is the number of episodes multiplied by the training time in each episode.

Table 3. Training time comparison between H-ELM and traditional stacked auto encoder

Method	Training time in each episode (s)	Total Training time (s)
Stacked Auto encoder [10]	71	3550
H-ELM	5	250

4. Conclusion and future work

This paper proposed a combination of Hierarchical Extreme learning machine and Reinforcement learning. This combination is fruitful to learn optimal policy directly from visual input. This combination outperforms other methods in terms of accuracy and learning speed.

The advantages of the proposed architecture are:

1) Suitable action is produced when observation is forwarded through the network. This is important to reduce the testing time.
2) The use of H-ELM doesn't require to fine tune the encoder's weights iteratively and as the consequence, it reduces the time of learning and training.

This paper only focuses on the static images and does not take the relationship between individual frames into consideration. Future work should focus on this point to use dynamic information of the system.

5. References

[1] Sutton R S and Andrew B G 1998 Reinforcement Learning: An Introduction (*MIT Press, Cambridge, MA).*

[2] Lange S and Riedmiller M 2010 Deep Auto-Encoder Neural Networks in Reinforcement Learning *The International Joint Conference on Neural Networks (IJCNN), Barcelona.*

[3] Lange S and Riedmiller M 2010 Deep Learning of Visual Control Policies *ESANN, 18th European Symposium on Artificial Neural Networks.*

[4] Caicedo J C and Lazebnik S 2015 Active Object Localization with Deep Reinforcement Learning *The IEEE International Conference on Computer Vision (ICCV).*

[5] Mattner J, Lange S and Riedmiller M 2012 Learn to Swing Up and Balance a Real Pole Based on Raw Visual Input Data *In Neural Information Processing* **7667** 126-133.

[6] Lange S, Riedmiller M A and Voigtländer A 2012 Autonomous reinforcement learning on raw visual input data in a real world application *the International Joint Conference on Neural Networks (IJCNN), Brisbane, Australia.*

[7] Mnih V and et al 2015 Human level control through deep reinforcement *Nature* **518** 529–533.

[8] Tang J, Deng C and Huang G B 2015 Extreme Learning Machine for Multilayer Perceptron *IEEE Transactions on Neural Networks and Learning Systems* **27** 809 - 821.

[9] Huang G B, Zhu Q Y and Siew C K 2006 Extreme learning machine: Theory and applications *Neurocomputing* **70** 489–501.

[10] Hinton G E and Salakhutdinov R R 2006 Reducing the Dimensionality of Data with Neural Networks *Science* **313** 504-507

Design and fabrication of a fixed-bed batch type pyrolysis reactor for pilot scale pyrolytic oil production in Bangladesh

Mohammad Abdul Aziz[1], Rami Ali Al-khulaidi[1], MM Rashid[1], M.R Islam[2], MAN Rashid[3]

[1]International Islamic University Malaysia, Department of Mechatronics Engineering Jalan Gombak, 53100 Kuala Lumpur, Selangor, Malaysia
[2]Rajshahi University of Engineering & Technology, Department of Mechanical Engineering, Bangladesh
[3]International Islamic University Malaysia, Department of Material and Manufacturing Engineering Jalan Gombak, 53100 Kuala Lumpur, Selangor, Malaysia

E-mail: abdul.aziz07me@gmail.com

Abstract. In this research, a development and performance test of a fixed-bed batch type pyrolysis reactor for pilot scale pyrolysis oil production was successfully completed. The characteristics of the pyrolysis oil were compared to other experimental results. A solid horizontal condenser, a burner for furnace heating and a reactor shield were designed. Due to the pilot scale pyrolytic oil production encountered numerous problems during the plant's operation. This fixed-bed batch type pyrolysis reactor method will demonstrate the energy saving concept of solid waste tire by creating energy stability. From this experiment, product yields (wt. %) for liquid or pyrolytic oil were 49%, char 38.3 % and pyrolytic gas 12.7% with an operation running time of 185 minutes.

1. Introduction

The civilization of human growing rapidly and transportation becomes an important part of human life. The faster the rate of population, it tends to increment of vehicles which leads higher transportation[1].Every year, more than 450 factories produce over 1.5 billion tons of tires around the world. In Bangladesh alone, 120000 tons are produced annually. Bangladesh's energy consumption by fuel type consists of natural gas (61.82 %), furnace oil (21.68%), diesel (7.75%), imported power (4.86%), coal and hydro power (1.86%). The total installed electricity generation capacity is 12339 MW. However, imported coal and petroleum oil, Bangladesh's annual gross domestic product disrupted 2% and 60% gas consumed from national grid gas production unit[2]. The precious land is being occupied by waste scrap tires which may affects environment seriously. Natural firing operation in bricks field that has been done by using waste tires. Moreover, burning of waste carried out emissions that are highly toxic to humans. Finally, rain water is known to carry waste to larger areas, causing disease. Advancements in technology have minimized the problem of solid waste tires in the environment. The process of pyrolysis involves feed stock being fed into an oxygen-free environment in a reactor and heated, with inlet and outlet temperatures being sufficiently high (between 550°C and 680°C) [3].

In recent years, many advances in research in the field of pyrolysis have occurred. The gross calorific value (GCV) of pyrolysis liquids is 41-44 MJ/kg. Past research groups examined the composition of pyrolysis gas fraction. Several research groups reported it to contain higher percentages of CH_4(methane) C_2H_6(ethane) C_4H_6(butadiene) and hydrocarbon gases. The gross calorific value (GCV) of approximately $37MJ/m^3$, a value sufficient to provide the energy required by the pyrolysis process. They found the (GCV) of pyrolysis oil to be 33.7 MJ/kg, 37 MJ/kg, 40 MJ/ kg[4] and 41 MJ/kg –44 MJ/kg[5] in the fixed-bed reactor. Where the temperature stayed within the range of 400°C-700°Cand the operations lasted for 4 hours. Solid products in these experiments made up 47% - 63%, liquid products 30% - 43% and gas products 2.2% - 4.4% [6].

Meanwhile, many different experimental procedures were done for the production of pyrolysis liquid products from solid tire waste including fixed-bed, fluidized–bed, vacuum and spouted bed reactors [5]. In the fixed-bed fire-tube heating pyrolysis system, the maximum liquid and char yield was 52 wt. % and 35 wt. %, for bicycles and rickshaws respectively. For truck tires, the liquid yield and char yield came to 60 and 23 wt. %, respectively. The heat value of liquid of rickshaw and truck tires was found to be 41 MJ/kg and 40.7 MJ/kg, respectively [11]. However, for car tires, the maximum liquid yield and char yield came to 50 and 40 wt. %, respectively. For car tires, the liquid yield and char yield came to 45 wt. % and 28 wt. %, respectively. The heat value of liquid and char was found to be 42.2 and 35.6 MJ/kg, respectively [12]. A burgeoning research area is the use of the pyrolytic oil in a single cylinder, 4-stroke, water-cooled DI diesel engine with varying load[7].

In this research, solid tire wastes are processed to pyrolysis oil, solid char and gases. This conversion system is done using a fixed-bed fire-tube heating reactor system. The techno-economic appraisement was conducted taking into account three different sizes of plants: medium commercial scale (144 tons /day), small commercial scale (36 tons /day), and pilot scale (3.6 tons /day). The liquid produced from the plant should be used for boiler operation and upgraded liquid production[8]. However, the solid waste tires are being used for energy production by pyrolysis oil production. The reaction pathway of pyrolysis is shown in Figure 1.

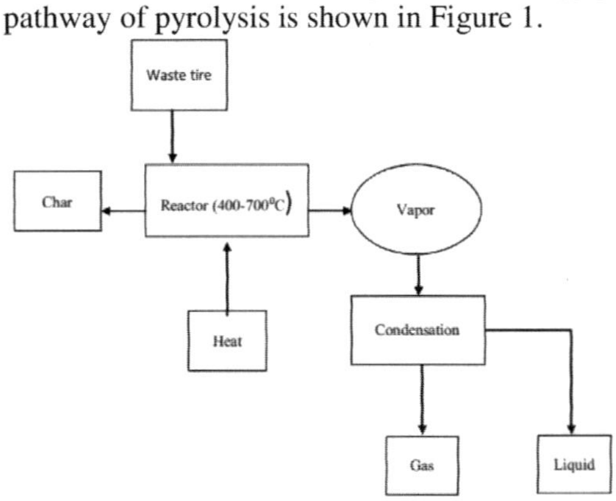

Figure 1. Reaction pathway of pyrolysis process.

2. Design & Fabrication of Fixed-Bed Reactor

The design and fabrication of a fire-tube heating reactor for the batch type fixed-bed pyrolysis unit was the major part of the project and thesis work. The fire-tube heating reactor system of pyrolysis unit was designed on the basis of the size of the system. It was designed so that sufficient amount of feed material could be taken into it, and in a way that there would be a short vapor residence time in the reactor, which would promote a high yield of pyrolysis liquid product.

IOP Conf. Series: Materials Science and Engineering **184** (2017) 012056 doi:10.1088/1757-899X/184/1/012056

The selection of the size of the fixed-bed reactor depended upon the amount of feed to be taken into the reactor. For maximum liquid production, the residence time of the mixture of volatile liquids and gases produced is very important. For simplicity of design, a cylindrical reactor was considered. Due to the corrosive nature of pyrolytic liquid, stainless steel was used as a reactor material. The reactor's height was 990 mm, while the diameter was 400 mm. The reactor had a through type feeder, 250 mm in diameter. This feeder was covered by a flat plate with the help of nut-bolts. Inside the reactor, there were three u-shaped spirals and two semi-hexagonal stainless steel pipes, 19 mm in diameter. The total length of each spiral pipe was 1370 mm and the bending radius was 76 mm. The pipes provided uniform and fast heating inside the reactor. At the bottom of the reactor, there was another throught of 200 mm diameter for char removal. A fixed-bed pyrolysis reactor is shown in Figure 3 before installation and after reactor installation. The annular shaped reactor shield was made while inside wall of reactor was made of 2 mm thick mild steel sheet and outer side being made by 2 mm thick GP sheet. The total length of the furnace 249 cm and length of reactor shield 60.96 cm and internal diameter of the reactor shield 4 cm. The dimensional view of fixed bed reactor shows by Figure 2.

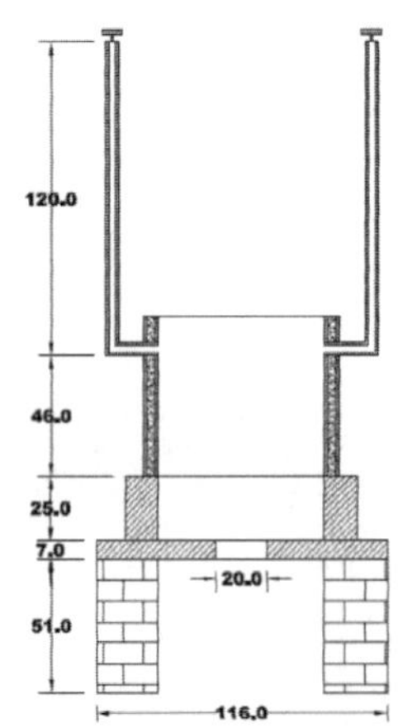

Figure 2. Dimensional view of fixed bed reactor.

Figure 3. Fixed-bed batch type pyrolysis reactor before installation (left) and after reactor installation (right).

2.1 Reactor shield

Reactor shield is a most important component in fixed bed pyrolysis reactor and consisting of four main parts. Two parts are portable and other two parts are fixed. The reactor shield was made of annular shaped consisting of 2 mm thick mild steel sheet while the outside was made of 2 mm thick GP sheet. Moreover, the hollow portion was filled with glass wool for insulating the material. Figure 4 represents the dimensional view of reactor shield.

2.2 Oil or gas burner

An oil burner is a very important part of furnace. The proper firing of burner can maintain the proper heating temperature as well as required temperature for reactor operation. Two burners were designed in such a way so that they can burn the fuel from both side of the furnace. They covered the proper heating inside the furnace. Burners allow flowing of both air and fuel to the furnace.

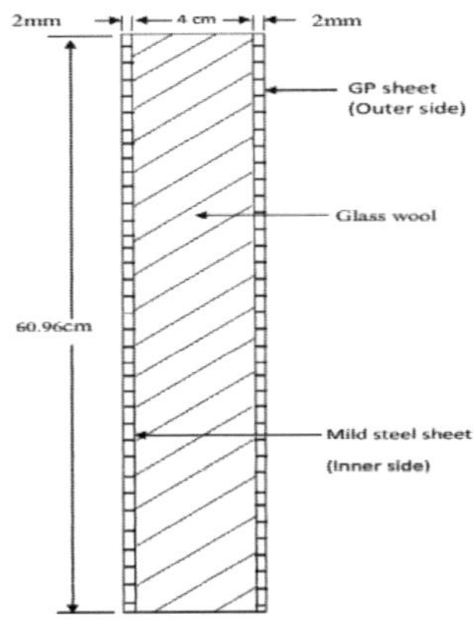

Figure 4. Dimensional view of reactor shield pyrolysis reactor.

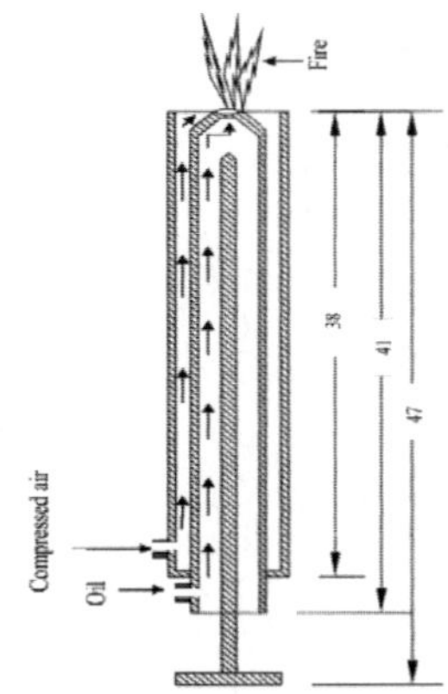

Figure 5. sectional view of burner for furnace heating.

Compressed air was passed through the outside pipe having 3.8 cm diameter. Fuel was supplied through the inner pipe having 2.8 cm diameter with a flow controlling screw. An orifice was introduced at the outer face of the burner to flow the air-fuel mixture at high speed. By controlling the flow of fuel through burner, required heating rate may be maintained. Figure 5 represents sectional view of burner for furnace heating.

3. Condensation principle

To yield liquid oil from pyro-gas, the condensing system is necessary. In condensation system the gas can be reduced to liquid by decreasing its temperature and increasing its pressure. Condensers are typically used as pretreatment devices. They are used ahead of incinerators or absorbers to reduce the total gas volume to be treated by these more expensive pieces of equipment. When hot vapor steam contacts a cooler surface, heat is transferred from the hot gases to the cooler surface. As the temperature of the vapor stream is lowered, the average kinetic energy of the gas molecules is reduced. Ultimately, the gas molecules are slowed down and crowded together so closely that the attractive forces between the molecules cause them to condense to liquid. Two conditions aid condensation: low temperature (so that the gas molecules are low) and high pressure (so that the gas molecules are brought closer). The actual conditions at which the particular gas will condense depend on its physical and chemical properties.

Condensers fall into two basic categories: contact condensers and surface condensers. In a contact condenser, the gas stream and coolant are physically mixed. The coolant and condensed vapors leave the device by separate exits. The surface condensers are also called shell and tube heat exchangers. The temperature of the coolant is increased, so that their devices also act as heaters. The diameter of condenser and length was 15 mm and 170 mm respectively. The outer shell diameter of condenser was 11.71 cm. In figure 6 and 7 presents thermal decomposition of organic solid wastes and dimensional view of condenser. A solid horizontal condenser with a pyro gas inlet, a pyro gas outlet, a hot water outlet and a cold water inlet is shown in Figure 8.

3.1 Selection of materials for separator and condenser

Due to low cost, availability and high operating temperature, a mild steel sheet was chosen to fabricate the separator and condensers. The chemical composition of the mild steel is: carbon 0.42% – 0.48%, silicon 0.15% – 0.35%, manganese 0.6% – 0.9%, phosphorus 0.030%, Sulphur 0.035%.

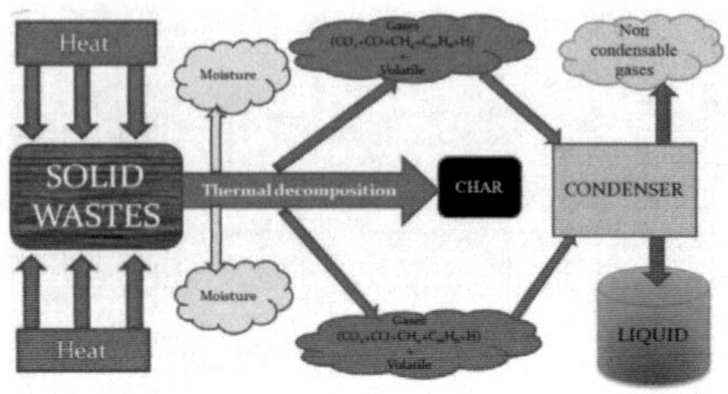

Figure 6. Thermal decomposition of organic solid wastes.

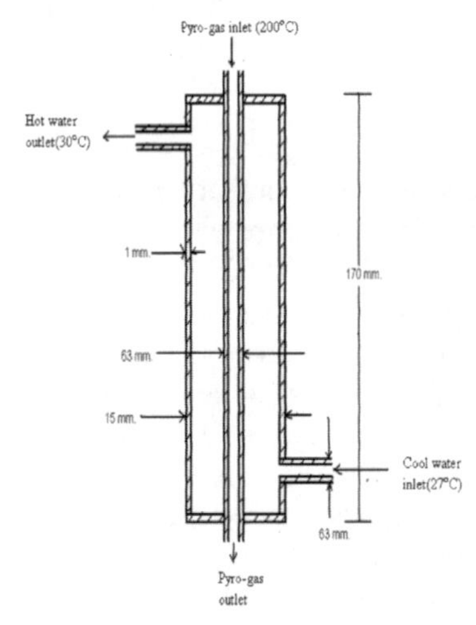

Figure 7. Section view of vertical condenser with dimensions.

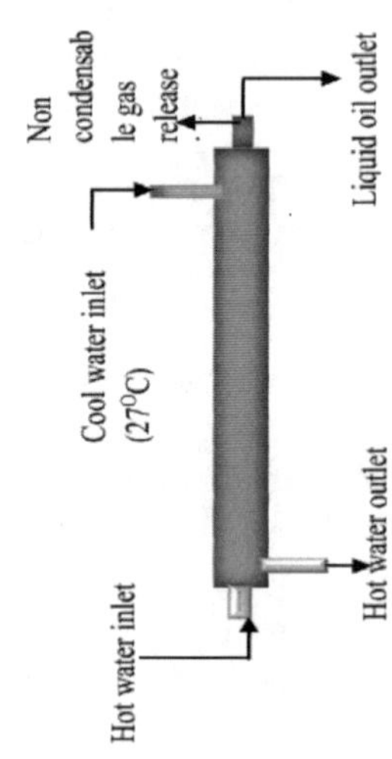

Figure 8. Solid horizontal condenser.

4. Pyrolysis unit operation process

In pyrolysis, the thermal decomposition of waste tires in the absence of oxygen causes them to undergo cracking into simpler organic constituents. Inside the pyrolysis plant, the breaking of chemical bonds occurs through the processes of decomposition and vaporization. The temperature range in which pyrolysis can occur is from 250°C to 500°C or 250°C to 900°C for some special pyrolysis operations. Shredding tires makes it easier to extract higher amounts of liquid oil products and gases, while temperatures higher than 450°C cause the yield of oil and solid tire-derived char to decrease relative to gas production[9].

The main factor that will allow accurate calculation of the batch type fixed-bed reactor's performance is the proper assembly of the pyrolysis unit. The generic features of the fixed-bed fire tube heating reactor system have been illustrated in research literature [10-13]. This pyrolysis unit contains a lot of components, such as: a fixed-bed fire-tube heating reactor chamber, a gravity feed type reactor feeder, two ice-cooled condensers, a N_2 gas cylinder, a N_2 gas pre-heater, an air compressor, a char collection bag, and thermocouples. Adequate assembly of these various components was the most important task needed to allow better performance of the plant. At first, a concrete slab supported by four pillars was placed inside underground holes at the base of the pilot plant. Then, the furnace was made on the concrete slab by using fire bricks. Two burners were connected at the bottom of the furnace. Then reactor chamber was placed on the furnace, supported with two m d steel rods. Next, the insulated side walls were placed around the reactor. A cover plate was attached to the top of the furnace. Two fixed parts of the side wall were attached to the exhaust pipes and the other two portable parts were connected with the fixed parts by means of nuts and bolts. Then, a gas flow line with a gate valve was connected to the fractionating column. The purpose of the fractionating column was to separate the heavy compounds from volatile liquids and gases. Thus, proper flow of gases through the pipes was maintained by separating the heavy compounds of high density as well as those with a high viscosity.

The fractionating column was attached to a vertical condenser, which was in turn connected to a horizontal condenser. A digital pyrometer probe was inserted into the reactor through the cover plate. The fractionating column and the gas pipe lines were supported by a metallic stand and insulated by glass wool with a thickness of 3 cm. Finally, a hole was made for the liquid collector at the outlet of the horizontal condenser. Figure 9 and 10 show the flow diagram and photograph of the completely assembled pilot scale pyrolysis unit, respectively.

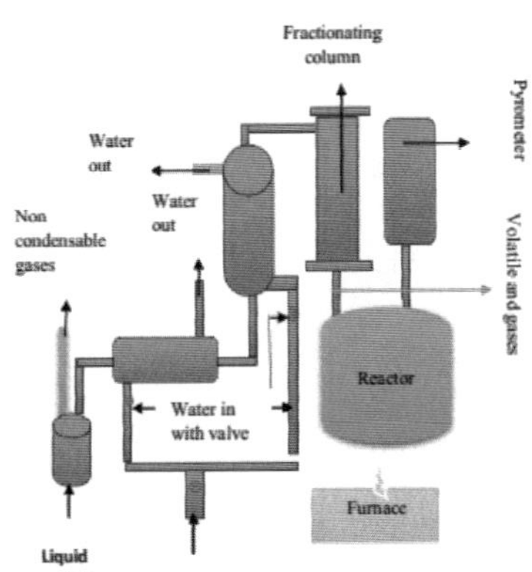

Figure 9. Flow diagram of the pyrolysis process.

Figure 10. Photograph of pilot scale pyrolysis unit.

5. Experimental procedure for the complete pyrolysis unit

In the production of tires, several materials and substances are used in tire processing plants[14]. The materials are: elastomer, which is natural or synthetic rubber; reinforcing agents such as carbon black; plastificants, which are hydrocarbon oils; vulcanizing agents, which are sulphur and sulphur components; accelerating agents to facilitate sulphur's action; and protective agents such as anti-oxidizing agents, stabilizers, etc.

Most primary raw materials are natural rubber, which is derived from the tree ***Hevea Brasiliensis***. Some parts of a processed tire cannot be separated and this waste is disposed in landfills. Process flow diagram of fixed bed pyrolysis reactor is shown in figure 11. Typical tire compositions for new and used tires are presented in Table 1.

Table 1. Typical tire compositions for new and used tires.

Tire compositions	New tires (wt %)	Used tires (wt %)
Elastomer	45	43
Carbon black	22	21
Steel	25	27
Zinc oxide	2	2
Sulphur	1	1
Other oils	5	6

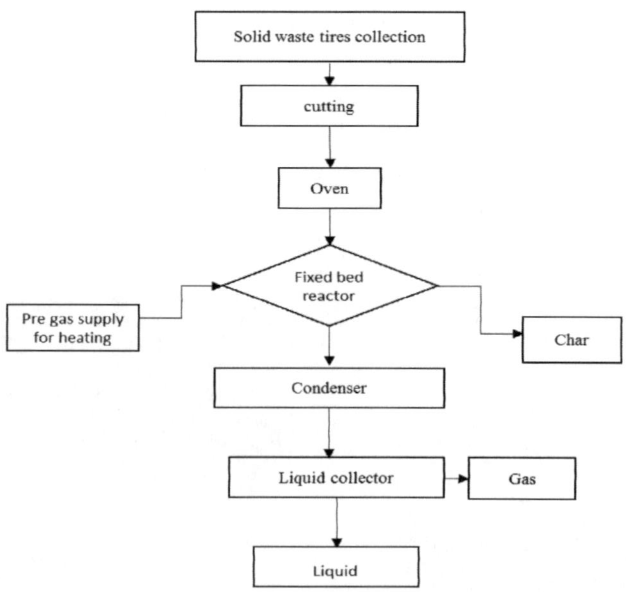

Figure 11. Fixed bed pyrolysis process flow diagram.

For this pyrolysis experiment, a sample of a feed size tire was heated in a fixed batch type pyrolysis plant at 300°C and at 410°C[14]. The yields of char and liquid were measured in comparison to the primary initial staring materials. Coke yield ($wt.$%) = $a/ W_S \times 100$, tar yield (wt. %) = $b/W_S \times 100$.

Where, a = weight of semi coke; b = weight of tar; W_S = initial weight of the sample.

A running pyrolysis unit for tire oil production was set up in the pyrolysis plant[15] Dry feed material was weighed and fed to the reactor and the experimental setup was checked properly. A high temperature adjustable gasket was used to seal the joints and fittings of the pipe lines and flange plates. A temperature measuring pyrometer probe was inserted into the reactor chamber. Subsequently, compressed air and fuel were supplied to the burner for burning the air fuel mixture into the furnace. For adjusting the prop heating, the flow of air and fuel was controlled by flow regulating valves. The temperature variations were measured by a digital pyrometer. The gas flow control valve was opened before starting gas production, and a water flow rine was opened to condensers for condensing the gases into liquid. Liquid was collected into a liquid collector at the outlet of the horizontal condenser. Moreover, the temperature of non-condensable gases and water was measured by thermometers.

The time calculation was done by means of a hand watch, and after the completion of pyrolysis oil production, the gas flow control valve was closed so that no air would enter into the reactor to bum the hot char. After all the portable side walls of the furnace were removed and cooled, the reactor's temperature rose to 50°C. Finally, char was removed from the reactor and both the liquid and char were weighed in order to get product distribution in weight percentage ($wt.\%$) of total feedstock.

6. Collection and Preparation of Feed Material

The collected scrap tires originally manufactured by birla and apollow, two Indian brands. The tires were originally designed for use in trucks or buses and weighed about 50 kg each. The waste tires were collected locally from a dump site in Rajshahi, Bangladesh. The tire was steel cord free and cut into pieces of size 16 x 11 x 3 cm. Pieces measuring 20 x 12 x 3cm and 20 x 11 x 2.5 cm were also made. Figures 4.4 and 4.5 show a dump site for solid tire wastes and prepared tire feed, respectively. A typical dump site of scrap tires and the feed size of waste tires are shown in Figure 12.

Figure 12. Typical dump site of scrap tires (left), and feed size of waste tires (right).

7. Heating and cooling operation

The heating and cooling operation is a very important factor for a pyrolysis plant. The reactor and furnace were designed in such a way that the produced heat was transferred into reactor chamber quickly and uniformly. Initially, the recorded temperature was only 30°C, but after only ten minutes the temperature rose to 77°C and the temperature reading was recorded. Finally, during the next 80 minutes of the reactor's operation, the reactor heated up at an increasing rate up to first 100 minutes of running maintained at constant temperature near about 430°C ($\pm 100^{O}$C). The portable parts of the insulated side wall of furnace were removed for cooling purposes, and the next 10 minutes saw a temperature decrease until the temperature reached 500° C. The total time for the reactor's heating and cooling operation was 360 minutes. The temperature distribution during the reactor's heating and cooling operation is presented in Figure 13 and 14.

8. Pyrolysis products from the pilot plant

From tire pyrolysis, three products were obtained: liquid, char and gases. Pyrolysis oil or liquid is the main product and char and gases are another two by-products containing higher gross calorific value which may be used as fuel for reactor heating purposes. In this experiment, a successful operation was carried out using four samples of waste tires.

IOP Conf. Series: Materials Science and Engineering **184** (2017) 012056 doi:10.1088/1757-899X/184/1/012056

The final results obtained from this pilot plant experiment are presented in Table 2. From the previous study, it was found that liquid production at the laboratory scale was slightly higher than that produced in a pilot plant For a particle feed size of 10 x 2 x 1cm, the maximum liquid, char and gas production was found to be 57, 26 and 17 *wt* %, respectively at an operating temperature of 475°C. Table 3 shows the pilot plant product yield distribution as compared to previous laboratory results. Table 4 shows the comparison of pyrolysis liquid with petroleum products. In figure 15 photograph of pyrolytic liquid (left), and photograph of char (right) are shown.

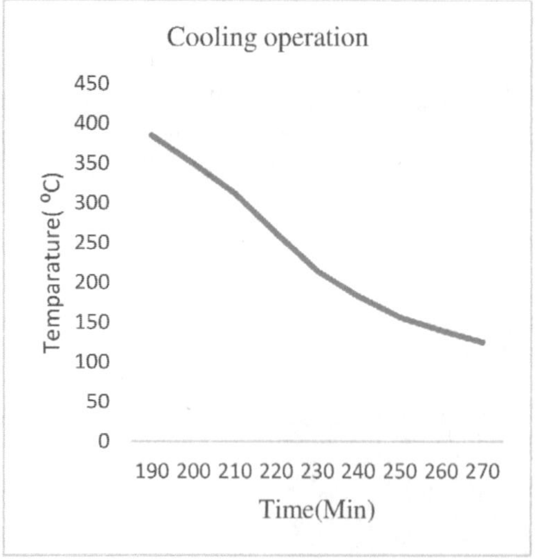

Figure 13. Temperature distribution during the reactor's cooling operation.

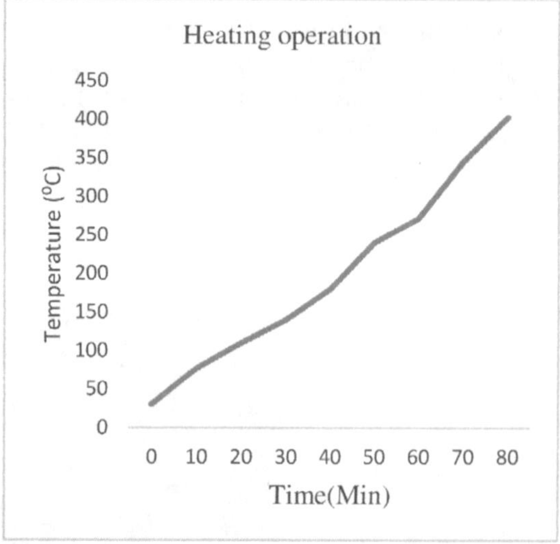

Figure 14. Temperature distribution during the reactor's heating operation.

Table 2. Product yield distribution for tire pyrolysis of different sizes.

Running time (min)	Feed size (cm)	Tire feed (kg)	Product yield (*wt* %)		
			Liquid	Char	Gas
180	16*11*3	35	48.5	37.7	13.8
210	20*12*3	33	48.4	38.1	15.3
185	20*11*3	30	49	38.3	12.7
180	19*12*3	32	46	37.5	16.5

Figure 15. Photograph of pyrolytic liquid (left), and photograph of char (right).

Table 3. Comparison of product yield distribution between a pilot plant and a laboratory unit.

Product yield (wt %)	**Pilot plant**	**Laboratory unit**
Liquid	48	53
Char	38	35
Gas	14	17

Some properties of the pyrolytic liquid produced were very close to petroleum products. The comparison between pyrolytic liquid and petroleum products is given in Table 4, while tire pyrolysis gas yields for different authors and experimental systems is shown in Table 5.

Table 4. Comparison between pyrolytic liquid and petroleum products.

Properties	Pyrolytic liquid	Diesel[7]	Furnace oil
Density (kg/m^3)	950	830	930
Viscosity (cSt)	4.0	2	6.2
HCV (MJ/kg)	42	46.5	46

Table 5. Tire pyrolysis gas yields taken from different authors and experimental systems.

Ref.	Liquid yield (wt %)	Temperature ($^\circ$C)	Reactor
[12]	46 – 55 wt%	475° – 575°C	Fixed-bed reactor
[16]	30 – 45 wt%	500° – 550°C	Fluidized bed reactors
This paper	49 wt%	430° – 500°C	Fixed-bed reactor
[17]	49.6 wt.%	450° – 650°C	Semi batch reactor

9. Environmental impact of pyrolysis plants and its remedy

During the process of pyrolysis, waste is disposed from the pyrolysis plant. This waste can be categorized by source of emission. Waste water in the tire pyrolysis plant is generated from general plant housekeeping works. A small amount of wastewater is produced from wet scrubber. The solid materials which adhere to the tires are cleaned with compressed air. Tires which have been exposed to road salt necessitate the use of water to remove the salt. For instance, it was reported that the amount of waste water produced from general housekeeping was 150 tons per month at a Shanghai plant[18]. The water is collected in a pond and is reused after removal of oil by a centrifugal separator.

The amount of oil averages 0.26 (about 390 liters per month). This oil is then sent to the furnace for combustion. In the pyrolysis plant, steel and gaseous wastes are produced and the composition of the gases is mostly hydrocarbons and hydrogen. The entire gaseous product is burned to provide heat energy for the pyrolysis process. In the case of pyrolysis, some non- condensable gases are emitted from the condenser: CO_2, CO, NO_X, and SO_X. Also, many fugitive and furan emissions are produced, as well as dioxins. To ensure pollution free pyrolysis plant fabric filters, more water and proper design is required.

10. Results and discussion

In this research, a few trial runs have been carried out at various operating conditions to reach maximum liquid production. The maximum liquid yield for tires was found to be 49 $wt.\%$ of feed. During condensation, it was found that the outlet temperature of the water from the condenser was 27°C, with the gas temperature reaching 68°C (Figure16). The outlet temperature was increased slowly and was raised up to 30°C. The pyro-gas inlet temperature in the condenser was 200°C and reduced to $38°C$. There was a temperature rise of the water in the condensers of 3°C.

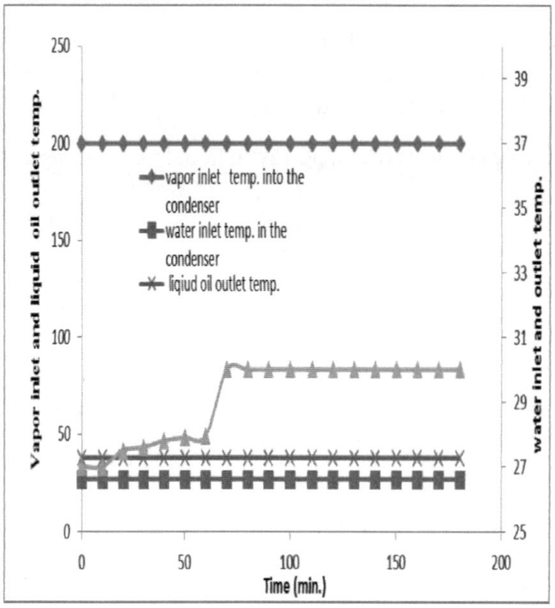

Figure 16. Temperature distribution of pyro-gas in the condensers, of the water inlet, and of the outlet of the condenser.

11. Conclusion

In this fixed-bed batch type pyrolysis system reactor, temperatures ranged from 430°C to 500°C and a maximum pyrolysis oil yield of 49 $wt.\%$ was achieved. At higher temperatures, a secondary cracking reaction takes place, causing an increase in gaseous product. It was found that some properties of the pyrolytic oil produced were close to diesel fuels and other conventional fuels. The obtained gross calorific value of pyrolytic oil was significantly high, meaning it could feasibly be used as an alternative to fossil fuel after proper treatment.

References

[1] Aziz, M., et al., DESIGN AND DEVELOPMENT OF A COMPRESSED AIR MACHINE USING A COMPRESSED AIR ENERGY STORAGE SYSTEM. 2006.

[2] Bangladesh power dovelopment board,Key Statistics. 2016; Available from: http://www.bpdb.gov.bd/bpdb/.

[3] Xue, Y., et al., Fast pyrolysis of biomass and waste plastic in a fluidized bed reactor. Fuel, 2015. **156**(0): p. 40-46.

[4] Rada, E.C., et al., Energy recovery from tyres waste through thermal option. Scientific Bulletin, Politehnica University of Bucharest, Series D, Mechanical Engineering, 2012. **74**: p. 201-210.

[5] Osayi, J.I., S. Iyuke, and S.E. Ogbeide, Biocrude Production through Pyrolysis of Used Tyres. Journal of Catalysts, 2014. **2014**.

[6] Aydın, H. and C. İlkılıç, Optimization of fuel production from waste vehicle tires by pyrolysis and resembling to diesel fuel by various desulfurization methods. Fuel, 2012. **102**(0): p. 605-612.

[7] Nabi, A.R., et al., Purification of TPO (Tire Pyrolytic Oil) and its use in diesel engine. IOSR Journal of Engineering, 2014. **4**(3): p. 1.

[8] Islam, M., et al., Feasibility study for thermal treatment of solid tire wastes in Bangladesh by using pyrolysis technology. Waste management, 2011. **31**(9): p. 2142-2149.

[9] Antoniou, N. and A. Zabaniotou, Features of an efficient and environmentally attractive used tyres pyrolysis with energy and material recovery. Renewable and Sustainable Energy Reviews, 2013. **20**(0): p. 539-558.

[10] Berrueco, C., et al., Pyrolysis of waste tyres in an atmospheric static-bed batch reactor: Analysis of the gases obtained. Journal of Analytical and Applied Pyrolysis, 2005. **74**(1–2): p. 245-253.

[11] Musale, H., Y. Bhattacharyulu, and R. Bhoyar, Design Consideration Of Pyrolysis Reactor For Production Of Bio-Oil.

[12] Islam, M.R., et al., Valorization of solid tire wastes available in Bangladesh by thermal treatment. 2011.

[13] Karthikeyan, S., C. Sathiskumar, and R.S. Moorthy, Effect of process parameters on tire pyrolysis: a review. Journal of Scientific and Industrial Research, 2012. **71**(5): p. 309-315.

[14] Fernández, A.M., C. Barriocanal, and R. Alvarez, Pyrolysis of a waste from the grinding of scrap tyres. Journal of Hazardous Materials, 2012. **203–204**(0): p. 236-243.

[15] Lopez, G., et al., A CONICAL SPOUTED BED REACTOR FOR THE VALORISATION OF WASTE TIRES. 2010.

[16] Ringer, M., V. Putsche, and J. Scahill, Large-Scale Pyrolysis Oil. Assessment, 2006.

[17] Pradhan, D. and R. Singh, Thermal pyrolysis of bicycle waste tyre using batch reactor. Carbon (C), 2011. **57**: p. 74.41.

[18] Li, S.-Q., et al., Properties of pyrolytic chars and activated carbons derived from pilot-scale pyrolysis of used tires. Journal of the Air & Waste Management Association, 2005. **55**(9): p. 1315-1326.

Validation of ESP Oil Wells Measured Parameters Using Simulation Olga Software

T A Ganat, M Hrairi*, MNA Hawlader

Department of Mechanical Engineering, International Islamic University Malaysia, P.O. Box 10,50728 Kuala Lumpur, Malaysia

*E-mail: meftah@iium.edu.my

Abstract. The significant challenge in the oil and gas industry is the concurrent measurement of commingled gas, oil and water production, either using three phase test separator or multiphase flow meter (MPFM). A major issue in these applications is the uncertainty of the measurements, due to different measurement operations conditions. A new computational approach has been generated to estimate oil well flow rate of 48 oil wells using Electrical Submersible pump (ESP) from D, G, and W oil fields located in North Africa. The idea is to close the wellhead wing valve and the ESP is kept running normally and the wellhead flowing pressure before shut-in the well and the build-up of wellhead flowing pressure after shut-in the well are measured. OLGA software has been used to make comparison with multiphase flow model available in the OLGA software against each nominated ESP oil well parameters obtained from measured field data. The objective was to verify the obtained shut-in wellhead pressure after closing the choke wing valve (WHPa) from the measured field data with the obtained shut-in wellhead pressure valve from the simulation model. In this paper the simulation results showed that the estimated WHPa are in agreement with the measured WHPa. The relative errors for individual oil field are within accuracy standard specification (typically +/- 10%). The overall relative errors are low and within acceptable uncertainty range, where the aggregate relative error for all wells was less than +/-4% which is considered acceptable. Therefore, the results have demonstrated that the new computational method can work under ESP oil wells conditions and has the ability to perform accurate results even when closing the wellhead wing valve for short time span.

1. Introduction

The need for multiphase flow measurements in the oil and gas production industry has been evident for many years. A number of such meters have been developed since the early eighties by research organizations, and meter manufacturers [1]. Currently the oil and gas industry use the three phase test separator or Multiphase Flow Meter (MPFM). A major issue in the application of using either of these techniques in the oil field is the uncertainty of the fluid measurements, due to different process and operations conditions (mainly at high water and gas fraction) [2].

In this paper, the use of multiphase flow models in OLGA software has been used. The purpose of using OLGA software is to compare multiphase flow models available in the OLGA software against the well parameters obtained from measured field data. This work was performed on a new computational method to estimate the flow rates of ESP oil wells, where the new method was applied on 48 ESP oil wells in North Africa oil fields [3]. The research idea is based on keeping the pump running and closing the wellhead wing valve for few minutes (continuously monitoring the limited

ESP shutting off pressure), and measure the wellhead flowing pressure before and after closing the well head valve, and record the total shut-in time period. Explicit physics concepts for estimation of the multiphase fluid flow rate in a vertical pipe were employed. The formulas deal with changes of fluid flow parameters along the vertical pipe in the well as a function of pressure and temperature variations with the depth.

Figure 1 shows the flow chart of the simulation model structure where the input parameters must contain many input fields. The latters are shown in the below OLGA board tabs (Figure 2).

Simulation model was performed for each individual ESP well from three different oil fields. The objective was to verify the obtained shut-in wellhead pressure after closing the choke valve from the measured field with the obtained shut-in wellhead pressure valve from the simulation model.

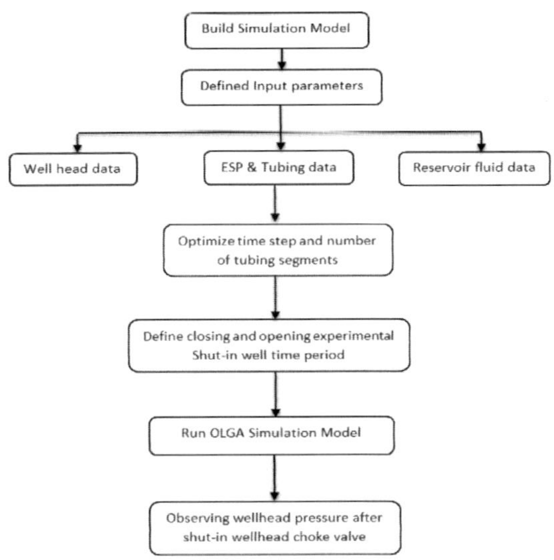

Figure 1. Flow chart of the simulation model structure

2. Software simulation

The OLGA simulator is one of the principal components in analysis of pipeline operations and production well operations. OLGA is a computational program developed to simulate multiphase flow in pipelines and pipelines networks, with processing equipment included. The program solves separate continuity equations for the gas, liquid bulk and liquid droplets, two momentum equations, one for the continuous liquid phase and one for the combination of gas and possible liquid droplets and one mixture energy equation, considering that both phases are at same temperature. The equations are solved using the finite volume method and semi-implicit time integration [4]. Successful design and operation of multiphase production systems rely on detailed understanding of flow behavior [5]. OLGA provides solutions through accurate modeling of true dynamics. Transient simulator provides an added dimension to steady-state analyses by predicting system dynamics such as time-varying changes in flow rates, fluid compositions, temperature, and operational changes [6].

From wellbore dynamics for any well completion to pipeline systems with all types of process equipment, simulator provides an accurate prediction of key operational conditions involving transient flow. The simulation model was run based on the field data gathered from D, G, and W oil fields. The aim of simulation was to find models that accurately predict the wellhead pressure after shut-in the wellhead valve for each single nominated ESP wells. The prediction results depend on the phase redistribution phenomenon and investigate the effect of changing simulation parameters. This was done by using same wells parameters and validating the modeling results against the measured well data.

2.1 The simulation parameter structure

The simulation model in terms of OLGA input file, file name of the optional restart file, specification of input variables, specification of output variables (trend/profile), and time interval between the major time steps are to be specified in an input parameter structure.

The simulator is controlled by defining a set of data groups consisting of a keyword followed by a list of keys with appropriate values. Each data group can be seen as either a simulation object, information object, or administration object. The input parameter structure must contain many input fields, these fields are shown in the below OLGA board tabs (figure 2).

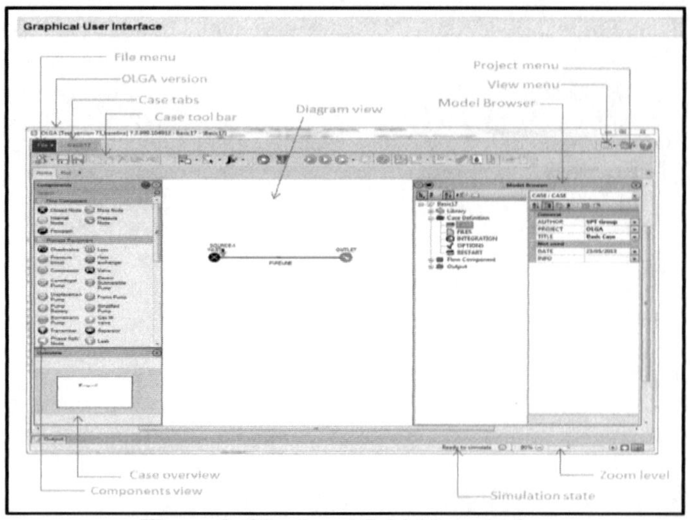

Figure 2. Olga input fields' board tabs.

The entire well path must be identified, (figure 3), using the input files such as well casing strings data and tubing string data, (figure 4), fluid properties data (PVT) using PVTI simulation, (figure 5), pump data, wellhead data, and reservoir data. Simulator requires a description of the fluid properties as a unique function of temperature and pressure. These can be given either as a fluid file, through specific component data given in a feed file (compositional tracking) or defined in the main file (black oil). Besides, a variable time step model is set up and simulator determines the major time steps. In this study, the well simulation design was built for each nominated ESP oil wells, using the measured field data available.

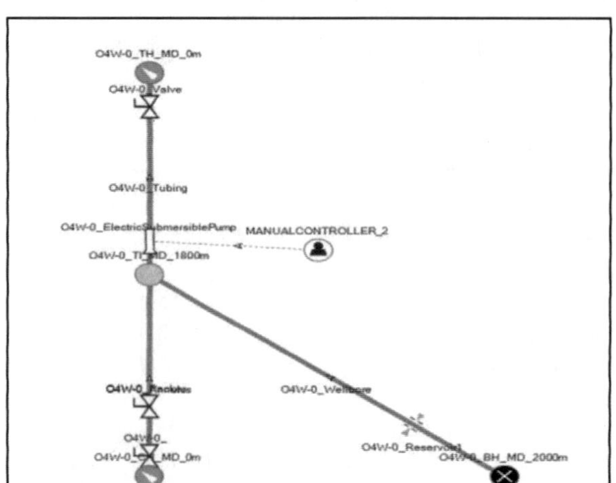

Figure 3. OLGA well structure model.

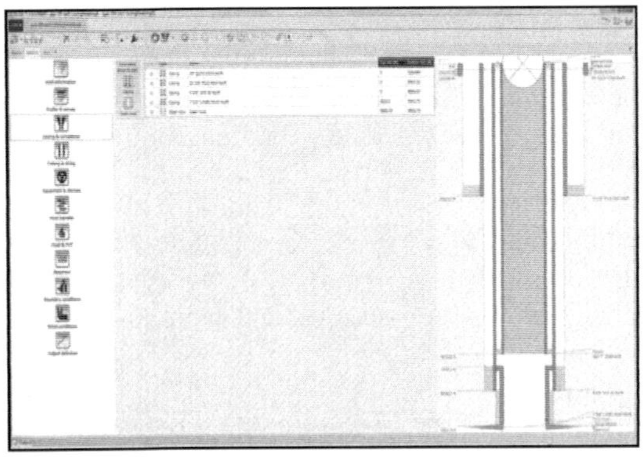

Figure 4.Well casing and tubing design profile.

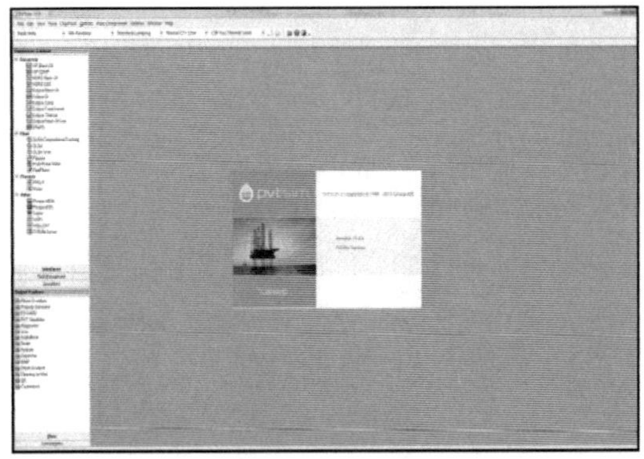

Figure 5. PVTI simulation model.

2.2 Simulation Viewer

The basic objective of multiphase flow visualization in the vertical well tubing is to imagine the multiphase flow phenomena and phase re-distribution during open and shut-in the wellhead choke valve and to prove that the new computational methodology process was made under same production conditions. Using OLGA software, one can evaluate and compare well result alternatives with the visualization of the fluid flow results with real time interaction. Also, one can understand the fluid flow inside the vertical riser pipe using the section plots. It will also allow getting values quickly and easily for critical parameters, such as pressure drop, liquid and gas volumes, and holdup fractions.

Figure 6 shows the 2D simulation viewer profile for multiphase flow in the vertical tubing string. The viewer shows the re-distribution phase phenomena when shut-in the wellhead choke valve where the gas phase dissolved back to the liquid phase, and shows the bubble point location depth before and after shut-in wellhead choke valve. It can be seen from figure 6 that the depth location of the bubble point pressure has been moved up to new bubble point location depth due to the dissolved gas and vapor phase into the liquid phase during shut-in the wellhead choke valve, and back to the original location depth after opening the choke valve again. This visualization demonstrated exactly what happened during shutting and opening wellhead choke valve process at the well site. The difference between the two bubble point location depths represents the amount of fluid volume pumped during shut-in the well as a function of total shut-in time period.

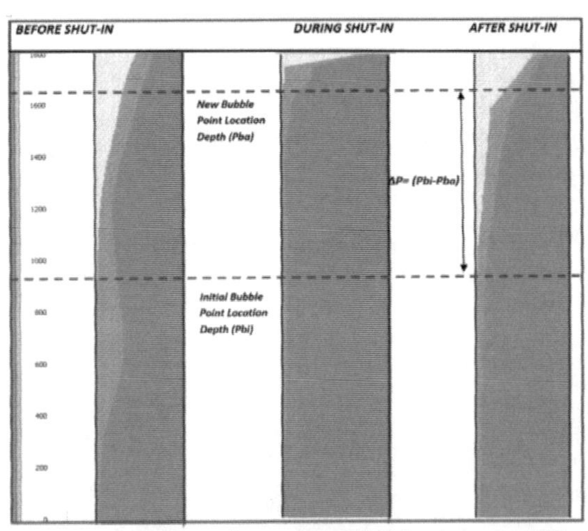

Figure 6. 2D vertical well cross section pipe using OLGA viewer profile showing bubble point location depth before and after shut-in the well.

2.3 Simulation Results

The series of simulation results that were carried out reflect the nominated wells production performance and yields many possible values for the outcomes of interest. The 48 ESP oil wells simulation results were compared and validated with field measured data. The results were very encouraging, where the verified WHPa was very close to the measured WHPa for each well. Tables 1, 2 and 3 and figures 7, 8, and 9 show the comparison results of OLGA simulation values of WHPa with experimental values of WHPa for all the ESP wells in D, G, and W oil fields, respectively. The result shows small values of average relative error of ± 10%, with a high R^2 of 99.06%, 99.44%, and 96.51%, respectively, indicating that the measured shut-in wellhead pressure values onsite for most of the ESP oil wells are satisfactory.

Table 1. WHPa Comparison between the measured and simulation data for D oil field.

well Name	Calculated WHPa (Psia)	Measured WHPa (Psia)	Absolute Error (Psia)	Relative Error (%)
B56	170	172.45	-2.45	-1%
B70	230	240	-10	-4%
B121	160	159.62	0.38	0%
B119	160	153.35	6.65	4%
B50	250	241.53	8.47	4%
B88	160	166.5	-6.5	-4%
B14	270	278.59	-8.59	-3%
B151	230	235	-5	-2%
B164	170	176.33	-6.33	-4%
B51	310	322.33	-12.33	-4%
Q89	150	146.36	3.64	2%
Q21	150	147.45	2.55	2%
Q53	150	148.05	1.95	1%
Q14	130	131.194	-1.194	-1%
Q100	130	130.65	-0.65	0%
Q12	150	146.126	3.874	3%
Q85	150	147.53	2.47	2%
Q82	150	156.38	-6.38	-4%
Q78	150	147.62	2.38	2%
Q76	160	153	7	5%

Table 2. WHPa comparison between the measured and simulation data for G oil field.

well Name	Calculated WHPa (Psia)	Measured WHPa (Psia)	Absolute Error (Psia)	Relative Error (%)
E89	200.53	190	10.53	6%
E210	134.12	120	14.12	12%
E211	122.28	120	2.28	2%
E286	106.86	100	6.86	7%
E192	114.63	110	4.63	4%
E327	119.52	110	9.52	9%
E325	112.54	110	2.54	2%
E197	117.55	110	7.55	7%
E208	107.06	110	-2.94	3-%
E226	115.88	110	5.88	5%
E284	132	120	12	10%
E258	92.64	90	2.64	3%
E326	104.49	100	4.49	4%
E227	158	150	8	5%
4_E3	310.71	300	10.71	4%

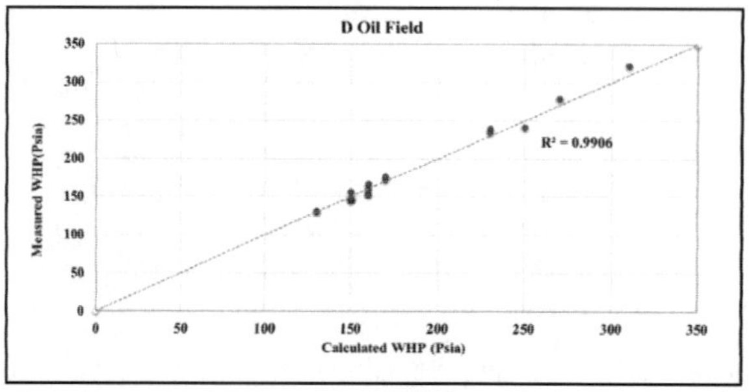

Figure 7. Comparison between the measured WHPa and simulated WHPa for D oil field.

Table 3. WHPa comparison between the measured and simulation data for W oil field.

Well Name	Calculated WHPa (Psia)	Measured WHPa (Psia)	Absolute Error (Psia)	Relative Error (%)
A33	202.34	200	2.34	1%
A125	209.86	200	9.86	5%
A64	242.64	250	-7.36	-3%
A29	251.5	270	-18.5	-7%
A23	335.35	300	35.35	12%
A135	303.36	300	3.36	1%
A126	176.32	170	6.32	4%
A12	305.15	300	5.15	2%
A108	225.12	210	15.12	7%
5J5	270.12	270	0.12	0%
J52	261.52	260	1.52	1%
5J4	249.52	250	-0.48	0%
5J7	312.17	300	12.17	4%

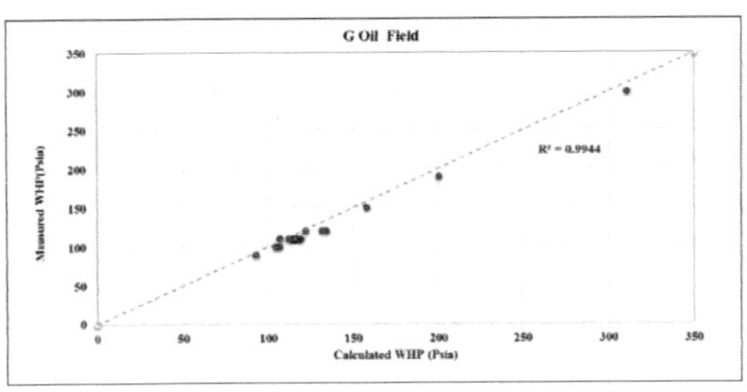

Figure 8. Comparison between the measured WHPa and simulation WHPa for G oil field.

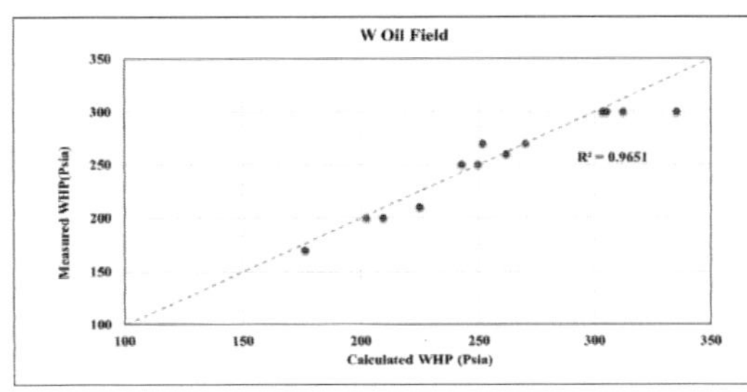

Figure 9. Comparison between the measured WHPa and simulation WHPa for W oil field.

At an average relative error band of ±10%, 100%, 87%, and 85%, of total number of ESP oil wells in D, G, and W, oil fields respectively, had correctly predicted shut-in pressure values.

Generally, it is important to know, how much the measured value is likely to deviate by using different method. All cross plot figures have showed a quite good correlation around the actual values. The results of the simulations were encouraging in terms of model performance.

Figures 10, 11 and 12 display the estimated relative error results in wellhead pressures after shut-in the choke valve from experimental measurements verses simulation calculations for all wells in D, G, and W oil fields respectively. These figures show the stability of estimating the wellhead shut-in pressure. It has been noticed that most investigated wells showed good relative error results.

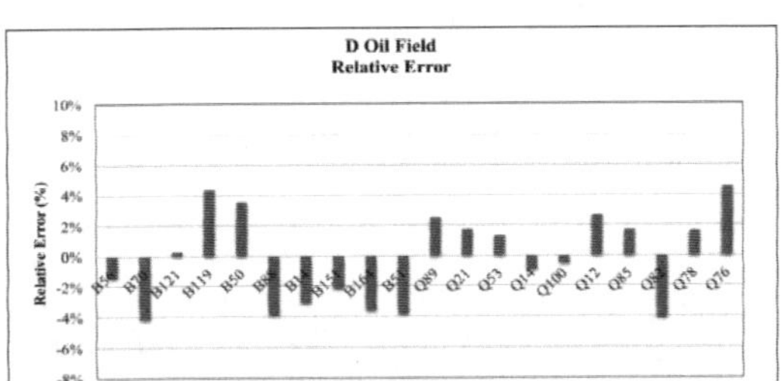

Figure 10. The measured WHPa accuracy for D oil field.

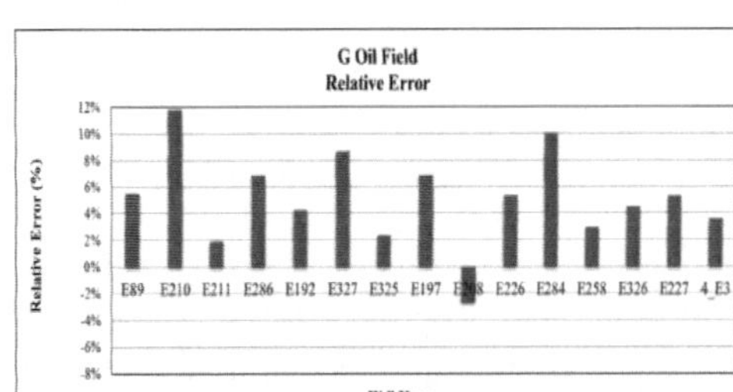

Figure 11. The measured WHPa accuracy for G oil field.

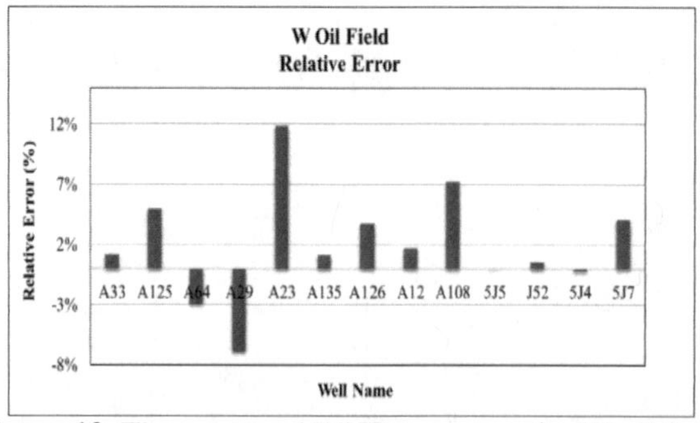

Figure 12. The measured WHPa accuracy for W oil field.

Figure 10 shows that 100% of the total number of oil wells in D oil field had correctly predicted shut-in pressure values where the error was within the specified target accuracy of ±10%. 95% of data had an accuracy of less than ±5%. However, there was a trending small spread of errors during verification of these accuracies.

Figure 11 shows that 87% of the total number of oil wells in G oil field, had correctly predicted shut-in pressure values where the error was within the specified target accuracy of ±10%, while 13% had shut-in pressure values are more than ±10%. Here, 60% of data had an accuracy of less than ±5%. However, there was a trending small spread of errors during verification of these accuracies.

Figure 12 shows that 85% of the total number of the oil wells in W oil field, had correctly predicted shut-in pressure values where the error was within the specified target accuracy of ±10%, while 15% had shut-in pressure values are more than ±10%. Here again, 69% of data had an accuracy of less than ±5%. However, there was a trending small spread of errors during verification of these accuracies.

In general, 96% of the total 48 oil wells had predicted the shut-in wellhead pressure correctly, within ±10% relative error, whereas 4% of the wells recorded shut-in pressure data accuracy of less than ±12%.

Figure 13 shows the histogram construction for all relative error measures (±), which records the overall shut-in wellhead pressure accuracy for all ESP oil wells. The histogram demonstrated that the inclusive average relative error of ± 4% is more accurate, and represents 25% of the data set, while the rest of the data are flocculating between 0% and 12% average relative error.

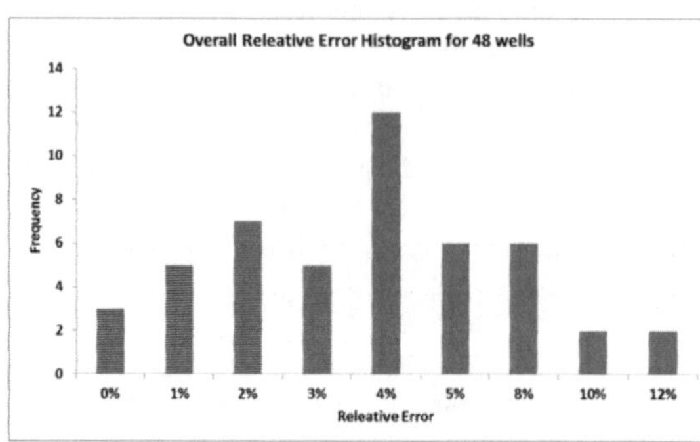

Figure 13. Overall relative Error Histogram for all ESP wells.

3. Conclusion

The estimation of the 48 ESP oil wells flow rate using the new computational approach has been generated and validated with measured flow rate data in the case study oil fields. The measured WHPa was investigated using OLGA simulation Software.

The simulation results showed that the estimated WHPa are in agreement with the measured WHPa. The relative errors for individual oil field are within the accuracy standard specification (typically +/-

10%). The overall relative errors are low and within acceptable uncertainty range, where the aggregate relative error for all wells was less than +/-4% this accuracy considered acceptable. Therefore, the results have demonstrated that the new computational method can work under ESP oil wells conditions and has the ability to perform accurate results even when closing the wellhead wing valve at short time span.

References

[1] Corneliussen S *et al.* 2005 Handbook of Multiphase Flow Metering (ed: NFOGM, Tekna, Oslo)
[2] Gryzlov A *et al.* 2013 *Proc. North Sea Flow Measurement Workshop (Tønsberg)* p 40
[3] Ganat T, Hrairi M, Hawlader M N A, and Farj O 2015 *J. Pet. Sci. Eng.* **135** 466
[4] Nemoto R H *et al.* 2010 *Proc. Brazilian Congress of Thermal Sciences and Engineering (Uberlândia)* p 9
[5] OLGA Dynamic Multiphase Flow Simulator, Schlumberger, Houston, TX, USA
[6] OLGA Software Integrated Solutions 2015 http://www.sis.slb.ru/sis/pdf/NL_June_ENG_2015.pdf

Impedance Based Detection of Delamination in Composite Structures

M Djemana and M Hrairi*

Department of Mechanical Engineering, International Islamic University Malaysia, P.O. Box 10,50728 Kuala Lumpur, Malaysia

*E-mail: meftah@iium.edu.my

Abstract. Nowadays commercial and military aircrafts are increasingly using composite materials to take advantage of their excellent specific strength and stiffness properties but impacts on composites due to bird-strike, hail-storm cause barely visible impact damage (BVID) that underscores the need for robust structural health monitoring methods. Hence, damage identification in composite materials is a widely researched area that has to deal with problems coming from the anisotropic nature of composites and the fact that much of the damage occurs beneath the top surface of the laminate. This paper focuses on understanding self-sensing piezoelectric wafer active sensors (PWAS) to conduct electromechanical impedance (EMI) in glass fibre reinforced polymer composite to perform structural health monitoring. With the aid of a 3D ANSYS finite element model, an analysis of different techniques for the detection of position and size of a delamination in a composite structure using piezoelectric patches had been performed. The real part of the impedance is used because it is known to be more reactive to damage or changes in the structure's integrity and less sensitive to ambient temperature changes compared to the imaginary part. Comparison with experimental results is presented to validate the FE results. The experimental setup utilizes as its main apparatus an impedance analyser HP4194 that reads the in-situ EMI of PWAS bonded to the monitored composite structure. A good match between experimental and numerical results has been observed for low and high frequencies. The analysis in this paper provides necessary basis for delamination detection in composite structures using EMI technique

1. Introduction

Structural health monitoring (SHM) method is very significant in the engineering world and every year new ideas and mechanisms have to be generated and developed. In engineering, exploration of new thing is a must, since from the 21st century many founding and discoveries of magnificent theories and innovative product have been created. Even though with the mind blowing theories and innovations, it is impossible to reach the understanding without further scientifically elaboration and explanation. In few years ago, a smart material has been discovered and it is not being expected that there is material which can react to external stimuli by itself without human intervention. Then, this smart material has been developed to the extent where it can use in the field of structural health monitoring (SHM) and non-destructive evaluation (NDE).

Damage detection on aircraft composite structures has been one of the major concerns of operators during the last two decades [1]. EMI method by using piezoelectric sensors and/or actuators incorporated into composite structures offer a promising option to fulfill such requirements and needs. This method can use finite element analysis combined to experimental results in order to detect damage. Thus, it is possible to identify, locate and estimate the damage events, comparing dynamic responses between damaged and health structures. The basic idea of the impedance based damage detection method consists on assuming that damage is a combination of different failure modes, which affect the local stiffness of the structure, and this modifies the dynamic characteristics of the structure, i.e., the modal frequencies, mode shapes and modal damping values. Structural Health Monitoring (SHM) can provide, all the time, during the aircraft life, a diagnosis of the material state and the structure situation, as well as a prognostic like the residual strength of the structure. Thus, SHM can reduce the maintenance costs, avoiding useless inspections. Many researchers have been done on the topic of structural health monitoring for composite structure such as Zou et al. [2] reviewed the model-based delamination detection methods for composite structures using vibrations. Firstly, it describes the most commonly, model-based methods for damage detection, used structural modeling techniques for delamination and the effects of delamination on dynamic parameters. Then it focuses on the application of vibration-based model-dependent damage detection methods in composite structures. Finally, the review is devoted to the development of methods with incorporated piezoelectric sensors and actuators for on-line delamination detection for composite structures.

Gresil et al. [3] have conducted, they compared experimental measurements on a beam structure which is the impedance is calculated using electromechanical impedance device such as AD5933 where the FEM model is mechanically coupled to PWAS and the calculated impedance from coupled-field multi-physics-based finite element method (MP-FEM) model where the PWAS is directly excited with electric

Pérez et al. [4] presented the results of an experimental campaign conducted to investigate the feasibility of using vibration-based methods to identify damages sustained by composite laminated due to low-velocity impacts. The experimental program include an evaluation of impact damage resistance and tolerance according to ASTM test methods, characterization of damage by ultrasonic testing and quantification of the effects on the vibration response. The experimental results indicate that impact-induced damages result in detectable changes in the vibration response of coupons.

Moreover, most of the works existing in the literature about composite structures are still limited. In practical, for the composite structures, it is difficult to detect damage. In order to detect damage throughout the whole structure, especially some composite structure, a methodology, which uses impedance based SHM, has been more adequate. Recently, this type of methodology has been used for composite structures. Therefore, this article presents experimental and finite element analyses on glass fiber reinforced polymer (GFRP) composite structures and focus of this paper will be on the analyses the effect of delamination on the EMI system.

2. Impedance based SHM

The electro-mechanical impedance method for structural health monitoring, damage detection, and NDE was explained in detail by Giurgiutiu et al. [5]. An overview of its principles is given next. Consider a piezoelectric transducer wafer intimately bonded to the surface of a structural member. When excited by an alternating electric voltage, the piezoelectric transducer applies a local strain parallel to the surface. Thus, elastic waves are transmitted into the structure. The structure responds by presenting to the transducer the drive-point mechanical impedance

$$Z_{str}(\omega) = i\omega m_e(\omega) + c_e(\omega) - ik_e(\omega)/\omega \qquad (1)$$

Through the mechanical coupling between the PZT transducer and the host structure, and through the electro-mechanical transduction inside the PZT transducer, the drive-point structural impedance directly reflects into the effective electrical impedance as seen at the transducer terminals (Figure 1).

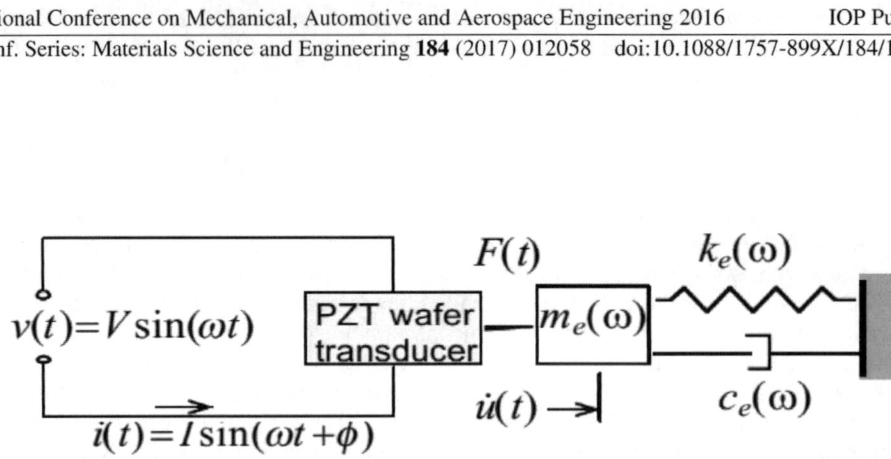

Figure 1. Electro-mechanical coupling between the PZT transducer and the structure.

The electro-mechanical (E/M) impedance technique for health monitoring, damage detection and NDE [6] utilizes the changes that take place in the drive-point structural impedance to identify incipient damage in the structure.

The apparent electro-mechanical impedance of the piezo transducer as coupled to the host structure is given by

$$Z(\omega) = \left[i\omega C \left(1 - k_{31}^2 \frac{k_{str}(\omega)}{k_{PZT} + k_{str}(\omega)} \right) \right]^{-1} \tag{2}$$

Here, $Z(\omega)$ is the electro-mechanical impedance as seen at the PZT transducer terminals, C is the zero-load capacitance of then PZT transducer, k_{31} is the electro-mechanical cross coupling coefficient of the PZT transducer ($k_{31} = d_{13} / \sqrt{s_{11}\varepsilon_{33}}$), $k_{str}(\omega)$ is the dynamic stiffness of the structure, and k_{PZT} is the stiffness of the PZT-wafer active sensor [7].

3. Experimental Analyses

The experimental analyses were carried on composite structure. The structures are made of glass fiber reinforced polymer (GFRP) with PZT transducers bonded to its surface. The beam considered is 60 mm x 10 mm x 1 mm whereas the piezoelectric patch has dimensions of 7mm diameter and 0.2mm thickness. There is also adhesive layer between the beam and piezoelectric patch with 30μm thickness. An impedance analyzer HP4294 was used to read the in-situ E/M impedance of piezoelectric wafer active sensors (PWAS) attached to the monitored structure. The experimental set up is shown in Figure 2.

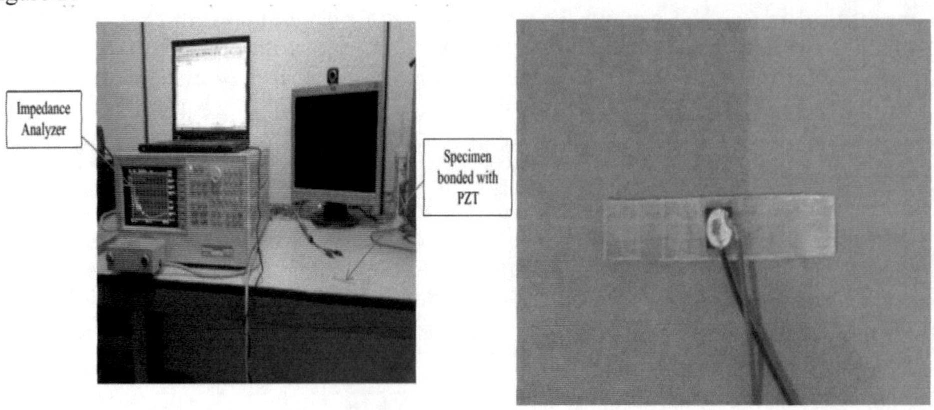

Figure 2. Experimental set up

4. Finite element analyses

All the specimens are numerically modeled with 3D mesh shown in Figure 3. The GFRP is modeled using 20 node SOLID186 layered structural solid in order to obtain a 3D five laminated layers beam with 0° orientation along x - axis and the adhesive layer is modeled with 20 node 3D SOLID95 element. Next for piezoelectric wafer active sensor transducer, it will be modeled with SOLID226 element which is for 3D 20 node coupled field analysis. The size of the mesh for all the elements is 0.5 mm for the frequency range between 0 to 5 MHz and 7MHz to 15MHz.

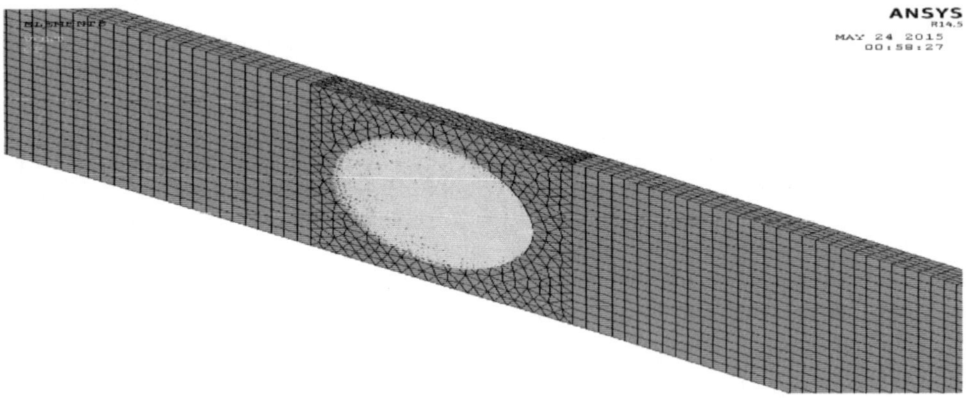

Figure 3.FE model of composite structure beam

In a finite element model of piezoelectric wafer active sensor, SOLID226 has degree of freedom (DOF) up to four DOFs per node. Electric voltage degree of freedom (DOF) will be used in order to obtain charge accumulation data for the model. PIC255 material properties will be used in the piezoelectric wafer active sensor. The material properties are given in Table 1.

Table 1.PIC255 Properties

Piezoelectric Stiffness Matrices

$$[C_p]=\begin{bmatrix}123.0 & 76.7 & 70.25 & 0 & 0 & 0\\ 76.7 & 123.0 & 70.25 & 0 & 0 & 0\\ 70.25 & 70.25 & 97.11 & 0 & 0 & 0\\ 0 & 0 & 0 & 23.15 & 0 & 0\\ 0 & 0 & 0 & 0 & 22.26 & 0\\ 0 & 0 & 0 & 0 & 0 & 22.6\end{bmatrix} GPa$$

Piezoelectric Relative Permittivity

$$[\varepsilon_p]=\begin{bmatrix}1650 & 0 & 0\\ 0 & 1650 & 0\\ 0 & 0 & 1750\end{bmatrix} 10^{-8} F/m$$

where $[c_p]$ and $[\varepsilon_p]$ are piezoelectric stiffness matrix and relative permittivity. For the adhesive layer, 0.00003m thick isotropic properties are considered. The Young's modulus, Poisson ratio and density of the adhesive layer are E = 5 GPa, =0.4 and ρ = 1700 , respectively (Gresil et al., 2012). A homogeneous orthotropic material with assumed density of 1960 is assigned to the GFRP composite beam. The orthotropic properties of the glass fiber reinforced polymer beam are listed in Table 2.

Table 2. Unidirectional Glass / Epoxy Properties.

Young's Modulus (GPa)			Poisson Ratio			Shear Modulus (GPa)		
E_{11}	E_{12}	E_{13}	$\mu_{12}\mu_{23}$		μ_{13}	G_{12}	G_{23}	G_{13}
21.0	2.779	2.779	0.33	0.4	0.33	4.481	4.9	4.481

The composite beam will be assigned with mass damping coefficient and stiffness damping coefficient of and, respectively.

5. Result and discussion

Several studies have provided the result of impedance response on piezoelectric patch on the composite structure. However, only two works, [3] and [5] are chosen because they have the most information regarding the material properties and simulation result on impedance response. Experimental results for pristine and damaged will be compared and also the impedance signal obtained from the experiment and simulation will be compared and discussed. The first simulation result of composite beam and circular piezoelectric patch are shown in Figure 4.

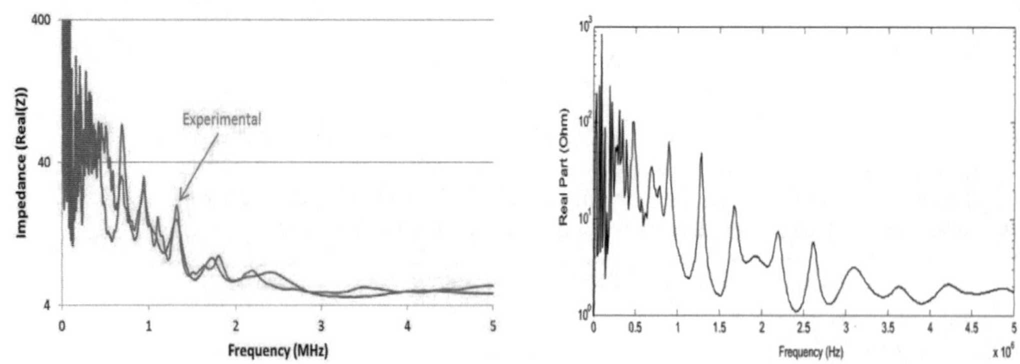

(a)

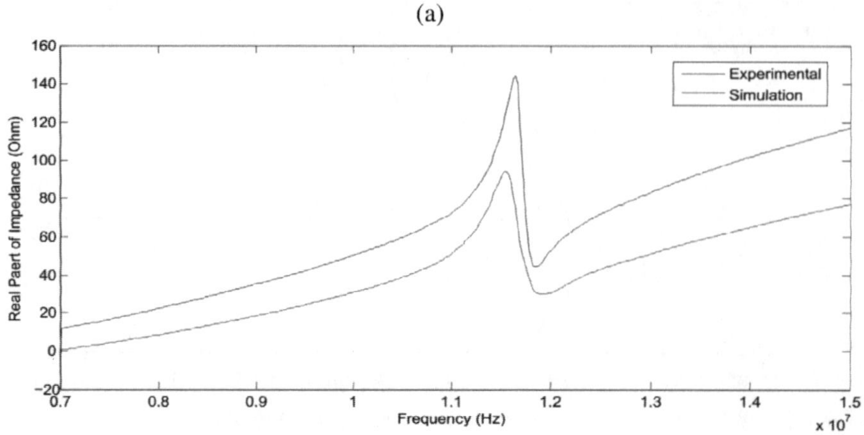

(b)

Figure 4. Comparison of experimental and FE simulation (a) for a frequency range 0–5 MHz; (b) for a frequency range 5–15 MHz.

The comparison between the simulation and experimental results show a good agreement It has revealed two different regions of behavior: (a) below 5 MHz, the result from a FE model with structural damping were taken as has a good matching with the experimental result from and (b) above 5 MHz, the experimental result matches better with a model with structural damping β= 10-9.

5.1 Delamination effects:

The composite structures have a tendency to delaminate which reduces the strength and stiffness and thus limits the life of a structure. This behavior of composites has caused concern amongst the designers to find the ways to delay or prevent delamination in order to increase the life and the load bearing capability of the structure. We studied the effect of a delamination introduced between the layers 4 and 5 as illustrated in Figure 5.

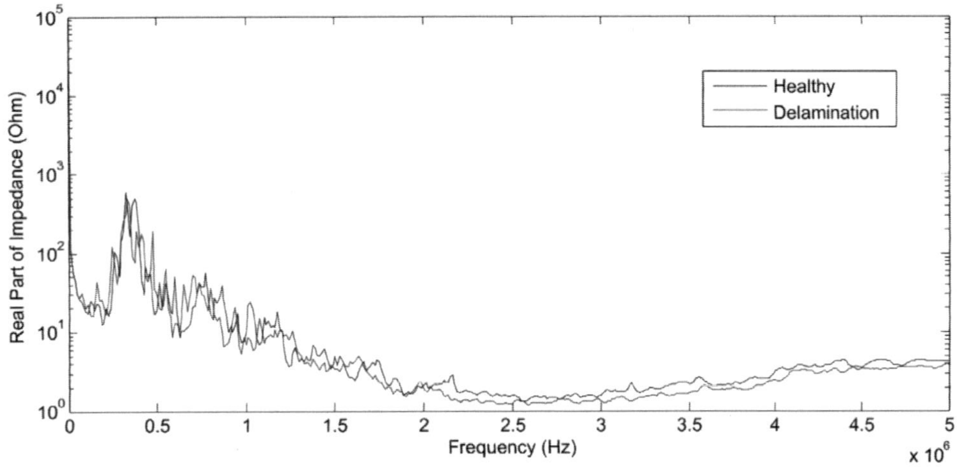

(a)

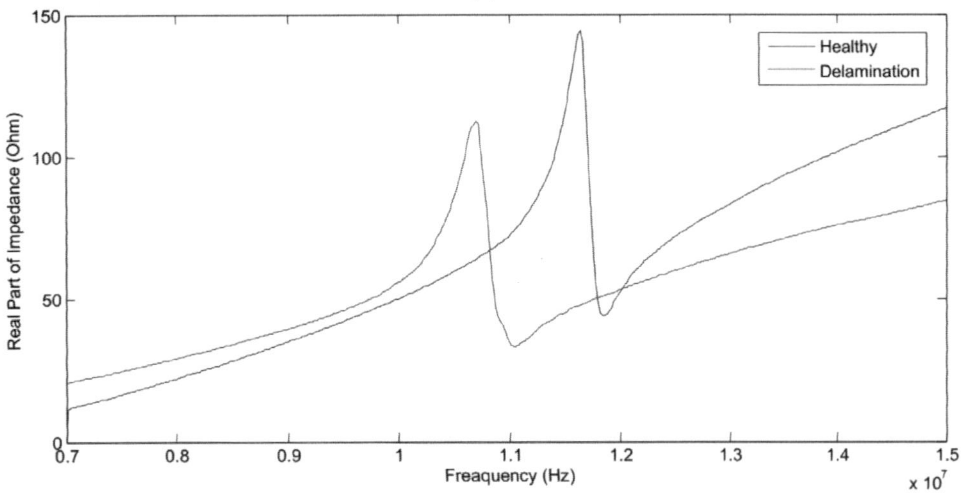

(b)

Figure 5. Comparison of experimental for Healthy and Delamination of laminate GFRP: (a) for a frequency range 10 kHz–5 MHz; (b) for a frequency range 7–15 MHz.

It can be seen that, when delamination is introduced in the composite, there will be shifts in response peaks and modification of peaks amplitude. From Figure 5(a), the four major resonance peaks that shift leftwards have higher amplitude. Splitting of the laminated beam between the third ply and the fourth ply caused the PWAS to respond in shorter distance compared to the respond of PWAS on healthy GFRP beam where longer respond is needed to detect non – gap plies of the composite beam. The gap that exists between the plies caused the PWAS to detect the thinner beam first resulting to higher amplitude of resonance peaks. For frequencies higher than 2.5 MHz, the impedance response between delaminated and healthy beam is significant as there is major resonance peak that exist in that range of frequencies except that the impedance response of the delaminated beam has a lower value of electrical impedance compared to the electrical impedance value of the healthy beam. This shows that significant changes of impedance response can observed clearly in Figure 5(b).These shifting of peaks and changes of the resonance amplitude occurred vividly because the nodes on the delaminated region are not directly connected with each other or disengaged with each other during meshing.

6. Conclusion:

Different methods are used to identify damage in composite structures. Impedance analyzer and piezoelectric transducers are used for data acquisition. The analyses were carried out to obtain modal parameters of both an undamaged and a damaged structure. Experimental and numerical analyses were presented. For the numerical analyses, finite element for piezoelectric active composite was implemented in Ansys 14.5. The analyses were performed to obtain modal parameters of both healthy and damaged structures, simulated in the structure by delamination introduced between the layers 4 and 5. The experimental and numerical results showed that this system can be used for continuous monitoring of structures .It gives information on the overall condition of the structures. The experimental analyses combined to the damage diagnose if the structure is damaged or not. Finally, it is possible to conclude that there is a great future perspective for the application of impedance-based SHM by using piezoelectric sensors for composite structures. However, there are still a lot of work that needs to be done numerically as well as experimentally to determine and investigate the effect of different variables introduced to the system.

Acknowledgments
The authors are grateful to the support of the Ministry of Higher Education Malaysia through the Grant No. FRGS 15-189-0430.

References:

[1] Medeiros R *et al* 2012 *Proc. Int. Conf. on Noise and Vibration (Leuven)* p 3285
[2] Zou Y, Tong L, and Steven G P 2000 *J. Sound Vibr.* **230** 357
[3] Gresil M, Lingyu Y, Giurgiutiu V, and Sutton. M 2012 *Struct. Hlth. Monit.* **11** 671
[4] Pérez M A, Gil L, and Oller S 2014 *Compos. Struct.* **108** 267
[5] Selva P *et al* 2013 *Eng. Struct.* **56** 794
[6] Giurgiutiu V and Rogers C A 1997 *Int. Workshop on Structural Health Monitoring* (Stanford CA) p 15.
[7] Giurgiutiu V and Rogers C A 1999 *Sympo. On Adaptive Structres and Material Systems* (Nashville TN) vol 59 p 39

IOP Conf. Series: Materials Science and Engineering **184** (2017) 012059 doi:10.1088/1757-899X/184/1/012059

Numerical study of heat transfer of embossed plate fined heat sink

Suliman Mohamed Mohamed Ali[1], Waleed Fekry Faris[2] and Ahmed Faris Ismail[3]
[1,2,3]Department of Mechanical Engineering, International Islamic University Malaysia

[1] solli22@yahoo.com, [2] waleed@iium.edu.my, [3]Faris@iium.edu.my

Abstract. In this article, a study of the thermal performance of embossed fined mini-channel heat sink numerically has been carried out using CFD. Commercial software STAR-CCM+ was used for the investigation. Embossed impressions were vertical to the heat sink base. The numerical results were presented in terms of flow visualization,a temperature distribution on the heat sink, velocity contour, and streamlines, Nusselt number for heat supply ranging from 18 to 70.4 W under natural convection. The existence of artificial geometry on the fin surface leads to greater Nusselt number due to the high energy fluid flow mixing near the heat sink wall by the separation and reattachment of the boundary layer. In addition, an embossed heat sink has shown better temperature distribution. The result showed a good agreement with available experimental data.

1. Introduction

The electronics industry is gradually developing. Light, small, and high-power electronic components are becoming more ubiquitous. Without efficient heat transfer, excessive temperatures may make the working performance of these components unstable, reducing their lifespan or even causing damage to them. Therefore, a major challenge in the design of electronic components is to enhance the efficiency of their heat transfer (Hung-Yi Li, 2013). Effective method to remove high rate heat is essential to control substrate temperatures under critical temperatures which causes devices to fail. Even though liquid is used widely in cooling system as the most important applicants for high flux applications, air refrigerating is still by far the most used cooling method to cool electronics devices for its reliability, simplicity, and low cost. Air has low heat transfer rate, due to its low thermal conductivity. As a result, the general method for heat transfer enhancement is through manipulation of smaller fin space to accommodate extra fin surface. The cannelure fin with the hollow/cavity configuration of heat sink could attain 25% increase of heat transfer with increasing in friction decrease of around 20% (Chi-Chuan Wang, 2011). Furthermore, opposite to interrupted fin or triangular vortex generator, it is approved that their suggested fin configuration leads to important increase in fully developed flow. Vortex generators shape, location, dimension have major contribution on the thermal and hydraulic ability of heat sink (Timothy Dake, 19-23 July 2009). They have shown that the angle of attack of generators play significant part in the heat transfer rise of heat sinks. Moreover, when the vortex generator height, the heat transfer improves and the greater improvement was registered when the generator and channel heights are equal. Finally, they denoted that the vortex generators could enhance the thermal ability of heat sink. High frontal velocity and large fin pitch lead to more useful vortex generators (K.S. Yang, 2010).

Starner at el (Starner & McManus, 1963) investigated four different fin groups and three different base plates to measure natural convective heat transfer rate considering fin pitch and the fin height. Welling and Wooldridge (Welling, 1965)) approved that the vertically fin array is the best among all

IOP Conf. Series: Materials Science and Engineering **184** (2017) 012059 doi:10.1088/1757-899X/184/1/012059

kinds of fin arrays comparing fin heights. The change of temperature with the variation in the fin height to fin pitch ratio was argued and an equation for best magnitude of the fin height to fin pitch ratio was suggested. Chaddock they changed the fin spacing and the fin height in the maximum vertically fin groups and explained that the radiation mode shares around 20% of the total heat transfer (Chaddock, 1970). Alhara studied the influence of fin geometry and temperature on average heat transfer rate obtained empirical correlation (Alhara, 1970). Singh and Patil studied experimentally the effect of semi-circle embossed impressions (Singh & Patil, 2015).

2. Geometry

A three-dimensional geometry has been drawn using SolidWorks and the simulation has been performed using starccm 8.1. The heat supply is assumed at the bottom of the mini-channel heat sink. The test heat sinks of 150 mm 150 mm 50 mm dimensions having aluminum fins of 150 mm length, 50 mm height and 1 mm thickness were used in the experiments. Three power supply t values, 18, 39.6, 70.4 W were used with each heat sink configuration. Impressions have dimensions of 40 mm length and 2 mm height were placed in the fins at different angles (45°, 90°) and at pitches 12 mm.

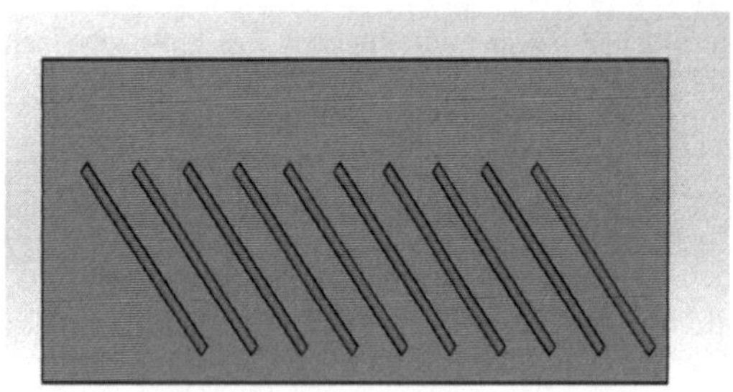

Figure 1. front view of heat sink geometry

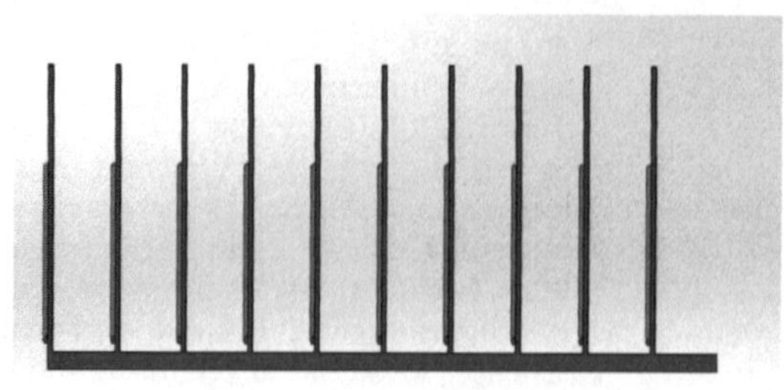

Figure 2. side view

3. Meshing Independence Study

To understand the sensitivity of the heat sink grid, four different grids, 34347 coarse grids cells, 399351 fine grids cells and 498976 and 589388 very fine grids cells were used to check the stability of the grids.

Best grids have been used within the heat sink in order to measure the fluid flow and heat transfer characteristics. Difference between surface temperature and ambient temperature has been calculated.

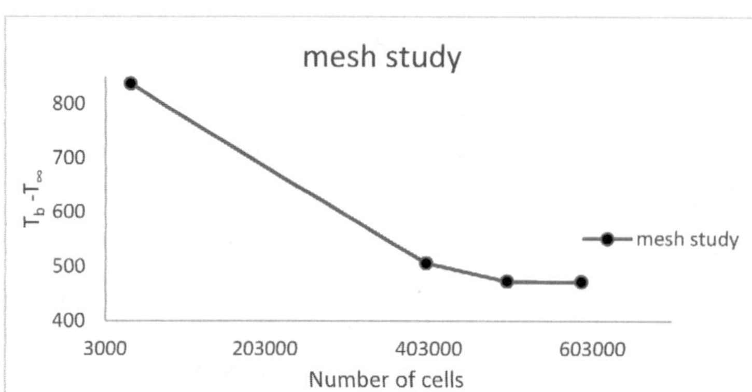

Figure 3. meshing independence study

4. Results and Discussions

Figure 4 presents the variant of the Nusselt number with different values of heat input for changed mini channel heat sink. The graphs visibly represent that the Nusselt number increases with the increase in heat input regardless of the shape of heat sink. It may be due to the fact that as the heat input to the sink higher the influence of convective rate of heat transfer increases more. The maximum value of the nusselt number is 47.647 correspond to embossed heat sink at angle 45°, while the minimum value of nusselt number is 24.5 correspond to smooth heat sink.

A significant improvement in the Nusselt number values can be seen by the use of inclined impressions on the fin surfaces for all the values of heat supply.

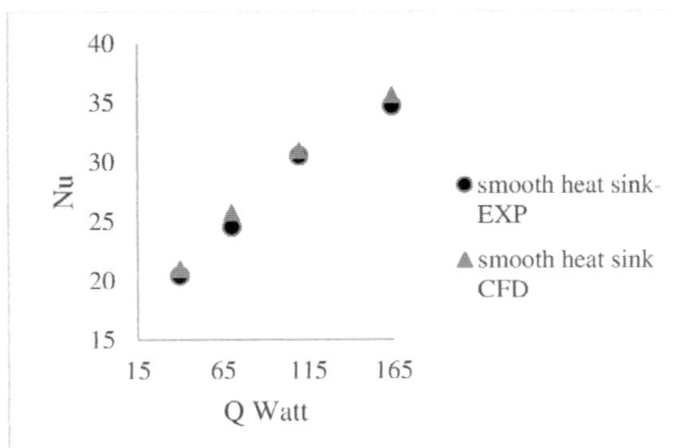

Figure 4. nusselt number and heat input of smooth heat sink.

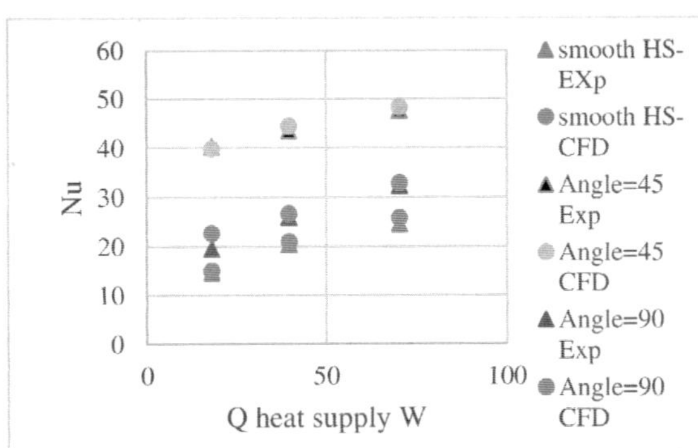

Figure 5. Nusselt number and heat input for different impression angles at Pi = 12 mm.

Figures 5 explains the consequence of impression angle on Nusselt number matching to the impression angle of 45 and 90 degree and smooth, for the different values of heat source. Same kind of trend is noticed in these graphs excluding the fact that the Nusselt number of embossed heat sink is more changeable in heat supply for the impression angle of 45 degree comparing to other values of angle. The existence of artificial geometry on the fin surface leads to greater Nusselt number due to the high energy fluid flow mixing near the heat sink wall by the separation and reattachment of the boundary layer. Also the restart of fluid film between the two successive impressions, the secondary fluid flow along the declined impressions interacts with the stream flow and so increases turbulence and heat transfer rate at the interface of fluid wall.

Figures 6, 7, and 8 represent the temperature distribution on the heat sink while figures 9 and 10 show the velocity contour at two different heat supplies. It can be seen from figures that embossed heat sink has better temperature than smooth heat sink. At heat supply 18 W, smooth heat sink, maximum temperature is 554.3 K, while it is 412.2 K at embossed heat sink. The minimum temperature is 551.76 K and 399.87 K at smooth and embossed heat sink, respectively. Temperature at heat sink wall increases with heat supply, it is seen clearly at Q=70.4 W, the maximum and minimum temperatures are 859.79, 850.35 and 838.72, 629.46 K at smooth and embossed heat sink respectively.

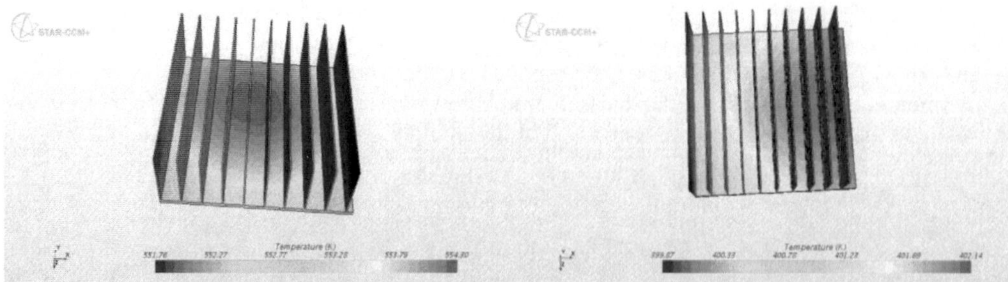

Figure 6. Temperature distribution on heat sink at Q=18 w, Smooth, Angle =45 respectively.

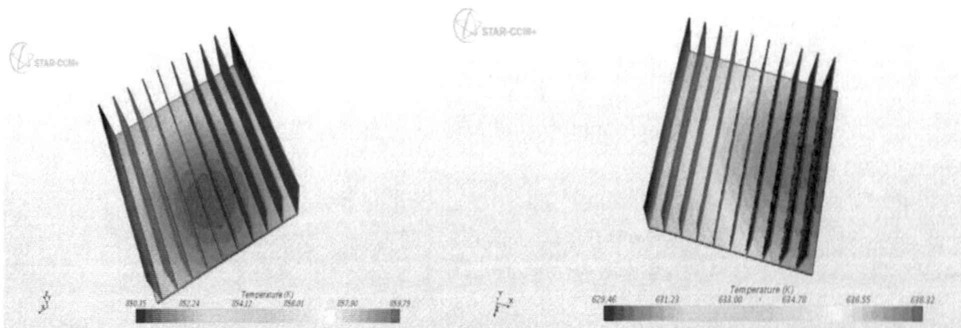

Figure 7. Temperature distribution on smooth heat sink, Angle of impression is 45 at Q=70.4 w.

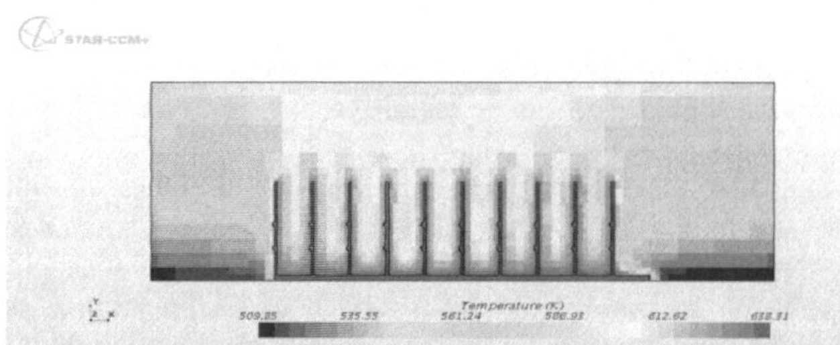

Figure 8. Temperature distribution at the middle of heat sink at Q=70.4 w, Angle of impression is 45.

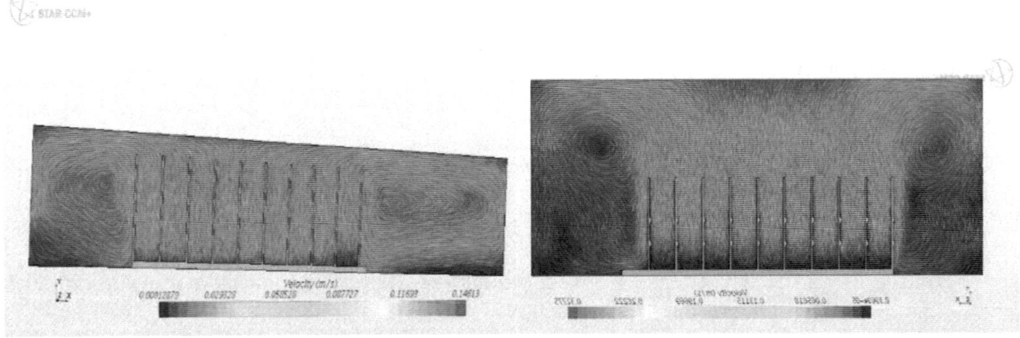

Figure 9. Velocity contour at Q=70.4 w, Smooth heat sink, Angle=45 at middle z=0.

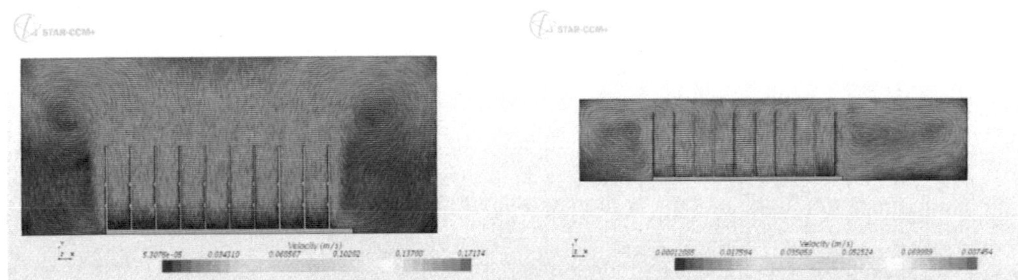

Figure 10. Velocity contour at Q=18 w, Angle=45, Smooth heat sink and at middle z=0.

5. Conclusion

The hotness dissipation capability of the heat sink under natural cooling has been found to increase by the adding surface geometry on the fin surface. Computational evidences validated with experimental results explain that the variation in the impression inclinations angle has important result on the heat sink thermal performance. Based on the numerical study, it can be conclude:The value of the Nusselt number raised directly with the increase of heat supply to the base for all heat sink shapes.Remarkable improvement in the Nusselt number values has been seen for the embossed fins heat sink, which confirms that the nusselt number is a strongly affected by impression angle. the Nusselt number increased and reaches the maxima at 45 angle with the lessening at impression angle 90.The improvements in Nusselt number between 1.12 and 1.76 matching to the impression pitch and angle of 12 mm and 45, correspondingly.

Embossed heat sink better temperature distribution than smooth het sink, 102 K differnce between embossed and smooth heat sink.

References

[1] Alhara T and et al 1970 Natural convection heat transfer from vertical rectangular-fin arrays: Part 2 heat transfer from fin-edges *Bull. JSME* 1108–1120.

[2] Chaddock B 1970 Free convection heat transfer from vertical rectangular fin arrays *ASHRAE J* 12 53–60.

[3] Chi-Chuan Wang K and et al 2011 Effect of cannelure fin configuration on compact aircooling heat sink. *Applied Thermal Engineering, 31* pp 1640-1647.

[4] Hung-Yi Li and et al 2013 Enhancing heat transfer in a plate-fin heat sink using delta winglet

vortex generators *International Journal of Heat and Mass Transfer* 67 pp 666–677.

[5] Yang J J and et al 2010 On the heat transfercharacteristics of heat sinks: with and without vortex generators *IEEE Trans.Compon. Packag. Technol 33 pp*. 391–397.

[6] Singh P and Patil A. K 2015 Experimental investigation of heat transfer enhancement through embossed fin heat sink under natural convection. *Experimental Thermal and Fluid Science* 61 24–33.

[7] Starner K and McManus H 1963 Experimental investigation of free convection heat transfer from rectangular fin arrays *Heat Transfer* 85 273–278.

[8] Timothy Dake J M and et al 2009 Improving Flow Circulation in Heat Sinks Using Quadrupole Vortices *ASME Summer Heat Transfer Conference.*

[9] Welling C W and et al 1965 Free convection heat transfer coefficient from rectangular vertical fins *Heat Transfer* 87 439–445.

Comparative Study of R134a and R744 Driven Solar Assisted Heat Pump Systems for Different Applications

Y. Baradey, M.N.A. Hawlader*, A.F. Ismail, M. Hrairi

Department of Mechanical Engineering, Faculty of Engineering, International Islamic University Malaysia, Jalan Gombak, 53100, Kuala Lumpur, Malaysia

*mehawlader@iium.edu.my

Abstract. A solar assisted heat pump (SAHP) system is considered in this study where space cooling, water heating, drying and desalination are carried out using solar energy, ambient energy and the waste heat available from air conditioning system. The integration of different applications with SAHP system leads to many benefits, but the most significant is the energy conservation issue where the prices of fossil fuels are increasing. In recent years, researches have been conducted worldwide to improve the performance of these systems. These systems usually use synthetic refrigerant, such as R134a, which contributes to many atmospheric problems. Currently, embracing natural refrigerants, such as, carbon dioxide in SAHP has become a trend specifically after Kyoto agreement to reduce global warming and improve the performance of SAHP systems. In this paper, comparative study using R134a and R744 in SAHP systems have been considered. The comparison includes the most significant results of previous work relevant to this issue, and the thermo physical properties of the two refrigerants when used in SAHP.

1. Introduction

Heat pump (HP) system is a conventional vapor compression air conditioning system which collects or absorbs the heat from a heat source, usually air conditioning room, and dissipates it to the atmosphere [1-3]. Air-conditioners in tropical countries, such as Malaysia, are used almost in all residential, industrial and office buildings. The amount of waste heat from air conditioners discharged to the environment is considered very large. Exploiting these energy resources instead of throwing it away is considered a viable option to reduce the global warming and other environmental issues. It is also considered an excellent solution for energy conservation issue where the price of oil and other non-renewable energy sources are increasing due to many reasons [2,3]. In recent years, researches have been conducted worldwide in order to modify the performance of heat pump systems and make it more reliable and efficient. The attempts include exploiting the waste heat from condenser to be used for different thermal applications such as drying, water heating, and desalination of seawater. Replacing the synthetic refrigerants, such as R134a, with natural refrigerants e.g. R744 has also been under study. Attempt to reduce the irreversibility of the cycle by replacing the expansion valve with turbine, ejector, and using internal heat exchanger has also been made.

Solar assisted heat pump (SAHP) is a conventional system where solar energy is used as the main or supplementary heat source. Integration of solar energy with conventional heat pump system is considered viable solution to reduce the dependancy on fossil fuel, and improve the performance of the system. SAHP systems usually use conventional synthetic refrigerants such as R134a. Recently, many researchers are moving toward embracing and using carbon dioxide because of environmental issues such as Ozone depletion and Global warming. Besides that, CO_2 has thermodynamic properties which enable it to be competitive and most promising alternative of conventional refrigerants, which will be discussed in this paper [4]. Using CO_2 in SAHP systems makes the system works under the transcritical thermodynamic cycle not the traditional vapor compression cycle [5-7].

In this paper, the focous is on using carbon dioxide as an alternative of synthetic refrigeratns, mainly R134a, in heat pump and solar assisted heat pump systems for space cooling, water heating, drying, and desalination of Seawater. Comparative study between using R744 and R134a have been conducted based on available literature and previous researches.

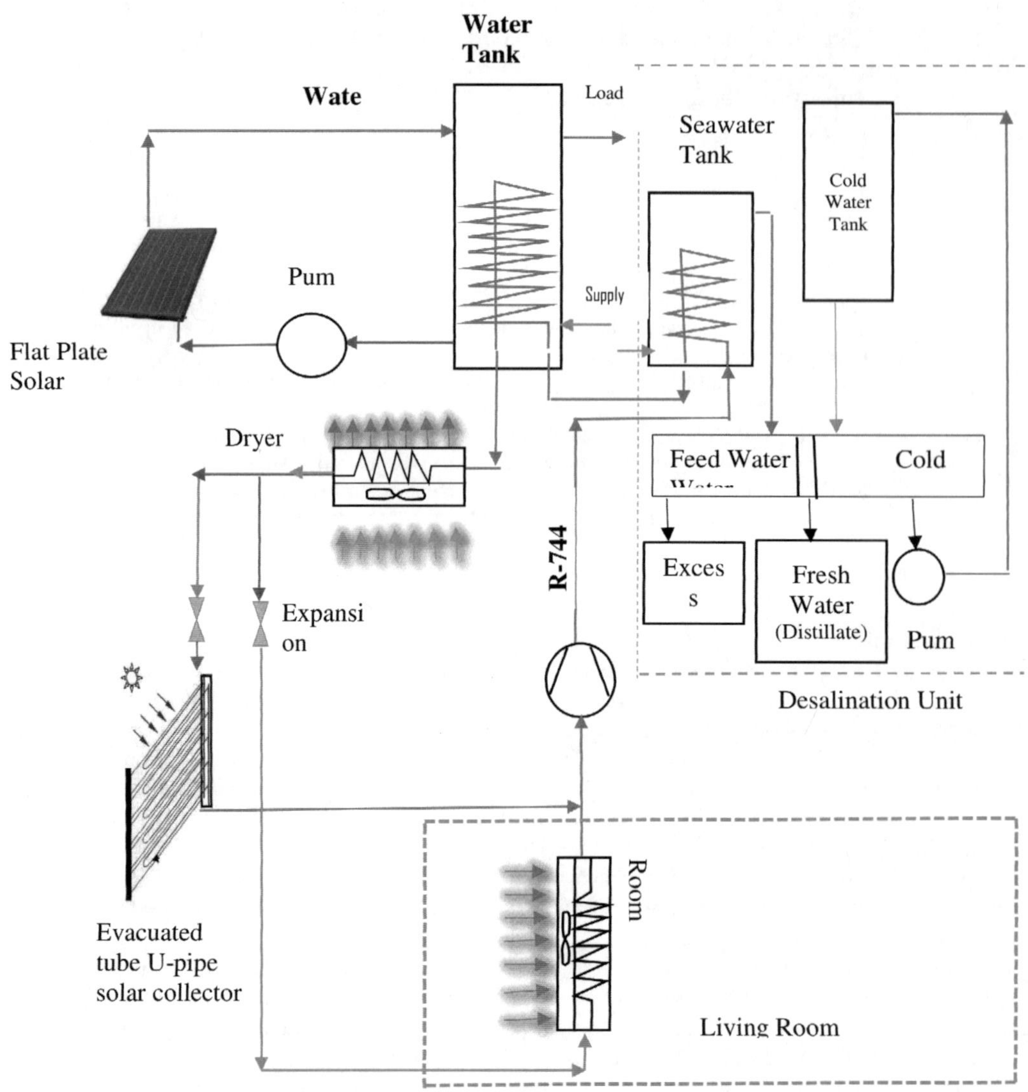

Figure 1: Schematic Diagram of Carbon Dioxide Solar Assisted Heat Pump System for Space Cooling, Water heating, Drying, and Desalination

2.0 The System

The schematic diagram of a Carbon Dioxide driven solar assisted heat pump system for space cooling, water heating, drying, and desalination of seawater is shown in Figure 1. The system comprises of mainly a R-744 compressor, water tank, room evaporator, evaporator collector, flat plat water collector, expansion valves, and a desalination unit. Air Gap Membrane Desalination (AGM) unit is used to convert seawater into fresh drinking water. Hot air drying unit is used in the system for clothes drying purpose.The system works under supercritical or transcritical cycle where heat rejection process (high pressure side) occurs at the supercritical region (above 73.8 bars), and the evaporation process (low pressure side) occures at the sub-critical region. Special design is required for the system components because the pressure inside the pipes is 5-10 times higher than those system which work using conventional synthetic refrigerants such as R-134a.

2.1 SAHP for Space Cooling, Water Heating

Lorentzen and Petersen [5] experimentally investigated the performance of a CO_2 system for automobile air conditioning. A comparison between CO_2 and R12 was conducted with equal heat exchanger dimensions. The results showed that CO_2 and R12 efficiencies were approximately equal. Based on this study, many projects were initiated in USA and European Union countries. Cecchinato et al. [6] conducted a detailed comparison between heat pump system using CO_2 as refrigerant and another one where HFC 134a was used. They found out that Carbon Dioxide showed better results and it is a potential alternative for synthetic refrigerants. In the same year, Sarkar [7] investigated, both theoretically and experimentally, the performance of carbon dioxide HP system for simultaneous cooling and heating. A simulation model of CO_2 heat pump dryer also was developed by him. It was concluded that carbon dioxide showed better performance than R134a in HP systems for drying applications, but poorer than R22. The effect of air mass flow rate, ambient temperature, recirculation air ratio, and dryer efficiency were significant on the system performance. The experimental and simulated results of the cooling and heating system showed that the effect of water inlet temperature of the gas cooler is significant on the system performance unlike the effect of mass flow rate.

Carmo et al. [8] designed and built a CO_2 heat pump system (Known as Thermal Battery "TB") for simultaneously heating and cooling. The experimental results showed that thermal battery system could supply water heating with temperature greater than 70 as well as cooling service with temperature less than 10 for commercial and residential building. They also performed optimization study for the system. The results of the optimization relied on the basic operational parameters, such as, water mass flow rate, and the inlet and outlet temperature in the gas cooler. They found out that the COP of the TB system could reach a value more than 4, and it was increasing with increasing mass flow rate in the evaporator. Hawlader and Schaochum [9] designed and experimentally investigated the performance of a solar assisted heat pump system for air conditioning, water heating, and drying using R134a as the refrigerant of the system. The experiments were conducted under the meteorological conditions of Singapore. The average COP of the systems for 24 hours operating was around 4, while the maximum value of it was 6. It is concluded that this system can ensure long term operation, and capable of reducing a lot of buildings thermal costs particularly buildings located in tropical region.

Kokila and Rajakumer [10] theoretically compared the performance of carbon dioxide driven solar assisted heat pump water heating system with conventional (Thermo syphon) solar water heating system. Numerical analysis of the direct expansion SAHP system was introduced as well. Stene [11] compared the performance of CO_2 HP water heating system with other systems using conventional refrigerants and concluded that the system achieved 20% higher COP at different evaporaing temperature, as shown in Figure 2.

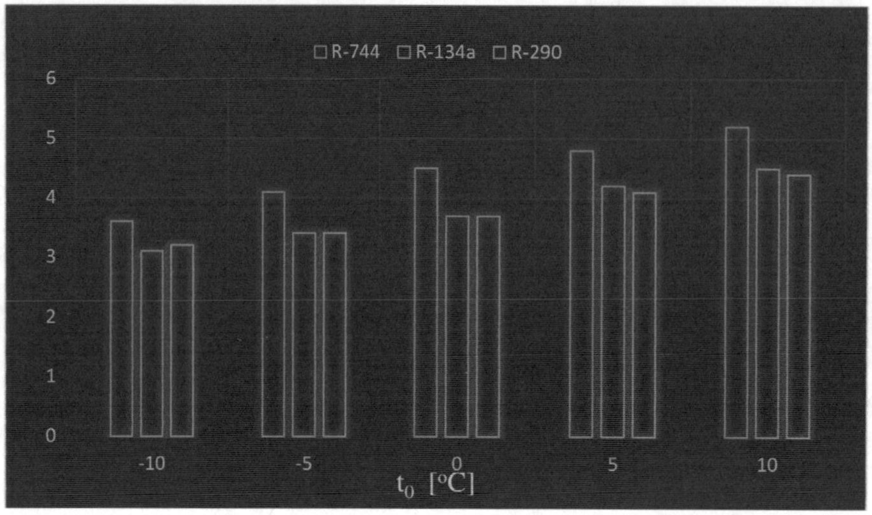

Figure 2: Caclulated COP as a function of Evaporation Temperature t_0 [°C]
(Source of Data: Stene, J [11])

2.1 SAHP Systems for Drying

Drying is an important process which is nowadays widely used in many industries, such as, chemical industry, timber, pharmaceuticals, food, and paper industry. This process usually require great amount of energy that sometime accounts for up to 15% of all industrial energy usage, often with relatively low thermal efficiency in the range of 20-25% [12]. Solar assisted heat pump system for drying has been widely used in industry since many years, and a lot of researches conducted on it. Manuel et al. [13] discussed the shortcomings and the economic issues related to the tumbler dryer technology, and they proposed a solution for the energy conservation in the laundry drying. Hawlader et al. [14-15] investigated the effects of hot air drying, vacuum drying, and freeze drying methods on fruits' quality. In this study, hot air was replaced by Carbon Dioxide and Nitrogen gases. Guava and Papaya were dried under different conditions. The experimental results showed that using inert gas such as Carbon Dioxide in heat pump drying system lead to a better physical appearance of the product, much better quality, better nutritional value of the product, and faster rehydration process. Sarkar et al. [16] developed a simulation model to study the performance of carbon dioxide heat pump dryer and compare its performance with conventional dryers operating with R-22 and R-134a refrigerants.

2.2 SAHP Systems for Desalination

The most abundant substance in the world is water and about 97% of the water is seawater while the rest is fresh water. According to some reports [17], about one billion people are not able to drink fresh water across the world. Billions of Dollars are currently spent worldwide on different desalination technologies for producing fresh water from brackish and seawater that could meet the demand of drinking water. Bahar and Hawlader [18] reported that about 60 Million RM have been spent on Seawater desalination project in Sarawak- Malaysia. According to Hawlader et al. [19] and Carter [20], integration of renewable energy in the conversion of seawater into fresh water by using thermal distillation process, such as air gap membrane distillation shows significant potential. Tjandra [21] conducted a study on solar assisted heat pump desalination system. The system consisted of a solar assisted heat pump working on R-134a, and a single effect desalination unit. Byrne et al. [22] investigated the performance of heat pump system for simultaneous cooling and desalination. Air Gap MD type was integrated with heat pump system.

3. Thermo physical properties of Carbon Dioxide (R-744)

Pressure drop and heat transfer process in solar assisted heat pump systems are strongly affected by the transport properties of the refrigerant, such as, thermal conductivity and viscosity. Thermal conductivity of the refrigerant influences the heat transfer coefficient in both single and two-phase flow. Viscosity affects the liquid flow behavior, two-phase heat transfer, and pressure drop in the system. It is reported that thermal conductivities of CO_2 in both saturated liquid and vapor are 20 and 60% higher than liquid and vapor thermal conductivity of R134a respectively. On the other hand R134a has liquid viscosity 40% lower than carbon dioxide. In the supercritical region, the high pressure and temperature are not coupled, so that they can be regulated independently to obtain the optimal operating conditions, as shown in Figure 3.

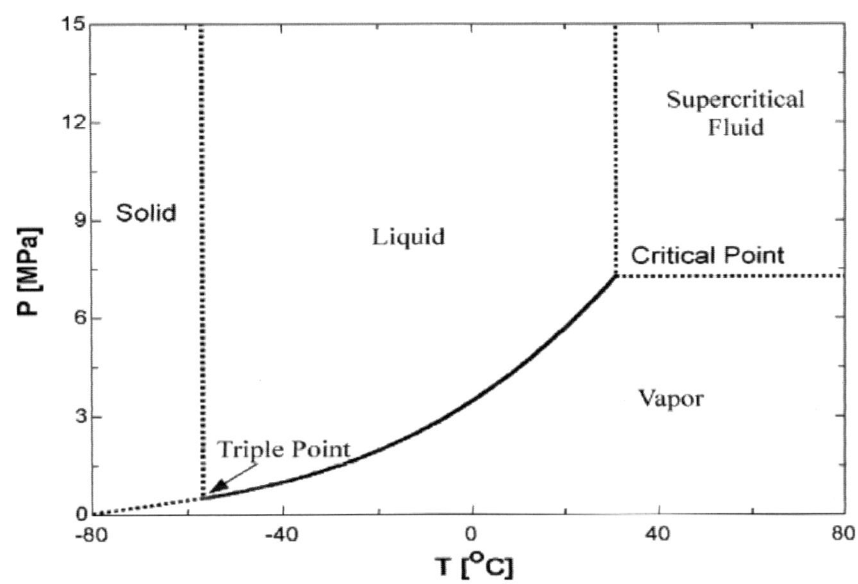

Figure 3: Pressure- Temperature Phase Diagram of Carbon Dioxide [23]

3.1 Heat Transfer Coefficient of R-744 compared to HFC R-134a

Mastrullo et al. [24] experimentally compared the heat transfer coefficients (HTC) of carbon dioxide and R134a during flow boiling in a 6mm diameter horizontal and circular tube at similar operating conditions. A comparison was performed to accurately evaluate the effect of the thermodynamic and transport properties, such as, vapor quality, the refrigerant mass flow flux, saturation temperature, and heat flux on the heat transfer coefficients. However, it is concluded that the HTC of the Carbon Dioxide is always higher than R134a at same operating conditions. Therefore, HTC of the R134a is dependent on mass flux and independent on saturation temperature whereas HTC of R744 is dependent upon saturation temperature and independent of mass flux.

4. Comparison between R-744 and R-134a Compressors

The pressure in CO_2 heat pump system is usually 5-10 time higher than other refrigerants which requires proper design for the system component especially compressor and gas cooler. For instance, suction and discharge pressures for R134a compressor at 0 °C and 80 °C are 3 and 27 bars, while suction and discharge pressures for CO_2 compressor at 0 °C and 80 °C are 35 and above 120 bars, respectively. The wall of Carbon Dioxide compressor is thicker, but the compressor itself is much smaller compared to conventional compressor of HFC refrigerants because of the high volumetric capacity of CO_2, as shown in Figure 4 [25].

R134a Compressor

CO_2 Compressor

Figure 4: Pictures of R-134a and Carbon Dioxide Compressors [25]

According to Kruse et al. [26] CO_2 compressors consume energy 18-37% less than compressors for conventional refrigerants (Mainly R-134a). Figure 5 shows comparison between CO_2 and R-134a Compressors without fans, motors and lights.

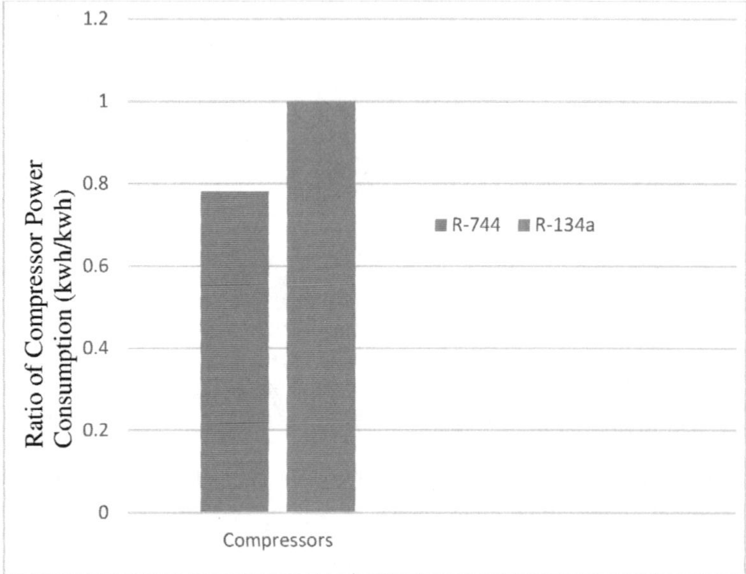

Figure 5: Comparison of Power Consumption from R-744 and R134 Compressors
(Source: Kruse et al. [26])

5. Conclusions

Using carbon dioxide in Solar assisted heat pump system for different applications could lead to a significant improvement in the htermal performance of the system and reduce the global warming issue. Comparative study between using R134a and R744 in SAHP system have been presented in this paper. Drying and desalination technolgies consume large amount of energy which usually contribute to more than 50% of these technolgoies costs. This energy usually obtained from electricity. Integration of drying and desalination technologies with solar assisted heat pump system could reduce the dependency on fossil fuel to operate them which will reduce the cost and make them envirementally friendly applications.

References

[1] Zakaria M A and Hawlader MNA 2013 A review on solar assisted heat pump systems in Singapore, Journal of Renewable and Sustainable Energy Reviews, 26, p286-293.

[2] Baradey Y, Hawlader MNA, Ismail, AF, Harairi M 2015 Waste Heat Recovery in Heat Pump Systems: Solution to reduce the Global Warming, IIUM Engineering Journal, Special Issue on Energy 16 (2).

[3] Hawlader MNA 2013 Utilization of solar energy and waste heat for water heating, drying and desalination of seawater, Proc ICMIME2013, 1-3 November, RUET, Rajshahi, Bangladesh.

[4] Austin BT, Sumathy K 2011 Transcritical Carbon Dioxide Heat Pump Systems: A review. Renewable and Sustainable Energy Reviews 15, p4013-4029.

[5] Lorentzen G and Pettersen J 1993 A new, efficient and environmentally benign system for car air conditioning, Int J Refrigeration, 16(1) p4-12.

[6] Cecchinato L, Corradi M, Fornasieri E and Zambon L 2005 Carbon Dioxide as refrigerant for tap water heat pumps: A comparison with the traditional solution, International Journal of Refrigeration, Vol 28(8), p1250-1258.

[7] Sarkar J 2005 Transcritical Carbon Dioxide Heat Pump for Simultaneously Cooling and Heating, A Thesis Submitted in Partial Fulfillment of the Requirements for the Award of the degree of Doctor of Philosophy in Engineering. Department of Mechanical Engineering Indian Institute of Technology.

[8] Carmo C, Blarke M, Yazawa K and Shakouri A 2012 Energy Optimization for Transcritical CO2 Heat Pumps for Combined Heating and Cooling and Thermal Storage Applications, International Refrigeration and Air Conditioning Conference, Paper 1168, http://docs.lib.purdue.edu/iracc/1168

[9] Hawlader MNA and Schaochum Y 2008 Performance of a Solar Assisted Heat Pump Air Conditioner, Water Heater and Dryer. Journal of Ambient Energy, Volume 29 Number 4, p189- 196.

[10] Kokila RN and Rajakumar S 2013 Performance Evaluation of Solar Assisted Heat Pump Water Heating System, IOSR Journal of Engineering Vol. 3, Issue 4, p12-17.

[11] Stene J 2008 High efficiency CO_2 heat pump water heater systems for residential and non-residential buildings. Retrived on 20 October from http://www.r744.com/assets/link/R744com_CO2-HPWH_2008.pdf

[12] Chua KJ, Mujumdar AS, Hawlader MNA, Chou SK and Ho JC 2001 Batch drying of banana pieces – Effect of stepwise change in drying air temperature on drying kinetics and product colour, Food Research International, 34, p721-731.

[13] Manuel R C 1997 Energy conservation with tumbler drying in laundries, Applied thermal engineering, 17, p1163-1172.

[14] Hawlader MNA, Chou SK, Jahangeer KA, Rahman SMA and Eugene Lau KW 2003 Solar-assisted heat-pump dryer and water heater, Applied Energy, 74(1-2), p185-193.

[15] Hawlader MNA, Perera CO and Tian M 2005 Influnce of Different Drying Methods on Fruits' Quality, 8th Annual IEA Heat Pump Conf, 30 May- 2 June, Las Vegas, Nevada, USA.

[16] Sarkar J, Bhattacharyya S and Gopal MR 2004 Transcritical Carbon Dioxide Based Heat Pumps: Process Heat Applications, International Refrigeration and Air Conditioning Conference, Paper 691.

[17] Selvi SR and Baskaran R 2014 Desalination of Well water by Solar Power Membrane Distillation and Reverse Osmosis and its Efficiency Analysis, International Journal of ChemTech Research, Vol.6, No.5, p2628-2636.

[18] Bahar R, Hawlader MNA 2013 Desalination: Conversion of Seawater to Freshwater, 2nd International Conference on Mechanical, Automotive and Aerospace Engineering (ICMAAE 2013), 2-4 July, Kuala Lumpur, Malaysia.

[19] Hawlader MNA, Dey PK, Sufyan D and Chung CY 2004 Solar Assisted Heat Pump Desalination system. Elsevier, desalination, vol. 168, p49 – 54.

[20] Carter NT 2015 Desalination and Membrane Technologies: Federal Research and Adoption Issues, Congressional Research Service. www.crs.gov.

[21] Tjandra TB 2007 The Study of a Solar Assisted Heat Pump Desalination System. A Thesis Submitted For the Degree of Master of Engineering, Department Of Mechanical Engineering National University Of Singapore.

[22] Byrne P, Oumeziane YA, Serres L and Mare T 2016 Study of a Heat Pump for Simultaneous Cooling and Desalination, Applied Mechanics and Materials, Trans Tech Publications, Advances in Heat Transfer, Flow Engineering and Energy Installations 819, p.152-159. <10.4028/www.scientic.net/AMM.819.152 >. <hal-01253254>

[23] Kim YM, Kim CG and Favrat D 2012 Transcritical or supercritical CO_2 cycles using both low- and high-temperature heat sources, Elsevier, Energy 43, p402-415.

[24] Mastrullo R, Mauro AW, Rosato A and Vanoli GP 2009 Comparison of R744 and R134a heat transfer coefficients during flow boiling in a horizontal circular smooth tube. International enewable energies and power quality (ICREPQ'09), 15 - 17 April,Valencia, Spain.

[25] Kim MH, Pettersen J and Bullard CW 2004 Fundamental Process and System Design Issues in CO_2Vapor Compression Systems, Progress in Energy and Combustion Science 30, p119–174

[26] Kruse H, Jakobs R and Russamann H 2005 On the Energy Efficiencyof Carbon Dioxide in Small Commercial Cooling Applications, Retrieved on 18 January 2016 from http://lms.i-know.com/mod/resource/view.php?id=572.

Numerical Investigation of Magnetic Nanoparticles Trajectories for Magnetic Drug Targeting

Syireen Shazri and Moumen Idres

Mechanical Engineering Department, Kulliyah of Engineering, International Islamic University Malaysia, 50728 Kuala Lumpur, Malaysia

midres@iium.edu.my

Abstract. In this work, the trajectories and capturing of magnetic nanoparticles coated with drug agent are investigated for drug targeting application. Nanoparticles are injected at the entrance of a microvessel and captured by a permanent magnet located at a specified location where tumor exists. The problem is divided into two steps; blood flow solution and nanoparticles trajectory solution. The blood flow in microvessel is obtained both analytically and numerically. Integration of nanoparticles equations of motion to obtain the trajectories is performed using both Matlab and ANSYS Fluent. Discrete Phase Model (DPM) in ANSYS Fluent is used for nanoparticles tracking. The dominant magnetization and drag forces acting on magnetic particles are incorporated to study the trajectories of magnetic particles. Parametric studies for steady flow with single point and multiple point injections are conducted. This includes varying particle diameter and magnet location. Critical minimum diameter for capturing is predicted. Capturing efficiency is reported for all cases. It is found that particle trajectories are strongly dependent on particle size and location of the magnet. The simulation can be used to determine the optimum particle size for treating a tumor, given its size and location.

1. Introduction

Magnetic Drug Targeting (MDT) provides a promising technique for improving the available conventional drug delivery techniques. During chemotherapy, for example, the drug is not only concentrated locally on the tumor tissue, but it is globally exposed to the other healthy tissues throughout the body. To avoid damages to healthy cells from the toxic drug, it is highly recommended to physically target the drug to the specific locations where the unhealthy cells exist. In MDT technique, ferrofluids consisting of suspended magnetic nanoparticles are used as a carrier for chemotherapeutic agents and for other medications. Then, they are targeted to a specified location by the help of an applied magnetic field coming from an external permanent magnet. The role of the external magnet is to provide the magnetic force that can attract and capture the magnetic particles. This allows the treatment to be targeted locally to the only necessary locations instead of going globally throughout the body, which can cause side effects. This technique can be very useful for treating many diseases such as cancer, stenosis, arterial occlusion, stroke, and much more.

The use of ferrofluids for magnetic drug targeting has become the primary carriers due to their high saturation magnetization and magnetic susceptibility. For general applications, ferrofluids in pharmaceutical and medical applications should be biocompatible, biodegradable, and nontoxic [1].

IOP Conf. Series: Materials Science and Engineering **184** (2017) 012061 doi:10.1088/1757-899X/184/1/012061

Magnetite nanoparticles (Fe_3O_4) that have excellent magnetic saturation (~78 emu/g) are desirable for these applications due to the strong ferromagnetic behavior, less sensitivity to oxidation and relatively low toxicity compared to many other materials (e.g., nickel and cobalt) [2]. Such nanoparticles can be driven to the specific target site by applying external magnetic fields. The effectiveness of this technique is critically dependent on the properties of magnetic nanoparticles [3-5].Ferrofluids are also dielectric (non conducting) and paramagnetic which means they are attracted by magnetic fields, and do not retain magnetization in the absence of an applied field [6].

There are many recent studies regarding this magnetic drug targeting technique. Furlani et al., 2009 presented an analytic model for transport and capture of magnetic nanoparticles in the human microvasculature for targeted drug delivery applications [7]. He also performed parametric studies by varying the size of magnetic particles and the distance of the external magnetic field. It is also been reported that the magnetic drug targeting is relatively safe and effective for targeting the drugs to a desired and specific location in patient's body [8-9]. Micron sized magnetic particles have been injected in a capillary of a blood vessel inside a tumor by locating an external magnetic field by 10 cm away from the tumor location [10].

During targeting, not all the particles will be trapped since part of them will be lost between the injection and the target point. The efficiency of the target and capture of the magnetic nanoparticles is very important in order to have a maximum-trapped drug in the desired location. Thus, it is vital to study about the optimal parameters such as magnetic nanoparticle size and external magnetic field that can significantly improve the efficiency of the magnetic drug targeting.

In this work, we presented several theoretical formulae and performed simulations through finite volume based ANSYS Fluent software under Discrete Phase Model (DPM) to predict the trajectories and captures of magnetic nanoparticles at the microvessel wall. Different magnetic particle radii and different external magnet locations are varied to observe the effect on particle trajectories. Two cases of injections are performed; single point injection and multipoint injection case. This study will help to design a better drug targeting technique with optimized size of magnetic nanoparticles and the relative location of the external magnet.

2. Methodology

Figure 1 depicts the schematic diagram or geometric configuration that simulated the particle trajectories in the two-dimensional straight cylindrical tube of microvessel with a radius (R_v). A cylindrical external permanent magnet with a radius of $R_{mag}$is located at a distance of y_{mag} from the axis and x_{mag} from the inlet of the vessel. This cylindrical magnet with the role of producing the requisite magnetic field is assumed to be infinitely long and oriented perpendicular to the direction of the blood flow.

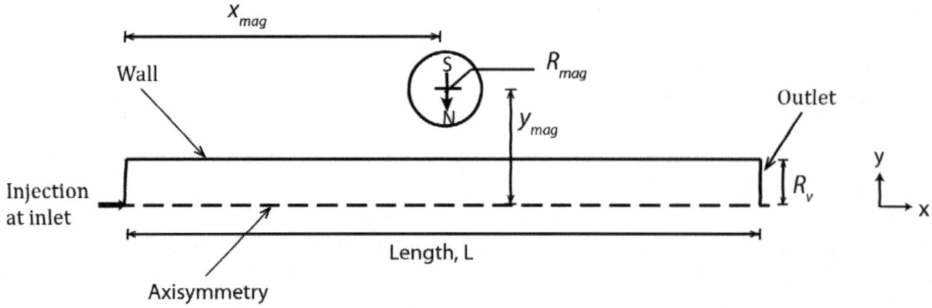

Figure 1. Schematic diagram for predicting particle trajectories in microvessel.

2.1. Theoretical Formulation

There are many forces acting simultaneously on the magnetic particles as they are transported in the microvessel. The forces are magnetic force, viscous drag force, gravitational force, and buoyancy force. The other factors that affect the motion of magnetic nanoparticle in microvessel are Brownian motion and particle inertial effect. However, in this model, only dominant forces which are the magnetic force and viscous drag force are considered, while the rest of the other forces are neglected.

The trajectory of the magnetic particles is predicted by integrating the Newton's law of equation of motion

$$m_p \frac{d\boldsymbol{u}_p}{dt} = \boldsymbol{F}_D + \boldsymbol{F}_{\boldsymbol{m}} \tag{1}$$

where m_p and $\boldsymbol{u}_p$ denote the mass and the velocity vector of the magnetic nanoparticle respectively, while$\boldsymbol{F}_D$and $\boldsymbol{F}_m$are theviscous drag force and magnetic force.

The viscous drag force was predicted by using spherical drag law from Stoke's approximation as following [13]

$$\boldsymbol{F}_D = -6\pi\eta R_p(\boldsymbol{u}_p - \boldsymbol{u}_f) \tag{2}$$

where R_p is the radius of the particle, η is the viscosity of the fluid, $\boldsymbol{u}_p$ and $\boldsymbol{u}_f$ are the velocities of the particles and fluid respectively. The flow is determined in terms of average speed of u_f

$$u_f(y) = 2u_{f,average}\left(1 - \frac{y^2}{R_v^2}\right) \tag{3}$$

where$u_f(y)$ is the velocity of fluid in the axial direction at a radial distance, y from its axis.

Next, for the magnetic force $\boldsymbol{F}_{\boldsymbol{m}}$, which also known as the Kelvin force on a magnetized particle of volume, V is given by the following equation,

$$\boldsymbol{F}_{\boldsymbol{m}} = \iiint_V \mu_0 \boldsymbol{M}\,.\,\nabla\mathbf{H}\,\mathrm{dV} \tag{4}$$

Where $\mu_0 = 4\pi \times 10^{-7}\mathrm{N}/A^2$ is the magnetic permeability of vacuum, $\boldsymbol{M}$ is the material's magnetization (magnetic dipole moment per unit volume), and $\mathbf{H}$ is the auxiliary magnetic field vector.

Furthermore, to determine the magnetic force, a model for the magnetic response of the particle is needed. We considered the particles are below saturation and the magnetization of such particles can often be assumed proportional to the applied magnetic field. In other words, we used a linear magnetization model with saturation.

$$\boldsymbol{M} = \chi\mathbf{H} \tag{5}$$

Here the proportionality constant χ is called the magnetic susceptibility. The magnetic susceptibility χ quantifies the tendency of a material to form magnetic dipoles. The effective susceptibility of the spherical particles is related to their intrinsic susceptibility. Since we have assumed the particle is below saturation, the intrinsic susceptibility, $\chi_{eff} = 3$.

After the integration of equation (4), the magnetic force term due to the presence of the magnetic field is expressed as following,

$$\boldsymbol{F}_{\boldsymbol{m}} = \mu_0\left(\frac{4}{3}\pi {R_p}^3\right)\chi_{eff}(\mathbf{H}\,.\,\nabla)\mathbf{H} \tag{6}$$

The magnetic field in the expression of the coordinate reference frame of the microvessel is derived by the following equations,

$$\boldsymbol{H} = H_x(x,y)\hat{\mathbf{x}} + H_y(x,y)\hat{\mathbf{y}} \tag{7}$$

Next, the magnetic field terms in equation (6) is reduced to the following equation,

$$\nabla \boldsymbol{H}_a^2 = H_x(x,y)\frac{\partial H_x(x,y)}{\partial y} + H_y(x,y)\frac{\partial H_y(x,y)}{\partial y} \tag{8}$$

This can be reduced after we have assumed **rot H** = 0. This results in $\frac{\partial H_x(x,y)}{\partial y} = \frac{\partial H_y(x,y)}{\partial x}$ and finally leads to this expression; $(\mathbf{H}\,.\nabla)\mathbf{H} = \nabla \boldsymbol{H}_a^2$.

The magnetic force in equation (6) can also be written in the x- and y-component as in the following equation,

$$\boldsymbol{F_m}(x,y) = \mathrm{F}_{mx}(x,y)\hat{\mathbf{x}} + \mathrm{F}_{my}(x,y)\hat{\mathbf{y}} \tag{9}$$

Substitute equation (8) into equation (6) in the form of equation (9),

$$\mathrm{F}_{mx}(x,y) = \mu_0\left(\frac{4}{3}\pi {R_p}^3\right)\chi_{eff}\left[H_x(x,y)\frac{\partial y(x,y)}{\partial x} + H_y(x,y)\frac{\partial H_y(x,y)}{\partial y}\right] \tag{10}$$

$$\mathrm{F}_{my}(x,y) = \mu_0\left(\frac{4}{3}\pi {R_p}^3\right)\chi_{eff}\left[H_x(x,y)\frac{\partial H_x(x,y)}{\partial x} + H_y(x,y)\frac{\partial H_x(x,y)}{\partial y}\right] \tag{11}$$

Next, the external source magnet at the horizontal distance, x_{mag} and vertical distance, y_{mag} with a radius ofR_{mag} (centered at its local origin) as shown in figure 1 is considered in the formulation of the magnetic force. The magnetization of the source magnet is assumed to be at a level of $M_s = 1 \times 10^6\ A/m$ through its cross section. Then, the final form of magnetic force can be written as the following,

$$\mathrm{F}_{mx}(x,y) = -\mu_0\left(\frac{4}{3}\pi {R_p}^3\right)\chi_{eff}M_s^2R_{mag}^4\frac{x - x_{mag}}{2\left[(y - y_{mag})^2 + (x - x_{mag})^2\right]^3} \tag{12}$$

$$\mathrm{F}_{my}(x,y) = -\mu_0\left(\frac{4}{3}\pi {R_p}^3\right)\chi_{eff}M_s^2R_{mag}^4\frac{y - y_{mag}}{2\left[(y - y_{mag})^2 + (x - x_{mag})^2\right]^3} \tag{13}$$

Lastly, the equation of motion shown in equation (1) is then written in its final form that comprises of x- and y-component as stated in the following equation (14) and (15). These two equations are also been used to make a comparison between the analytical and the numerical solutions (from Fluent) for single injection point. Meanwhile, equation (16) shows the formula for determining the capture efficiency in multipoint injection case.

$$m_p \frac{du_{px}}{dt} = -\mu_0 \left(\frac{4}{3}\pi R_p{}^3\right) \chi_{eff} M_S^2 R_{mag}^4 \frac{x - x_{mag}}{2\left[(y - y_{mag})^2 + (x - x_{mag})^2\right]^3} - 6\pi\eta R_p \left[u_{px} - 2u_f \left(1 - \frac{y^2}{R_v^2}\right)\right] \quad (14)$$

$$m_p \frac{du_{py}}{dt} = -\mu_0 \left(\frac{4}{3}\pi R_p{}^3\right) \chi_{eff} M_S^2 R_{mag}^4 \frac{y - y_{mag}}{2\left[\left(y - y_{mag}\right)^2 + \left(x - x_{mag}\right)^2\right]^3} - 6\pi\eta R_p u_{py} \quad (15)$$

$$\text{Capture Efficiency} = \frac{\text{Trapped particle}}{\text{Injected particle}} \quad (16)$$

2.2. Solution and Numerical Methods

In this work, there are two ways of solving the equations of motion of magnetic particles for predicting their trajectories. For single injection case, the trajectories of particles are predicted through simulations using ANSYS Fluent under DPM. Meanwhile, for multipoint injection case, the trajectories of the particles are predicted by solving the equations of motion analytically and numerically using Matlab.

Discrete Phase Model plays an important role in tracking particles hence the trajectories of particles can be predicted. Discrete Phase Model in ANSYS Fluent follows the Euler-Lagrange approach. The fluid phase, which is the continuous phase, is treated as a continuum by solving the time-averaged Navier-Stokes equations in the Eulerian reference frame. Meanwhile, the dispersed phase is solved by tracking the number of particles through the calculated flow field of the continuous phase in Lagrangian reference frame [11].

For continuous phase, the pressure-based solver is selected and under this, pressure-velocity coupling method with selecting SIMPLE segregated algorithm using the semi-implicit method is been done. The SIMPLE algorithm uses a relationship between velocity and pressure corrections to enforce the conservation of mass and to obtain the pressure field. The pressure corrections can be obtained by specifying the under-relaxation factor for pressure which usually around 0.03.For spatial discretization scheme, Least Square Cell-Based scheme is used to discretize the gradient, whereas central differencing scheme of second order accuracy is used to discretize the pressure and second order upwind to discretize momentum.

For discrete phase, ANSYS Fluent predicted the trajectory of a particle by integrating the force balance on the particle as in equation (1), which is written in a Lagrangian reference frame [11]. In this case, the particle is only subjected to a fluidic drag force and magnetic force as stated previously. The magnetic force in equation (12) and equation (13) are compiled into Fluent using a User Defined Function(UDF) option. They are programmed using the C language based on the particle position, not the cell position [12].They are also being hooked in the "body force" options under DPM.

The trajectory equation is solved by stepwise integration over discrete time steps using the trapezoidal discretization scheme. The maximum number of time steps used to compute a single particle trajectory via integration of equation (1) used in the simulation is 100000. This should be large enough so that the particles can travel from entrance to exit. When the maximum number of steps is exceeded; ANSYS Fluent abandons the trajectory calculation for the current particle injection and reports the trajectory fate as "incomplete" [12]. Step Length Factor is inversely proportional to the integration time step. It is roughly equivalent to the number of time steps required to traverse the current continuous phase control volume. In this simulation, the step length factor is chosen as 5.

For single injection, a particle was released from the inlet at one point, which is exactly at the axis of the vessel. The fluid flow is assumed as laminar and axisymmetric flow with a no-slip boundary condition at the wall. For DPM boundary condition, the inlet and outlet were set to allow particles to escape while the wall of the vessel is set to trap the particles. The particle is being tracked using only for "one-way coupling" and steady particle tracking in DPM.

Next, for multipoint injection case, the prediction of magnetic particles trajectories is done by solving equation (14) and (15) analytically and numerically. Stiff Ordinary Differential Equation (ODE23s) in Matlab is used to solve the numerical parts instead of using ANSYS DPM. For this case, since the particles are injected at many different points through the inlet, the axisymmetric flow option in Fluent cannot be selected because the magnetic force is unsymmetrical across the entire domain of the microvessel. This results in the unsymmetrical particles trajectories. This is the reason for why we could not simulate the trajectories in Fluent since only the fluid flow is axisymmetric but not the particle trajectories.

Hence, in equation (14) and (15), the magnetic force terms are solved numerically using ODE23s, while the viscous drag force terms are solved analytically. From these two solutions, the particles trajectories are been predicted for the multipoint injection case.

2.3. Mesh Independency

Mesh independence study is very important in computational fluid dynamics to achieve a fair accurate solution. The size of mesh needs to be tested and specified for different mesh densities in order to have the best mesh independency within a reasonable CPU time. Mesh is refined until the solution is independent of the variation of the mesh size.

The range of sizes of the mesh consisting of fine, medium, and coarse sizes are varied in the radial direction (orthogonal to the flow) and in the axial direction (parallel to the flow). After the calculation is done, the convergence histories for all the meshes are plotted. Then, the accuracy of each mesh is analyzed. In order to identify which mesh has better accuracy, the velocity profile of the flow solution at the outlet of microvessel for each mesh is plotted. It is then been compared with the analytical velocity profile formula for fully developed flow as presented in equation (3).

3. Results and Discussion

The magnetic nanoparticles with the density of 5000 kg/m^3 are injected into the microvessel (L = 16 cm and R_v = 75μm). The cylindrical external permanent magnet with a radius of R_{mag} = 2 cm is placed at a distance of y_{mag} and $x_{mag} = 8\,cm$. The magnetic nanoparticles with different radii ranging from 150 nm to 350 nm are injected parallel with the same fluid flow velocity (x-direction). The flow is assumed to be a laminar artificial blood flow with the velocity, u_f of 0.015 m/s, density, ρ of 1056 kg/m^3 and viscosity, η of 0.0036 N s/m^2.

3.1. Mesh Independence

Figure 2 depicts 10 velocity profiles solutions from different mesh sizes. All the velocity profiles are plotted together with the analytical solution. It is observed that the higher number of nodes, which means finer mesh, has higher accuracy when compared with the analytical solution. It shows that the higher the number of nodes or elements, the smaller the error is.

However, we should also consider the time taken for CPU to solve the simulation. To that extent, we will observe at which mesh the solution will be constant and not dependent anymore on the refinement of mesh. From figure 3, the maximum velocity at the centerline of the vessel is plotted. From this figure, approximately at node number 16000, the solution started to have a constant value even though the mesh is refined more. We chose this mesh size because it is reasonable since the fully developed flow at the outlet and the error for the maximum velocity at the centerline is showing the least error when compared with the analytical data as shown in figure 4. It is also reasonable if compared to the other finest meshes as those meshes need more time to have a converged solution.

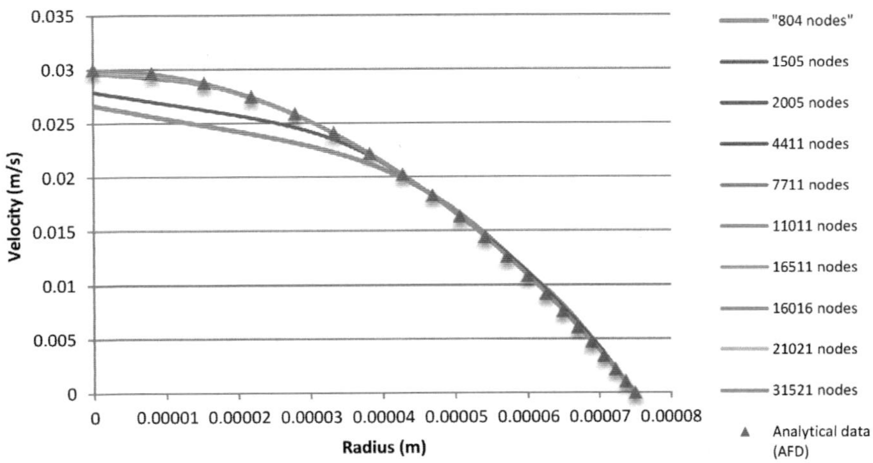

Figure 2. Velocity profiles at outlet for different mesh sizes.

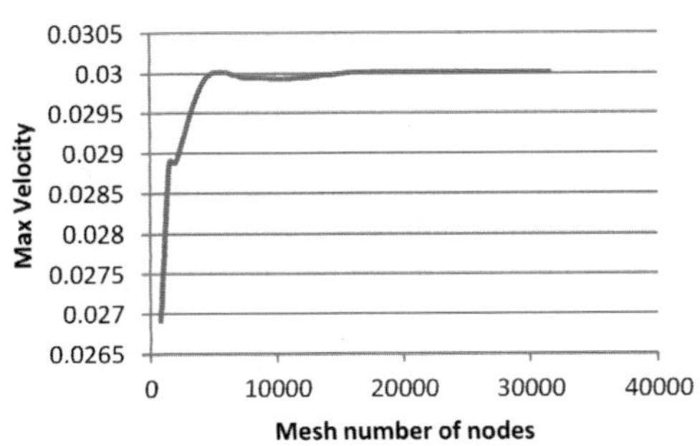

Figure 3. Mesh number of nodes against the maximum velocity at centerline.

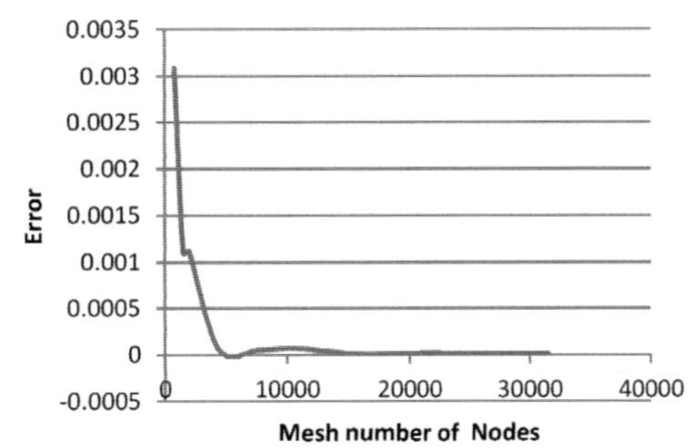

Figure 4. Mesh number of nodes against the error of maximum velocity with AFD.

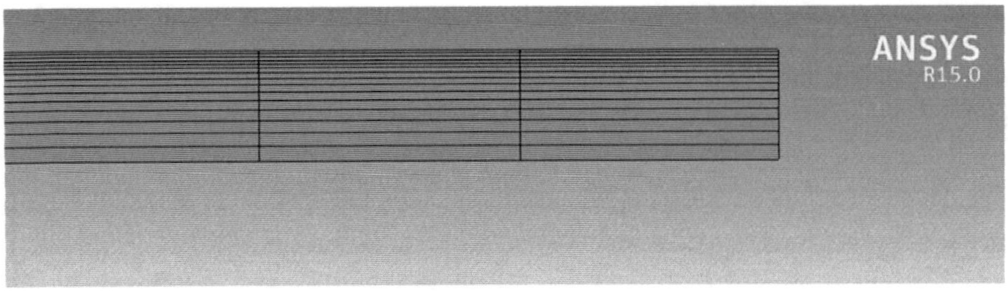

Figure 5. Structured non-uniform mesh.

Furthermore, in this project, the domain was discretized into a large number of quadrilateral structured meshes, which has been specified through mesh independence study. 15000 number of elements. Since the velocity is changing in the y-direction (orthogonal to the flow), thus the mesh resolution needs to be higher in that direction or in other words, the mesh size needs to be finer at the wall of the domain. In order to make it finer or higher resolution only at the wall of the domain, bias

IOP Conf. Series: Materials Science and Engineering **184** (2017) 012061 doi:10.1088/1757-899X/184/1/012061

factor of 6is specified. Finer meshes are not recommended to be used through the entire domain since it will take longer time for solver to run. This means that the non-uniform mesh will be applied since the refinement of the mesh only recommended on the region of interest.

3.2. Single Point Injection Case

We predicted the trajectories of five different sizes of particles which are $R_p = 150, 200, 250, 300, 350$ nm. These five particles are being injected at different locations of magnet-to-vessel distances which are $y_{mag} = 0.025, 0.03, 0.035$ m as shown in Figure 1. Each particle is being injected at the origin point ($x = 0$ and $y = 0$), which is far enough upstream to have a negligible magnetic force acting on the particle during injection.

The data of each simulated particle trajectory are being exported from ANSYS Fluent to Microsoft Excel so that they can be normalized easier. The particle trajectory needs to be normalized by since the length of the domain is extremely long if compared to its radius resulting in unclear and invisible particle trajectory. Therefore, the radial position of the particle is normalized with respect to the microvessel radius (R_v), while the axial position (x), is normalized with respect to the magnet radius (R_{mag}). After normalization, the x_{mag}location at the x-axis is at point 4 in figure (6), (7) and (8).

From the plots also, it can be shown that for y_{mag}=0.025 m as in figure (6), only the smallest particle with a radius of 150 nm is escaped and the rest four particles are being captured at the wall of the vessel. For y_{mag}=0.030 m as shown in figure (7), the particle with a radius of 150 nm and 200 nm are escaped and the rest three particles are being captured at the wall of the vessel. Lastly, for y_{mag}=0.035 m in figure (8), only the largest particle is being captured and the rest four are escaped. From these, we can conclude that the magnetic force acting on the particle is directly proportional to the volume of the particle. It has been shown that fewer particles are being captured as $y_{mag}$is increasing. Only a larger particle experienced a stronger capture force and can be captured at a higher value of y_{mag} (farther distance).

Furthermore, the minimum capture radius particle for each ofy_{mag}distance also being predicted. Fory_{mag} = 0.025 m, the minimum captured radius is 150 nm, for y_{mag} = 0.0250 m, the minimum capture radius is 250 nm, and lastly, for y_{mag} = 0.035 m, the minimum capture radius is 350 nm.

From this parametric study, it can be shown that the magnetic particle trajectories are strongly dependent on the particle size and the location of the source magnet that will give the magnetic force on the particles.

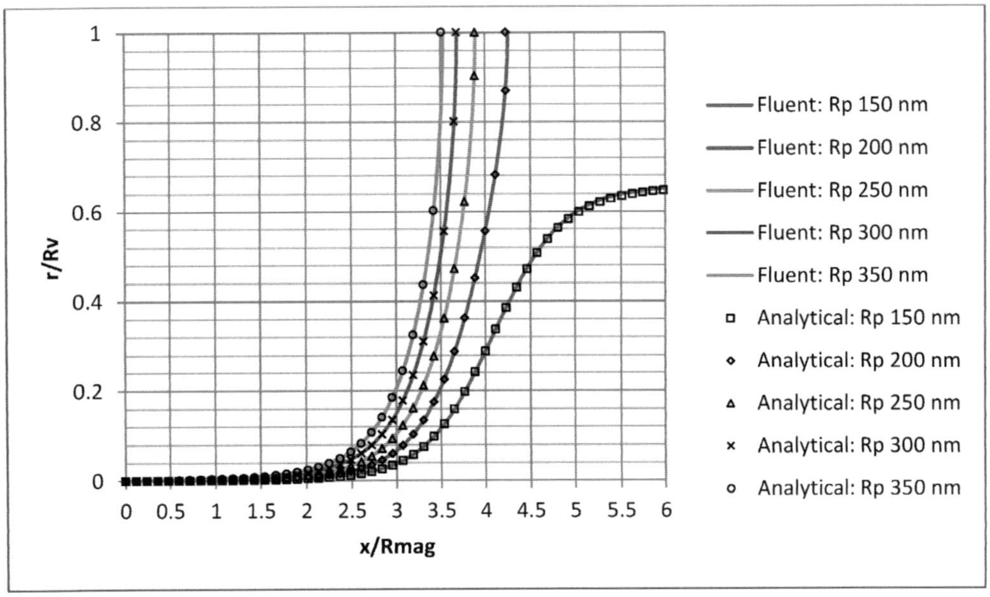

Figure 6. Trajectories for different particles radii for single injection case (y_{mag} = 0.025 m).

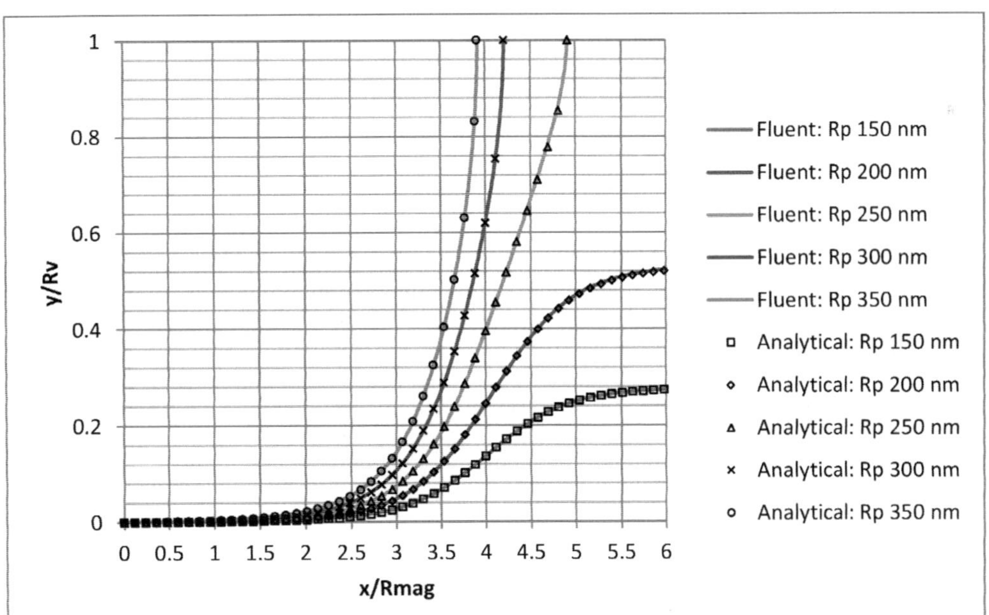

Figure 7. Trajectories for different particles radii for single injection case (y_{mag} = 0.030 m).

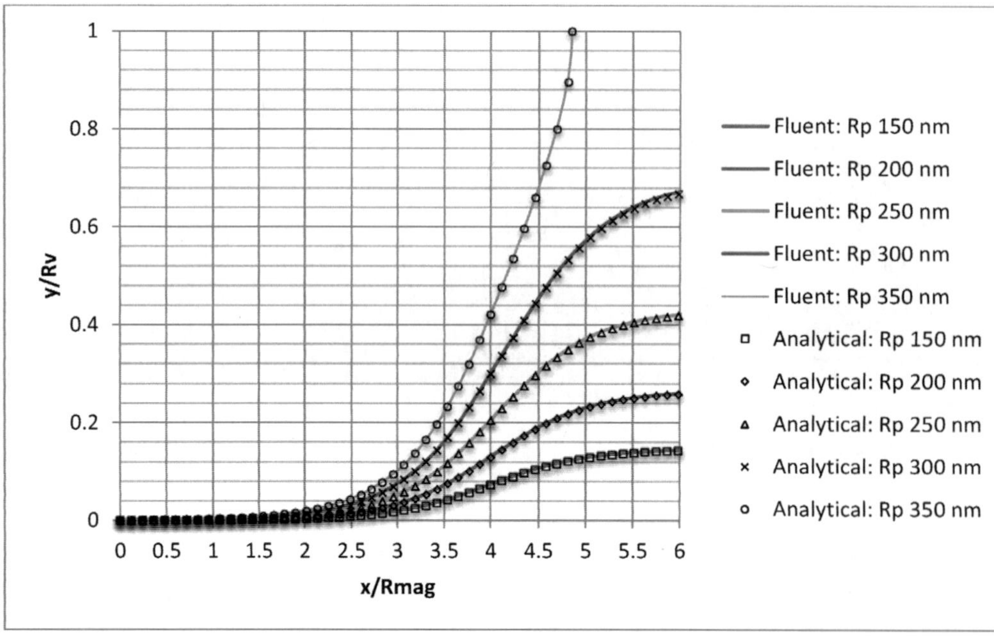

Figure 8. Trajectories for different particles radii for single injection case (y_{mag} = 0.035 m).

3.3. Multipoint Injection Case

For this case, the two components of equation of motion of the particle in the equation (14) and equation (15) are being solved to observe their trajectories fory_{mag} = 0.035 m. 19 particles for each radius of $R_p$for 150 nm, 200 nm, 250 nm, 300 nm and 350 nm are injected at the inlet. Next, the capture efficiency for each radius of the particle is calculated as shown in table 1.

From the five particles radii above, it is observed that the magnetic force is not symmetric. This is because, when the particles are injected closer to the external magnet (closer to the upper wall), it will go towards the left side position because of the higher value of magnetic force in both components of x and y there. Therefore, the particle will be captured more as it moves closer to the external magnet position. From the capture efficiency in table 1 below, it shows that the higher the particle radius, the higher the capture efficiency. It has been shown that the higher size of particle has the higher tendency to be captured by the external magnet because of the proportionality between the size of particle and magnetic force.

Furthermore, by simulating the particle trajectories using line injection method, the Region of Interest (ROI) of the particles can also be determined. As shown in figure 9, larger particle has larger and wider ROI because it has higher capture efficiency.

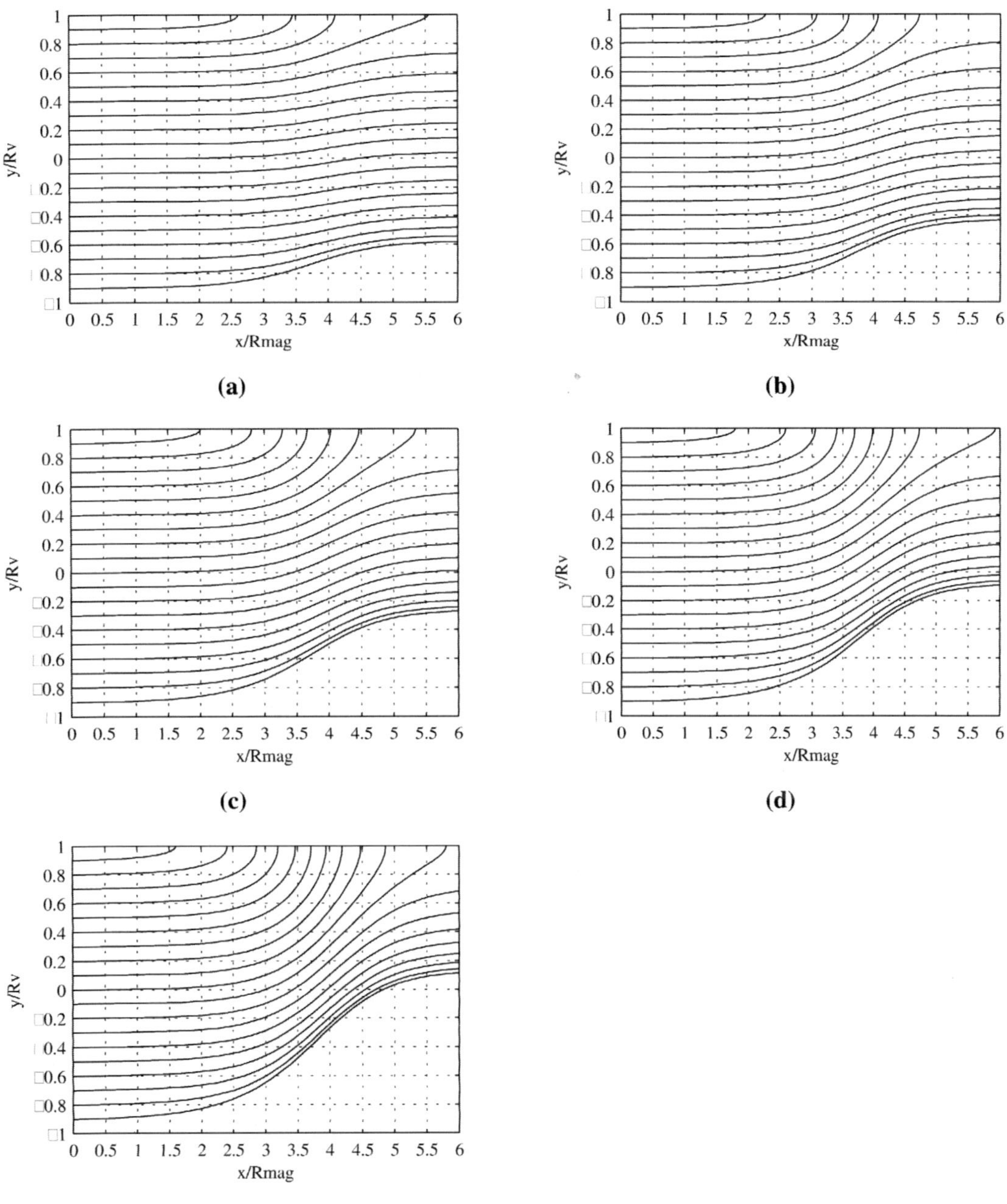

Figure 9. Trajectories for different particles radii for multipoint injection caseat (y_{mag} = 0.035m).
(a) R_p = 150 nm, (b) R_p = 200 nm, (c) R_p = 250 nm, (d) R_p = 300 nm,(e) R_p = 350 nm.

Table 1 Capture Efficiency for multipoint injection case.
(19 injected particles at inlet and $\boldsymbol{y_{mag}} = \mathbf{0.035\ m}$)

Radius (nm)	Trapped	Escaped	Capture Efficiency
150	4	15	21 %
200	5	14	27 %
250	7	12	37 %
300	9	10	48 %
350	11	8	58 %

4. Conclusion

In summary, the simulations using DPM ANSYS Fluent are successfully performed to study the effect on particle trajectories in microvessel when varying the particle radii and the external magnetic field locations. By only considering the dominant magnetic forces and fluid drag forces, it is found that the particle trajectories are strongly dependent on the particle size and the location of the external magnetic source. For single injection case, the minimum captured radius is $R_p = 150$ nm, 250 nm, and 350 nm for $y_{mag} = 0.025$ m, 0.030 m, and 0.035 m respectively. The simulation then can be used to determine the optimum particle size for magnetic drug targeting technique, given its size and location. Meanwhile, for multipoint injection, the capture efficiency is predicted for injected particles at $y_{mag} = 0.035$ and it is increasing as the size of particle increases. The capture efficiency is 21%, 27%, 37%, 48% and 58% for $R_p = 150$ nm, 200 nm, 250 nm, 300 nm, and 350 nm respectively.

It can be concluded that the concept of the minimum captured radius in single point injection is not valid for multipoint injection case. This is because there are particles will be captured even the radius of the particle is smaller than the minimum captured radius in multipoint injection case. Even for minimum captured radius, not all particles are captured since there are some particles lost because the injection points are everywhere through the inlet, not only at the centerline. However, the capture efficiency in multipoint injection will increase as the radius of the particle is close to the minimum captured radius in single injection point. At the end, this study can give some information about the size of the magnetic nanoparticles and the location of the magnetic field.

Acknowledgments
Authors gratefully thank Kulliyah of Engineering, International Islamic University Malaysia for financial support to publish this work.

References

[1] Ramchand C, Pande P, Kopcansky P, Mehta R 2001 Application of magnetic fluids in medicine and biotechnology *Indian Journal of Pure and Applied Physics* **39**(10) pp 683– 686
[2] Asmatulu R, Zalich M, Claus R, Riffle J 2005 Synthesis, characterization and targeting of biodegradable magnetic nanocomposite particles by external magnetic fields *J. Magn. Mag. Mater* **292** pp 108–119
[3] Kayal S and Ramanujan R V 2010 Anti-cancer drug loaded Iron-Gold core-shel nanoparticles (Fe@Au) for magnetic drug targeting *J. Nanosci. Nanotechnol.* **10** pp 5527-5539
[4] Mahmoudi M, Simchi A, Imani M, Milani A S and Stroeve P 2008 Optimal design and characterization of superparamagnetic iron oxide nanoparticles coated with polyvinyl alcohol for targeted delivery and imaging *Journal of Physical Chemistry* B **112** pp 14470-14481
[5] Lee J, Isobe T and Senna M 1996 Preparation of ultrafine Fe_3O_4 particles by percipitation in the presence of PVA at high PH *Journal of Colloid and Interface Science* **117** pp 490-494

[6] Behrens S, Bonemann H, Matoussevitch N, Modrow H, Kempter V, Riehemann W, Wiedenmann A, Odenbach S, Will S, Eberbeck D, Hergt R, Landfester R K, Schmidt A, Schiller D, and Hempelmann R 2009 Synthesis and characterization *Colloidal Magnetic fluids:Basics, Development and application of Ferrofluids*, ed S Odenbach (Berlin: Springer-Verlag) chapter 1 pp 1- 69

[7] Furlani EJ, Furlani E P 2007 A model for predicting magnetic targeting of multifunctional particles in the microvascular *J. Magn. Mag. Mater.* **312** pp 187–193

[8] Aviles M O, Ebner A D and Ritter J A2008 *In vitro* study of magnetic particle seeding for implant-assisted-magnetic drug targeting*J. Magn. Magn. Mater.* **320**(21) pp 2640-2646

[9] Ritter J A, Ebner A D, Daniel K D, and Stewart K L 2004 Application of high gradient magnetic separation principles to magnetic drug targeting*J. Magn. Magn. Mater.* **280**(2) pp 184-201

[10] Rotariu O and Strachan N J C 2005 Modelling magnetic carrier particle targeting in the tumor microvasculature for cancer treatment *J.Magn. Magn. Mater*. **293** pp 639-646

[11] ANSYS® Academic Research, Release 15.0, Help System, Fluent Theory Guide, ANSYS, Inc.

[12] ANSYS® Academic Research, Release 15.0, Help System, Fluent User's Guide, ANSYS, Inc.

[13] Batchelor G K 1970 An Introduction in Fluid Dynamics (London:Cambridge University Press), p 233

Investigation of aluminum heat sink design with thermoelectric generator

Mohiuddin A K M[1,a], Muhammad Yazid Ameer[1], Ataur Rahman[1], Ahsan Ali Khan[2]

[1]Department of Mechanical Engineering, [2]Department of Manufacturing & Materials Engineering,
Kulliyyah of Engineering, International Islamic University, Malaysia
P.O.Box 10, 50728 Kuala Lumpur, Malaysia

E-mail: [a]mohiuddin@iium.edu.my

Abstract. This paper presents an investigation of aluminium heat sink designs with thermoelectric generator. Basically, for thermoelectric generator (Peltier module), the thermal conversion uses Peltier effect. Two heat sinks with different design, with thermoelectric module of Bismuth Telluride, Bi_2Te_3 were used in this investigation. The simulation and experimental studies were conducted with two different heat sinks attached with thermoelectric generator (TEG). System modelling was used to collect data and to predict the behaviour and performance of thermoelectric modules. Experiment was conducted in exhaust system at muffler section since the temperature at muffler section meets the requirement of thermoelectric generator.The result of the experiments shows that rectangular fin heat sink is more efficient in heat transfer compared to circular tube fin heat sink due to its geometry and properties.

1. Introduction

It has been quite some time that the researchers have developed means to generate power and generate voltage through temperature difference from thermal alternative energy. Recent development leads to generate energy by using thermoelectric generator. Basically, when there is a temperature difference at the module, energy will be converted to electricity. It is similar as the concept of pipeline system where potential energy is converted to kinetic energy. Thermoelectric generator produces power output due to its movement of electron in the semi-conductor material. When there is hot and cold side temperature, the electron will move from the hot side to the cold side, thus electricity will be produced.

There are many ways to get the renewable energy, such as wave energy, solar energy, wind energy, nuclear energy and many more. Therefore, in a creation and innovation, it must have boundaries and limitations. The main problem in dealing with the thermoelectric generator is the cold side also become hotter as the temperature at the hot side increases. So, the power output will be small as the temperature different small. In order to prevent this problem, the thermoelectric generator is attached with heat sink as a medium of heat dissipation. Recent research shows that heat sink plays the main role for optimizing the performance of thermoelectric generator as a coolant mechanism.

The main objective of this work is to determine the relationship between power output of the thermoelectric generator with the difference in temperature and the efficiency in heat transfer of various designs of heat sink used for test with thermoelectric generator. Heat sink is a cooling mechanism that has been used to cool the cold side of the TEG.

Zhang, Chau and Chan (2008) mentioned that, "Based on the Seebeck effect, the Thermoelectric Generator system has no part that is moving. It works in silent operation" [1]. According to Martins, Brito, Goncalves and Antunes (2011), the Seebeck effect is also used for the thermocouple operation. The electric voltage and electric current are produced by temperature difference between the junctions of two different materials when the electric circuit is closed [2]. The Seebeck effect is due to electromotive force generated in the contact of the closed circuit consisting of dissimilar conductors, when the temperature difference exists (Khripach, Papkin and Korotkov, 2014) [3].

Stobart and Weerasinghe (2006) carried out research on thermoelectric regeneration in vehicles. They had found that the output of the TEG has same potential as the alternator for a small vehicle [4]. Stobart et al. (2010) stated about the possibility of saving fuel by using thermoelectric devices for vehicles. These show that hybrid engine with the thermoelectric generator will increase the performance and also can reduce pollution [5].

2. Thermoelectric working principle

According to Shameer and Christopher (2013), thermoelectricity implies the immediate convert from heat into electricity, or the other way around. As indicated by Joule's law, a conductor conveying a current creates heat at a rate relative to the result of the resistance (R) of the conductor and the square of the current (I). A circuit of this type is called a thermocouple; a number of thermocouples connected in series are called a thermopile [6].

Thermoelectric devices are attached to the exhaust pipe to convert waste heat to electrical energy. This thermoelectric generator is placed at exhaust as shown in Fig.1. It is easy to implement and will cause less influence on the operation of engine. The thermoelectric device has two different semiconductors that are subjected to a heat source and heat sink. A voltage is created between the two conductors. The heat from cold side of thermoelectric devices is removed by liquid-cooling cold plate. Thermoelectric generator have several advantages such as, it is free of maintenance, silent in operation, system is independent on weather and has no moving and complex mechanical parts [6].

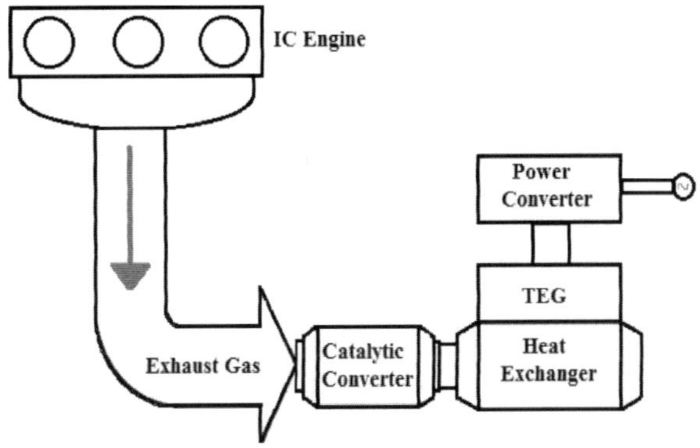

Figure 1. Position of the thermoelectric generator

2.1. Thermoelectric material
Sharan and Felix (2014) explained that by using Bismuth telluride, thermoelectric will act as an efficient material that is good for power generation [7]. Bismuth telluride is a compound of tellurium and bismuth. It will become semiconductor when it is alloyed with antimony or selenium. Therefore, it will produce electricity through the direct conversion of heat when it is joined with other by metal conductor, and placed between two thermal conductive surfaces. According to Khripach, Papkin and Korotkov (2014), "However, a thermoelectric material intended for use in vehicles must satisfy additional requirements, such as environmental safety and accessibility, and low cost. Currently, few thermoelectric materials that meet all these requirements are in mass production. One such example is bismuth telluride (III) Bi_2Te_3 [3]." They claimed that bismuth telluride satisfies all the requirements needed as a part of thermoelectric generator material.

2.2. Basic idea of heat sink
Heat sinks (Fig. 2) are devices that improve heat dissipation from a hot side, usually the case of a heat generating component, to a cooler ambient, usually air. For the following discussions, air is assumed to be the cooling fluid.

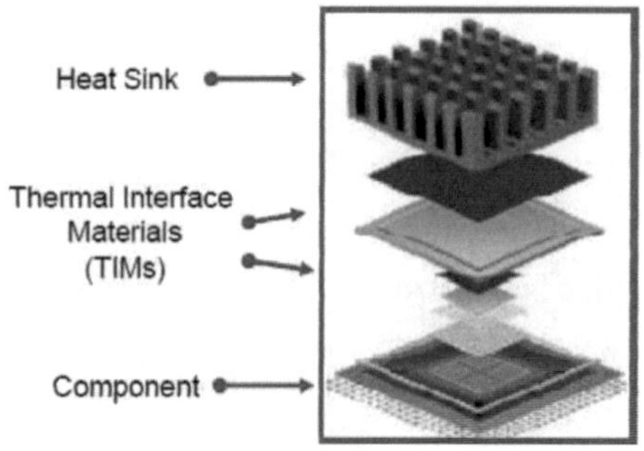

Figure 2. Heat sink

In order to have more prominent surface area for heat transfer, the heat sink is firmly packed (Fig. 3). However, it will give small coefficient of heat transfer due to the resistance. The extra fins restrict fluid to flow through the inter-fin passages. Consequently, heat sinks with sufficient space between fins will have a higher heat transfer coefficient; however they will have small surface area.

Figure 3. Designs of heat sink, flat plate, rectangular fin and circular tube fin.

To get a better surface area for heat transfer, the fins of heat sink must be closely packed. But it will have a low coefficient of heat transfer. This is a direct result of extra resistance of the supplementary fins that influences a fluid moving through the fins.

3. Modelling and experiment

3.1. Mathematical modelling of heat sink

The design is based on orientation, size and number of Peltier module.
Size of Peltier module = $40 \times 40 \times 3\ mm$
Length (Peltier module), $L = w_{mod} \times N_{modseries} = 40 \times 2 = 80\ mm$
Width (Peltier module), $W = w_{mod} \times N_{modparallel} = 40 \times 1 = 40\ mm$
Surface area (Peltier module), $A_{mod} = L \times W = 80 \times 40 = 3200\ mm^2$
Ratio of the length (Peltier module), $\beta_{lw} = \frac{L}{W} = \frac{80}{40} = 2$
Surface area of zone, $A_{zone} = \beta_{lw} \times A_{mod} = 2 \times 3200 = 6400\ mm^2$
Insulation area, $A_{ins} = A_{zone} - A_{mod} = 6400 - 3200 = 3200\ sq.mm$
Heat sink's length = $\sqrt{\beta lw \times A_{zone}} = \sqrt{2 \times 6400} \approx 100\ mm$
Final dimension of heat sink = $100\ mm \times 100\ mm$.

3.2. Experiment of heat sink without and with thermoelectric generator

The experiment was carried out to determine the best geometry of the heat sink. In the experiment the temperature at which heat dissipates from the heat sink was measured. The apparatus and device like thermocouples are connected to G.U.N.T. WL350 TEST UNIT. The heater is connected with the heat sink, and then placed it to the heater position at G.U.N.T. WL350 TEST UNIT (Fig. 4).

All results obtained were recorded. The experiments were carried out for both rectangular fin heat sink and circular tube fin heat sink.

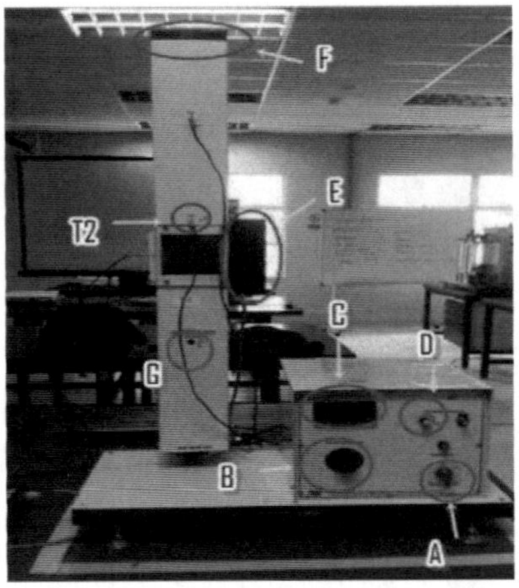

Figure 4. The Experiment setup

D = Speed of fan controller
E = Heat sink
F = position of fan
G = Anemometer port
T2 = Thermocouples

After analyzing the result, the best heat sink was chosen to proceed with the next experimentation with the Peltier modules.

Thermoelectric module TEC-12706 was used as shown in Figure 5. The dimension of the module is 40 mm × 40 mm ×3.5 mm.

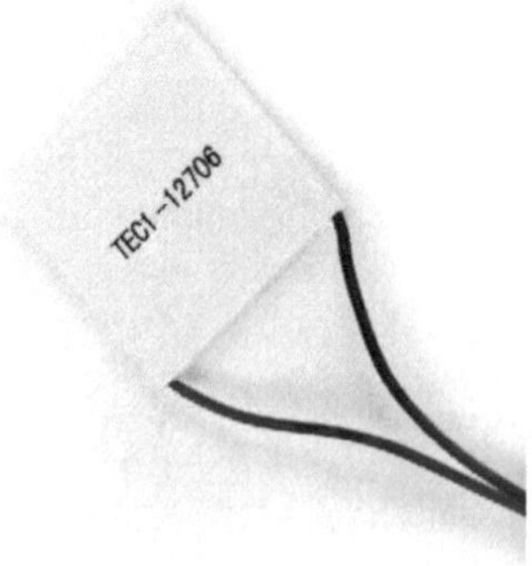

Figure 5. TEG Modules

In this experiment, four Peltier modules were used. All the modules were connected in series. The contact surfaces of the modules are enchanted by using thermal grease for efficient heat transfer.

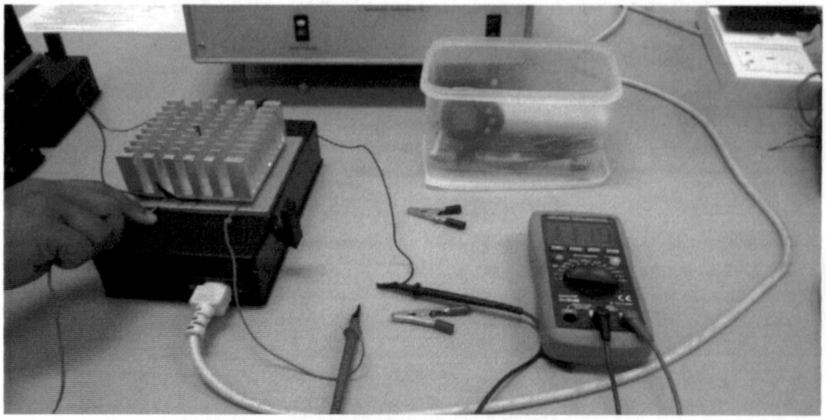

Figure 6. Experiment of heat sink with TEG

Figure 6 shows the heat sink attached with the TEG modules to conduct experiment. The hot section is maintained with a constant high temperature through heating and the cold section is supplied with the surrounding air at environmental temperature.

3.3. Application of thermoelectric generator (TEG) at exhaust system

The experiment continues with the application of the TEG at exhaust system. Surrounding air is used in cooling mechanism and heat supply is from the exhaust system. With the running engine data were recorded according to varying rpm. The temperature was recorded by using thermocouples and the current and voltage were measured by using a multimeter.

A 4G15 engine was used to complete this experiment. The waste heat from the exhaust was the heat source for the TEG. Basically, the heat source will transfer heat from the exhaust to aluminium plate and then to the TEG hot side.

4. Simulation and experimental results

The outcome of this project is based on the experiments which are the results of heat sink simulation, experiment of aluminum heat sink, experiment of Peltier modules with the heat sink and application of TEG at exhaust system.

4.1. Heat sink simulation

The thermal analysis of the heat sink simulation was carried out by using ANSYS software (Figs. 7 and 8)

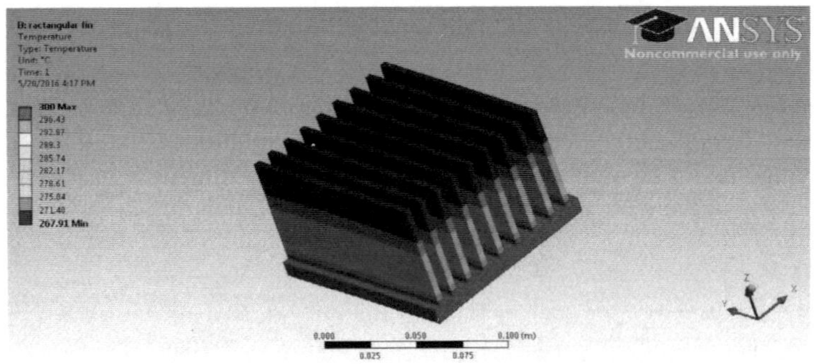

Figure 7. Heat sink (rectangular fin type)

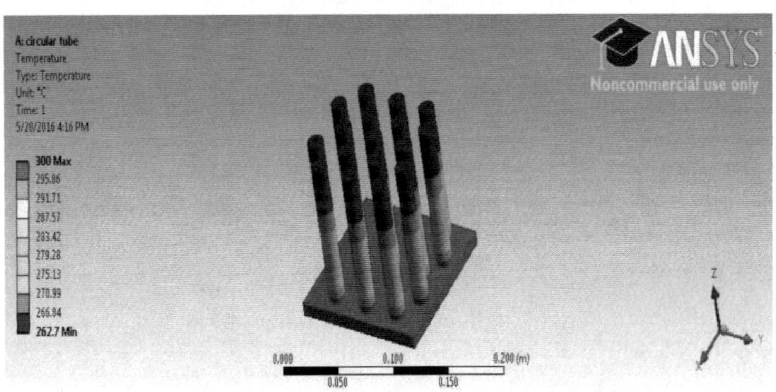

Figure 8. Heat sink (circular tube fin type)

Based on figures 7 and 8, the thermal analysis was carried out by using ANSYS software. From the figures it can be observed that the hot section is at the base plate and flow out through the fins. The top section of the fins has low temperature compared to lower section of the fins.

Table 1. Heat flux and temperature of heat sink (rectangular fin type)

	Results	
Minimum	484.23 W/m²	267.91 °C
Maximum	6.9874e+005 W/m²	300. °C

Table 2. Heat flux and temperature of heat sink (circular tube fin type)

	Results	
Minimum	3551.3 W/m²	262.7. °C
Maximum	2.1351e+005 W/m²	300. °C

From the results in Table 1 and Table 2, it can be seen that the heat sink with rectangular type of fin has higher rate of heat energy transferred compared to circular tube type of fin heat sink. The execution of rectangular fin heat sink is better compared to circular tube heat sink. Heat sink with rectangular fin design enhances the rate of heat transfer.

4.2. Experiment of aluminum heat sink (Rectangular and circular tube fin)

In this experiment, two types of heat sink were compared in terms of their temperature after dissipation of heat. This is to find out which one is more effective in heat transfer.

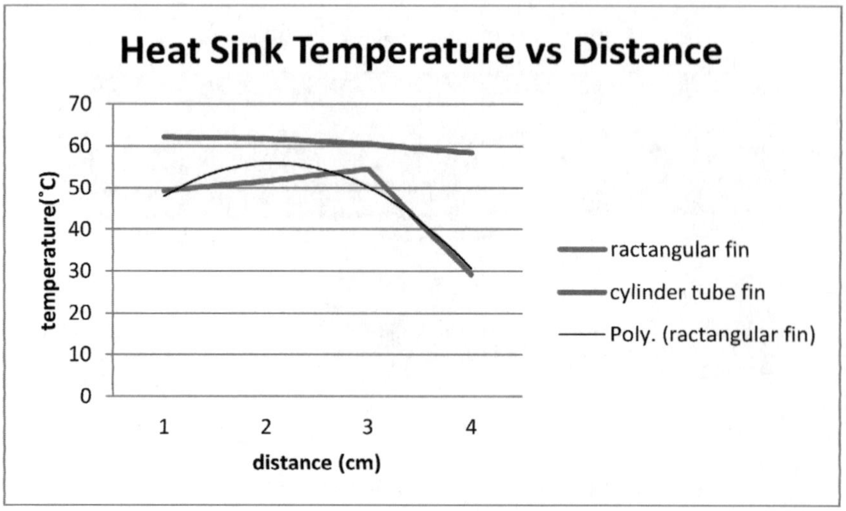

Figure 9. Heat sinks temperature

From the graph in Figure 9, it can be seen that rectangular fin heat sink has lower temperature compared to circular tube heat sink. Therefore, rectangular fin type has higher temperature difference compared to circular tube type. The differences in temperature near the heat exchanger (T2) with the temperature farthest away from the surface (T5) for free convection in this experiment is quite large which is 20.2 °C for rectangular fin compared to cylinder tube fin which is only 3.7 °C. This shows that rectangular fin type of heat exchanger transfers heat efficiently.

4.3. Experiment of Peltier module with the heat sink

The experiment results of performance for thermoelectric generator (TEG) are presented in Table 3.

Table 3. Experiment result

T_{hot} (˚C)	T_{cold}(˚C)	T_{hot}-T_{cold}(˚C)	Q (Watt)
67.6	60	7.6	0.06
68	50	18	0.221
76	55	21	0.312
80	61	19	0.253
85	69	16	0.194
94.6	71	23.6	0.323
98	70	28	0.401
105.8	78	27.8	0.372
115.7	80	35.7	0.623
118.9	87	31.9	0.48

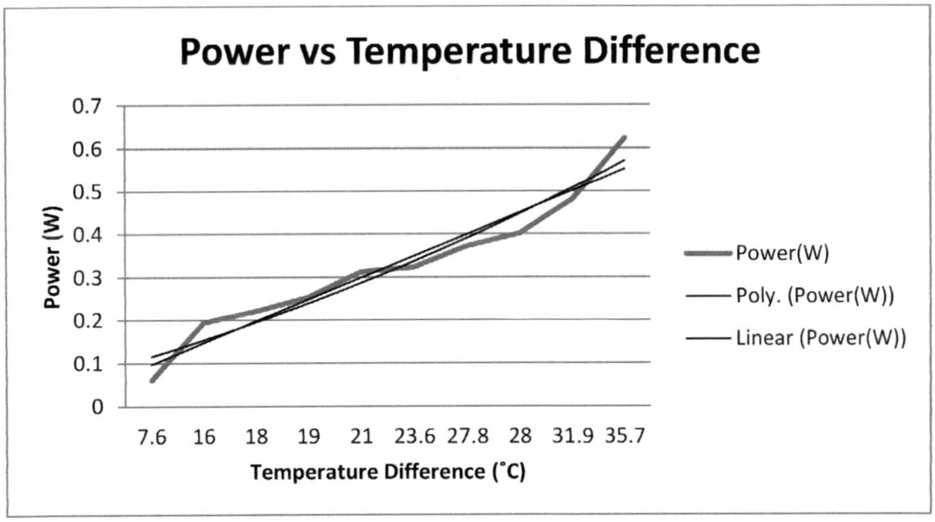

Figure 10. Power generated vs. Temperature Different

Figure 10 shows the power produced by TEG. It is clear from the figure that the power is steadily increased from 0-35.7°C temperature difference. When the temperature difference reached 35.7°C, the TEG produces 0.623 W of power.

4.4. Application of TEG at exhaust system

Table 4 shows the data collected at different rpm from the TEG attached with the heat sink at the exhaust system.

IOP Conf. Series: Materials Science and Engineering **184** (2017) 012062 doi:10.1088/1757-899X/184/1/012062

Table 4. Result (Application of TEG at Exhaust System)

RPM	Voltage (V)	Current (A)	ΔT (°C)	Q (Watt)
1000	1.8	0.12	17	0.216
1500	2.9	0.18	32	0.522
2000	3.95	0.23	38	0.9085
2500	4.8	0.3	54	1.44
3000	5.1	0.38	63	1.938
3500	5.5	0.4	65	2.2
4000	5.8	0.43	67	2.494

Figure 11 illustrates the relationship between the temperature differences with the engine rotation (RPM). At 4000 rpm, the temperature difference is 67°C. It can be seen that the temperature difference increases with the increase in rpm. So the power output of the TEG also increases.

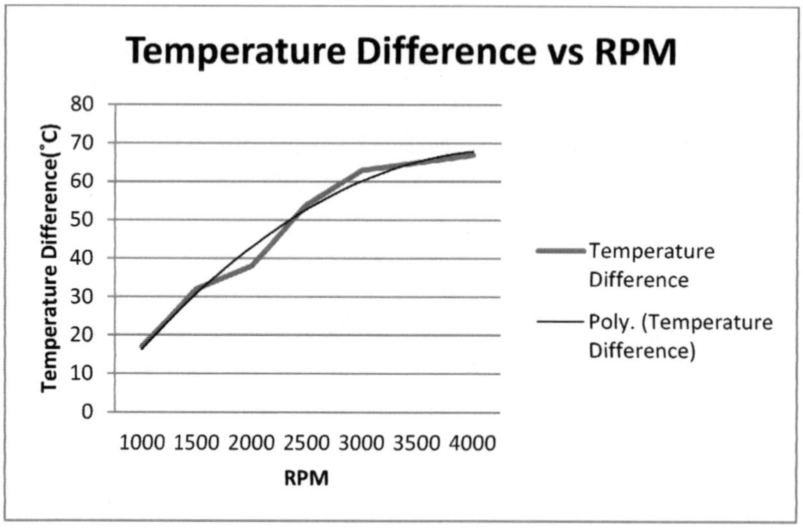

Figure 11. Temperature Difference vs. RPM

Figures 12 and 13 show that the maximum voltage obtained is 5.8V and maximum current produced is 0.43A respectively. This is due to the electrons of the material of semiconductor which will excite through the band gap when the temperature is sufficiently high enough. This will enhance the voltage and current.

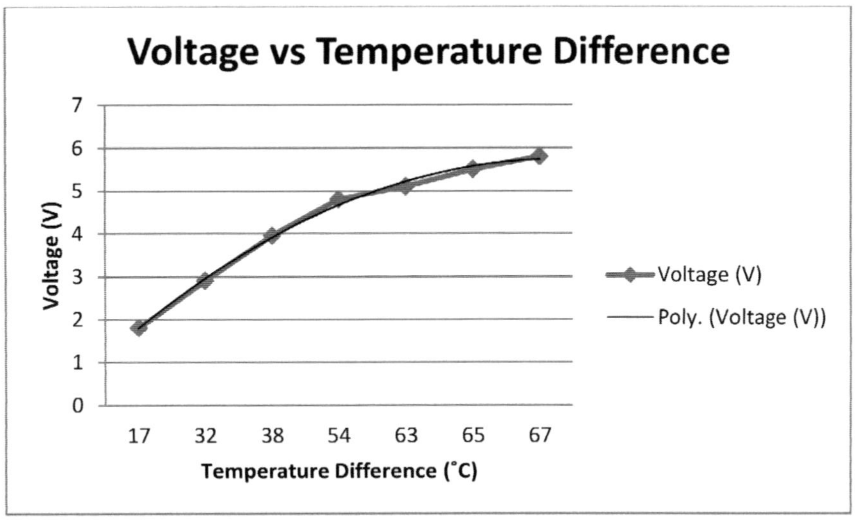

Figure 12. Voltage vs. Temperature Difference

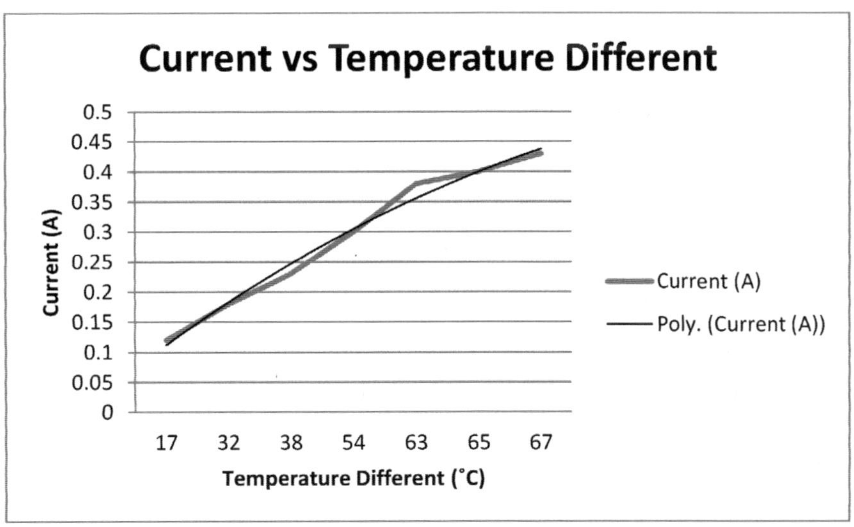

Figure 13. Current vs. temperature difference

Figure 14 shows the power produced by this TEG on the exhaust system. It can be seen that the same trend prevails as the other experiment. As the temperature difference increases, the power also increases. The maximum power output achieved is 2.494W at 4000 rpm.

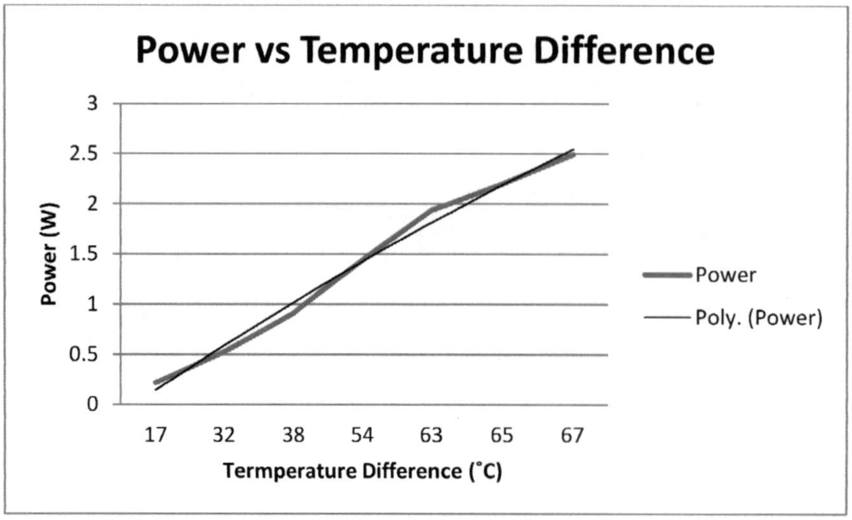

Figure 14. Power generated vs. Temperature Different

5. Conclusion

The following conclusion can be drawn from the present study:

1. The relationship between the power output of TEG and temperature different has been determined.
2. As the temperature difference increases, the power output also increases.
3. The result of the experiments shows that rectangular fin heat sink is more efficient in heat transfer compared to circular tube fin heat sink due to its geometry and properties.
4. For the application of thermoelectric generator in the exhaust system, the selection of material must be suitable as it will affect the performance of the thermoelectric generator.
5. The contact of surfaces of the module, the heat sink and the exhaust must be ensured. Smooth complete contact to each other is essential for boosting its performance.

Acknowledgements

The authors would like to acknowledge the RIGS of International Islamic University Malaysia for the support to carry out this work.

References

[1] Xiaodong Zhang, Chau K T, and Chan C C 2008 *Journal of Asian Electric Vehicle,* Overview of Thermoelectric Generation for Hybrid Vehicles, **6**(2).

[2] Jorge Martins, Francisco P B, Goncalves L M, Joaquim Antunes 2011 Thermoelectric Exhaust Energy Recovery with Temperature Control through Heat Pipes, *MIT-Pt/EDAM-SMS/0030/2008*.

[3] Nikolay Khripach, Boris Papkin, Viktor Korotkov 2014 *Life Science Journal*, Thermoelectric Generators of motor vehicle powertrains, problems and prospects, 11(12) p.505.

[4] Stobart R K, Weerasinghe R 2006 Heat recovery and bottoming cycles for SI and CI engines—a perspective, SAE Paper no. 2006-01-0662, Presented at *SAE 2006 World Congress & Exhibition*, April 2006, Detroit, MI, USA, Session : Advanced Hybrid Vehicle Powertrains(Part 2 of 5).

[5] Stobart R K, Wijewardane A, Allen C 2010 The potential for thermo-electric devices in passenger vehicle applications, SAE Paper no.2010-01-0833, Presented at *SAE 2010 World Congress & Exhibition*, April 2010, Detroit, MI, USA, Session: Advanced Hybrid Vehicle Powertrains (Part 2 of 3).

[6] P Mohamed Shameer and Christopher D 2013 Design of Exhaust Heat Recovery Power Generation System Using Thermo-Electric Generator, *International Journal of Science and Research (IJSR)* ISSN (Online): 2319-7064 Index Copernicus Value (2013): 6.14 | Impact Factor (2013): 4.438, pp. 2.

[7] S Karthik Sharan and D Giftson Felix 2014 *IOSR Journal of Mechanical and Civil Engineering (IOSR-JMCE)*, Thermoelectric Materials in Hybrid Cars, e-ISSN: 2278-1684,p-ISSN: 2320-334X, Volume **11**, Issue 5 Ver. VI (Sep-Oct. 2014), pp.73-76.

Intelligent Control Electromagnetic Actuated Continuously Variable Transmission System for Passenger Car

Ataur Rahman*, Sazzad Sharif, Mohiuddin AKM, Ahmed Faris Ismail, Sany Ihsan Izan
Department of Mechanical Engineering, Faculty of Engineering,
International Islamic University Malaysia,
50728 KL, Malaysia
*Corresponding author: arat@iium.edu.my

Abstract. Continuously variable transmission (CVT) system transmits the engine /battery power to the car driving wheel smoothly and efficiently. Cars with CVT produces some noise and slow acceleration to meet the car power demand on initial start-ups and slow speed. The car noise is produced as a result of CVT adjustment the engine speed with the hydraulic pressure. The current CVT problems incurred due to the slow response of hydraulic pressure and CVT fluid viscosity due to the development of heat.The aim of this study is to develop electromagnetic actuated CVT (EMA-CVT) with intelligent switching controlling system (ICS). The experimental results of ¼ scale EMA shows that it make the acceleration time of the car in 3.5-5 sec which is 40% less than the hydraulic CVT in the market. The EMA develops the electromagnetic force in the ranged of 350 –1200 N for the supply current in the range of 10-15 amp. This study introduced fuzzy intelligent system (FIS) to predict the EMA system dynamic behaviour in order to identify the current control for the EMA actuation during operation of the CVT. It is expecting that the up scale EMA-CVT would reduce the 75% of vehicle power transmission loss by accelerating vehicle in 5 sec and save the IC engine power consumption about 20% which will makes the vehicle energy efficient (EEV) and reduction of green house gas reduction.

1. Introduction

Acceleration stays in a "sweet spot" to minimise wasted power, thus improving a vehicle's fuel efficiency. In the most advanced form, the CVT and its control forms an actuation system for the transmission which determines the best operating speed and torque of the engine given a set of driver inputs and car operating conditions. The CVT transmission has 15% less fuel consumption than that of a conventional gearbox along with a reduction in harmful emissions of about 30%. Comparisn has been made with the simulation study based on different CVT concepts with a manual transmission for gasoline and diesel engines and show up to 10% less fuel consumption for gasoline engines and up to 19% for diesel engines [1, 2]. The author [3] has conducted a study on the performance for a car of 3000cc by simulation on the manual transmission, automatic transmission and the continuously variable transmission and he reported that time taken to accelerate to 100 km/h is 10.20 s for manual transmission, 10.76 s for automatic transmission, 7.85 s for CVT. CVT is preferred over other automatic transmission because of the comfortability, reliability, durability as well as the efficiency [4]. The power flow through the power train does not need to be interrupted during acceleration as in conventional transmissions and this makes it possible to gain a smooth, rapid and steeples response to

IOP Conf. Series: Materials Science and Engineering **184** (2017) 012063 doi:10.1088/1757-899X/184/1/012063

drivers' demand without disturbing jerk. Technically, oil hydraulic system is preferred over the pneumatic due to leakage and compressible fluid when high accuracy and great amount of pressure are needed [5]. Hydraulic control high gear and low gear mechanism for CVT is quite effective to develop the sufficient pressure but the problem is to hold the movable sheave of the pulley as it is desired. Though spring can be used, owing to spring oscillation there will be slip, which results torque losses. High pressures and flow rates are required in the hydraulic control system to maintain the high torque at start-ups and slow speed. However, fluid pumping losses are the major causes of torque loss in modern CVT [6] . The half toroidal CVT could not replace replace the belt-push drive CVT because of its bulk deformation of contact component resulting imperfect contact angle. This problem was exacerbated by the comparative delay controller strategies which caused the engine speed to go to its maximum speed range following a full pedal input and it was only after a delay that the car speed increased as well [7].

Clamping force on the pulley produces a normal force, which results in traction force between a V-belt's element and the pulley (Tarutani et al. 2005, Nishizawa et al., 2005). Insufficient clamping force causes an excessive slipping of the V-belt and causes the engine to develop more power for the hydraulic pressure of the CVT to maintain the clamping force [8,9,10]. Therefore, to address these issues, some researchers have explored a novel idea of replacing the application of the existing electro-hydro-mechanical (EHM) system for the clamping force with an electro-mechanical (EM) system in a pulley-based CVT (EM CVT). The author [11] has studied on the disk spring based EM-CVT and reported that the gear can be achieved 3.0 for reduction and 0.6 for overdrive. The researcher have not reported about the actuation time for the movable sheave towards the fixed sheave. The actuation time is the core objective for this study. It could be mentioned that if the actuation time is more the vehicle will be in no traction force, which is the main reason for the customers of CVT to feel 'No-Traction" on hilly track.

The market for belt type CVT systems is growing rapidly. Since belt production began at Van Doorne's Transmissie (VDT) in 1985, more than 5 million vehicles have been equipped with a push-belt CVT. Today, approximately 1 million push-belt CVTs are manufactured annually for the Japanese, USA, European and Korean markets. More than 45 different vehicle models are currently available with push-belt CVT systems. The push-belt performance is illustrated by the CVT system installed in the Nissan Murano, which has a ratio coverage of 5.4 and operates with a 3.5 liter V6 180 kW/350 Nm engine and a torque convertor, applying drive side torque levels on the belt over 500 Nm. Further extension of the CVT application range can be achieved, among other things, by continuously increasing the push-belt fatigue strength [2].

The current state-of-the-art of push-belt performance can be illustrated by using the EMA-CVT, which could have fast actuation to accelerate the vehicle in 5-7 sec without incurring fatigue strength and transfer the power to the tires in any road conditions with targeted CVT ratio and without making noise and jarking, good drivability, and better fuel economy.

The objective of this paper is to present a intelligent control electromagnetic actuated CVT system for the vehicle to make the vehicle energy efficient by reducing the acceleration time, ensurining optimal traction in hilly track, reducing trasmission losses and noises.

2. Methodology

The EMA-CVT is developed with a small battery pack of 12V with 2 sub module of lithium ion battery (each module is built with 4 cells 2 in series and 2 in parallel; nominal cell voltage 3.5 V and capacity 43 Ah), 1.6 mm diameter copper coil of 900 m length and a high carbon steel plunger. The alternator always keeps battery full charging mode to power the EMA as required. This EMA-CVT system's development is conducted with incorporating the car traction torque dynamic analysis. The EMA design has been performed in such a way that the drive mechanism of the actuator be able to develop the electromagnetic force to actuate the CVT for the maximum gear ratio at start-ups or slow speed (reduction) and lowest gear ratio for maximum speed (overdrive or cruising speed).

2.1 Kinematics of EMA-CVT

Kinematics of the EMA-CVT is developed based on the traction torque of the car and the clamping force of the CVT. The electromagnetic force analysis of the EMA has been made based on the clamping force of the pulley for the different number of windings and the length of the EMAs. The electromagnetic force would be greater than and equals to the clamping force of the pulley to develop the highest gear ratio for start-ups or low speeds and the lowest gear ratio for the highest speed. The mathematical model has been developed for estimating the driving torque of the primary pulley based on the engine of the full-scale car. The transmission dynamics of the EMA-CVT could be modeled as:

$$T_{inw} = [T_e - J_t \alpha_e] GR \tag{1}$$

with $\alpha_e = \frac{1}{J_e}[T_e - T_{in}]$

where, T_{inw} is the torque applied to the wheels by the EMA-CVT, J_t is the inertia of the transmission unit and GR is the gear ratio, α_e engine angular acceleration, T_e is the torque generated by the engine and T_{in} is the torque applied by the engine to the CVT and J_e is the inertia of the engine. For electromagnetic actuated CVT system, CVT speed ratio as well as instantaneous torque is changed with the axial (back and forth) movement of pulley movable sheave through attraction and repulsion of solenoids. The instantaneous torque incremental for the primary pulley of the EMA-CVT is calculated by using the equation:

$$\Delta T_{ins} = \frac{T_{out(\max)} - GR_{ins} T_{in(\min)}}{GR_{ins} + 1} \tag{2}$$

with $GR_{ins} = 1 + \frac{\tan\theta(L_s - 2ds)}{r + \tan\theta ds}$

where, $T_{out(max)}$ is the maximum output torque in the secondary pulley and $T_{in(min)}$ is the minimum input torque in the primary pulley from the engine and GR_{ins} represents the instantaneous gear ratio, L_s is the stroke length in m, and ds is the instantaneous displacement of the movable sheave in m, r is the radius of pulley in m and θ is the pulley angle in deg. The traction torque for initial condition of the car is computed [12]:

$$T_{initial} = T_{t(\theta_g)} = m_c g \left[\mu_R \frac{(l_f - f_r h)/L_w}{1 + \mu_R h / L_w} \right] (r_{wheel}) \tag{3}$$

where, μ_R is the adhesion coefficient of the road, m_c is the mass of car in kg, f_r is the rolling motion resistance coefficient, h is the height of the centre of gravity in m, L_w is the wheel base in m, l_f is distance of the CG from the front wheel in m, θ_R is the slope angle of the road in deg, g is the acceleration due to gravity in m/s^2, r_{wheel} is for radius of the drive wheel in m. The power transferred from the engine to the primary pulley and from the primary to secondary pulley can be computed as:

$$P_{in} = \left[\frac{\Delta v_p + v}{R_p}\right](T_{in}) \; and \; P_{out} = \left[\frac{v - \Delta v_s}{R_s}\right](T_{out}) \tag{4}$$

where, P_{in} is the power generated by the engine or input power of the primary pulley in Watt and P_{out} is the output power of the primary pulley in Watt and Δv is slip velocity in m/s. The slip velocity for the primary pulley and the secondary pulley can be calculate as, $\Delta v_p = \omega_p R_p - v$ and $\Delta v_s = v - \omega_s R_s$, where $\omega_p R_p$ and $\omega_s R_s$ is the speed of the primary and secondary pulley, v is the belt speed, Δv_p and Δv_s are the slip velocity respectively.

2.2 Clamping Force

The EMA gets advantage to pull the movable sheave whereas the difficulties for pushing the sheave against the rotating belt which is considered as the clamping force (Figure 1). Two sets of solenoids with common plunger are used in this study to overcome the clamping force with developing the electromagnetic force. In each sets of solenoids one develops the pulling force and other one develops pushing force.

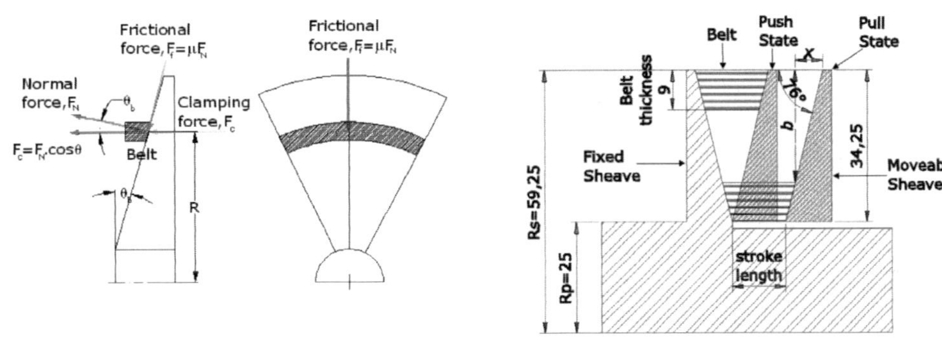

Figure 1. Geometry of pulley [13]

The clamping force for the primary and secondary pulley of the EMA-CVT system can be figured out form the modified formula [14]:

$$F_p = \frac{T_{in}\cos(90-\theta_b)}{2\mu_p R_p} \tag{5}$$

$$F_s = \frac{T_{out}\cos(90-\theta_b)}{2\mu_s R_s} \tag{6}$$

where, T_{in} and T_{out} is transmission torque of primary and secondary pulley in Nm, θ_b is the belt angle. μ_p and μ_s belt frictional coefficient of primary and secondary pulley. R_p and R_s are radiuses in m of primary and secondary pulley respectively.

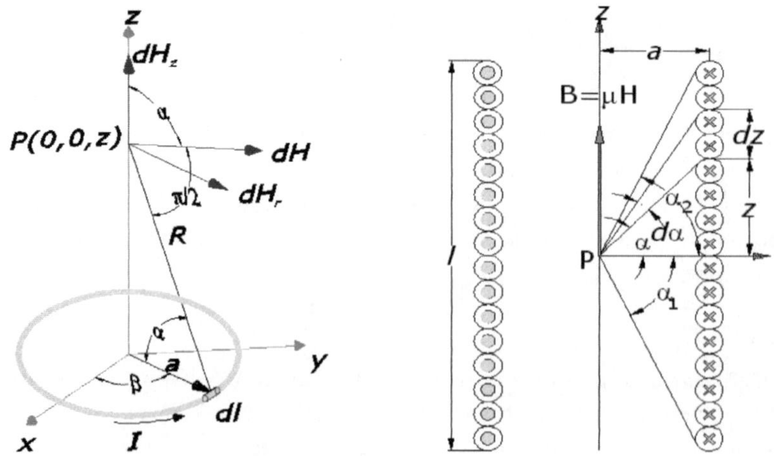

Figure 2. Magnetic field of single and multiple coil winding [13].

The developed electromagnetic actuator (EMA) produces magnetic force must be higher than that of clamping force to push the movable sheave of the fully towards the fixed sheave to maintain the gear ratio of the CVT in start-ups or on the hilly traction. The acceleration time of EMA-CVT system for ¼ scale car Nm is estimated by using the equation, $t = \dfrac{m_c r_{wheel} \, {}^{R_p}\!/_{R_s}}{T_e}(v_c)$, where, m_c is the mass of the car in kg, r_{wheel} is wheel radius in m, v_c is car velocity in m/s and t is the acceleration time.

2.3 Electromagnetic Force

The pushing and pulling the movable sheaves of the pulleys by EMA to meet the load demands of the car is indispensable. The actuator solenoid has been designed to develop the maximum electromagnetic force to overcome the maximum clamping force. Each set of the solenoids has been equipped with individual pulley which one has been used for pushing and the other one for pulling along with a common plunger.

The mathematical models are developed for the EMA by considering the dynamic behavior of the magnetic flux, density, strength, electromagnetic force and energy according to the Faraday's Law, Ampere's Law and Lenz's law, Maxwell's dynamic condition, and the modified equations [15] (Hayt & Buck, 2006). The magnetic concentration is gathered in the centre of the solenoid P. Therefore; overall magnetic flux density can be presented [13]:

$$\vec{B}_z = \hat{z} \sum_{a=-h/2}^{h/2} \sum_{z=-L_{solenoid}/2}^{L_{solenoid}/2} \left(\frac{\mu \left(\iint_{S(rJ_{segment})} J\, dS_{wire} \right) a}{4\pi \sqrt[3]{a^2+z^2}} \int_{l(a,\beta)} dl \right) \tag{7}$$

$$= \hat{z} \sum_{a=-h/2}^{h/2} \sum_{z=-L_{solenoid}/2}^{L_{solenoid}/2} \left(\frac{\mu \left(\iint_{S(rJ_{segment})} J\, dS_{wire} \right) a^2}{2.\sqrt[3]{a^2+z^2}} \right)$$

where, μ is magnetic permeability, dH magnetic flux intensity along the z direction α is the angle between resultant magnetic flux intensity and z axis. J is the current density across wire segment cross section S_{wire}, a is the radius of the wire segment from the centre of the solenoid and dl is the length of the wire segment split. μ is magnetic permeability, dH magnetic flux intensity along the z direction α is the angle between resultant magnetic flux intensity and z axis. J is the current density across wire segment cross section S_{wire}, a is the radius of the wire segment from the centre of the solenoid and dl is the length of the wire segment split.h_2, h_1 is the outer and inner radius of the solenoid from the centre P. $L_{solenoid}$ is the length of the solenoid. It is noted that the development of F_{em} is the function of supplied current (i.e. $F_{em} = f^{n}\left(\iint_{S(r,l_{segment})} JdS_{wire}\right)$). The supplied current of the solenoid is controlled with the controlling the voltage for the desired F_{em}. The main purpose of controlling current is to prevent the actuator from temperature spike. Figure 2 shows the development of electromagnetic force. The magnetic flux (B) develops with supplying current to each loop of the EMA solenoid. The magnetic flux squeezes to the solenoid x direction and expands it along the z direction. The largest F_{em} is generated in the middle of the solenoid due to magnetic field concentration, which causes the plunger to attract (push) and to repel (push).

Simplifying Eq.7, the total B at point P can be estimated by using the equation [13]:

$$B = \hat{z} \sum_{a=h_1/2}^{h_2/2} \sum_{z=-L_{solenoid}/2}^{L_{solenoid}/2} \left(\frac{\mu\left(\iint_{S(r,l_{segment})} JdS_{wire}\right)a^2}{2.\sqrt[3]{a^2+z^2}} \right) = \hat{z}N_{loop}\frac{\mu N_{wind\,perlength} I}{2L_{solenoid}}(\sin\alpha_2 - \sin\alpha_1) \tag{8}$$

where, μ is magnetic permeability (degree of magnetization of a material in response to magnetic field), $N_{wind\ per\ length}$ is the number of turns in a single loop, N_{loop} is the number of loop in solenoid housing, $N{=}N_{wind\ per\ length}\,.N_{loop}$ is the total number of turns in the solenoid. $I = \iint_{S(r,l_{segment})} JdS_{wire}$ is the total supply current, $L_{solenoid}$ is the solenoid length and $\alpha = \tan^{-1}\left[\frac{L_{solenoid}/2}{a/2}\right]$ is the limiting angle with radius (a) for the first loop of the solenoid depends on solenoid inner dimension (h_1). Therefore, the electromagnetic force:

$$F_{em} = N\frac{\mu I^2}{L_{solenoid}}(\sin\alpha_2 - \sin\alpha_1)\, l_{sin\,gle winding} \tag{9}$$

If the solenoid length is much larger than its radius, then $\alpha_2 \cong 90^\circ$ and $\alpha_1 \cong (-)90^\circ$ in which Eq.9 reduced to:

$$F_{em} = N\frac{\mu I^2}{L_{solenoid}} l_{sin\,gle winding} = N\frac{\mu I^2}{L_{solenoid}} 2\pi a \tag{10}$$

with $\mu = \mu_o \mu_r$

where a is the radius of the wire segment from the centre of the solenoid and $L_{solenoid}$ is the length of the solenoid and number of windings N are defined for satisfying the desired magnetic force.

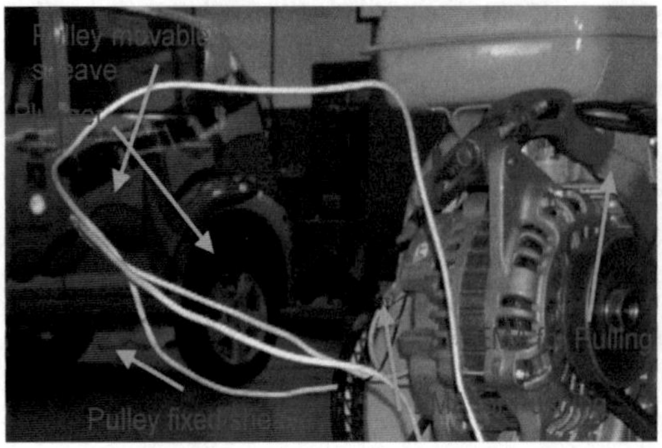

Figure 3. Assembly of Developed EMA and plunger

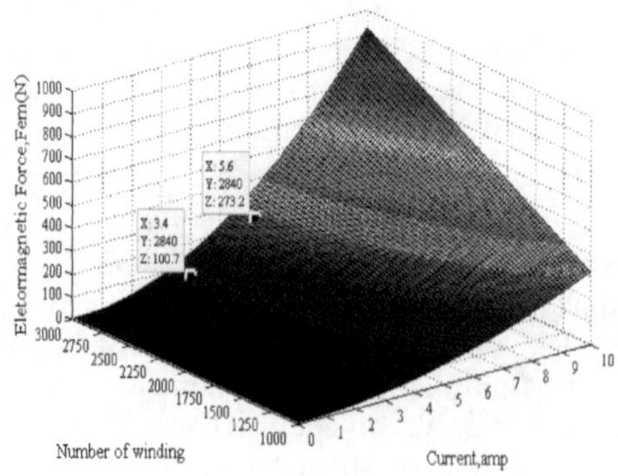

Figure 4. Electromagnetic force for the variation of current and coil windings

Table 1. Specification of EMA

Parameter	unit	Value
Coil diameter (d_w)	mm	1.6
Cross sectional area of wire	mm^2	2.01
Length of solenoid ($L_{solenoid}$)	mm	200
Solenoid Minimum diameter (h_1)	mm	31
Solenoid Maximum diameter (h_2)	mm	171
Cross-sectional surface of the Solenoid	mm^2	755
Packing factor, $f_p = \frac{105(Actual\ windings)}{200(Theoretical\ windings)}$		0.525
Number of turns (N)= $\frac{L_{solenoid}(h_2 - h_1)}{2d_w^2} \times f_p$	turns	2843
Length of coil (L_w)= $\pi \frac{L_{solenoid}(h_2^2 - h_1^2)}{4d_w^2} \times packingfactor$	mm	90232
Volume of the Solenoid coil(Vol^m)= $\pi \frac{(h_2^2 - h_1^2)}{4} L_{solenoid} \times packingfactor$	mm^3	2.3×10^6
Mass of EMA	kg	20.65

IOP Conf. Series: Materials Science and Engineering **184** (2017) 012063 doi:10.1088/1757-899X/184/1/012063

3. Development Of EMA

The EMA-CVT in this study is designed and developed for a car of mass 633 kg. It is developed with different size of solenoid for pushing and pulling the movable sheave towards and away from the fixed sheave is shown in Figure 3. The bigger solenoid has been used to push the movable sheave towards the fixed sheave to develop the maximum gear ratio (GR) of 3.6 to develop the maximum torque (reduction) or the vehicle on uphill gradient of 10%. While, the smaller solenoid has been used to take away (pull) the movable sheave from the fixed sheave to develop the minimum GR of 0.65 for the car is in overdrive. EMA for generation of electromagnetic force is designed in such a way that it can push and pull the rotating pulleys. Figure 4 shows the magnetic force produces by the EMA for current.
The dynamic analysis of EMA with rear spring and plunger by applying the motion equation,

$$\begin{aligned} &\rightarrow + \sum F = m\ddot{x} \\ &F_{em} - F_{clampingforce} = m\ddot{x} + kx \end{aligned} \tag{11}$$

where, m is total mass of pulley and plunger which close to 1.5kg, k is the spring stiffness. x is the displacement of pulley or plunger here it is known as stroke length., $x = \dfrac{T_{out}}{2F_{em}\tan(90-\theta_b)}$ with

$$dx = \frac{dr}{\tan\theta} \text{ and } \ddot{x} = \frac{1}{2F_{em}\tan\theta}\frac{d^2T_{out}}{dt^2}$$

The equation (11) can be rewritten as,

$$\mu NI^2 - \frac{T_{out}\cos\theta_b}{2\mu_x R_x} = m\ddot{x} + kx$$

$$\mu NI^2 - \frac{T_{out}\cos\theta_b}{2\mu_x R_x} = \frac{m}{2F_{em}\tan\theta}\frac{d^2T_{out}}{dt^2} + k\frac{T_{out}}{2F_{em}\tan\theta}$$

$$\left(\frac{m}{2F_{em}\tan\theta}\right)\frac{d^2T_{out}}{dt^2} + \left(\frac{k}{2F_{em}\tan\theta} + \frac{\cos\theta_b}{2\mu_x R_x}\right)T_{out} = \mu NI^2 \tag{12}$$

where, μ is magnetic permeability, N is number of windings, I is the supplied or consumed current, T_{out} is the torque output, θ_b is the belt angel, μ_x is the frictional coefficient, and R_x is the pulley radius. The analytical mathematical model of the EMA-CVT can be fitted with the Laplace transfer function for getting the gain on vehicle traction torque (T_{out} with the current control) and can be modeled as,

$$aS^2T_{out}(S) + bT_{out}(S) = \frac{2c}{S^3}I(S) \tag{13}$$

where, $a = \left(\dfrac{m}{2F_{em}\tan\theta}\right), b = \left(\dfrac{k}{2F_{em}\tan\theta} + \dfrac{\cos\theta_b}{2\mu_x R_x}\right), c = \mu N$

$$\textit{Therefore, Gain} = \frac{T_{out}(S)}{I(S)} = \frac{2c}{S^3(aS^2+b)}$$

where, T_{out} is the input torque which defines the traction torque of the vehicle and can be modeled as, $T_w = n_d T_{out}$, where, T_w is the vehicle traction torque and n_d is the differential constant.

3. RESULT AND DISCUSSION

A ¼ scale car is considered for this study both for the theoretical and experimental performance investigation of EMA. Figure 3 shows the EMA with plunger for controlling the primary pulleys operating diameter.

3.1 Theoretical Performance

The EMA-CVT is able to develop 274 N force equivalent to clamping force on the 10% with current supplying 6 A. Figure 5 shows the ¼ scale car model with EMA-CVT. Figure 6 shows the performance of the CVT movable sheave operation with applying 100 N load to the shaft of the secondary pulley's fixed sheave. The larger EMA develops the electromagnetic force to move the movable sheave towards toe fixed sheave to start-ups, which is called reduction. Total force push the movable sheave, $F_{pushing} = F_{em} + k_s\delta$, where, F_{em} is the electromagnetic force in N, k_s is the spring stiffness in N/m, and δ_s is the spring deflection in m. While, the smaller EMA develops the electromagnetic force to pull the movable sheave towards the EMA and against the spring force, $F_s = k_s\delta_s$ for the vehicle overdrive. Gear ratio of the CVT has made 1.0 by controlling the same diameter for both of the pulleys.

Result (Figure 6a) shows that the pushing time is greater than the pulling time. However, both of the actions of the pulley is equalized with current supplying 4 A and more. This happens because of setting-up the gear ratio where the pulley does not require any force to oppose the clamping force and releasing force. The responding time for static condition is less compared to with load and without load (Figure 6b). It is obvious that during pulling EMA exercises some benefits due to mechanical advantage. In general, the travelling time (responding time) of the plunger of the EMA for pulling mechanism lower than the travelling time of pushing mechanism. Therefore, with a proper development of EMA, a typical car alternator with 40 Ah gives optimal gear ratio in any load condition within the range.

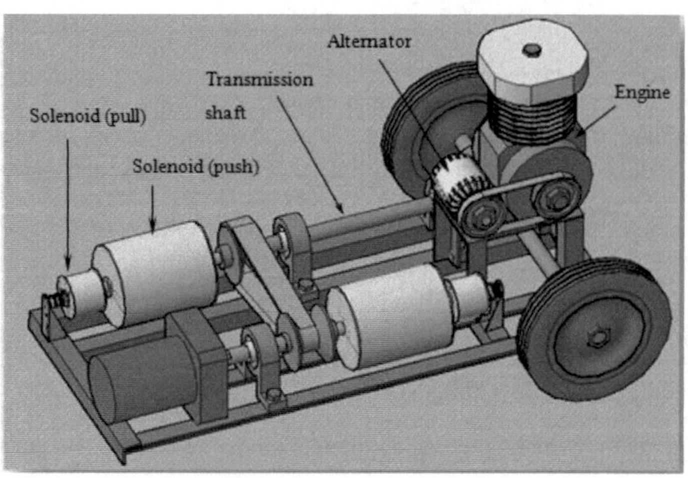

Figure 5a. CAD drawing of proposed quarter scale car with EMA-CVT system [12].

Figure 5b. Developed ¼ scale car with EMA-CVT system [12].

. The EMA of the CVT is mainly contributes on the actuation with developing electromagnetic force. The electromagnetic force develops due to the development of magnetic field for the supplying to the EMA. The electromagnetic force increases with increasing the current supply to the solenoid as shown in Figure 7. The electromagnetic force is the function of current supply and number of windings, which can be presented as, $F_{em} = \oint(I,N)$.

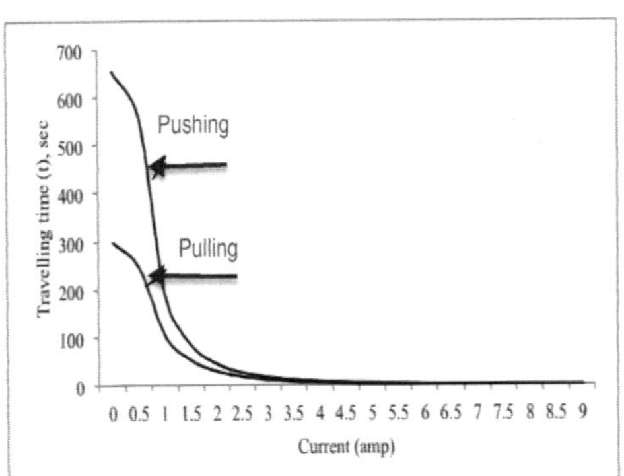

(a)

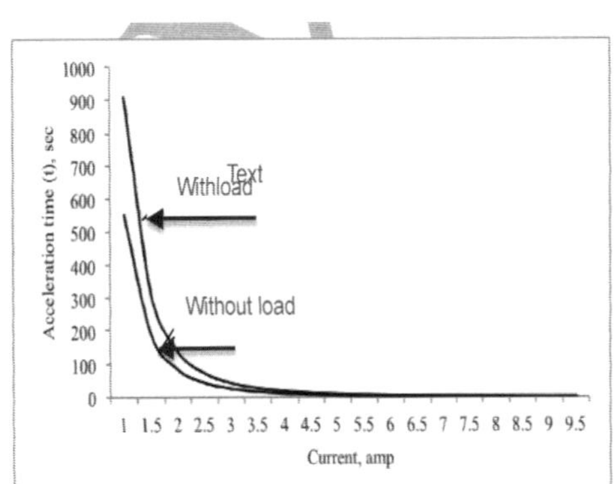

(b)

Figure 6. EMA performance on moveable sheave operation.

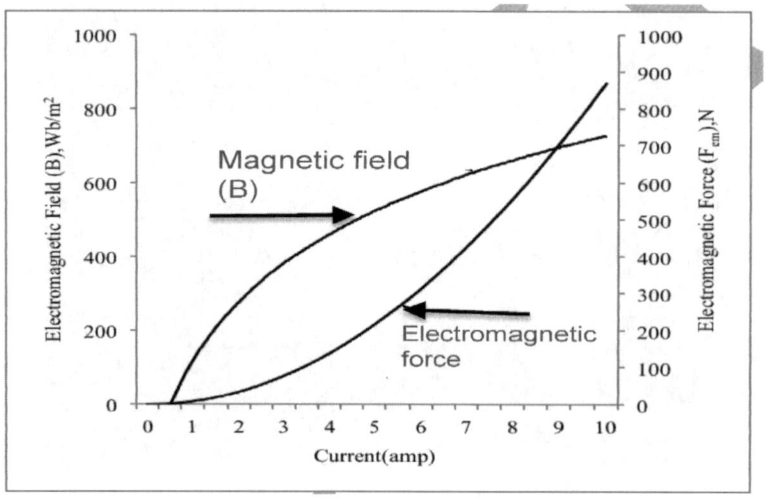

Figure 7. Performance of EMA for different current.

3.2 *EMAs' Power Control*

A movable sheave position control system with fuzzy logic controller design is proposed to realize the electromagnetic force targets and thus minimize the total power consumption and eliminate unnecessary power loss based on the rpm of drive wheel only on simulation. It has the advantage of fuzzy controller being simple (relations between input and output variables can be explained in a linguistic-based rule base), robust (performance is not depending on training and new input variables and rules can be easily added) and not requiring precise mathematical model (Carman, 2008). In the control system of the solenoid, wheel torque (*Y*) is selected as controlled variable and power supply (*P*) as regulated variable through the change in variable resister knob position. The Fuzzy-Proportional-Derivative-Integrator (FPID) controller acts as a self PID tuner. The structure of the FPID controller is shown in Figure 8. The fuzzy logic controller (FLC) with FPID is preferred to control the EMA of the proposed study to control the EMA-CVT to meet the load demands of the car. The FLC is used in this study to control power supply to the EMA for maintaining the desired traction torque of the car in different road conditions.

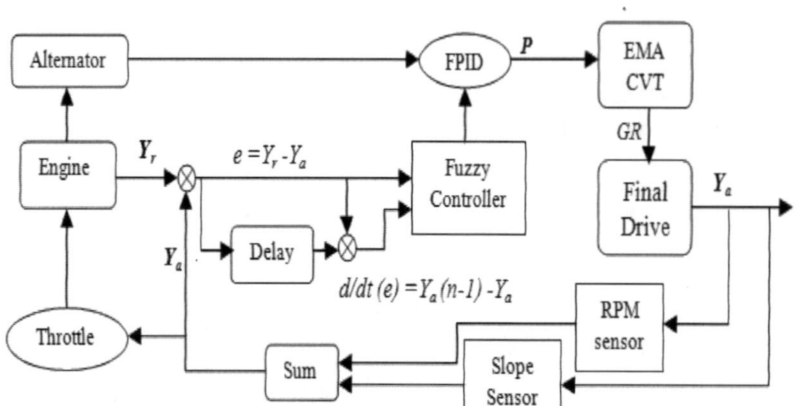

Figure 8. Block diagram of the control system.

IOP Conf. Series: Materials Science and Engineering **184** (2017) 012063 doi:10.1088/1757-899X/184/1/012063

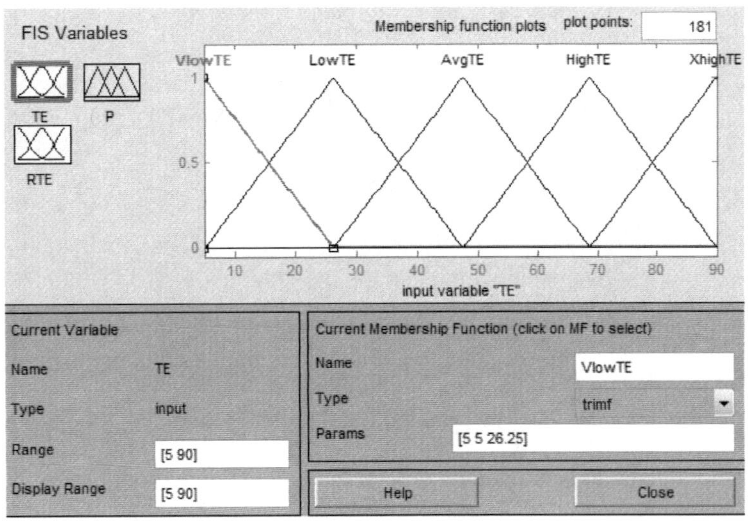

Figure 9. Prototype membership functions of input variable TE.

Based on the difference between measured value of torque (Y_a) and reference value (Y_r), the plunger position i.e. movable sheave displacement is controlled by a regulation of variable, i.e., power supply (*P*). The reference torque is calculated based on the maximum allowable (maximum boundary) generated torque by the motor and then is compared with the measured torque numeric values using dynamometer (rpm sensor) and slope sensor. Hence, the resultant deviation, i.e., torque error (*TE*), and differential torque or rate of torque error (*RTE*) are continuously calculated in operation (Figure 9).

The controlled variable is considered as Torque Error (***TE***), Rate of change of Torque Error (***RTE***) and regulated variable as Power consumption (***P***) by EMA actuator. It is noted that the values of membership functions (***TE***, ***RTE***) change their values with respect to time. For instance, the degree of ***TE*** changes in Nm from 5 to 90, ***RTE*** is fluctuates in Nm/s from -5 to 5, and ***P*** is measured in watt from 50 to 200, respectively. It is noticed that different membership functions acquire on zero and nonzero values indicating the degree to which the linguistic value suitably illustrates the present value of ***TE***. For example, at ***TE*** = 5; it is certain that the torque error is "very low (Vlow)", and as the value of ***TE*** moves toward 26.25 it is become less certain that it is "Vlow" and more certain that it is "Low" as shown in Fig.9. The membership function of the fuzzy inputs from fuzzy logic expert system where *i* stand for input from into Eq. (14), (15) and (16).

$$\mu_{Vlow}(TE(i_1)) = \begin{Bmatrix} 1; & (TE) \le 5 \\ \dfrac{26.25-(TE)}{21.25}; & 5 < (TE) < 26.25 \\ 0; & (TE) \ge 26.25 \end{Bmatrix} \tag{14}$$

$$\mu_{Low}(TE(i_1)) = \begin{cases} 0; & (TE) \leq 5 \\ \frac{(TE)-5}{21.25}; & 5 < (TE) \leq 26.25 \\ \frac{47.5-(TE)}{21.25}; & 26.25 < (TE) < 47.5 \\ 0; & (TE) \geq 47.5 \end{cases} \quad (15)$$

$$\mu_{Pos}(RTE(i_2)) = \begin{cases} 0; & (RTE) \leq 1 \\ \frac{(RTE)-1}{2}; & 1 < (RTE) \leq 3 \\ \frac{5-(RTE)}{2}; & 3 < (RTE) < 5 \\ 0; & (RTE) \geq 5 \end{cases} \quad (16)$$

To comprehend fuzzification, an example is considered. It is assumed that at a particular point at time t, ***TE*** = 20 Nm and ***RTE*** = 2.5 Nm/s. These are the crisp inputs directly from the sensors.

Determining applicability of each rule is called "firing" [17]. The fuzzy inference system (FIS) seeks to determine which rules fire, to find out which rules are relevant to the current situation. For crisp input ***TE*** (i_1) = 20 Nm, and ***RTE*** (i_2) = 2.5 Nm/s, the rules 3 and 6 are satisfied to be fired. The firing strength of truth values of membership function for individual rules is obtained as:

$$\alpha_3 = \min\{\mu_{Vlow}(TE), \mu_{Pos}(RTE)\} = \min\left(\frac{26.25-20}{21.25} = 0.294, \frac{2.5-1}{2} = 0.75\right) = 0.294$$

$$\alpha_6 = \min\{\mu_{Low}(TE), \mu_{Pos}(RTE)\} = \min\left(\frac{20-5}{21.25} = 0.705, \frac{2.5-1}{2} = 0.75\right) = 0.705$$

3.3 Experimental Result

Electromagnetic force is the heart of this current CVT system, which offers desired gear ratio by placing the pulley shave to its exact position. Figure 10 shows thing the behavior of electromagnetic force response of the system with varying the current supply. It supposed to be linear change with respect to power followed gear ratio. However, due to slip in belt and pulley during the change of gear ratio the electromagnetic force trend is showing its fluctuating behavior. Sufficient Electromagnetic force is indispensible for perfect gripping the belt and pulley.

Point 'A' (Figure10) shows the initial start-ups of the car, which is called reduction, the ¼ scale car needs gear ratio 2.33 for the supply current to the EMA of primary pulley's EMA. While, point 'B' indicates the overdrive of the car at gear ratio 1.1 and the current required 2.85A to EMA to push the primary pulley's movable sheave towards the fixed pulley. Figure 11 shows the electromagnetic force developed by the EMA by the supplying current. The electromagnetic force 144 N develops by the EMA for the current 1.35 A for the initial start-ups while electromagnetic force develops 248 N for the current supply 2.89 N. It is noted that the secondary pulley movable sheave's initially placed at maximum position for the reduction or initial start-ups. Beyond the point 'B' the movable sheave needs to push towards the fixed to get the reduction gear ratio for developing the maximum traction force for the vehicle. In this case the EMA needs current supply 3.37 A to develops the electromagnetic force of 320 N. The develops traction force for the car by the engine is estimated as, $F_t = \frac{n_d \eta_t T_e}{r_w}$, where, n_d is the differential speed ratio, η_t is the transmission efficiency in percentage, T_e is the engine torque in Nm, and r_w is the wheel radius in m. By referring to the current supplied to the coil, the variety of responding (travelling) time of the movable sheave is recorded. It is difficult to adjust infinite number of gear ratio by manual control. In order to ensure this intelligent control system would be a better solution. Fuzzy logic intelligent system has been considered for this study to maintain the gear ratio as required for the car to maintain its traction power and cruising speed.

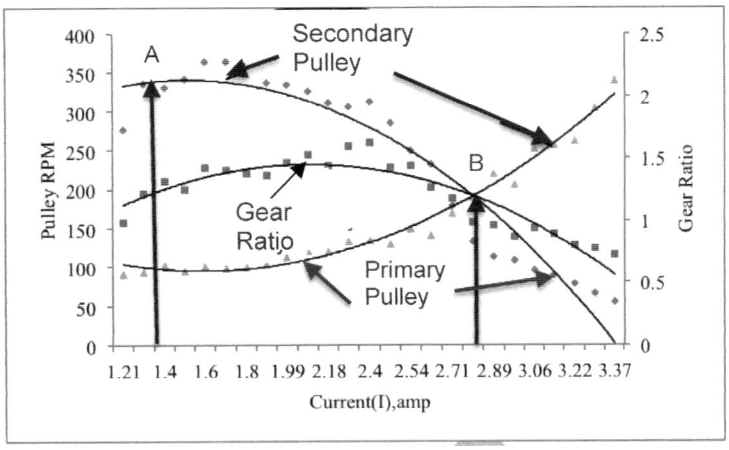

Figure 10. Current requirement for CVT gear ratio.

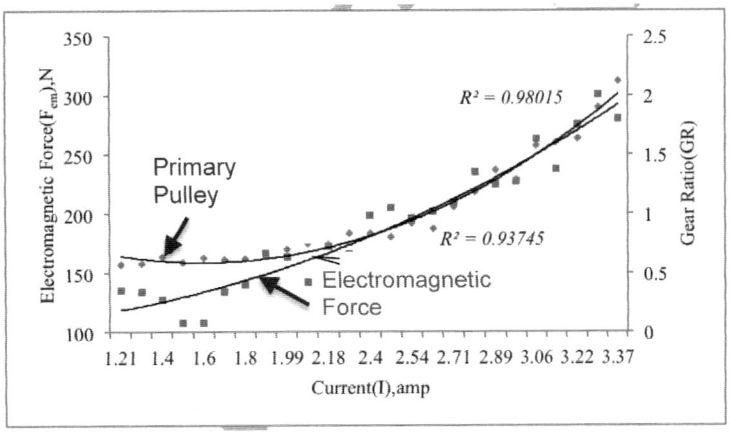

Figure 11. Electromagnetic force, Gear Ratio VS Supply current.

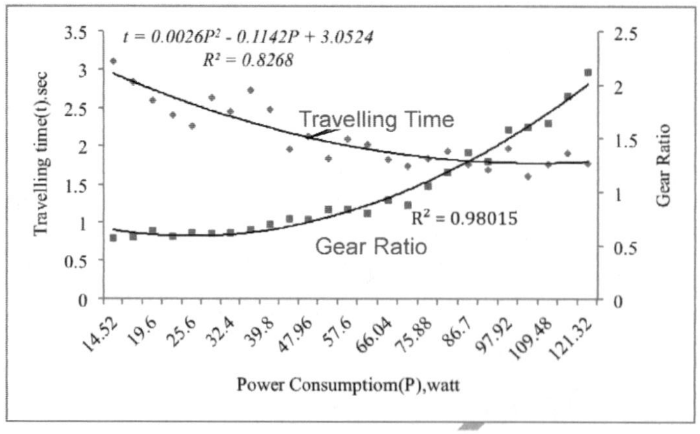

Figure 12. Travelling time vs Power Supply.

Figure 12 illustrates the gear ratio with respect to power where higher power was observed during higher gear ratio under any pushing condition of pulley sheaves. It seems to be that pushing and pulling time is equal. However, in practice the travelling time (responding time) of the plunger of the EMA for pulling mechanism was found significantly lower than that of pushing mechanism. The basic reason was the load on matching area of the sheaves is higher and the pulling of the sheave was found easier than the pushing. Furthermore, clamping releasing force always tends to pull the movable sheave of the pulley towards the solenoid of the EMA while, clamping force push the movable sheave against the rotating belt. Spring could be enhancing the pulling time even further but it would consume more power during pushing. It should be noted that the movable sheave of the pulley during push needs to press against the rotating belt, and spring to create the higher GR for the higher torque of the car in starting and inclined road. Hence, the electromagnetic force required to push the movable sheave, $F_{push} = F_{em} = \frac{n_g \eta_t T_e}{r_w} - k_s \delta_s$ while the pulling force, $F_{pull} = F_{em} = k_s \delta_s - x \frac{n_g \eta_t T_e}{r_w}$, where, k_s is the spring constant in N/m, δ_s is the spring deflection in m, and x is the fraction of traction force. It is noticed during experiment that for low rpm (in start-ups or climbing slope) acceleration time is higher compared with the other kinds of rpm. High torque means high clamping force of the CVT, which needs to overcome by the solenoid. Once the car get its full inertia need less time to be accelerate. This minimum acceleration time or travelling time otherwise; is the consequence of fast response characteristics of solenoid.

3.3.1 Fuzzy Simulation Verification

Dynamic torque behavior in between 30-90 Nm is observed in both experimental and fuzzy simulation method shows in Figure 13. The fuzzy simulation can be verified by experimental data taken form the ¼ scale model. Fuzzy data has been extracted for the MATLAB workspace. Fig.13 represents the torque characteristics of the final drive. Transmitted torque to the final drive has only 2.74% of average deviation with respect to gear ratio because of belt slippage as well as electromagnetic force. Enhanced electromagnetic force could be a better solution for avoiding belt slip. It is noticed that 108 N electromagnetic forces is required in the primary pulley solenoid to transmit the initial torque when

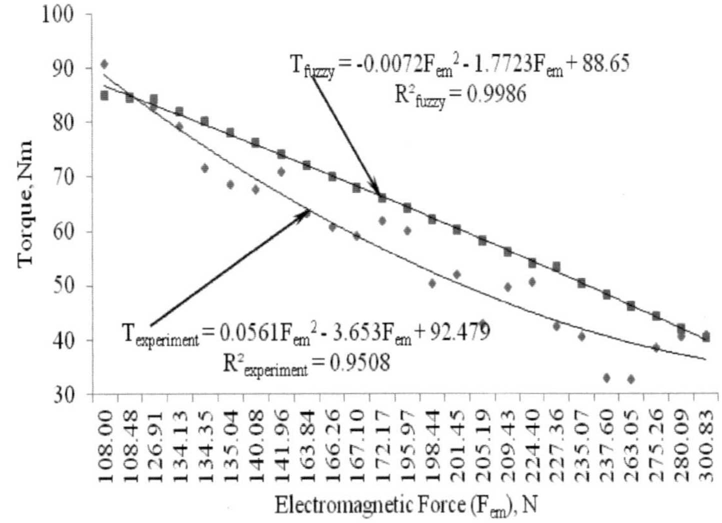

Figure 13. Torque verification with experimental and fuzzy simulation.

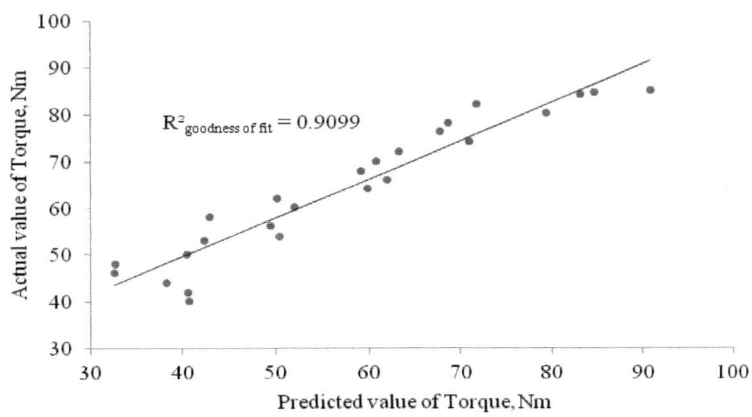

Figure 14. Correlation between actual and predicted value of Torque.

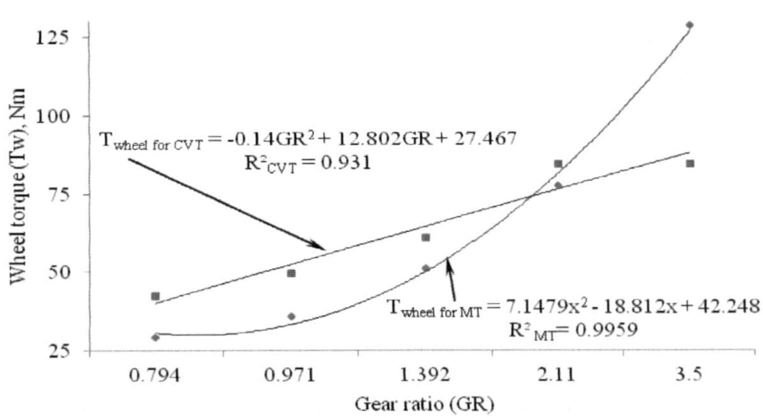

Figure 15. EMA-CVT performance over manual transmission.

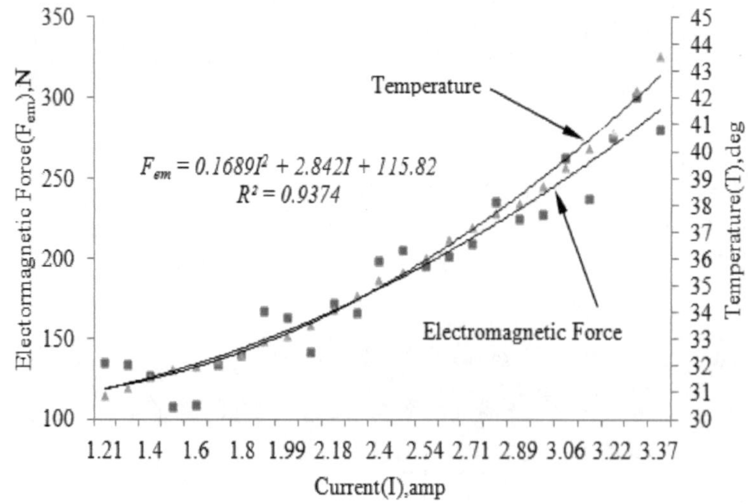

Figure 16. Electromagnetic force, Temperature VS Supply current.

the gear ratio is 1.8. It is also observed that to keep the car in motion the minimum torque needed 32.6Nm for this current set up. The correlations between experimental values and predicted (FLES) values of traction torque have been illustrated in Figure 14. The correlation coefficient of traction torque or goodness of fit is found as 0.9099. It indicates that the predicted data over the measured data have a closed agreement and thus, validity of the fuzzy simulation results. However, the mean relative error of measured and predicted values from the FLES model on traction torque is found as 9.01% since this study neglecting some transmission losses. Relative error of predicted values is in the acceptable limits of 10%. The goodness of fit gives the ability of the developed system and its highest value is 1 according to statistical method [18,19].

The final torque in the wheel is compared with the down scaled conventional VIVA manual car in Figure 15. The available five transmission ratio is considered as it is discrete transmission unit for same final gear ratio 3. Similar gear ration has been chosen from a number of gear ratio in CVT for EMA in order to observe torque transmission performance of CVT compared with manual. CVT perform better in the region of low torque of gear ratio. Afterward, it became stable at gear ratio 2 and leveled off. It represents the up to certain limit CVT performed much better to its counterpart manual transmission. However, for high torque this system is not applicable. This is why the concern of this study to implement this system in small car not in heavy vehicle.

Figure 16 shows after an hour operation of the solenoid with maximum current flow the temperature jumps to 44°C from the ambient temperature. To some extend in thermal management point of view this heat generation is a matter of deep concern. Heat sink could be better solution; However, current experiment overlook this constrain of the system which could be the further scope of research.

4. Conclusion

i. The EMA able to develop the electromagnetic forces 101.22 N to 274.82N equivalent to the clamping forces by supplying current in the range of 1- 4 amps with 36 volts.
ii. The EMA is able to develop the electromagnetic forces 101.22–274.82 N and dynamic torque 31.38 - 85.19Nm at transmission gear ration 0.820 - 2.369.

iii. The solenoid able to pull and push the plunger in the desired distance when current supply is 3 amp in almost same time. While, it is different when the current supply to the EMA is less than 3 amp.
iv. The responding time of the plunger for kerb weight of the ¼ car in static condition is less compared to with-load of 15kg.
v. The fuzzy simulation represents the time required reaching the desired maximum torque 85.19 Nm is 3.75sec.and
vi. The corresponding current is 4.5 or the maximum desired torque 85.19 Nm.

This virtual control of the EMA operated CVT with fuzzy controller helps the system to operate manually when the real fuzzy controller is not attached with EMA in practice. Experimentally, it is observed that:

i. The EMA develops the electromagnetic force in the ranged of 108–301 N which has been maintained with supply current maximum 3.37 amp.
ii. The traction torque has been estimated maximum 90 Nm by using the electromagnetic force 301N which has been found with the corresponding supplied current 3.37 amp for the maximum gear of 1.8.
iii. Travelling time of the plunger is in the range of 1.8– 3.1sec without loading. It shows higher with the incremental loads. However, prime mover was unable to transmit torque when the applied load increased more than 3kg. This could be due to the lower rating of prime mover.
iv. The correlations between measured (experimental) and predicted (FLES) values of traction torque has found 90.99%, which closely verify the fuzzy simulation model.

ACKNOWLEDGEMENT
With the financing of Research Management System, International Islamic University, an Electro-magnetic actuated CVT system has been developed for PREVE PROTON car and it was patent which is granted on 31 July 2013 (Ref. Patent No. MY-149226-A by Ataur Rahman).

References

[1] W. Kriegler et al. IC-Engines and CVTs in Passenger Cars: A System Integration Approach.Advanced Vehicle Transmissions and Powertrain Management. *IMechE*, Birdcage Walk, London, 1997.
[2] (http://community.carfax.com/t5/Car-Technology-Forum/Pros-and-Cons-of-Continuously-Variable-Transmisssions/m-p/88, Retrieved date 15 February 2016)].
[3] K. K. Ang et al. MCMAC-CVT: a novel on line associative memory based CVT transmission control system. Neural Networks, 15 (2002) 219-236, 2002.
[4] B. Marcus et al. Elastic modelling of bodies and contacts in continuous variable transmissions. *Multibody System Dynamics, 13*, 175-194, 2005.
[5] H.L. Stewart. *Hydraulic and Pneumatic Power for Production* (4th Edn.). New York: Industrial Press Inc., 1997
[6] M. Pesgens et al. Control of a hydraulically actuated continuously variable transmission', *Vehicle System Dynamics*, 44:5, 387 – 406, 2006.

[7] H. Tanaka and H. Machida. Half Toroidal Traction-Drive Continuously Variable Power Transmission pp 205-212,Vol 120. *Journal of Engineering Tribology.IMechE* 1996.
[8] T. W.G.L Klaassen. The Empact CVT; Dynamics and Control of an Electromechanically Actuated CVT. PhD Thesis. Eindhoven, 2007.
[9] S. Akehurst et al. Modelling of loss mechanisms in a pushing metal V-belt continuously variable transmission. Part 1: Torque losses due to band friction. Proc. Of the Inst. of Mech. Eng., Part D: J. of Auto. Eng. 218: 1269-1281, 2004.
[10] E. Kirchner. Leistungsübertragung in Fahrzeuggetrieben; Grundlagen der Auslegung, Entwicklung und Validierung von Fahrzeuggetrieben und deren Komponente. Heidelberg: Springer-Verlag, 2007.
[11] I. Izhari et alApplication Of Disc Spring In Clamping Force Mechanism For Electro-Mechanical Continuously Variable Transmission. Jurnal Teknologi, Vol. 77:22 (2015) 129-135.
[12] A. Rahman et al. Energy efficient electromagnetic actuator for CVT system. Journal of Mechanical Science and Technology, 28(4), pp: 1153-1160 (2014).
[13] A. Rahman et al. Kinematics and nonlinear control of an electromagnetic actuated CVT, 26 (7), pp. 1-9 (2012).
[14] T. Toshie. Electromagnetic actuator design technology using electromagnetic coupled with motion analysis, Mitsubishi Electric.,116, 2-4.a.
[15] W. H. Hayt et al. Engineering electromagnetics. 7th Ed. McGraw-Hill International Edition, (2006).
[16] F. T. Ulaby. Electromagnetics for engineers, Pearson Inter-national Edition, 2005.
[17] M. Gopal. Digital control and state variable methods: con- ventional and intelligent control systems (3rd ed.), Tata McGraw-Hill Education Pvt. Ltd, 2009.
[18] K. Carman, K. Prediction of soil compaction under pneumatic tires a using fuzzy logic approach. *Journal of Terramechanics*, 45, 103-108, 2008.
[19] A. Hossain et al. Cushion pressure control system for an intelligent air-cushion track vehicle, Journal of Mechanical Science and Technology, 25 (4) (2011) 1253-1260.

Estimation of Stress Concentration Factor of Plate with Hole using Piezoelectric Actuator and Finite Element Method

A Abuzaid, M Hrairi* and M.S.I. Shaik Dawood

Department of Mechanical Engineering, International Islamic University Malaysia, P.O. Box 10,50728 Kuala Lumpur, Malaysia

*E-mail: meftah@iium.edu.my

Abstract. In this paper, the effect of piezoelectric actuators placed above a circular hole of a rectangular plate subjected to uniform uniaxial tension is studied. The core idea is to investigate the stress (compression/tension) produced by the piezoelectric actuators on the stress distribution around the hole and along the width of the host plate. For this purpose, Finite Element Analysis (FEA) was carried out through parametric study in ANSYS software. The results demonstrated that the positive electric field would decrease and change the state of the stress distribution along the width of the host plate in contrast to the negative applied electric filed which increases the stress distribution smoothly without affecting its behaviour. The results also indicated that the reduction of the stress concentration factor increases with the decrease of the ratio (D/W) for the same applied positive electric field.

1. Introduction

Holes location in engineering structures will produce stress or strain concentrations and hence will decrease the mechanical properties and may cause a crack initiation. The amount of strains or stresses concentration due to the presence of the hole is known as the stress-concentration factor (SCF). The stress concentration factor is the ratio of the maximum stress at the edge of the hole to the nominal (reference) stress [1].

Recently, structures incorporating smart materials such as piezoelectric transducers and actuators are becoming attractive for many researchers. These smart structures could be applied in many engineering applications, such as aircraft structures, satellites, large space structures, etc. [2-5]. The piezoelectric materials are widely used in different engineering structures as actuators and sensor due to their coupling behavior. This coupling behavior is attributed to the interaction effects of the mechanical and electrical properties.

Many theoretical and experimental investigations show that the piezoelectric actuator affects the strain and stress in isotropic materials [6]. The crack tip stress concentration reduction by using piezoelectric actuator to generate counteracting bending moment has been investigated by Wang et al [7]. The piezoelectric actuator stress has been employed by Liu [8] to reduce the stress intensity factor of a crack beam using the fracture mechanics approach. The compression stress induced by the piezoelectric actuator near the crack tip has been used to reduce the fatigue damage propagation in thin aluminum panels [9]. The stress produced by the piezoelectric actuator was used to control the edge debonding in beams strengthened with composite materials [10]. Mode-I active repair using the piezoelectric behaviour restricted to the square actuator without accounting for the singularity of the stress at the crack tip and the influence of adhesive properties in the repair process was carried out by Abuzaid et al. [11-12]

The control of the stress concentration for plate with a hole by using piezoelectrics actuator has been studied by Sensharma et al. [13]. Shah et al. [14] demonstrated that the stress concentration around the hole in a plate under tension can be reduced effectively by using the stress produced by the piezoelectric actuators. They investigated numerically two locations for the piezoelectric actuator around the hole. Their results show that the stress flow in the plate can reduce smoothly.

Lately, the effect of stiffness and thickness ratio of host plate and piezoelectric patches on reduction of the stress concentration factor has been investigated by Fesharaki et al. [15]. Their results indicated the importance of the stiffness ratio on the piezoelectric best placement for the SCF reduction. Fesharaki and Golabi [16] furthered the state of knowledge by specifying the actuators pattern recognition for the optimization of the stress concertation reduction.

Most of the previous works analyzed the piezoelectric actuator effects near the discourteous locations for the purpose of actively controlling the high stress concentration of the host structure. However, only limited numbers of studies have investigated the effects of the piezoelectric actuator when employed above the discontinuous locations. In this paper the effects of the piezoelectric actuators placed above the circular hole of a rectangular plate under uniform uniaxial tension were considered. The core idea is to investigate the stress (compression/tension) generated by the piezoelectric actuators on the stress distribution around the hole. For this purpose, the finite element analysis (FEA) was employed and two cases for the applied electric filed (negative/positive) were studied. The analysis of the two cases was carried out through parametric study in ANSYS finite element software.

2. Formulation of the problem

The rectangular plate containing a central hole with integrated piezoelectric actuator (placed above the hole) and loaded in a uniform uniaxial tension shown in figure 1 is considered. When the uniform tension is applied, the theoretical stress concentration factor K_t in case without actuator is [1]:

$$K_t = \frac{\sigma_{max}}{\sigma_{nom}} \tag{1}$$

where σ_{max} is the maximum stress and σ_{nom} is the nominal stress. For the linear piezoelectric materials, the strain is linearly related to the electric field. Therefore, the displacement produced by an actuator can be easily controlled by changing the applied electric field. Applying electric field to the piezoelectric actuator would induce stress on the host structure which affects the host plate stress concentration factor and the stress distribution.

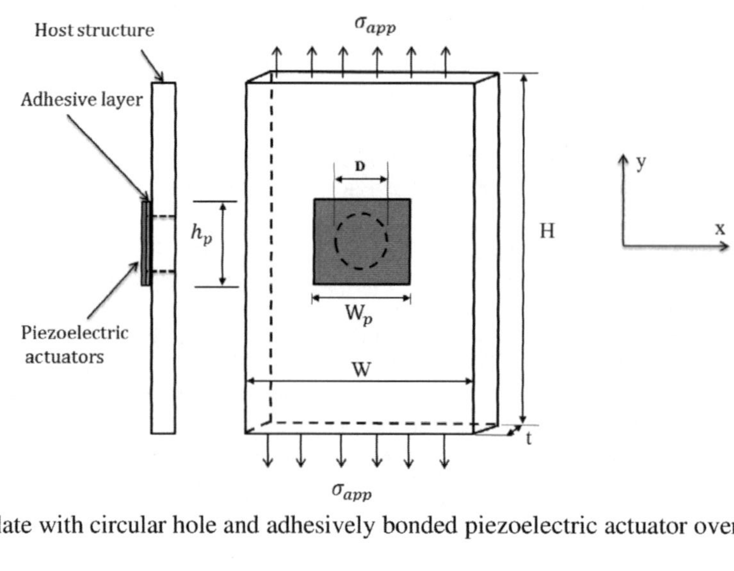

Figure 1. Plate with circular hole and adhesively bonded piezoelectric actuator over the hole.

Two different applied electric fields were considered in this study: the first case is a positive applied electric filed (produces compression stress in the piezoelectric actuator) while the second case is a negative applied electric filed (produces tension stress in the piezoelectric actuator)

For the purpose of comparison, explicit relationship between the applied mechanical stress (σ_{app}) and the applied electric field (E_0) was computed. The applied electric field (E_0) was found such that it would produce equal stress in the infinite piezoelectric actuator to the applied mechanical stress (σ_{app}) [14-15].

3. Piezoelectric Constitute Equations

The constitutive equations governing the linear electro-mechanical piezoelectric effect are given as below [17].

$$S_1 = s_{11}^E T_1 + d_{31} E_3 \quad (2)$$

$$D_3 = d_{31} T_1 + \varepsilon_{33}^T E_3 \quad (3)$$

where S_1 is the mechanical strain, T_1 is the stress, D_3 is the electrical displacement, E_3 is the electric field, s_{11}^E is the mechanical compliance at zero field, ε_{33}^T is the dielectric constant at zero stress, d_{31} is the piezoelectric coefficient.

4. Finite Element Analysis

Finite Element model

In ANSYS, the structural element SOLID186 is used to model both the host structure and the bonding layer. This element is appropriate for the analysis of solid structures (3D) and recommended for linear application. Fine resolution is used to model the plate near the circular area. Piezoelectric actuator is molded with the structural element SOLID226.

The dimensions of the host plate considered in this work are $W = 100$ mm, $H = 200$ mm, with a central hole of diameter $D = 0.2W$ and thickness t of 1mm. The PIC151 piezoelectric actuator and the adhesive have the dimensions of $h_P = h_{ad} = 1.5D$, $w_P = w_{ad} = 1.5D$, and thickness of $t_P = 0.5t$, $t_{ad} = 0.03$mm. The material used in this simulation for host structure has the parameters: density $\rho = 2715\ \text{kg m}^{-3}$, Young's modulus $E = 200$ Gpa and Poisson ratio $\nu = 0.3$. For the piezoelectric actuator PIC151 was used. Some of the important material and electrical data for the piezoelectric actuator are: density $\rho_P = 7800\ \text{kg m}^{-3}$, elastic compliance constant $S_{11} = 19 \times 10^{-12}\ \text{m}^2\ \text{N}^{-1}$, piezoelectric constant $d_{31} = -2.10 \times 10^{-10}$ m/V and electric permittivity coefficient $\varepsilon_{33}^T = 2400$.

Validations of the finite element model

In order to validate the finite element model, the plate with a hole and piezoelectric actuator placed on the top/bottom of the hole in figure 1 of Ref. [15] is considered. By comparing the results obtained by Ref. [15] and the present finite element results, there is a good agreement as shown in table 1.

Table 1. Verification of the present finite element model.

Condition	Applied electric field	SCF Ref. [11]	SCF present work
Without actuator	-	3.17	3.16
With actuator	-E0	1.7	1.68

5. Results and Discussion

In this section, the FEA was employed to study the effect of the piezoelectric actuators placed above the circular hole of a rectangular plate under uniform uniaxial tension for the problem shown in figure 1. In this study the stress σ_y is normalized by the applied stress σ_{app}.

The normalized stresses σ_y around the circular hole in host plate with positive and negative applied electric field are depicted in figures 2 and 3, respectively. The results show that the normalized stress distribution σ_y reduced with the application of the positive electric filed $(+E_0)$ and this reduction increases with increasing the positive applied electric field (figure 2). Alternatively, the application of the negative electric field $(-E_0)$ leads to increase of the SCF and at the same time the stress distribution (figure 3). This could be attributed to the strain produced by the piezoelectric actuator which is positive in the case of the $(+E_0)$ and hence lead to reduction of the effective stress around the hole. It is noted from figure 2 that applying electric field of $(2E_0)$ would change the state and the location of the maximum stress around the hole. It is also observed that at the compression zone of the stress around the hole ($\theta \simeq 90$), the curves meet each other for both cases of the applied electric field (negative/positive).

The normalized stresses σ_y along the width of the host plate with various applied electric fields (positive/negative) are illustrated in figures 4 and 5, respectively. The results show that the application of the positive electric filed decreases the SCF and the stress distribution σ_y along the host plate width (figure 4). The results also show that there is a dip in the curves at the end location of the piezoelectric actuator.

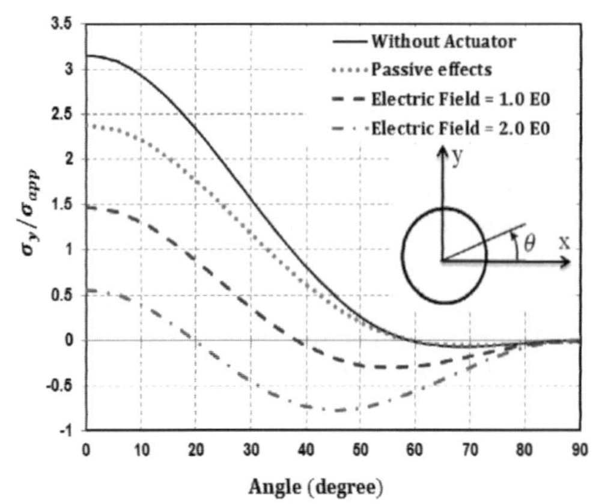

Figure 2. The stress σ_y around hole in host plate (Case 1).

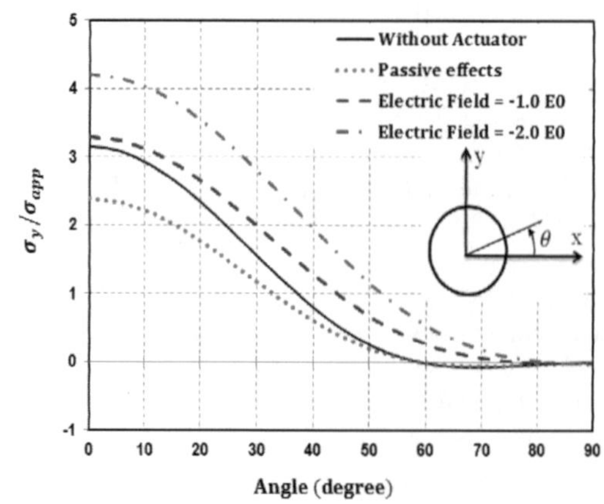

Figure 3. The stress σ_y around hole in host plate (Case 2).

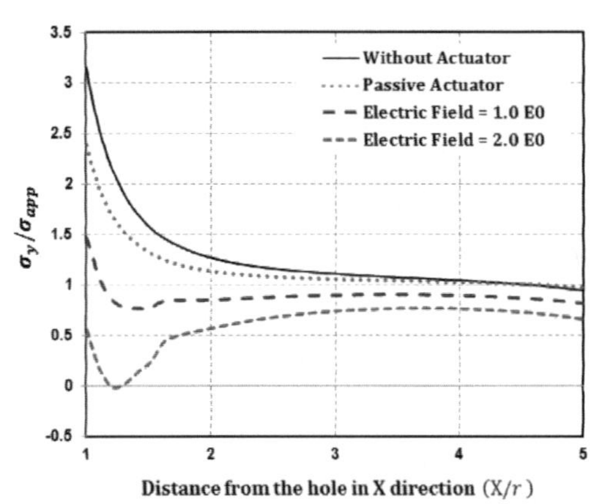

Figure 4. The distributions of the stress σ_y of the plate along the x −axis (Case 1).

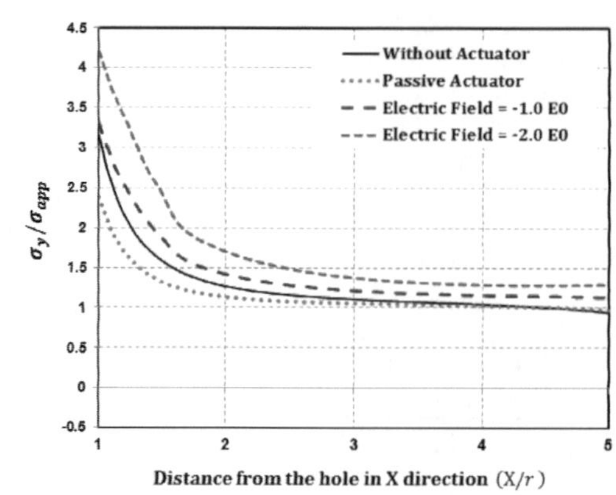

Figure 5. The distributions of the stress σ_y of the plate along the x −axis (Case 2).

It is also observed that the reduction of the stress distribution at far distance from the hole and the piezoelectric actuator edge are slightly less compared to stress reduction near the vicinity of the hole. In contrast, the negative applied electric field smoothly increases the SCF and also the stress distribution along the width of the host plate (figure 5).

Figures 6 and 7 show the normalized stress distribution σ_{ypiezo} in piezoelectric patch with positive and negative applied electric field, respectively. It is clear from the results that increasing the applied negative electric filed increases the tension stress (figure 6) while increasing the positive applied electric field increases the compression stress (figure 7).

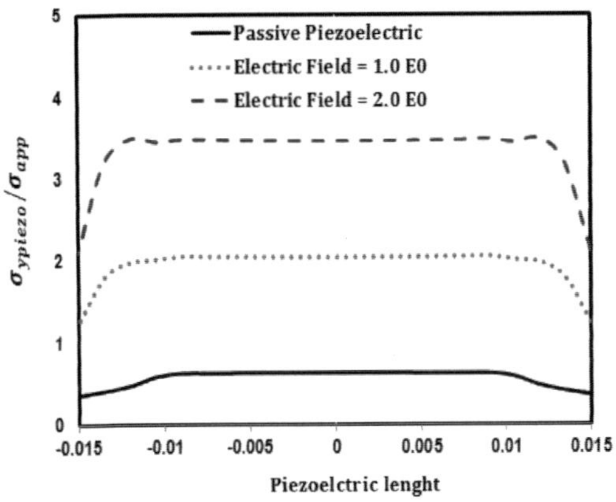

Figure 6. The distributions of the stress σ_{ypiezo} of the piezoelectric actuator along the x −axis (Case 1).

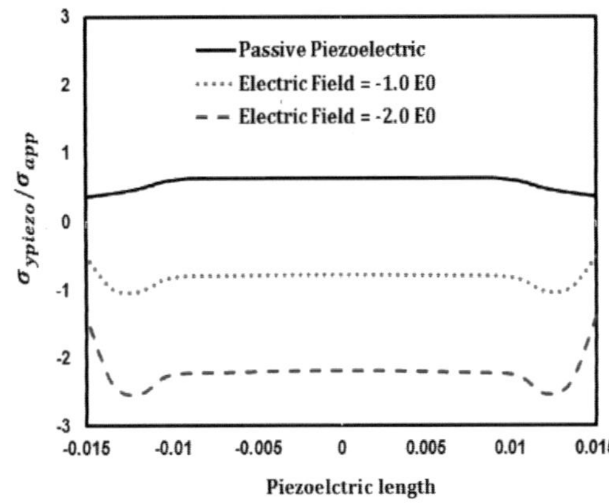

Figure 7. The distributions of the stress σ_{ypiezo} of the piezoelectric actuator along the x −axis (Case 2).

The passive effects of the piezoelectric actuator with different actuator thicknesses on the normalized stress distribution and the stress σ_y around the hole in host plate are given in figures 8 and 9, respectively. The results show that increasing the actuator thickness increases the reduction of the stress distribution. In fact, this is due to the increase of the integrated structure stiffness.

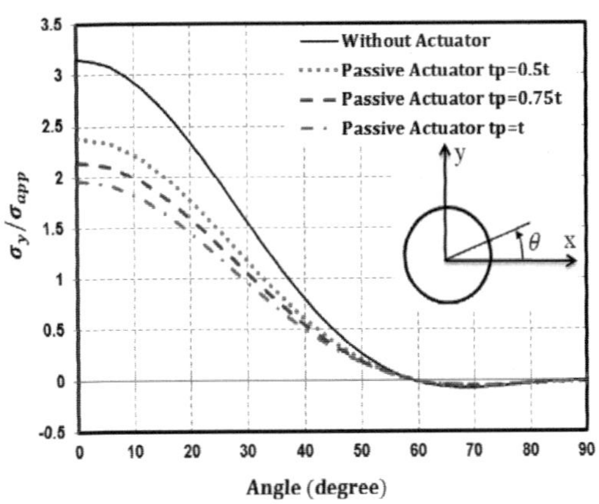

Figure 8. Passive effects on the Stress σ_y around hole in host plate (Case 1).

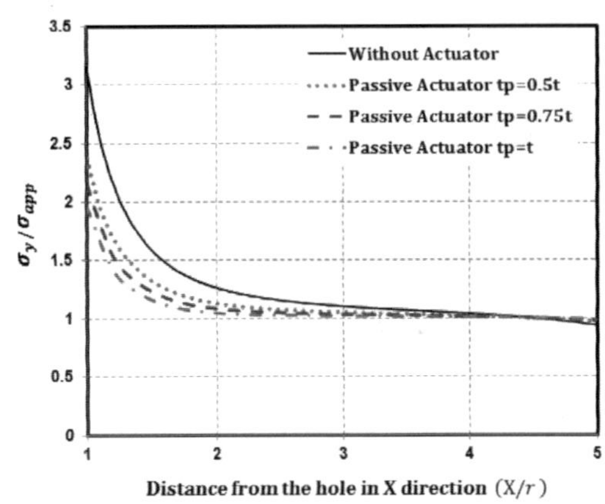

Figure 9. Passive effects on the distributions of the stress σ_y along the x −axis (Case 1).

For the purpose of observing the effect of the piezoelectric actuator thickness on the plate stress distribution, the normalized stress σ_y around the circular hole in host plate with negative/positive applied electric field is depicted in figure 10. From the figure, it is clear that thicker actuator with positive electric field relatively gives higher decrease of the SCF and the stress distribution compared to thin actuators. It is important to mention that the increase of the actuator thickness will not affect the location of the maximum stress around the hole of the host plate with application of the electric filed.

Figure 11 demonstrates the effect of the piezoelectric actuator thickness on normalized stress σ_y along the width of the host plate with various applied electric fields. The figure shows that applying electric field of (E_0) would change the state of the stress distribution along the width of the host plate in contrast to the negative applied electric filed which increases the stress distribution smoothly without affecting the behaviour of the stress distribution.

The effects of the ratio (D/W) with fixed piezoelectric actuator size on the normalized stress σ_y along the width of the host plate with negative and positive applied electric fields are shown in figures 12 and 13, respectively. The results in figure 12 show that the reduction of the stress concentration factor increases with decrease of the ratio (D/W) with the application of the same positive electric filed (E_0). It is also observed that the stress drop in the curves at the end location of the piezoelectric actuator increases with decrease of the ratio (D/W). It is clear from figure 13 that the reduction of the stress distribution at far distance from the hole and the piezoelectric actuator edge are slightly less compared to stress reduction near the vicinity of the hole.

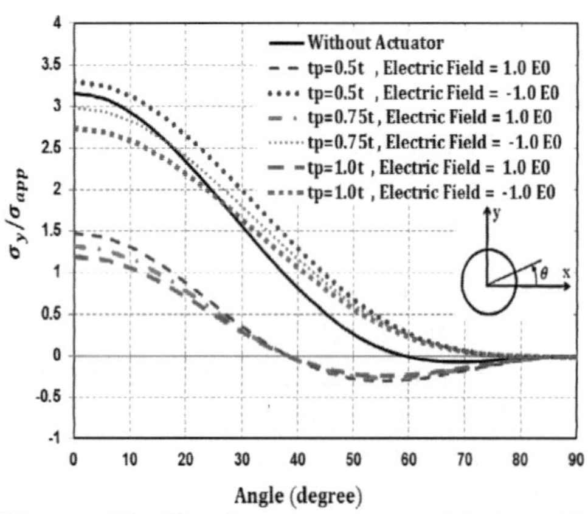

Figure 10. The Stress σ_y around hole with different piezoelectric actuator thickness and applied electric filed.

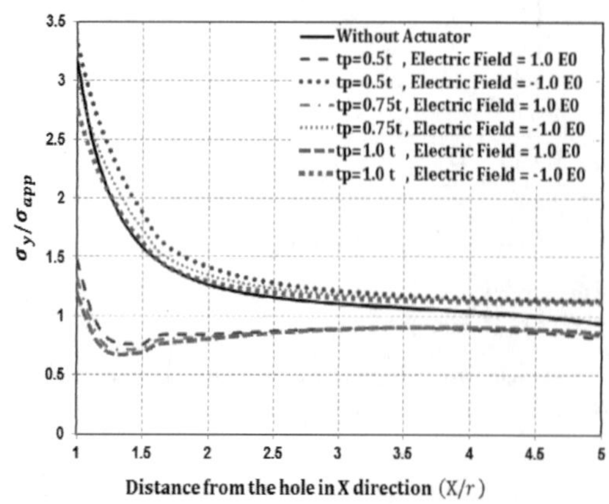

Figure 11. The stress distributions σ_y of the plate along the x −axis with different piezoelectric actuator thickness.

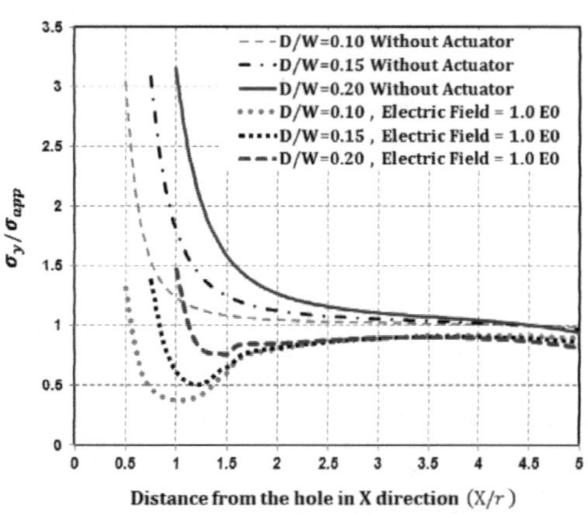

Figure 12. The stress distributions σ_y of the plate along the x −axis with different D/W ratio (Case 1).

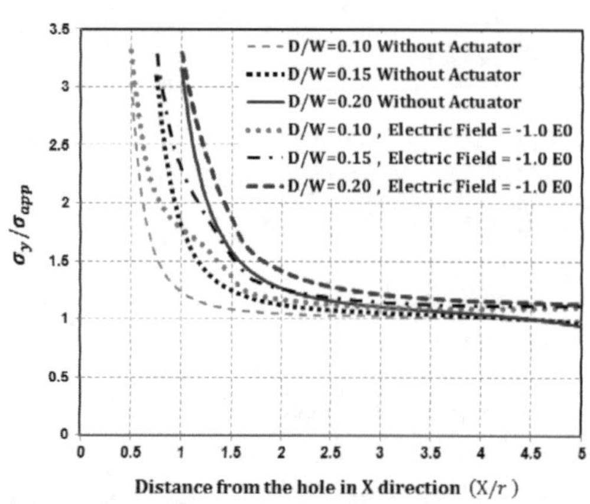

Figure 13. The stress distributions σ_y of the plate along the x −axis with different D/W ratio (Case 2).

6. Conclusion

This paper investigated the effect of piezoelectric actuators placed above a circular hole of a rectangular plate under uniform uniaxial tension. The methodology of this work is based on the investigation of the electromechanical behaviour of the piezoelectric actuator on the stress distribution around the hole and along the width of the host plate. The results showed that the normalized stress distribution reduces with the application of positive electric field and this reduction increases with increasing the positive applied electric field. Alternatively, the application of a negative electric field would lead to increase of the SCF and at the same time the stress distribution. The results also indicated that the ratio (D/W) affects the reduction of the stress concentration factor and the stress distribution at the vicinity of the hole.

Acknowledgments

The authors are grateful to the support of the Ministry of Higher Education Malaysia through the Grant No. FRGS 15-189-0430.

References

[1] Walter D F P and Pilkey D 2008 Stress Concentration Factors (ed: John Wiley & Sons), p. 560.
[2] Crawley E F and De Luis J 1987 *AIAA J.* **25** 1373
[3] Srinivasan A V and McFarland D M 2001 *Smart structures: analysis and design* (Cambridge University press).
[4] Frecker M I 2003 *J. Intel. Mat. Syst. Str.* **14**. 207
[5] Chopra I 2002 *AIAA J.* **40** 2145
[6] Abuzaid A, Hrairi M and Dawood M S I S 2015 *Actuators* **4** 77
[7] Wang Q, Quek S T and Liew K M 2002 *Smart Mater. Struct.* **11** 404
[8] Liu T J C 2007 *Theor. Appl. Fract. Mech.* **47** 120
[9] Platz R, Stapp C, and Hanselka H 2010 *Smart Mater. Struct.* **20** 085009
[10] Rabinovitch O 2007 *J. Compos. Mater.* **41** 525
[11] Abuzaid A, Hrairi M, and Dawood M S I 2015 *Adv. Mater. Res.* **1115** 517
[12] Abuzaid A, Dawood M S I, and Hrairi M 2015 *Appl. Mech. Mater.* **799-800** 788
[13] Sensharma P K, Palantera M J, and Haftka R T 1993 *J. Intel. Mater. Syst. Str.* **4** 509
[14] Shah D, Joshi S, and Chan W 1994 *Smart Mater. Struct.* **3** 302
[15] Fesharaki J J, Madani S G 2016 *Int. J. Adv. Struct. Eng.* **8** 229
[16] Fesharaki J J and Golabi S I 2016 *Smart Struct. Syst.* **17** 725
[17] IEEE 1988 IEEE Standard on Piezoelectricity ANSI/IEEE Std 176-1987.

Numerical simulation of non-isothermal thin liquid film flow on inclined plane using an implicit finite difference scheme

Mohammad Ameer Hamza[1*], **Ahmad Tariq Jameel**[2], **Waqar Asrar**[1] **and Asif Hoda**[3]

[1]Dept. of Mechanical Engineering, [2]Dept. of Biotechnology Engineering,
International Islamic University Malaysia, P.O. Box 10,
Gombak, 50728 Kuala Lumpur, Malaysia

[3]Dept. of Mechanical Engineering, Jubail University College, P. O. Box 10074
Jubail Industrial City 31961, Kingdom of Saudi Arabia

Corresponding author's e-mail address: mahamza2000@gmail.com

Abstract. The classical problem of the stability and dynamics of thin liquid films on solid surfaces has been studied extensively. Particularly, thin liquid films subjected to various physico-chemical effects such as thermocapillarity, solutal-Marangoni and evaporative instabilities at the film surface has been the focus of research for more than two decades. Various flow configurations of thin film such as thin film on plane, inclined, and wavy surfaces has been the subject of recent investigations. An inclined film compared to a horizontal film, also experiences the gravity force which may significantly influence the nonlinear dynamics of the film coupled with other forces. In this research, we attempt to study the stability and dynamics of thin liquid films subjected to thermocapillarity and evaporative instabilities at the free surface besides instability owing to ubiquitous van der Waals attraction, using numerical simulations. For a Newtonian liquid, flow in thin liquid film on a planar support and bounded by a passive gas, is represented by Navier-Stokes equation, equation of continuity and appropriate boundary conditions. The external effects are incorporated in the body force term of the Navier-Stokes equation. These governing equations are simplified using the so called long-wave approximation to arrive at a nonlinear partial differential equation, henceforth called equation of evolution (EOE), which describes the time evolution of the interfacial instability in the film caused by internal and/or external effects. Efficient numerical method is required for the solution of the equation of evolution (EOE) in order to comprehend the nonlinear dynamics of the thin film. Here we present the results of our numerical simulation using Crank-Nicholson implicit finite difference scheme applied to the thin film model incorporating instabilities owing to gravity, evaporation and thermo-capillarity. Comparison of our results with those obtained from Spectral method, show remarkable agreement for most of the cases investigated.

Keywords: Thin liquid film flow; Nonlinear Dynamics; Implicit Finite Difference; Spectral Method.

IOP Conf. Series: Materials Science and Engineering **184** (2017) 012065 doi:10.1088/1757-899X/184/1/012065

1. Introduction

The stability and dynamics of thin liquid films on solid substrates has been studied extensively for the past three decades. Particular attention has been focused on the case of thin liquid films subjected to various physical-chemical effects such as thermocapillarity, solutal Marangoni and evaporative instabilities at the surface. Furthermore, if the film is inclined at an angle to the horizontal, it experiences the gravitational force which might influence the nonlinear dynamics of the film coupled to other forces. Heat transfer in thin liquid film flow on an inclined plane occurs in numerous chemical and biochemical processes such as solid coating, food processing and in systems with evaporative cooling as well as high throughput exchange devices.

In the case of Newtonian fluid flow of thin liquid film on a solid support and bounded by a passive gas, the dynamics are represented by Navier Stokes equations and equation of evolution together with appropriate boundary conditions. The external effects are generally incorporated in the body force term of the Navier Stokes equations. These governing equations can then be simplified using the so called long wave approximation to arrive at fourth order nonlinear partial differential equation, henceforth called the equation of evolution which describes the time evolution of the interfacial instability caused by internal and/or external effects. The details of the derivation of the equation of evolution are available in the literature.

The linear stability characteristics can be obtained by the solution of the linearized equation of evolution. However, for the accurate simulations of the nonlinear dynamics and surface morphology of thin-film, we require efficient numerical methods for the solution of the equation of evolution (EOE). The extent of nonlinearity and the stiffness of the resulting differential equation depend upon the nature of various physico-chemical effects incorporated in the thin-film model [1-11]. Numerous works on numerical simulations of various liquid thin-film models have been reported in the literature [e.g., 12, 13, 14, 15]. Burelbach et al. (1988) [14] has solved the problem of evaporating/condensing liquid film on horizontal plane using an implicit finite difference scheme, Crank-Nicholson mid-point rule, by employing centered difference for space and forward difference for time derivatives. The resultant nonlinear difference equations are solved by Newton-Raphson iterations. Ali et al. (2005) [13] have employed similar discretization scheme for their isothermal liquid thin-film on inclined plane, however IMSL subroutines were employed for the solution of resulting difference equations which essentially uses an adoption of Newton-Raphson method. Fourier spectral method was used by Joo et al. (1991) [12] for numerical simulations of heated falling films under the influence of thermocapillarity and evaporative effects at the surface, however, they have ignored van der Waals effects in their formulation. There are many other works that have been reported in the literature.

The present paper illustrates some results from the numerical simulation of the flow of the thin liquid layer on inclined plane and subjected to evaporative and thermocapillarity instabilities. An implicit time averaged Crank Nicholson finite difference scheme formulation of the EOE has been used for numerical simulations. To validate our numerical codes of our implicit Crank Nicholson finite difference scheme we apply them to the solution of the equation of evolution of long-wave instabilities of heated falling film discussed by Joo et al. (1991) [12]. The results generated using our code are compared with the results reported by Joo et al. (1991) [12] which uses Fourier spectral method. The results from our implicit finite difference code closely match with the Fourier spectral thereby proving the viability of the implicit scheme as a reliable alternative to an explicit finite difference method. The validated solution procedure has thereafter been employed to study the time evolution of surface instabilities in a non-isothermal thin-film by incorporating the thermocapillarity and evaporation effects.

2. Thin film model

We assume a thin viscous Newtonian liquid film. The film is bounded above by its vapour and below by a uniformly heated rigid plane inclined at an angle θ to the horizontal. The film is sufficiently thick so that a continuum theory of the liquid is valid. The liquid film consists of an incompressible Newtonian fluid with constant material properties and is laterally unbounded. Evaporation occurs at the layer resulting in mass loss, momentum transfer and energy consumption. Plate has a fixed constant temperature T_H. Liquid temperature T_F on the free surface is controlled by losses to the passive gas above. The surface tension σ_o depends on temperature so that the thermocapillarity effects are present.

We consider a 2-D Cartesian coordinate system as shown in figure 1. The vapor=liquid interface is located at z=h(x,t). Liquid is assumed to evaporate in a direction normal to the interface. The Navier-Stokes and continuity equations in a non-dimensional Cartesian co-ordinate system (x, z) defined as in figure 1 are:

$$x: u_t + uu_x + wu_u = -p_x - \phi_x + Gsin\theta + u_{xx} + u_{zx} \quad (1)$$

$$z: w_t + uw_x + ww_z = -p_z - Gcos\theta + w_{xx} + w_{zz} \quad (2)$$

$$EOC: u_x + w_z = 0 \quad (3)$$

Where, the spatial coordinates are scaled by mean film thickness h_0; time by viscous time scale h_0^2. u and w are the components of velocity in x- and z- directions, respectively and p is the pressure. The subscripts denote the derivatives.

2.1 Equation of Evolution

The length scale of the initial instability in a thin film is much larger than the mean film thickness h_0. For flows with a characteristic length in the x-direction typically proportional to the disturbance wavelength and much larger than the thickness of the liquid layer, the long wave analysis of Benney (1966) [15] can be applied to the problem at hand. Following the derivation in Joo et. al. (1991) [2], the long wave approximation of our system of equations result in an equation of evolution (in space & time for the film interface h(x, t) in original coordinates (x, z) given by

$$\begin{aligned} h_t &+ \frac{E}{h+K} + Gh^2h_x sin\theta \\ &+ [\frac{2}{15}G^2h^6h_x sin^2\theta + \frac{KM}{P}\frac{h^2h_x}{(h+K)^2} + \frac{E^2}{D}\frac{h^3h_x}{(h+K)^3} - \frac{1}{3}Gh^3h_x cos\theta + \frac{A}{h}h_x \\ &+ Sh^3h_{xxx}]_x + \frac{5}{24}GE(\frac{h^4}{h+K})_x sin\theta \\ &+ EP(\frac{h}{h+K})^3\left[\frac{E}{3(h+K)} + \frac{G}{120}(7h-15K)hh_x sin\theta\right] = 0 \end{aligned} \quad (4)$$

2.2 EOE by Joo et. al. (1991)

The EOE of Joo et. al. is [12]

$$\begin{aligned} h_t &+ \frac{\bar{E}}{h+K} + Gh^2h_\xi sin\theta \\ &+ \varepsilon[\frac{2}{15}G^2h^6h_\xi sin^2\theta + \frac{KM}{P}\frac{h^2h_\xi}{(h+K)^2} + \frac{E^2}{D}\frac{h^3h_\xi}{(h+K)^3} - \frac{1}{3}Gh^3h_\xi cos\theta + \frac{A}{h}h_\xi \\ &+ \bar{S}h^3h_{\xi\xi\xi}]_\xi + \frac{5}{24}\varepsilon GE(\frac{h^4}{h+K})_\xi sin\theta \\ &+ \varepsilon EP(\frac{h}{h+K})^3\left[\frac{E}{3(h+K)} + \frac{G}{120}(7h-15K)hh_\xi sin\theta\right] = 0 \end{aligned} \quad (5)$$

The following scale factors are used: $\xi=\varepsilon x$, $\zeta=z$, $\tau=\varepsilon t$

$(E, D, S) = (\varepsilon\bar{E}, \varepsilon^2 D, \varepsilon^{-2}\bar{S})$. Joo et. al. have employed ε=0.2 and $\bar{S}$=0.1 throughout their simulations.

3. Numerical method

We have employed an implicit finite difference numerical discretization scheme discussed below.

3.1 Implicit Finite Difference Scheme-Crank Nicholson Mid Point Rule

We use the conservative form of the rescaled equation of evolution. This equation of evolution is discretized using forward difference in time and central difference in space with time averaging, resulting into a set of difference equations for the film thickness h. These equations are then solved as a set of nonlinear coupled algebraic equations by an iterative procedure of the type of Newton's method. Let subscript i denotes space discretization and superscripts n and n + 1 denote values of variables at nth and (n + 1)th times. Thus, Eq. (5) after discretization appears as:

$$
\begin{aligned}
&\frac{h_i^{n+1}}{\Delta\tau} + \frac{S}{2A^2}\left[\left(\frac{E}{h_i + k}\right)^{n+1} + \left(\frac{E}{h_i + k}\right)^{n}\right] + \left(\frac{S}{A}\right)^{1/2}\frac{Re}{A}\frac{1}{4\Delta x}[\{h_i{}^2(h_{i+1} - h_{i-1})\}^{n+1} \\
&+ \frac{1}{2x}\left[\frac{2}{15}\frac{Re^2}{A}\frac{1}{4\Delta x}[\{h_{i+1}^6(h_{i+2} - h_i) - h_{i-1}^6(h_i - h_{i-1})\}^{n+1}\right. \\
&+ \left.\{h_{i+1}^6(h_{i+2} - h_i) - h_{i-1}^6(h_i - h_{i-1})\}^{n}\right] \\
&+ \frac{KM}{PA}\frac{1}{4\Delta x}\left[\left\{\frac{h_{i+1}^2}{(h_{i+1} + K)^2}(h_{i+2} - h_i) - \frac{h_{i-1}^2}{(h_{i-1} + K)^2}(h_i - h_{i-2})\right\}^{n+1}\right. \\
&+ \left.\left\{\frac{h_{i+1}^2}{(h_{i+1} + K)^2}(h_{i+1} - h_i) - \frac{h_{i-1}^2}{(h_{i-1} + K)^2}(h_i - h_{i-2})\right\}\right]^{n} \\
&+ \frac{E^2}{DA}\frac{1}{4\Delta x}\left[\left\{\frac{h_{i+1}^3(h_{i+2} - h_i)}{(h_{i+1} + K)^3} - \frac{h_{i-1}^3(h_i - h_{i-2})}{(h_{i-1} + K)^3}\right\}^{n+1}\right. \\
&+ \left.\left\{\frac{h_{i+1}^3(h_{i+2} - h_i)}{(h_{i+1} + K)^3} - \frac{h_{i-1}^3(h_i - h_{i-2})}{(h_{i-1} + K)^3}\right\}^{n}\right. \\
&- \frac{1}{3}\frac{G\cos\theta}{A}\frac{1}{4\Delta x}\left[\{h_{i+1}^3(h_{i+2} - h_i) - h_{i-1}^3(h_i - h_{i-2})\}^{n+1}\right. \\
&+ \left.\{h_{i+1}^3(h_{i+2} - h_i) - h_{i-1}^3(h_i - h_{i-2})\}^{n}\right] \\
&+ \frac{1}{4(\Delta x)^3}\left[\{h_{i+1}^3(-h_{i-1} + 2h_i - 2h_{i+2} + h_{i+3})\right. \\
&- h_{i-1}^3(-h_{i-3} + 2h_{i-2} - 2h_i + h_{i+1})\}^{n+1} \\
&+ \left.\{h_{i+1}^3(-h_{i-1} + 2h_i - 2h_{i+2} + h_{i+3}) - h_{i-1}^3(-h_{i-3} + 2h_{i-2} - 2h_i + h_{i+1})\}^{n}\right] \\
&+ \frac{1}{4\Delta x}\left[\{(h_{i+1})^{-1}(h_{i+2} + h_i) - (h_{i-1})^{-1}(h_i + h_{i-2})\}^{n+1}\right. \\
&- \left.\{(h_{i+1})^{-1}(h_{i+2} + h_i) - (h_{i-1})^{-1}(h_i + h_{i-2})\}^{n}\right] \\
&+ \frac{5}{24}\left(\frac{S}{A^3}\right)^{1/2}\frac{Re}{4\Delta x}\left[\left\{\frac{h_{i+1}^4}{h_{i+1} + K} - \frac{h_{i-1}^4}{h_{i-1} + K}\right\}^{n+1} + \left\{\frac{h_{i+1}^4}{h_{i+1} + K} - \frac{h_{i-1}^4}{h_{i-1} + K}\right\}^{n}\right. \\
&+ \left(\frac{S}{A^3}\right)^{1/2}\frac{EP}{2}\left[\left\{\left(\frac{S}{A}\right)^{1/2}\frac{Eh_i^3}{3(h_i + K)^4} + \frac{Re}{120}\frac{(7h_i - 15K)h_i^4}{(h_i + K)^3}\left(\frac{h_{i+1} - h_{i-1}}{2\Delta x}\right)\right\}^{n+1}\right. \\
&+ \left.\left\{\left(\frac{S}{A}\right)^{1/2}\frac{Eh_i^3}{3(h_i + K)^4} + \frac{Re}{120}\frac{(7h_i - 15K)h_i^4}{(h_i + K)^3}\left(\frac{h_{i+1} - h_{i-1}}{2\Delta x}\right)\right\}^{n}\right] = 0
\end{aligned}
\qquad (6)
$$

IOP Conf. Series: Materials Science and Engineering **184** (2017) 012065 doi:10.1088/1757-899X/184/1/012065

4. Results and discussions

The results of the numerical solution of the evolution equation of Joo et. al. (1991) using implicit method is compared with the numerical results of Joo et. al. (1991) obtained using Fourier spectral method. We have incorporated the various thermophysical effects from Joo et. al. (1991) model in our thin film model, with the exception of van der Waals effects. Figure (2) shows the evolution of film interface for an isothermal film. Our implicit finite difference method gives almost identical results with Joo et. al. (1991) with only a slight deviation. Figure (3) shows the evolution of liquid layers with thermocapillarity in the absence of gravity. Significant deviation is observed both in the film rupture time as well as film profile.

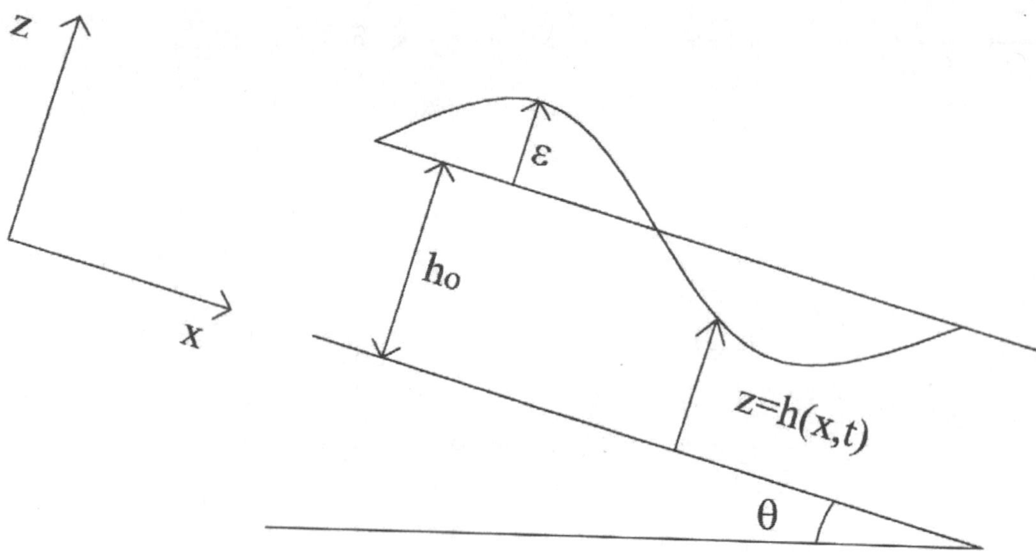

Figure 1: The physical configuration of thin film flowing down an inclined plane

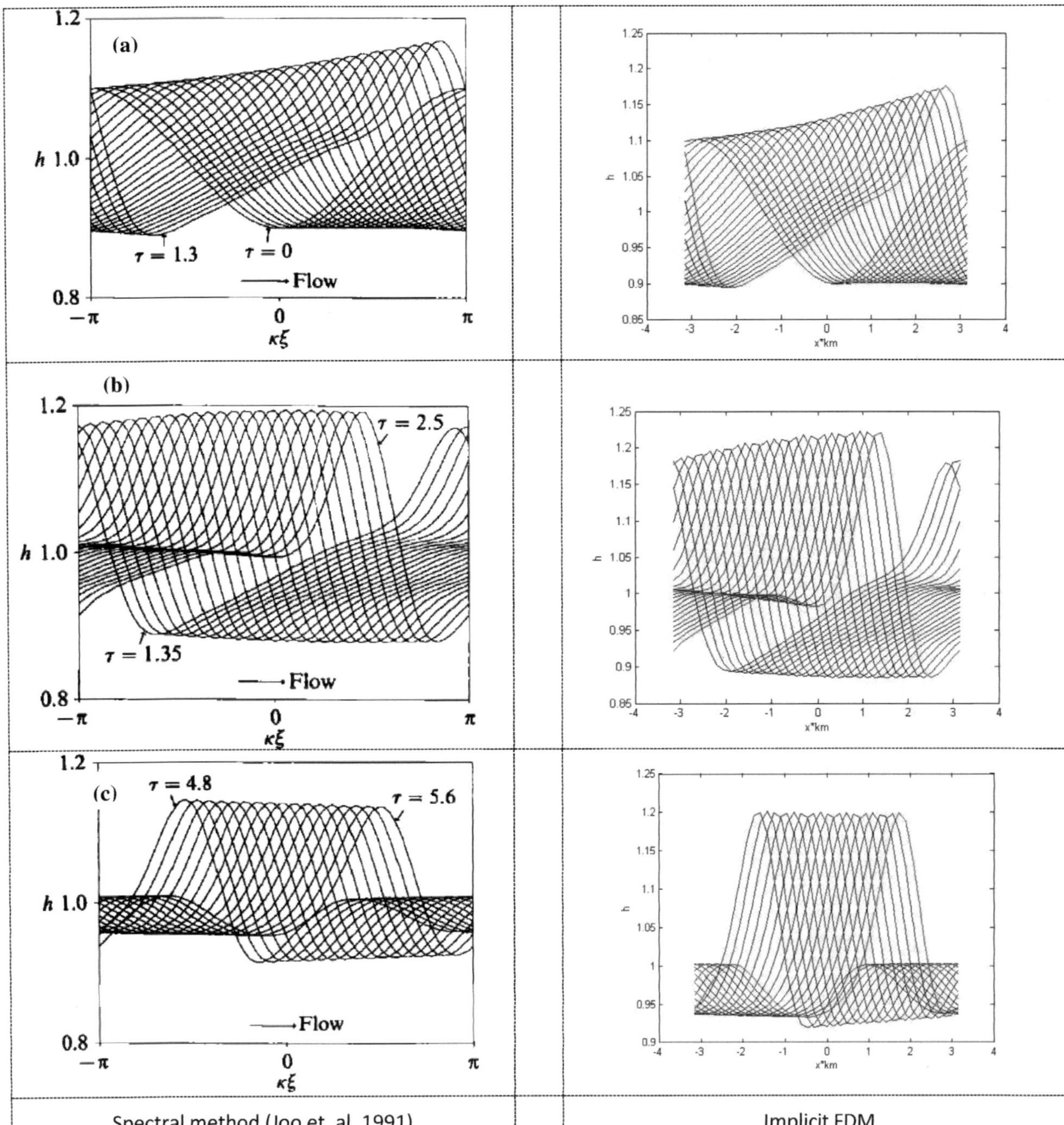

Figure 2: Surface wave instability for an isothermal layer with k=1.11, G=5, β=45^0, $\bar{S}$=0.1. **(a)** Free surface configuration for $0 \leq \tau \leq 1.3$ with Δτ=0.05; **(b)** free surface configurations for $1.35 \leq \tau \leq 2.5$ with Δτ=0.05; **(c)** free surface configurations for $4.8 \leq \tau \leq 5.6$;

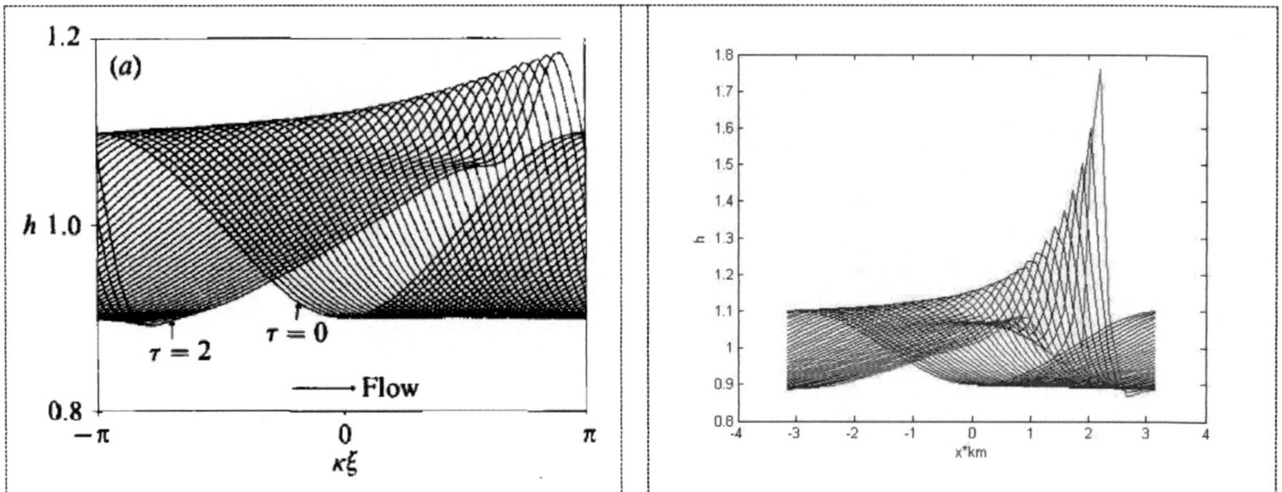

(a) Spectral method (Joo et. al. 1991) **(b)** Implicit FDM

Figure 3: Free surface configuration of gravitational instability for an isothermal layer with k=0.7, G=5, $\beta=45^0$ and $\bar{S}=0.1$ with $\Delta\tau=0.05$ up to $\tau=2$.

5. Conclusion

The results obtained by our implicit Crank Nicholson finite difference method demonstrate a close conformity to the results obtained by Fourier spectral method of Joo et. al. (1991), with some slight deviation in some cases and significant deviation in one of the cases investigated. This indicates that the implicit finite difference method is a reliable method for investigating the highly nonlinear thin film model to a high degree of accuracy.

References

[1] Huppert, H. E., and J. E. Simpson, "The slumping of gravity currents," 1980, J. Fluid Mech. **99,** 785-789

[2] Huppert, H. E., "The propagation of two-dimensional and axisymmetric viscous gravity currents over a rigid horizontal surface," 1982a, J. Fluid Mech. **121,** 43-58.

[3] Grotberg, J. B., "Pulmonary flow and transport phenomena," 1994, Annu. Rev. Fluid Mech. **26,** 529-571.

[4] Sharma, A., and E. Ruckenstein, "An analytical nonlinear theory of thin film rupture and its application to wetting films," 1986a, J. Colloid Interface Sci. **113,** 456-479.

[5] Wong, H., I. Fatt, and C. J. Radke, 1996, "Deposition and thinning of the human tear film," J. Colloid Interface Sci. **184,** 44-51.

[6] Gaines, G. L., 1966, *Insoluble monolayers at liquid-gas interfaces* (Interscience, New York).

[7] Bikerman, J. J., 1973, *Foams* (Springer, New York).

[8] Edwards, D. A., H. Brenner, and D. T. Wasan, 1991, *Interfacial Transport Phenomena* (Butterworth Heineman, Boston).

[9] Schramm, L. L., and F. Wassmuth, 1994, "Foams: basic principles," in *Foams: Fundamentals and Applications in the PetroleumIndustry*, edited by L. L. Schramm (American Chemical Society, Washington), pp. 3-46.

[10] Wasan, D. T., K. Koczo, and A. D. Nikolov, 1994, "Mechanisms of aqueous foam stability and antifoaming action with and without oil: a thin film approach," in *Foams: Fundamentalsand Applications in the Petroleum Industry*, edited by L. L. Schramm (American Chemical Society, Washington), pp. 47-114.

[11] Wong, H., C. J. Radke, and S. Morris, "The motion of long bubbles in polygonal capillaries. Part 1. Thin films," 1995, J. Fluid Mech. **292,** 71-94.

[12] S. W. Joo, S. H. Davis, S. G. Bankoff, "Long wave instabilities of heated falling films: Two-dimensional theory of uniform layers," 1991, J. Fluid. Mech. **230,** 117-146.

[13] M. A. Ali, A. T. Jameel, F. R. Ahmadun, "Stability and rupture of nano-liquid film (NLF) flowing down an inclined plane," 2005, Computers and Chemical Engineering, **29**, 2144–2154.

[14] J. P. Burelbach, S. G. Bankoff, "S. H. Davis, Nonlinear stability of evaporating/condensing liquid films," 1988, J. Fluid Mech. **195**, 463–494.

[15] M. B. Williams, S. H. Davis, "Nonlinear theory of film rupture," 1982, Journal of Colloid and Interface Science. **90**, 220-228.

IOP Conf. Series: Materials Science and Engineering **184** (2017) 012066 doi:10.1088/1757-899X/184/1/012066

Adoption of JMM practices – A key to performance improvement of a local automotive industry

Mohiuddin A K M[1,a], Julia Bt. Haji A. Jabbar[2], Ahsan Ali Khan[2], A N Mustafizul Karim[2]

[1]Department of Mechanical Engineering, [2]Department of Manufacturing & Materials Engineering,

Kulliyyah of Engineering, International Islamic University, Malaysia

P.O.Box 10, 50728 Kuala Lumpur, Malaysia

E-mail: [a]mohiuddin@iium.edu.my

Abstract. It is imperative for a manufacturing company all over the world to constantly look for ways to increase productivity and at the same time to lower cost to secure a competitive position. It is recognized that practices associated with Japanese Manufacturing Management (JMM) can yield a superior competitive advantage in terms of productivity, quality and provide overall successful business performance. This paper discusses the transfer of the best practices of the JMM locally and analyses the impact of adoption and adaptation of the management system as an in-depth case study conducted in a Malaysian automotive company. This study is to identify what are the changes in terms of the philosophy and practices undertaken by the company and ascertain the impact of the JMM on its manufacturing and financial performances. The elements of business performance from the viewpoint of manufacturing are based on safety, Parts Per Million (PPM), in-line Defect per Unit (DPU), First Time Quality (FTQ), cycle time, productivity, efficiency and stock level. The results show a positive impact to the automotive plant manufacturing performance. For example, safety index has reduced to 0 major accident occurrences. The PPM and In-line DPU have improved by 98% and 91% respectively whereas the FTQ has improved by 167%. Cycle time has reduced from 20 to 6 minutes and productivity increased up to 43% whilst the efficiency reached at 99.9%. The stock level was reduced from half month to 3 days after the adoption of the JMM. The revenue has increased up to 92%, the percentage of expenses has reduced from 11.04% to 3.06% giving an improvement of 72% whilst the net profit has increased from 5.33% to 8.15%. The Return of Asset (RoA) and Return of Equity (RoE) also showed slight improvement despite the effects from the restructuring exercise, Tsunami calamity and fluctuation in Japanese exchange rate.

1. Introduction

Automotive companies all over the world are constantly looking for ways to increase productivity and at the same time to reduce cost to remain competitive. They are looking for ways and means to produce quality and competitive products and services to the market. Currently, Toyota Motors Corporation is the largest auto maker in the world. They exceeded the sales records of the giant automotive companies such as General Motors (GM) and Ford. The title of number one automotive manufacturer formerly held by GM since 1950s for more than seven decades and was claimed by Toyota in 2008. Toyota lost the title when sales declined in 2011 due to natural disasters of tsunami

IOP Conf. Series: Materials Science and Engineering **184** (2017) 012066 doi:10.1088/1757-899X/184/1/012066

and Thailand flood. But it managed to recover it in the following year [1]. In 2012, Toyota sold 9.75 million vehicles in the United States compared with 9.29 million for GM [2]. There remains a wide gap in profitability between the giant American automotive manufacturers and their Japanese automotive counterparts. Toyota continues to be the industry benchmark of automakers. But the American automakers have been facing with various problems including bankruptcy, reorganization, declining demand, high operational costs, and overproduction [3].

The success for Toyota does not happen overnight. The success for Toyota starts since post World War II (WWII). Toyota has developed its own principles in excellent manufacturing practices and philosophy. After WWII, with scarce materials and resources, Toyota has utilized different techniques of manufacturing rather than mass production introduced by Ford in the USA. Unlike mass production with high inventory level to cater for big production volume to take advantage of economy of scale, the manufacturing practices by Toyota emphasizes on Lean Production. Lean Production focuses on producing high quality finished goods satisfactory to customer demands by eliminating waste (muda). The Lean Production approach includes a wide variety of management practices, including just-in-time (JIT), continuous improvement (Kaizen), quality systems, quality circles, work teams, cellular manufacturing, supplier management, etc. in an integrated system [4,5]. These practices, which later referred to the best practices when applied to manufacturing scenarios, are continuously improved and tied to form a management system such as Toyota Production System (TPS). The effectiveness of this technique motivated most of the Japanese automotive companies in the 70s to implement this technique and later it was adapted and implemented in the USA, Europe and other parts of the world. Isuzu also practices Lean Production and the Japanese way of manufacturing with its philosophy and practice called Isuzu Manufacturing Management (IMM). IMM is implemented by its parent company in Japan and similar best practices are adopted and shared by its overseas plants throughout the world. The TPS, IMM and JIT have been adopted and adapted in various developed and developing countries to the extent they form what is known as "The Japanese Way" [6]. Until now, the Japanese has continuously perfected their manufacturing strategy in automotive industry [7]. Japanese Manufacturing Management (JMM) has evolved over the last thirty years and it continues to evolve. A lot of studies examined the transfer of Japanese production practice and its impact on the company performance in the US [6, 7, 8]. Also, in the UK, Japanese manufacturing practices and its relationship with performance have been studied extensively. Phan et al. [9] conducted two surveys with twenty seven Japanese manufacturing companies in the 1990s and 2000s to compare the degree of implementation of eleven quality management practices and their impact on different dimensions of competitive performance during the period. This paper provides an insight of the performance improvement of a local automotive plant by adoption and adaptation of JMM.

2. Overview of automotive manufacturing in Malaysia

The manufacturing sector in Malaysia plays a significant role in the financial health of the nation. In 2011, the manufacturing sector contributed 27.9% of Malaysia Gross Domestic Product (GDP) compared to other sectors i.e. agriculture (7.4%), mining (7.1%), construction (3.2%) and services (57.3%). The percentage had also steadily increased from 26.6% to 27.6% for respective years of 2009 and 2010 [10]. The automotive industry is viewed as one of the most pertinent contributors for the economy since it is directly correlated with manufacturing and service sectors. The industry is also significant because more than half of the population in Malaysia owns a vehicle with the rate of car utilization stood at 67%. The number of new car registration has also increased every year [11].

The presence of Japanese influence in Malaysian automotive industry has existed for quite some time. The 'Look East' policy by the former Prime Minister Tun Mahathir Mohamad has resulted in the establishment of the first national car company Proton in Malaysia, in 1983 which jointly collaborated with Mitsubishi Motors Corporation and Mitsubishi Corporation. The second national car company of

Malaysia was established in 1992 under the name PERODUA. The company is a joint venture between a Malaysian company and Daihatsu Motor Corporation.

3. Japanese manufacturing management (JMM) and Toyota production system (TPS)

The objective of manufacturing management is to manage resources effectively and contribute significantly to the achievement of corporate objectives. The main target of manufacturing management is productivity, meaning manufacturing or production effectiveness. Manufacturing requires integration with other management functions to ensure that the manufacturing objectives can be accomplished. This includes the running of production processes, quality management, logistics, maintenance, industrial engineering, procurement as well as human resources.

For effective operation, top management must acknowledge that the organization as a whole needs to develop an integrated approach to the business and its activities. This involves recognizing and understanding the interactions between different functions that are necessary for the organization to be effective. Different management functions must be involved in setting the corporate strategy of the organization. Each function will then set its own functional objectives and develop a functional strategy to achieve those objectives.

JMM places a great emphasis on mission and vision of a company. In the Executive Summary of the 14 Toyota Way Principles extracted from the book 'The Toyota Way' written by Liker [12], it is stated that companies should base the management decisions on a long-term philosophy, even at the expense of short-term financial goals. Company should also have a philosophical sense of purpose that supersedes any short-term decision making. The company's philosophical mission is the foundation for all the other principles [12]. Liker and other researchers believe that the basis of Toyota's success stems from the business philosophy that underlies its production system.

TPS is the most vital component for understanding Japanese production management. It can be seen as a technology-based, comprehensive production management system with the primary goals of increasing productivity and reducing costs. This is achieved by reducing cycle time, increasing flexibility, reducing stock levels and shortening machine changeover times [13]. The best word to describe TPS is Lean Production. The use of the term "Lean", in a business or manufacturing environment, describes a philosophy that incorporates a collection of tools and techniques into the business processes to optimize time, human resources, assets, and productivity, while improving the quality level of products and services to their customers.

Shingo (1989) in his book 'The Study of the TPS from an Industrial Engineering Perspective' [14] identified the basic features of TPS as follows:

1. It achieves cost reductions by eliminating waste, be it staff time, materials, or other resources.
2. It reduces the likelihood of overproduction by maintaining low inventories ("non-stock") and keeps labor costs low by using minimal manpower.
3. It reduces production cycle time drastically with innovations like the Single-Minute Exchange of Die (SMED) system, which cuts downtime and enables small-lot production.

The driving force behind the Japanese system of production is eliminating waste, thereby maximizing process efficiency and the returns on resources. A wide number of principles and practices can be employed to achieve this goal. While low inventory levels are a key outcome of the TPS, an important element of the philosophy behind its system is to work intelligently and eliminate waste so that inventory is no longer needed. The act of imitating without understanding the underlying concept or motivation may have led to the failure of those projects.

4. Adoption of JMM

The findings by Laosirihongthong et al. [6] on Japanese auto companies established in Thailand, Indonesia, Malaysia, Singapore, Philippines, Vietnam and Australia show that there are six pillars to support JMM implementation. These are the key enablers to success in adopting JMM. These six pillars are summarized as follows:

Cultural change: local employees are encouraged to adapt and be familiar with JMM by providing on job training and education.

Quality at the shop floor: Assembly line employees practice quality check point to ensure they do not receive defects, make any defects or transfer the defects to the next workstation. Defects are identified using root cause analysis through PDCA – problem solving cycle to detect defects in dimensions of Man (skill and training), Machine (preventive maintenance), Material (Suppliers development program) and Method (Standard Operating Procedure and Work Instruction). Morning briefing is held daily as part of quality circle to improve production process.

Consensus: Top Down and Bottom Up approaches present in JMM companies.

Incremental continuous improvement: Implementation of Kaizen at shop floor including open concept office layout with no partitions for easy communication and access. Gemba (Go see problem at source) are conducted every day by managers to better understand issues at workplace and to develop relations with subordinates. Visual Management System and Boards are effectively used in the companies under study. Also a large number of Kaizen program are implemented.

Benchmarking: All cases studied use benchmarking techniques for quality, delivery, cost, flexibility and innovation metrics.

Backward-forward integration: The relationship integration exists among the suppliers, the auto company itself and customers.

Laosirihongthong et al. [6] outlined that manufacturing practices should describe the contributions it makes to the cost, quality, flexibility, delivery and future objectives of the business. Manufacturing practice / strategy which is developed and implemented through a set of activities into output measurable in terms of the competitive priorities of cost, quality, delivery, flexibility and innovation, can be further elaborated as follows:

Quality: Manufacture of products with high quality or performance standards.

Delivery: Meet delivery schedules and respond quickly to customer orders.

Cost: Production and distribution of product at low cost.

Innovation: Number of new products and processes that are developed to fit customers' requirement and differ from competitors.

5. Implementation of JMM

The local company, upon adopting the JMM, had taken massive improvement activities under JMM umbrella during its restructuring exercise in late 2007. In line with the implementation of JMM, the improvement activities can be listed as follows:

Plant Layout Improvement: the initial improvement in plant layout was conducted by rearranging the original layout to fit the lean requirement. All of the plant reengineering and relocation activities are performed under a project called J-Project.

Quality Improvement: Among the quality improvement activities conducted are workshop quality improvement, open space concept for the offices and supplier quality management. The workshop on quality improvement is dedicated to minimize wrong assembly and defects in production line and most detection are performed by production operators. This self-check activity has empowered operators to ensure the quality of the vehicle manufactured to be in-line with the standard required by JMM.

With the implementation of the J-Project and quality management adopted from JMM in 2008 onwards, all operations personnel including manufacturing supervisors, technical specialist, engineers, managers and in particular production operators were directly involved in the Quality Improvement Activities of the JMM. Their skill and energy are the sources of continuous improvement as well as standardization and has become the biggest asset in the organization today. Their competence and commitment are an invaluable capital to the organization.

In quality control for manufacturing, the JMM adopted by the local company emphasizes on quality management of not passing defect and being unable to make defects. For JMM, quality must be measured, controlled and charted at shop floor where the process and manufacturing strategy is executed. The reason is because the "process owner" is a person who is responsible for providing the resources to a process to perform activities and tasks thereby intimately knowing where exact problems occur. Therefore, such operational personnel should be given the opportunity to suggest the options to solve any problems at source through Gemba and Kaizen activities.

The local company has been practicing three main quality-check-point activities for operational personnel

1). Do not receive any defects;
2). Do not make any defects; and
3). Do not transfer any defects".

The JMM adopted by the local company necessitates the changes in quality activities implemented in accordance to JMM requirement. The activities for workshop improvement consist of daily quality audit meeting; two (2) minutes parts check; two (2) minutes tips check; defect parts management; work stop rule; torque analyser usage; daily inspection of torque wrench and torque meter; daily checking of inspection tools; fallen bolt, torque wrench and meter; 4M change control point at QC inspection points; QC in-line inspection and process matrix; QC quality agreements; feedback rule among production lines and from QC to production; abnormalities guidelines; maintenance agreements; joint operation manual; purchase of new torque; colour code etc. [15].

6. Results and discussions

6.1. Manufacturing performance

The results of the implementation of the JMM in the local company, gauged in the process performance metrics based on safety, quality, cost and delivery (SQCD). The performance metrics are listed as follows:

a) Safety - Safety index
b) Quality – Parts Per Million (PPM), In-Line Defect Per Unit (DPU), First Time Quality (FTQ)
c) Cost – cycle time and productivity
d) Delivery - efficiency

6.1.1. Safety. Implementation of JMM shows a significant improvement of the safety index. In financial year 2007/2008, the local company registered 6 occurrences of recordable major accidents with medical leave and 8 minor accidents as shown in Figure 1. The accident rate dropped sharply in FY2011/2012 with no occurrences of recordable major accident and 6 occurrences of minor accident. This is the best accident record ever achieved for the major accident. Although in early implementation of JMM, the minor accidents were above the plant target of 8 occurrences per year, the rate dropped below the target of 8 occurrences in FY2008/2010 onwards.

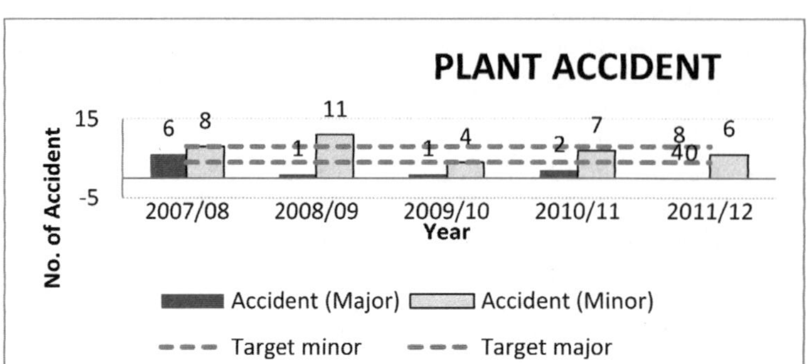

Figure 1. Plant accidents

6.1.2. Parts per Million. PPM (Parts per Million) is a measurement used by the local company to measure quality performance of parts received from suppliers. One PPM means one defect in a million or 1/1,000,000. In the past a good supplier would have a defect rate of less than 1%, (10,000 PPM). However, production and quality control procedures have improved with an expected defect rate of as low as 300 PPM.

Since the adoption of the JMM, the defective parts shipment to the plant by suppliers had been reduced significantly over the last few years. The plant originally recorded 6,500 PPM in FY2007/2008 and then improved by 97.8% at 141 PPM and 98.1% at 121 PPM for FY2010/2011 and FY2011/2012 respectively. The downward trend for the PPM at the local company is depicted in Figure 2.

Even though PPM is only one part of the overall quality picture, it is still a very large part. At the local company, quality awards are hand out to suppliers based on a number of criteria, including low PPM. Supplier quality programs are geared to make PPM as low as possible.

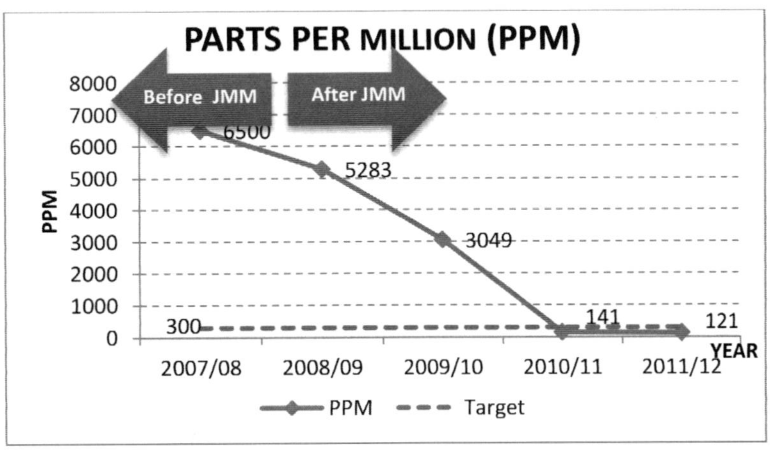

Figure 2. Quality trend of incoming parts measured in Parts per Million

6.1.3. In-Line Defect Per Unit (DPU) and First Time Quality (FTQ). Defect per unit (DPU) is the average number of defects per unit. The ratio of defects to unit is the universal measure of quality simply defined by the following equation:

$$\text{In-Line Quality Defect Per Unit (DPU)} = \frac{Total\ number\ of\ defects}{Total\ number\ of\ units}$$

The In-line DPU is simply the number of defects found in assembly line divided by the number of units inspected. Since the implementation of the JMM, the In-line DPU had shown a consistent downward trend. From FY2007/2008 to FY2011/2012, the In-line DPU was reduced from 5.7 DPU to 0.49 DPU giving 91% improvement for the DPU.

First Time Quality (FTQ) is a lean metric that indicates to what extent parts are manufactured correctly the first time without need for inspection, rework, or replacement. FTQ is calculated by dividing the number of good units produced by the number of total units going into the process. FTQ is an important tool in any quality management system to deliver a high quality product: built right, the first time. This is a key element in the effort to exceed the expectations of both internal and external customers. A well-strategized FTQ system offers a multitude of benefits, among them includes the opportunity to identify opportunities for continuous improvement, reduce costs, and monitor process stability. It also provides added value to the customers by demonstrating and documenting measurable performance.

The FTQ in percentage also indicated an improvement trend beginning FY2009/2010. The upward trend for FTQ in FY2009/2010 was slightly dropped due to the effect of Tsunami and Thailand flood which contributed to disruption of operation and parts supply to the plant. In FY2011/2012, the FTQ improved at 84.97% as compared to 31.8% before the JMM implementation. This shows improvements of 167% for the FTQ as compared to those before the JMM adoption.

The trend for the DPU and FTQ shows that the quality has substantially improved due to the adoption of JMM. The In-line DPU and FTQ are depicted in Figure 3.

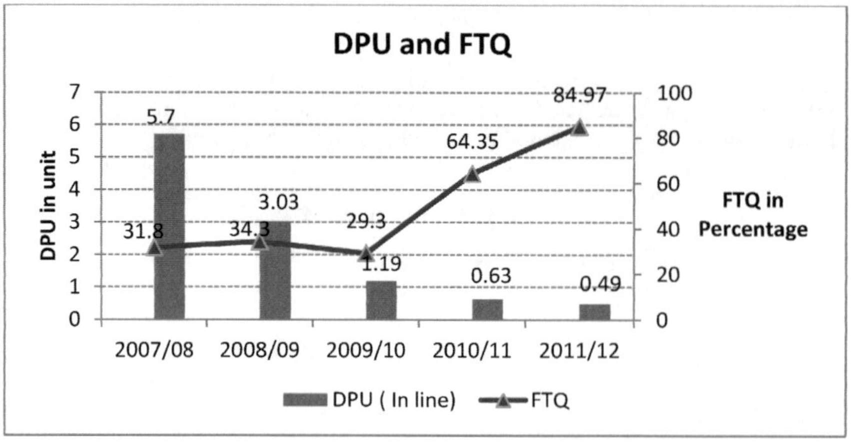

Figure 3. First Time Quality and In-line Defect Per Unit

6.1.4. Cycle time. Understanding the concept of cycle time is a useful method in continuous improvement efforts. Cycle time describes how long it takes to complete a specific task from start to finish. Based on lean approach, cycle time should be considered a viable option when an organization is trying to improve efficiency, productivity, cost base, customer responsiveness and flexibility.

Cycle time reduction is identifying and implementing more efficient and effective ways to do things. Reducing cycle time requires elimination or reduction of non-value-added tasks and activities, which do not provide any real value to the customer. Reduction in cycle time has a significant impact on a company's profitability and ultimately customer satisfaction. The benefits of cycle time reduction include lower costs, elimination of waste and non-value activity, streamlined processes, improved communications, improved process consistency, schedule integrity and improved on-time delivery.

Over the years of JMM implementation, the cycle time for general assembly had been reduced significantly as shown in Table 1. Beginning with more than 20 minutes prior to the adoption of JMM, the cycle time was further improved to 6 minutes in financial year 2012/2013. The cycle time had improved by 70% as compared to the value prior to JMM Implementation.

Table 1. Cycle time for general assembly TF2

	Prior to JMM	After JMM				
Financial Year	**2007/08**	**2008/09**	**2009/10**	**2010/11**	**2011/12**	**2012/13**
Cycle time	20	16	12	10	8	6

The improvement was also due to the improved processes, skill upgrading, kaizen activities undertaken by all, not only the engineering personnel but also all process owners i.e. supervisors, foremen and operators, and improved manpower skill. It is also to be noted that the cycle time reduction is an ongoing process that takes strong management commitment.

6.1.5. Productivity. The standard measurement of productivity is output per worker-hour, or the ratio of output to inputs in production. The productivity is simply defined by the following equation:

Productivity = Output / Manpower x working hours

During the early year of JMM in FY2008/2009, Body Shop and General Assembly employees collectively produced an average of 0.23 units per hour and 0.05 units per hour respectively which were below the plant target. In FY2011/12, over 3 years after JMM, Body Shop productivity improved by 43% at 0.33 units per hour. The General Assembly performance is quite stable at 0.07 units due to conveyorized system. Therefore, both shops had shown improvement in subsequent years after the introduction of JMM which exceeded or at par with the targets. The productivity for Body Shop and General Assembly is illustrated in Table 2.

Table 2. Productivity for body shop and general assembly

FY/Shop	Target	2008/09	2009/10	2010/11	2011/12
Body Shop	0.28	0.23	0.32	0.31	0.33
Gen. Assembly	0.07	0.05	0.07	0.07	0.07

6.1.6. Efficiency. Manufacturing efficiency determines how well a factory operates in production. To avoid wasting money, all processes in manufacturing must be as efficient as possible. Calculating a numerical value for the efficiency helps to identify if improvements to the production process need to be made. The calculation of percentage efficiency used by the plant is as follows:

Efficiency = [(Available working time – Downtime) / Available working time] x 100

The plant recorded an increase in efficiency for the past few years as compared to the first year of the JMM implementation in FY2008/2009. Although the efficiency performance at Body Shop recorded a mixed trend due to introduction of new models, new manpower and etc., the performance was above the target of 96% set by the plant. The highest achievement was efficiency at 99.9% accomplished in 2009/2010, an improvement of 4% against the target. The efficiency for the General Assembly had shown an increasing trend although a lot of improvement initiative are needed to reduce downtime in terms of machine breakdown, absenteeism, parts delays etc. in order to meet the target efficiency. The efficiency for Body Shop and General Assembly is presented in Table 3.

Table 3. Efficiency for body shop and general assembly

FY/ Shop Efficiency (%)	Target (%)	2008/09	2009/10	2010/11	2011/12
Body Shop	96	97.3	99.9	98.7	97.9
Gen. Assembly	89	73.9	80.1	85.4	85.6

6.1.7. Stock level. The control of stock level is one of the important aspects of lean manufacturing. It was found that the CKD stock level for the local company was steadily controlled from 1.7 months prior to the JMM to 1.5 months after the JMM implementation. The stock level for the local parts also was reduced from 0.5 months to 7 days before delivery from the local suppliers. With the recent implementation of distribution center and soon milk-run activities, the local parts stock level for the plant was further reduced to 3 days.

6.2. Financial performance

The financial results were taken in FY2007/2008, a year before the full implementation of the JMM in FY2007/2008 and assess throughout the adaptation period for the following 4 years. The local company began with RM189.7 million business turnovers during the midst of the JMM implementation in FY2007/2008 but had significantly increased by 92% to RM364 million in a 4-year period by FY2011/2012. The financial statement of the local company is shown in Table 4.

Table 4. Financial statement

	2007/2008	JMM started 2008/09	2009/10	2010/11	2011/12
	RM'000	**RM'000**	**RM'000**	**RM'000**	**RM'000**
Revenue	189,726	280,189	232,211	316,327	364,059
Total cost of sales	(163,275)	(254,041)	(207,097)	(269,323)	(324,796)
Gross profit	26,451	26,148	25,114	47,004	39,263
Operating expenses	20,954	15,519	8,511	10,807	11,146
Profit before taxation	5,497	10,629	16,603	36,197	28,117
Taxation	4,614	11	546	2,256	1,553
profit after taxation	10,111	10,640	17,149	38,453	29,670
Percentage Revenue VS Expenses (%)	11.04	5.54	3.67	3.42	3.06

Source: Audited report of the local company

Specifically, the revenue had increased by more than 47.6% from RM189.7 million in FY2007/2008 to RM 280,189 million in FY2008/2009. The revenue however had decreased to RM232.2 million in FY2009/2010 as a result of Tsunami calamity in Japan and Thailand flood which had disrupted the parts supply. Even so, the figure was still more than the revenue before the JMM implementation. The revenue continued to improve by 36% at RM316.3 million in the following financial year due to the increase in the volume of vehicles sold to the market, manufacturing efficiency and the acquisition of product ownership for the heavy-duty model. In FY2011/2012, the plant recorded an increase in revenue by 15% at RM364 million compared to the previous financial year.

It is noted that for FY2007/2008, the expenses was high at RM20.9 million mainly due to company restructuring cost of RM2.40 million and Japanese technical assistance fee for the J-Project totaling RM1.86 million. The percentage expenses stood at 11.04% in FY2007/2008 prior to the JMM, however, after full implementation of the JMM in FY2008/2009, percentage expenses sharply fell by 50% at 5.54% as compared to the previous financial year and reduced further by 34% at 3.67% in FY2009/2010. The expenses had been steadily reducing for the following 2 years. For example, in FY2011/12, the percentage of expenses was at 3.06%, an improvement of 72% in comparison before JMM in FY2007/2008.

Due to restructuring exercise, the net profit went slightly down from 5.33% to 3.8% giving a reduction of 29% in FY2008/2009, however, the net profit increased tremendously by 94% at 7.39% net profit in FY2009/2010, the second year after the JMM adoption. The net profit increased further by 65% at 12.16% net profit in FY2010/2011 whereas for FY2011/2012 it showed a reduction of 33%

at 8.15%. Although the figure is lower than previous year, the net profit had improved by 52% as compared to those before JMM. The reduction of net profit although with higher revenue was due to higher Japanese yen exchange rate for the CKD purchase from Japan. The percentage however, is still higher than the net profit before the JMM adoption. The financial performance of the local company is indicated in Table 5.

Table 5. Financial performance

	2007/08 RM'000	2008/09 (JMM started) RM'000	2009/10 RM'000	2010/11 RM'000	2011/12 RM'000
Revenue	189,726	280,189	232,211	316,327	364,059
Profit after taxation	10,111	10,640	17,149	38,453	29,670
Total assets	72,974	142,228	137,059	206,042	248,822
Inventory	27,454	36,531	22,007	30,084	77,517
Total s/equity	49,285	59,925	77,074	113,373	143,043
Net profit margin (%)	5.33	3.80	7.39	12.16	8.15
Return on asset ROA (%)	13.86	7.48	12.51	18.66	11.92
Return on equity ROE (%)	20.52	17.76	22.25	33.92	20.74
Inventory Turnover Ratio	6.91	7.67	10.55	10.51	4.70

Source: Audited report of the local company

The RoA (Return on Assets) reduced by 46% at 7.48% in FY 2008/2009 as compared to earlier financial year following to the restructuring exercise. The RoA increased by 67% at 12.51% in FY2009/2010 as compared to the previous financial year and then continued to increase by 49% at 18.66% in FY2010/2011. The RoA however, went down by 36% at 11.92% in FY2011/2012 as compared to earlier financial year due to the fluctuation in Japanese exchange rate.

The Return on Equity (RoE) which was at 20.52% in FY 2007/2008 reduced by 13% at 17.76% in FY2008/2009 after the restructuring of the plant. The RoE went up by 25% at 22.25% in the year FY2009/2010 and continuously increased by 52% at the percentage of 33.92 in FY2010/2011 before it went down by 39% at 20.74% for FY2011/2012 again due the fluctuation in exchange rate although the percentage RoE was still above the RoE prior to JMM implementation.

Following to the restructuring exercise and early stages of growth of the plant, the calculation of return on assets (RoA), return on equity (RoE) and net profit margin of the local company showed some reduction at the beginning of JMM introduction in FY2008/2009. The net profit margin, RoA and RoE however, had been steadily on the rise for the following three consecutive financial years before the performance was slightly declined due to the higher Japanese exchanged rate in FY2011/2012. The increasing financial performance for FY2008/2009 to FY2010/2011 after the JMM implementation had indicated that the profitability of the plant was improving.

The upward trend for the Inventory Turnover Ratio (ITR), which measures the rate at which the company manufactures and sells products to customers, indicates the company was enjoying strong sales and practicing JIT inventory methods throughout the 3-year period of JMM implementation. In FY2011/2012, however, the inventory turnover ratio dropped to 4.70 as compared to 10.51 in the previous year mainly due to the company was carrying too much inventory in preparation for special promotions and new model introductions.

When compared the revenue and expenses during the financial years under review, the revenue increased significantly while the expenses had downward trend. This means that since the adoption of the JMM, the local company was able to run production efficiently as the revenues increased and expenses decreased or were steadily under control. The revenue versus expenses trend is illustrated in Figure 4. From this it can be concluded that the financial performance of the local company has improved after the JMM adoption.

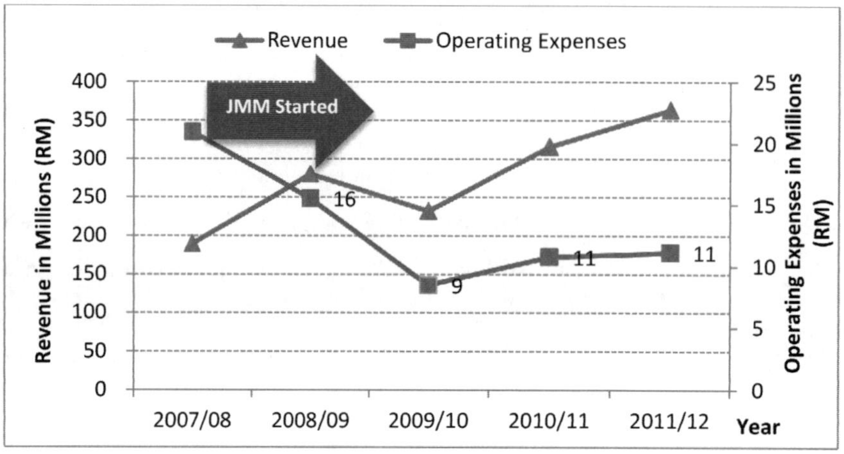

Figure 4. Revenue and expenses

When viewed over time since the adoption of the JMM, the revenue and the profit for the local company had shown an increasing trend. This indicates that the local company had effective control over cost in converting revenue into profit. The trend is shown in Figure 5.

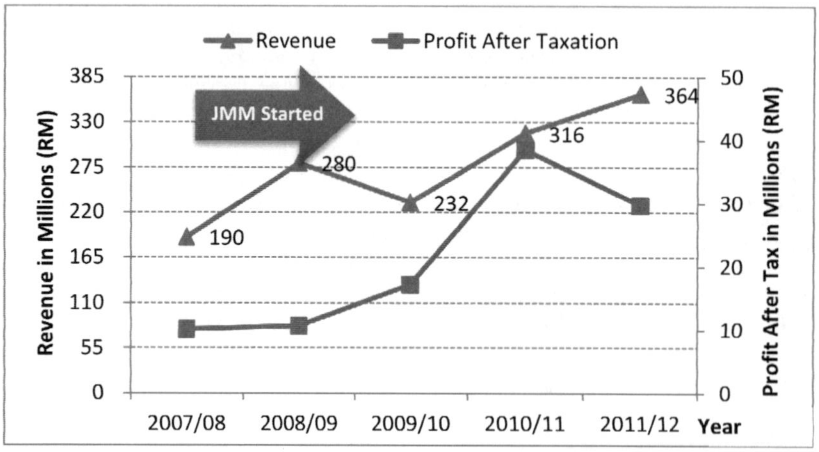

Figure 5. Revenue and profit

7. Conclusion

The research shows that through the adoption of JMM, the manufacturing performance improved in terms of the following aspects:

a) Safety index reduction from 6 major accidents to 0 accident,
b) PPM reduction of 98%,
c) In-line DPU improvement of 91%,
d) FTQ up from 31.8% to 84.97% giving an improvement of 167%
e) Cycle time improvement from 20 minutes to 6 minutes.
f) Productivity improvement up to 43% at Body Shop and at par for General Assembly

g) Efficiency improvement up to 4% against the target for Body Shop.
h) Stock level of local parts from half month to 3 days and CKD from 1.7 months to 1.5 months.

The results of this study also show that the implementation of JMM increases the firm's profitability and revenue besides maintaining the stability of expenses. The revenue has increased from 22% to 92% after the JMM adoption and the net profit has increased from 5.33% to 8.15%.

Acknowledgements
The authors would like to acknowledge the RIGS of International Islamic University Malaysia for the support to carry out this work.

References

[1] Krisher T 2012 Toyota Now Top Selling Automaker, Claims Title From GM. Huff Post Business, Retrieved March 2013. http://www.huffingtonpost.com/2013/01/14/toyota-top-selling-automakern_2471068.html.
[2] Dawson C 2013 Toyota Takes Top Car-Maker Title Back from GM. *Wall Street Journal*, Retrieved March 2013, http://blogs.wsj.com/drivers-seat/2013/01/28/toyota-takes-top-car-maker-title-back-from-gm.
[3] Bergenwall A L, Chen C and White R E 2012 TPS's process design in American Automotive Plants and its Effects on the Triple Bottom Line and Sustainability, *Int. Journal of Production Economics*, **140**(1) 374-384.
[4] Shah R and Ward P T 2003 Lean Manufacturing: Context, Practice Bundles, and performance, *Journal of Operation Management*, **21** pp 129-149.
[5] Fullerton R R, McWatters C S and Fawson C 2003 An Examination of the Relationship Between JIT and Financial Performance, *Journal of Operations Management*, **21**(4) 383-404.
[6] Laosirihongthong T, McLean M W, Leabsuetrakool P, Chongjareonjai T and Leedhirakul S 2007 Technology Strategy: Findings from Adoption and Adaptation of Japanese Manufacturing Management (JMM) to ASEAN and Australasian Automotive Manufacturers, *4th IEEE International Conference on Management of Innovation and Technology*, (2008) 1325-1330.
[7] Schonberger R J 2007 Japanese Production Management: An Evolution-With Mixed Success, *Journal of Operations Management*, **25**(2) 403-419.
[8] Pil F K 2000 Thrive: Managing the Transfer of "Best Practice" at Japanese Auto Plants in North America, *Journal of World Business*, **34**(4) 372-390.
[9] Phan A C, Abdallah A B and Matsui Y 2011 Quality Management Practices and Competitive Performance: Empirical Evidence from Japanese Manufacturing Companies, *International Journal of Production Economics*, **133**(2) 518-529.
[10] Economic Report 2010/2011. (2010). Official Website of Ministry of Finance Malaysia. Retrieved March, 2013. http://www.treasury.gov.my.
[11] Haji Alias M, Harris A and Ramli M A 2009 Penglibatan Bumiputera Dalam Industri Otomotif Sokongan: *Satu Penilaian Kritis*, **2** pp 199–207.
[12] Liker J K 2004 The Toyota Way: 14 Management Principles from the world's greatest manufacturer, McGraw- Hill, USA.
[13] Monden Y 1983 Toyota Production System: A Practical Approach to Production Management. Industrial Engineers and Management Press, Norcross, GA.
[14] Shingo S 1989. A Study of the Toyota Production System, revised ed. Productivity Press, Cambridge, MA.
[15] Mohiuddin A K M, Rahman M A and Julia Bt Haji A Jabar 2015 Adoption and Adaptation of Japanese Manufacturing Management in an Automotive company of Malaysia, *Advanced Materials Research* **1115** pp 589-595.

CFD Analysis of a SD 7003 Airfoil with a Local Correlation Based Transition and Turbulence Model

Mazharul Islam[1a], **Felix Langfeldt** [2b], **Jiří Fürst** [3c], **David H Wood** [1d]
[1] Department of Mechanical Engineering, Kulliyah of Engineering, International Islamic University Malaysia, P.O. Box 10, 50728 Kuala Lumpur, Malaysia
[2] Department of Automotive and Aeronautical Engineering, Hamburg University of Applied Sciences, Berliner Tor 9, 20099 Hamburg, Germany
[3] Department of Technical Mathematics, Faculty of Mechanical Engineering, Czech Technical University in Prague, Karlovo nam. 13, 121 35 Praha, Czech Republic
[4] Department of Mechanical and Manufacturing Engineering, Schulich School of Engineering, University of Calgary, 2500 University Dr NW, Calgary T2N1N4, Canada

E-mail: [a] mazharul@iium.edu.my, [b] felix.langfeldt@gmail.com, [c] Jiri.Furst@fs.cvut.cz, [d] dhwood@ucalgary.ca

Abstract.
The SD 7003 has been investigated quite extensively in the past at different Reynolds numbers. At moderate Reynolds numbers and low angles of attack its upper surface is dominated by a large laminar separation bubble which provides a demanding test case for RANS-based transition models. In this article, a widely used open-source Computational Fluid Dynamics (CFD) toolbox has been used mainly to investigate the (i) aerodynamics coefficients and (ii) boundary layer transitional behaviors of a SD 7003 airfoil using a local correlation-based $\gamma - Re_{\theta t}$ SST [1] transition and turbulence model. Several researchers have proposed different correlations which can be used in the $\gamma - Re_{\theta t}$ SST. In this article, five such correlations have been investigated for the SD 7003 at low Reynolds number which is particularly challenging for CFD analysis.

1. Introduction

Airfoils are used in different engineering applications like wind turbines, airplanes etc. In the operating conditions, airfoils typically encounter transitions from laminar to turbulent flow in the boundary layer (BL). The SD 7003 airfoil has been investigated quite extensively in the past at different Reynolds numbers through diversified computation techniques and experimental investigations. It has been found that at moderate Reynolds numbers and low angles of attack, the upper surface of SD 7003 is dominated by a large laminar separation bubble (LSB). According to [2] - "*At these low Reynolds numbers, the flow may remain laminar over a significant portion of the airfoil rendering it susceptible to separation from even mild adverse pressure gradients. For moderate incidence, separation leads to the formation of a closed laminar separation bubble (LSB) which reattaches through transition into turbulence.*"

CFD analysis of airfoils at low Reynolds number that is associated with LSB is a challenging task for computationally inexpensive RANS-based transition models. Diversified approaches can be used to model different types of BL transitions. Two popular transition and turbulence models in Computational Fluid Dynamics (CFD) analysis are:

(i) The local correlation-based transitional model called $\gamma - Re_\theta$ which was originally proposed by Menter et al. [1]; and
(ii) The phenomenological $k - k_L - \omega$ model [3].

Over the years, different researchers [4–8] have proposed different correlations for calculating the two undisclosed proprietary parameters required by the original $\gamma - Re_\theta$ model. In the present work, CFD analysis of SD 7003 airfoil has been performed at a low Reynolds number of $60,000$. To tackle the transition modeling, $\gamma - Re_\theta$ model is used in OpenFOAM [9] using five different correlations, proposed by [4–8]. It should be noted that the present work is an extension of the earlier research activities with $k - k_L - \omega SST$ model which was presented in [10]. Selected final results from the CFD analysis have been compared with other computational (DNS & LES) and experimental datasets for verification and validation.

2. The OpenFOAM Case

To perform the CFD analysis, a widely used open-source CFD toolbox called OpenFOAM has been used in the present work. To verify the results, mesh sensitivity analysis has been conducted with 3 mesh sizes (termed as Coarse, Medium and Fine). In Table 1, the salient features of these three mesh sizes are presented. It should be noted that the same three mesh sizes were previously investigated in the previous research [10] using the $k - k_L - \omega$ model. It should be noted that the SD 7003 has sharp trailing edges for which the C-type mesh is preferred and thus used in the present work. More discussions about the computational domain and the mesh were included in [10].

In Figure 1, the mesh around SD 7003 is illustrated. Whereas, the mesh distribution around the leading and trailing edges of the airfoil are depicted in Figures 2 and 3 respectively. For the present analysis, a steady-state solver called "simpleFoam", which is available in OpenFOAM and employs the SIMPLE algorithm to solve the incompressible Navier-Stokes equations, was used. The Reynolds Number (Re) for all the cases is about 60,000 and all the CFD analyses were conducted at 4° angle of attack.

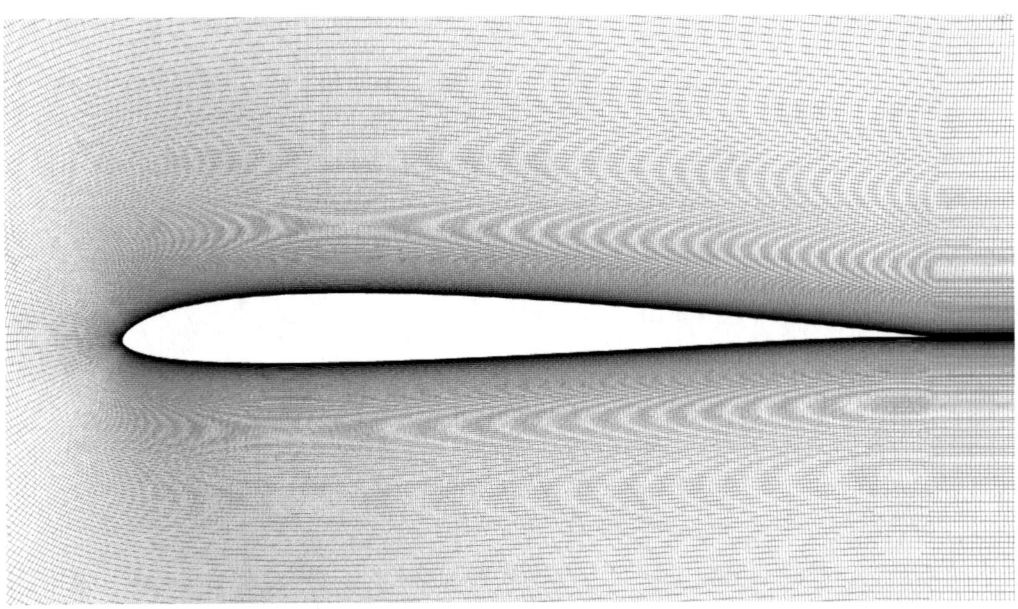

Figure 1: Mesh Around the SD 7003 Airfoil

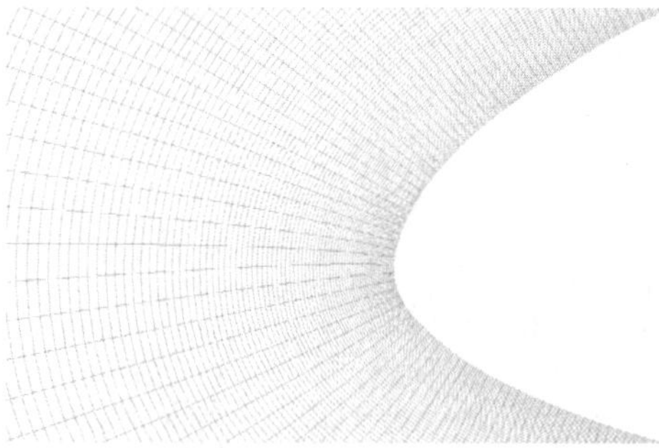

Figure 2: Mesh Around the Leading Edge of the SD 7003 Airfoil

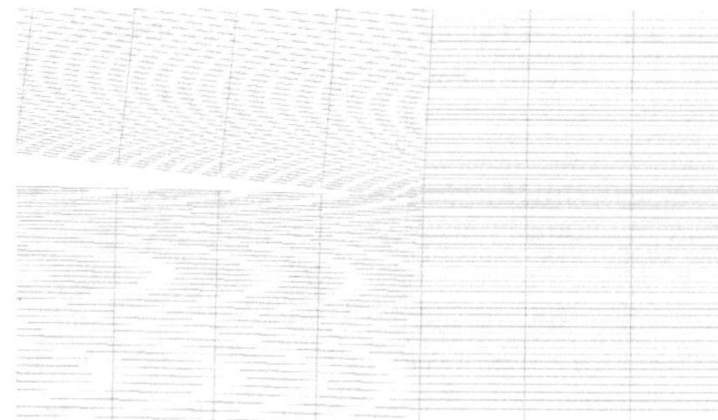

Figure 3: Mesh Around the Trailing Edges of the SD 7003 Airfoil

Table 1: Features of 3 Meshes

Parameter	Coarse	Medium	Fine
Number of Cells	223571	396506	519707
Number of points over the airfoil surface	200	300	450
Number of points in the direction normal to the airfoil surface	80	120	180
Maximum y+ for the final iteration	0.9118	0.9126	0.9132

3. Results and Discussions

3.1. Mesh Sensitivity Analysis

To verify the CFD results, the variation of the pressure coefficients (C_p) around the surface of SD 7003 airfoil has been investigated using the $\gamma - Re_{\theta t}\ SST$ [1] model using the correlation proposed by Tomac et al. [8], which is shown in Figure 4. The C_p distributions for the three mesh sizes are quite close to each other for the entire surface. These results have been compared with the ILES results obtained by [2] and it can be seen that there are noticeable differences around certain locations. The variation of aerodynamic coefficients for 3 mesh sizes are presented in Table 2, which shows that further refinement of mesh sizes is not needed.

Table 2: Moment, Drag and Lift Coefficients for 3 Mesh Sizes

Mesh	Cm	Cd	Cl
$\gamma - Re_{\theta t}\ SST$ with Tomac et. al - Coarse	-0.0329	0.0203	0.4772
$\gamma - Re_{\theta t}\ SST$ with Tomac et. al - Medium	-0.0328	0.0202	0.4773
$\gamma - Re_{\theta t}\ SST$ with Tomac et. al - Fine	-0.0328	0.0202	0.4773

3.2. Validation of Aerodynamic Coefficients

In Table 3, aerodynamic coefficients (C_l,C_d & C_m) obtained from this research work with five correlations are compared with - (1) CFD results of $k - \omega\ SST$, (2) $k - k_L - \omega$ (Falkner-Skan) - Fine [11] (3) ILES results from [12], (4) Experimental results from [13] [1]. It can be seen from

[1] Interpolated for $\alpha = 4°$ from the results at Re=61,400

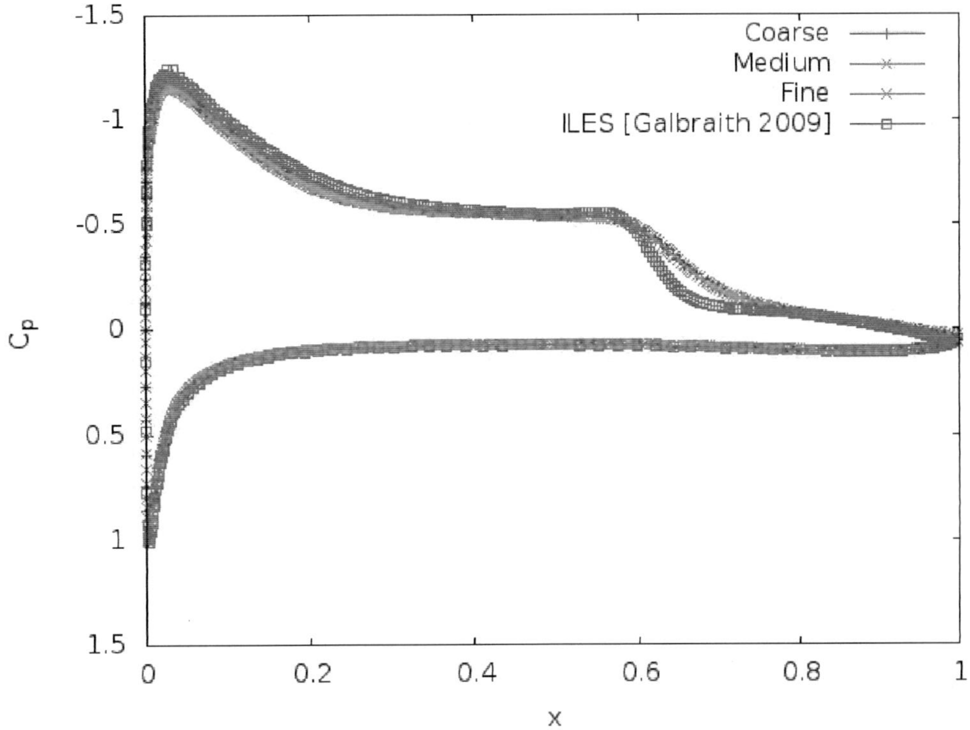

Figure 4: Variation of Pressure Coefficient (C_p) around the SD 7003 Airfoil

this table that C_l and C_d values obtained from [10] are better than the five correlations [4–8] with the $\gamma - Re_{\theta t}$ SST [1] model and closer to the experimental results.

Table 3: Comparison of Moment, Drag and Lift Coefficients

	Cm	Cd	Cl
$k - k_L - \omega$ (Falkner-Skan) - Fine	-0.0387	0.0188	0.5150
$\gamma - Re_{\theta t}$ SST with Tomac et. al - Fine	-0.0328	0.0202	0.4773
$\gamma - Re_{\theta t}$ SST with Langtry and Menter - Fine	-0.0477	0.0255	0.4782
$\gamma - Re_{\theta t}$ SST with Malan et. al - Fine	-0.0453	0.0245	0.4801
$\gamma - Re_{\theta t}$ SST with Sorensen - Fine	-0.0346	0.0208	0.4780
$\gamma - Re_{\theta t}$ SST with Suluksna et. al - Fine	-0.0394	0.0225	0.4807
XFOIL (Ncrit = 9)	-0.0355	0.0199	0.6253
ILES [GalbraithAndVisbal2010]	-	0.021	0.59
Experimental [Selig et al. 1995] (Interpolated for $\alpha = 4°$ from the results at Re=61,400)		0.0166	0.6038

3.3. Skin Friction Coefficients

In Figure 5, the variations of skin friction coefficients (C_f) around the SD 7003 airfoil that are obtained from the CFD analysis using the five correlations [4–8] with the $\gamma - Re_{\theta t}$ SST [1] model are shown. These results are compared with Implicit Large Eddy Simulation (ILES) [2] and XFOIL [14].

Both the transition from laminar to turbulent flow and the laminar separation bubble are situated between the separation point $x_{separation}$ (where the C_f becomes negative near $x/c \approx 0.23$ for the ILES result) and reattachment point $x_{reattachment}$ (where the C_f value becomes positive again near $x/c \approx 0.65$ for the ILES result). So, the distance between $x_{separation}$ and $x_{reattachment}$ is indicative of the length of the bubble size. The two major observations from this figure are:

- All the RANS-based correlations are over-predicting the reattachment points significantly when compared with the ILES result.
- Among the five correlations [4–8], the the distance between $x_{separation}$ and $x_{reattachment}$ is maximum for [7] and minimum for [8].

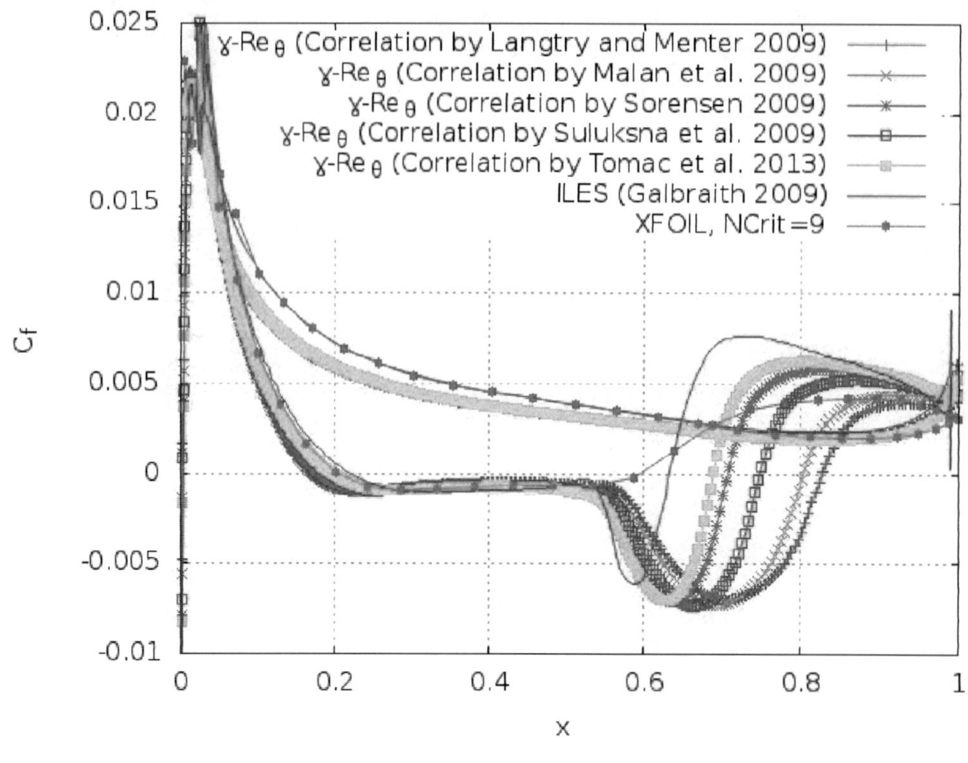

Figure 5: Variation of Skin Friction Coefficient (C_f) around the SD 7003 Airfoil

4. Concluding Remarks

Five different correlations for $\gamma - Re_\theta Semicond.Sci.Technol.$ model [1], proposed by [4–8], have been investigated through CFD analysis of SD 7003 airfoil at low Reynolds number in the present study. It has been found that all the correlations have under-predicted the C_l values and over-predicted the C_d values. However, aerodynamic coefficients (C_m,C_l, and C_d) obtained from the $k - k_L - \omega$ model [3] using the Falkner-Skan based modification [10] are closer to the experimental results than Tomac et al. [8]. Also, the correlation by Tomac et al [8] is predicting the separation and reattachment points, for the flow around SD 7003 at low Reynolds number, better than the other four correlations [4–7].

5. References

[1] Menter F R, Langtry R B, Likki S R, Suzen Y B, Huang P G and Völker S 2004 *Journal of Turbomachinery* **128** 413–422 ISSN 0889-504X URL http://dx.doi.org/10.1115/1.2184352

[2] Galbraith M C 2009 *Implicit Large Eddy Simulation of Low-Reynolds-Number Transitional Flow Past the SD7003 Airfoil* Master's thesis University of Cincinnati URL http://rave.ohiolink.edu/etdc/view?acc_num=ucin1242939414

[3] Walters D K and Cokljat D 2008 A Three-Equation Eddy-Viscosity Model for Reynolds-Averaged Navier–Stokes Simulations of Transitional Flow

[4] Sørensen N N 2009 *Wind Energy* **12** 715–733 ISSN 1099-1824 URL http://dx.doi.org/10.1002/we.325

[5] Malan P, Suluksna K and Juntasaro E 2009 Calibrating the γ-Reθ transition model for commercial CFD *47th AIAA Aerospace Sciences Meeting including the New Horizons Forum and Aerospace Exposition*

[6] Suluksna K, Dechaumphai P and Juntasaro E 2009 *International Journal of Heat and Fluid Flow* **30** 66–75 ISSN 0142-727X URL http://www.sciencedirect.com/science/article/pii/S0142727X0800146X

[7] Langtry R B and Menter F R 2009 Correlation-based transition modeling for unstructured parallelized computational fluid dynamics codes

[8] Tomac M, Pettersson K and Rizzi A 2013 Calibration and Verification of a $\gamma - \widetilde{Re}_{\theta t}$ Transition Prediction Method for Airfoil Computations *51st AIAA Aerospace Sciences Meeting including the New Horizons Forum and Aerospace Exposition* Aerospace Sciences Meetings (American Institute of Aeronautics and Astronautics) URL http://dx.doi.org/10.2514/6.2013-407

[9] OpenFOAM ® - the open source computational fluid dynamics (cfd) toolbox

[10] Islam M, Fürst J and Wood D 2015 CFD Analysis of SD 7003 Airfoil at Low Reynolds Number with a Laiminar Kinetic Energy based Transition Model *International Conference on Computational Fluid Dynamics in Research & Industry 2015, Kuala Lumpur, Malaysia*

[11] Fürst J, Islam M, Příhoda J and Wood D 2015 Modifications to the k-kL-ω Transition Model based on Pohlhausen and Falkner-Skan Profiles URL http://www.manager.cpe.vt.edu/contributionDisplay.py?contribId=59&confId=18

[12] Galbraith M and Visbal M 2010 Implicit Large Eddy Simulation of Low-Reynolds-Number Transitional Flow Past the SD7003 Airfoil *40th Fluid Dynamics Conference and Exhibit* Fluid Dynamics and Co-located Conferences (American Institute of Aeronautics and Astronautics) URL http://dx.doi.org/10.2514/6.2010-4737

[13] Selig M S, Guglielmo J J, Broeren A P and Giguère P 1995 *Summary of low speed airfoil data* vol 1 (SoarTech Publications) ISBN 0964674718

[14] Drela M 1989 *Lecture Notes in Engineering* 1–12 ISSN 01765035

IOP Conf. Series: Materials Science and Engineering **184** (2017) 012068 doi:10.1088/1757-899X/184/1/012068

Dynamic response for structural health monitoring of the Penang (I) cable-stayed bridge

M I Mohammed[1], E Sulaeman[2], F Mustapha[3]

[1]PhD candidate, Department of Aerospace Engineering, Faculty of Engineering, University Putra Malaysia, 43400 UPM Serdang, Selangor, Malaysia
[2]Associate Professor, Department of Mechanical Engineering, Faculty of Engineering, International Islamic University Malaysia, 53100 Kuala Lumpur, Malaysia
[3]Associate Professor, Department of Aerospace Engineering, Factualty of Engineering, University Putra Malaysia, 43400 UPM Serdang, Selangor, Malaysia

[1]Email: mohd.idris.mohd@gmail.com
[2]Email: esulaeman@iium.edu.my
[3]Email: faizalms@upm.edu.my

Abstract. The paper discusses the dynamic response of the Penang (I) cable stayed bridge structure under various moving load representing typical traffic load of the bridge. The bridge has a total span of 440 m excluding the transition bridge that assumed to be not connected structurally to the main bridge structure. The bridge that links the fast growing Pinang Island and the Malaysian Mainland Peninsula has been known to be fully utilized which leads to the construction of Penang (II) bridge and now the third one. Due to highly traffic use of the bridge that may lead to reduction of the bridge design life, the dynamic response of the bridge becomes important to predict critical part of the bridge structure elements including the main girder and the 144 stay cables. The present study reveals that, due to flexible nature of the cable stayed bridge, the moving load that interacts with the natural dynamic characteristics of the bridge, gives significant stress increment compare to proportional static load especially when the moving load is un-symmetric. For this reason, several classes of typical vehicle passing the bridge with various vehicle speeds are investigated to demonstrate their effect on the bridge displacement, internal forces and stresses. The results can be used for further fatigue assessment of the bridge.

1. Introduction

Rapid growth of economy and societies demand for highly functional transportation infrastructures such as bridges, which are designed to carry loads and expected to serve effectively. Bridges infrastructures are supposedly function to operate safely and work for decades [1]. However, they may expose to exponentially heavy traffic. In addition, despite the satisfactory in design, they may experience deterioration over a period due to unpredictable environmental conditions. In order to control such conditions and to maintain the reliability as well as integrity of these infrastructures, continuous monitoring of the structures is of a paramount importance.

The bridge monitoring system is vital as a technique to measure operation, loading environment and the critical responses to track and evaluate the symptoms operation incidents, anomalies and deterioration indicator that may affect the serviceability or safety reliability of a structure [2]. In this context, the monitoring means to distinguish the condition of the bridge where the parameters to be investigated and bridge responses are simulated with finite element (FE) method, then analyzed by modeling and equated with calculation for accuracy [3]. The monitoring aspects can be assessed based on two behavior of bridge infrastructures i.e. static and dynamic loads. The static behavior verifies the changes to static parameters of its components and material while the dynamic verifies to the vibration mode [4]. Ref [5 - 9] explored the areas i.e. bridge monitoring based on static and dynamic load. They conducted static load field test of the cable stayed bridge in Taiwan. Ref. [5] developed FE model to compare the bridge responses at the field. The structure load test revealed the linear superposition features and gave considerable agreement with model analysis [5]. In another investigation, numerical analysis and field dynamic tests of a 605m main span, cable-stayed bridge in China is conveyed by [6]. The study has disclosed that FE modeling gives description of the bridge physical and modal characteristic, whereas the experimental modal analysis at the dynamic field tests provides information to validate drawing-based idealized FE model. Also, the developed model shows the structure displacements and achieves a good correlation with measured modal parameters identified from the dynamic filed tests. Such model can be used to serve as baseline in structural dynamic for long duration health monitoring, damage detection, dynamic predictions and other condition assessment of the cable stayed bridge.

The static dead and traffic, wind, temperature effects and their combinations on long cantilever truss bridge in United States is investigated in [7]. The study is to evaluate how structural health monitoring can be used to reduce the uncertainties of phenomena that are difficult to model. The inquiry is focused in particular on the temperature changes to determine the reliability of the entire bridge members. The analysis at [7] has revealed that temperature-induced strains is vital factor to be included to determine the reliability of the bridge components since the temperature levels changes during the yearly seasons. From the perspective of improving the design and maintenance, dynamic stress response of three spans of highway composite steel girder bridge is assessed in [8]. Field measurement was carried out to monitor the dynamic response developed due to passing by truck of weight 25ton. Then numerical model to predict the vibration is setup. The model is verified through the comparison between the measured data and the calculated results. The investigation has revealed that the bridge natural frequencies and the global dynamic stress response are in good agreement with the measurement. Another study of dynamic analysis is carried out in [9]. The investigation is based on three selected parameter indexes including vehicle load, speed and road surface condition effects. The FE Model of the concrete – filled steel tube arch bridge [9] and vehicle coupled system is developed and validated with field test. The results have disclosed that the dynamic impact factors differ between the selected structure members / locations and are greatly altered by the proposed parameters [9]. The results also show that the well noticeable level of the bridge vibration is highly affected by the surface road state and vehicle loading situation. In addition, the rid comfort level of the bridge is reduced as the road surface state worsen [9].

Other researchers have employed monitoring instruments to assess the reliability of bridge components [10]. Wind loading field test is conducted at one of the three towers cable bridge in China. The study adopted Global Positioning System (GPS), accelerometer and anemometer to acquire the quasi-static and dynamic responses of the middle tower displacement [10]. At first, Fast Fourier Transform (FFT) algorithm is utilized to identify the dominant frequencies. Next the displacement measurements of the noisy GPS and data of the accelerometer are de-noised by utilizing the Vondrak filter. Then noise-mitigated accelerations are used to the low frequency noise elimination and displacement reconstruction scheme. Meanwhile, the adaptive recursive least squares (RLS) filter is employed to the total horizontal displacement of the pylon which improved measurement accuracy so that they can be reconstructed. The initial natural frequency of 0.17Hz is determined from the data of the GPS and is validated and compared with that of the accelerometer data. The method of [10] improves the accuracy of the pylon displacement under the influence of the extreme winds. The stability of a bridge is assessed under loading condition.

Another research [11] has investigated the structural stability under buckling failure of cable stayed bridge decks. The study has presented that decks sensitivity is depended on several parameters such as span and deck stiffness, cable spacing, towers height, pattern of live load at deck and side intermediate piers. Subsequently, statistical methods have become significant to identify automatically the damage at a bridge structure for structural health monitoring. Multivariate statistical analysis approach is explored to determine the abnormality at the main cables of suspension bridge model for structural health monitoring. one main cable of the bridge is damaged and subjected to wind loading [12]. The study considered the natural frequencies as a mean to detect abnormalities of the structure. The technique of Principle component analysis is used and applied to the response data of long-duration pseudo-experimental buffeting. The results have shown feasibility of wind-excited to bridges creating relative changes at the most sensitive natural frequency lower than 0.1%. Consequently, [13] has carried out vertical vibration-based damage identification utilizing time-series analysis technique to the three spans bridge supported by five steel girders. The study has measured vibration (velocity) before and after damage. The method is employed for detecting damage. FE models are developed to determine the bridge behavior at undamaged and damaged state. Frequencies and modes shapes from numerical simulations are compared with that obtained from fields tests. FE models are adjusted utilizing field tests vibration data and compared with each damage scenario. The damage detection approach is considerably reliant on the damage location and vibration measurement noise.

The concept of multi-loading is carried out in [14] to assess the condition of a structure for structural health monitoring purpose. The research conducted fatigue analysis using multiple loadings of railway, highway and wind to Tsing Ma suspension bridge in Hong Kong. Their work used integrated computer simulation and measured data from wind and SHM system (WASHMS). The study has conducted computationally efficient method for the dynamic stress analysis under the discussed loading scenario. Then establish the critical locations of the fatigue at the bridge elements. The time histories of the dynamic induced stresses of singular loading can be created based on databases in their work [14].

In the present paper the dynamic response of one unit moving load applied on a cable-stayed bridge is investigated. The objective is to determine the displacement, internal forces and stress under the influence of dynamic effect due to moving vehicle along one lane. Comparison between the responses of the bridge structure having undamaged and damaged cables is conducted. Several cases involving the failure of critical cables having highest stress loading are selected to assess the influence of the damage cables to the overall response of the bridge structure.

2. Brief description of Penang (I) cable-stayed bridge

Penang (I) bridge is a cable-stayed, concrete girder bridge with harp configuration as presented in figure 1. The construction was began in 1982 and completed in 1985. The main bridge has three span segments, two flanking spans at the approach level of 107. 5m and main span of 225m. The deck is carried by 148 cables attached to two twin towers with the height of 101.5 m as shown in figure 2. The width of the deck is 29.7 m in total, narrows up at the main piers to 24.7m so that it can accommodate the inside tower faces. The carriageway consist of three lanes of each side. Figure 3 portray the superstructure which is constructed of 98 edge girder segments and 147 floor-beams [15].

Figure1. The oblique view of Penang Bridge

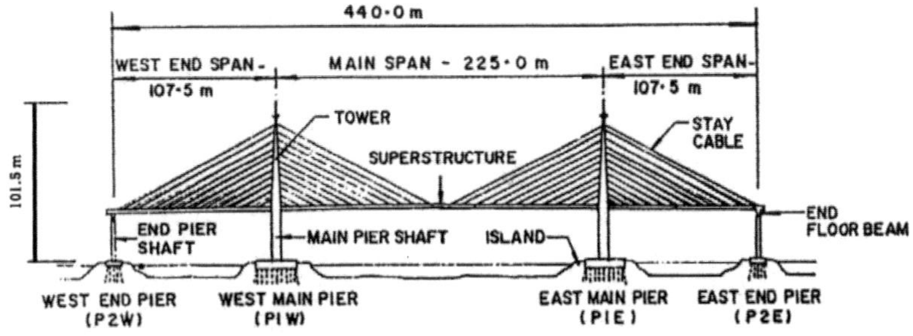

Figure 2. Penang (I) cable bridge longitudinal profile [15]

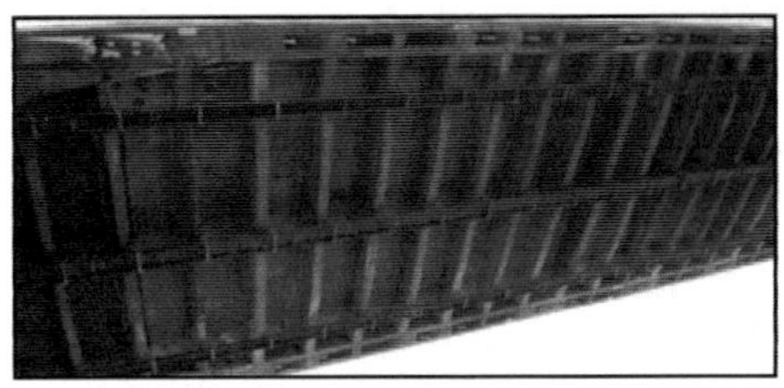

Figure3. Edge girder segments and floor-beams

3. Finite element modelling of the bridge

According to [16] cable stayed bridges show nonlinear performance under either static or dynamic loads. Such characteristic is the result of: (I) sagging consequences in the geometrical alteration of cables, (II) interaction between the forces of bend-moment and axial of towers and deck;(III) alterations of bridge geometry resulted from the structure deflects. The bridge is exposed to traffic loading which causes significant changes in its mechanical properties. The effect of traffic flow has been investigated to evaluate the ultimate bridge displacement /strain, internal forces and stresses in an effort to preserve the structure integrity from replacement [17, 18]. To exercise structural analysis finite element model is created using Nastran and Patran program as depicted in Figure 4.

The structural model consists of 554 grid points, 671 CBAR beam elements, 192 CQUAD4 plate elements and 144 CROD truss elements. The structure has fixed supports at the bottom of the bridge tower piles. The static analysis of the model has been reported in [19].

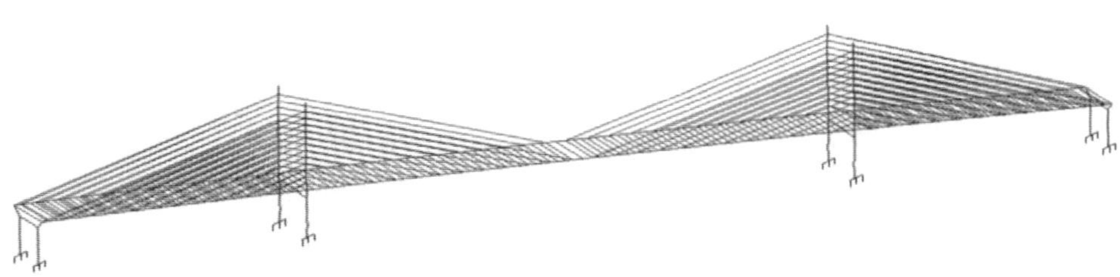

Figure 4. FE model of Penang (I) Bridge.

4. Modal analysis of the bridge

Modal analysis is carried out using FE model of Penang (I) Bridge in similar manner as [20]. The natural frequencies and resembling mode shapes of the first 8 vibration modes of 15 mode shapes are provided in Table 1 and the related shapes are portrayed in figure 5.

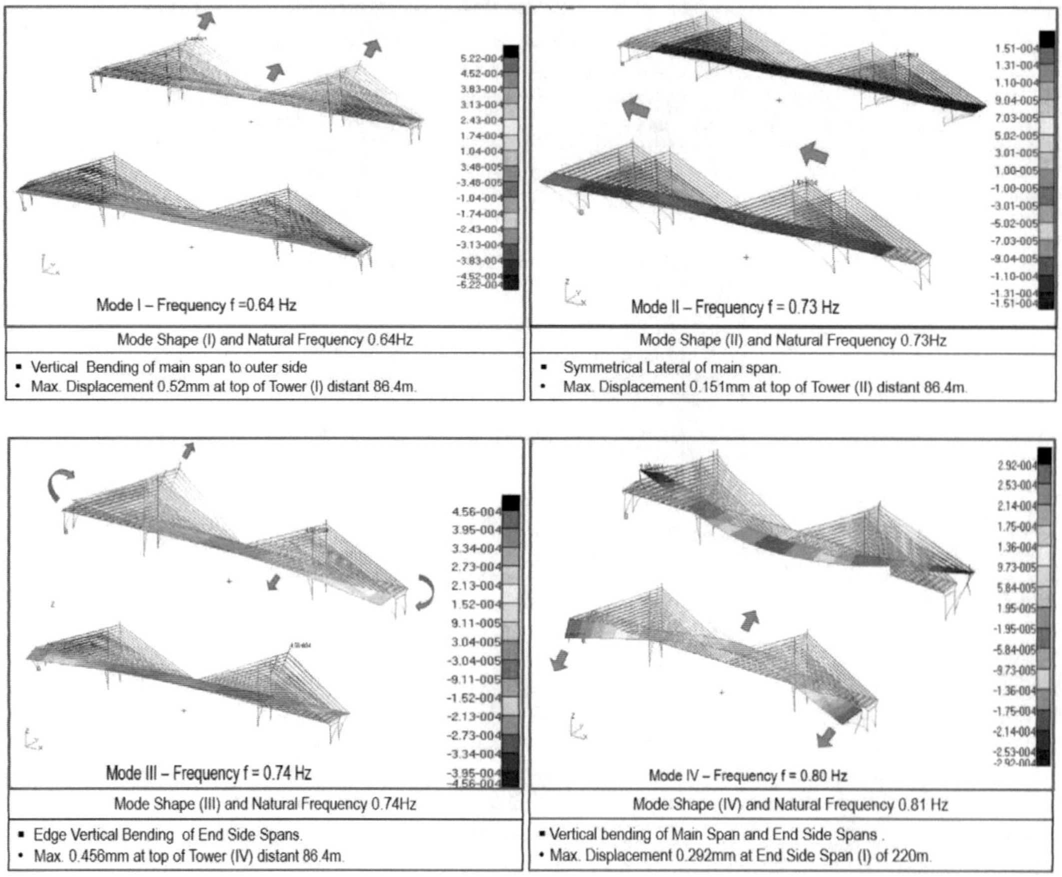

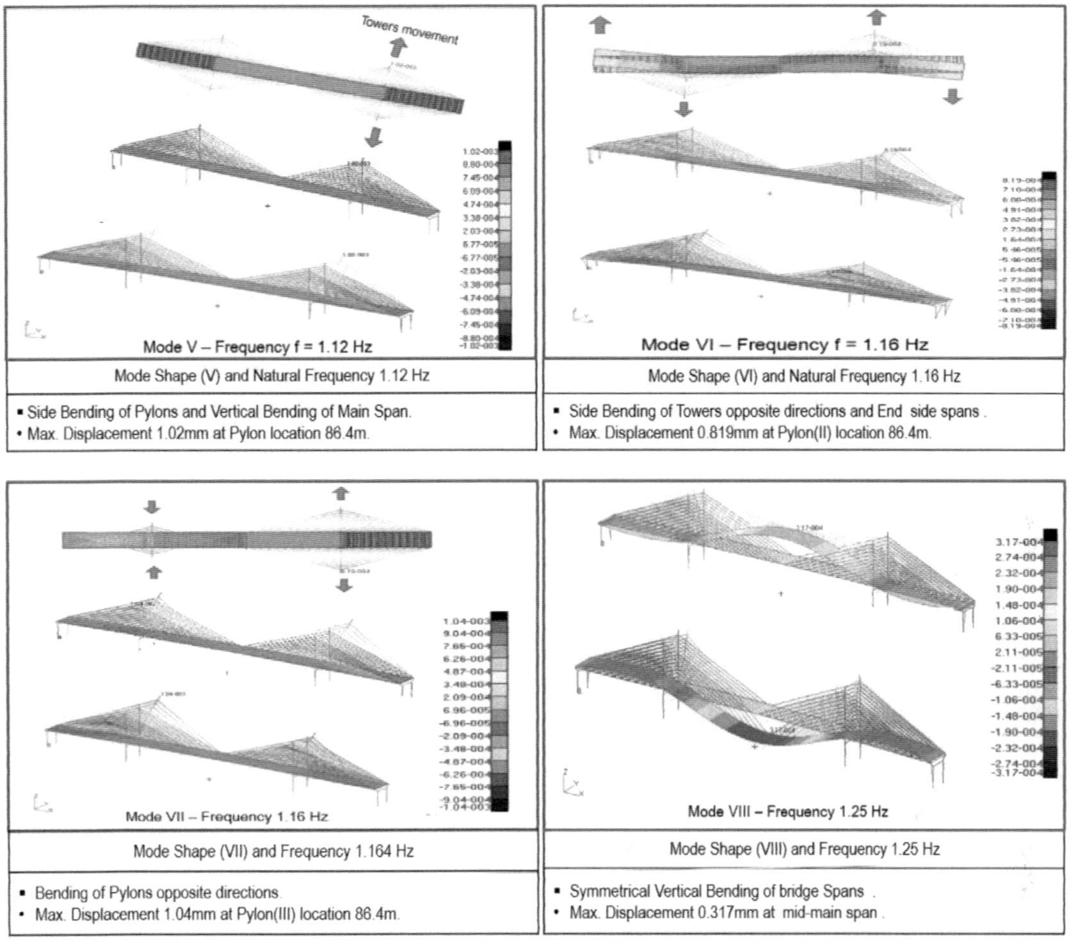

Figure 5. Mode shapes of the bridge displacements

Mode	**1**	**2**	**3**	**4**	**5**	**6**	**7**	**8**
Frequency (Hz)	0.64	0.73	0.74	0.81	1.12	1.16	1.16	1.25

Table 1. Natural frequencies at of undamaged bridge structure

5. Dynamic Analysis of One Unit Moving Load

The motion of a one-unit load is illustrated in figure 6. The impact of the unit load moving with certain velocity is evaluated through the displacement, inner force and stress levels on the bridge such as performed by [21]. The study is carried out on mid-main span grid 14 and element 9156 located at 112m from the left end side span support for assessment.

IOP Conf. Series: Materials Science and Engineering **184** (2017) 012068 doi:10.1088/1757-899X/184/1/012068

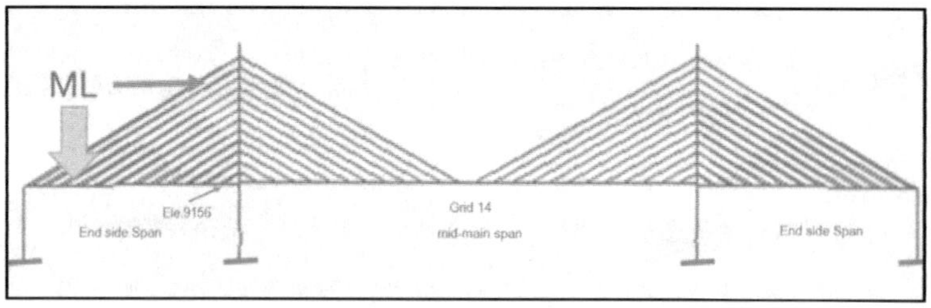

Figure 6. Moving one-unit load along bridge spans

The results obtained are portrayed in figure 7. The dynamic response due to this one unit moving load can be further used to simulate the dynamic response due to more realistic load such as truck, bus, car or other vehicles by using superposition methods.

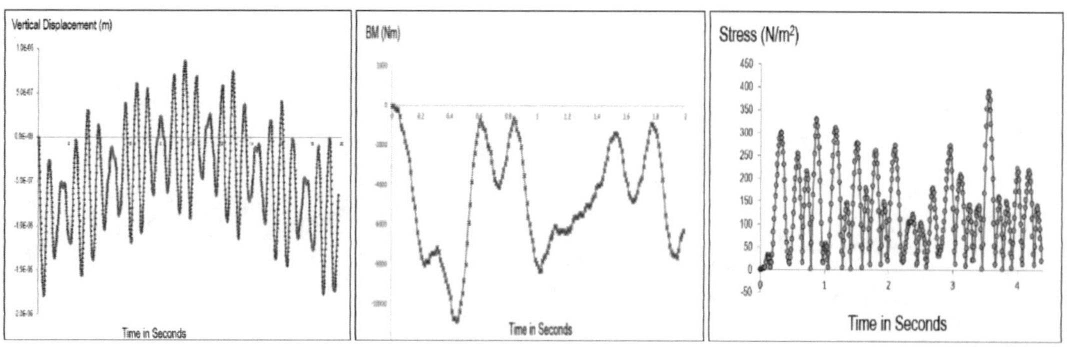

Displacement at G14 mid-main span | Bend-Moment in Element 9156 at 112m | Stress in Element 9156

Figure 7. Dynamic responses of the parametric study

6. Dynamic response of the bridge structure having damaged cables

The dynamic response is further investigated to find the influence of any failure on the bridge structure. Particularly, in the present work, the focus is on the influence of the bridge cables. The static loading analysis is first conducted and the axial cable force is established as portrayed in figure 8. The data shows that cable C9 experiences the highest stress level compare to others. The study introduces failure at the cable to determine their dynamic response. The second highest stress level after C9 is C8 cable such that the failures of both C8 and C9 are also important to be investigated.

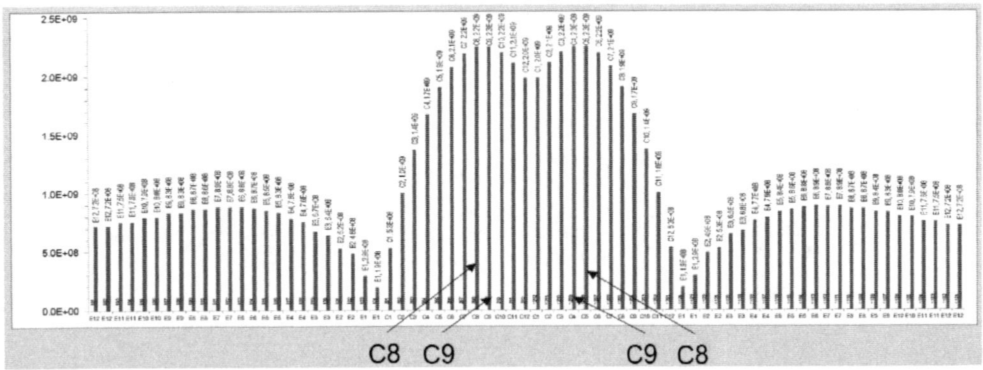

Figure 8. Maximum Stress of Cables at static loading

In case (I) the inactive of cable C9 has resulted an increase of the displacement at Grid 14 which is the mid span point and increase of the stress level in Cable E7 /112 as presented in Figure 9 and 10 respectively. It should be noted that the bridge structure is still can carry the load with the damaged of cable C9.

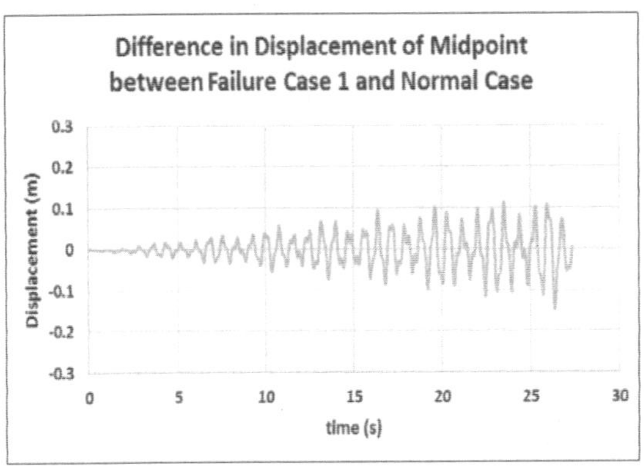

Figure 9. Case (I) Displacement at Grid 14

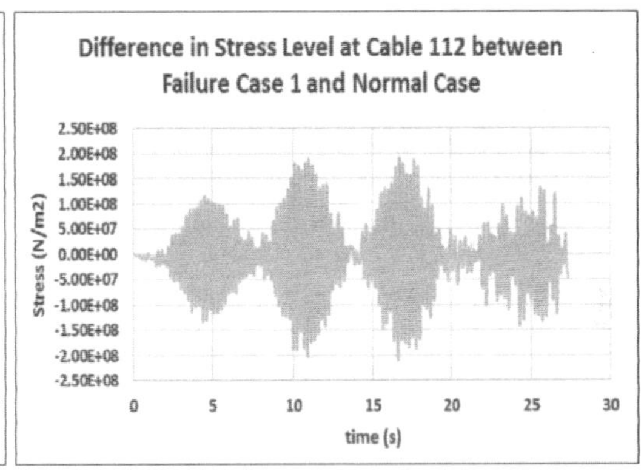

Figure 10. Case (I) Stress at Cable E7/112

Furthermore, in case (II) where cables C8 and C9 are both fail, the response in the form of increment of displacement at Grid 14 and stress at Cable E7/112 are shown in Figures 11 and Figure 12 respectively. Similarly, the bridge is still intake with the loss of the cables C8 and 9. Both cases clearly show different pattern of dynamic responses which can be used to further identify the failure of the structural components.

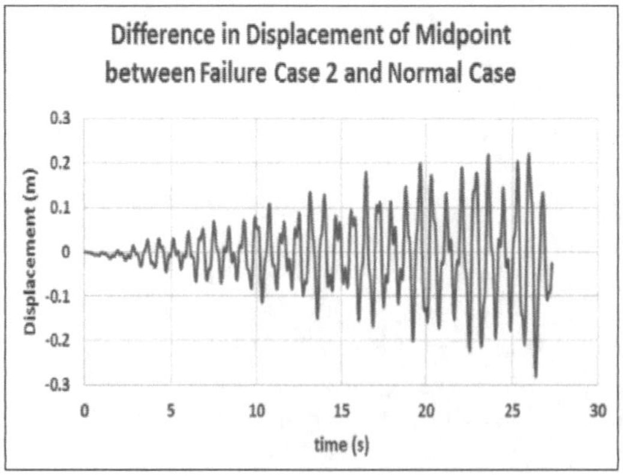

Figure 11. Case (II) Displacement at Grid 14

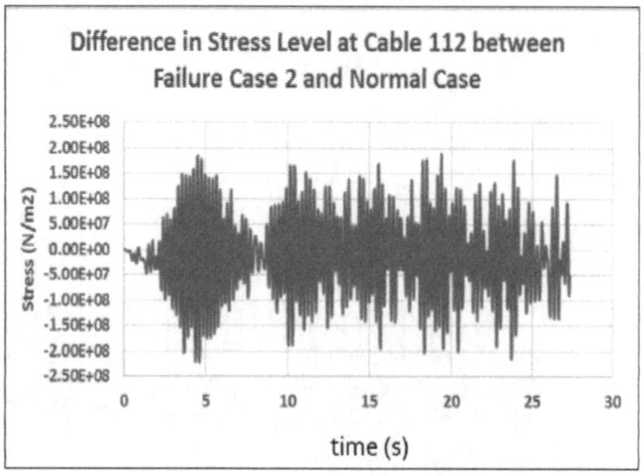

Figure 12. Case (II) Stress at Cable E7/112

7. Structure Health Monitoring (SHM)

The analytical dynamic analysis has indicated that each cable failure contributes different dynamic response characteristics. Therefore, by analyzing the response of several parameters such as displacement and stress level in several parts of the structure, it is possible to estimate which part of the structure is inactive. This procedure can be conducted if the bank data of the dynamic response of normal and failure structures is available.

For example, if the SHM equipment is planted in Grid 14 and Cable E7/112, the dynamic response at these positions can be compared with the data of figures 9 - 12 to check if the possible failures are due to Cable 9 or combination of Cables 8 and 9. This checking will involve a huge number of cases which can be performed using digital computer.

8. Conclusion

The study has proposed several scenarios of loading effect on Penang (I) bridge to assess the structural behavior for SHM. The dynamic response in each scenario is validated by the corresponding displacement, bending-moment force and the induced stress. The work has demonstrated that SHM can be conducted by detecting the dynamic response of the bridge due to particular moving load, such as truck, bus, or car, provided that their data are known. The dynamic response that can be monitored includes structural stress, force or displacement. These stress, force and displacement response data due to the failure of each structural member and their combination should be stored for SHM.

Reference

[1] Estes A and Frangopol D 2001 Bridge lifetime system reliability under multiple limit states *J. Bridge Eng.* **6** pp523-8.

[2] Aktan A, Catbas F, Grimmelsman K and Tsikos C 2000 Issues in infrastructure health monitoring for management *J. Eng. Mech.* **126** pp 711-24.

[3] Chang, P. C., Flatau, A., & Liu, S. C. (2003). Review paper: health monitoring of civil infrastructure. Structural health monitoring, **2** pp 257-267.

[4] Brownjohn, J. M. (2007). Structural health monitoring of civil infrastructure. Philosophical Transactions of the Royal Society of London A: Mathematical, Physical and Engineering Sciences, **365** pp 589-622.

[5] Fang I, Chen C and Chang I 2004 Field static load test on Kao-Ping-Hsi cable-stayed bridge *J. Bridge Eng.* **9** pp 531-540.

[6] Ren W, Lin Y and Peng X 2007 Field load tests and numerical analysis of Qingzhou cable-stayed bridge *J. Bridge Eng.* **12** pp 261-270.

[7] Catbas, N, Susoy M, and Frangopol M 2008 Structural health monitoring and reliability estimation: Long span truss bridge application with environmental monitoring data *J Eng. Structures* **30** pp2347-2359.

[8] Su D, Laomenit J and Fujino Y 2011 Traffic-induced response prediction of highway bridges *J. Procedia Eng.* **14** pp1071-1078.

[9] Wang W, Yan W, Deng L and Kang H 2014 Dynamic analysis of a cable-stayed concrete-filled steel tube arch bridge under vehicle loading *J. Bridge Eng.* **20** pp 04014082.

[10] Han H, Wang J, Meng X and Liu H 2016 Analysis of the dynamic response of a long span bridge using GPS/accelerometer/anemometer under typhoon loading *J. Eng. Structures* **122** pp 238-50.

[11] Pedro J J O, and Reis A, 2011 Stability of Composite Cable-Stayed Bridges *Proc. of 6th European conf. on steel and composite structures* pp 1263-68.

[12] Comanducci G, Ubertini F and Materazzi A 2015 Structural health monitoring of suspension bridges with features affected by changing wind speed *J. Wind Eng. and Industrial Aerodynamics* **141** pp 12-26.

[13] Farahani R V and Penumadu D, 2016 Damage identification of a full-scale five-girder bridge using time-series analysis of vibration data *J. Eng. Structures* **115** pp 129-39.

[14] Chen Z, Xu Y, Xia Y, Li Q and Wong K 2011 Fatigue analysis of long-span suspension bridges under multiple loading: Case study *J. Eng. Structures.* **33** pp 3246-56.

[15] Kee C F 1988 *The Penang Bridge Planning, Designing, and Construction* (Kuala Lumpur: Lembaga Lebuhraya Malaysia)

[16] Chang C, Chang T and Zhang Q 2001 Ambient vibration of long-span cable-stayed bridge *J. Bridge Eng.* **6** pp 46-53.

[17] Caprani C, OBrien E and McLachlan G 2008 Characteristic traffic load effects from a mixture of loading events on short to medium span bridges *J. Structural Safety.* **30** pp 394-404.

[18] Freire A, Negrao J and Lopes A 2006 Geometrical nonlinearities on the static analysis of highly flexible steel cable-stayed bridges *Computers and Structures* **84** pp2128-2140.

[19] Mohammed I M, Mustapha F, Hrairi M, Sulaeman E, Khairol A, Dayang L and Hojazi F 2013 Penang Bridge 1 Loading Analysis using British Standard and Finite Element Method for Structural Health Monitoring *Proc. of 2nd Int. Conf. on Mechanical, Automotive and Aerospace Engineering.*

[20] Ren W, Peng X and Lin Y 2005 Experimental and analytical studies on dynamic characteristics of a large span cable-stayed bridge *Eng. Structures* **27** pp 535-548.

[21] Yu Y, Zhao X, Shi Y and Ou J 2013 Design of a real-time overload monitoring system for bridges and roads based on structural response *Measurement* **46** pp 345-352.